STATISTICAL THERMOPHYSICS

Harry S. Robertson

University of Miami

P T R Prentice Hall
Englewood Cliffs, New Jersey 07632

Library of Congress Cataloging-in-Publication Data

Robertson, Harry S.
    Statistical thermophysics / Harry S. Robertson.
       p.   cm.
    Includes bibliographical references (p.   ) and index.
    ISBN 0-13-845603-8
    1. Statistical physics.   2. Statistical thermodynamics.
  I. Title.
  QC174.8.R63   1993
  530.1'3—dc20

92-17052
CIP

Editorial/production supervision: *Ann Sullivan*
Cover design: *Ben Santora*
Prepress buyer: *Mary McCartney*
Manufacturing buyer: *David Dickey*
Acquisitions editor: *Betty Sun*
Editorial assistant: *Maureen Diana*

© 1993 by P T R Prentice-Hall, Inc.
A Simon & Schuster Company
Englewood Cliffs, New Jersey 07632

The publisher offers discounts on this book when ordered
in bulk quantities. For more information, write:

Special Sales/Professional Marketing,
Prentice-Hall, Inc.
Professional Technical Reference Division
Englewood Cliffs, New Jersey 07632

Printed in the United States of America

10 9 8 7 6 5 4 3 2 1

ISBN 0-13-845603-8

Prentice-Hall International (UK) Limited, *London*
Prentice-Hall of Australia Pty. Limited, *Sydney*
Prentice-Hall Canada Inc., *Toronto*
Prentice-Hall Hispanoamericana, S.A., *Mexico*
Prentice-Hall of India Private Limited, *New Delhi*
Prentice-Hall of Japan, Inc., *Tokyo*
Simon & Schuster Asia Pte. Ltd., *Singapore*
Editora Prentice-Hall do Brasil, Ltda., *Rio de Janeiro*

TABLE OF CONTENTS

PREFACE

If I thought that modern statistical mechanics is a tranquil, noncontroversial subject, I should not have written about it. It is, in fact, an academic battle field, and has long been so. The Gibbs-Boltzmann differences in outlook and method have further subdivided and balkanized the thermal-physics community (to use the term loosely), with the result that workers in one camp are often unaware of the substantial progress being made in another, and the benefits of cross-fertilization are lost. In this book, I discuss some of the controversy, and I hope to effect some degree of reconciliation. I also discuss some of the outstanding problems and challenges.

The title *Statistical Thermophysics* was chosen to indicate as accurately as two words will allow that the subject matter is chosen in the area of thermal phenomena, and the treatment is statistical. The older title, *Statistical Mechanics*, is evidently inadequate because the subject matter is not primarily within the present-day connotation of the word *mechanics*, although certainly the classical and quantum-mechanical foundations underlie the subject. The more modern title *Statistical Physics*, on the other hand, is too broad in that it properly should include many other topics not treated here, such as statistical optics, quantum electronics, radioactivity, the statistics of electromagnetism in matter, and a variety of other topics that are the province of treatises on so-called stochastic processes. The term *thermophysics* is meant to imply that the concepts of temperature and entropy are of central importance to the subject matter treated.

I have tried to display the statistical foundations of thermodynamics soundly, to expose techniques for solving problems, and to provide a pathway to some of the frontiers of thermal physics.

While I recognize the great contributions to kinetic theory by Boltzmann and his students, my approach to understanding the foundations of the subject follows Gibbs, enhanced by the insights and contributions of Callen, Jaynes, and a number of others cited within the text. The treatment I have chosen, which I regard as offering the greatest conceptual clarity and postulational economy, often evokes a surprisingly emotional response (usually a violent rejection) from established scientists whose philosophy causes them to discredit the concept that our information about a system has anything to do with its real physical state. I quite sympathize with their attitude, and I certainly agree that our ignorance of the molecular structure of a block of ice cannot result in its melting. But no such claim is ever even implied. Moreover, the standard assertion that molecular chaos exists is nothing more than a poorly disguised admission of ignorance, or lack of detailed information about the dynamic state of a system.

Molecular chaos is an ambiguous term, especially since modern studies of phenomena that have come to be called *chaotic* have pre-empted the word of more general meanings. In dynamics, the state of a system of particles could be specified by the instantaneous position and momentum vectors of all particles in the system. Surely no subset of all possible dynamical states could be widely accepted as states of molecular chaos. The only attitude I have found productive in this context is that if I am able to perceive order, I may be able to use it to extract work from the system, but if I am unaware of internal correlations, I cannot use them for macroscopic dynamical purposes. On this basis, I shall distinguish heat from work, and thermal energy from other forms.

This book was written in keeping with my observation that most graduate students in physics keep their textbooks for use as references throughout their careers. Therefore, I have included more material than can usually be covered in a two-semester course. The book contains a core of subject matter suitable for a first graduate course in more-or-less traditional statistical mechanics, with a one-chapter reminder of basic thermostatics. It also contains a number of rather advanced topics; these can be disregarded if the book is used as an unadorned introduction to the subject, or they can be used as needed for students who are prepared to bypass, or hurry through, the more elementary material.

The first chapter introduces the ideas of information theory, surprise, entropy, probabilities, and the evolution of entropy in time, quite independently of any thermal considerations. A brief description of the intuitive relationship of these concepts to statistical thermophysics is included as a means of orientation and as a plausibility argument for the treatment of equilibrium macroscopic thermophysics in Chapter 2.

Since one of the conventional objectives of statistical thermophysics is the derivation of macroscopic thermal properties and relationships (equations state, heat capacities, phase transitions) from the microscopic dynamical properties of the particles, and since these results are often expressed in terms of what is usually called *thermodynamics*, Chapter 2 is included as a short summary of macroscopic thermostatics. In keeping with the later statistical development, a compact postulational approach is given, based on similar treatments by H. B. Callen and L. Tisza, which are directly descended from the Gibbs formulation of the subject. Although the approach is not conventional, the results are the standard ones, and the entire chapter may be omitted by readers who are already well acquainted with macroscopic thermostatics.

The basic concepts of equilibrium statistical thermophysics are presented on Chapter 3, together with simple applications to systems of noninteracting particles. More advanced topics in this chapter include treatments of the statistical operator, and Feynman path-integral methods. Much of this can be ignored on first reading.

Because of the conceptual importance and computational detail, fermion and boson systems of noninteracting particles are treated in their own chapter rather than as a part of Chapter 3. In addition to the basic material, Chapter 4 treats white dwarfs and neutron stars, and it includes a long commentary on generalizations of quantum mechanics that permit particles to obey parafermi or parabose statistics. Although such particles may never be encountered, they seem not to be forbidden, and the possibility of other statistics should be kept in mind, even if not studied in detail.

Chapter 5 treats some of the thermodynamics of electric and magnetic systems. These topics are often confusing, because so many different choices are possible regarding the placement of energies attributable to the fields that would be present even in the absence of matter. At least one puzzle remains to be resolved. Much of this chapter can be omitted, if necessary.

The next three chapters examine the thermodynamics and statistical behavior of systems of interacting particles. Chapter 6 concentrates on the thermodynamics of critical phenomena, and the remarkable existence of universal critical exponents that characterize the behavior of a system's properties and response functions in the neighborhood of a critical point, independent of details of microscopic interaction for systems in the same universality class. Exact solutions of model systems, such as the Ising model of interacting spins on a lattice, are derived in Chapter 7. Generalized treatments of such systems, based on the renormalization techniques that originally appeared in quantum field theory, are employed in Chapter 8 to provide insight about the theoretical basis for the universality found in Chapter 6. Each of these chapters introduces its subject at a level that should be accessible without difficulty in a first course, and each ends at a much higher level of complication, but with the intermediate steps included. It is therefore possible to cover enough of each of these chapters in a basic course to provide some of the flavor and methodology of its subject, or to extend the coverage to the more advanced topics. The initial treatment in each of these chapters is independent of the more advanced topics in previous chapters.

Chapters 9 and 10 return to nonequilibrium systems. I have tried to present a representative sampling of the various ways to calculate the behavior of these systems, both in terms of thermodynamics and of statistical physics. The study of nonequilibrium systems emphasizes matters of principle. The long-standing controversy about choice of the Gibbs or Boltzmann entropy, or perhaps some other contender, is addressed in Chapter 9, with some degree of clarification as a result. (The rather imposing harmonic oscillator system presented in Section 1.7 could be delayed until deemed appropriate in connection with Chapter 9.) The underlying importance of fluctuations in the behavior of thermal systems is explored in Chapter 10. Again, both of these chapters begin at a low-enough level for any graduate course in the subject, and they build to the point that current literature should be accessible.

A set of problems has been included at the end of each chapter, followed by the solutions. The inclusion of these sections as parts of their chapters, and not relegated to an appendix, is intended to emphasize that solving problems is an essential part of learning. The presented solutions exhibit useful techniques and applications that enhance the material without lengthening the presentation of basic topics in the main body of the text.

The commentary at the end of each chapter supplies background information and flesh to the rather skeletal body of basic subject matter. The references are intended to encourage further study by offering easy access to good sources of additional information. When the problems, solutions, commentary, and references are subtracted from the total volume, there results a relatively compact coverage of a considerable body of statistical thermodynamics and related topics. Some of the commentaries contain sermons intended to convert the heathen to the truth of the information-theoretic approach to statistical thermophysics. By placing these sermons in the commentaries, I hope to keep them out of the way of anyone who wishes to avoid them, while providing ready access for readers who are still in the process of examining arguments on both sides. While my own point of view is decidedly biased, I have quoted and referred to authors who disagree with me.

No textbook of acceptable size can cover all topics that have ever been considered appropriate for courses in statistical mechanics. A number of traditional subjects have been omitted, such as the Mayer cluster-integral theory of nonideal gases, the interminable discussions of coarse-graining, ergodic theory, and detailed solutions of the Boltzmann equation for gases, including realistic collision terms. A number of other topics that could have been included were omitted on the assumptions that they are standard material in condensed-matter or quantum mechanics texts and that they do not provide background material needed in the development of the more timely topics. The two topics I wanted to include, but lacked both time and space, are (1) dynamic, time-dependent critical phenomena, and (2) the statistical thermophysics of chaotic systems. These promise to be a source of interesting problems for a long time.

I am indebted to my colleagues James C. Nearing, Manuel Huerta, and Lawrence C. Hawkins for helpful criticism, additional references, and error correction, and to James C. Nearing and Marco Monti for TeXnical assistance. Thanks also are due to a succession of students who have detected typographical errors and ambiguities, and to Judy Mallery and Melody Pastore for transferring early versions of some chapters to computer memory. Finally, I thank my wife, Mary Frances, for her patience and support.

The remaining errors are entirely my responsibility. The manuscript was generated by me in TeX, and submitted camera-ready to the publisher, who therefore cannot be blamed for typographical errors, and certainly not for conceptual ones. I thank the staff at Prentice Hall, especially Betty Sun, Maureen Diana, and Ann Sullivan, for their continued courtesy, support, and attention to detail.

Comments, corrections, and suggestion for improvement will be welcome.

Harry S. Robertson

1. BASIC CONCEPTS

1.1 INTRODUCTION

Most of our decisions, predictions, and expectations are based on incomplete or inexact information. We may sit down confidently on a chair we have never seen before, walk without hesitation on an unfamiliar pathway, or assert that a car is red after having glanced at only one side of it. Ventures in the stock market and bets on horse races are much more likely to cause surprise.

In statistical physics, one considers the behavior of a number (usually very large) of microscopic and essentially unobservable entities, each in principle describable in terms of such dynamical variables as coordinates and momenta. No attempt is made to base predictions on detailed knowledge of these dynamical variables, however, since they are, at least for practical purposes, unknowable. Instead, one chooses to describe the collection of entities in terms of known or measurable macroscopic properties that are related to averages of functions of the microscopic variables. On the basis of the information obtained from the macroscopic observations, one then attempts to predict other macroscopic properties.

As a simple numerical illustration of the nature of the problem, suppose it is known that an urn contains balls each numbered 0, 1, or 2; that the average of the numbers, $\langle n \rangle$ (their sum divided by the number of balls), is some definite number, say 2/7; and nothing else is known. On the basis of this information, what is the expected value of some other function of the ball numbers, say $\langle n^3 \rangle - 2 \langle n \rangle$? Obviously the relative numbers of the three balls cannot be uniquely determined from the information given. There could be 5/7 numbered 0, 2/7 numbered 1, and none numbered 2, as a possibility, in which case $\langle n^3 \rangle - 2 \langle n \rangle = -2/7$. Other evident choices are $n_0 : n_1 : n_2 = 6 : 0 : 1$ or $11 : 2 : 1$, giving $\langle n^3 \rangle - 2 \langle n \rangle$ to be 4/7 and $-1/7$, respectively. It is possible to choose the relative numbers of balls to give any value in the range $-2/7 \leq \langle n^3 \rangle - 2 \langle n \rangle \leq 4/7$, and we are asked to select the most probable result, based on the only information at hand.

The most evident point of difference between this example and the problems of statistical thermophysics is that the preparation of the urn was the act of an intelligent and perhaps capricious being, so that we might well expect a distribution of balls differing significantly from the most probable one, whereas we have no such expectation when treating the behavior of a very large number of molecules. If our simple problem were rephrased as the number of light-bulb replacements per month in a certain set of desk lamps or the number of children per couple per year in a particular community, then the element of caprice is diminished, and abstract mathematical treatment would seem intuitively acceptable.

The systematic treatment of prediction from insufficient or incomplete data is a part of *information theory*. The ideas of information theory form an excellent foundation for the characteristic attitudes of statistical thermophysics, and the calculational techniques are often identical. The development to be presented here is essentially that of Shannon [1948], though several later mathematical treatments differ somewhat in detail. The presentation of further historical and philosophical details is deferred to Sec. 1.9.

1

1.2 PROBABILITIES

The possible outcomes of an experiment, measurement, contest, or trial are called, for want of a better word, *events*. The events are numbered 1 to N, where it is assumed at first that N is finite, and it is also assumed that a numerical (perhaps vector) value can be assigned for each event, say y_i for the ith event.

Associated with each y_i is a probability p_i, such that $0 \le p_i \le 1$, with the p_i's satisfying the condition

$$\sum_{i=1}^{N} p_i = 1. \tag{1.2.1}$$

Each p_i represents, in some sense, the probability that the event y_i will occur. Sometimes the p_i can be assigned on the basis of a series of tests. For example, an unbiased die should show each of its six faces equally probably, so $p_i = 1/6$ for $i = 1 \cdots 6$. On the other hand, the outcome of a horse race can hardly be treated on the same basis; yet probabilities are assigned, often by highly fanciful processes, to many kinds of one-shot or unrepeatable occurrences.

A familiar problem in physics – radioactive decay – combines the *objective* probabilities assignable to the die with the *subjective* probabilities often assigned to unique happenings. A radioactive nucleus can decay only once, after which it is regarded as a different nucleus. Yet it is meaningful to assign probabilities to its decay in, say, a sequence of one-second intervals.

The probabilities that are important in statistical physics are seldom the strictly objective ones obtainable from counting or from repeated measurements. Usually additional assumptions, extrapolations, or logical analyses are necessary before a working set of probabilities is finally attained. One such procedure is that of information theory, presented in Section 1.3. More appropriate procedures in other contexts are used later in this chapter and elsewhere throughout the book.

It is now assumed that a probability p_i is assigned to each event y_i. The average value of y_i is given by

$$\langle y \rangle = \sum_i p_i y_i, \tag{1.2.2}$$

and in fact the average of any power of y_i is given by

$$\langle y^n \rangle = \sum_i p_i y_i^n. \tag{1.2.3}$$

The *variance* of the distribution is

$$\langle (y - \langle y \rangle)^2 \rangle = \langle y^2 \rangle - \langle y \rangle^2, \tag{1.2.4}$$

a quantity often used as a measure of dispersion in statistics, and often related to temperature in statistical thermophysics.

More generally, any function of the events, $f(y_i)$, may be averaged to give

$$\langle f(y) \rangle = \sum_i p_i f(y_i), \tag{1.2.5}$$

and the events themselves may be more complicated entities than single numbers; they may be vectors or matrices, for example, and the operations of raising to powers, as in Eq. (1.2.3), or of forming functions, as in Eq. (1.2.5), must be appropriately defined. In any case, the probabilities themselves are simple non-negative-definite scalars. Their use as a measure of information will be discussed in Section 3.

1.3 INFORMATION THEORY

It has long been realized that the assignment of probabilities to a set of events represents information, in a loose sense, and that some probability sets represent more information than do others. For example, if we know that one of the probabilities, say p_2, is unity, and therefore all the others are zero, then we know that the outcome of the experiment to determine y_i will give y_2. Thus we have complete information. On the other hand, if we have no basis whatever for believing that any event y_i is more or less likely than any other, then we obviously have the least possible information about the outcome of the experiment.

A remarkably simple and clear analysis by Shannon [1948] has provided us with a quantitative measure of the uncertainty, or missing pertinent information, inherent in a set of probabilities. Let $S(\{p_i\})$ denote the desired quantitative measure of uncertainty, where $\{p_i\}$ denotes the set of p_i's. With Shannon, we require the following:

(a) $S(\{p_i\})$ is a continuous, differentiable, single-valued function of the p_i (with $0 \leq p_i \leq 1$; $\sum_{i=1}^{N} p_i = 1$).

(b) When all the p_i's are equal ($p_i = 1/N$), the function $A(N) = S(\{p_i = 1/N\})$ is a monotone increasing function of N.

(c) The uncertainty in a set of probabilities is unchanged if the probabilities are grouped into subsets, each with its own uncertainty. Specifically, if we write

$$w_1 = \sum_{i=1}^{n_1} p_i; \ w_2 = \sum_{i=n_1+1}^{n_2} p_i; \ w_k = \sum_{i=n_{k-1}+1}^{n_k} p_i;$$

where $n_k \equiv N$, this combinatorial property may be written as

$$S(\{p_i\}) = S(\{w_j\}) + \sum_{j=1}^{k} w_j S(\{p_i/w_j\}_j), \tag{1.3.1}$$

where the notation $\{ \ \}_j$ signifies the set of p_i's that make up a particular w_j. Each of the p_i's in the second term must be divided by its appropriate w_j, as indicated, so that the sum of each of the sets is unity, since S is defined only for such a set. The w_j multiplier in the second term represents the probability that the event is in the jth subset, *i.e.*, the uncertainty associated with the jth subset should contribute only in proportion to the probability that the event is *in* that subset.

The requirements (a), (b) and (c) are sufficient to determine the form of S. We choose, at first, all $p_i = 1/mn$, where m and n are integers, and all $w_j = 1/m$, so that $p_i/w_j = 1/n$. By use of the A's defined in (b), Eq. (1.3.1) becomes

$$A(mn) = A(m) + A(n), \tag{1.3.2}$$

since the sum term can be easily summed. But the continuity requirement now permits m and n to be treated as continuous variables, and partial differentiation of Eq. (1.3.2) with respect to n gives, with $mn \equiv p$,

$$m\frac{dA(p)}{dp} = \frac{dA(n)}{dn}.$$

Multiplication of both sides by n gives

$$p\frac{dA(p)}{dp} = n\frac{dA(n)}{dn} = C,$$

where C must be constant since p may be changed while n is kept constant. Therefore it follows immediately that

$$A(n) = C \ln n + d.$$

Substitution of this result in Eq. (1.3.2) gives $d = 0$, and

$$A(n) = C \ln n, \tag{1.3.3}$$

where C is as yet unspecified.

Now we return to Eq. (1.3.1), choose all the $p_i = 1/n$, where n is an integer, and let $w_j = \alpha_j/n$, where α_j is an integer and $\sum \alpha_j = n$. With these choices, Eq. (1.3.1) becomes

$$A(n) = S(\{w_j\}) + \sum (\alpha_j/n) A(\alpha_j),$$

or

$$S(\{w_j\}) = \sum (\alpha_j/n)[A(n) - A(\alpha_j)].$$

The use of Eq. (1.3.3) leads to

$$S(\{w_j\}) = -C \sum (\alpha_j/n)(\ln \alpha_j - \ln n),$$
$$= -C \sum (\alpha_j/n) \ln(\alpha_j/n),$$

or

$$S(\{w_j\}) = -C \sum w_j \ln w_j. \tag{1.3.4}$$

Except for the constant C, then, the form of the set function S has been determined uniquely. Note that since $w_j \leq 1$, the $\ln w_j$ term is nonpositive, and the negative sign in front leads to a non-negative measure of uncertainty when C is positive. (In nonthermal contexts, C is usually set equal to unity, or else set equal to $1/\ln 2$, so that effectively $C \ln w_i$ is replaced by $\log_2 w_i$.)

The uncertainty $S(\{p_i\})$ associated with a set of probabilities is seen to be proportional to the average value of $-\ln p_i$, by Eqs.(1.2.5) and (1.3.4). The quantity $-\ln p_i$ appears so often that it has been given the quite descriptive name of *surprise* (Watanabe [1969]). When any $p_i = 1$, the outcome of the experiment is certain; there is no surprise, and $-\ln p_i = 0$, as it should to be so named. For $p_i < 1$, the outcome is not certain, and we may regard the surprise as positive, given by $-\ln p_i$. The smaller p_i, the larger the surprise if the ith event occurs, and $-\ln p_i$ therefore seems aptly named.

It should be noted the $p_i \ln p_i$ is always taken to be zero when $p_i = 0$, in accord with the usual l'Hospital's-rule result. The inclusion of events for which the probability is zero in the sum of Eq. (1.3.4) does not change S.

For a given number of possible events n, it is easily shown that S is maximum when the events are equally probable ($p_i = 1/n$ for all i), and $S(\{p_i\}) = A(n)$, in accord with an earlier remark. The next problem to be examined is that of assigning the p_i's on the basis of our knowledge. This problem then leads to the conceptual foundation of the approach to statistical mechanics used in the present development.

1.4 MAXIMUM UNCERTAINTY

The uncertainty associated with a probability set $\{p_i\}$ has been found in Eq. (1.3.4) to be

$$S(\{p_i\}) = -C \sum_i p_i \ln p_i, \tag{1.4.1}$$

where C is positive constant. (At von Neumann's suggestion, Shannon called this function the *entropy* (Denbigh and Denbigh [1985]).) In statistical physics (and elsewhere) we are often asked to assign probabilities to events based on a few significant bits of information, but far too little to yield objective probabilities. We may, for example, know only one thing, such as the average value of y, given by Eq. (1.2.5), to be

$$\langle y \rangle = \sum_i y_i p_i. \tag{1.4.2}$$

The fact that the probabilities must sum to unity,

$$1 = \sum_i p_i \tag{1.4.3}$$

provides one more equation, but if there are more than two values of i, then there are not sufficient equations to yield the p_i's uniquely.

The procedure to be followed (Jaynes [1957a]) is that of maximizing the uncertainty, Eq. (1.4.1), subject to the limiting conditions on the p_i's [Eqs.(1.4.2) and (1.4.3), in this case]. The underlying attitude is that we are not entitled to assume more knowledge (less uncertainty) than we are given by the subsidiary conditions, and any assignment of the p_i's that satisfies the limiting conditions but yields a value of S less than its maximum is unjustified on the basis of the known data. For example, if one of the y_i's, say y_k, happens to be equal to $\langle y \rangle$, then the assignment $p_k = 1$, $p_{i \neq k} = 0$ will give $S = 0$ and will satisfy Eq. (1.4.2) and Eq. (1.4.3), but no such choice is warranted without further data.

The most convenient technique for maximizing S subject to the subsidiary conditions is by use of the Lagrange undetermined multipliers, so that all the p_i's may then be regarded as independent variables. We vary the quantity

$$S/C - \alpha' \sum p_i - \beta \sum p_i y_i,$$

to get

$$-\sum_i \delta p_i (\ln p_i + 1 + \alpha' + \beta y_i) = 0. \qquad (1.4.4)$$

Let $\alpha' + 1 = \alpha$, regard the δp_i as independent, so that the coefficient of each must vanish independently, and solve for the p_i to obtain

$$p_i = e^{-\alpha - \beta y_i}. \qquad (1.4.5)$$

The multipliers α and β can now be determined by means of Eqs. (1.4.2) and (1.4.3). Substitution of p_i from Eq. (1.4.5) into Eq. (1.4.3) gives

$$e^\alpha = \sum_i e^{-\beta y_i} \equiv Z, \qquad (1.4.6)$$

where the Z so defined is called the *partition function*. Similarly, Eq. (1.4.2) gives

$$\langle y \rangle = \sum_i y_i e^{-\alpha - \beta y_i}$$

$$= \sum_i y_i e^{-\beta y_i} \bigg/ \sum_i e^{-\beta y_i}$$

$$= -\frac{\partial \ln Z}{\partial \beta}, \qquad (1.4.7)$$

where the last line follows by inspection from the previous one. While Eq. (1.4.7) can in principle be solved for β, it usually cannot be done easily, and other procedures are used for discovering the nature of β.

The second derivative of $\ln Z$ with respect to β yields the variance,

$$\langle y^2 \rangle - \langle y \rangle^2 = \frac{\partial^2 \ln Z}{\partial \beta^2}, \qquad (1.4.8)$$

a quantity also used to indicate uncertainty, but not suitable in several respects for the uses to which S is put.

If other subsidiary conditions are known, say average values of functions denoted by $f_k(y_i)$, with $f_1(y_i) = y_i$, then Eq. (1.4.5) becomes

$$p_i = \exp\left[-\alpha - \sum_k \beta_k f_k(y_i)\right], \qquad (1.4.9)$$

and the resultant value of S is either reduced or left unchanged. To show this quite generally, assume that the p_i's are given by Eq. (1.4.5); then examine some other set of probabilities, P_i, that also satisfy Eqs.(1.4.2) and (1.4.3). Let $S_0 = S(\{p_i\})$, and $S_1 = S(\{P_i\})$. Now form the difference $(S_0 - S_1)/C$, add and subtract the quantity $\sum_i P_i \ln p_i$, and obtain

$$(S_0 - S_1)/C = -\sum_i p_i \ln p_i + \sum_i P_i \ln P_i - \sum_i P_i \ln p_i + \sum_i P_i \ln p_i$$

$$= \sum_i (P_i - p_i) \ln p_i - \sum_i P_i \ln(p_i/P_i)$$

$$= \sum_i (P_i - p_i)(-\alpha - \beta y_i) - \sum_i P_i \ln(p_i/P_i)$$

$$\geq -\sum_i P_i \left(\frac{p_i}{P_i} - 1\right) = 0, \qquad (1.4.10)$$

where the first sum in the third line vanishes because both sets of probabilities satisfy Eqs.(1.4.2) and (1.4.3), and the inequality is merely $\ln x \leq x - 1$ for all real x. It has been shown, then, that any choice of the probability set that differs from that of Eq. (1.4.5), including of course refined sets such as those of Eq. (1.4.9), will either reduce the uncertainty or else leave it unchanged. (It will be unchanged only if all of the new probabilities are identical with the original ones, because the inequality becomes an equality only if the argument of the logarithm is unity, in which case the additional information is redundant, and the extra β's are zero.)

The quantity S, called the *information entropy*, will be shown later to correspond to the thermodynamic entropy, with $C = k$, the Boltzmann constant, and y_i usually an energy level ϵ_i while β becomes $1/kT$, with T the thermodynamic temperature. The probabilities given by Eq. (1.4.5) are then called *canonical*, and Z is called the canonical partition function. A thermodynamic system is characterized by a microscopic structure that is not observed in detail, although it is assumed to be describable in principle in terms of dynamical variables. We observe macroscopic properties such as volume, energy, temperature, pressure, and magnetic moment. We attempt to develop a theoretical description of the macroscopic properties of the system in terms of its underlying microscopic properties, which are not precisely known. We attempt to assign probabilities to the various microscopic states of the system based upon a few scattered macroscopic observations that can be related to averages of microscopic parameters. Evidently the problem that we attempt to solve in statistical thermophysics is exactly the one just treated in terms of information theory. It should not be surprising, then, that the uncertainty of information theory becomes a thermodynamic variable when used in proper context.

Useful relationships may be developed between Lagrange multipliers, β_k, in Eq. (1.4.9) and the constraints denoted by $\langle f_k \rangle$. Note that the partition function is

$$Z = e^{\alpha} = \sum_i \exp\left[-\sum_k \beta_k f_k(y_i)\right].$$

From this expression, it follows that

$$-\frac{\partial \ln Z}{\partial \beta_j} = Z^{-1} \sum_i f_j(y_i) \exp\left[-\sum_k \beta_k f_k(y_i)\right] = \langle f_j \rangle. \qquad (1.4.11)$$

From $S/C = -\sum_i p_i \ln p_i$, it follows that

$$S/C = \ln Z + \sum_k \beta_k \langle f_k \rangle, \qquad (1.4.12)$$

and, if we regard S as a function of the set of constraints $\{\langle f_k \rangle\}$, we have

$$\frac{\partial (S/C)}{\partial \langle f_m \rangle} = \frac{\partial \ln Z}{\partial \langle f_m \rangle} + \beta_m + \sum_k \frac{\partial \beta_k}{\partial \langle f_m \rangle} \langle f_k \rangle$$

$$= -Z^{-1} \sum_i \sum_r f_r(y_i) \frac{\partial \beta_r}{\partial \langle f_m \rangle} \exp\left[-\sum_k \beta_k f_k(y_i)\right] + \beta_m + \sum_k \frac{\partial \beta_k}{\partial \langle f_m \rangle} \langle f_k \rangle$$

$$= -\sum_r \langle f_r \rangle \frac{\partial \beta_r}{\partial \langle f_m \rangle} + \beta_m + \sum_k \frac{\partial \beta_k}{\partial \langle f_m \rangle} \langle f_k \rangle$$

$$= \beta_m. \qquad (1.4.13)$$

From Eq. (1.4.11), differentiation with respect to β_r gives

$$\frac{\partial \langle f_j \rangle}{\partial \beta_r} = -\frac{\partial^2 \ln Z}{\partial \beta_r \partial \beta_j} = \frac{\partial \langle f_r \rangle}{\partial \beta_j} = A_{jr} = A_{rj}, \qquad (1.4.14)$$

and similarly from Eq. (1.4.13),

$$\frac{\partial \beta_m}{\partial \langle f_k \rangle} = \frac{\partial^2 (S/C)}{\partial \langle f_k \rangle \partial \langle f_m \rangle} = \frac{\partial \beta_k}{\partial \langle f_m \rangle} = B_{mk} = B_{km}. \qquad (1.4.15)$$

The matrices A and B are easily seen to be reciprocal, since

$$\frac{\partial \langle f_j \rangle}{\partial \langle f_k \rangle} = \delta_{jk} = \sum_m \frac{\partial \langle f_j \rangle}{\partial \beta_m} \frac{\partial \beta_m}{\partial \langle f_k \rangle} = \sum_m A_{jm} B_{mk},$$

or

$$A = B^{-1}.$$

As will be shown in Chapter 2, the entities S/C and $\ln Z$ contain the same amount of information about the system, and every result that can be calculated from a knowledge of one can also be calculated from a knowledge of the other. The relationship given in Eq. (1.4.12) is just a Legendre transformation.

Thermodynamics is treated on a postulational basis in Chapter 2, with the principal postulate being that the entropy of a thermodynamic system of constant energy is maximum for equilibrium. In present terms, the postulate means that we expect to find the equilibrium properties of the system consistent with the distribution of microstates that maximizes our uncertainty, subject to the existing constraints. The uncertainty, it should be emphasized, is in our knowledge of the microstate of the system, and the uncertainty is maximum (roughly) when the number of different microstates consistent with the observed macrostate is as large as possible. This point is discussed in more detail in Chapter 3, when samples from thermophysics become appropriate. The idea, however, can be studied in terms of heuristic models that can be understood without any previous knowledge of thermal physics.

1.5 TIME-DEPENDENT SYSTEMS

The statement that the uncertainty, or the information entropy, of a system in equilibrium is some sort of maximum, to be defined more precisely when necessary, raises the question: "How did it get to be a maximum?" The evolution of the uncertainty in time requires that the probabilities be time dependent, and in some cases equations of motion for these probabilities can be exhibited and solved. Several examples of this kind will be examined in this section, both in terms of probabilities and in other ways.

1.5.1 Radioactive Decay. The familiar problem of radioactive decay is treated here first in its most elementary fashion, and then in terms of time-dependent probabilities, not as an approach to equilibrium, which does not occur here, but as an example of the treatment of such a problem in terms of probabilities. Typically it is stated that one has n_0 radioactive nuclei at time $t = 0$. The change in the number n in the time interval dt is given by

$$dn = -\alpha n \, dt, \qquad (1.5.1)$$

where αdt is the probability that any one nucleus decays in the time interval dt. The solution to Eq. (1.5.1) and its initial condition is easily seen to be

$$n = n_0 e^{-\alpha t} \tag{1.5.2}$$

where n is the population at time t. But since each nucleus decays at random, a realistic expectation is that even if n_0 were known precisely (which it usually is not), at least $n(t)$ should be somewhat uncertain.

In order to treat the problem in terms of probabilities, let $p(n;t)$ be the probability that n nuclei exist at time t. An equation for $p(n;t)$ may immediately be written as

$$p(n;t+\delta t) = (1 - n\alpha\delta t)p(n;t) + (n+1)\alpha\delta t p(n+1;t), \tag{1.5.3}$$

where the first term represents the probability that if there are n nuclei at time t, all will survive to $t+\delta t$, and the second term represents the probability that if there are $n+1$ nuclei at time t, one of them will have decayed in the interval δt. Expansion of the left-hand side to first order leads to

$$\dot{p}(n;t) = -n\alpha p(n;t) + (n+1)\alpha p(n+1;t). \tag{1.5.4}$$

The set of these coupled equations can be solved easily by a variety of methods, some of which are exhibited in the solutions to problems. If it is known that at $t=0$, $n_0 = N$, then the solution is

$$p(n;t) = \binom{N}{n} \left(1 - e^{-\alpha t}\right)^{N-n} \left(e^{-\alpha t}\right)^n, \tag{1.5.5}$$

a result that is almost evident by inspection, since the decays are independent. If $p_s(t)$ is the probability that one nucleus survives to time t, and $p_d(t)$ the probability that it decays before t, then $p_s + p_d = 1$, and the terms in the generating function $(p_s + p_d)^N$ give the probabilities that of N particles initially, n will have survived and $N-n$ decayed. The individual probabilities are evidently $p_s = e^{-\alpha t}$ and $p_d = 1 - e^{-\alpha t}$; hence Eq. (1.5.5). It is also evident, on the basis of this discussion, that $\sum_{n=0}^{N} p(n;t) = 1$ for all t. (See Appendix A on generating functions and Appendix C on combinatorial problems.) The average value of n is given by

$$\langle n(t)\rangle = \sum_{n=0}^{N} np(n;t) = \sum_{n=0}^{N} n\binom{N}{n} e^{-n\alpha t}\left(1 - e^{-\alpha t}\right)^{N-n}$$

$$= Ne^{-\alpha t}\sum_{n=1}^{N}\binom{N-1}{n-1} e^{-(n-1)\alpha t}\left(1-e^{-\alpha t}\right)^{(N-1)-(n-1)}$$

$$= Ne^{-\alpha t}, \tag{1.5.6}$$

in agreement with the simple, deterministic theory.

More generally, when N is not known precisely at $t=0$, the initial probability set must be incorporated in the solution to Eq. (1.5.4), to give

$$p(n;t) = \sum_{N=0}^{\infty} p(N;0)\binom{N}{n} e^{-n\alpha t}\left(1-e^{-\alpha t}\right)^{N-n} \tag{1.5.7}$$

a result that reduces to Eq. (1.5.5) when, for some N, $p(N;0) = 1$. The result corresponding to Eq. (1.5.6), for any initial probability set, is simply

$$\langle n(t) \rangle = \langle N \rangle e^{-\alpha t}. \tag{1.5.8}$$

This development exhibits several of the characteristics of statistical thermophysics in that even if we know the state of the system very well at $t = 0$, its free evolution without further input of information leads to a distribution of probabilities and a growing uncertainty of the actual number of particles. The temporal increase in $S = -C \sum p_n \ln p_n$ is a consequence of the state of our knowledge. There is, presumably, a definite number of undecayed nuclei at any time, so the growth of uncertainty is not a property of the system itself, but of our knowledge of it. Our inability or failure to incorporate enough dynamic detail in the theory to permit precise calculation of the decay time of each nucleus has led to a statistical theory in which uncertainty increases as we move in time away from the initial observations.

1.5.2 The Dog-Flea Model.
The same technique can be used in an example of the approach to equilibrium. This is a classic example put forth by the Ehrenfests in their attempt to clarify the meaning of the Boltzmann H-theorem and the approach of a system to equilibrium (Ehrenfest and Ehrenfest [1959]). It was written about formally in terms of numbered balls in urns, but informally in lectures it was presented as a collection of N fleas on two equally attractive dogs, with the fleas hopping back and forth at random. The object of the analysis was to show that as time goes on, each dog acquires half the fleas as an equilibrium state, no matter what the initial distribution, and that small fluctuations persistently return to this state of equipartition.

The Ehrenfests took, in effect, a quantity $n_1 \ln n_1 + n_2 \ln n_2$, corresponding to Boltzmann's H (which differs both in sign and in conceptual content from the S used here as a measure of uncertainty) where n_i is the flea population of the ith dog, and $n_1 + n_2 = N$. They tried to show that as time progressed this quantity tended to settle down to its minimum value of $N \ln(N/2)$, with only minor fluctuations, and they then regarded this demonstration as support for the idea that random processes lead to equilibrium. A surprising experimental paper by Schrödinger, in collaboration with Kohlrausch (Kohlrausch and Schrödinger [1926]), patiently verifies the Ehrenfests' rationalizations without contributing any improvement in the conceptual basis of the Boltzmann H-theorem.

One problem that disturbed the Ehrenfests and continues to cause confusion is that of time reversal. If the fleas, in a sequence of random hops, could start out from any initial distribution and eventually become equipartitioned, then why could they not with equal probability reverse each hop separately and return to the initial state? The answer, despite attempted rationalizations, is that the two sequences of hops are equally probable when the fleas are regarded as identifiable individuals. Thus for every sequence of hops leading toward equipartition, there is an equally probable sequence leading away from it. No amount of rationalization could contrive the state of equilibrium to be other than approximate equipartition with continual fluctuations, although with the probability for these fluctuations diminishing as their size increases.

The view to be presented here is that equilibrium is not dependent upon the populations themselves, but only on our expectation of these populations. The populations should properly be regarded as random variables, and we should ask the probability of finding particular populations. The calculation of the probabilities proceeds as in the case of radioactive decay, with $w_N(n_1, n_2; \tau)$ the probability of n_1 fleas on dog 1, $n_2 = N - n_1$ on dog 2 at time $\tau = \alpha t$, where αdt is the probability that one flea hops to the other dog in the time dt. The equations for the probabilities are

$$\dot{w}_N(n_1, n_2; \tau) = - N w_N(n_1, n_2; \tau) + (n_1 + 1) w_N(n_1 + 1, n_2 - 1; \tau)$$
$$+ (n_2 + 1) w_N(n_1 - 1, n_2 + 1; \tau) \tag{1.5.9}$$

and the solution again may be found, compounding the $N = 1$ solution by means of a generating function (See Appendix A), to be, with $w_N(n_1, N - n_1; \tau) = w_N(n_1; \tau)$,

$$w_N(n_1; \tau) = e^{-N\tau} \sum_{k=0}^{N} \sum_{m=0}^{k} \binom{k}{m}\binom{N-k}{n_1 - k + m}(\cosh \tau)^{\gamma}(\sinh \tau)^{N-\gamma} \times w_N(N - k; 0),$$
$$\tag{1.5.10}$$

where $\gamma = n_1 - k + 2m$. This rather complicated expression evolves from any initial probability distribution, including the extreme case where, say, dog 1 has all the fleas, smoothly in time to a time-independent (at $t \to \infty$) distribution given by

$$w_N(n_1, N - n_1; \infty) = 2^{-N}\binom{N}{n_1}, \tag{1.5.11}$$

which is just a binomial (or binary, or Bernoulli) distribution for equally probable events. This distribution is, of course, sharply peaked for large N in the neighborhood of equipartition, with diminishing probabilities in either direction away from balance. The quantity

$$S = -C \sum_{n_1=0}^{N} w_N(n_1, N - n_1; \tau) \ln w_N(n_1, N - n_1; \tau)$$

evolves smoothly, continuously and monotonously toward its maximum value, which occurs when the probabilities are given by Eq. (1.5.11).

The result is that after a long wait we may expect to find any possible distribution of the fleas, but with the probabilities given by Eq. (1.5.11). There is no concern about fluctuations, because it is not the populations themselves, but rather the probabilities, that evolve to the time-independent state called *equilibrium*.

There is no question of time reversibility in the probabilistic analysis, since Eq. (1.5.9) and its solution, Eq. (1.5.10), are not time-reversible. An attempt to study the evolution of the probabilities objectively implies, in effect, an ensemble of identical dog-flea systems. Each such system could be started with a definite number of fleas on each dog, say with a_1 fleas on each dog 1 and $a_2 = N - a_1$ on each dog 2, so that $w_N(a_1, N - a_1; 0) = 1$ for each system. A later observation would produce data from which objective (measured) probabilities could be computed. If these probabilities are used as new initial conditions in Eq. (1.5.10), there is still no realistic expectation that the populations will evolve so as to produce the objective measure $w_N(a_1, N - a_1; 2\tau) = 1$, *i.e.*, a certain return to the initial value. An intrinsic irreversibility is evident here, and it has led some physicists and philosophers to regard the growth of entropy as pointing the direction of time. One should not agree too hastily with this point of view, however, since a retrodictive set of equations equivalent to Eq. (1.5.9) but with τ replaced by $-\tau$ could be written to give the probabilities at earlier times based on a measured distribution of probabilities at $t = 0$. The result would lead to the same time-independent probability set, Eq. (1.5.11), at $t = -\infty$, which merely says that as we move in time away from measured data, the pertinence of our information deteriorates, and we become increasingly uncertain of the state of the system, no matter which way we go in time. Thus, any profound conclusions on entropy evolution and time's arrow may be specious (Gal-Or [1974a]; Gal-Or [1974b]; Hobson [1966]; Ne'eman [1974]; Penrose and Lawrence [1974]).

Another interesting aspect of this problem concerns the equilibrium value of S, given in terms of the probabilities of Eq. (1.5.11), as

$$S_N/C = -\sum w_N \ln w_N = N \ln 2 - 2^{-N} \sum \binom{N}{n} \ln \binom{N}{n}. \qquad (1.5.12)$$

But since there are N independent fleas, each with the probability $1/2$ of being on each of the dogs, the expected S for each flea is given by

$$S_1/C = \left(-\frac{1}{2}\ln\frac{1}{2}\right) \times 2 = \ln 2$$

and one might expect the value of S_N/C to be simply N times the single-flea value, or $N \ln 2$. The explanation of this apparent paradox is that we choose to treat the fleas as indistinguishable. If they *were not* so regarded, there would be 2^N equally probable configurations of fleas on the two dogs, each with probability 2^{-N}. This value gives the quantity $A(2^N)$ of Eq. (1.3.3), which is seen to be the naïvely expected result already obtained. By regarding the individual identities of the fleas as not being pertinent information, one can group the ways of having n fleas on dog 1 and $N - n$ on dog 2, to get a probability $w_N(n, N-n) = \binom{N}{n}\left(2^{-N}\right)$, which is just the value of w_N in Eq. (1.5.11). The basic equation used for deriving the form of S, Eq. (1.3.1), can be written in this context as

$$A(2^N) = S_N(\{w_j\}) + 2^{-N}C \sum_n \binom{N}{n} \ln \binom{N}{n},$$

where the last term results from using $p_i/w_n = (2^{-N})/w_n$ in Eq. (1.3.1). This is easily seen to give the correct value for S_N. The point here is that since the identities of the individual fleas are not regarded as pertinent, the correct S_N is less than $A(2^N)$ because the locations of the individual fleas are not counted as missing pertinent information.

1.5.3 The Wind-Tree Model. Another famous example of the Ehrenfests, called the "wind-tree model," suffers from many of the same defects in their treatment of it, but the correct probabilistic analysis is too lengthy to be included in detail. Since the useful conceptual understanding can be developed in terms of the results, rather than the calculation of them, and since the details of the calculation are not particularly important in thermophysics, only the principal results will be presented, along with a résumé of the Ehrenfests' treatment.

The wind-tree model involves specular collisions of the point-particle "wind molecules" with the fixed "tree" molecules. The former all move with the same speed initially in

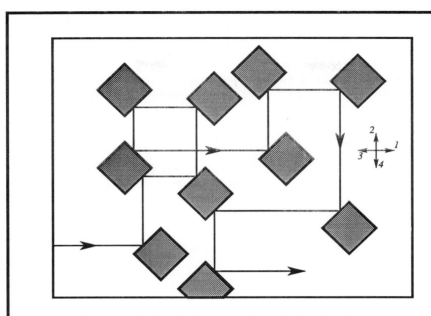

Figure 1. Path of a wind molecule through Ehrenfest forest.

directions parallel and antiparallel to the x and y axes. The tree molecules are smooth squares with their diagonals parallel to the x and y axes, as shown in Fig. 1, placed without apparent order in the $x - y$ plane. Edge effects are ignored. Since the reflections are perfectly elastic from smooth surfaces, the wind molecules continue to have only four possible directions of motion.

The Ehrenfests took n_1, n_2, n_3 and n_4 to be the time-dependent populations of the four groups of wind molecules, with the directions as shown in Fig. 1. The collisions with tree molecules are taken to be random, leading to the set of equations

$$\dot{n}_1 = -2\alpha n_1 + \alpha\left(n_2 + n_4\right)$$
$$\dot{n}_2 = -2\alpha n_2 + \alpha\left(n_1 + n_3\right)$$
$$\dot{n}_3 = -2\alpha n_3 + \alpha\left(n_2 + n_4\right)$$
$$\dot{n}_4 = -2\alpha n_4 + \alpha\left(n_2 + n_3\right) \tag{1.5.13}$$

where $\sum n_i = N$, and α is a constant containing data on the speed of the wind molecules, the geometry of the tree molecules, and their density. With $\tau = \alpha t$, $\cosh\tau \equiv c$, and $\sinh\tau \equiv s$, the solution to this set of equations may be written as

$$\begin{pmatrix} n_1(\tau) \\ n_2(\tau) \\ n_3(\tau) \\ n_4(\tau) \end{pmatrix} = e^{-2\tau} \begin{pmatrix} c^2 & cs & s^2 & cs \\ cs & c^2 & cs & s^2 \\ s^2 & cs & c^2 & cs \\ cs & s^2 & cs & c^2 \end{pmatrix} \begin{pmatrix} n_1(0) \\ n_2(0) \\ n_3(0) \\ n_4(0) \end{pmatrix}. \tag{1.5.14}$$

It is easily seen that as $\tau \to \infty$, $n_1(\infty) = n_2(\infty) = n_3(\infty) = n_4(\infty) = N/4$, according to Eq. (1.5.14). The quantity $\sum n_i \ln n_i$, taken as equivalent to Boltzmann's H, evolves smoothly to its minimum value of $N \ln(N/4)$ as $\tau \to \infty$. But time reversal presents itself more interestingly here than in the dog-flea model, because here the collisions are clearly time reversible. Evidently, if we start with an initial set of populations $\{n_i(0)\}$ and allow the system to evolve to the set $\{n_i(\tau)\}$, then an instantaneous reversal of velocities would form a new set $\{n_i^*(\tau)\}$ given by

$$\{n_1^*(\tau), n_2^*(\tau), n_3^*(\tau), n_4^*(\tau)\} = \{n_3(\tau), n_4(\tau), n_1(\tau), n_2(\tau)\} \qquad (1.5.15)$$

Since the collisions are reversible, if we put the set $\{n_i^*(\tau)\}$ as the initial conditions in Eq. (1.5.14), we should expect, from the dynamics, a return to a population at 2τ that, again with the velocities reversed, is identical with the initial set $\{n_i(0)\}$. Schematically,

$$\{n_i(0)\} \quad \text{evolves to} \quad \{n_i(\tau_1)\};$$
$$\{n_i(\tau_1)\} \quad \text{reverses to} \quad \{n_i^*(\tau_1)\}, \quad \text{given by Eq. (1.5.15)};$$
$$\{n_i^*(\tau_1)\} \quad \text{evolves to} \quad \{n_i^*(\tau_1 + \tau)\};$$
$$\{n_i^*(2\tau_1)\} \quad \text{reverses to} \quad \{n_i^{**}(2\tau_1)\}, \quad \text{as in Eq. (1.5.15)}.$$

For every value of τ_1, then, the set $\{n_i^{**}(2\tau_1)\}$ should equal the initial set $\{n_i(0)\}$. The results of Eq. (1.5.14), however, give

$$
\begin{pmatrix} n_1^{**}(2\tau_1) \\ n_2^{**}(2\tau_1) \\ n_3^{**}(2\tau_1) \\ n_4^{**}(2\tau_1) \end{pmatrix}
= e^{-4\tau_1}
\begin{pmatrix} c^2 & cs & s^2 & cs \\ cs & c^2 & cs & s^2 \\ s^2 & cs & c^2 & cs \\ cs & s^2 & cs & c^2 \end{pmatrix}_{2\tau_1}
\begin{pmatrix} n_1(0) \\ n_2(0) \\ n_3(0) \\ n_4(0) \end{pmatrix}
\qquad (1.5.16)
$$

which agrees with the set $\{n_i(0)\}$ only at $\tau_1 = 0$. The subscript on the 4×4 matrix indicates that the functions are to be evaluated at $2\tau_1$, and the result is the same as if there had been no reversal. The Ehrenfests took this as evidence that something was wrong with the so-called "Stosszahlansatz," the assumption about collisions that underlies the factor α and the set of relations given in Eq. (1.5.13). They argued, in effect, that since Eq. (1.5.16) indicates the failure of the system to return to its original state after time reversal, in contradiction to its obvious dynamic behavior, then the collisions clearly could not be random in both the forward and reversed situations. (The term "random" seems to deny the existence of order or pattern, either evident or not. How can one ever assert that a sequence of numbers, for example, contains no order, either observable or to be perceived later? The Ehrenfests apparently did not consider the irreversible behavior of Eq. (1.5.13) as an argument that no possible placement of the tree molecules could be random, but more likely that even if a sequence of events is truly random (whatever that means), its inverse cannot be so regarded.) See Ford [1983], Ford [1986] for useful discussions of randomness.

The point of view that equilibrium is a physical condition of the system (here with $n_1 = n_2 = n_3 = n_4 = N/4$) continued to be strongly presented by the Ehrenfests and others. The view that equilibrium is a state of information is much freer of worrisome problems than the older attitude, as may be seen from the following treatment of the wind-tree model.

As in the dog-flea model, we define $w_N(n_1, n_2, n_3, n_4; \tau)$ as the probability that there are n_i wind molecules in direction i, with $\sum n_i = N$, as time $t = \tau/\alpha$. The equation satisfied by these functions is

$$\dot{w}_N(n_1, n_2, n_3, n_4; \tau) = -2N w_N(n_1, n_2, n_3, n_4; \tau) +$$
$$+ (n_1 + 1)[w_N(n_1 + 1, n_2 - 1, n_3, n_4; \tau) + w_N(n_1 + 1, n_2, n_3, n_4 - 1; \tau)] +$$
$$+ (n_2 + 1)[w_N(n_1 - 1, n_2 + 1, n_3, n_4; \tau) + w_N(n_1, n_2 + 1, n_3 - 1, n_4; \tau)] +$$
$$+ (n_3 + 1)[w_N(n_1, n_2 - 1, n_3 + 1, n_4; \tau) + w_N(n_1, n_2, n_3 + 1, n_4 - 1; \tau)] +$$
$$+ (n_4 + 1)[w_N(n_1 - 1, n_2, n_3, n_4 + 1; \tau) + w_N(n_1, n_2, n_3 - 1, n_4 + 1; \tau)] \quad (1.5.17)$$

When $N = 1$, the four possible equations reduce to a set of the same form as Eq. (1.5.13), with solution equivalent to Eq. (1.5.14). Since the collisions are independent, the solution to these equations can be used to form a generating function from the one-molecule solutions. The result is, again with $c \equiv \cosh \tau$ and $s \equiv \sinh \tau$,

$$w_N(n_1, n_2, n_3, n_4; \tau) =$$
$$\left(e^{-2\tau} cs\right)^N \sum_{q_i=0}^{n_i} \sum_{r_i=0}^{q_i} \sum_{a_i=0}^{r_i} (N - k_1; a_1, a_2, a_3, a_4)(k_1 - k_2; r_1 - a_1, r_2 - a_2, r_3 - a_3, r_4 - a_4)$$
$$\times (k_2 - k_3; q_1 - r_1, q_2 - r_2, q_3 - r_3, q_4 - r_4)(k_3; n_1 - q_1, n_2 - q_2, n_3 - q_3, n_4 - q_4)$$
$$\times (c/s)^{(\gamma_1 - \gamma_2)} w_N(N - k_1, k_1 - k_2, k_2 - k_3, k_3; 0), \quad (1.5.18)$$

where $k_1 = N - \sum a_i$; $k_2 = N - \sum r_i$; $k_3 = N - \sum q_i$; $\gamma_1 = a_1 - a_2 + r_2 - r_3 + q_3 - q_4 + n_4$; $\gamma_2 = a_3 - a_4 + r_4 - r_1 + q_1 - q_2 + n_2$; the expression $(N; a_1, a_2, a_3, a_4)$ is the multinomial coefficient $N!/\prod(a_i!)$, with $\sum a_i = N$, and in all summations, i goes from 1 to 4.

Note that there are $\binom{N+3}{3}$ – the tetrahedral numbers – possible w_N's for each N. For a rather small number of wind molecules, say 200, there are 1,373,701 w_N's, and the matrix corresponding to the 4×4 one in Eq. (1.5.14), where $N = 1$, has well over 10^{12} elements! The probabilities evolve smoothly to $w_N(m_1, m_2, m_3, m_4; \infty) = 4^{-N}(N; m_1, m_2, m_3, m_4)$, the equivalent of Eq. (1.5.11) for the dog-flea model. It may be noted that the wind-tree model is identical to a dog-flea model for four dogs standing at the corners of a square and interchanges forbidden along the diagonals. Evidently a wide variety of such models can be devised, and they are perhaps useful in biological, social, economic, or political contexts, but they have little further use in elucidating the basic ideas of statistical thermophysics. In their relationship to random walks on a lattice, however, they recur in the study of phase transitions.

As in the radioactive decay problem, the average populations follow the predictions of the deterministic equations, but in both the dog-flea and the wind-tree models, the probabilities evolve to a time-independent set that maximizes the uncertainty $S = -c \sum w_i \ln w_i$. The probabilities are greatest in the region of equipartition, as is intuitively anticipated, and the expectation values are exactly those of equipartition. The probability of the system's being in a state far from equipartition becomes very small, decreasing as the departure from equipartition increases.

Thus the treatment of the approach to equilibrium in terms of probabilities does everything that is intuitively expected of it – giving time independence and equipartition –

as well as providing a conceptually different quantity, S, the information entropy, that does what Boltzmann's H was supposed to do (except for direction), and evolves smoothly in time to its maximum value. The next test for the information-theoretic S function is to see what it will do in a physically meaningful example.

1.6 CONTINUOUS PROBABILITY DISTRIBUTIONS

The probability sets of the previous examples were summable over discrete indices. In many problems of interest, however, it is convenient to replace discrete indices by continuous variables and sums by integrals. The probability density of a continuous variable x, denoted by $\rho(x)$, is defined such that the probability that x is found in the infinitesimal interval dx around x is $\rho(x)dx$ with

$$\int \rho(x)\, dx = 1 \qquad (1.6.1)$$

being equivalent to $\sum p_i = 1$, where the integral is taken over the permitted values of x. The counterpart of Eq. (1.2.5) is

$$\int f(x)\rho(x)\, dx = \langle f \rangle, \qquad (1.6.2)$$

the expectation value, or average, of some function of x.

In physics a common function of this type is the Liouville function, defined over the phase space of a system of particles. The phase space of an N-particle system is the $6N$-dimensional space with N orthogonal x axes, one for each particle, and similarly for y, z, p_x, p_y and p_z. A single point in the space represents the locations of all the N particles in the system. The probability density $\rho_N(\{\mathbf{r}_i, \mathbf{p}_i\})$ represents our knowledge of the locations of the particles. The probability that the particles of the system have their coordinates and momenta in infinitesimal intervals around the values $\{\mathbf{r}_i, \mathbf{p}_i\}$ is given by

$$\rho_N(\{\mathbf{r}_i, \mathbf{p}_i\}) \prod_{i=1}^{N} \mathbf{dx}_i\, \mathbf{dp}_i. \qquad (1.6.3)$$

The integral of this expression over the entire phase space gives unity, and the average of a function of any, or all, of the coordinates and momenta may be found by direct integration in accord with Eq. (1.6.2).

Since ρ_N is a conserved quantity (its integral over phase space is always unity), it must satisfy a continuity equation

$$\frac{\partial \rho_N}{\partial t} + \sum_{i=1}^{N} \frac{\partial(\rho_N \dot{\mathbf{x}}_i)}{\partial \mathbf{x}_i} + \sum_{i=1}^{N} \frac{\partial(\rho_N \dot{\mathbf{p}}_i)}{\partial \mathbf{p}_i} = 0, \qquad (1.6.4)$$

where the $\dot{x}_i$ and $\dot{p}_i$ are the local velocity components of the probability fluid, and the independent variables are the coordinates, momenta, and time. This equation is analogous to the ordinary continuity equation

$$\frac{\partial \rho}{\partial t} + \nabla \cdot (\rho \mathbf{V}) = 0$$

for a conserved fluid with no sources or sinks. Every particle is assumed to move according to the equations of motion:

$$\dot{\mathbf{p}}_i = -\frac{\partial \mathcal{H}\left(\{\mathbf{p}_i, \mathbf{x}_i\}, t\right)}{\partial \mathbf{x}_i},$$

$$\dot{\mathbf{x}}_i = \frac{\partial \mathcal{H}\left(\{\mathbf{p}_i, \mathbf{x}_i\}, t\right)}{\partial \mathbf{p}_i},$$

where $\mathcal{H}$ is the Hamiltonian of the system. The motions of the representative points are almost everywhere regular and diffusionless.

The second and third terms of Eq. (1.6.4) contain the expressions

$$\sum_{i=1}^{N} \rho_N \left(\frac{\partial \dot{\mathbf{x}}_i}{\partial \mathbf{x}_i} + \frac{\partial \dot{\mathbf{p}}_i}{\partial \mathbf{p}_i}\right) = \sum_{i=1}^{N} \rho_N \left(\frac{\partial^2 \mathcal{H}}{\partial \mathbf{x}_i \partial \mathbf{p}_i} - \frac{\partial^2 \mathcal{H}}{\partial \mathbf{p}_i \partial \mathbf{x}_i}\right) = 0. \qquad (1.6.5)$$

Therefore Eq. (1.6.4) can be written

$$\frac{\partial \rho_N}{\partial t} + \sum_i \left(\dot{\mathbf{x}}_i \frac{\partial \rho_N}{\partial \mathbf{x}_i} + \dot{\mathbf{p}}_i \frac{\partial \rho_N}{\partial \mathbf{p}_i}\right) \equiv \frac{d\rho_N}{dt} = 0,$$

or

$$\frac{\partial \rho_N}{\partial t} + \{\rho_N, \mathcal{H}\} = 0, \qquad (1.6.6)$$

where $\{\rho_N, \mathcal{H}\}$ is the classical Poisson bracket.

The Liouville theorem, Eq. (1.6.6), states that the total time derivative of the probability density in phase space, or Liouville function, is zero; ρ_N is therefore a constant during the motion of the system.

In order to form the continuous-function analog of the uncertainty S, as given by Eq. (1.4.1), we note that the argument of the ln is a probability, not a probability density. It is therefore convenient to multiply ρ_N by the volume of a small arbitrary cell in phase space to obtain a true probability for use in the ln factor. Usually this cell volume is denoted by h^{3N}, where h has the units of action, $\delta x \delta p$. The choice of the letter h was not accidental; it will turn out that Planck's constant of action is almost always either the proper choice for h or else an acceptable choice in situations where the cell size is unimportant. In quantum systems, h is the basic cell size determined by the uncertainty principle. In coordinate transformations, it is necessary that the new variables scale according to $\Delta x \Delta p \geq h$ in order for S to be invariant. (This is not the full story. See Section 1.9.4 for further comments.)

The analog to Eq. (1.4.1) then becomes

$$S = -c \int \rho_N \ln\left(h^{3N}\rho_N\right) d\Gamma_N, \qquad (1.6.7)$$

where $d\Gamma_N$ is the $6N$-dimensional phase space volume element.

Because of the Liouville theorem, Eq. (1.6.6), it is evident that S in Eq. (1.6.7) does not change in time; therefore it cannot evolve to an equilibrium maximum. Partly on this basis, the Gibbs expression for entropy (essentially Eq. (1.6.7)) (Gibbs [1960]) was rejected by a number of leading physicists of his time. In terms of information theory, the constancy of S

in Eq. (1.6.7) says that if we know the dynamics exactly, then there is no loss of information as the system evolves in time. Since the equations of motion are time reversible, there is evidently no state of *dynamics* that corresponds to thermodynamic equilibrium, and there is no reason to regard any particular set of coordinates and momenta (a particular point in Γ space) as being representative of equilibrium while another is not. It is for this reason that we seek equilibrium outside the formalism of dynamics.

Gibbs offered an analogy to the behavior of the Liouville function for an isolated system in terms of an immiscible black dye in a white background fluid of the same density. Initially the dye drop may be well localized and easily defined, but as time goes on the shape of the drop becomes increasingly filamentary until it becomes so drawn out and tangled that the eye can no longer separate the dye from the background fluid. There is no diffusion of the dye into the surrounding fluid, but a "coarse-grained" examination fails to reveal the two as separate. The reversal of time untangles the dye filaments and returns the drop to its original form.

A mathematically different and physically more suitable picture, to be offered in Section 1.7, may again be described in terms of Gibbs's dye drop, but this time miscible. As time goes on, the drop moves with the fluid but also diffuses into it, ultimately spreading so as to color it uniformly. Since this is really the picture wanted, it is certainly a proper objective to develop a mathematical formalism that is compatible with this picture and free of the objections to the formulation attributed to Gibbs. A careful reading of his book (Gibbs [1960, pp. 143-144]) shows that he clearly understood the problem. (See Section 9 for further comments.)

1.7 COUPLED HARMONIC OSCILLATORS

A typical situation in thermodynamics is that of a finite, uniform, bounded body of matter, called the *system*, in contact (so that energy can be transferred) with an infinite and unchanging body of matter called a *heat bath*. The system may be initially at any temperature whatever, but ultimately it reaches that of the heat bath. The problem to be treated in this section is an entirely mechanical analog of this thermodynamic situation, chosen to show the approach to equilibrium, equipartition of energy, and the evolution of the uncertainty (later, the entropy) to a value determined entirely by the heat bath.

As shown in Fig. 2, the system is a segment of an infinite chain of identical masses m, coupled to nearest neighbors by identical springs of constant k, each anchored to its home position by a leaf spring of constant K. In the treatment of the dynamics, nothing distinguishes the system from the heat-bath, and the entire infinite chain is regarded as an entity. The equation of motion for the nth mass is

$$m\ddot{x}_n = -Kx_n + k\left(x_{n+1} - x_n\right) + k\left(x_{n-1} - x_n\right),$$

or with

$$\Omega^2 = \left(K + 2k\right)/m, \ \omega^2 = k/m, \ \text{and} \ \gamma = k/\left(K + 2k\right),$$

$$\ddot{x}_n = -\Omega^2 x_n + \gamma\Omega^2\left(x_{n-1} + x_{n+1}\right). \tag{1.7.1}$$

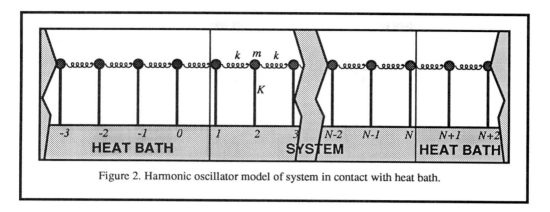

Figure 2. Harmonic oscillator model of system in contact with heat bath.

The easiest way to solve this set of equations is by means of a generating function (See Appendix A), defined as

$$G(t, \ s) = \sum_{n=-\infty}^{\infty} x_n(t)s^n. \tag{1.7.2}$$

The second partial derivative of G with respect to t is seen to be

$$\frac{\partial^2 G}{\partial t^2} = -\Omega^2 \left[1 - \gamma \left(s + \frac{1}{s} \right) \right] G, \tag{1.7.3}$$

which can be solved by inspection to give

$$G(t,s) = G(0,s) \cos \left[\Omega t \sqrt{1 - \gamma(s + s^{-1})} \right] + \frac{\partial G(0,s)}{\partial t} \frac{\sin \left[\Omega t \sqrt{1 - \gamma(s + s^{-1})} \right]}{\Omega \sqrt{1 - \gamma(s + s^{-1})}}. \tag{1.7.4}$$

$G(0, s)$ and $\partial G(0, s)/\partial t$ are seen from Eq. (1.7.2) to contain the initial coordinates and velocities. In order to recover a particular $x_n(t)$ from this expression, a simple contour integral produces the required result as

$$x_r(t) = \frac{1}{2\pi i} \oint \frac{G(t,s)\,ds}{s^{r+1}}$$

With the substitution $s = e^{i\theta}$ and the use of Eq. (1.7.2) in Eq. (1.7.4), it follows that

$$x_r(t) = \frac{1}{2\pi} \int_0^{2\pi} d\theta \sum_{n=-\infty}^{\infty} \left\{ x_n(0)e^{i(n-r)\theta} \cos \left[\Omega t \sqrt{1 - 2\gamma \cos\theta} \right] \right.$$

$$\left. + \dot{x}_n(0)e^{i(n-r)\theta} \frac{\sin \left[\Omega t \sqrt{1 - 2\gamma \cos\theta} \right]}{\Omega \sqrt{1 - 2\gamma \cos\theta}} \right\}. \tag{1.7.5}$$

By changing n to $n + r$, Eq. (1.7.5) may be rewritten more conveniently as

$$x_r(t) = \sum_{n=-\infty}^{\infty} [x_{n+r}(0)f_n(t) + p_{n+r}(0)g_n(t)/m\Omega],$$

and

$$p_r(t) = m\dot{x}_r(t), \tag{1.7.6}$$

where

$$f_n(t) = \frac{1}{\pi} \int_0^\pi d\theta \cos n\theta \cos\left[\Omega t\sqrt{1 - 2\gamma\cos\theta}\right], \tag{1.7.7}$$

and

$$g_n(t) = \Omega \int_0^t f_n(t')\,dt'. \tag{1.7.8}$$

The integrals were modified from Eq. (1.7.6) to Eq. (1.7.7) by retaining only the even part of the integrand, since the odd part vanishes, and using twice the integral over half the total angle.

An exact treatment of the problem in terms of the functions f_n and g_n defined by Eqs.(1.7.7) and (1.7.8) is possible (Huerta, Robertson and Nearing [1971]). For present purposes, however, the choice of very weak coupling ($k \ll K$, or $0 < \gamma \ll 1/2$) is made to permit each oscillator to have, very nearly, its own energy. Therefore an approximation to f_n and g_n based on the small-γ expansion of $\sqrt{1 - 2\gamma\cos\theta}$ is both valid and considerably simpler. Thus when $\sqrt{1 - 2\gamma\cos\theta}$ is replaced by $1 - \gamma\cos\theta$ in the integrand of Eq. (1.7.7), the resulting approximation is

$$f_n(t) \approx J_n(\gamma\Omega t)\cos(\Omega t - n\pi/2). \tag{1.7.9}$$

A similar treatment of Eq. (1.7.8) yields

$$g_n(t) \approx J_n(\gamma\Omega t)\sin(\Omega t - n\pi/2). \tag{1.7.10}$$

These equations show oscillation at the angular frequency Ω, with $\pi/2$ phase difference between the coordinate function $f_n(t)$ and its corresponding momentum function $g_n(t)$, as expected, with a slowly varying modulation function $J_n(\gamma\Omega t)$, that exhibits the dynamically interesting features. In order for an individual oscillator to reach equilibrium with the others in any sense, it is necessary that the influence of its own initial conditions on its behavior must ultimately vanish. The contribution of the nth oscillator's initial conditions to its position is seen to be contained in the terms

$$x_n(0)f_0(t) + p_n(0)g_0(t)/m\Omega = x_n(0)J_0(\gamma\Omega t)\cos\Omega t + [p_n(0)/m\Omega]\,J_0(\gamma\Omega t)\sin\Omega t,$$

both of which die out at large t, as is well known from the behavior of the Bessel function. Even more strongly, since $J_n(x)$ goes as $|x|^{-1/2}$ for $|x| \gg n$, it is apparent that any finite segment of the chain must evolve to a state that is determined entirely by the initial conditions of increasingly remote parts of the surrounding oscillator chain, so that at least the dynamic requirements for equilibration are met.

The one other feature of note in the dynamics is the time reversibility of Eq. (1.7.6), which of course follows from that of Eq. (1.7.1), when it is noted that the $p_n(0)$ terms in Eq. (1.7.6) change sign on time reversal. A film of the motion of the chain would make sense when run in either direction. Apparently, then, the approach to equilibrium of this system cannot lie entirely in the dynamics, though the loss-of-memory feature discussed in the previous paragraph is essential in order that equilibration is not dynamically precluded.

The dynamics has now said all it has to say. Note that Eq. (1.7.6) is not statistical; it gives a definite value for $x_n(t)$ in terms of all the (presumably known) initial conditions. At this point it becomes necessary to introduce the state of our knowledge of the system into the calculation in terms of the initial conditions. The only statistical input to this development is the description of our knowledge of the initial conditions. The simplest choice, and the one to be used here, is that each initial coordinate and momentum in the heat bath is specified by its own independent Gaussian distribution, while those of the system may be chosen as displaced Gaussians with small variances to indicate that the system initial conditions are quite well known.

It is convenient to present the initial distributions of the system variables by writing

$$x'_n(0) = x_n(0) - u_n,$$

$$p'_n(0) = p_n(0) - v_n, \tag{1.7.11}$$

where u_n and v_n are the initial expectation values of the system variables. In terms of these variables, the initial value of the Liouville function, or probability density of the entire phase space of the chain, is chosen to be

$$
\begin{aligned}
\rho_0 &\equiv \rho[\{x_n(0)\}, \{p_n(0)\}] \\
&= \prod_{n=1}^{N} \frac{\exp\{-[x'_n(0)/\alpha]^2/2\}}{\alpha\sqrt{2\pi}} \prod_{n=1}^{N} \frac{\exp\{-[p'_n(0)/\delta]^2/2\}}{\delta\sqrt{2\pi}} \\
&\times \prod{}' \frac{\exp[-x_n^2(0)/2\epsilon^2]}{\epsilon\sqrt{2\pi}} \prod{}' \frac{\exp[-p_n^2(0)/2\zeta^2]}{\zeta\sqrt{2\pi}},
\end{aligned}
\tag{1.7.12}
$$

where α^2 and δ^2 are the initial variances of the system coordinates and momenta respectively, $\prod'$ denotes the product over all members of the chain outside the system (numbered 1-N), and ϵ^2 and ζ^2 are the initial variances of the heat-bath variables.

Any function of the dynamical variables $\{x_n(t)\}$ and $\{p_n(t)\}$ can be averaged by expressing the variables in terms of their initial values by the use of Eq. (1.7.6), and averaging as

$$\langle F[\{x_n(t), p_n(t)\}] \rangle = \int \rho_0 F[\{x_n, p_n\}] \prod_{n=-\infty}^{\infty} dx_n(0)\, dp_n(0), \tag{1.7.13}$$

with the understanding that the system variables are usually most conveniently primed. An immediate application of Eq. (1.7.13) yields, for all n,

$$\langle x_n(t) \rangle = \sum_{r=1}^{N} \left[u_r f_{n-r}(t) + (v_r/m\Omega)\, g_{n-r}(t) \right].$$

and

$$\langle p_n(t) \rangle = m \sum_{r=1}^{N} \left[u_r \dot{f}_{n-r}(t) + (v_r/m\Omega) \dot{g}_{n-r}(t) \right].$$
(1.7.14)

The primed variables may now be defined for all t as

$$x_n'(t) = x_n(t) - \langle x_n(t) \rangle,$$

$$p_n'(t) = p_n(t) - \langle p_n(t) \rangle.$$
(1.7.15)

The objectives of this section are to calculate a probability density for the system variables and to show that this function leads to a correct and useful state of equilibrium. The Liouville function of the entire chain is independent of time, as shown in Section 6, but the *reduced* Liouville function $\rho_N(\{x_n, p_n\}, t)$, which is a function only of the system variables, is time dependent. It could, in principle, be calculated simply by integration of the Liouville function for the entire chain over the heat bath variables, but in practice this calculation is usually not convenient.

If we define a column vector $Y(t)$ in the centered system variables, such that its transpose is $\tilde{Y} = (x_1' x_2' \ldots x_N' p_1' \ldots p_N')$, with $x_i' = x_i'(t)$, we can denote the reduced Liouville function as $\rho_N(Y)$.

The characteristic function for the system variables is defined as

$$\Phi_N(Z) = \int \exp(i\tilde{Y}Z) \rho(\{x_n(t), p_n(t)\}) \prod_{n=-\infty}^{\infty} dx_n(t)\, dp_n(t),$$
(1.7.16)

where Z is a column vector of $2N$ Fourier-transform variables, $\tilde{Z} = (k_1 \ldots k_N\, q_1 \ldots q_N)$, that soon disappear from the calculation. As Eq. (1.7.16) shows, $\Phi_N(Z)$ is nothing more than the average of $\exp(i\tilde{Y}Z)$, an average that can be found by means of Eq. (1.7.13) as

$$\Phi_N(Z) = \int \exp(i\tilde{Y}Z) \rho_0 \prod_{n=-\infty}^{\infty} dx_n(0)\, dp_n(0),$$
(1.7.17)

where it is necessary to express the components of $\tilde{Y}$ in terms of the initial conditions by means of Eq. (1.7.6). The reason for calculating $\Phi_N(Z)$ is that its Fourier inverse is $\rho_N(Y)$, the function sought, or

$$\rho_N(Y) = \int \exp(-i\tilde{Z}Y) \Phi_N(Z) \prod_{j=1}^{2N} \left(\frac{dZ_j}{2\pi} \right).$$
(1.7.18)

✳ Details of the calculation, included here for reference, may be deferred indefinitely, and the conceptual development taken up with Eq. (1.7.51). The terms needed in $\tilde{Y}Z$, Eq. (1.7.17) are

$$\sum_{i=1}^{N} k_i x_i'(t) = \sum_{n=-\infty}^{\infty} \{ x_n'(0) \sum_{i=1}^{N} k_i f_{n-i}(t) + [p_n'(0)/m\Omega] \sum_{i=1}^{N} k_i g_{n-i}(t) \},$$
(1.7.19)

and

$$\sum_{n=1}^{N} q_i p_i'(t) = m \sum_{n=-\infty}^{\infty} \{x_n'(0) \sum_{i=1}^{N} q_i \dot{f}_{n-i}(t) + [p_n'(0)/m\Omega] \sum_{i=1}^{N} q_i \dot{g}_{n-i}(t)\}. \tag{1.7.20}$$

With these expressions and the standard formula

$$\int_{-\infty}^{\infty} \exp\left(-ax^2 + bx\right) dx = \left(\frac{\pi}{a}\right)^{1/2} \exp\left(\frac{b^2}{4a}\right),$$

the characteristic function can be calculated to be

$$\Phi_N(Z) = \prod_{r=1}^{4} P_r, \tag{1.7.21}$$

where

$$P_1 = \exp\left([(\epsilon^2 - \alpha^2)/2] \sum_{n=1}^{N} \{\sum_{j=1}^{N} [k_j f_{n-j}(t) + m q_j \dot{f}_{n-j}(t)]\}^2\right), \tag{1.7.22}$$

$$P_2 = \exp\left([(\xi^2 - \delta^2)/2m^2\Omega^2] \sum_{n-1}^{N} \{\sum_{j=1}^{N} [k_j g_{n-j}(t) + m q_j \dot{g}_{n-j}(t)]\}^2\right), \tag{1.7.23}$$

$$P_3 = \exp\left([-\epsilon^2/2] \sum_{n=-\infty}^{\infty} \{\sum_{j=1}^{N} [k_j f_{n-j}(t) + m q_j \dot{f}_{n-j}(t)]\}^2\right), \tag{1.7.24}$$

and

$$P_4 = \exp\left([-\zeta^2/2m^2\Omega^2] \sum_{n=-\infty}^{\infty} \{\sum_{j=1}^{N} [k_j g_{n-j}(t) + m q_j \dot{g}_{n-j}(t)]\}^2\right). \tag{1.7.25}$$

The double finite sum in Eq. (1.7.22) can be replaced by

$$\sum_{i,j=1}^{N} [k_i k_j A_{ij}(t) + 2 k_i q_j B_{ij}(t) + q_i q_j C_{ij}(t)],$$

where

$$A_{ij}(t) = \sum_{n=1}^{N} f_{n-i}(t) f_{n-j}(t) = A_{ji}(t), \tag{1.7.26}$$

$$B_{ij}(t) = m \sum_{n=1}^{N} f_{n-i}(t) \dot{f}_{n-j}(t), \tag{1.7.27}$$

and

$$C_{ij} = m^2 \sum_{n=1}^{N} \dot{f}_{n-i}(t) \dot{f}_{n-j}(t) = C_{ji}(t). \tag{1.7.28}$$

The sum in Eq. (1.7.24) is replaced in exactly the same way by the expression

$$\sum_{i,j=1}^{N} [k_i k_j a_{ij}(t) + 2 k_i q_j b_{ij}(t) + q_i q_j c_{ij}(t)],$$

where a_{ij}, b_{ij}, and c_{ij} differ from A_{ij}, B_{ij}, and C_{ij} only by having $\sum_{n=-\infty}^{\infty}$. Also, since $f_j = f_{-j}$, it may be shown that $b_{ij} = b_{ji}$, a result not true for B_{ij}. From well-known properties of the Bessel functions, it is not difficult to obtain

$$2a_{ij}(t) = \delta_{ij} + J_{i-j}(2\gamma\Omega t)\cos[2\Omega t - (i-j)\pi/2], \tag{1.7.29}$$

$$2b_{ij}(t) = -m\Omega J_{i-j}(2\gamma\Omega t)\sin[2\Omega t - (i-j)\pi/2], \tag{1.7.30}$$

and

$$2c_{ij}(t) = (m\Omega)^2\delta_{ij} - (m\Omega)^2 J_{i-j}(2\gamma\Omega t)\cos[2\Omega t - (i-j)\pi/2]. \tag{1.7.31}$$

The double finite sum in Eq. (1.7.23) is replaced by

$$\sum_{i,j=1}^{N}[k_1 k_j D_{ij}(t) + 2k_i q_j E_{ij}(t) + q_i q_j A_{ij}(t)],$$

where Aij appears again because $\dot{g}_j = \Omega f_j$, and

$$D_{ij}(t) = (m\Omega)^{-2}\sum_{n=1}^{N} g_{n-i}(t)g_{n-j}(t) = D_{ji}(t), \tag{1.7.32}$$

$$E_{ij}(t) = m^{-1}\Omega^{-2}\sum_{n=1}^{N} g_{n-i}(t)\dot{g}_{n-j}(t). \tag{1.7.33}$$

Similarly, the double sum in Eq. (1.7.25) is replaced by

$$\sum_{i,j=1}^{N}[k_i k_j d_{ij}(t) + 2k_i q_j e_{ij}(t) + q_i q_j a_{ij}(t)]$$

again with finite sum replaced by an infinite one for d_{ij} and e_{ij}. The Bessel function sums yield

$$d_{ij}(t) = (m\Omega)^{-4} c_{ij}(t), \tag{1.7.34}$$

and

$$e_{ij}(t) = -(m\Omega)^{-2} b_{ij}(t). \tag{1.7.35}$$

These results in Eq. (1.7.21) now give

$$\Phi_N(Z) = \exp\left[-\frac{1}{2}\sum_{i,j=1}^{N}(k_i k_j M_j + k_i q_j G_{ij} + q_i q_j Q_{ij})\right], \tag{1.7.36}$$

where

$$M_{ij} = (\alpha^2 - \epsilon^2) A_{ij} + \epsilon^2 a_{ij} + (\delta^2 - \zeta^2) D_{ij} + \zeta^2 d_{ij} = M_{ji}, \tag{1.7.37}$$

$$G_{ij} = (\alpha^2 - \epsilon^2) B_{ij} + \epsilon^2 b_{ij} + (\delta^2 - \zeta^2) E_{ij} + \zeta^2 e_{ij} \neq G_{ji}, \tag{1.7.38}$$

and

$$Q_{ij} = (\alpha^2 - \epsilon^2) C_{ij} + \epsilon^2 c_{ij} + (\delta^2 - \zeta^2) A_{ij} + \zeta^2 a_{ij} = Q_{ji}. \tag{1.7.39}$$

The quantities M_{ij}, G_{ij} and Q_{ij} may be seen by direct calculation to be the covariances:

$$M_{ij} = \langle x'_i(t) x'_j(t) \rangle, \tag{1.7.40}$$

$$G_{ij} = \langle x'_i(t) p'_j(t) \rangle, \tag{1.7.41}$$

and

$$Q_{ij} = \langle p'_i(t) p'_j(t) \rangle, \tag{1.7.42}$$

and the matrix

$$W = \begin{pmatrix} M & G \\ \tilde{G} & Q \end{pmatrix}, \tag{1.7.43}$$

with $M = (M_{ij})$, etc., is called the covariance matrix. Details about covariances may be found in Cramér [1946], Kullback [1968], and Miller [1964]. In terms of W, $\Phi_N(Z)$ may be written as

$$\Phi_N(Z) = \exp[-\tilde{Z} W Z / 2].$$

The reduced Liouville function, from Eq. (1.7.18), is

$$\rho_N(Y) = \int \exp[-i\tilde{Z}Y - \tilde{Z}WZ/2] \prod_{n=1}^{N} (dk_n/2\pi)(dq_n/2\pi). \tag{1.7.44}$$

Since W is a real, symmetric matrix, it can be diagonalized by an orthogonal transformation, say L, so that

$$L^{-1} W L = \Lambda,$$

where Λ is a diagonal matrix, with elements $\lambda_1, \ldots, \lambda_{2N}$. We can then write

$$\tilde{Z} W Z = \tilde{Z} L L^{-1} W L L^{-1} Z = \tilde{Z} L \Lambda L^{-1} Z = \tilde{Z} L \Lambda \tilde{L} Z, \tag{1.7.45}$$

since $L^{-1} = \tilde{L}$. Let $\tilde{L}Z = J$, a new column vector that simplifies Eq. (1.7.45) to

$$\tilde{Z} W Z = \tilde{J} \Lambda J.$$

Similarly write $\tilde{Z}Y = \tilde{Z} L \tilde{L} Y = \tilde{J} X$, where $\tilde{L}Y = X$, so that Eq. (1.7.44) becomes

$$\rho_N(Y) = \int \exp[-i\tilde{J}X - \tilde{J}\Lambda J/2] \prod_{n=1}^{N} (dk_n/2\pi)(dq_n/2\pi). \tag{1.7.46}$$

The variables of integration can be changed from $\prod_{n=1}^{N} dk_n dq_n$ to $\prod_{n=1}^{2N} dJ_n$. The Jacobian of the transformation is simply $|\det L| = 1$. Since Λ is diagonal, Eq. (1.7.46) becomes

$$\rho_N = \prod_{n=1}^{2N} \int \exp[-iJ_n X_n - J_n^2 \lambda_n / 2](dJ_n/2\pi)$$

$$= \prod_{n=1}^{2N} (2\pi\lambda_n)^{-1/2} \exp[-X_n^2 / 2\lambda_n]. \tag{1.7.47}$$

But since we know that

$$\prod_{n=1}^{2N} \lambda_n = \det W, \tag{1.7.48}$$

and

$$\prod_{n=1}^{2N} \exp[-X_n^2/2\lambda_n] = \exp\left[-\sum_{n=1}^{2N} X_n^2/2\lambda_n\right] = \exp[-\tilde{X}\Lambda^{-1}X/2], \tag{1.7.49}$$

and since

$$\tilde{X}\Lambda^{-1}X = \tilde{Y}L\Lambda^{-1}L^{-1}Y = \tilde{Y}W^{-1}Y, \tag{1.7.50}$$

we may finally write

$$\rho_N(Y) = (2\pi)^{-N}(\det W)^{-1/2}\exp[-\tilde{Y}W^{-1}Y/2]. \tag{1.7.51}$$

The uncertainty S from Eq. (1.6.7), with k written in place of c, is

$$S_N = -k\int\rho_N(Y)\ln\left(h^N\rho_N\right)\prod_{j=1}^{N} dx'_j\, dp'_j, \tag{1.7.52}$$

which, after straightforward calculations, becomes

$$S_N = Nk + k\ln[\hbar^{-N}(\det W)^{1/2}], \tag{1.7.53}$$

where $\hbar = h/2\pi$, and h so far is only dimensionally equal to Planck's constant. Thus the uncertainty is a function only of the covariance matrix W. The quantity S_N will be shown later to evolve into the equilibrium-thermodynamic entropy of a classical system of harmonic oscillators.

Even at this point, it is evident that S_N does not contain the initial expectation values of system variables *at any time*. The initial expectation values are part of the dynamic description of the system. If the values are well known, with small variances $\left(\alpha^2 \text{ and } \delta^2\right)$, the motion of the system may be described dynamically for some time after $t = 0$, but finally the dynamical picture will begin to blur; our knowledge of the coordinates and momenta will become increasingly fuzzy; expectation values all become zero-centered, independent of time, as may be seen from Eq. (1.7.14) and properties of finite sums of Bessel functions; and it will now be shown that the properties of the system ultimately become indistinguishable from those of the heat bath.

From ρ_0, Eq. (1.7.12), it may be seen that the initial kinetic energy of an oscillator in the heat bath is

$$\langle p_n^2(0)\rangle/2m = \zeta^2/2m,\ n \ni 1, N, \tag{1.7.54}$$

and the potential energy is approximately (for $k \ll K$),

$$m\Omega^2\langle x_n^2(0)\rangle/2 = m\Omega^2\epsilon^2/2. \tag{1.7.55}$$

The expectation values of the kinetic and potential energies of system variables may be computed from Eq. (1.7.51) or read off from Eqs.(1.7.37) through (1.7.42) as

$$\langle p'^2_n(t)\rangle/2m = Q_{nn}/2m, \tag{1.7.56}$$

and

$$m\Omega^2\langle x'^2_n(t)\rangle/2 = M_{nn}m\Omega^2/2. \tag{1.7.57}$$

As $t \to \infty$, the finite sums of Bessel functions, given by the upper-case letters in Eqs.(1.7.37) to (1.7.42), vanish, as do the average values that distinguish x'_n, p'_n. Thus it may be seen that as $t \to \infty$, these energies become

$$\langle p_n^2 \rangle / 2m = \left(m\Omega^2 \epsilon^2 / 4 \right) \left[1 - J_0(2\gamma\Omega t) \cos(2\Omega t) \right] + \left(\zeta^2 / 4m \right) \left[1 + J_0(2\gamma\Omega t) \cos 2\Omega t \right]$$
$$\to (1/2)[\zeta^2/2m + m\Omega^2\epsilon^2/2], \tag{1.7.58}$$

and

$$m\Omega^2 \langle x_n^2 \rangle / 2 = \left(m\Omega^2 \epsilon^2 / 4 \right) \left[1 + J_0(2\gamma\Omega t) \cos 2\Omega t \right] + \left(\zeta^2 / 4m \right) \left[1 - J_0(2\gamma\Omega t) \cos 2\Omega t \right]$$
$$\to (1/2)[\zeta^2/2m + m\Omega^2\epsilon^2/2]. \tag{1.7.59}$$

We have found, entirely from considerations of dynamics and our knowledge of initial conditions, equipartition of the kinetic and potential energies of all the system variables; moreover the equilibrium kinetic (or potential) energy of each oscillator *in the system* is the arithmetic mean of the *initial* kinetic and potential energies *in the heat bath*. Since the system could have been any part of the infinite chain, the conclusion is that the expectation value of the energy of every oscillator in the chain is the same. Initially, the kinetic and potential energies outside the system were not equipartitioned, as shown by Eqs.(1.7.54) and (1.7.55), and the energies inside the system could be chosen at will. The initial kinetic energy of a system mass is

$$\langle p_n^2(0) \rangle / 2m = \left(\delta^2 + v_n^2 \right) / 2m, \ 0 \le n \le N, \tag{1.7.60}$$

and the potential energy is

$$m\Omega^2 \langle x_n^2(0) \rangle / 2m = m\Omega^2 \left(\alpha^2 + u_n^2 \right) / 2, \ 0 \le n \le N. \tag{1.7.61}$$

The terms in u_n and v_n are strictly dynamical, *i.e.*, they are the normal mechanical ones. The terms that arise from the variances α^2 and δ^2 are to be identified conceptually as *thermal* energies.

In thermodynamics, a basic distinction is made between energy transfers called *heat* and those called *work*. Although the distinction is often poorly presented, or given in terms of circular definitions, or couched in 19th century terms that refuse to recognize the underlying microstructure of matter, the conceptual basis for distinction is that work involves calculable changes in macroscopic dynamic variables and/or is expressible in recognized macroscopic mechanical or other work terms, whereas heat is energy transfer that cannot be so identified. Heat comprises transfers of energy to or from the microscopic motions of the system, with the typical result that heat leads to changes in the variances of system microvariables. These variances are conveniently represented in terms of temperature.

We can tentatively define temperatures for the coupled-oscillator system and verify that the resulting thermal behavior is in accord with expectations. Accordingly, we can write the initial thermal kinetic and potential energies in terms of kinetic and potential temperatures, as

$$\begin{aligned} &{}^1\!/_2 k T_{ks}(0) = {}^1\!/_2 \delta^2/m, \ \text{and} \\ &{}^1\!/_2 k T_{ps}(0) = {}^1\!/_2 m\Omega^2 \alpha^2, \end{aligned} \tag{1.7.62}$$

for the system, and

$$\begin{aligned} &{}^1\!/_2 k T_{kb}(0) = {}^1\!/_2 \zeta^2/m, \ \text{and} \\ &{}^1\!/_2 k T_{pb}(0) = {}^1\!/_2 m\Omega^2 \epsilon^2, \end{aligned} \tag{1.7.63}$$

for the heat bath, where k is Boltzmann's constant. If the final temperatures are written similarly as

$$\begin{aligned} {}^{1}\!/_{2}kT_{ks}\left(\infty\right) &= \langle p_n^2\rangle_\infty/2m, \quad \text{and} \\ {}^{1}\!/_{2}kT_{ps}\left(\infty\right) &= m\Omega^2\langle x_n^2\rangle_\infty/2, \end{aligned} \tag{1.7.64}$$

then Eqs.(1.7.58) and (1.7.59) give the equilibrium temperature T_e as

$$T_e = T_{ks}\left(\infty\right) = T_{ps}\left(\infty\right) = ({}^{1}\!/_{2})\left[T_{kb}(0) + T_{pb}(0)\right], \tag{1.7.65}$$

showing that at equilibrium the chain has reached a uniform temperature throughout. This is precisely the result expected for thermal equilibrium.

The behavior of S_N in Eq. (1.7.53) as a function of time is interesting and instructive. At $t = 0$, the covariance matrix W is diagonal, and we have

$$\begin{aligned} (\det W)^{1/2} &= \langle x_n'^2(0)\rangle^{N/2}\langle p_n'^2(0)\rangle^{N/2} \\ &= (\alpha\delta)^N = \left[k\sqrt{T_{ks}(0)T_{ps}(0)}/\Omega\,\right]^N, \end{aligned} \tag{1.7.66}$$

and

$$S_N(0)/Nk = 1 + \ln[kT_0/\hbar\Omega], \tag{1.7.67}$$

where

$$T_0 = \sqrt{T_{ks}(0)T_{ps}(0)}. \tag{1.7.68}$$

At $t = \infty$, since W is once again diagonal, we obtain

$$(\det W)^{1/2} = \left[k\sqrt{T_{ks}\left(\infty\right)T_{ps}\left(\infty\right)/\Omega}\,\right]^N, \tag{1.7.69}$$

and

$$S_N(\infty)/Nk = 1 + \ln[kT_e/\hbar\Omega], \tag{1.7.70}$$

where the equilibrium temperature T_e is that of Eq. (1.7.65). For intermediate times, $S_N(t)$ may be written as

$$S_N(t)/Nk = 1 + \ln[kT_N(t)/\hbar\Omega], \tag{1.7.71}$$

where $T_N(t)$ can be easily related to temperatures for $t = 0$ or $t \to \infty$, but at other times it is merely a notational convenience. There seems to be no apparent general expression for $T_N(t)$, but for the first few N's, with $T_{ks}(0) = T_{ps}(0) = T_0$ and $T_{kb}(0) = T_{pb}(0) = T_b$, the results are

$$T_1(t) = T_b + (T_0 - T_b)\, J_0^2(\gamma\Omega t), \tag{1.7.72}$$

$$T_2(t) = T_b + (T_0 - T_b)\left(J_0^2 + J_1^2\right), \tag{1.7.73}$$

$$\begin{aligned} T_3(t) = \Big[&T_b^3 + T_b^2(T_0 - T_b)(2a_3 + b_3) + T_b(T_0 - T_b)^2(a_3^2 + 2a_3 b_3 - 2c_3^2 - d_3^2) \\ &+ (T_0 - T_b)^3\left[b_3(a_3^2 - d_3^2) - 2c_3^2(a_3 + d_3)\right]\Big]^{1/3}, \end{aligned} \tag{1.7.74}$$

and

$$T_4(t) = \left[T_b^4 + 2T_b^3(T_0 - T_b)(a_4 + b_4) + T_b^2(T_0 - T_b)^2 \left[(a_4 + b_4)^2 - 2(c_4^2 + d_4^2 - a_4 b_4) \right] \right.$$

$$\left. + 2T_b(T_0 - T_b)^3(a_4 + b_4)(a_4 b_4 - c_4^2 - d_4^2) + (T_0 - T_b)^4(c_4^2 + d_4^2 - a_4 b_4)^2 \right]^{1/4}, \quad (1.7.75)$$

with $a_3 = J_0^2 + J_1^2 + J_2^2$, $b_3 = J_0^2 + 2J_1^2$, $c_3 = J_1 J_2$, $d_3 = 2J_0 J_2 - J_1^2$, $a_4 = a_3 + J_3^2$, $b_4 = b_3 + J_2^2$, $c_4 = c_3 + J_2 J_3$, and $d_4 = d_3 + J_1 J_2$, where the arguments of all Bessel functions are $\gamma\Omega t$. These four temperature variables are shown in Fig. 3 for the initial condition $T_b/T_0 = 2$. The behavior of $S_N(t)$ is similar to that of $T_N(t)$. The single-particle temperature function, which could reasonably be called the temperature of the system, starts at T_0, increases to T_b when $\gamma\Omega t = 2.405$, and bounces back to a lower temperature, returning to T_0 at successive zeros of J_0. These pre-equilibrium swings to $T = T_b$, with subsequent bounces, seem at odds with the ideas of the H theorem and the simplistic "time's arrow" concept of entropy. The reason for this behavior is clear, however, when one realizes that for $J_0 = 0$, the system's initial conditions have (at those instants) no influence on its behavior, and its motion is determined entirely by the initial conditions of heat-bath variables. At other times, the system's influence *on* the heat bath returns as an echo *from* the heat bath to reduce its temperature, but ever more feebly.

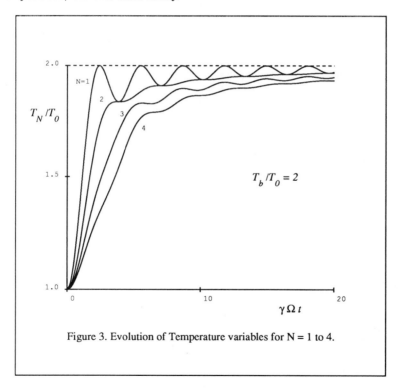

Figure 3. Evolution of Temperature variables for N = 1 to 4.

The two-particle system shows no bounces in temperature, since $J_0^2 + J_1^2$, is a monotone nonincreasing function of its argument (with $t \geq 0$), but the slope of $T_2(t)$ is zero at values corresponding to the zeros of J_1. At no pre-equilibrium time is the two-particle system completely determined by heat-bath variables, and internal influence is shared. The three-particle system shows an almost monotonous temperature function, $T_4(t)$ is even smoother, and presumably $T_N(t)$ and $S_N(t)$ becomes increasingly structureless as N increases.

There is no necessity for the heat bath to be at higher temperature than the system, and both $T_N(t)$ and $S_N(t)$ can decrease in time if the initial variances of the system are greater than those of the heat bath.

The structure of the temperature function should be noted from Eqs.(1.7.68) and (1.7.69), and their relationship to Eqs.(1.7.53) and (1.7.43). At $t = 0$, W is diagonal, and $T_N(0)$ can be written as a geometric mean, as in Eq. (1.7.68), of kinetic and potential temperatures. For t in general, the temperature function involves terms from the entire matrix W, including cross terms from G, but as $t \to \infty$, all elements of G vanish, W becomes diagonal, and $T_N(\infty)$ is once again the geometric mean of kinetic and potential temperatures, this time at equilibrium, equal to each other and to the arithmetic mean of initial heat-bath temperatures.

An expression for S_N that is sometimes useful as a working approximation in more general equilibrium situations is patterned from Eqs.(1.7.66) and (1.7.53) as

$$S_N \approx (Nk/2)\ln[\langle x'^2\rangle\langle p'^2\rangle/\hbar^2] \tag{1.7.76}$$

where $\langle x'^2\rangle$ and $\langle p'^2\rangle$ are the equilibrium variances in coordinate and momentum. (For a three-dimensional system, a similar term must be written for each dimension.) One extra refinement needed for freely moving indistinguishable particles confined to an area or volume is that the argument of the ln properly contains, not the spatial variance *per se*, but the spatial variance divided by N^2, or the standard deviation per particle, squared. The reasoning for such a modification will become evident when proper calculations for these systems are made in subsequent chapters.

It should be noted for future application that S_N is proportional to N in Eq. (1.7.76), as it is in most systems of practical importance.

1.8 QUANTUM SYSTEMS

The basic concepts and some of their applications have been presented so far in classical terms, whereas most of the systems of interest in statistical thermophysics have an underlying microstructure that properly should be described quantum mechanically. There is no fundamental difficulty in doing this, and it will be seen that many of the results are identical in form or easily recognized.

If a quantum system is known to be in a particular state $|\phi\rangle$, then the expectation value for some dynamical variable f is

$$\bar{f} = \langle\phi|\mathcal{F}|\phi\rangle, \tag{1.8.1}$$

where $\mathcal{F}$ is the operator corresponding to f. This expectation value is not the one of primary importance in statistical physics. When the system is not so well known, it might be described as being in one of a number of states, with the probability p_ϕ that it is in the state $|\phi\rangle$, with $0 \le p_\phi \le 1$, $\sum_\phi p_\phi = 1$, where $\{|\phi\rangle\}$ is an orthonormal basis.

The density operator, or statistical operator, ρ is defined as

$$\rho = \sum_\phi |\phi\rangle p_\phi \langle\phi|, \tag{1.8.2}$$

where the possibly time-dependent c-numbers p_ϕ can, of course, be put anywhere in the symbolic product. The set of state vectors need not be complete at this point; it need only include the states for which $p_\phi \neq 0$. The density operator has the following properties:

(a) ρ is self adjoint
(b) $\mathrm{Tr}\,\rho = 1$
(c) ρ is positive definite
(d) $\langle \bar{f} \rangle = \mathrm{Tr}\,(\rho \mathcal{F})$,

where the last one says that the statistical expectation value of the dynamical variable f, now a combination of the suitably weighted single-state expectation values, is given by the trace of the operator corresponding to f with the density operator. This is the quantum mechanical equivalent of Eq. (1.6.2).

To prove these properties, note that the adjoint of ρ is

$$\rho^\dagger = \left(\sum_\phi |\phi\rangle p_\phi \langle\phi| \right)^\dagger = \sum_\phi |\phi\rangle p_\phi \langle\phi| = \rho, \tag{1.8.3}$$

since the p_ϕ are real, so ρ is self-adjoint (the term *Hermitian* is often used in this context, and there is no difference for bounded operators). Since $\langle\phi|\phi\rangle = 1$, and since

$$\mathrm{Tr}\,|\alpha\rangle\langle\beta| = \sum_\gamma \langle\gamma|\alpha\rangle\langle\beta|\gamma\rangle = \langle\beta|\alpha\rangle,$$

where $\{|\gamma\rangle\}$ is any complete, orthonormal basis for the state space, it follows that

$$\mathrm{Tr}\,\rho \equiv \mathrm{Tr}\sum_\phi |\phi\rangle\, p_\phi\, \langle\phi| = \sum_\phi p_\phi \langle\phi|\phi\rangle$$

$$= \sum_\phi p_\phi = 1. \tag{1.8.4}$$

The statement that ρ is positive definite (more properly, non-negative definite) means that in any representation the diagonal matrix elements of ρ are real and non-negative. For any ket $|\gamma\rangle$, the corresponding diagonal matrix element is

$$\langle\gamma|\rho|\gamma\rangle = \sum_\phi \langle\gamma|\phi\rangle\, p_\phi\, \langle\phi|\gamma\rangle = \sum_\phi p_\phi |\langle\gamma|\phi\rangle|^2 \geq 0$$

which completes the proof and shows that the diagonal elements of ρ in any representation are real and lie between 0 and 1.

The statistical average, property (d) follows from

$$\mathrm{Tr}\,(\rho\mathcal{F}) = \mathrm{Tr}\left(\sum_\phi |\phi\rangle p_\phi \langle\phi|\mathcal{F} \right) = \sum_\gamma \sum_\phi \langle\gamma|\phi\rangle p_\phi \langle\phi|\mathcal{F}|\gamma\rangle$$

$$= \sum_{\phi,\gamma} p_\phi \langle\phi|\mathcal{F}|\gamma\rangle\langle\gamma|\phi\rangle = \sum_\phi p_\phi \langle\phi|\mathcal{F}|\phi\rangle$$

$$= \sum_\phi p_\phi \bar{f}_\phi = \langle\bar{f}\rangle, \tag{1.8.5}$$

a result that is independent of the particular representation of ρ.

The temporal evolution of ρ may be computed directly from the basic Schrödinger equations

$$\mathcal{H} \mid \phi \rangle = i\hbar \frac{d}{dt} \mid \phi \rangle, \tag{1.8.6}$$

and

$$\langle \phi \mid \mathcal{H} = -i\hbar \frac{d}{dt} \langle \phi \mid, \tag{1.8.7}$$

where $\mathcal{H}$ is the Hamiltonian and the Schrödinger picture is being used. When it is assumed that the p's are not explicitly time dependent, direct differentiation and substitution from Eqs. (1.8.6) and (1.8.7) gives

$$i\hbar \frac{d\rho}{dt} = i\hbar \sum_\phi \left[\frac{d\mid\phi\rangle}{dt} p_\phi \langle\phi\mid + \mid\phi\rangle p_\phi \frac{d\langle\phi\mid}{dt} \right] + i\hbar \frac{\partial\rho}{\partial t}$$

$$= \sum_\phi \left[\mathcal{H} \mid \phi \rangle p_\phi \langle \phi \mid - \mid \phi \rangle p_\phi \langle \phi \mid \mathcal{H} \right] + i\hbar \frac{\partial\rho}{\partial t}$$

$$= (\mathcal{H}\rho - \rho\mathcal{H}) + i\hbar \frac{\partial\rho}{\partial t},$$

where $\partial\rho/\partial t = \sum \mid\phi\rangle \dot{p}_\phi \langle\phi\mid$, in case the p's are time dependent, or

$$i\hbar \frac{d\rho}{dt} = [\mathcal{H}, \rho] + i\hbar \frac{\partial\rho}{\partial t}, \tag{1.8.8}$$

a result that is the quantum relative of Eq. (1.6.6), since the quantum equivalent of the Poisson bracket is $(1/i\hbar)$ times the commutator. This is the statistical-operator expression of the Liouville theorem, known as the von Neumann equation. Note that if ρ commutes with the Hamiltonian, and the probabilities are constant in time, then ρ is constant. As in the classical case, it is possible to devise a statistical operator that permits evolution to equilibrium, either by providing a means for explicit time dependence to enter through the p_ϕ or by using a reduced density operator for a system in contact with a heat bath (Agarwal [1971]). Chapter 9 returns to time-dependent statistical operators.

The uncertainty (or information entropy) is defined in analogy to Eq. (1.4.1) as

$$S = -k \operatorname{Tr}(\rho \ln \rho). \tag{1.8.9}$$

A typical information-theoretic procedure is the calculation of ρ at equilibrium by a variational principle, the maximization of Eq. (1.8.9), subject to Eq. (1.8.4) and any further information available, just as in Section 1.4. For an explicit example, suppose the expectation value of the energy is known as

$$\langle \bar{E} \rangle = \operatorname{Tr}(\rho\mathcal{H}).$$

Then with the undetermined multipliers α and β, as in Eq. (1.4.4), we vary the quantity

$$\operatorname{Tr}(\rho \ln \rho + \alpha'\rho + \beta\rho\mathcal{H}),$$

to get, with $\alpha = \alpha' + 1$,

$$\ln \rho = -\alpha - \beta \mathcal{H},$$

or

$$\rho = e^{-\alpha - \beta \mathcal{H}}, \qquad (1.8.10)$$

where $\mathcal{H}$ is the Hamiltonian operator but α and β are c-numbers.

The trace of Eq. (1.8.10) yields

$$e^{\alpha} = \operatorname{Tr} e^{-\beta \mathcal{H}} \equiv Z, \qquad (1.8.11)$$

and ρ may then be written as

$$\rho = e^{-\beta \mathcal{H}} / Z, \qquad (1.8.12)$$

where Z is again the partition function. It is evident from Eqs. (1.8.11) and (1.8.12) that $\operatorname{Tr} \rho = 1$ and that

$$\langle \bar{E} \rangle = \frac{\operatorname{Tr} \mathcal{H} e^{-\beta \mathcal{H}}}{\operatorname{Tr} e^{-\beta \mathcal{H}}} = -\frac{\partial}{\partial \beta} \ln Z, \qquad (1.8.13)$$

in accord with Eq. (1.4.7).

In the representation of Eq. (1.8.12), ρ commutes with $\mathcal{H}$, and when $\mathcal{H}$ is the Hamiltonian it follows from Eq. (1.8.8) that ρ is not explicitly time dependent, thereby satisfying a requirement for equilibrium.

If S from Eq. (1.8.9) is evaluated in a representation in which ρ is diagonal, the result is

$$S = -k \sum_{\phi} p_{\phi} \ln p_{\phi}, \qquad (1.8.14)$$

where $p_{\phi} = \langle \phi \mid \rho \mid \phi \rangle$, from which it follows that the general fundamental results of Section 1.4 are also valid for quantum systems.

Applications of these basic techniques will be developed in subsequent chapters; further applications and related material are to found in the references, as are fundamental concepts and basic background material.

1.9 COMMENTARY

The purpose of this section is to provide some background, historical perspective, and enrichment to the rather starkly presented developments in this chapter. I should say at the outset that it is not possible, in an acceptable allocation of space, to present fairly and unobjectionably the fully qualified opinions of every author who is cited. I therefore apologize in advance to those who are misrepresented, and, which may be even worse, to those who are not represented at all. I have attempted to be fair and accurate; for oversights and inaccuracies, I take full responsibility.

Some of the comments here are beyond the scope of this chapter, and they should become more meaningful after coverage of the next two chapters. But since it is, ultimately, of more use to collect the pertinent background material into one essay, rather than fragmenting it on the basis of chapter divisions, there may result some puzzlement at first reading.

Ever since Maxwell [1871] invoked his demon to separate gas molecules in a divided chamber, it has been evident that our knowledge of a physical system determines what we can do with it. The demon was postulated to be able to use the knowledge he obtained by observing the gas molecules. He could operate a workless trap door in the wall dividing the chamber, allowing selected molecules to pass from one side to the other, thereby causing a separation based on his selection criterion. If he allows only fast molecules to enter side A from side B, and only slow ones to enter B from A, he effects a thermal separation, with the gas in side A becoming warmer than the equilibrium temperature, and the gas in side B becoming cooler. This separation could be used to operate a heat engine, with the result that the demon could use his information of the microscopic motions of the system to produce a perpetual motion machine. He could similarly produce a pressure difference, simply by allowing all approaching molecules to enter one of the chambers, but none to leave. The nonexistence of Maxwell's demon is generally accepted as a basic truth, but the implied connection between information and thermodynamics continues to be debated (Ehrenberg [1967]; Prigogine [1967]; Rowlinson [1970]; Tribus [1979]).

1.9.1 Demonology. In the course of scientific evolution, several demons have contributed to our understanding of the nature of physical laws. The oldest of the modern ones, Laplace's demon, knows the state of every particle in the universe at some particular time, and therefore, since he knows dynamics, he can determine the state of the universe at any other time, past or future. His is a completely deterministic, fatalistic universe, in which every move we make, no matter how much thought is involved, is preordained by deterministic mechanics. His contribution to thermodynamics will be examined later.

Maxwell's demon has been treated in detail by Brillouin [1962], who established, by analysis of the physics that must underlie the demon's operation, that there is an energy cost in observing a molecule such that the overall entropy of the system cannot decrease as a consequence of the demon's observations. Brillouin offered compelling physical arguments for a relationship between information and a quantity he called *negentropy*, a term already used by Schrödinger [1945] He did not then proceed to develop a useful formalism relating information and statistical physics.

Leo Szilard [1929] proposed a simplified version of Maxwell's demon, in that Szilard's needed to monitor only one gas molecule, but by so doing, he could cause the operation of a perpetual motion device (Denbigh and Denbigh [1985]; Poundstone [1985]; Rothstein [1959]; Skagerstam [1975]; Yu [1976]). The principal significance of Szilard's work is that he offered the first modern analysis of the operation of Maxwell's demon. His procedure for analyzing that paradoxical being was a precursor of Brillouin's work, and the beginning of a proliferation of papers that created a fertile climate for the development of the information-theoretic approach to statistical mechanics. Szilard demonstrated that the observer becomes an influential part of the system, a fact that was amplified by his successors (Brillouin [1956]; Gabor [1950]; Gabor [1961]).

As Harvey Leff and Andrew Rex have pointed out in their excellent book on the subject, Maxwell's Demon lives on. (Leff and Rex [1990]) They provide a more detailed introduction to demonology than I have space for, and they include reprints of many of the classic and modern papers, including interesting analyses of the demon's adventures in the field of computation.

1.9.2 Information Theory. In 1948, two mathematicians independently published the same quantitative measure of information (Shannon [1948]; Wiener [1948]). Both authors went on to develop the theory, primarily in the context of communication theory. Wiener and Shannon both concerned themselves with communication in the presence of noise, and the problems related to the fidelity of recovering a signal that has been distorted by noise. Wiener's approach was in the direction of processing the received signal, using techniques of filtering, correlation, and prediction to reduce the effects of noise, while Shannon considered, in addition, techniques of preprocessing the signal before transmission to optimize the received signal-to-noise ratio. While these topics are not of direct concern in the context of statistical mechanics, they are important in the functioning of the modern world. Information theory has become a fully developed discipline, studied for its own sake, and closely allied with mathematical statistics (Khinchin [1957]; Kullback [1968]; Tribus [1979]; Wiener [1956]). It has become a powerful tool in inductive reasoning, as a consequence of the Bayes theorem and the work of Laplace and Jeffreys, and of Cox and Jaynes (Cox [1946]; Cox [1961]; Cox [1979]; Hobson [1972]; Jaynes [1984]). But more than that, it has pervaded many other disciplines, including statistical physics, communication theory, optics, cryptology, linguistics, artificial intelligence, pattern recognition, various areas of biology, and a wide range of activities requiring the extraction of reliable information from less-than-ideal data (Baierlein [1971]; Gabor [1961]; Grandy, Jr. [1987]; Hobson [1971]; Jumarie [1986]; Kapur and Kesavan [1987]; Katz [1967]; Kesavan and Kapur [1989]; O'Neill [1963]; Patera and Robertson [1980]; Rothstein [1951]; Rothstein [1979]; Shannon [1949]; Shannon and Weaver [1949]; Singh [1966]; Watanabe [1969]; Watanabe [1985]; Weiss [1977]; Wiener [1956]; Yu [1976]). (It is of passing interest to note that both Shannon and Kullback have made professional contributions to cryptology, the former in the early years after his original papers (Shannon [1949]; Shannon [1951]), while the latter was in the Army Security Agency, and then in the 1950's he was head of the National Security Agency's Office of Research and Development. The problem of extracting deliberately concealed information from intercepted messages is similar to some aspects of the recovery of signals from noise.)

There is little need for a student to acquire the full armament of the discipline of information theory in order to learn statistical mechanics, but he will find the introductory treatment in this context a useful entrée to further studies as are deemed appropriate. Some idea of the scope and pervasiveness of information theory can be found in a book edited by Zurek [1990], and that of Kapur [1989] gives a coherent account. Jumarie [1986] presents a highly individualistic view of the applicability of information theory to a variety of subjects, including linguistics, biology, and psychology.

1.9.3 The Maximum-Entropy Formalism. The significant developments in statistical physics based on the maximum-entropy formalism stem from the two papers published by Edwin T. Jaynes (Jaynes [1957a]; Jaynes [1957b]). In these, he proposed that in problems of statistical inference, including those of statistical physics, the probabilities should be assigned so as to maximize the Shannon entropy. He clearly expressed the attitude underlying this procedure – that to do otherwise is to assume more information than is justifiable by the data at hand. (A similar attitude was expressed by Grad [1961]. Harold Grad's paper on entropies is a classic, and should be read at some point by all students of statistical physics.) This attitude does not require or imply that the Shannon (information) entropy (Eq. (1.4.1)) must be equated *a priori* to that of thermodynamics. Jaynes, however, did so identify the two entropies for equilibrium systems, based on a demonstration that the probabilities derived to maximize the information entropy are the standard Boltzmann values, as in Eq. (1.4.9). He then proceeded to show that the standard results of equilibrium statistical mechanics followed directly (Baierlein [1971]; Grandy, Jr. [1987]; Hobson [1971]; Katz [1967]; Tribus [1961]).

As Jaynes and others advanced and polished the conceptual foundations of the formalism of entropy maximization, most physicists working in thermodynamics either ignored the work or objected to it. In a typical objection, Jaynes [1979] quotes Uhlenbeck as follows: "Entropy cannot be a measure of 'amount of ignorance,' because different people have different amounts of ignorance; entropy is a definite physical quantity that can be measured in the laboratory with thermometers and calorimeters." I would answer objections of this sort by agreeing that entropy is a property of any thermodynamic system, but our information about the system defines the way we specify its state. If we agree on its mass, composition, volume, temperature, and other such properties, then we have identical information about the system, and, as a consequence, we agree on its entropy. In most cases, we are content to measure an entropy change as

$$\Delta S = \int \frac{dQ}{T}$$

from some initial state to a final state, along a reversible path, without trying to extrapolate the measurement back to $T = 0$. But there must always be contact with physics. If our information-theoretic calculation yields a result different from that of a careful experimental determination, then our presumed knowledge of the microstates of the system must be wrong or else macroscopic constraints must exist that have not been included in the calculation. The information-theoretic entropy is correct for the system that was the subject of the calculation, but the real system must differ from the presumed one. A wrong description of the system yields the wrong entropy for that system, as should be expected, and experimental evidence that the result is incorrect provides a means for revising the set of constraints, leading to a revised set of calculated probabilities and a revised entropy. Jaynes's delayed answer to Uhlenbeck is, "Certainly, different people have different amounts of ignorance. The entropy of a thermodynamic system is a measure of the degree of ignorance of a person *whose sole knowledge about its microstate consists of the values of the macroscopic quantities* X_i *which define its thermodynamic state.* This is a perfectly 'objective' quantity, in the sense that it is a function only of the X_i, and does not depend on anybody's personality. There is no reason why it cannot be measured in the laboratory."

Other objections have been expressed and answered by Jaynes and others (Dresden [1961]; Landsberg [1984]; Rowlinson [1970]). Careful replies to most of the objections are to be found in Jaynes [1979], Jaynes [1982a], Jaynes [1984], and Grandy, Jr. [1987]. These are recommended reading for any serious student of statistical thermophysics. One recent scholarly work (Denbigh and Denbigh [1985]) examines much of the literature on thermodynamic entropy as related to information theory. (See Price [1991].) The authors conclude that "although information theory is more comprehensive than statistical mechanics, this very comprehensiveness gives rise to objectionable consequences when it is applied to physics and chemistry. Since the subjective interpretation of probability, on which information theory has been based, is more general than is the objective interpretation, the language in which the theory is presented necessarily takes on a subjective character. A probability distribution is thus said to represent *our* state of knowledge, whereas in physico-chemical contexts it is usually justifiable to go much further in an objective direction and to regard a probability distribution as being characteristic of the nature of the system in question and of the physical constraints upon it." Although they quoted the Jaynes reply to Uhlenbeck's objection, they apparently did not fully agree with it, because theirs is the same objection as Uhlenbeck's. There seems to be a breakdown in communication that leads sincere and thoughtful people to disagree, even when they try to understand each other.

But perhaps a breakthrough is yet possible. In a review of the Denbighs' book, the reviewer summarizes these objections so bluntly that clarification becomes explicit. He says "The extra information one gains from molecular measurement indeed reduces our ignorance, but it leaves the system's entropy unaffected. (No? Imagine that you know the location and momenta of all particles to within uncertainty limits. So you have the entropy near zero? Now cool the system. You haven't left room for the entropy to decrease.)" (Parsegian [1987])

I could attempt to answer him as Szilard or Brillouin might have done, in terms of the entropy cost of acquiring the information, but this approach is obviously not accepted by either him or the Denbighs. Instead, I offer a different scenario. Suppose a system of particles is cooled to our lowest convenient temperature, and thus to low entropy. Now allow this refrigerated system to move past at some velocity **v**, where the speed is not so excessive that one must be concerned with relativistic formulations of thermodynamics. The macroscopic motion of the system does not change its temperature or entropy. Our monitors allow us to keep track of this system, and we expand our endeavors by adding several more such frigid systems, all moving at arbitrary but known velocities. We monitor all of them, and we continue to regard them as being cold, low-entropy systems. Our model, so far, is no worse in concept than that existing all day long in the control tower at any major airport. In principle, we can increase the number of these systems substantially, and they continue to be cold. Also, in principle, since there is no specified size for these systems, we can allow them to be single atoms, each moving at its own velocity. And there we have it. We have devised a system of atoms at very low temperature, with non-negligible kinetic energy, each moving along a path that we can follow. The entropy is low because of our information about the positions and momenta of the individual atoms. Our description of the system is in dynamic terms, not thermal ones. How do we cool such a system? There is no effective or meaningful way to do so. The individual systems are already as cold as we can conveniently make them, and there is no reason for us to ascribe a temperature to the known dynamical motions of the system. If our monitors fail, and we lose the ability to describe the system dynamically, then we must regress to a thermal description, with consequent temperature in terms of the mean kinetic energy of the atoms (if this is the appropriate measure), and increased entropy as a result of our loss of information. (Think of what could happen at a major airport if all monitors in the control tower fail.)

Clearly, then, the entropy of a system *does* depend upon our information about it. If two observers have different descriptions of the state of a system, they will ascribe different values to its entropy, as well as to other variables. However, if two observers perform equivalent experiments on equivalent systems, they should obtain equivalent results, including their values for the entropy. Laplace's demon should see a zero-entropy universe (except for quantum effects), and as a consequence a zero-temperature universe. How, then, do we perceive every temperature to be higher than that? Because we treat energy transfers to the microscopic motions of our systems, including thermometers, as thermal rather than dynamic. If we return to our low-temperature, low-entropy system of the previous paragraph and carry out the crude operation of inserting a thermometer into our carefully monitored collection of atoms, the thermometer will read a temperature corresponding to its own equilibrium state, as it acquires energy from the colliding atoms and loses energy by radiation or other processes. (I have used an atomic system, instead of a molecular one, to preclude the objection that vibrational and rotational temperatures are still measurably the result of the original cooling, and only the translational temperature is subject to two interpretations. I have also assumed that the atomic motions are collisionless. None of this should violate the spirit of Parsegian's comments.)

My example starts with our knowing the positions and momenta of all the particles, to within uncertainty limits, and concedes that it is already as cool as conveniently possible. By loss of this information, we are forced to describe the system as having an increased temperature, if we regard it as being in equilibrium, or even if we insert a thermometer. The thermodynamic entropy has increased because of lost information. My claim is that the inverse sequence, the acquisition of detailed data on the positions and momenta of all the particles, allows us to reduce the temperature and the entropy to the values that existed at the beginning of the example.

In another recent objection to the use of information theory in statistical mechanics, Ma [1985] seems to regard as a basis for rejection the observation that there are many more possible trajectories in phase space than the small fraction that is realized by the system in any short-term observation, and therefore many unrealized states must be grouped with any actual one in order to obtain measurable probabilities. He concludes that "if we use information theory ... to discuss statistical mechanics, we would be missing the point." Perhaps his point suffered in translation, but as it is stated, I have missed it.

In an entirely different treatment, Callen [1985], in an excellent and highly recommended book, renames the uncertainty, or entropy function, $S = -\sum p_i \ln p_i$, the *disorder*, thereby making quantitative the often-heard statement that entropy is a measure of the disorder of the system. (Boltzmann's disorder is the quantity $\ln \Omega$, where Ω is the number of different, equally probable ways the system can have a particular energy, and the two definitions agree for microcanonical systems, in which all accessible states have the same energy, and there are no other constraints. This topic is expanded in Chapter 3.) Callen's formal development is equivalent to the present one, or to that of Jaynes, but his interpretation of the probabilities is different. He points out that the concept of probabilities has two distinct interpretations. *Objective probability* refers to a frequency, or a fractional occurrence, and is always obtained from experiment. *Subjective probability* is a measure of expectation based on less than complete information. Therefore the disorder, which is a function of the probabilities, has two corresponding interpretations. He says, "The very term *disorder* reflects an objective interpretation, based upon objective fractional occurrences. The same quantity, based upon a subjective interpretation (of the probabilities) is a measure of the uncertainty of a prediction that may be based upon the (probabilities)." And further, "There is a school of thermodynamicists who view thermodynamics as a subjective science of prediction. If the energy is known, it constrains our guess of any other property of the system. If *only* the energy is known the most valid guess as to the other properties is based on a set of probabilities that maximize the residual uncertainty. *In this interpretation the maximization of the entropy is a strategy of optimal prediction.*"

"To repeat," he continues, "we view the probabilities ... as objective fractional occurrences. The entropy is a measure of the objective disorder of the distribution of the system among its microstates. That disorder arises by virtue of random interactions with the surroundings or by other random processes (which may be dominant)."

My first response to this is that I cannot think of a case, in the context of thermodynamics, in which anyone has made an objective determination of the probabilities associated with the microstates of the system, except in the almost trivial cases of chemical or mass-spectrometric measurements of the mole fractions of the constituents of systems. One of the reasons that we describe systems thermodynamically is that we are unable to determine which microstates they occupy. And if we are unable to ascertain that a system is in a particular microstate, I fail to see how we can expect to make an objective determination of the probabilities of occupation, at least in the sense that Callen defines. I therefore suggest that a source of disagreement is in the definitions and meanings of terms.

My next response, in support of this suggestion, is to quote a commentary by Jaynes that precedes the third article in Jaynes [1982b]. "The reaction of some readers to my use of the word 'subjective' in these articles was astonishing....There is something patently ridiculous in the sight of a grown man recoiling in horror from something so harmless as a three-syllable word. 'Subjective' must surely be the most effective scare word yet invented. Yet it was used in what still seems a valid sense: 'depending on the observer'."

He continues: "In Euclidean geometry the coordinates of a point are 'subjective' in the sense that they depend on the orientation of the observer's coordinate system; while the distance between two points is 'objective' in the sense that it is independent of the observer's orientation. That is all I ever meant by the term; yet twenty-five years later the shock waves are still arriving."

As I have tried to make evident, the entropies and the probabilities that we assign to our thermodynamic systems do, in fact and in practice, depend upon the macroscopic coordinates that we use to define their thermodynamic states. Given these macroscopic coordinates, however, there is nothing ambiguous about the method for computing the entropy or the probabilities. Furthermore, the terms *subjective* and *ambiguous* are not synonymous. On the other hand, *subjective* can mean *having its source in the mind*, or *existing only in the mind*, or *depending on the personality of the observer*. Since these are the connotations that are regarded as objectionable, it is evident that a different choice of terminology could have averted some of the controversy. I therefore propose that the terms *measured* and *calculated* be used henceforth, to replace the offending ones. I define the term *measured* in this context to mean that the $\{p_i\}$ are determined experimentally as fractional occurrences, and the term *calculated* to mean that the $\{p_i\}$ are obtained as the least-biased set of probabilities compatible with the known macrostate of the system. I reiterate that if the calculated probabilities yield expectation values or predictions that do not agree with subsequent macroscopic measurements, then these measurements provide further input data that should be used to refine the unsatisfactory probability set. It is by this procedure, and not by obtaining measured probabilities, that we generate acceptable probability sets by which we describe our thermodynamic systems.

One further objection to the use of information-theoretic entropy in thermal physics is sometimes heard. The point is raised that in the context of information theory, an observer must be present to establish a set of probabilities so that the entropy can exist. Only then can the entropy evolve to an equilibrium value. But what about systems that are not being observed? This objection seems almost too trivial to warrant a serious reply, but apparently it is not so regarded by its advocates. As a partial answer, I point to the equations for the evolution of the probabilities in the dog-flea model, Eq. (1.5.9). The evolution in time of the probabilities in these equations is independent of the existence of an observer, and the final, equilibrium set of probabilities is not at all dependent upon there being an earlier set of observed probabilities. The microprocesses that underlie the temporal evolution of the system are unaffected by the presence or absence of an unobtrusive observer. The equations that describe the evolution of the probability set represent our understanding of these microprocesses, but they do not require that there be any observation of the probabilities at any time. Further, it is clear that the entities that execute the microprocesses are not aware that we might have written equations to describe their behavior; they merely exhibit their intrinsic behavior, and a system of them evolves unselfconsciously. If we choose, at some time, to observe the system and ascribe an entropy to it, based on our observations (what else?), then, except for the inevitable perturbation caused by the process of observation, the system continues to evolve as before. It should be noted that an initially observed flea population, with observed transition probabilities, is expected to evolve according to Eq. (1.5.9). The equilibrium probability set that evolves, independent of the initial observation or the continued presence of the observer, is exactly the one determined by the maximum-entropy postulate of information theory, and even at equilibrium, the underlying microscopic processes are random fluctuations. The same comment applies to the coupled-oscillator system of Section 1.7, with the additional result that the equilibrium state of a finite system of oscillators in an infinite heat bath turns out to have the same description, whether arrived at by the time-dependent evolution of Section 1.7, by direct use of the maximum-entropy principle, as shown in Sections 3.2.1 and 3.5.1., or by the usual methods set forth in most other books on statistical mechanics.

As a final, and noncontroversial, topic in this subsection, I call attention to an interesting paper (Montroll and Shlesinger [1983]) in which the authors invert the usual maximum-entropy problem. They assume that a probability set is known and ask what constraints will lead to such a set in the maximum-entropy formalism. They are particularly concerned with long-tailed distributions that sometimes occur in statistics, and like all of Montroll's papers, the presentation is both informative and culturally enlightening.

1.9.4 Continuous Probability Distributions. As noted in Section 1.6, the expression in Eq. (1.8.5) is not the most general one for S when ρ is a probability density. A general, mathematically rigorous treatment of systems that are described by probability densities requires the full apparatus of mathematical statistics (Cramér [1946]), including measure theory and Lebesgue integration. (Cramér's first twelve chapters offer a good, readable introduction to these topics.) Both Jaynes [1982b] and Kullback [1968] give generalizations of Eq. (1.8.5) that replace the h^{3N} by the reciprocal of the measure function in the appropriate space. This kind of generalization is necessary when the invariance of S under transformation from one noncartesian space to another is required, but for elementary applications to statistical thermophysics, the expression given in Section 1.6 is usually all that is needed.

1.9.5 Approaches to Equilibrium. The equations of motion of the particles that make up our thermodynamic systems are time-reversible. How, then, is it possible to explain irreversible phenomena, such as equilibration, on a molecular basis? This is the question Boltzmann tried to answer with his H-theorem, and it is a question that is not yet cleared of controversy (Boltzmann [1872]).

Boltzmann defined the function $f(\mathbf{v})$ such that the number, dn, of molecules with velocities between $\mathbf{v}$ and $\mathbf{v} + d\mathbf{v}$ is given by

$$dn = f(\mathbf{v})\, d^3 v.$$

He then defined the functional H as

$$H = \int f(\mathbf{v}) \ln f(\mathbf{v})\, d^3 v, \qquad (1.9.1)$$

where the integral is taken over the entire velocity space. The statement of his famous H-theorem is that for an isolated system the quantity H will never increase; it will decrease as the system approaches equilibrium, and then remain at its minimum value. His proof of the theorem, however, implicitly invoked an assumption about collision rates, the so-called *Stosszahlansatz*, that led to two paradoxes, and to the realization that the H-theorem in its unrestricted form could not be valid.

If, at some instant of time, t_i, the positions and momenta of all of the particles in the isolated system are represented by a single vector, $\mathbf{R}_i$, in Γ-space, then H should be given as $H_i = H(\mathbf{R}_i)$. Every such point that satisfies the integral constraints on the system, such as the fixed total energy, momentum, angular momentum, and perhaps others, is an allowed point, and each point lies on a unique trajectory in this space, determined by the equations of motion, which are time reversible. This trajectory was assumed to pass through every point allowed by the constraints, since all restrictions are presumed to be incorporated in the constraints. One such system, for which the only constraint with value different from zero is the total energy, was called by Boltzmann and others at that time *ergodic*. The meaning of the term has been modified in the literature, as the related subject matter became increasingly a subject for mathematicians and its physical roots became more and more remote. (Fomin, Kornfeld and Sinai [1982]; Halmos [1956]; Kitchens [1981]; Sinai [1976]; Smorodinsky [1971]; Walters [1982].) Suppose the system vector for such a system at a series of times ...., t_1, t_2, t_3,..., leads to corresponding values of H_i given by the sequence

$$... \geq H_1 \geq H_2 \geq H_3 \geq .... \geq H_n \geq ....,$$

where the equality holds only if the system is in equilibrium. Now consider another set of system vectors, at another sequence of times, starting at R'_n, where all the momenta of R_n have been reversed, but all the coordinates are unchanged. Then it follows that

$$... \geq H'_n \geq ... \geq H'_3 \geq H'_2 \geq H'_1 \geq ....,$$

where $H_i = H'_i$ for all points. Evidently, then, there are inconsistencies in the theory. This one was called the *Umkehreinwand*, or reversibility paradox, pointed out by Loschmidt.

Another paradox, discussed by Zermelo, is that a theorem of Poincaré establishes that for systems enclosed in a finite volume, any sequence of points on a trajectory in Γ-space will be repeated to any prescribed accuracy after a finite – usually very long – time, a result that converts the Boltzmann H-theorem into an analog of M.C. Escher's lithograph, *Waterfall*, in which water apparently flows downhill around a complete, closed circuit. This paradox was called the *Wiederkehreinwand*, or recurrence paradox.

The Ehrenfests studied the dog-flea and wind-tree models in order to look at these paradoxes in a simpler context. Their treatments in terms of populations examine the analogs of the Boltzmann H as a function of time. It is informative to study their treatments of these problems, if only to appreciate the struggle going on at that time with the conceptual foundations of the subject (Ehrenfest and Ehrenfest [1959]). In the treatments given in this chapter, the flea jumps and wind-molecule collisions are treated as first-order Markov processes, which means that the future of any system is determined entirely by its present state, and not by its history. The growth of entropy in these cases is simply a consequence of the evolution of the calculated probability set towards a state of maximum possible uniformity, consistent with the existing constraints. It is often observed that if one waits long enough, either of these systems will return to its initial state, thus restoring the entropy to its original value. The point about the given calculations is that the entropy we assign to the system depends upon our knowledge of it. Whenever the flea populations are measured, it becomes possible to assign all probabilities the value zero, except the one that corresponds to the observed populations. The entropy at time of observation is then $S(0) = 0$. But until an observation is next made, the entropy evolves monotonously toward its equilibrium value, and if the populations happen to return temporarily to their initial values, we are unaware of it, unable to act on it, and therefore not entitled to say that the *entropy* has returned to its original value.

Willard Gibbs published his careful construction of statistical mechanics (a term he was the first to use) in 1902, ten years before the first appearance of the Ehrenfests' cited work. (Gibbs [1960]) Gibbs found the entropy to be exactly equivalent to that of Eq. (1.6.7). He wrote it in terms of his *average index of probability*, $\overline{\eta}$, which he found to be the negative of the entropy. His average index of probability was to be minimum at equilibrium, just as was Boltzmann's H, but the Gibbs index was to be evaluated as an integral over the entire Γ-space, not just the phase space of a single particle. Although the Gibbs entropy was demonstrated to yield the correct value for equilibrium systems, it was rejected, in accord with the Liouville theorem, as being unable to evolve in time toward higher values as a system approaches equilibrium. (Ehrenfest and Ehrenfest [1959]). Gibbs was aware of this difficulty, and he attempted to resolve it, but not in his usual clear style. He says, in his Chapter XII, words that indicate the necessity for maximizing the entropy of Eq. (1.6.7) in order to obtain a distribution that will remain constant in time. He then goes on to discuss his well-known example of the stirring of two colored, immiscible liquids. But he did not offer the kind of mathematically evident evolution that Boltzmann had attempted, or that is presented in Sections 1.5 and 1.7 of this chapter. (It should be noted that the probabilistic calculations for the dog-flea and wind-tree models of Section 1.5 are in terms of probability sets that will yield the analogs of the Gibbs entropy, rather than that of Boltzmann, but in these cases there is no Liouville theorem to restrict evolution. Similarly, the treatment of Section 1.7 is again in terms of a Gibbs entropy, but defined on a reduced space to permit interaction with the outside world.)

Gibbs seems to be saying, in his discussion of the stirring of colored liquids, that while his average index of probability is conserved as the system evolves (because of the Liouville theorem), it is conserved only because the Liouville function contains information about correlations that, in practical situations, becomes unusable. Therefore, the Gibbs expression sets a lower bound for the entropy as the system evolves, and the useful measure for the entropy emerges as a modified Gibbs index in which the useless information in the Liouville function is discarded. Tolman [1938] expressed this idea quantitatively by defining first a quantity P that is a coarse-grained average value of ρ, with the average taken over a small, but finite, volume element of Γ-space, such that the filamentary structure depicted by Gibbs is smoothed out (as if we regard the strands of different-colored dyes as being of a uniform local color). He then defined a function $\overline{\overline{H}}$ as

$$\overline{\overline{H}} = \int \rho \ln P \, d\Gamma.$$

He showed that this modified form of the H-function evolves in time as Boltzmann expected his H to do. Tolman's function is nothing more than a mathematical formulation of the idea expressed by Gibbs. (Tolman's entire discussion of the H-theorem is carefully presented and informative, but not always rigorous. It is interesting to note that, in this discussion, any reference to a later time could, with equal validity, be made to an earlier time; *i.e.* when he says that his $\overline{\overline{H}}$ diminishes with time later than that of the initial observation, he could just as well speak of times earlier than the initial observation.) He states, "In concluding this chapter on the H-theorem it is evident that we must now regard the original discovery of that theorem by Boltzmann as supplemented in a fundamental and important manner by the deeper and more powerful methods of Gibbs."

In a later chapter, Tolman notes "that the ensemble which we choose to represent a mechanical system of interest is determined by the observations we have made on the condition of that system. Thus also the value of $\overline{\overline{H}}$ will be determined by the nature of our knowledge of the condition of the system. Since the value of $\overline{\overline{H}}$ is lower the less exactly the quantum mechanical state of the system is specified, this provides the reason for the statement sometimes made that the entropy of a system is a measure of our ignorance as to its condition."

Further details of this kind of treatment are to be found in Tolman [1938], Ehrenfest and Ehrenfest [1959], ter Haar [1954], Jaynes [1982a], and Grünbaum [1974]. A lengthy analysis of the dog-flea problem is given by Kac [1959], and discussed by Dresden [1962]. These discussions include details of fine-graining *vs.* coarse-graining, the quasi-ergodic theorem, time reversibility, and the relative merits of the Gibbs and Boltzmann entropies. Most of this material is of interest in providing an appreciation for the difficulties that have been surmounted in our struggle to understand entropy and equilibration. It is not essential, however, to the understanding of the conceptual development presented in these pages. The information-theoretic point of view provides a convenient and efficient means of transcending this material, and for this reason alone it should be welcomed as a significant step forward.

A careful modern treatment of the dynamical foundations of the evolution of entropy to maximal states, by Mackey [1989], examines quite generally the conditions under which a dynamical system may evolve to thermodynamic equilibrium. He concludes, with much more detail than I can provide here, that equilibrium can result from ergodicity, from nontrivial coarse graining of the phase space, or from interaction with a heat bath. Specific examples of all three possibilities are mentioned in this commentary.

1.9.6 Ergodic Theory and Chaos. The idea that Boltzmann sought to formalize in the ergodic hypothesis is simply that of being able to use time-averages and ensemble-averages interchangeably. Boltzmann visualized the state of a gas of N atoms as represented by a point in the $6N$-dimensional Γ space, with three coordinates and three components of momentum for each atom. This point moves through the Γ space on a trajectory that is taken to be completely determinate, assuming that the Hamiltonian for the system is known and the system is isolated. With constant total energy, the representative point is confined to move on a surface in the Γ space corresponding to that energy. How much of that constant-energy surface is covered by the trajectory, as it evolves in time? Boltzmann and others conjectured that if no other constraints on the motion existed, then the trajectory would, in the course of time, pass as closely as anyone cared to specify to any designated point. If a Gibbs ensemble of identically prepared systems could each be represented by a point in Γ space, then the probability that the representative point of any one system of the ensemble is in a particular infinitesimal interval $d\Gamma$ is given by $\rho(\{\mathbf{r}_i, \mathbf{p}_i\})\, d\Gamma$. Thus the average value of any function of the coordinates and momenta, taken over the ensemble, can be written as

$$\langle f \rangle = \int_\Gamma f(\{\mathbf{r}_i, \mathbf{p}_i\})\rho(\{\mathbf{r}_i, \mathbf{p}_i\})\, d\Gamma,$$

the *ensemble average*. But as the representative point of any one system moves along its trajectory, the values of the arguments that determine the function change. A time average can be written as

$$\overline{f} = \frac{1}{T} \int_0^T f(\{\mathbf{r}_i(t), \mathbf{p}_i(t)\})\, dt,$$

where the time dependencies have been made explicit. The original Boltzmann statement was, in effect, that in the limit as $T \to \infty$, $\langle f \rangle = \overline{f}$. By the Liouville theorem, the probability density ρ is time invariant, as it must be for the ergodic theorem to be meaningful. The weighting function, or measure, by which ensemble averages are computed is an invariant function, and the transformation that carries the representative point on its trajectory is called *measure preserving*.

This is the property of transformations that mathematicians study in the current branch of mathematics called *ergodic theory*. It is an important and interesting area of mathematics, related to probability theory and mathematical statistics. To read the modern mathematics books on ergodic theory, it is necessary to know the basic ideas of measure theory, Lebesgue integration, and Hilbert spaces. But since the field has moved so far from its roots in physics, it is not necessary for the beginning student of statistical thermophysics to be conversant with this material. Furthermore, since it is usually more convenient to do the calculations of statistical thermophysics with systems that are not isolated, but instead are in equilibrium with a reservoir at constant temperature, the basic motivation for investigating ergodicity is no longer of overriding importance. The material included here is largely for cultural background, since it is not needed in order to understand the subject matter of statistical thermophysics as presented herein. There is, nevertheless, a significant body of interesting modern work in the area of ergodic theory and its relationship to statistical physics. This material is accessible only to those who have sufficient background, but several articles help considerably in providing the necessary preparation. (Arnol'd and Avez [1968]; Caldirola [1961]; Crawford, Lott and Rieke [1982]; Eckman and Ruelle [1985]; Falkoff [1961]; ter Haar [1954]; Helleman [1980]; Rasetti [1986]; Reichl [1980]; Truesdell [1961])

Rasetti devotes more than half of his book to the classical-mechanical background that underlies the ergodic-theory approach to statistical mechanics. His book, while quite rigorous and mathematical, is written to be read by students of physics. It offers an excellent pathway to this important and currently active approach. The older paper by Truesdell is carefully done and rich in both historical detail and mathematical nicety. Arnol'd and Avez have written what is generally regarded as the current classic development of the new mechanics. It is presented somewhat more in the style used by mathematicians than are the previous two references, but it is clearly written, and it should become accessible with no more than modest help from Rasetti. Reichl's book is another for physicists, but with less coverage on the topic of ergodic theory than the others offer. The review papers by Eckman and Ruelle, and by Helleman are also written so that physicists can read them. The former is a good review of the theory of chaos and its relationship to dimensionality and attractors. Helleman's paper points out that ergodicity and the approach to equilibrium do not hold for most Hamiltonian systems. Crawford *et al.* have written a pedagogical paper that is pertinent to statistical mechanics. This is a fascinating area of study in its own right. The fact that the information-theoretic approach to the study of statistical mechanics is able to bypass the need for this material in the development of the subject should not be taken as implication that these studies are unimportant in the overall background of a physicist. This material is of great current interest, and the insights it provides about the foundations of Newtonian mechanics are profound. Laplace's demon comes to a chaotic demise.

The important concept that has appeared in the new studies of ergodic theory is that of *chaos*. When motion of a system is chaotic, its coarse-grained future is not determined by its coarse-grained past, whereas non-chaotic systems do exhibit coarse-grained determinism. Neighboring points in the phase space of a chaotic system do not remain as neighboring points; they may diverge exponentially with time. Furthermore, chaotic systems appear to be the rule, rather than the exception, in the sense that almost all dynamic systems exhibit chaotic orbits. Orbits in chaotic systems become uniformly distributed over the appropriate energy surface, and the only integral of the equations of motion is the total energy. If additional such constants of the motion were to exist, the orbit of a representative point in phase space would be confined to a surface of dimension lower than $6N - 1$, the dimension of the energy surface. The non-existence of constants of the motion, other than the energy, is necessary in order for the system to be ergodic (Arnol'd [1986]; Crutchfield, Farmer and Packard [1986]; Ford [1986]; Holden [1986]; Mayer-Kress [1986]).

From the point of view of information theory, the overwhelming dominance of chaotic systems reinforces the position that the only appropriate description of complex mechanical systems is a statistical one, and that description should be based only on the information known (and knowable) about these systems. The adoption, at the outset, of this attitude allows the subject matter of statistical thermophysics to be presented without the necessity for detailed discussion of ergodic theory or chaos.

1.9.7 Coupled Harmonic Oscillators. Since the only nontrivial many-body problem we can solve analytically is that of coupled harmonic oscillators, numerous authors have studied these systems, in one to many dimensions, with finite and infinite chains of oscillators. One should note at the outset that since these systems are so unrepresentative of many-body systems in general, conclusions based solely on their behavior should be generalized only with caution. But it also should be noted that all integrable systems can be nonlinearly transformed into each other and are thus equivalent to systems of coupled harmonic oscillators (Helleman [1980]). There exists an extensive literature on systems of coupled oscillators, not always in the context of thermalization, and usually treating motions with nearest-neighbor couplings. Several of the more pertinent references are included in the bibliography. (Agarwal [1971]; Ford [1970]; Ford and Lunsford [1970]; Hemmer [1959]; Maradudin, Montroll and Weiss [1963]; Mazur [1965]; Phillipson [1974]) There has long been the urge to formalize the connection between classical mechanics and classical statistical mechanics by exhibiting a mechanical system that evolves, *via* prescriptions provided by the equations of motion, to a state that can be regarded as equilibrium. Coupled-harmonic-oscillator systems were examined early and found unsuitable. As is well known, a system of N coupled harmonic oscillators, when set into motion, is conveniently described as a superposition of the motions of N fictitious independent, uncoupled harmonic oscillators, each oscillating at a frequency determined by the masses and spring constants of the system oscillators, with amplitudes and phases determined by the initial conditions. For nondissipative systems, the amplitude and phase of each of these *normal modes* remains constant in time, not evolving at all. Therefore, in order to devise a solvable system that can evolve in time, it was natural to try to modify the coupled-harmonic-oscillator system by introducing small nonlinear couplings as perturbations. Analytic solutions become virtually impossible, even for finite systems; perturbation calculations are difficult to interpret; and the only hope seemed to be in numerical analysis. The basic idea is that each normal mode of the system represents a degree of freedom, and at equilibrium the total energy should be equipartitioned among these degrees of freedom. The small nonlinear couplings are supposed to provide a means for transfer of energy among the approximately normal modes, thereby permitting the system energy to become equipartitioned. In the summer of 1953, on the old MANIAC computer at Los Alamos, Fermi, Pasta and Ulam [1955] studied a one-dimensional chain of 64 harmonically coupled particles with additional quadratic, cubic, or broken linear coupling terms. To their surprise, they found that these nonlinear couplings led to very little sharing of energy; their behavior was far from the physicist's intuitive notion of ergodic. Much subsequent work made it evident that sharing of energy becomes important in these cases, only among oscillators with unperturbed frequencies almost satisfying a resonance condition

$$\sum_i n_i \omega_i = 0,$$

where the n_i are integers. For a careful discussion of this point and the related KAM theorem (for Kolmogorov, Arnol'd, Moser), see Rasetti [1986], and for earlier papers, see Ford and Lunsford [1970], Hemmer [1959].

   The harmonic oscillator system chosen in Section 1.7 has strictly linear coupling, and it therefore should not equilibrate in the sense that it equipartitions energy among its normal modes. The point of the development is that equilibrium is not sought within the context of dynamics, which gives the appearance of being deterministic, but rather in the context of information theory. Our knowledge of the system, no matter how precisely it is measured initially, is seen to vanish with time, and our description of it ultimately must be given in terms of our (necessarily statistical) information about the initial conditions of increasingly remote oscillators in the heat bath. Thus we are forced by the state of our knowledge to content ourselves with a statistical, or thermal, description of the system (Huerta and Robertson [1969]; Huerta and Robertson [1971]; Huerta, Robertson and Nearing [1971]; Robertson and Huerta [1970a]; Robertson and Huerta [1970b]). Formally, these infinite harmonic oscillator chains provide a continuing influx of thermally described information about remote members of the chain, thus simulating a system in contact with a heat bath. Dynamic information about the system propagates outward and is eventually lost – a behavior that is necessary for equilibration, since otherwise traces of the initial conditions would persist forever. In agreement with the conceptual basis of the equilibration process exhibited by this model, Blatt [1959] regards statistical mechanics as the mechanics of limited, not-completely isolated systems, rather than of large complicated ones. Accordingly, interactions between the system and the outside world are invoked to govern the approach to equilibrium, not the interactions within the system itself.

Evidently there can be other ways of obliterating any artifact of initial conditions, but these other mechanisms are not as easily handled analytically. Northcote and Potts [1964] have achieved *time-reversible* dynamic ergodicity and complete energy equipartition in a finite, isolated system of harmonically coupled hard spheres that can redistribute their normal-mode energies by elastic collision. Casati *et al.* (Casati et al. [1984]; Ford [1986]) model a system of bound, one-dimensional harmonic oscillators similar to those of Fig. 2, except that (1) the system is finite, and (2) the coupling springs are replaced by free masses, so that the oscillators interact, as in the Northcote-Potts model, by elastic collisions, in this case with the free masses. This model, called by its originators the *ding-a-ling* model, equipartitions energy among the masses and springs, provided the amplitude of the initial excitation is large enough for the collisions to be chaotic.

The primary mechanisms for the loss of initial information with time are (1) chaos, and (2) imprecision. Two quotations (Ford [1983]) express what must be regarded as fundamental truths in statistical physics. First, "...nonchaotic systems are very nearly as scarce as hen's teeth, despite the fact that our physical understanding of nature is largely based on their study. On the other hand, for the much more common, overwhelmingly dominant class of chaotic systems, initial observational error grows exponentially, in general, and determinism and the continuum become meaningless in an impressively short human time scale." And second, " Newtonian dynamics has, over the centuries, twice foundered on assumptions that something was infinite when in fact it was not: the speed of light, c, and the reciprocal of Planck's constant, $1/h$. Reformulations omitting these infinities led first to special relativity and then to quantum mechanics. Complexity theory now reveals a third tacitly assumed infinity in classical dynamics, namely the assumption of infinite computational and observational precision. In consequence, Newtonian dynamics now faces a third reformulation whose impact on science may be as significant as the first two." The procedures of information theory are already in place for use in calculations that exhibit the evolution of a physical system in terms of our changing information about it, as expressed by a time-dependent probability set.

Although the system of bound, coupled oscillators treated in Section 1.7 evolves in time from any initial state to a final state that is characterized by the parameters of the heat bath, its evolution is not a good model of diffusive heat transfer. In a classical analog to this model, a finite-length bar of metal, say, is placed in thermal end-to-end contact, at $t = 0$, with two semi-infinite bars of the same composition and cross-section at some heat-bath temperature T_0. The temperature of the midpoint of the finite bar (the system) evolves in time to that of the heat-bath bars asymptotically as $t^{-1/2}$, whereas the model of Section 1.7 evolves as t^{-1}. This result is valid also in a two-dimensional system, showing that the nondiffusive energy transport of these harmonically coupled systems is not merely a consequence of the dimensionality (Linn and Robertson [1984]). The ding-a-ling model, on the other hand, does exhibit proper diffusive energy transport, even in one dimension. This result reinforces the above-quoted *caveat* about basing our understanding of nature on the behavior of nonchaotic systems. Transport processes in solids are governed by the scattering of electrons or phonons by defects in the lattices. The prevalence of diffusive transport of heat or electric charge is further evidence for the importance of chaotic processes in nature.

1.10 REFERENCES

A few of the principal references are listed here; all references are listed in the Bibliography at the end of the book.

GRANDY, JR., W. T. [1987], *Foundations of Statistical Mechanics V.I: Equilibrium Theory.* D. Reidel, Dordrecht–Boston–London, 1987.

JAYNES, E. T. [1957a], Information Theory and Statistical Mechanics. *Physical Review* **106** (1957), 620–630.

JAYNES, E. T. [1957b], Information Theory and Statistical Mechanics II. *Physical Review* **108** (1957), 171–190.

KATZ, A. [1967], *Principles of Statistical Mechanics: The Information Theory Approach.* W. H. Freeman and Company, San Francisco, CA, 1967.

ROBERTSON, H. S. AND HUERTA, M. A. [1970b], Information Theory and the Approach to Equilibrium. *American Journal of Physics* **38** (1970), 619–630.

SHANNON, C. E. [1948], The Mathematical Theory of Communication. *Bell System Technical Journal* **27** (1948), 379–424, 623-657.

SHANNON, C. E. AND WEAVER, W. [1949], *The Mathematical Theory of Communication.* Univ. of Illinois Press, Urbana, 1949.

1.11 PROBLEMS

1. Show that for a given n, with

$$\sum_{i=1}^{n} p_i = 1$$

the function $S(\{p_i\})$ of Eq. (1.4.1) takes its maximum value when $p_i = 1/n$ for all i, *i.e.*, when $S(\{p_i\}) = A(n)$.

2. Corresponding to the set of events $\{y_i\}$ and knowledge of $\langle y \rangle$, a probability set $\{p_i\}$ is assigned. It is then found that each y_i can be sorted into subcategories, a and b, with $y_{ia} = y_{ib} = y_i$ for each i, but with $p_{ia} = 2p_{ib}$ for each i. Show that a new uncertainty function $S(\{p_{ia}, p_{ib}\})$ is larger that the original $S(\{p_i\})$, and calculate a numerical value for the increase. Why does the uncertainty appear to be larger when further detailed information is introduced?

3. Consider the problem mentioned in the introduction to Chapter 1. An urn is filled with balls, each numbered 0, 1, or 2, such that the average number is 2/7. Calculate the maximum uncertainty estimate of the probabilities p_0, p_1, and p_2. Find the expectation, based on these probabilities, of $\langle n^3 \rangle - 2\langle n \rangle$.

4. Problem 3 is revised to include the additional information that $\langle n^2 \rangle = 3/7$, with $\langle n \rangle = 2/7$ as before.

a. Calculate the revised probabilities.

b. Show that the revised S is smaller than that of problem 3.

5. Calculate $\langle (n - \langle n \rangle)^2 \rangle$ for the radioactive decay of Section 1.5.1. Interpret the behavior of the ratio $\langle (n - \langle n \rangle)^2 \rangle^{1/2} / \langle n \rangle$ as a measure of the validity of Eq. (1.5.8).

6. Use the Vandermonde convolution

$$\binom{n}{m} = \sum_{k=0}^{p} \binom{p}{k}\binom{n-p}{m-k}$$

which follows from $(1+x)^n = (1+x)^{n-p}(1+x)^p$ to derive Eq. (1.5.11) from Eq. (1.5.10).

7. Solve the dog-flea problem as set up in Eq. (1.5.9) for the case of one flea. Show that the result agrees with Eq. (1.5.10) for $N = 1$.

8. Use the binomial expansion for $[w_1(1,0;t) + w_1(0,1;t)]^N$ and the appropriate expressions for $w_N(N-k,k;0)$ obtained from the solution to problem 7 to derive Eq. (1.5.10)).

9. Show that if the initial flea probabilities are given by

$$w_N(N-k,k;0) = 2^{-N}\binom{N}{k},$$

as in Eq. (1.5.11), then the probabilities $w_i(n_1, N-n,;\tau)$ calculated from Eq. (1.5.11) are independent of τ. This result further emphasizes that the binomial distribution is the equilibrium one for this problem.

10. Verify that Eq. (1.5.16) for the doubly reversed populations at time $2\tau_1$ follows from the solution to the wind-tree problem, Eq. (1.5.14), and the population interchanges at τ_1 and $2\tau_1$.

11.a. Write the equations of motion for the simple harmonic oscillator in terms of the variables $y_1 = \omega x \sqrt{m}$ and $y_2 = p/\sqrt{m}$.

b. Solve these equations for $y_1(t)$ and $y_2(t)$ in terms of the initial conditions.

c. Write the initial probability density as a product of Gaussian, y_{10} centered at u_1, and y_{20} centered at u_2, with variances α^2 and δ^2 respectively.

d. Calculate the characteristic function.

e. Calculate $\rho(y_1, y_2)$ from the characteristic function.

f. Calculate the entropy, and show that it is a constant, dependent only on the covariance matrix.

g. Now consider ω to have a Gaussian probability density, centered at ω_0, with variance ω_1^2. Replace the elements of the covariance matrix with their ω averages, and show that the entropy evolves to a reasonable equilibrium value. Discuss the final result.

12. Derive the equilibrium probability set of Eq. (1.5.11), $w_N(n, N-n)$, directly from information theory. Note that $w_N(n)$ is composed of a number of equivalent flea-population distributions, because the fleas are individual and localizable, though considered indistinguishable. Therefore there are $\binom{N}{n}$ ways of having a state for which there are n fleas on dog 1. (See Appendix C.)

1.12 SOLUTIONS

1. With $\sum p_i = 1$, use the method of undetermined multipliers to write

$$\delta\left[S/c - \alpha' \sum_i p_i \right] = -\sum \delta p_i \left(1 + \ln p_i + \alpha'\right) = 0.$$

Let $\alpha = \alpha' + 1$. Since the δp_i are now independent, we have

$$p_i = e^{-\alpha}$$

for all i, and it follows that

$$p_i = 1/n.$$

To check for maximum, examine the second variation of S

$$\delta^2 (S/c) = - \sum_i (\delta p_i)^2 / p_i \leq 0.$$

Therefore S/c is a convex function and its extremum must be a maximum.

2.

$$S_1 (\{p_i\}) = -c \sum_i p_i \ln p_i$$

$$S_2 (\{p_{ia}, p_{ib}\}) = -c \sum_i p_{ia} \ln p_{ia} - c \sum_i p_{ib} \ln p_{ib}.$$

Since

$$p_i = p_{ia} + p_{ib} \text{ and } p_{ia} = 2p_{ib},$$

it follows that

$$p_{ib} = p_i/3,$$

and

$$S_2 = -c \sum_i (2p_i/3) \ln (2p_i/3) - c \sum_i (p_i/3) \ln (p_i/3)$$

$$= S_1 + c \ln 3 - (c/3) \ln 4 > S_1.$$

Although further information is introduced, the number of possible events is doubled. The new information does not help in the choice of original events y_i. The new uncertainty function is greater than the original but less than it would have been if no further information had been supplied about p_{ia} and p_{ib}.

3.

$$p_0 = 1/z; \quad p_1 = e^{-\beta}/z; \quad p_2 = e^{-2\beta}/z; \quad z = 1 + e^{-\beta} + e^{-2\beta}$$

Let $e^{-\beta} = x$. Then $z = 1 + x + x^2$, and

$$2/7 = x/z + 2x^2/z, \quad \text{or} \quad x = 1/4; \quad p_0 = 16/21; \quad p_1 = 4/21; \quad p_2 = 1/21.$$

Finally, since $\langle n^3 \rangle = 4/7$, it follows that

$$\langle n^3 \rangle - 2 \langle n \rangle = 0.$$

4. This problem is now determinate; no variational calculation is necessary (but it is possible and leads to the correct results).

$$p_0 + p_1 + p_2 = 1,$$

$$p_1 + 2p_2 = 2/7,$$

and

$$p_1 + 4p_2 = 3/7.$$

By simple algebra, it follows that a) $p_0 = 11/14$; $p_1 = 1/7$; $p_2 = 1/14$.
b) For problem 3,

$$S/c = (16/21)\ln(21/16) + (4/21)\ln(21/4) + (1/21)\ln 21$$
$$= 0.668.$$

For problem 4,

$$S/c = (11/14)\ln(14/11) + (1/7)\ln 7 + (1/14)\ln 14$$
$$= 0.656 < 0.668.$$

5.

$$\left\langle \left(n - \langle n\rangle\right)^2 \right\rangle = \langle n^2\rangle - \langle n\rangle^2.$$

From Eq. (1.5.6), $\langle n(t)\rangle = Ne^{-\alpha t}$. By Eq. (1.5.5),

$$\langle n^2\rangle = \sum_{n=0}^{N} n^2 \binom{N}{n} \left(1 - e^{-\alpha t}\right)^{N-n} \left(e^{-\alpha t}\right)^n$$
$$= N(N-1)e^{-2\alpha t} + Ne^{-\alpha t}$$
$$= \langle n\rangle^2 + \langle n\rangle - \langle n\rangle^2 /N.$$

From this it may be seen that

$$\frac{\left(\langle n^2\rangle - \langle n\rangle^2\right)^{1/2}}{\langle n\rangle} = \left(\frac{1}{\langle n\rangle} - \frac{1}{N}\right)^{1/2}.$$

The result, called the *fractional standard deviation* or *coefficient of variation*, is particularly convenient for very large N so that only the first term need be considered. Evidently the simple result in Eq. (1.5.8) is quite accurate for large $\langle n\rangle$.

6. From Eq. (1.5.10) at $t \to \infty$,

$$w_N(n_1, N - n_1, \infty) \to 2^{-N} \sum_{k=0}^{N} \sum_{m=0}^{k} \binom{k}{m}\binom{N-k}{n_1 - k + m} w_N(N - k, k, 0)$$
$$= 2^{-N} \sum_{k=0}^{N} w_N(N - k, k, 0) \sum_{m=0}^{k} \binom{k}{m}\binom{N-k}{N - n_1 - m}$$
$$= 2^{-N} \sum_{k=0}^{N} w_N(N - k, k, 0) \binom{N}{n_1} = 2^{-N}\binom{N}{n_1}$$

7. Let $w_1(1, 0;\ \tau) \equiv w_a(\tau)$; $w_1(0, 1;\ \tau) \equiv w_b(\tau)$. Then

$$\dot{w}_a = -w_a + w_b$$
$$\dot{w}_b = w_a - w_b,$$

and

$$\ddot{w}_a(\tau) = -\dot{w}_a + \dot{w}_b = -\dot{w}_a + w_a - w_b = -\dot{w}_a + w_a - (\dot{w}_a + w_a) = -2\dot{w}_a.$$

$$\dot{w}_a(\tau) = \dot{w}_a(0)e^{-2\tau} = \left[-w_a(0) + w_b(0) \right] e^{-2\tau}$$

$$w_a(\tau) = -\left[-w_a(0) + w_b(0) \right] e^{-2\tau}/2 + c,$$

$$w_a(0) = -\left[-w_a(0) + w_b(0) \right]/2 + c,$$

or

$$c = \left[w_a(0) + w_b(0) \right]/2.$$

It follows that

$$w_a(\tau) = e^{-\tau} \left[w_a(0) \cosh\tau + w_b(0) \sinh\tau \right],$$

and

$$w_b(\tau) = e^{-\tau} \left[w_a(0) \sinh\tau + w_b(0) \cosh\tau \right].$$

Second method: Let

$$P = \begin{pmatrix} w_a(\tau) \\ w_b(\tau) \end{pmatrix}.$$

Then the two d.e.'s become $\dot{P} = AP$, where

$$A \equiv \begin{pmatrix} -1 & 1 \\ 1 & -1 \end{pmatrix}$$

The formal solution is $P(\tau) = e^{A\tau}P(0)$. Note that $A^n = (-2)^{n-1}A$. Therefore $e^{A\tau}$ can be written

$$e^{A\tau} = \sum_{n=0}^{\infty} A^n \tau^n / n! = I + A/2 - e^{-2\tau}A/2 = e^{-\tau} \begin{pmatrix} \cosh\tau & \sinh\tau \\ \sinh\tau & \cosh\tau \end{pmatrix},$$

giving the same result as the previous method.

8. Let $w_1(1, 0; \tau) \equiv w_a(\tau)$; $w_1(0, 1; \tau) \equiv w_b(\tau)$. Then

$$w_N(n_1, N - n_1; \tau) = \binom{N}{n_1} w_a^{n_1} w_b^{N-n_1},$$

where

$$w_a(\tau) = e^{-\tau} \left[w_a(0) \cosh\tau + w_b(0) \sinh\tau \right], \text{ and}$$

$$w_b(\tau) = e^{-\tau} \left[w_a(0) \sinh\tau + w_b(0) \cosh\tau \right].$$

Note that, with $c \equiv \cosh\tau$ and $s \equiv \sinh\tau$,

$$w_a^{n_1}(\tau) = e^{-n_1\tau} \sum_i \binom{n_1}{i} w_a^i(0) w_b^{n_1-i}(0) c^i s^{n_1-i},$$

$$w_b^{N-n_1}(\tau) = e^{-(N-n_1)\tau} \sum_j \binom{N-n_1}{j} w_a^j(0) w_b^{N-n_1-j}(0) s^j c^{N-n_1-j}.$$

Then, as in Eq. (1.5.10), with $w_N(n_1, N - n_1; \tau) = w_N(n_1; \tau)$,

$$w_N(n_1; \tau) = \binom{N}{n_1} e^{-N\tau} \sum_{j,i} \binom{n_1}{i}\binom{N-n_1}{j} w_a^{i+j}(0) w_b^{N-(i+j)}(0) c^{N+i-n_1-j} s^{n_1+j-i},$$

$$\text{(with } i+j=r), = e^{-N\tau} \sum_{r,i} \binom{N}{n_1}\binom{n_1}{i}\binom{N-n_1}{r-i}\binom{N}{r}^{-1} w_N(r;0) c^{N+2i-n_1-r} s^{n_1+r-2i}$$

$$\text{(with } r=N-k), = e^{-N\tau} \sum_{k,i} \binom{N-k}{i}\binom{k}{n_1-i} w_N(N-k;0) c^{k+2i-n_1} s^{N+n_1-k-2i},$$

$$\text{(with } m=k-n_1+i) = e^{-N\tau} \sum_{k,m} \binom{N-k}{n_1-k+m}\binom{k}{m} w_N(N-k;0) c^{n_1+2m-k} s^{N-(n_1+2m-k)}.$$

9.

$$w_N(n_1,n_2;\tau) = (2e^{\tau})^{-N} \sum_{k,m} \binom{k}{m}\binom{N-k}{n_1-k+m}\binom{N}{k} c^{\gamma} s^{N-\gamma}, \text{ with } \gamma = n_1 - k + 2m,$$

$$= (2e^{\tau})^{-N} \sum_{k,m} \binom{N}{n_1}\binom{n_1}{k-m}\binom{N-n_1}{m} c^{\gamma} s^{N-\gamma}$$

$$= (2e^{\tau})^{-N} \binom{N}{n_1} \sum_{q,m} \left[\binom{n_1}{q} c^{n_1-q} s^{q}\right]\left[\binom{N-n_1}{m} c^{m} s^{N-n_1-m}\right]$$

$$= (2e^{\tau})^{-N} \binom{N}{n_1} (c+s)^{n_1} (c+s)^{N-n_1} = 2^{-N}\binom{N}{n_1}, \text{ as given,}$$

independent of τ.

10. Write Eq. (1.5.14) as $N(\tau) = e^{-2\tau} A(\tau) N(0)$, where $N(\tau)$ is the column matrix of the $n_i's$ and $A(\tau)$ is the 4×4 matrix of hyperbolic functions. Write the first transformation as $N^*(\tau_1) = BN(\tau_1)$, where B is given by

$$B = \begin{pmatrix} 0 & 0 & 1 & 0 \\ 0 & 0 & 0 & 1 \\ 1 & 0 & 0 & 0 \\ 0 & 1 & 0 & 0 \end{pmatrix}.$$

Note that $B^2 = I$ and $BA = AB$. For $\tau_1 < \tau < 2\tau_1$, the evolution of $N^*(\tau)$ is given by $N^*(\tau) = e^{-2\tau} A(\tau - \tau_1) N^*(\tau_1)$; at $\tau = 2\tau_1$ this expression becomes $N^*(2\tau_1) = e^{-2\tau_1} A(\tau_1) N^*(\tau_1)$. The second transformation is

$$N^{**}(2\tau_1) = B^{-1} N^*(2\tau_1) = e^{-4\tau_1} B^{-1} A(\tau_1) BA(\tau_1) N(0),$$

or

$$N^{**}(2\tau_1) = e^{-4\tau_1} A^2(\tau_1) N(0) = e^{-4\tau_1} A(2\tau_1) N(0) = N(2\tau_1),$$

where the conversion $A^2(\tau_1) = A(2\tau_1)$ may be obtained either by direct matrix multiplication and use of the hyperbolic-function identities or more generally from the iterated solution:

$$N(\tau_1 + \tau_2) = e^{-2\tau_2} A(\tau_2) N(\tau_1) = e^{-2(\tau_1+\tau_2)} A(\tau_2) A(\tau_1) N(0),$$

from which it follows that

$$A(\tau_1 + \tau_2) = A(\tau_2) A(\tau_1).$$

11.a.

$$H(p,x) = p^2/2m + m\omega^2 x^2/2; \quad \omega^2 = k/m.$$

$$\frac{\partial H}{\partial p} = \dot{x} = p/m,$$

$$\frac{\partial H}{\partial x} = -\dot{p} = m\omega^2 x,$$

the equations of motion.

Let $x = y_1/\omega\sqrt{m}$ and $p = y_2\sqrt{m}$. With these substitutions, the equations of motion become

$$\dot{y}_1 = \omega y_2; \quad \dot{y}_2 = -\omega y_1, \text{ and}$$

b.

$$\ddot{y}_1 = \omega\dot{y}_2 = -\omega^2 y_1, \text{ and } \ddot{y}_2 = -\omega\dot{y}_1 = -\omega^2 y_2.$$

$$y_1(t) = y_1(0)\cos\omega t + y_2(0)\sin\omega t,$$
$$y_2(t) = -y_1(0)\sin\omega t + y_2(0)\cos\omega t.$$

c.

$$\rho_0 = \frac{\exp\left\{-[y_1(0) - u_1]^2/2\alpha^2\right\}}{\alpha\sqrt{2\pi}} \cdot \frac{\exp\left\{-[y_2(0) - u_2]^2/2\delta^2\right\}}{\delta\sqrt{2\pi}}$$

d.

$$\Phi(k_1, k_2) = \int_{-\infty}^{\infty} \int_{-\infty}^{\infty} \rho_0 e^{ik_1 y_1(t)} e^{ik_2 y_2(t)} \, dy_1(0) \, dy_2(0).$$

Let $y_1'(t) = y_1(t) - \langle y_1(t)\rangle$, etc., where

$$\langle y_1(t)\rangle = \int\int y_1(t)\rho_0 \, dy_1(0) \, dy_2(0)$$
$$= u_1 \cos\omega t + u_2 \sin\omega t.$$

Similarly,

$$\langle y_2(t)\rangle = -u_1 \sin\omega t + u_2 \cos\omega t.$$

Then write

$$\Phi(k_1, k_2) = \int\int \rho_0 \exp\left\{ik_1\left[y_1(0)\cos\omega t + y_2(0)\sin\omega t\right]\right.$$
$$+ ik_2\left[-y_1(0)\sin\omega t + y_2(0)\cos\omega t\right]\left.\right\} dy_1(0)dy_2(0)$$
$$= \int\int \rho_0 \exp\left\{ik_1\left[(y_1'(0) + u_1)\cos\omega t + (y_2'(0) + u_2)\sin\omega t\right] + \right.$$
$$+ ik_2[\text{corresponding terms}]\left.\right\} dy_1'(0)dy_2'(0),$$

where ρ_0 is now to be expressed in terms of $y_1'(0)$ and $y_2'(0)$. By means of the standard formula following Eq. (1.7.20),

$$\Phi(k_1, k_2) = \exp\left[ik_1(u_1\cos\omega t + u_2\sin\omega t) + ik_2(-u_1\sin\omega t + u_2\cos\omega t)\right] \times$$
$$\times \exp\left[-(\alpha^2/2)(k_1\cos\omega t - k_2\sin\omega t)^2 - (\delta^2/2)(k_1\sin\omega t + k_2\cos\omega t)^2\right].$$

e. The Fourier inverse of $\Phi(k_1, k_2)$ is given by

$$\rho(y_1(t), y_2(t)) = \int_{-\infty}^{\infty} \int_{-\infty}^{\infty} e^{-i(k_1 y_1 + k_2 y_2)} \Phi(k_1, k_2) \left[\frac{dk_1}{2\pi}\right] \left[\frac{dk_2}{2\pi}\right]$$

$$= \int \int e^{-i(k_1 y_1' + k_2 y_2')} \exp\left[-\left(\frac{\alpha^2}{2}\right)(k_1 \cos \omega t - k_2 \sin \omega t)^2 + \right.$$

$$\left. -\left(\frac{\delta^2}{2}\right)(k_1 \sin \omega t + k_2 \cos \omega t)\right] \frac{dk_1}{2\pi} \frac{dk_2}{2\pi}.$$

Let $K_1 = k_1 \cos \omega t - k_2 \sin \omega t$, $K_2 = k_1 \sin \omega t + k_2 \cos \omega t$, and note that

$$\frac{\partial(K_1, K_2)}{\partial(k_1, k_2)} = 1.$$

It follows that, with $s = \sin \omega t$ and $c = \cos \omega t$,

$$\rho(y_1, y_2) = \int \int \exp\left[-iy_1'(K_1 c + K_2 s) - iy_2'(-K_1 s + K_2 c)\right] \times$$

$$\times \exp\left[-\left(\frac{\alpha^2 K_1^2}{2}\right) - \left(\frac{\delta^2 K_2^2}{2}\right)\right] \frac{dK_1}{2\pi} \frac{dK_2}{2\pi}$$

$$= \frac{\exp\left[-(y_1' c - y_2' s)^2 / 2\alpha^2\right]}{\alpha\sqrt{2\pi}} \frac{\exp\left[-(y_1' s + y_2' c)^2 / 2\delta^2\right]}{\delta\sqrt{2\pi}} = \rho_0.$$

Observe that the Liouville theorem permits $\rho(y_1, y_2)$ to be derived from ρ_0 merely by replacement of $y_1(0)$ and $y_2(0)$ by the expressions

$$y_1(0) = y_1(t) \cos \omega t - y_2(t) \sin \omega t, \text{ and}$$

$$y_2(0) = y_1(t) \sin \omega t + y_2(t) \cos \omega t,$$

obtained from part b. The derivation given here is intended to illustrate the techniques of Section 1.7 and to reinforce the content of the Liouville theorem by actual calculation. It is one thing to know intellectually the content of the Liouville theorem, but quite another to verify it by direct calculation.

f. In $y_1 y_2$ space, the density ρ_0 must be multiplied by $\hbar\omega$ to yield the dimensionless quantity required as the argument of a logarithm, where $\hbar$ has the units of action. In this case, then,

$$S = -k \int \int \rho_0 \ln(\rho_0 \hbar\omega) \, dy_1'(0) \, dy_2'(0) = k\left[1 + \ln(\alpha\delta/\hbar\omega)\right]$$

by direct calculation. Since ρ is constant, the entropy does not change in time. Note that the covariance matrix is

$$W = \begin{pmatrix} \langle y_1^2 \rangle - \langle y_1 \rangle^2 & \langle y_1 y_2 \rangle - \langle y_1 \rangle \langle y_2 \rangle \\ \langle y_1 y_2 \rangle - \langle y_1 \rangle \langle y_2 \rangle & \langle y_2^2 \rangle - \langle y_2 \rangle^2 \end{pmatrix} = \begin{pmatrix} \langle (y_1')^2 \rangle & \langle y_1' y_2' \rangle \\ \langle y_2' y_1' \rangle & \langle (y_2')^2 \rangle \end{pmatrix}$$

$$= \begin{pmatrix} \alpha^2 \cos^2 \omega t + \delta^2 \sin^2 \omega t & (\delta^2 - \alpha^2) \sin \omega t \cos \omega t \\ (\delta^2 - \alpha^2) \sin \omega t \cos \omega t & \alpha^2 \sin^2 \omega t + \delta^2 \cos^2 \omega t \end{pmatrix},$$

from which it follows that $\det W = a^2 \delta^2$, and S is given by Eq. (1.7.53).

g.

$$\rho_\omega = \frac{\exp\left[-(\omega - \omega_0)^2 / 2\omega_1^2\right]}{\omega_1 \sqrt{2\pi}}$$

Use the formula

$$\int_{-\infty}^{\infty} e^{i\omega t} \rho_\omega \, d\omega = e^{i\omega_0 t - (\omega_1 t)^2 / 2}$$

to calculate all the integrals necessary for ω-averaging. The elements of the ω-averaged covariance matrix, found from the individual matrix terms *before* subtraction, are

$$\langle \bar{y}_1^2 \rangle - \langle \bar{y}_1 \rangle^2 = \frac{\alpha^2 + \delta^2}{2} + \left[\frac{\alpha^2 - \delta^2}{2}\right] e^{-2\omega_1^2 t^2} \cos 2\omega_0 t +$$

$$+ \frac{(1 - e^{-\omega_1^2 t^2})}{2} \left[(u_1^2 + u_2^2) - (u_1^2 - u_2^2)e^{-\omega_1^2 t^2} \cos 2\omega_0 t - 2u_1 u_2 e^{-\omega_1^2 t^2} \sin 2\omega_0 t\right],$$

$$\langle \bar{y}_2^2 \rangle - \langle \bar{y}_2 \rangle^2 = \frac{\alpha^2 + \delta^2}{2} + \left[\frac{\alpha^2 - \delta^2}{2}\right] e^{-2\omega_1^2 t^2} \cos 2\omega_0 t +$$

$$+ \frac{(1 - e^{-\omega_1^2 t^2})}{2} \left[(u_1^2 + u_2^2) + (u_1^2 - u_2^2)e^{-\omega_1^2 t^2} \cos 2\omega_0 t + 2u_1 u_2 e^{-\omega_1^2 t^2} \sin 2\omega_0 t\right],$$

and

$$\langle \bar{y}_1 \bar{y}_2 \rangle - \langle \bar{y}_1 \rangle \langle \bar{y}_2 \rangle = \left[\frac{e^{-\omega_1^2 t^2}}{2}\right] \left[(\delta^2 - \alpha^2)e^{-\omega_1^2 t^2} \sin 2\omega_0 t +\right.$$

$$\left.+ (u_1^2 - u_2^2)(1 - e^{-\omega_1^2 t^2}) \sin 2\omega_0 t - 2u_1 u_2 (1 - e^{-\omega_1^2 t^2}) \cos 2\omega_0 t\right],$$

where $^-$ denotes ω-averaging, $\langle \ \rangle$ denotes phase-space averaging, and the order of the two is commutative. At $t = 0$, these matrix elements equal the initial values of the matrix elements in part (f) and lead to an initial entropy equal to the entropy of part (f). As $t \to \infty$, the covariance matrix becomes

$$W(\infty) = \frac{1}{2} \begin{pmatrix} \alpha^2 + \delta^2 + u_1^2 + u_2^2 & 0 \\ 0 & \alpha^2 + \delta^2 + u_1^2 + u_2^2, \end{pmatrix}$$

and the entropy at the equivalent of equilibrium is

$$S = k\left[1 + \ln(kT/\hbar\omega)\right],$$

where

$$kT = (\alpha^2 + \delta^2 + u_1^2 + u_2^2)/2.$$

The important point about this result is that when phase information is lost, as a consequence of the uncertainty in ω, all of the energy evolves into what can be described only as *thermal* energy. This means that it becomes impossible to extract work from the system except by using it as a heat source, subject to the limitations imposed by the Carnot theorem. Further examination of this problem appears in Chapter 9, beginning on p. 456.

12. Suppose the fleas are regarded as identifiable individuals. Let p_n be one of the $\binom{N}{n}$ equivalent ways of having n fleas on dog 1. If the entropy were defined in terms of the $\{p_n\}$, such that each configuration equivalent to p_n contributes $-p_n \ln p_n$ to the entropy, then

$$S/k = -\sum_n \binom{N}{n} p_n \ln p_n,$$

which, together with

$$1 = \sum_n \binom{N}{n} p_n,$$

and

$$\langle n \rangle = N/2 = \sum_n n \binom{N}{n} p_n,$$

provides an equation set that leads to the variational equation

$$0 = \sum_n \delta p_n \binom{N}{n} \left(\ln p_n + \alpha' + 1 + \beta n \right).$$

This gives $p_n = \exp(-\alpha - \beta n)$. Let $x = e^{-\beta}$, so that

$$1 = e^{-\alpha} \sum_n \binom{N}{n} x^n = e^{-\alpha} (1 + x)^N,$$

or

$$p_n = x^n / (1 + x)^N.$$

The equation for $\langle n \rangle$ now becomes

$$\frac{N}{2} = \frac{1}{(1+x)^N} \sum_n n \binom{N}{n} x^n = \frac{x}{(1+x)^N} \frac{d}{dx} (1+x)^N = \frac{Nx}{1+x}.$$

This gives $x = 1$, $\beta = 0$, $p_n = 1/2^N$, and the weighted probabilities are $w_N(n) = 2^{-N} \binom{N}{n}$, as required.

2. *THERMOSTATICS*

2.1 INTRODUCTION

Statistical mechanics has often been presented as providing the microphysical foundation for thermodynamics. On the other hand, some thermodynamicists have insisted that their subject is self-contained and based firmly on experiment. They have seemed to regard the development of thermodynamics as a sort of game, the object being to avoid any appeal to microphysics for support, guidance, or understanding, on the grounds that such an appeal is not helpful and that the edifice of thermodynamics should stand undamaged even if the laws of microphysics turn out to be wrong.

The attitude adopted herein is that the physicist's objective is to understand his subject matter in whatever ways are open to him. He will examine macroscopic and microscopic concepts and treatments, mathematical models and experimental results, logical analysis and heuristic description, all as legitimate tools for their proper purposes. Therefore, thermodynamics will be presented in a manner that relates it closely to statistical thermophysics and in particular to the information-theoretic approach. The successes of the approach in yielding the conventional results of thermodynamics, including the identification of the defined temperature with Carnot and ideal-gas temperatures, will be regarded as satisfactory evidence, not rigorous proof, that the development of thermodynamics is sound in itself and that the tie with statistical thermophysics is viable, without any insistence that the subject of thermodynamics is unable to stand without the support of its underlying microphysics.

2.2 THE STATISTICAL MODEL

In this section a statistical model of thermostatics will be explored, based on the information entropy of Sections 1.4 and 1.8. The properties of this model will be used in the next section to suggest a postulational basis for thermostatics. The postulates will be shown to yield the standard results of macroscopic thermostatics, insofar as they are needed for later developments. No attempt is made to present the full context of thermodynamics, which includes not only equilibrium thermostatics, but also nonequilibrium processes with time dependence, heat and matter flows, and other processes that justify the usage of *dynamics*, in this chapter. In the remainder of this chapter, the term *thermodynamics* is taken to mean equilibrium thermodynamics, or thermostatics.

A macroscopic system, taken as a quantum system with a countable set of states (with the energy level of the nth state denoted by E_n) is known from Section 1.8, to have an energy average, or expectation value, given by

$$\langle \bar{E} \rangle = \sum_n E_n p_n, \qquad (2.2.1)$$

and an information entropy

$$S = -k \sum_n p_n \ln p_n, \qquad (2.2.2)$$

where a representation is used in which the statistical operator ρ is diagonal. The equilibrium probabilities, for Eq. (1.8.12) are

$$p_n = e^{-\beta E_n}/Z, \tag{2.2.3}$$

where

$$Z = \sum_n e^{-\beta E_n}. \tag{2.2.4}$$

The use of (2.2.3) in (2.2.2), with the aid of (2.2.1), gives

$$S = -k \sum_n p_n \left[-\beta E_n - \ln Z \right]$$
$$= k\beta \left\langle \bar{E} \right\rangle + k \ln Z. \tag{2.2.5}$$

The differential of S is given by

$$dS/k = \beta d \left\langle \bar{E} \right\rangle + \left\langle \bar{E} \right\rangle d\beta + dZ/Z. \tag{2.2.6}$$

From (2.2.4), dZ/Z is found to be

$$dZ/Z = -\left(\sum_n E_n e^{-\beta E_n}/Z \right) d\beta - \beta \sum_n dE_n e^{-\beta E_n}/Z. \tag{2.2.7}$$

The first term is easily seen to be $- \left\langle \bar{E} \right\rangle d\beta$, but the second requires more examination. The energy levels E_n are, say, N-body energy levels of a collection of (perhaps interacting) particles in a definite bounded region of space. These may be simply N noninteracting particles in a box of volume V, but they may also be the atoms of a block of metal or a container of liquid. The states identified by the index n are N-body eigenstates of the energy, not necessarily reducible to single-particle terms. These energy eigenstates are the complete set of such N-body states for the system; *i.e.*, the system may not ever be in a particular eigenstate, but the state nevertheless exists. The differential dE_n represents a shift in an energy *level*, and it therefore has nothing to do with temperature. The energy levels depend upon geometry and total population, N, as well as, perhaps, such properties as total magnetic or electric moment. For present purposes, it is sufficient to choose, in addition to N, just one geometric variable, the volume V, to represent the E_n dependencies, giving

$$dE_n = \left(\frac{\partial E_n}{\partial V} \right)_N dV + \left(\frac{\partial E_n}{\partial N} \right)_V dN. \tag{2.2.8}$$

Somewhat more generally, the first term here could be replaced by $\sum_{i=1}^{6} \left(\frac{\partial E_n}{\partial a_i} \right) da_i$, where the da_i are the six independent components of the (symmetric) deformation tensor. Terms of this sort can always be introduced when needed, but they are unnecessarily awkward to carry along in the course of a conceptual development. Equation (2.2.7) can now be written

$$dZ/Z = -\left\langle \bar{E} \right\rangle d\beta - \frac{\beta dV}{Z} \sum_n \left(\frac{\partial E_n}{\partial V} \right)_N e^{-\beta E_n} - \frac{\beta dN}{Z} \sum_n \left(\frac{\partial E_n}{\partial N} \right)_V e^{-\beta E_n}. \tag{2.2.9}$$

The second term must have the dimensions of βE to match the first term; therefore the coefficient of βdV must have the units of pressure. It may be argued that since the average of $(\partial E_n/\partial V)_N \, dV$ can be understood only as the work of deformation in changing the energy levels of the system, the term must be $-PdV$ (or an appropriate tensor generalization); this identification requires the further remark that the probability factors in the term ensure that it represents a change in system energy and not merely a shifting of unoccupied levels. (A powerful and interesting generalization of this paragraph is to be found in Section 9.10.2.)

Similarly, but less familiarly, the final term of Eq. (2.2.9) represents energy change as a consequence of change in the number of particles in the system, and it presumably can be written in terms of the chemical potential μ, giving the tentative expression for Eq. (2.2.9) as

$$dZ/Z = -\left\langle \bar{E} \right\rangle d\beta + \beta PdV - \beta\mu dN. \tag{2.2.10}$$

Just as the second term could be generalized to include other forms of work, so the third term can be generalized to allow for the presence of different kinds of molecules, becoming $-\beta \sum_i \mu_i dN_i$. If N_i is strictly the number of molecules of the ith species, then it is an integer and therefore not differentiable. If it is regarded as an expectation value, however, no conceptual difficulty arises from use of dN_i as a differential. For brevity, μdN will be written unless a sum is needed. Equations (2.2.10) and (2.2.5) combine to give

$$dS/k\beta = d\left\langle \bar{E} \right\rangle + PdV - \mu dN, \tag{2.2.11}$$

where further evidence is still needed for a working identification of $k\beta$. The interpretation of P as pressure and μ as chemical potential is shown to behave as expected in the subsequent development.

Suppose two independent systems exist, each satisfying Eqs.(2.2.1) through (2.2.4), with generally different sets of energy levels. The two sets of independent probabilities, say p_n and w_m may be combined to form a single set of probabilities W_{mn} given by

$$W_{mn} = p_n w_m. \tag{2.2.12}$$

The combined entropy is then found to be

$$
\begin{aligned}
S/k &= -\sum_{m,n} W_{mn} \ln W_{mn} \\
&= -\sum_{m,n} W_{mn} \ln (p_n w_m) \\
&= -\sum_{m,n} p_n w_m (\ln p_n + \ln w_m) \\
&= -\sum_n p_n \ln p_n - \sum_m w_m \ln w_m \\
&= (S_1 + S_2)/k, \tag{2.2.13}
\end{aligned}
$$

since $\sum_n p_n = \sum_m w_m = 1$. Thus the entropies of the two systems are additive. If the systems are put into weak thermal contact, so that their energy level structures are essentially unperturbed, exchange of energy should take place until the combined entropy attains its maximum value. Two points here require amplification. First, "weak" contact implies that each system has an identifiable energy of its own, as, for example, the coupled oscillators of Section 1.7. Second, the phrase "weak thermal contact" seems to imply a knowledge of the meaning of "thermal," which will be considered by some readers as a violation of the rules. The point is that thermal contact defines a mode of energy transfer that is not identifiable as any kind of macroscopic work. There can be, for a strictly thermal contact, no macroscopic deformations or displacements of boundaries, no known processes involving work, including change of polarization, by externally applied fields, and no exchange of matter across the boundaries. By "thermal contact," the physicist means an interaction permitting the transfer of energy into and out of the unobservable microscopic modes of the system, usually associated with the exchange of kinetic energy in molecular collision processes. If, however, other hidden processes exist that might be called work and used to transfer energy to macroscopic variables *if they were known*, they too can be included in the so-called thermal, or heat transfer, process as long as they remain unavailable for the transfer of energy by (macroscopic) work processes. The basic and essential difference between heat and work, between thermal and nonthermal processes, is in the state of availability of the knowledge necessary to define and permit the transfer of energy by means of macroscopic variables.

Historically the distinction between heat and work was made for entirely practical reasons and without any very consistent or objection-free prescription for distinguishing the two. Heat was transferred to a boiler by building a fire under it, and work was extracted from the steam engine *via* a movable piston, clearly and without subtleties, doubts, or confusion. But the realization of the true nature of the distinction between heat and work has been rather slow in arriving, and the logical structure of the subject has suffered because authors have attempted distinctions based on spurious reasoning or circular definitions.

We now return to the two systems in weak thermal contact, with the further proviso that no energy is to be transferred by work processes or matter flow. The total energy U is taken to be constant, given by

$$U = U_1 + U_2, \tag{2.2.14}$$

where U_i has been written for $\langle \bar{E} \rangle_i$, the expectation value of the energy of system i, and the thermodynamic U is used to avoid confusion with a particular energy level E_i of one of the systems. The entropy from Eq. (2.2.13) is

$$S = S_1 + S_2, \tag{2.2.15}$$

and the objective is to maximize S, subject to the constraint that U is constant and the symbolic work and matter-transfer terms of (2.2.11) are zero. When S is maximum, its variation δS should be zero, where

$$\begin{aligned} \delta S &= \delta S_1 + \delta S_2 \\ &= k\beta_1 \delta U_1 + k\beta_2 \delta U_2, \end{aligned} \tag{2.2.16}$$

from Eq. (2.2.11) with $\delta V_1 = 0 = \delta V_2$ and $\delta N_1 = 0 = \delta N_2$. But from Eq. (2.2.14) and the constancy of U, we may write $\delta U_2 = -\delta U_1$, and Eq. (2.2.16) becomes

$$\delta S = k\left(\beta_1 - \beta_2\right)\delta U_1. \tag{2.2.17}$$

When S is maximum, $\delta S = 0$ for small, nonzero variations δU_1 of either sign, from which it follows that

$$\beta_1 = \beta_2 \qquad\qquad (2.2.18)$$

at entropy maximum, or equilibrium. We attempt to incorporate into the theory the expectation that this purely thermal equilibrium should result in equality of the two system temperatures, so the equality of the β's should be tantamount to equality of temperature. Energy levels are, as a rule, bounded from below but not from above, implying that if the partition function is to be finite in Eq. (2.2.4), the parameter β should be positive. Furthermore, we expect that if the two systems are not in equilibrium, the energy should flow in such a direction as to increase S toward its maximum value; *i.e.*, the energy should flow from the hotter to the cooler subsystem. In Eq. (2.2.17), with $\delta S > 0$ and δU_1 chosen positive, it follows that $\beta_1 > \beta_2$. But if δU_1 is positive, energy must flow to system 1 from system 2, and the temperature of system 2 then should be greater than that of system 1. This argument leads to the identification

$$T \equiv 1/k\beta, \qquad\qquad (2.2.19)$$

with the residual possibility that T is not the thermodynamic temperature *per se*, but is some function of it. It will turn out, however, that the T so defined is the correct temperature and that it is also the ideal gas temperature, provided k is chosen as Boltzmann's constant and the energy levels of Eq. (2.2.3) are the correct ones.

The constraints on the interaction of the two subsystems can now be relaxed to permit not only weak thermal contact but also energy transfer mechanically. In this case, a movable piston as a common wall separating the two subsystems provides a convenient means of mechanical energy transfer, and it can also be the means of weak thermal contact. In addition to Eq. (2.2.14) and Eq. (2.2.15), which are still valid, there is added

$$V = V_1 + V_2, \qquad\qquad (2.2.20)$$

where the total volume V is taken to be constant, and $\delta V_2 = -\delta V_1$. The new equation corresponding to Eq. (2.2.17) is

$$\delta S = k\,(\beta_1 - \beta_2)\,\delta U_1 + k\,(\beta_1 P_1 - \beta_2 P_2)\,\delta V_1. \qquad\qquad (2.2.21)$$

Since U_1 and V_1 can be changed independently, it is necessary that both coefficients be zero for equilibrium, so that in addition to and because of (2.2.18), we have

$$P_1 = P_2, \qquad\qquad (2.2.22)$$

a result that agrees with the expected behavior of pressure. When the system is approaching equilibrium and, for convenience, the β's are equal, an increase in entropy δS corresponding to an increase in V_1 ($\delta V_1 > 0$) implies $P_1 > P_2$, as expected of pressure for a spontaneous increase in V_1.

A further relaxation of the constraints could permit the exchange of matter between subsystems. If both have the same composition, with only one constituent, a similar and obvious treatment gives, for equilibrium,

$$\mu_1 = \mu_2, \qquad\qquad (2.2.23)$$

and for the approach to equilibrium the result that matter flows from larger μ to smaller, the expected result for chemical potential. For a system containing a number of components, each with its own chemical potential, equilibrium implies equality of subsystem chemical potentials of each component for which interchange is possible. Thus a semipermeable membrane could permit certain components to equilibrate while others remain separated.

In terms of the identifications so far, Eq. (2.2.11) can now be written as

$$dS = \frac{1}{T}dU + \frac{P}{T}dV - \frac{\mu}{T}dN, \tag{2.2.24}$$

implying that S is a function of U, V and N in this simplified treatment. For such a function, we have

$$dS = \left(\frac{\partial S}{\partial U}\right)_{V,N} dU + \left(\frac{\partial S}{\partial V}\right)_{U,N} dV + \left(\frac{\partial S}{\partial N}\right)_{U,V} dN, \tag{2.2.25}$$

from which it follows, with reference to Eq. (2.2.24), that

$$\left(\frac{\partial S}{\partial U}\right)_{V,N} = \frac{1}{T}, \tag{2.2.26}$$

$$\left(\frac{\partial S}{\partial V}\right)_{U,N} = \frac{P}{T}, \tag{2.2.27}$$

and

$$\left(\frac{\partial S}{\partial N}\right)_{U,V} = -\frac{\mu}{T}. \tag{2.2.28}$$

Somewhat more generally, Eq. (2.2.28) could be written

$$\left(\frac{\partial S}{\partial N_i}\right)_{U,V,\ N_j \neq N_i} = -\frac{\mu_i}{T}, \tag{2.2.29}$$

for the chemical potential of the ith constituent in the presence of any number of others.

In summary, then, the information-theoretic maximum entropy for a system of known energy expectation value U has been used to suggest the identification of macroscopic thermodynamic variables. The entropy is found to be a function of U, V, and N, and the additive property that came out in Eq. (2.2.13) suggests that the entropy might well be an *extensive* function of its basic variables.

Extensive functions are those that are proportional to the size of a homogeneous system. Doubling the size (volume, total energy, and number of particles) results in doubling the value of an extensive function. Extensive variables are those that are direct measures of the size of the system (V, N, U, for example), and an extensive function can best be defined as one that is a homogeneous first-order function of the extensive variables, *i.e.*, if each of the extensive variables is multiplied by a constant, λ, the function itself is in effect multiplied also by λ.

Boundary effects and interactions can destroy the strict additivity of the entropy function by introducing terms that are proportional to areas of surfaces, lengths of edges, or other nonextensive parameters. The fraction of the total entropy that comes from these nonextensive terms usually diminishes as the size of the system increases, leading physicists whose objective is mathematical rigor to treat systems in the so-called thermodynamic limit, *i.e.*, the limit approached when the size of the system becomes infinite but the density, energy density, and concentrations of constituents remain constant. The present attitude is that infinite systems are physically objectionable, even though they do offer mathematical advantages; nonextensive terms are usually negligible and can be ignored for most purposes; sometimes they really are important and should properly appear in both the thermodynamics and the statistical mechanics; and, finally the modification of the basic formalism to accommodate nonextensive terms is usually straightforward and need not be carried along in what is intended to be a conceptual development. Therefore the entropy will be regarded as an extensive function of the extensive parameters of the system. This topic is discussed further in Section 2.11, and the thermodynamic limit is examined in more detail in Section 3.8.1.

On the basis of this model of the relationship of statistical and macroscopic thermophysics, an axiomatic development of thermostatics is outlined in the next section.

2.3 MACROSCOPIC EQUILIBRIUM THERMODYNAMICS (THERMOSTATICS)

In this section, the structure and content of standard equilibrium thermodynamics will be developed to the extent necessary to meet the needs of the statistical theories to be presented later. The statistical–microscopic theories are considered to underlie and support the macroscopic theory, and indeed the point of view is sound, but thermodynamics is a sturdily founded subject in its own right, and the integrity of the thermodynamic forest should not be obscured by the dendritic details of the underlying microphysics.

2.3.1 Preliminary Statements. The objectives of an axiomatic presentation of thermostatics in this context are threefold: First, the intuitive understanding of the macroscopic and microscopic theories is mutually reinforcing when they are presented in closely allied conceptual frameworks; second, the mathematical structure of thermostatics seems more easily related to the underlying physical ideas in the approach; and third, the possibility of faults in the logical structure of thermostatics needs wider exposure.

The presentation herein is based largely on the excellent and detailed work of Callen [1985], though it is by no means identical. Aside from being generally briefer than Callen's careful work, the present emphasis is more heavily on the information going in to such constructs as the entropy function, and on the existence of the underlying microstructure of matter. A physicist's set of axioms (including both Callen's and the present ones) differs from a mathematician's in that the physicist tolerates a considerable body of background and supplementary discussion in a relatively informal style, and he may also exhibit less concern about axiomatic economy. It may therefore be more stylistically proper to consider the presentation as based on a set of carefully stated fundamental concepts that form the foundations upon which the theory is built, rather than as a set of axioms that in themselves imply the entire theoretical structure.

Thermostatics will be developed in terms of closed systems rather than with reference to the universe as a whole. A **closed system** is regarded as a bounded region of real space that contains the physical entities under study and that is closed to the transfer of matter and energy across its boundaries. Closed systems do not exist. They may be approximated by sealed containers that are insulated from thermal contact, say by means of a Dewar flask, shielded from electric and magnetic fields and, as well as possible, from cosmic rays and ambient radiation. Nevertheless, gravitational fields are not easily screened out, nor are cosmic rays completely. (A calculation by Borel (Borel [1913]; Borel [1914]) notes that the motion of one gm. of matter through a distance of 1 cm at the distance of α Centauri, ~ 4 light years, would change the gravitational field on earth by about 10^{-100}, and that a change of this order of magnitude in the initial positions and momenta of the molecules of a macroscopic gas would make it impossible to compute the trajectories for more than about one microsecond!) The Liouville theorem, Eq. (1.6.6), is proved for a closed system with known (in principle) Hamiltonian; its pertinence in the proof that the Gibbs entropy does not change in time is evidently highly questionable.

Nevertheless, the concept of a closed system is useful in macroscopic thermostatics. It is clearly possible to construct a good approximation to a closed system in that the rate at which energy crosses its boundaries can be reduced to the point of negligibility, so that macroscopic processes are, to a very good approximation, not affected by external events. The real importance of the outside world as seen by a quasi-closed system is in our knowledge of its microstructure, since very little energy is required to alter slightly a phase space trajectory, with the result that minute perturbations will lead quickly to complete loss of dynamic information. But as long as we assume ignorance of detailed dynamic information, the circumstance that we should soon be restored to a state of ignorance if we were not already there is of little importance. (Further discussion of this point may be found in Section 1.9.) Practically, then, a closed system is one for which energy transfer across its boundaries is negligible when compared with the energies of the macroscopic processes being studied.

A **simple system** is a bounded region of space that is macroscopically homogeneous. That we regard a system as simple may imply nothing more than that we have not examined it on a fine enough scale. The failure to do so may be one of choice, or of lack of it. We may usually choose to regard an obviously complex system as simple if its internal behavior is not involved in the problem at hand. For example, if part of a complex system is a heat reservoir made up of water and stones in a copper tank, we may elect to write down a rather large and undetailed entropy function that satisfactorily expresses all we need to know about it. Usually, however, the simple systems that are treated in terms of statistical thermophysics are made up of atoms or molecules, the spatial distribution of which is described by probability densities that are constant (or periodic on the scale of a crystal lattice) over the volume of the system.

The boundary of a simple system may be material, or it may be described by a set of mathematical surfaces in space. Boundaries may be restrictive to matter flow, as closed vessels; they may be open to some molecules and not to others, as semipermeable membranes, or they may be completely open; they may allow the transfer of energy *via* a work process, typified here by a moveable piston; and they may allow the transfer of energy by *thermal contact*, already discussed in Section 2.2. Boundaries that allow thermal contact are called **diathermal** and those that prevent it are called **adiabatic**.

These preliminary statements and definitions now permit a relatively uncluttered display of the basic ideas in the next section.

2.3.2 Fundamental Statements. The following five statements (which may be regarded as physicists' axioms, with inherent shortcomings) are presented as a basis for the development of classical macroscopic equilibrium thermodynamics.

1. The macroscopic thermodynamic equilibrium states of simple systems are completely specified in terms of the extensive parameters $(U, V, \{N_i\})$ in the present context), where $\{N_i\}$ is the set of component populations, with all the information contained in the entropy function $S(U, V, \{N_i\})$, called the *fundamental relation.*

2. The entropy of a simple system is a continuous, differentiable, homogeneous-first-order function of the extensive parameters, monotone increasing in U.

3. The entropy of a composite system $S(\{U_i, V_i, \{N_i\}\})$ is the sum of the entropies of the constituent subsystems:

$$S(\{U_i, V_i, \{N_i\}\}) = S_1(U_1, V_1, \{N_1\}) + S_2(U_2, V_2, \{N_2\}) + \cdots.$$

4. The equilibrium state of a composite system when a constraint is removed maximizes the total entropy over the set of all possible constrained equilibrium states compatible with the constraint and the permitted ranges of extensive parameters.

5. The entropy of a simple system approaches zero when $(\partial U/\partial S)_{V,\{N\}} \to 0$.

The first statement is based on the statistical model developed in Section 1.2. We expect to be able to represent our knowledge of a simple system in terms of an entropy function obtained from the maximization of the information-theoretic uncertainty, subject to the information available. If we know the correct energy levels, then the fundamental relation $S = S(U, V, \{N\})$ so obtained should correctly specify the state of our knowledge of the system. The requirement that S be a function of the extensive parameters is suggested by the analysis of Section 1.2, where it was pointed out that boundary effects also can contribute nonextensive terms. When such terms arise from the statistical analysis, their physical significance is easily recognizable and they offer no fundamental difficulty. Since they are usually negligible, however, especially for large thermodynamic systems, the wording of statement 1 is much simplified by their omission.

Statement 2 is primarily for mathematical convenience, although it does specify S as extensive, since $S(\lambda U, \lambda V, \{\lambda N\}) = \lambda S(U, V, \{N\})$. The requirement that S be a monotone increasing function of U has the useful feature that it permits the **entropic fundamental relation,** $S = S(U, V, \{N\})$ to be solved uniquely for the **energetic fundamental relation,** $U = U(S, V, \{N\})$. The statistical basis for this requirement is that the energy levels of a simple system are usually bounded from below but not from above. Therefore an increase in the energy expectation value permits increased relative probabilities of the upper energy levels, a greater number of non-negligible probabilities, and a consequent increase in S. The existence of simple systems for which the energy levels are bounded from above as well as from below (such as spin systems with their moments aligned in a magnetic field) leads to a behavior of S in violation of the requirement of monotonous increase with U, as will be seen in Chapter 3, but these systems are quite exceptional, and they are always treated by means of an S derived from the underlying statistical theory. For most of the domain of thermodynamics, including all of the conventional classical systems, there is no contradiction of this requirement. In cases where there is a contradiction, one consequence is a negative absolute temperature, much to the consternation of classical thermodynamicists.

The sum rule of statement 3 is suggested by Eq. (2.2.13) and the development that follows. As already pointed out, non-extensive terms may exist in the entropy function, in violation of this statement, but they are usually negligible. Furthermore, for most purposes the requirement may be stated in terms of differentials, as

$$dS(\{U_i, \ V_i, \ \{N_i\}\}) = dS_1(U_1, \ V_1, \ \{N_1\}) + dS_2(U_2, \ V_2, \ \{N_2\}) + \cdots,$$

which is usually valid. One of the effective uses of statement 3 is in the treatment of thermodynamic stability, for which it is convenient to consider a simple equilibrium system as composed of two smaller systems in equilibrium internally and with each other.

Statement 4 is the principal working rule in this development of thermostatics. The underlying idea is still the one from information theory that equilibrium represents the state of maximum uncertainty subject to the existing constraints. A composite system consisting of two simple subsystems in weak thermal contact, as treated in Section 2.2, is expected to redistribute energy internally until the composite entropy is maximum, leading to equality of subsystem temperatures. If the subsystems were not in thermal contact at all, but energy could be transferred from one to the other by some other process that left them otherwise unchanged (say by heating one and refrigerating the other, with careful control of the individual energy transfers), then the partition of energies that maximizes the total entropy is the same as the one resulting from direct thermal contact.

In analogy with the development in Section 2.2, we may take a composite system comprising two simple systems, each with its own fundamental relation. With only thermal contact through a diathermal boundary, there is no change of volume or component numbers. Conservation of energy (the composite system is closed) implies

$$\delta U = \delta U_1 + \delta U_2 = 0, \tag{2.3.1}$$

and the change in total entropy is

$$\begin{aligned} \delta S &= \delta S_1 + \delta S_2 \\ &= \left(\frac{\partial S_1}{\partial U_1} \right)_{V_1, \{N_1\}} \delta U_1 + \left(\frac{\partial S_2}{\partial U_2} \right)_{V_2, \{N_2\}} \delta U_2. \end{aligned} \tag{2.3.2}$$

The requirement of entropy maximum for equilibrium, so that $\delta S = 0$, and the use of Eq. (2.3.1), give

$$0 = \left[\left(\frac{\partial S_1}{\partial U_1} \right)_{V_1, \ \{N_1\}} - \left(\frac{\partial S_2}{\partial U_2} \right)_{V_2, \ \{N_2\}} \right] \delta U_1 = 0 \tag{2.3.3}$$

for all small δU_1 around equilibrium, implying that

$$\left(\frac{\partial S_1}{\partial U_1} \right)_{v_1, \{N_1\}} = \left(\frac{\partial S_2}{\partial U_2} \right)_{V_2, \{N_2\}}, \tag{2.3.4}$$

in direct analogy to Eq. (2.2.18). Repetition of the discussion following Eq. (2.2.18) leads to the temperature identification of Eq. (2.2.19) and Eq. (2.2.26), *i.e.*

$$\frac{1}{T} = \left(\frac{\partial S}{\partial U} \right)_{V, \{N\}}. \tag{2.3.5}$$

Replacement of the diathermal wall by a diathermal movable piston, with conservation of total volume $V = V_1 + V_2$, leads to the pressure identification of Eq. (2.2.27), *i.e.*,

$$\frac{P}{T} = \left(\frac{\partial S}{\partial V}\right)_{U, \{N\}}. \tag{2.3.6}$$

Similarly, the removal of matter-flow constraints leads to equality of the terms

$$\left(\frac{\partial S_1}{\partial N_{1j}}\right)_{U_1, V_1, \{N_1 \neq N_{1j}\}} = \left(\frac{\partial S_2}{\partial N_{2j}}\right)_{U_2, V_2, \{N_2 \neq N_{2j}\}} \tag{2.3.7}$$

and identification of chemical potentials, in agreement with Eq. (2.2.29), as

$$\frac{\mu_j}{T} = -\left(\frac{\partial S}{\partial N_j}\right)_{U, V, \{N \neq N_j\}}. \tag{2.3.8}$$

The mathematical formalism appears straightforward and free of ambiguity or circularity. The identifications made in Eqs. (2.3.5), (2.3.6), and (2.3.8) can be regarded as definitions and their consequences examined to show that they do indeed exhibit the precise properties required of them. Accordingly, the differential of entropy for a simple system may be written

$$dS = \frac{1}{T}dU + \frac{P}{T}dV - \sum_i \frac{\mu_i}{T}dN_i,$$

or, when solved for the differential of energy,

$$dU = TdS - PdV + \sum_i \mu_i dN_i. \tag{2.3.9}$$

When the entire energy change can be accounted for in terms of work and matter transfer, conservation of energy permits the writing of dU as

$$dU = -PdV + \sum_i \mu_i dN_i, \tag{2.3.10}$$

where of course other identifiable work terms should be included in Eqs. (2.3.10) and (2.3.9) if they are pertinent. Also, when other energy transfers are present that are not identifiable as work, they must be called heat, and the additional term in dU is TdS, which is often written dQ. The use of Eq. (2.3.10) in Eq. (2.3.9) gives

$$dS = 0, \tag{2.3.11}$$

a result closely akin to the results expected for a reversible, adiabatic process in conventional treatments. Identification of Eq. (2.3.11) with a real physical process requires some care, however, because the fundamental relation has been postulated to define states of equilibrium, and dS in Eq. (2.3.8) refers to a general differential change in S at equilibrium resulting from a change in the extensive parameters. In order to relate the equilibrium changes that are represented in the formalism to something resembling real physical processes, even if idealized, it is customary to speak of quasi-static processes; *i.e.*, processes for which departures from equilibrium are very slight because rates of change are slow compared to rates of internal relaxation. The term *internal relaxation* refers to the physical idea that a localized disturbance will propagate throughout the system, for example as an acoustic wave, reflecting from boundaries, attenuating as it spreads and scatters, until it is ultimately indistinguishable from thermal noise. When a simple system is internally relaxed, it is in equilibrium, with a single temperature, pressure, and set of chemical potentials throughout. A quasi-static process, then, is one for which localized departures from uniformity are negligibly small for the intensive parameters. Therefore a quasi-static process may be represented, for example, by a line in a *P-V* diagram, with pressure regarded as being defined and uniform. In calculations, the equilibrium intensive parameters are used; in the laboratory, moving averages in time of such variables as pressure and temperature are read by measuring instruments that are designed to be unresponsive to small, rapid fluctuations. Comparison of theory and experiment requires, of course, a one-to-one correspondence.

Any theoretical presentation based on abstract postulates must make contact with the real physical world somewhere if it is to be called physics. Two principal points of contact are offered in this development; first, the fundamental relation may be derived from the underlying microphysics, the procedure to be exposed in the next and later chapters; and second, the thermodynamicist may require that Eq. (2.3.11) be satisfied by real systems for all quasi-static processes in which all energy transfers can be accounted for in terms of work and matter flow. The necessity of the second procedure in the absence of underlying microphysics may be seen from the observation that any fundamental relation $S(U, V, N)$ may be modified by the addition of an arbitrary function $f(U, V, N)$ that satisfies the fundamental statements. Thus

$$S^*(U, \ V, \ N) = S(U, \ V, \ N) + f(U, \ V, \ N)$$

could be used to calculate quantities T^*, P^*, and μ^* that behave formally in the same way as T, P and μ. Since real work and matter-flow processes can be identified, however, one can construct correct fundamental relations thermodynamically.

The fundamental relations themselves contain all of the thermodynamic information about a system, but comparisons with experiment are often made by use of **equations of state**; these are just the relations already presented as Eqs.(2.3.5), (2.3.6), and (2.3.8). As an example, the Sackur-Tetrode equation, Eq. (3.2.30) developed in Chapter 3 as the fundamental relation for a monatomic ideal gas, may be written here as

$$S(U, \ V, \ N) = (5Nk/2) + Nk \ln\left((V/V_0)(U/U_0)^{3/2}(N/N_0)^{-5/2}\right), \tag{2.3.12}$$

where V_0, U_0, and N_0 are constants. Successive use of Eqs. (2.3.5), (2.3.6), and (2.3.8) gives

$$\frac{1}{T} = \frac{3Nk}{2U},$$

or

$$U = (3/2)NkT; \tag{2.3.13}$$

$$\frac{P}{T} = \frac{Nk}{V},$$

or

$$PV = NkT; \tag{2.3.14}$$

and

$$\frac{\mu}{T} = -k \ln\left((V/V_0)(U/U_0)^{3/2}(N/N_0)^{-5/2}\right). \tag{2.3.15}$$

The so-called ideal gas law, Eq. (2.3.14), is often called *the* equation of state, with no such generic name given to the energy equation, Eq. (2.3.13), which is valid only for monatomic gases. The final equation of state, Eq. (2.3.15), is less likely to be a familiar one.

Fundamental statement 5, not previously discussed, and often called *the Third Law of Thermodynamics*, is violated by Eq. (2.3.12). Since $(\partial U/\partial S)_{V,\{N\}} = T$ by Eq. (2.3.5), statement 5 says that entropy is zero at $T = 0$. It will be seen in Chapter 4 that the ideal gas law, Eq. (2.3.14), is valid only at temperatures well above $T = 0$ and the violation of fundamental statement 5 by a fictitious substance should cause no great concern. There may be violations by real substances, however; these are mentioned in Chapter 7.

2.4 FURTHER DEVELOPMENTS

Since U, V, and N are extensive, the total energy and volume may be written

$$U \equiv Nu \tag{2.4.1}$$

and

$$V \equiv Nv, \tag{2.4.2}$$

where u and v are so defined as the energy and volume per particle when N is the number of particles, and as the energy and volume per mole when N represents the number of moles. (When several species of particle coexist in the system, let $N = \sum N_i$, and $N_i = f_i N$, where f_i is the fractional population of the ith species, usually called the *mole fraction*, and the extension of the development to multispecies systems is evident.) The use of statement 2 permits the entropy to be written as

$$S(U,\ V,\ N) = S(Nu,\ Nv,\ N) = Ns(u,\ v). \tag{2.4.3}$$

Partial differentiation of Eq. (2.4.3) with respect to N gives

$$\frac{\partial S(Nu, Nv, N)}{\partial N} = \left(\frac{\partial S}{\partial U}\right)_{V,N}\left(\frac{\partial U}{\partial N}\right)_u + \left(\frac{\partial S}{\partial V}\right)_{U,N}\left(\frac{\partial V}{\partial N}\right)_v + \left(\frac{\partial S}{\partial N}\right)_{U,V}$$

$$= \frac{u}{T} + \frac{Pv}{T} - \frac{\mu}{T} = s(u,v), \tag{2.4.4}$$

or

$$u = Ts - Pv + \mu. \tag{2.4.5}$$

Multiplication of (2.4.5) by N restores the extensive equation;

$$U = TS - PV + \mu N. \tag{2.4.6}$$

The differential form of Eq. (2.4.6) is

$$dU = TdS + SdT - PdV - VdP + \mu dN + Nd\mu,$$

or, by use of Eq. (2.3.9),

$$SdT - VdP + Nd\mu = 0. \qquad (2.4.7)$$

Equation (2.4.7) and its generalizations to multiple species are known as Gibbs-Duhem relations; they are of prime importance in chemical thermodynamics and in some aspects of the study of phase transitions.

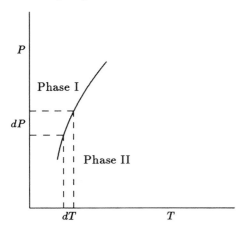

Figure 2.1 T-P plot of phase boundary

As an illustration of the use of Eq. (2.4.7) in phase transitions, consider a phase boundary in the T-P plane in Fig. 2.1, where Phase I may be typically liquid or solid and Phase II is then vapor or liquid. At equilibrium the two phases are at the same temperature and pressure; furthermore the chemical potentials are equal, so that if the volume is fixed, the average number of molecules in each phase is constant in time. A Gibbs-Duhem equation can be written for each phase as

$$d\mu_i = v_i dP - s_i dT, \qquad (2.4.8)$$

where $i = I$ or II. But since $\mu_I(T, P) = \mu_{II}(T, P)$, it follows that $d\mu_I = d\mu_{II}$, and

$$\frac{dP}{dT} = \frac{s_{II} - s_I}{v_{II} - v_I}. \qquad (2.4.9)$$

It is usual to write $N_0(s_{II} - s_I) = \ell/T$, where ℓ is the (molar) *latent* heat for the phase transition, and $N_0(v_{II} - v_I) = \Delta v$, the difference in the molar volumes of the two phases, with $N_0 = $ Avogadro's number. In these terms, Eq. (2.4.9) is written as

$$\frac{dP}{dT} = \frac{\ell}{T\Delta v}, \qquad (2.4.10)$$

giving the slope of the phase boundary in P-T space in terms of the latent heat and the molar volume change. Equation (2.4.10) is called the *Clapeyron relation*, as is Eq. (2.4.9). For fusion or vaporization, the numerator of Eq. (2.4.9) or Eq. (2.4.10) is always positive, but Δv in the denominator may sometimes be negative as it is for melting ice. Thus the melting temperature of ice is decreased by an increase of pressure, where as for most substances the phase transition occurs at higher temperatures for increased pressures.

A general relation, first derived by Gibbs, expresses the number of degrees of freedom, f, of a system made up of c components with p phases in equilibrium. The number of degrees of freedom is the number of independent intensive parameters. For equilibrium a total of $c + 2$ intensive parameters, T, P, μ_1, μ_2, $\cdots$, μ_c is needed, since T, P, and the μ's have the same values in each phase.

But there are p Gibbs-Duhem relations, one for each phase; these are independent because the molar volumes and entropies differ from phase to phase. Therefore the number of degrees of freedom is

$$f = c + 2 - p, \qquad (2.4.11)$$

the well-known Gibbs phase rule. As an example, a one-component system such as water when entirely in one phase, say liquid, can have independent values of temperature and pressure. A two phase system, say liquid water and its vapor, has the pressure (called *vapor pressure*) determined by specification of the temperature alone, and three-phase system such as ice, liquid water, and water vapor, can exist in equilibrium at only one value of P, T, and μ. For more detailed discussion of the phase rule and the thermodynamics of phase transitions, a number of good books are listed in the references for this chapter. (ter Haar and Wergeland [1966]; Landsberg [1961]; Wetmore and LeRoy [1951])

2.5 ADDITIONAL CONCEPTS AND DEFINITIONS

There are several derived and measurable quantities, usually called *response functions*, that are of interest in thermodynamics, and the calculation of these quantities is often an objective of statistical thermophysics. It is usual to write Eq. (2.3.9), for systems with all $dN_i = 0$, as

$$dU = TdS - PdV. \tag{2.5.1}$$

Since the entropy change arises from an energy transfer that is not attributable to mechanical work, and since the discussion following Eq. (2.3.10) indicates that Eq. (2.5.1) must refer to quasi-static (or reversible) processes, for which any point on the trajectory differs but infinitesimally from the equilibrium coordinates, it is customary to define $dQ = TdS$ as the reversible, or quasi-static, heat. Similarly, the reversible work is $dW = \pm PdV$, where the choice of sign varies, depending upon the author's taste. In the present case, both dQ and dW are regarded as positive when the energy flow into the system is positive, so $dW = -PdV$. The notation dQ (and similarly dW) indicates that dQ is not an exact differential, *i.e.*, that there is no function Q (a property of the system) such that dQ is its differential. *Work* and *heat* refer to energy transfer *processes*, and not to states of the system.

The use of $dQ = TdS$ leads to the definition of heat capacity as

$$C = dQ/dT = T(dS/dT). \tag{2.5.2}$$

but since dQ pertains to a process, it is necessary to restrict the derivative in Eq. (2.5.2) to the appropriate process. If, further, a molar system is considered, then $S = N_0 s$, where N_0 is the Avogadro number and s is the entropy per particle. In this case Eq. (2.5.2) may be written

$$C_r = N_0 T \left(\frac{\partial s}{\partial T} \right)_r, \tag{2.5.3}$$

where the subscript r indicates a particular process, such as one at constant volume or constant pressure.

Other measured and tabulated response functions are the *thermal expansivity*, given by

$$\alpha = \frac{1}{V} \left(\frac{\partial V}{\partial T} \right)_{P,N}, \tag{2.5.4}$$

and the *isothermal compressibility*, given by

$$\kappa_T = -\frac{1}{V} \left(\frac{\partial V}{\partial P} \right)_{T,N}. \tag{2.5.5}$$

Equation (2.5.1) is a relationship among exact differentials, found directly from Eq. (2.3.9). Since dU is exact and may be written

$$dU(S,V) = \left(\frac{\partial U}{\partial S}\right)_V dS + \left(\frac{\partial U}{\partial V}\right)_S dV = TdS - PdV,$$

and since the order of taking second derivatives is assumed not to matter, the equality of $\partial^2 U/\partial S \partial V$ and $\partial^2 U/\partial V \partial S$ implies

$$\left(\frac{\partial T}{\partial V}\right)_S = -\left(\frac{\partial P}{\partial S}\right)_V. \tag{2.5.6}$$

Equation (2.5.6) is one of many so-called Maxwell relations, all derived by application of the criterion of exactness to the differential of a thermodynamic potential, such as the energy, and all used in the reduction of calculated expressions to forms that contain tabulated or readily observable properties. As an example of such a reduction, suppose one wished to calculate the change in temperature of a substance that expands reversibly and adiabatically (*i.e.*, S = const.) from an initial volume V_1 to a final volume V_2. Evidently one can write

$$\Delta T = \int_{V_1}^{V_2} \left(\frac{\partial T}{\partial V}\right)_S dV. \tag{2.5.7}$$

An expression for the derivative is needed, since no such quantity is ordinarily tabulated or measured. The Maxwell relation Eq. (2.5.6) and the use of several elementary manipulations with partial derivatives permit Eq. (2.5.7) to be written

$$\Delta T = \frac{-1}{N} \int_{V_1}^{V_2} \frac{\alpha T}{C_v \kappa_T} dV, \tag{2.5.8}$$

where here N is the number of moles of the substance, α and κ_T are defined in Eqs.(2.5.4) and (2.5.5), C_v is the molar heat capacity at constant volume defined generally in Eq. (2.5.3), and all quantities in the integrand are standard thermodynamic properties or variables. No further reduction of Eq. (2.5.8) is worthwhile in general, but for a specific material with known equations of state the integral can be evaluated either analytically or numerically.

The properties of partial derivatives most frequently needed for such calculations are:

$$\left(\frac{\partial z}{\partial x}\right)_y = 1 \bigg/ \left(\frac{\partial x}{\partial z}\right)_y, \tag{2.5.9}$$

$$\left(\frac{\partial z}{\partial x}\right)_y = \left(\frac{\partial z}{\partial t}\right)_y \bigg/ \left(\frac{\partial x}{\partial t}\right)_y, \tag{2.5.10}$$

and

$$\left(\frac{\partial z}{\partial x}\right)_y \left(\frac{\partial x}{\partial y}\right)_z \left(\frac{\partial y}{\partial z}\right)_x = -1, \tag{2.5.11}$$

all well-known results available in any intermediate calculus text. Evidently any number of other variables u, $v, \cdots, w$ could be retained as constant subscripts on all the partial derivatives; *i.e.*, variables that remain constant throughout the manipulations, such as N or $\{N_i\}$ in (2.5.6) *et seq.* The calculation leading to Eq. (2.5.8) is, starting with Eq. (2.5.6):

$$\left(\frac{\partial T}{\partial V}\right)_S = -\left(\frac{\partial P}{\partial S}\right)_V = -\left(\frac{\partial P}{\partial T}\right)_V \bigg/ \left(\frac{\partial S}{\partial T}\right)_V, \qquad \text{by Eq. (2.5.10)}$$

$$= \left(\frac{\partial P}{\partial V}\right)_T \left(\frac{\partial V}{\partial T}\right)_P \bigg/ \left(\frac{\partial S}{\partial T}\right)_V, \qquad \text{by Eq. (2.5.11)}$$

$$= (-1/V\kappa_T)(\alpha V)/(NC_v/T), \qquad \text{by Eqs. (2.5.9), (2.5.3), and (2.5.5)}$$

where $NC_v = T\left(\partial S/\partial T\right)_v$, and N is the mole number.

Other potentials are derived from $U(S, V, N)$ by Legendre transformations. A Legendre transformation permits the information content of a fundamental relation to be preserved while one (or more) of the independent extensive variables in U is replaced by an intensive one. In general, suppose $y = y(x)$ represents a fundamental relation, with the independent variables other than x to remain unchanged by the transformation. The variable complementary to x is $m = \partial y/\partial x$, just as $T = \partial U/\partial S$ and $U = TS + \cdots$. But the relation $m = \partial y/\partial x$ contains considerably less information than is in the original expression $y = y(x)$, since integration of the expression for m is indeterminate to the extent of the arbitrary constant (or function of the other independent variables) of integration. If, however, the original curve $y = y(x)$ is replaced by the set of all lines tangent to it, then the y-intercept, b, of each tangent must be specified, so that $(y - b)/x = m$ is an equation relating the intercept b to the slope m of the original curve. It follows that $b = y - mx$, and

$$db = dy - m\,dx - x\,dm$$
$$= -x\,dm, \tag{2.5.12}$$

since $dy = m\,dx$. Therefore the new curve $b = b(m)$, which gives the intercept b in terms of the slope m and thus specifies the family of tangents, contains all of the information in the original $y = y(x)$. Except for sign, the Legendre transformation is exactly the procedure used in mechanics to transform the Lagrangian $\mathcal{L}(q_i, \dot{q}_i)$ to the Hamiltonian $\mathcal{H}(q_i, p_i)$, where $p_j \equiv \partial L/\partial \dot{q}_j$, and

$$\mathcal{H}(q_i, p_i) = \sum_i p_i \dot{q}_i - L(q_i, \dot{q}_i).$$

The common Legendre transformations used in thermodynamics are those that yield (1) the Helmholtz free energy

$$F = U - S\left(\frac{\partial U}{\partial S}\right)_{V,N} = U - TS, \tag{2.5.13}$$

(2) the enthalpy

$$H = U - V\left(\frac{\partial U}{\partial V}\right)_{S,N} = U + PV, \tag{2.5.14}$$

(3) the Gibbs free energy

$$G = U - S\left(\frac{\partial U}{\partial S}\right)_{V,N} - V\left(\frac{\partial U}{\partial V}\right)_{S,N} = U - TS + PV, \tag{2.5.15}$$

and (4) the grand potential

$$\Phi = U - S\left(\frac{\partial U}{\partial S}\right)_{V,N} - N\left(\frac{\partial U}{\partial N}\right)_{S,V} = U - TS - \mu N \qquad (2.5.16)$$

It is evident from Eq. (2.4.6) that $G = \mu N$ (or more generally $\sum_i \mu_i N_i$) and $\Phi = -PV$.

Differentiation of these expressions and use of Eq. (2.3.9) gives

$$dF = -S\,dT - P\,dV + \mu\,dN, \qquad (2.5.17)$$

$$dH = T\,dS + V\,dP + \mu\,dN, \qquad (2.5.18)$$

$$dG = -S\,dT + V\,dP + \mu\,dN, \qquad (2.5.19)$$

and

$$d\Phi = -S\,dT - P\,dV - N\,d\mu, \qquad (2.5.20)$$

which are comparable to

$$dU = T\,dS - P\,dV + \mu\,dN. \qquad (2.5.21)$$

The grand potential reduces trivially to $d\Phi = -d(PV)$, either directly or by means of Eq. (2.4.7); therefore Maxwell relations will not be developed for it. The evident independent variables in Eq. (2.5.17) - Eq. (2.5.21) show that the potentials are $U(S, V, N)$, $F(T, V, N)$, $H(S, P, N)$, $G(T, P, N)$ and $\Phi(T, V, \mu) = \Phi(P, V)$. Three Maxwell relations may be constructed by additional Legendre transformations. For purposes of illustration, only those that do not involve the N's and μ's will be examined, and only the P, V coordinates related to external work will be used here. Evidently other work-related coordinates will lead to additional potentials and relations among derivatives.

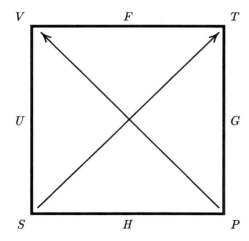

Figure 2.2 Mnemonic diagram for obtaining the Maxwell relations.

A mnemonic diagram, Fig. 2.2, may be used to summarize the Maxwell relations that were derived from U, F, G, and H. The potentials in clockwise alphabetical order label the sides of a square, the extensive variables in clockwise alphabetical order at the left hand corners, and the complementary intensive variables at the diagonally opposite right-hand corners. Each potential is now a function of the independent variables at adjacent corners; thus $H = H(S, P, N)$, with the N-dependence implicit and not contained in the diagram. Moreover, the differential expression Eq. (2.5.18) has both $T\,dS$ and $V\,dP$ positive as implied by the diagonal arrows in the diagram pointing away from S and P. In Eq. (2.5.21), the $P\,dV$ term is negative and arrow points toward V; thus both dU and dF contain $-P\,dV$.

The left hand side of Eq. (2.5.6), $(\partial T/\partial V)_S$, involves the three consecutive variables T, V, and S going counterclockwise from the top right, with the arrow pointing **toward** V.

The diagram yields the Maxwell relation Eq. (2.5.6) if one starts at the corner not involved in T, V, S; *i.e.* at P, and proceeds clockwise, to get P, S, V, or $(\partial P/\partial S)_V$. But since the arrow points *away* from S, the middle variable of this sequence, the minus sign appears in Eq. (2.5.6). As another example, the relation

$$\left(\frac{\partial P}{\partial T}\right)_V = \left(\frac{\partial S}{\partial V}\right)_T \tag{2.5.22}$$

can be read from the diagram or obtained directly from the requirement that dF in Eq. (2.5.17) is an exact differential.

In the reduction of thermodynamic expressions to forms involving P, V, T, N, α, κ_T, C_p, and C_v, it should be noted that the entropy appears only in C_p and C_v, and then always as a derivative with respect to T. Therefore any derivative with S held constant must be expressed in other terms, either by the use of Eq. (2.5.11) or by a Maxwell relation. If a derivative of S appears with T constant, then Eq. (2.5.22) or its counterpart derived from Eq. (2.5.19) can be used to eliminate S altogether. If S appears with anything else constant, Eq. (2.5.10) is usually helpful; for example

$$\left(\frac{\partial S}{\partial V}\right)_P = \left(\frac{\partial S}{\partial T}\right)_P \Big/ \left(\frac{\partial V}{\partial T}\right)_P$$
$$= (NC_P/T)/(\alpha V).$$

For more complicated reductions, Jacobian techniques are often helpful. A brief resume of pertinent properties of Jacobians appears in Appendix E. As an example, consider the problem of expressing C_p in terms of C_v and the usual thermodynamic variables. The Jacobian procedure is as follows:

$$C_P = T\left(\frac{\partial s}{\partial T}\right)_P = T\frac{\partial(s,P)}{\partial(T,P)} = T\frac{\partial(s,P)}{\partial(T,v)}\frac{\partial(T,v)}{\partial(T,P)}$$

$$= T\begin{vmatrix} \left(\dfrac{\partial s}{\partial T}\right)_v & \left(\dfrac{\partial s}{\partial v}\right)_T \\ \left(\dfrac{\partial P}{\partial T}\right)_v & \left(\dfrac{\partial P}{\partial v}\right)_T \end{vmatrix} \cdot \begin{vmatrix} 1 & \left(\dfrac{\partial T}{\partial P}\right)_T \\ \left(\dfrac{\partial v}{\partial T}\right)_P & \left(\dfrac{\partial v}{\partial P}\right)_T \end{vmatrix}$$

$$= T\left[\left(\frac{\partial s}{\partial T}\right)_v\left(\frac{\partial P}{\partial v}\right)_T - \left(\frac{\partial s}{\partial v}\right)_T\left(\frac{\partial P}{\partial T}\right)_v\right]\cdot\left(\frac{\partial v}{\partial P}\right)_T$$

$$= C_v - T\left(\frac{\partial P}{\partial T}\right)_v^2\left(\frac{\partial v}{\partial P}\right)_T$$

$$= C_v + \frac{Tv\alpha^2}{\kappa_T}, \tag{2.5.23}$$

where the last step makes use of Eqs.(2.5.9), (2.5.11), (2.5.4), and (2.5.5). The derivation of Eq. (2.5.23) by direct methods is much longer and more troublesome.

2.6 THE CARNOT THEOREM

Although no direct use will be made of Carnot's theorem on the ideal efficiency of heat engines, the development is relatively brief and it serves as a further bridge between abstract theory and physical reality. We consider the compound closed system of Fig. 2.3, with a real source of energy s, assumed to be able to transfer energy across its boundaries as heat and/or work; an ideal engine, which is itself unaffected by energy transfers but merely serves as a transfer agent; a reversible work source w; and a reversible heat reservoir, q. The reversible work source must satisfy $dU_w = dW_w$, transferring energy in and out entirely by macroscopic work processes with no loss, dissipation, degradation, or entropy production. The reversible heat reservoir satisfies $dU_q = T_q dS_q$, transferring energy in and out only as heat, with no macroscopic work possible. It is sometimes useful to require such heat reservoirs to be of very large heat capacity so the the temperature remains constant for finite energy transfers: *i.e.*,

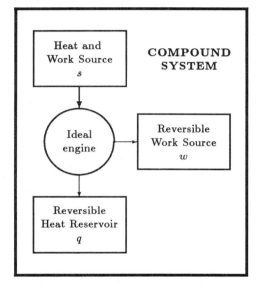

Figure 2.3 Compound closed system containing a real heat and work source, s, an ideal engine, an ideal reversible work source, w, and an ideal reversible heat reservoir, q.

$$\frac{dT}{dU} = \frac{dT}{T dS} = \frac{1}{C},$$

where C is the total heat capacity. Thus if C is very large, $dT/dU \to 0$. The ideal engine takes whatever kind of energy it receives from the real source, stores as much as possible in the reversible work source, and passes the rest on to the reversible heat source. The only restrictions are those implied by the closed system; that the total energy is constant and that any real process that takes place must be in the direction of equilibrium, which is toward entropy maximum. Formally, then, conditions are

$$dU = dU_s + dU_q + dU_w = 0, \tag{2.6.1}$$

and

$$dS = dS_s + dS_q + dS_w \geq 0, \tag{2.6.2}$$

with

$$dS_w \equiv 0, \ dU_q = T_q dS_q, \ dU_w = dW_w,$$

and

$$dU_s = T_s dS_s + dW_s.$$

The use of these in Eq. (2.6.1) gives

$$T_s dS_s + dW_s + T_q dS_q + dW_w = 0,$$

and combination with Eq. (2.6.2) leads to

$$dW_w = -dW_s - (T_s - T_q)dS_s - T_q dS \qquad (2.6.3)$$

where Eq. (2.6.3) gives the total work transferred to the reversible work source as the negative of the work done on the source s, as it should be, plus some heat transfer terms. The conventional Carnot theorem follows easily if $dW_s = 0$, so the real source is regarded as strictly a heat source. The heat transferred from s to the ideal engine is $-T_s dS_s$, and the efficiency of conversion to work is

$$\eta = \frac{dW_w}{-T_s dS_s} = \frac{T_s - T_q}{T_s} + \frac{T_q dS}{T_s dS_s}, \qquad (2.6.4)$$

when $dW_s = 0$. The second term is non-positive, since both temperatures are positive and $dS \geq 0$ by Eq. (2.6.2), but $dS_s < 0$ in order that energy is transferred from s to the engine. Therefore the conversion efficiency η is maximum when $dS = 0$, and it follows that

$$\eta_{\max} = \frac{T_s - T_q}{T_s}, \qquad (2.6.5)$$

the well-known expression for the efficiency of a Carnot engine.

The generality of this development should be particularly satisfying to physicists, since Eq. (2.6.5) represents the maximum conversion efficiency of thermal energy to work, independent of engine cycles. The result is equivalent to the statement that for an ideal energy conversion in a closed system there is no net change in the total entropy. Any reduction in entropy of a heat source must be compensated by the increase in entropy of a heat sink. If the heat source is at high temperature, then a heat transfer dQ_s corresponds to relatively small entropy change $dS_s = dQ_s/T_s$, while if the heat sink is at low temperature, the heat transfer dQ_q required to compensate for the entropy loss of the heat source is the relatively small amount $dQ_q = T_q dS_q$, with $dS_s + dS_q = 0$. Thus maximum the amount of heat convertible into work is the total heat extracted from the source, $-T_s dS_s$, minus the heat required by the necessity to compensate for the entropy reduction of the source, $T_q dS_q$, for a net available thermal energy $-T_s dS_s - T_q dS_q = -(T_s - T_q)dS_s$. This energy gives $\eta_{\max}$ when divided by the heat transferred from s, $-T_s dS_s$. Therefore the objectives in efficient ideal energy conversion are (1) to obtain energy at the least possible entropy cost, *i.e.*, at highest possible T, and (2) to repay the entropy debt with as little energy as possible. (In real devices, the optimum practical efficiency involves numerous compromises and trade-offs, recognizing the existence of nonideal devices and cost limitations. Usually, practical engines attain their optimum efficiencies by means of engine cycles that differ considerably from Carnot cycles, although the limitations imposed by the Carnot theorem always apply. These matters are properly the concern of engineers.)

A thermodynamic temperature scale can be established experimentally in terms of near-ideal energy conversion processes. Equation (2.6.5) implies

$$\eta_{\max} = 1 - \frac{T_q}{T_s} = 1 - \frac{|dQ_q|}{|dQ_s|},$$

or

$$\frac{T_q}{T_s} = \frac{|dQ_q|}{|dQ_s|}. \qquad (2.6.6)$$

Thus a measurement of near-ideal efficiencies will establish a scale of temperature ratios, and the assignment of a particular $T \neq 0$, such as the ice point, will determine every temperature uniquely. It may be seen that the thermodynamic temperatures in Eq. (2.6.6) are identically those introduced in Eqs. (2.2.26) and (2.3.5). The further identification of the T in Eq. (2.6.6) with the ideal gas temperature is left as an exercise, although it follows implicitly from the fundamental relation for ideal gases derived in Chapter 3.

The derivation leading to Eq. (2.6.3) is satisfactory from yet another point of view. If information is available that permits the extraction of work directly from the energy source s, so that $-dW_s \neq 0$, then such work may be completely transferred, in the ideal case, to the reversible work source with no production of entropy. It becomes evident that energy may be transferred to the work source at an efficiency greater than that given by Eq. (2.6.5), in apparent violation of Carnot's theorem. This development sometimes confuses students who encounter it for the first time.

The use of information to permit the extraction of work directly from an energy source at higher than thermal efficiency is neither a violation of Carnot's theorem nor an exercise in Maxwellian demonology. Windmills and waterwheels have long existed; flywheels, springs, and weights can be used for energy storage, as can batteries and resonant circuits. Carnot's theorem evidently applies only to the conversion of thermal energy to work; energy is regarded as thermal only when there is insufficient information available to treat it otherwise. Thus a tank of gas at high pressure can be used efficiently to drive a gas turbine, provided (1) it is recognized as such, and (2) the technology is available. Otherwise its energy can be treated only as thermal. Even at this level of thermodynamic reasoning, the connection between thermodynamics and information is manifest.

2.7 STABILITY OF EQUILIBRIUM

Fundamental statement No. 4 in Section 2.3.2 requires that the equilibrium entropy of a composite system at constant energy is the maximum possible, subject to the existing constraints. No use has been made so far of the requirement that the equilibrium entropy be a true maximum, not merely *any* extremum or stationary value. In this section, the entropy maximum postulate is used as the requirement for stability. The entropy as a function of its several extensive variables may be regarded as a surface in a multidimensional space. The entropy-maximum postulate can be applied locally, to a point on the entropy surface, or globally, to the entire surface that is accessible without violating the existing constraints. These aspects are discussed separately.

2.7.1 Local Stability. In principle, any simple system can be regarded as a composite system made up of two or more subsystems that are partitions of the original simple system. For intrinsic stability of a simple system, we require that the total entropy of any two complementary partitions of the system be maximum at constant total energy. In this section, we examine local maxima.

The entropy-maximum calculation, however, is rather lengthy because of the necessity to express such quantities as $\left(\partial(P/T)/\partial V\right)_{U,N}$ in more recognizable terms, but since the calculation is direct and free of tricks, it is left as an exercise, and an equivalent calculation is presented here. Then, based on this treatment, the stability criterion is formulated in several other ways near the end of this section.

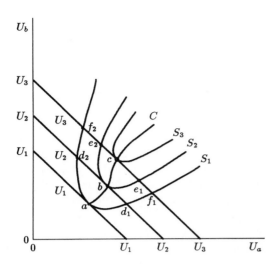

Figure 2.4 Topographic map showing isentropes in $U_a - U_b$ space.

The entropy was postulated to be a monotone increasing function of the energy. We re-examine a composite system comprising two simple systems, a and b, each with its own fundamental relation. The total energy is $U = U_a + U_b$, and, as in Eq. (3.3.1), $\delta U = \delta U_a + \delta U_b = 0$. Figure 2.4 shows *isoergs* of the composite system for different total energies, with $U_1 < U_2 < U_3$. A dot at point a on U_1 represents the point for which the entropy is maximum at that energy, *i.e.* the equilibrium point. Similarly, dots b and c represent the equilbruim points at the energies U_2 and U_3. The successively larger dots signify that the equilibrium entropy of the composite system increases with U, in accord with fundamental statement 2, and the curve C passes through the equilibrium points, along the entropy crest. The points d_1 and d_2 on the U_2 isoerg show the same size dots as that at a, indicating the same entropy, as do the dots f_1 and f_2 on U_3. A curve through these points is drawn to represent an *isentrope*, a contour of constant entropy, here designated as S_1. Similar isentropes are shown as S_2 and S_3. Evidently the points of entropy maximum for fixed energies correspond to the points of energy minimum for fixed entropy. Therefore the condition of intrinsic stability of a simple system may be expressed as that of minimum total energy at constant entropy.

The total energy of the simple system is now written as the sum $U = U_a + U_b$ of arbitrary subsystems a and b, with similar expressions for the entropy $S = S_a + S_b$, the volume $V = V_a + V_b$, and the populations $N = N_a + N_b$. For present purposes, the subsystems will be regarded as having fixed populations N_a and N_b, so $\delta N_a = \delta N_b = 0$. The first-order variation of U gives

$$\begin{aligned}
\delta U &= \delta U_a + \delta U_b \\
&= \left(\frac{\partial U_a}{\partial S_a}\right)\delta S_a + \left(\frac{\partial U_a}{\partial V_a}\right)\delta V_a + \left(\frac{\partial U_b}{\partial S_b}\right)\delta S_b + \left(\frac{\partial U_b}{\partial V_b}\right)\delta V_b \\
&= (T_a - T_b)\delta S_a - (P_a - P_b)\delta V_a.
\end{aligned} \qquad (2.7.1)$$

Thus energy minimum at equilibrium gives $T_a = T_b$ and $P_a = P_b$, in agreement with the results of Sections 2.2 and 2.3.

In order that the equilibrium point be identified as a true energy minimum, it is necessary that $\delta^2 U$ be positive. The second variation is

$$\begin{aligned}
\delta^2 U &= \left(\frac{\partial^2 U_a}{\partial S_a^2}\right)(\delta S_a)^2 + 2\left(\frac{\partial^2 U_a}{\partial S_a \partial V_a}\right)\delta S_a \delta V_a + \left(\frac{\partial^2 U_a}{\partial V_a^2}\right)(\delta V_a)^2 + \text{similar } b \text{ terms} \\
&= \left(\frac{\partial T_a}{\partial S_a}\right)_{V_a}(\delta S_a)^2 + 2\left(\frac{\partial T_a}{\partial V_a}\right)_{S_a}\delta S_a \delta V_a - \left(\frac{\partial P_a}{\partial V_a}\right)_{S_a}(\delta V_a)^2 + \text{similar } b \text{ terms} \quad (2.7.2)
\end{aligned}$$

By the definitions and reduction techniques of Section 2.5, Eq. (2.7.2) becomes

$$\delta^2 U = \left(\frac{T_a}{N_a} + \frac{T_b}{N_b}\right)\frac{(\delta S_a)^2}{C_v} - \frac{2\alpha}{C_v\kappa_T}\left(\frac{T_a}{N_a} + \frac{T_b}{N_b}\right)\delta S_a\delta V_a + \frac{C_p}{vC_v\kappa_T}\left(\frac{1}{N_a} + \frac{1}{N_b}\right)(\delta V_a)^2.$$
(2.7.3)

If subsystem a is chosen very much smaller than subsystem b, then all b terms in Eq. (2.7.3) can be disregarded, giving

$$\delta^2 U = \frac{T_a}{N_a C_v}(\delta S_a)^2 - \frac{2\alpha T_a}{N_a C_v\kappa_T}\delta S_a\delta V_a + \frac{C_p}{N_a C_v v_a\kappa_T}(\delta V_a)^2 > 0.$$
(2.7.4)

The conditions for $\delta^2 U$ to be positive for all arbitrary δS_a and δV_a are just the conditions for the quadratic form $z = ax^2 + 2bxy + cy^2$ to be positive definite; *i.e.*, $a > 0$, $c > 0$, and $b^2 - ac < 0$. Application of these conditions to Eq. (2.7.4) gives $T_a/N_a C_v > 0$, or

$$C_v > 0;$$
(2.7.5)

the not particularly informative result from $c > 0$ that $C_p/N_a v C_v\kappa_T > 0$; and, from $b^2 - ac$,

$$0 > \frac{\alpha^2 T^2}{N^2 C_v^2 \kappa_T^2} - \frac{TC_p}{N^2 C_v^2 v\kappa_T} = -\frac{T}{N^2 C_v v\kappa_T},$$
(2.7.6)

by Eq. (2.5.23). From Eqs. (2.7.5) and (2.7.6) it follows that

$$\kappa_T > 0.$$
(2.7.7)

The intrinsic local stability conditions (2.7.5) and (2.7.7) say that in order for a pure substance to be able to exist in a state of thermodynamic equilibrium it must have the properties that (1) its temperature rises when it is heated at constant volume ($C_v > 0$), and (2) its volume decreases when it is squeezed at constant temperature ($\kappa_T > 0$). These are conditions on the behavior of matter that are in reassuring agreement with our normal expectations and experiences. They are specific examples of *Le Chatelier's Principle*, which is most often presented in the context of chemical equilibration, but is of wide validity. The principle is examined in some detail in the next subsection. They are also examples of general convexity relations developed by Gibbs, and examined in Chapter 9. (Gibbs [1961], Falk [1970])

The condition for the mutual stability of the equilibrium of two systems follows immediately from examination of Eqs. (2.7.2) and (2.7.4). Evidently the intrinsic stability of the simple component systems is sufficient to ensure the mutual stability of equilibrium in a composite system. Stability can also be expressed in terms of the other potentials F, H, G, and Φ of Eqs. (2.5.13)–(2.5.16). In terms of F, we consider a system of energy U, entropy S, in contact with a reservoir at constant temperature T_r. The total energy of the composite system is $U + U_r$, and energy exchanged by the reservoir is only in the form of heat. Mutual stability requires, at $S + S_r = $ constant,

$$\begin{aligned}
0 &= \delta(U + U_r) \\
&= \delta U + T_r\delta S_r \\
&= \delta U - T_r\delta S \\
&= \delta(U - TS); \quad T = T_r,
\end{aligned}$$
(2.7.8)

and

$$0 < \delta^2(U + U_r) = \delta^2 U + \delta^2 U_r$$
$$= \delta^2 U + \delta(T_r \delta S_r)$$
$$= \delta^2 U - \delta(T_r \delta S)$$
$$= \delta^2(U - TS); \quad T = T_r, \tag{2.7.9}$$

or

$$0 < \delta^2 F; \quad T = T_r. \tag{2.7.10}$$

Thus the Helmholtz potential of a system in contact with a heat bath at constant temperature is minimum for equilibrium.

Similarly, equilibrium conditions for the other potentials are given by

$$\delta H = 0, \quad \delta G = 0, \quad \text{or} \quad \delta \Phi = 0,$$

with

$$0 < \delta^2 H; \quad P = P_r, \tag{2.7.11}$$

$$0 < \delta^2 G; \quad T = T_r \quad \text{and} \quad P = P_r, \tag{2.7.12}$$

and

$$0 < \delta^2 \Phi; \quad T = T_r \quad \text{and} \quad \mu = \mu_r. \tag{2.7.13}$$

Thus, for example, (2.7.13) says that for a system in equilibrium with a heat reservoir at T_r and a matter reservoir at μ_r, the grand potential is a minimum.

Other stability conditions and other criteria for equilibrium can be developed when different work terms are present. Especially interesting are the criteria that arise when magnetic work is treated; further details are to be found in Chapter 5. An application of stability criteria to equations of state for nonideal gases is given in Section 2.8.

The one further set of stability criteria that will be mentioned at this point pertains to the behavior of the chemical potential, which was omitted in the development beginning with Eq. (2.7.1). After this much of the theory has been presented, the inclusion of variations in the subsystem populations can be seen to proceed in the same way. The principal condition on the chemical potential that arises from the energy-minimum formulation is easily found to be

$$\left(\frac{\partial \mu}{\partial N}\right)_{S,V} > 0. \tag{2.7.14}$$

Other relationships involving chemical response functions may be derived, either directly as stability conditions from other thermodynamic potentials, or from general relationships among the response functions that are analogs of Eq. (2.5.23), derived in the same way by means of Jacobian transformations. The appropriate Maxwell relations may be read from the equations for the differentials of the potentials, Eqs. (2.5.21), (2.5.17), (2.5.18), and (2.5.19), respectively. From these and the analogs of Eq. (2.7.14), which say $(\partial \mu / \partial N)_W \geq 0$, where the subscript W stands for whatever one chooses (almost), numerous relationships can be derived, following the procedures of Eq. (2.5.23). Two examples of these are:

$$\left(\frac{\partial \mu}{\partial N}\right)_{V,S} = \left(\frac{\partial \mu}{\partial N}\right)_{P,S} + \frac{V \kappa_T C_v}{C_P} \left(\frac{\partial \mu}{\partial V}\right)_{S,N}^2; \tag{2.7.15}$$

and

$$\left(\frac{\partial \mu}{\partial N}\right)_{P,S} = \left(\frac{\partial \mu}{\partial N}\right)_{P,T} + \frac{T}{NC_P}\left(\frac{\partial \mu}{\partial T}\right)_{P,N}^2. \tag{2.7.16}$$

The most general result obtained in this way is that the chemical potential of a substance increases with the concentration of the substance, as is expected, so that equilibration with respect to matter flow requires the direction of flow to be from higher concentration to lower. It should be noted that for systems with more than one component, a Maxwell relation always says $(\partial \mu_i/\partial N_j)_W = (\partial \mu_j/\partial N_i)_W$, for whatever W is selected, but with only one component, $(\partial \mu/\partial N)_{P,T} = 0$, one of the few examples for which the inequality required for stability becomes an equality. This simply states that systems with identical intensive parameters can by combined, with neutral consequences for stability.

2.7.2 The Le Chatelier Principle. An often useful path toward the understanding of thermodynamic behavior is provided by the Le Chatelier Principle, which may be stated quite generally as follows:

Any local departures from equilibrium (fluctuations) that arise in a stable system induce changes in the intensive parameters that diminish these fluctuations.

This statement includes Le Chatelier's original principle and the extension attributable to Braun. As an example, consider the effect of a local compression of part of the system and local expansion of another part, such as in a two-part system with a movable piston. The pressure in the compressed region increases, and that in the expanded region decreases. The induced pressure difference acts to move the piston in the direction of pressure equality, which represents expansion of the compressed section and compression of the expanded one. The pressures respond to the compression and expansion in the proper sense to restore homogeneity because the compressibility, κ_T, is positive for a stable system.

The principle can be examined in general, abstract terms. The extensive parameters of a system are denoted by X_i, and the corresponding intensive parameters by Z_i. In a development similar to that beginning with Eq. (2.7.1), choose a small subsystem, a, in equilibrium with a large surrounding system, b, here treated as a reservoir for whatever exchanges a requires to compensate for its internal fluctuations. The statement that equilibrium corresponds to an energy minimum implies that if a fluctuation disturbs the equilibrium state, then the natural direction of the ensuing processes is toward decreasing the energy to its equilibrium value. When the system has fluctuated, it is governed by the process of equilibration to respond by minimizing the energy, or $\delta(U_a + U_b) \leq 0$. With only two extensive variables exhibited in the analysis, the governing inequality becomes, as a general statement that does not depend upon the existence of fluctuations,

$$\begin{aligned}
\delta(U_a + U_b) &= \frac{\partial U_a}{\partial X_{a1}}\delta X_{a1} + \frac{\partial U_a}{\partial X_{a2}}\delta X_{a2} + \frac{\partial U_b}{\partial X_{b1}}\delta X_{b1} + \frac{\partial U_b}{\partial X_{b2}}\delta X_{b2} \\
&= Z_{a1}\delta X_{a1} + Z_{a2}\delta X_{a2} + Z_{b1}\delta X_{b1} + Z_{b2}\delta X_{b2} \\
&= (Z_{a1} - Z_{b1})\delta X_{a1} + (Z_{a2} - Z_{b2})\delta X_{b2} \\
&= \delta Z_1 \delta X_{a1} + \delta Z_2 \delta X_{a2} \leq 0,
\end{aligned}$$

since $\delta X_{ai} + \delta X_{bi} = 0$. The reservoir, subsystem b, is considered so large relative to a that the intensive-parameter differences, the δZ_i, can be thought of as the changes in the parameters of subsystem a, because the fluctuations in a do not significantly affect the intensive parameters of b. The subscripts designating subsystem a can now be dropped and the treatment continued as that of a small system interacting with an appropriate reservoir.

Since it is true in general that the extensive parameters can change independently, the inequality requires that each term be nonpositive, or $\delta Z_i \delta X_i \leq 0$ for $i = 1, 2, \cdots$.

Once these inequalities are established, the Le Chatelier Principle and its relatives can be understood formally in several ways. First, denote the change in the intensive parameter Z_i, in response to the extensive-parameter fluctuation δX_i^f, by δZ_i^r. This response, in turn, is attributable to an intensive-parameter fluctuation δZ_j^f, or

$$\delta Z_i^r \frac{\partial X_i}{\partial Z_j} \delta Z_j^f \leq 0, \tag{2.7.17}$$

where i and j are not necessarily different. As an example, let $X_1 = S$ and $X_2 = V$, so $Z_1 = T$ and $Z_2 = -P$. The inequalities of (2.7.17) give

$$\delta T^r \left(\frac{\partial S}{\partial T} \right)_P \delta T^f = \frac{N C_P}{T} \delta T^r \delta T^f \leq 0, \tag{2.7.18}$$

and

$$-\delta P^r \left(\frac{\partial V}{\partial T} \right)_P \delta T^f = -\alpha V \delta P^r \delta T^f \leq 0. \tag{2.7.19}$$

The first of these says, since $C_P > 0$, that the subsystem's response to a temperature fluctuation is a change in temperature of the opposite sign. This is the original content of the Le Chatelier Principle. The second inequality says that when $\alpha > 0$ (or < 0), temperature fluctuation leads to a pressure response of the same (or opposite) sign , which induces a volume change that is accompanied by a temperature change that opposes the original fluctuation, illustrating the amplification designated as the Le Chatelier-Braun Principle. For P as the fluctuating intensive parameter, the resulting inequalities are

$$-\delta P^r \left(\frac{\partial V}{\partial P} \right)_T \delta P^f = V \kappa_T \delta P^r \delta P^f \leq 0, \tag{2.7.20}$$

and

$$\delta T^r \left(\frac{\partial S}{\partial P} \right)_T \delta P^f = -\alpha V \delta T^r \delta P^f \leq 0. \tag{2.7.21}$$

The first of these again illustrates the Le Chatelier Principle, since $\kappa_T > 0$, and the second is the equivalent of (2.7.19), with pressure and temperature interchanged.

In another view of this rich and often ignored principle, denote a fluctuation in the ith extensive parameter of subsystem a by δX_i^f, without assigning a particular intensive parameter as the causative agent. Then regard δZ_i^r as the response to the fluctuation, and write it as

$$\delta Z_i^r = \sum_j \frac{\partial Z_i}{\partial X_j} \delta X_j^r, \tag{2.7.22}$$

where the δX_j^r are the extensive-variable responses to the fluctuation δX_i^f. The separate inequalities then become

$$\sum_j \frac{\partial Z_i}{\partial X_j} \delta X_j^r \delta X_i^f \leq 0.$$

But since the δX_j are presumably independent, this inequality spawns as many independent inequalities as there are j's, of the form

$$\frac{\partial Z_i}{\partial X_j} \delta X_j^r \delta X_i^f \leq 0. \tag{2.7.23}$$

For example, with S and V as the independent extensive variables, these inequalities are found to be $(T/NC_P)\delta S^r \delta S^f \leq 0$; $(1/\alpha V)\delta V^r \delta S^f \leq 0$; $(1/\alpha V)\delta S^r \delta V^f \leq 0$; and $(1/V\kappa_T)\delta V^r \delta V^f \leq 0$. These are evidently complementary to those that are expressed in terms of the intensive variables, showing that fluctuations in either intensive or extensive parameters lead to equivalent system responses that tend to restore the system to its equilibrium state, provided the stability criteria are satisfied. It is of interest that the Le Chatelier-Braun responses, involving $\delta P \delta T$ or $\delta V \delta S$ products are not explicitly dependent upon a stability criterion, since α can have either sign. The response to a fluctuation is nevertheless in the proper direction to diminish the departure from equilibrium produced by the fluctuation. (The principle was known to Gibbs, and examined by him in quite general terms. (See his Equation 171 and its consequences.)) (Falk [1970]; Gibbs [1878])

Other Le Chatelier relationships can be derived, both by using potentials other than U to minimize for equilibrium, and by including other extensive parameters. In particular, when $\{N_i\}$ are allowed to change, as in chemical reactions, much of the content of chemical thermodynamics emerges in this context. This subject will be examined further in Section 2.10. In the study of phase transitions, particularly in states near the critical point, another look at stability will become necessary, as will be seen in Chapter 6. For nonequilibrium and time-dependent systems, additional techniques are required. Some of these are developed in Chapters 9 and 10.

2.7.3 Global Stability.
As shown in Section 2.7.1, thermodynamic equilibrium is a state of minimum thermodynamic potential, U, F, G, H, subject to the appropriate constraints. Stability can be examined in terms of second-order variations of these potentials, and the results lead to requirements that certain response functions, C_v and κ_T, for example, are necessarily positive for stability. These requirements, and the related Le Chatelier Principle, are expressions of the local shape of the potential surfaces. The general result is that the thermodynamic potentials are convex functions of their extensive variables and concave functions of their intensive variables. But these conditions are restricted by the differential-based analysis to apply only to the local shape of the potential. If the full potential surface resembles a hilly terrain, it is entirely possible that there could be neighboring higher peaks or deeper valleys that offer improved stability.

Fluctuations in a system, as a result of the microscopic behavior of the constituent particles, enable the system to explore the surrounding terrain and probe for these remote regions of improved stability. If such a region is found, a transition is made. If the local curvature of the appropriate potential is large, fluctuations are small, and local equilibrium may appear stable, even when the potential at a remote point offers improved stability. Near a phase transition, however, it is typical for the local potential minima to become shallow and flattened, so that fluctuations can be large, and the transition becomes increasingly probable. Considerations of global stability are particularly important in the study of phase transitions, and they will be developed in more detail in Chapter 6. A brief view of a phase transition is presented in the next section, where the van der Waals equation of state is examined, and the phase transition is described by means of the Gibbs potential.

2.8 NONIDEAL GASES

For purposes of calculation and analysis it is often convenient to write analytic expressions (formulas) that summarize the properties of real substances in the same manner as Eqs. (2.3.13)–(2.3.15) do for the ideal gas. If a fundamental relation such as Eq. (2.3.12) were known, the equation of state would follow directly. For substances made up of real molecules that interact with each other, analytic expressions for the fundamental relations usually are not accurate, and the best representations of the properties of these systems appear as tables of data. For example, engineers use steam tables, which give the specific volume, internal energy, entropy, and enthalpy as functions of temperature and pressure, and similar tables are used for the condensable vapors used in refrigeration systems. The accuracy required in many of the calculations is not usually attainable in any convenient analytic expression.

Equations of state may be highly accurate for nearly ideal gases, and series expansions can yield as high accuracy as can be justified by the experimental data. The ultimate test of any proposed equation of state lies in its agreement with observation, but thermodynamic requirements of stability are often useful in eliminating nonsense and in establishing possible ranges of validity for proposed equations of state. As an example, we examine the celebrated van der Waals equation of state,

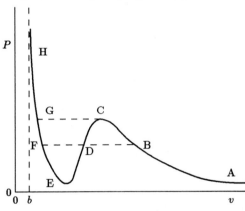

Figure 2.5 Van der Waals isotherm for $T < T_c$.

$$\left(P + \frac{a}{v^2}\right)(v - b) = RT, \qquad (2.8.1)$$

written for one mole of gas, with v the molar volume, b a constant that recognizes the highly compressed state as having a finite volume, and the a/v^2 term representing the interaction of the molecules. For a positive, the molecules attract each other and augment the external pressure in holding the gas together. Qualitative arguments can be given for the form of the a/v^2 term, and indeed many other interaction terms have been proposed, but no attempt to justify any of them will be given here.

It was once fashionable to evaluate a and b in terms of the critical constants. At the critical point, the temperature T_c is such that there is no condensation to the liquid state at any pressure for $T > T_c$, and the isotherm in $P - v$ space exhibits an inflection point at $v = v_c$, $P = P_c$. If we set $\partial P/\partial v = 0$ and $\partial^2 P/\partial v^2 = 0$ at the critical point, the constants are found to be

$$b = v_c/3, \text{ and } a = 9RT_c v_c/8 = 3P_c v_c^2, \qquad (2.8.2)$$

which says $P_c v_c/RT_c = 3/8$, a result not usually in accurate numerical agreement with critical data.

No matter how the constants are evaluated, the pressure is seen to be negative in the range

$$\frac{a}{2RT} - \frac{1}{2}\sqrt{\frac{a^2}{R^2T^2} - \frac{4ab}{RT}} < v < \frac{a}{2RT} + \frac{1}{2}\sqrt{\frac{a^2}{R^2T^2} - \frac{4ab}{RT}}, \qquad (2.8.3)$$

provided the quantity under the radical is positive. For a and b as given by Eq. (2.8.2), P can be negative for $T < (27/32)T_c$. Thus, the van der Waals equation is not expected to be valid for temperatures much below T_c, in the volume range given by (2.8.3). The range $v < b$ is likewise excluded because P is always negative there. For T and v such that P is positive and $T < T_c$, a typical van der Waals isotherm has the general shape shown in Fig. 2.5.

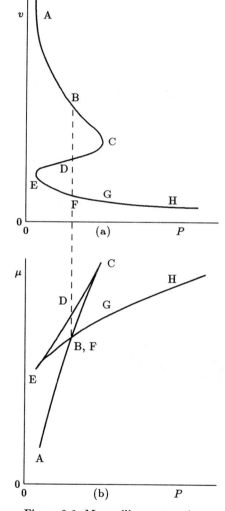

The stability criterion $\kappa_T > 0$ excludes the positive-slope portion of the curve between C and E, leaving as yet unanswered the question of how the sectors H-E and C-A are connected. One might expect a hysteresis loop, with the path H G F E A for increasing volumes and A B C G H for v decreasing, and indeed indications of such behavior are observed experimentally. But also it is easily possible to observe stable equilibrium at a point with $v = v_D$, which would be excluded by the hysteresis loop. The thermodynamically correct equilibrium path F D B can be determined by examination of the Gibbs-Duhem relation Eq. (2.4.7), written here as

$$d\mu = v\,dP - s\,dT. \qquad (2.8.4)$$

Along an isotherm, $dT = 0$, and $\mu(P, T)$ may be calculated by integration of Eq. (2.8.4), up to an arbitrary function of T that is constant along an isotherm. Thus

$$\mu(P, T) = \int_{P_i}^{P} v\,dP' + C(T) \qquad (2.8.5)$$

gives the chemical potential as shown in Fig. 2.6b as the area under the curve in Fig. 2.6a. The area under the sector of C D E is subtracted from the area obtained under A B C, giving the path C D E in Fig. 2.6b, after which the area is again additive along E F G H. By Eq. (2.5.19), $d\mu = dg$, or $\mu = g$, the molar Gibbs potential, and by (2.7.12) g is a minimum for equilibrium at fixed P and T. The points E and A in Fig. 2.6b are at the same temperature and pressure; evidently A is the equilibrium point since $\mu = g$ is minimum there.

Similarly it is seen that the equilibrium path in Fig. 2.6b is A B F G H where points B and F in Fig. 2.6a are such that the area B C D B is numerically equal to the area D E F D. Thus the

Figure 2.6. Maxwell's construction.
(a) Isotherm of Fig. 2.5 redrawn.
(b) Chemical potential from integration of Eq. (2.8.4).

Maxwell construction assigns the stable isotherm in Fig. 2.5 to the path F D B such that the area $B\ C\ D\ B$ equals the area $D\ E\ F\ D$. This portion of the isotherm is understood to be a two-phase mixture, with a fraction x of the substance in the vapor state with molar volume v_B and the fraction $1 - x$ in the liquid state with molar volume v_F. The total molar volume in the mixed state is

$$v = xv_B + (1 - x)v_F. \tag{2.8.6}$$

Similar results are valid for all extensive properties of the system, the entropy, for example, is given by

$$s = xs_B + (1 - x)s_F, \tag{2.8.7}$$

with the same x as in Eq. (2.8.6). Thus if the temperature is known, so that presumably v_B and v_F are known, and molar volume is measured, then x can be calculated and used to evaluate other properties of the mixture.

The evaluation of properties of the mixed state, as given in Eqs. (2.8.6) and (2.8.7), is a valid procedure independent of the quantitative correctness of the Maxwell construction applied to some approximate equation of state. The representation of the isotherm if Fig. 2.5 between F and B as an isobar is in agreement with the Gibbs phase rule Eq. (2.4.11), which assigns a single degree of freedom to a one-component two-phase system. Thus the specification of temperature fixes both P and μ, in accord with path $F\ D\ B$ in Fig. 2.5 and the identity of points F and B in Fig. 2.6b.

Many proposed equations of state for real gases resemble Fig. 2.5 in general appearance. Nothing in the discussion of stability is restricted to the van der Waals equation of state. A widely used equation of state originally proposed by Clausius, called the *virial expansion* is given by

$$\frac{P}{N_0kT} = \frac{1}{v} + \frac{B_2(T)}{v^2} + \frac{B_3(T)}{v^3} + \cdots \tag{2.8.8}$$

where N_0 is Avogadro's number, v is the molar volume, and the quantities $B_2(T)$, $B_3(T)$, ... are called the second, third, ... *virial coefficients*. Corrections to the ideal gas law are expected to diminish as the intermolecular spacing increases, and the temperature dependence of the B's permits an accurate representation of experimental data. A Maxwell construction is necessary even with a virial expansion for the two-phase regime.

2.9 ENTROPY OF MIXING

An interesting thermodynamic analysis offers support for the relationship between information theoretic and thermodynamic entropy. Suppose a partitioned volume contains a different ideal gas in each of its subvolumes, all at the same temperature and pressure. Removal of the partitions leads to free mixing and an increase in entropy as each of the different gases expands to occupy the entire volume. The condition that the gases are ideal ensures that they do not interact or affect each other in any way. For simplicity, assume that there are two different gases, each obeying the law $PV = NkT$. From Eq. (2.3.6), we may write

$$\left(\frac{\partial S}{\partial V}\right)_{U,\ N} = \frac{P}{T} = \frac{Nk}{V}, \tag{2.9.1}$$

from which it follows that

$$S = Nk\ln V + \mathcal{F}(U,\ N), \tag{2.9.2}$$

where $\mathcal{F}(U, N)$ is an undetermined function of energy and N. From Eqs. (2.3.5) and (2.9.2), it follows that

$$\left(\frac{\partial S}{\partial U}\right)_{V, N} = \frac{\partial \mathcal{F}(U, N)}{\partial U} = \frac{1}{T}, \tag{2.9.3}$$

so that we may write $\mathcal{F}(U, N)$ as a new function $\mathcal{G}(T, N)$. Then the total entropy of the two gases, a and b, before mixing is

$$\begin{aligned} S_0 &= S_a + S_b \\ &= N_a k \ln V_a + \mathcal{G}_a(T, N_a) + N_b k \ln V_b + \mathcal{G}_b(T, N_b), \end{aligned} \tag{2.9.4}$$

and the entropy after mixing, with $V = V_a + V_b$, is

$$S_f = N_a k \ln V + \mathcal{G}_a(T, N_a) + N_b k \ln V + \mathcal{G}_b(T, N_b). \tag{2.9.5}$$

The entropy increase is

$$\Delta S = S_f - S_0 = N_a k \ln(V/V_a) + N_b k \ln(V/V_b). \tag{2.9.6}$$

From the gas law, with $N = N_a + N_b$ and no change in T or P, the volume ratios in Eq. (2.9.6) are replaced by population ratios to give

$$\Delta S = -N_a k \ln(N_a/N) - N_b k \ln(N_b/N). \tag{2.9.7}$$

The entropy change per particle, rewritten with $N_a/N = p_a$ and $N_b/N = p_b$, becomes

$$\Delta S/N = -k \left[p_a \ln p_a + p_b \ln p_b \right], \tag{2.9.8}$$

which is just the expected surprise (in thermodynamic units) associated with the selection of a single gas molecule from the mixture. Evidently if there were n species instead of two, Eq. (2.9.8) would become

$$\Delta S/N = -k \sum_{i=1}^{n} p_i \ln p_i, \tag{2.9.9}$$

where $p_i = N_i/N = $ the relative frequency of the ith species. Thus the entropy of mixing is just like the information-theoretic entropy of Eq. (2.2.2), except that here the probabilities are actual relative frequencies instead of the derived quantities usually used. Further aspects of the entropy of mixing are to be found in the problems and in commentary on the Gibbs paradox, Section 3.8.2.

A sometimes puzzling situation occurs when the two gases are identical. Apparently Eqs. (2.9.9) and (2.9.10) show the same entropy increase as before, even though the removal of the partition, and its subsequent reinsertion, can cause no measurable change in any thermodynamic variable. This point of confusion is related to the Gibbs paradox, to be reconsidered in the next two chapters. It is of interest that it can be resolved, at least in the context of macroscopic thermostatics, without recourse to the underlying quantum theory that is usually invoked in more fundamental explanations.

The resolution lies, of course, in the fact that the functions $\mathcal{G}(T, N)$ need not be linear in N. When the gases are identical, there is no need for a subscript on $\mathcal{G}$. The final entropy in this case is

$$S_f = (N_a + N_b)k \ln V + \mathcal{G}(T, N) = Nk \ln V + \mathcal{G}(T, N), \qquad (2.9.10)$$

and the entropy change becomes

$$\Delta S = N_a k \ln(N/N_a) + N_b k \ln(N/N_b) + \mathcal{G}(T, N) - \mathcal{G}(T, N_a) - \mathcal{G}(T, N_b) = 0, \qquad (2.9.11)$$

since the reversible removal of a partition cannot result in an entropy change. It is now possible to refine the function $\mathcal{G}$ by using the requirement that the entropy change must be zero. The ΔS equation can be rewritten as

$$Nk \ln N + \mathcal{G}(T, N) = N_a k \ln N_a + \mathcal{G}(T, N_a) + N_b \ln N_b + \mathcal{G}(T, N_b). \qquad (2.9.12)$$

When this equation is differentiated partially with respect to N_a, the result is

$$k \ln N + k + \frac{\partial \mathcal{G}(T, N)}{\partial N} = k \ln N_a + k + \frac{\partial \mathcal{G}(T, N_a)}{\partial N_a} = f(T), \qquad (2.9.13)$$

where $f(T)$ is an as-yet-undetermined function of T alone, since the left side of the equation is dependent upon N_b, but the right side is not. Solution for $\mathcal{G}(T, N)$ gives

$$\mathcal{G}(T, N) = N f(T) - Nk \ln N, \qquad (2.9.14)$$

so that S may be rewritten as

$$S(V, N, T) = Nk \ln V/N + N f(T). \qquad (2.9.15)$$

The definition of C_V permits evaluation of $f(T)$:

$$T \left(\frac{\partial S}{\partial T} \right)_{V,N} = \frac{N C_V}{N_0} = N T f'(T), \qquad (2.9.16)$$

or

$$f'(T) = C_V / N_0 T. \qquad (2.9.17)$$

When C_V is constant, $f(T)$ is easily evaluated, and the entropy becomes

$$S(V, N, T) = Nk \ln \left(\frac{V}{\alpha N} T^{C_V/R} \right), \qquad (2.9.18)$$

where α is a dimensional constant, inserted for cosmetic reasons. From $U = (N/N_0) C_V T$, the temperature can be replaced by $T = N_0 U / N C_V$, making S a homogeneous first order function of the extensive parameters, as postulated. It is of interest to note that the correct form of S is, in this case, imposed by the requirement that the reversible mixing of two containers of an identical gas at the same temperature and pressure can cause no change in the total entropy. This same situation, when considered by Gibbs in the context of the emerging field of statistical mechanics, resulted in a paradoxical, *ad hoc* correction to his calculated partition function, of evident necessity but of dubious fundamental validity. This topic is reexamined in Chapters 3 and 4.

2.10 CHEMICAL EQUILIBRIUM

It is useful at this point to examine the basic ideas of chemical thermostatics, since the macroscopic equations are available to do so. The usual chemical reactions to be considered take place at constant temperature and pressure, so that equilibrium corresponds to a Gibbs-function minimum by Eq. (2.7.12), or at constant temperature and volume, so that equilibrium corresponds to a Helmholtz potential minimum by Eq. (2.7.10). In either case, chemical equilibrium requires that the thus-far-neglected terms in the chemical potentials sum to zero, or

$$\sum_0^c \mu_i dN_i = 0, \qquad (2.10.1)$$

where c is the number of different chemical constituents.

It is easiest to explain chemical equilibration by going first to a specific example. Suppose a chamber at constant volume and temperature contains a mixture of hydrogen, deuterium, and a hybrid of the two, all as diatomic molecular gases, denoted by H_2, D_2, and HD, respectively. For notational convenience, denote their mole numbers respectively as N_1, N_2, and N_3, and their chemical potentials by the same subscripts. No matter how the atoms of hydrogen and deuterium are distributed among the molecules, the number of atoms, or of moles, of each *atomic* species is conserved. The number of moles of hydrogen atoms, N_H, is given by

$$N_H = 2N_1 + N_3,$$

and the number of moles of deuterium atoms, N_D, is

$$N_D = 2N_2 + N_3.$$

These equations give $dN_H = 0 = 2dN_1 + dN_3$, and $dN_D = 0 = 2dN_2 + dN_3$, or $dN_3 = -2dN_1$ and $dN_2 = dN_1$. Substitution of these results into Eq. (2.10.1) gives

$$(\mu_1 + \mu_2 - 2\mu_3) dN_1 = 0. \qquad (2.10.2)$$

Since dN_1 is arbitrary in sign and magnitude, Eq. (2.10.2) requires that $\mu_1 + \mu_2 = 2\mu_3$, for equilibrium. Moreover, since the equilibrium state is to be a minimum of the Gibbs or Helmholtz function, as is appropriate, the left-hand side of Eq. (2.10.2) must be negative for nonequilibrium states. Therefore, if the term in parentheses is positive, then dN_1 must be negative, as must dN_2, since they are equal, and *vice versa*. The reaction must proceed down the chemical-potential hill, in this case by reducing the populations of H_2 and D_2, and increasing that of HD. (The chemical potentials were shown in Eqs. (2.7.15), (2.7.16), and related discussion, to be increasing functions of the concentrations of the constituents, in systems at constant volume or pressure.)

The equation for the chemical reaction may be read directly from the relationships among the dN_i's. The results that $dN_1 = dN_2$ and $dN_3 = -2dN_1$ say that for every molecule of H_2 created in the chemical reaction, there must be a molecule of D_2 created and two molecules of HD destroyed, as every beginning chemistry student knows who writes

$$H_2 + D_2 \Leftrightarrow 2HD.$$

The coefficients that appear in these so-called *stoichiometric* equations can be read directly from Eq. (2.10.2) and its counterparts.

There are always at least as many chemical potentials (or kinds of molecules) as there are different atoms that make up the molecules. If there are a kinds of atoms and c possible molecules, there must be a independent relationships among the dN_i. It follows that there must be $c - a$ independent mole-number changes dN_i, and therefore $c - a$ independent chemical potentials, with the same number of independent stoichiometric equations. One more illustration should provide sufficient clarity to preclude the need for a general formulation of this aspect of chemical equilibrium.

Consider a gaseous mixture of H_2, O_2, H_2O, and H_2O_2, with four molecular and two atomic species. Label the molecules with the subscripts 1, 2, 3, and 4, respectively. The conservation-of-atoms equations are

$$dN_H = 0 = 2dN_1 + 2dN_3 + 2dN_4,$$

and

$$dN_O = 0 = 2dN_2 + dN_3 + 2dN_4.$$

These may be solved for, say, dN_1 and dN_3 in terms of dN_2 and dN_4, to give $dN_3 = -2dN_2 - 2dN_4$, and $dN_1 = 2dN_2 + dN_4$. These relationships used in Eq. (2.10.1) yield, for equilibrium

$$0 = (2\mu_1 + \mu_2 - 2\mu_3)dN_2 + (\mu_1 - 2\mu_3 + \mu_4)dN_4, \qquad (2.10.3)$$

where dN_2 and dN_4 are regarded as independent. Two independent stoichiometric equations can be read off as

$$2H_2 + O_2 \Leftrightarrow 2H_2O,$$

and

$$H_2 + H_2O_2 \Leftrightarrow 2H_2O.$$

The choice of dN_2 and dN_4 as the independent variables was arbitrary, and another choice, say dN_1 and dN_3, gives a different representation of the equilibrium condition, and different stoichiometric equations:

$$H_2 + O_2 \Leftrightarrow H_2O_2,$$

and

$$2H_2O + O_2 \Leftrightarrow 2H_2O_2.$$

One set of equations is derivable from the other, in either order. The two equations in a set are independent, and it is evident that other pairs of independent stoichiometric equations can be constructed, corresponding to the two independent conditions on the chemical potentials that follow from Eq. (2.10.3).

Further details of chemical equilibrium are examined in the next chapter, where the partition function can be used to derive basic results and to extend the treatment begun here. (Tribus [1961b])

2.11 COMMENTARY

Since the commentary section of Chapter 1 is pertinent to this chapter as well, only a few additional points require further treatment. Some of these can be developed better in later chapters (such as the thermodynamic limit, boundary effects, and extensive parameters, discussed in the next chapter). Others require only a passing nod, such as the use of μ as the chemical potential in two senses: (a) when N is the number of molecules, then μ is the chemical potential per molecule, and $N\mu$ is the total contribution to the internal energy from the chemical-potential terms; (b) when N is the mole number, then μ is the chemical potential per mole, and $N\mu$ has the same meaning and value. The meaning is usually evident in the context. It is standard in macroscopic thermodynamics, including thermostatics, to treat N as the mole number when R is the molar gas constant, and in statistical physics to treat N as the particle number with k as the molecular gas constant.

Most of the space in this commentary is given to a discussion of the various treatments of thermodynamics. The attitude of writers about thermophysics, including the advocates of information theory, often seems apologetic or defensive; some authors write as if they feel compelled to offer something better than a physicist's proof that the statistical theories are related in the posited manner to macroscopic thermodynamics, and some authors act as if they have indeed done so. Much time and frustration can be spared by the dropping of any pretense of rigorous proof. Therefore the connections between information theory, statistical thermophysics, and thermodynamics will be made plausible and will be shown to lead to standard results, but the lingering uncertainty should remain that perhaps there is another way to do it with better results, sharper predictions, improved conceptual clarity, or greater postulational economy.

2.11.1 Approaches to Thermodynamics. Although I have restricted the subject matter of this chapter to equilibrium thermodynamics, or more properly, thermostatics, it should be realized that most books with titles that imply a broader coverage are similarly restricted. Authors who profess to be concerned with the foundations of thermodynamics turn out to have confined their treatments to the foundations of thermostatics, mostly as a consequence of limitations imposed by Clausius, although they may discuss some nonequilibrium formalisms. Despite this prevalent mistitling of their works, there are a number of carefully written and worthwhile books that can be studied for further enlightenment. (The fact that I do not mention a book in this commentary does not imply that I regard it as other than carefully written or worthwhile. So many books exist that there is not space to comment on all of them, even if I have examined them.)

Several authors are convinced that their subject should exist as an independent discipline, firmly founded within itself, with no need to seek its basis in microphysics (Adkins [1983]; Buchdahl [1966]). Others do not express this attitude as a conviction, but proceed to develop the subject with little, if any, reference to the underlying microphysics (Fermi [1956]; ter Haar and Wergeland [1966]; Landsberg [1961]; Planck [1945]; Zemansky and Dittman [1981]). Still others use the microphysical description of a system whenever it seems to illuminate the topic being presented, not necessarily as a foundation for the subject, but as a useful teaching tool (Guggenheim [1967]; Morse [1969]). I recommend these books as serious, enlightening works, even though I do not agree with everything they say, and even though no one of us is always correct in our presentations or perceptions.

Buchdahl specifically denies that his treatment is axiomatic, but it is carefully done, with its basis in the well-known inaccessibility theorem of Carathéodory. Tisza [1966] regards the theoretical structure developed by Carathéodory as an axiomatization of the Kelvin-Clausius formulation of thermodynamics, and, in fact, the first such axiomatization of the subject. (I view the Kelvin-Clausius formulation as an axiomatic development, proceeding as it does from the stated three laws of thermodynamics.) The book by Adkins is in the same spirit as that of Buchdahl, but somewhat less abstract and with a somewhat different selection of topics. Both of these books have relatively readable sections expounding Carathéodory's statement of the second law, which is: *In the neighborhood of any arbitrary state of a thermally isolated system, there are states that are inaccessible from the original state.*

Pfaffian equations, which are of the form $dY = \sum_i X_i dx_i$, where the $X_i's$ are given by $X_i = X_i(x_1, x_2, \cdots)$, were studied by Carathéodory, who professed to derive the Kelvin temperature scale and apparently every other consequence of the Kelvin-Planck statement of the second law, which is: *No process exists such that its only effect is the removal of heat from a reservoir and the conversion of all of this heat into work.* Born has asserted that the Kelvin-Planck statement covers the widest possible range of impossible processes, whereas the advantage of the Carathéodory statement is that it is necessary to know only that *some* local processes are impossible in order to develop the machinery of thermostatics. Needless to say, there seem to be no correct consequences of the Carathéodory statement that are not also consequences of the Kelvin-Planck statement. (Born [1964]; Carathéodory [1909]) Truesdell offers a devastating criticism of Carathéodory's work, with several references, dismissing his work both for inadequate mathematics and for faulty physics, in that he tries to reduce thermodynamics to mechanics. (Truesdell [1984]) Landsberg has refined the mathematics and provided a conceptual extension of the method that allows a natural place for the third law of thermodynamics and for quasistatic adiabatic changes. The details are too lengthy for inclusion here, particularly since this topic is well-removed from my principal subject matter, and also because of Truesdell's criticism that the Carathéodory theory is not worth the effort to fix it. (Landsberg [1961])

The account of thermostatics in this chapter closely follows that of Callen [1985], which is a modification of the work of Tisza [1966] on "MTE," the macroscopic thermodynamics of equilibrium. As Callen points out in the opening sentence of his second edition, his first edition was the most frequently cited thermodynamic reference in the physics literature, and the postulational formulation introduced therein is now widely accepted. I have attempted, in the early sections of the chapter, to offer plausible reasons, based on information theory, for the postulates that Callen introduced. I have shown that the thermodynamic developments are carried out in the same way, whether the maximum entropy is obtained through information theory or postulated, as by Callen. In the next chapter, I shall propose one additional postulate, that of Jaynes, connecting information theory to statistical mechanics and thermodynamics, thereby presenting a postulational package that seems to unify the entire subject, and also to permit extension to nonequilibrium situations, *i.e.*, to thermo*dynamics*. (In Chapter 1, most of the examples given were of time-dependent systems, and in Section 1.7, the concept of a time-dependent temperature was used, but without much discussion. Further nonequilibrium systems are presented in later chapters.) In Chapter 21 of Callen's second edition, a preliminary treatment of the foundation of thermostatics is offered, based on quite general symmetry principles. I recommend its study.

It should be noted that neither Callen nor Tisza originated the entropy-maximum postulate; Gibbs did. (Gibbs [1878]; Gibbs [1961]) In his 1876-78 paper *On the Equilibrium of Heterogeneous Substances*, he states two equivalent criteria for equilibrium:
I. *For the equilibrium of any isolated system it is necessary and sufficient that in all possible variations of the state of the system which do not alter its energy, the variation of its entropy*

shall either vanish or be negative, and

II. *For the equilibrium of any isolated system it is necessary and sufficient that in all possible variations of the state of the system which do not alter its entropy, the variation of its energy shall either vanish or be positive.*

He showed that the two statements are equivalent, and offered examples of their use. His paper was a long one (pp.108-248; 343-524), with much carefully presented material, but it was largely ignored by most of the established authorities. (But not by Duhem [1886].) His phase rule, the Gibbs-Duhem relation, and the Gibbs potential G did find their way into the literature, but the entropy-maximum principle seems not to have been recognized as a useful means of developing thermodynamics, except by Gibbs, until Tisza and Callen did so in the late 1950's. (It was cited by Duhem, who hails the *fécondité* of the new methods in thermodynamics, but it seems to have been ignored for some time thereafter.) Additional similar formulations exist (Bowers [1975]).

Gibbs used a *convexity property* as the basis for his derivations of stability. It is equivalent to the following procedure: The so-called first law of thermodynamics, the conservation of energy, is written $\delta U = T\delta S - P\delta V + \sum \mu_i dN_i$. Under the same conditions that led to Eq. (2.7.4), the second variation of U can be written

$$\delta^2 U = \delta T \delta S - \delta P \delta V + \sum \delta \mu_i \delta N_i \geq 0. \tag{2.11.1}$$

This powerful version of the stability criterion is used in a generalized form in Section 9.10.2.

The writings of Tribus on thermostatics and thermodynamics are significant in their use of a maximum-entropy principal, but they are based from the outset on information theory, whereas I introduced the information-theory formalism of Chapter 1 at the beginning of this chapter as a plausible background for the maximum-entropy axiomatization of thermostatics. (Costa de Beauregard and Tribus [1974]; Tribus [1961a]; Tribus [1961b]; Tribus, Shannon and Evans [1966]) I comment again on these papers in Chapter 3.

Finally, in this subsection, I come to the most difficult topic: the monumental work of Truesdell and his colleagues in the emerging field of *Rational Thermodynamics*. (Coleman and Noll [1963]; Coleman and Noll [1964]; Coleman and Owen [1974]; Day [1987]; Day [1988]; Gurtin [1975]; Noll [1973]; Truesdell [1969]; Truesdell [1984]; Truesdell [1986]; Truesdell and Bharatha [1977]) They are not cited in any of the references I have used so far in this subsection, and Truesdell does not cite Callen, among others. He treats true thermodynamics as an essential complement to continuum mechanics, properly ignoring, in that context, the microstructure of matter. He presents a mathematically sound and coherent unification of thermodynamics and the version of continuum mechanics that has come to be known as *rational mechanics*. He defines and uses the concepts of local temperature and pressure as time-dependent variables. He objects to the policy, apparently imposed by Clausius, of excluding time as a proper thermodynamic variable by writing $du = Tds - Pdv$, in terms of differentials, rather than writing $\dot{u} = T\dot{s} - P\dot{v}$, on the grounds that the intensive parameters are not definable except at equilibrium, and the changes implied by the differentials must therefore be made quasistatically. (But it is not enough simply to replace differentials by time derivatives, since the Clausius-Duhem inequality states that $\dot{s} \geq q/T$, where q is the rate of adding heat, per unit mass, to the system. The correct restatement of the time-dependent First Law is $q = \dot{u} + P\dot{v}$. This can be written, when u is regarded as a function of v and T, as

$$q = \left[\left(\frac{\partial u}{\partial v} \right)_T + P \right] \dot{v} + \left(\frac{\partial u}{\partial T} \right)_v \dot{T} = \Lambda_v \dot{v} + c_v \dot{T}, \qquad (2.11.2)$$

where c_v is the specific heat at constant volume, and Λ_v is called by Truesdell the *latent heat with respect to volume*.

Much of what he says is clearly correct; I have found nothing clearly wrong. So why is not this new formulation of thermodynamics hailed by all as a long-awaited improvement? Because, I believe, it is easier to ignore than to understand. (This was probably a significant part of the reason that Gibbs was ignored.) Truesdell uses nonstandard notation and terminology, sometimes with good reason, but sometimes, I believe, without any real justification. For example, he uses the term *latent heat*, as I have shown in Eq. (2.11.2), not just for the heat transferred per unit mass in a phase transition at constant temperature, but more generally for the heat transfer processes associated with expansion and contraction at constant temperature, even with no change of phase. He proposes the term *caloric* for specific entropy, and its integral over the mass of the system, the total entropy, is designated by him as the *calory*, because he regards *entropy* as a meaningless term. (He later withdrew the proposal, but did not revise his writing. (Truesdell [1984]) These notational divergences are not insurmountable, but they preclude the typical hurried scholar's inclination to look up a particular topic, somewhere in the middle of the book, without reading everything in the book up to that point. Therefore, apparently, everyone is waiting for someone else to translate Truesdell and his school into more or less standard jargon. I am not going to do that here, but some of his work will be examined in Chapter 9 in the context of nonequilibrium thermodynamics and statistical physics. I note in passing that one frequent objection to this body of work, particularly by the more traditional school of thermodynamicists, is that since a unique nonequilibrium entropy function does not exist, Truesdell's entropy and temperature variables are undefined, and thus of no practical value. Truesdell does not regard the entropy as undefined, and a more accurate expression of this objection is that the several definitions of nonequilibrium entropy do not necessarily agree with each other, so that the nonequilibrium entropy function is indeed not unique. Nevertheless, some of these functions have demonstrable practical value. (Woods [1975])

The rational thermodynamicists have had considerable success in attacking difficult and interesting problems, such as the thermodynamics of materials with (usually fading) memory, (Coleman and Noll [1963]; Day [1988]); wave propagation in dissipative materials, and in materials with memory; and the thermodynamics of multiphase flowing mixtures (Truesdell [1984]). These topics are too far afield to justify a more detailed commentary, but knowledge of their existence might lead to some future unification of the separate schools of thermodynamics.

2.11.2 Mnemonic Diagrams. Apparently the use of mnemonic diagrams, like that of Fig. 2.2, for easy access to information about thermodynamic potentials and Maxwell relations was first used by Max Born in Göttingen in the late 1920's. (Tisza [1966]) Other such diagrams are to be found in Callen [1985].

2.11.3 Systems with Nonextensive Entropies. There are circumstances for which fundamental statement 3 is violated, in which case the entire formal structure and its consequences are expected to be invalid. These are systems subject to long-range forces, such as gravitation or Coulomb interactions, and for which the potential energies associated with these forces cannot be neglected. Systems for which fundamental statement 3 holds are called *normal* systems, and those that are expected to violate it are then said to be *abnormal*. (Landsberg [1984]) Most thermodynamic systems that are commonly encountered are normal; they are electrically neutral; gravitation is either an unimportant uniform background or negligible in comparison with other energies; and all other significant interactions are short range. But astrophysical systems and non-neutral aggregates of electrically charged particles are not normal, in the sense that some terms in the energy and entropy are nonextensive. Landsberg has proposed an added law of thermodynamics: *In the absence of long-range forces, the principal variables are either extensive or intensive.* This topic is mentioned again in the commentary of Chapter 3.

2.11.4 The Phase Rule. The phase rule (Eq. (2.4.11) can be rewritten as $p = c + 2 - f$, and interpreted to say that the maximum number of phases that can coexist in equilibrium is $c + 2$, corresponding to no degrees of freedom (no independent intensive parameters). Thus a one-component system can have, at most, three coexisting phases. An interpretation of this result follows from the idea that the molar Gibbs potential for a one-component system is $g = u - Ts + Pv$, and $u = u(s, v)$ is a fundamental relation, *i.e.* a constitutive relation for the material. Therefore, at fixed temperature T_0 and pressure P_0, the molar Gibbs potential $g = u(s, v) - T_0 s + P_0 v$ is a two-dimensional surface in usv-space. The planes $g_i = u - T_0 s + P_0 v$, where u, v, and s are regarded as independent coordinates, are level planes for g at the fixed values of temperature and pressure. There is a value g_0 that minimizes g at T_0, P_0; this is the equilibrium value. The plane $g_0 = u - T_0 s + P_0 v$ is called a support plane for g at T_0, P_0, and the surface $g = u(s, v) - T_0 s + P_0 v$ must be tangent to its support plane at one or more points, and have no points on the surface below the support plane. In a three-dimensional space, three points determine a plane.

There are normally no more than three distinct points at which the g-surface is tangent to the support plane at level g_0, corresponding to the phase rule's maximum three phases. But it is possible to construct mathematical surfaces, corresponding to fundamental relations, with any arbitrary number of tangent points, and therefore that number of phases in equilibrium. How can this be? The phase-rule derivation is unambiguous, and there can be no less than zero degrees of freedom, so there can be no more than three phases in equilibrium for a single-component system, unless there are more than $c + 2 = 3$ intensive parameters (T, P, and μ).

It is, in fact, possible to have additional intensive parameters, such as components of the stress tensor, or electric and magnetic fields; these have associated extensive parameters—strains or electric and magnetic moments—which should therefore become variables in the constitutive relation and in the expression for the g surface. Then g becomes a surface in a space of more than three dimensions, and the number of points to determine a plane in n dimensions is n, so there can be a normal maximum of n tangent points, and n coexisting phases in thermal equilibrium.

In the absence of additional extensive parameters, the normal phase-rule limit of three phases for a one-component system must be valid, and the mathematical possibility of devising constitutive relations that can have more than three distinct tangents to the support plane cannot be realizable in actual substances. It would be comforting to derive this result in an independent way, rather than as a consequence of the phase rule, because the phase rule does not show what must be wrong with the artificially constructed fundamental relations.

2.12 REFERENCES

These are the principal references for Chapter 2. The complete reference list is to be found in the Bibliography at the end of this book.

ADKINS, C. J. [1983], *Equilibrium Thermodynamics, 3rd ed.*. Cambridge University Press, Cambridge, England, 1983.

CALLEN, H. B. [1985], *Thermodynamics and an Introduction to Thermostatistics*. John Wiley & Sons, New York, NY, 1985.

FALK, H. [1970], Inequalities of J. W. Gibbs. *American Journal of Physics* **38** (1970), 858–869.

GIBBS, J. W. [1876, 1878], On the Equilibrium of Heterogeneous Substances. *Transactions of the Connecticut Academy* **3** (1876, 1878), 108–248, 343-524.

GUGGENHEIM, E. A. [1967], *Thermodynamics, 5th ed.*. North-Holland, Amsterdam, 1967.

LANDSBERG, P. T. [1961], *Thermodynamics*. Interscience Publishers, A division of John Wiley & Sons, London, 1961.

TISZA, L. [1966], *Generalized Thermodynamics*. MIT Press, Cambridge, MA, 1966.

TRIBUS, M. [1961b], *Thermostatics and Thermodynamics*. D. van Nostrand Co., Inc., New York, 1961.

TRUESDELL, C. A. [1984], *Rational Thermodynamics, 2nd Ed.*. Springer-Verlag, New York–Heidelberg–Berlin, 1984.

2.13 PROBLEMS

1. Which of the following, offered as fundamental relations, are acceptable, and which violate one or more of the five fundamental statements? (R, N_0, V_0, U_0 and S_0 are constants.)

a. $S = NR + NR \ln \left[\left(\frac{U}{U_0} \right)^{3/2} \left(\frac{V}{V_0} \right) \left(\frac{N}{N_0} \right)^{-5/2} \right]$.

b. $S = R \left(UVN/U_0 V_0 N_0 \right)^{1/3}$.

c. $S = R \left(\frac{UV}{U_0 V_0} \right)^2 \left(\frac{N}{N_0} \right)^{-3}$.

d. $S = NR \exp \left[\left(\frac{U}{U_0} \right)^2 \left(\frac{V}{V_0} \right)^{-2} \right]$.

e. $S = NR \tan \left[\left(\frac{U}{U_0} \right) \left(\frac{V}{V_0} \right)^2 \left(\frac{N}{N_0} \right)^{-3} \right]$.

f. $S = NR \coth \left[\left(\frac{U_0}{U} \right)^2 \left(\frac{VN}{V_0 N_0} \right) \right]$.

g. $S = R \left[\frac{UN}{U_0 N_0} \right]^{1/2} \exp \left[- \left(\frac{VN_0}{V_0 N} \right)^2 \right]$.

h. $U = U_0 \left(\frac{S}{S_0} \right) \left(\frac{V_0}{V} \right) \exp \left(\frac{SN_0}{S_0 N} \right)$.

i. $U = U_0 \left[\frac{NV}{N_0 V_0} \right] \left[1 + \frac{S N_0}{S_0 N} \right] \exp \left[-\frac{SN}{S_0 N_0} \right].$

j. $S = R \left[\left(\frac{NU}{N_0 U_0} \right)^2 \frac{V}{V_0} \right]^{1/5}.$

2. The fundamental relation of a material is given by

$$s = s_0 + R \ln \left[(v - b) / (v_0 - b) \right] + (3/2) R \ln \sinh \left[c \left(u + a/v \right) \right].$$

a. Derive the equation of state $f(P, v, T) = 0$ for this material.

b. Show that $C_v = (3R/2)/[1 - (3RcT/2)^2]$. (Note that s_0, R, b, v_0, a, and c are constants.)

3. Show that when $\alpha = 1/(T + T_0)$, where T_0 is a constant, it follows that C_p is independent of pressure.

4. A condensable vapor undergoes a throttling process (for which the final enthalpy is the same as the initial enthalpy) from an initial state at pressure P_1 and molar volume v_1 to a final state at pressure P_2 and molar volume v_2.

a. Derive a general differential expression for the change in temperature of the vapor in terms of v, T, dv, and such tabulated properties as α, κ_T, and C_v.
b. Show that there is no temperature change for an ideal gas.

5. The fundamental relation for what may be regarded as a gas of photons in thermal equilibrium is $S = A U^{3/4} V^{1/4}$, where A is a constant.
a. Find the energy density (energy per unit volume) in terms of the temperature.
b. Show that the pressure equals one-third the energy density.
c. Find the entropy density in terms of temperature.

6. Suppose that $C_v = C_v(v, T)$. Show that for a van der Waals gas, C_v is a function of temperature only.

7. Consider a composite system, made up of two interacting subsystems, which fails to satisfy Eq. (2.2.12), but

$$p_n = \sum_m W_{mn} \text{ and } w_m = \sum_n W_{mn}, \text{ with } \sum_{mn} W_{mn} = \sum_n p_n = \sum_m w_m = 1.$$

Show that

$$S - (S_1 + S_2) = k \sum_{mn} W_{mn} \ln(p_n w_m / W_{mn}) \leq 0.$$

8. The adiabatic volume expansivity may be defined as

$$\alpha_s = \frac{1}{v} \left(\frac{\partial v}{\partial T} \right)_s.$$

a. Evaluate α_s in terms of α, C_v, κ_T, and the state variables of the system. Show that α_s is negative for an ideal gas, and explain the reason. In general, could α_s ever be positive? How?

9. The fundamental relation for a condensable vapor is assumed to be given by

$$s = s_0 + R \ln \left[(v - b) \left(u + a/v \right)^{5/2} / \left(v_0 - b \right) \left(u_0 + a/v_0 \right)^{5/2} \right].$$

(The Maxwell construction is used for the equilibrium state of phase mixtures, and the fundamental relation applies to both pure phases.) Show that the latent heat of vaporization at a temperature T is given by

$$\ell = RT \ln \left[(v_g - b) / (v_c - b) \right],$$

where v_g and v_c are the molar volumes of the pure saturated vapor and liquid respectively.

10. A material is found to have a volume expansivity given by

$$\alpha = (R/Pv) + \left(a/RvT^2 \right),$$

and the compressibility is of the form

$$\kappa_T = \left[Tf(P) + (b/P) \right] /v.$$

a. Find $f(P)$. b. Find $v(P,T)$.

c. Does the material appear to be stable? Under what conditions?

11. The fundamental relation for a new secret mob-control gas is reported to be

$$s = s_0 + R \ln \left[(v - b) / (v_0 - b) \right] + R \ln \left[\cosh c \left(u + a/v \right) \right] - R \ln \left[\cosh c \left(u_0 + a/v_0 \right) \right],$$

where a, b, c, s_0, u_0 and v_0 are constants.
a. Calculate the P/T equation of state for this gas.
b. Show that the gas is intrinsically unstable at all temperatures.

12. An empirical expression for the vapor pressure of cesium in equilibrium with liquid cesium at temperature T is found to be $\ln P = A - B \ln T - CT^{-1} + DT$. Find a theoretical basis for this formula, if possible, and evaluate the constants $A, B, C,$ and D in terms of the following data: Cesium vapor pressures are so low that cesium vapor is approximately an ideal gas. Therefore, since $v_g \gg v_c$, it follows that $v_g - v_c \simeq v_g$. The latent heat of vaporization varies with the temperature approximately as $\ell = a \left(T_c - T \right)^2$. The empirical formula is valid only up to a temperature $T_0 = (3/4) T_c$, where the pressure is P_0, and the volume approximation is still regarded as valid.

13. Show that in general the isothermal and adiabatic compressibilities are related by

$$\kappa_T = \kappa_s + \left(Tv\alpha^2/C_p \right).$$

14. An attempt to generalize the ideal gas law leads to an equation of state of the form

$$\left[P + f(v) \right] (v - b) = RT.$$

a. Assume that $C_v = C_v(v,T)$ and show that in this case C_v is a function only of T.
b. Measurements of the compressibility are found to fit the expression $\kappa_T = (v - b)/vP$. Use this result to find $f(v)$.

15. A composite isolated system consists of 1 mole of a monatomic ideal gas $[U_1 = (3/2)N_1 RT]$ and 2 moles of a diatomic ideal gas $[U_2 = (5/2)N_2 RT]$, separated into subsystems of equal volume by a fixed, impermeable, diathermal wall.

a. Given the total energy U and the total volume V, derive expressions for the equilibrium temperature and pressure of each subsystem in terms of known quantities.

b. The wall separating the two subsystems is released, allowing the systems to reach a new equilibrium. Calculate the consequent change in total entropy.

16. Measurements on a gas show that $\alpha = (Rv/P) + (Av/T^2)$ and $\kappa_T = Tvf(P)$, where A and R are constants. Find $f(P)$ and $v(P,T)$.

17. Calculate $C_v, C_p, \alpha,$ and κ_T for the substance of problem 1.b. Verify that for this substance $C_p = C_v + (Tv\alpha^2/\kappa_T)$.

18. The heat source and heat sink available in a composite thermodynamic system are composed of material with molar heat capacity $C_v = RT/\theta$, where R is the gas constant and θ is a (constant) characteristic temperature. The source consists of N_1 moles, initially at temperature T_1, and the sink is N_2 moles and initial temperature T_2. Show that the maximum work attainable from this system goes as the square of the initial temperature difference.

19. The fundamental relation of a gaseous material is found to be

$$S = Ns_0 + NR\ln\left[AU^n V N^{-1-n}\right]$$

where s_0 and A are constants, R is the gas constant, and n is a number constant. Show that this material is an ideal gas with $C_v = nR$.

20. Show that for an ideal gas $C_p = C_v + R$.

21. Two identical ideal gases, initially at the same pressure P and number of moles N, but at different temperatures T_1 and T_2 are in two containers with volumes V_1 and V_2. The containers are then connected. Calculate the change in entropy and show that $\Delta S > 0$, except in trivial cases.

22. The fundamental relation of a substance is given by

$$s = s_0 + R\ln\left[(v/v_0)\sinh^{3/2}(au/u_o)\right],$$

where $\sinh a = 1$.

a. Show that for this substance $C_p - C_v = R$.

b. Calculate C_v.

23. (This uncharacteristic and elementary problem is included only because the widespread and recurrent student surprise at its outcome indicates a common conceptual gap that should have been closed in earlier courses.) A home is to be maintained at $27°C$ while the outside temperature is $-3°C$. If an electric heater is used to supply the necessary heat, the cost for this purpose is \$30.00. If, instead, an ideal Carnot engine is used in reverse to pump heat from the outside (at $-3°C$) to the house, it is necessary to pay for the power to drive the heat pump, which must supply the same heat as before. If the power rate remains constant and the electric motor that drives the heat pump is 90% efficient and located inside the house, what is the cost of supplying the amount of heat that previously cost \$30.00? (The solution to this problem presumes prior knowledge of the Carnot engine, its reversibility, and the relationship of thermodynamic and common temperatures.)

24. A polytropic process, or polytrope, for an ideal gas is represented by the equation $Pv^{\Gamma} = K$, where K is a constant, $\Gamma = 1 + 1/n$, and n is called the *index* of the polytrope. A characteristic of polytropes is that the molar heat capacity for the process, C_{Γ}, is constant, and can be expressed in terms of Γ, γ, and C_v, where $\gamma = C_P/C_v$. Derive an expression for C_{Γ}, and examine the results for $\Gamma = 0$, 1, γ, and ∞.

25. The Gibbs paradox, discussed in Sec. 9, arises upon consideration of the entropy change that results from the mixing of two ideal gases. When they are identical, there is no change in entropy; when they are different, mixing results in an entropy increase. In this problem, the gases are formulated to allow the difference to be a continuous variable, so that the resulting entropy change is also continuous. Two equal containers, each of volume V, contain an ideal gas at pressure P and temperature T. The gas in chamber 1 consists of N_1 molecules of gas α and M_1 molecules of the distinctly different gas β. In chamber 2, there are N_2 molecules of gas α and M_2 molecules of gas β. (Note that $N_1 + M_1 = N_2 + M_2$.) The gases are allowed to mix. Derive an expression for the entropy of mixing, and show that it goes continuously from $\Delta S = 0$ when the mixtures are identical, to $\Delta S = 2kN \ln 2$ when the chambers each originally contain N molecules of a different gas.

2.14 SOLUTIONS

1. By inspection, the following relations are not homogeneous, first order, and are therefore not acceptable: c, h, i. In addition, e and i do not give S as a monotone increasing function of U, provided only positive values of U and S are allowed. The following violate the condition that $S = 0$ when $\partial U/\partial S = 0$: a, c, d, e, f, h. Therefore only b, g, and j are free from objection.

2.a.

$$\frac{1}{T} = \left(\frac{\partial s}{\partial u}\right)_v = \frac{3}{2} Rc \coth\left[c\left(u + a/v\right)\right].$$

$$\frac{P}{T} = \left(\frac{\partial s}{\partial v}\right)_u = \frac{R}{v - b} + \frac{3Rca}{2v^2} \coth\left[c\left(u + a/v\right)\right].$$

$$= \frac{R}{v - b} + \frac{a}{Tv^2} \text{ (the van der Waals equation)}$$

b. $C_v = (\partial u/\partial T)_v$. Differentiate the $1/T$ equation implicitly, keeping v constant.

$$-dT/T^2 = -(3/2) Rc^2 \operatorname{csch}^2\left[c\left(u + a/v\right)\right] du$$

$$C_v = \left(\frac{\partial u}{\partial T}\right)_v = \frac{2}{3Rc^2T^2\operatorname{csch}^2\left[c\left(u + a/v\right)\right]} = \frac{2}{3Rc^2T^2\left(4/9R^2c^2T^2 - 1\right)}$$

$$= (3R/2)\left[1 - (3RcT/2)^2\right]^{-1}.$$

3.

$$\left(\frac{\partial C_P}{\partial P}\right)_T = \left(\frac{\partial}{\partial P}T\left(\frac{\partial s}{\partial T}\right)_P\right)_T = T\frac{\partial^2 s}{\partial T \partial P} = T\left(\frac{\partial}{\partial T}\left(\frac{\partial s}{\partial P}\right)_T\right)_P.$$

But

$$\left(\frac{\partial s}{\partial P}\right)_T = -\left(\frac{\partial v}{\partial T}\right)_P = -\alpha v.$$

Then

$$\left(\frac{\partial}{\partial T}\left(\frac{\partial s}{\partial P}\right)_T\right)_P = -\left(\frac{\partial(\alpha v)}{\partial T}\right)_P = -v\left(\frac{\partial \alpha}{\partial T}\right)_P - \alpha\left(\frac{\partial v}{\partial T}\right)_P = 0$$

when $\alpha = (T + T_0)^{-1}$. Therefore

$$\left(\frac{\partial C_P}{\partial P}\right)_T = 0, \text{ and } C_P \neq C_P(P).$$

4.a.

$$dT = \left(\frac{\partial T}{\partial v}\right)_h dv = -\left(\frac{\partial h}{\partial v}\right)_T \left(\frac{\partial h}{\partial T}\right)_v^{-1} dv.$$

But $dh = Tds + vdP$; therefore

$$\left(\frac{\partial T}{\partial v}\right)_h = \frac{-\left(T\left(\frac{\partial s}{\partial v}\right)_T + v\left(\frac{\partial P}{\partial v}\right)_T\right)}{T\left(\frac{\partial s}{\partial T}\right)_v + v\left(\frac{\partial P}{\partial T}\right)_v} = \frac{-\left(T\left(\frac{\partial P}{\partial T}\right)_v - 1/\kappa_T\right)}{C_v + v\left(\frac{\partial P}{\partial T}\right)_v}$$

$$= \frac{-(\alpha T/\kappa_T) + (1/\kappa_T)}{C_v + (\alpha v/\kappa_T)} = \frac{1 - \alpha T}{C_v \kappa_T + \alpha v}, \text{ or}$$

$$dT = \left[(1 - \alpha T)(C_v \kappa_T + \alpha v)^{-1}\right] dv.$$

b. For an ideal gas, $\alpha = 1/T$; therefore $dT = 0$.

5.a.

$$\frac{1}{T} = \left(\frac{\partial S}{\partial U}\right)_V = \frac{3}{4} A (V/U)^{1/4}, \text{ or } U/V = (3A/4)^4 T^4.$$

b.

$$\frac{P}{T} = \left(\frac{\partial S}{\partial V}\right)_U = (A/4)(U/V)^{3/4}.$$

$$P = \frac{P/T}{1/T} = \frac{(A/4)(U/V)^{3/4}}{(3A/4)(U/V)^{-1/4}} = (1/3)(U/V).$$

c.

$$S/V = A(U/V)^{3/4} = A(3A/4)^3 T^3 = (3/4)^3 A^4 T^3.$$

6.

$$(\partial C_V/\partial v)_T = \left(\frac{\partial}{\partial v}\left(T\frac{\partial s}{\partial T}\right)_v\right)_T = T\frac{\partial^2 s}{\partial v \partial T}$$

$$= T\left(\frac{\partial}{\partial T}\left(\frac{\partial s}{\partial v}\right)_T\right)_v = T\left(\frac{\partial}{\partial T}\left(\frac{\partial P}{\partial T}\right)_v\right)_v = T\left(\frac{\partial^2 P}{\partial T^2}\right)_v.$$

For a van der Waals gas,

$$P = \frac{RT}{v - b} - \frac{a}{v^2}, \text{ and } \left(\frac{\partial^2 P}{\partial T^2}\right)_v = 0$$

Therefore $C_v(v, T) = C_v(T)$.

7.

$$S = -k \sum_{m,n} W_{mn} \ln W_{mn}; \quad S_1 = -k \sum_n p_n \ln p_n; \quad S_2 = -k \sum_m w_m \ln w_m.$$

$$S - (S_1 + S_2) = -k \sum_{m,n} [W_{mn}(\ln W_{mn} - \ln p_n - \ln w_m)]$$

$$= k \sum_{m,n} W_{mn} \ln(p_n w_m / W_{mn})$$

$$\leq k \sum_{m,n} W_{mn}\left(\frac{p_n w_m}{W_{mn}} - 1\right) = 0.$$

8.a.
$$\left(\frac{\partial v}{\partial T}\right)_s = -\left(\frac{\partial s}{\partial T}\right)_v \left(\frac{\partial s}{\partial v}\right)_T^{-1} = -\left(\frac{C_v}{T}\right)\left(\frac{\partial P}{\partial T}\right)_v^{-1}$$
$$= -\left(C_v/T\right)\left(\alpha/\kappa_T\right)^{-1} = -C_v\kappa_T/\alpha T.$$

As T increases, the entropy associated with the momentum-space distribution increases because of increased uncertainty about the momenta of the particles. Therefore the uncertainty in position space must decrease if the entropy is to remain constant. The volume must decrease, since the probability that a particle is in a particular volume element dV is dV/V.

b. For the ideal gas, $\alpha = 1/T$, and $\kappa_T = 1/P$, giving $\alpha_s = -C_v/P < 0$.

c. α_s can be > 0 when $\alpha < 0$, as in the case of water between the ice point and the temperature of maximum density.

9. Since $1/T = (\partial s/\partial u)_v = 5R/2(u + a/v)$, it follows that $u + a/v = (5/2)RT$. Then along an isotherm, s is conveniently written

$$s = s_0 + R\ln\left[(v - b)\left(5RT/2\right)^{5/2} / (v_0 - b)\left(5RT_0/2\right)^{5/2}\right],$$

and

$$s_g - s_c = R\ln\left[(v_g - b) / (v_c - b)\right].$$

The latent heat is

$$\ell = T\left(s_g - s_c\right) = RT\ln\left[(v_g - b) / (v_c - b)\right].$$

10.a. From α,
$$\left(\partial v/\partial T\right)_P = (R/P) + \left(a/RT^2\right), \text{ or}$$
$$v = (RT/P) - (a/RT) + C\left(P\right).$$

From this and κ_T,
$$\left(\frac{\partial v}{\partial P}\right)_T = -\left(RT/P^2\right) + C'\left(P\right) = -Tf\left(P\right) - b/P.$$

Therefore
$$f(P) = R/P^2; \quad C'\left(P\right) = -b/P; \quad \text{and } C(P) = -b\ln P + \text{Const.}$$

b.
$$v\left(P, T\right) = (RT/P) - (a/RT) - b\ln P + v_0$$

c.
$$\kappa_T = \left(RT/P^2 v\right) + \left(b/Pv\right).$$

The material satisfies the stability criterion $\kappa_T > 0$ for all $b > 0$ and even for $b < 0$ in the P, T range such that $(RT/P) + b > 0$.

11.a.
$$\frac{P}{T} = \left(\frac{\partial s}{\partial v}\right)_u = \frac{R}{v - b} - \frac{Rca}{v^2}\tanh c\left(u + a/v\right).$$
$$\frac{1}{T} = \left(\frac{\partial s}{\partial u}\right)_v = Rc\tanh c\left(u + a/v\right).$$

Therefore
$$\frac{P}{T} = \frac{R}{v - b} - \frac{a}{Tv^2} \text{ (the van der Waals equation)}$$

b. Since a van der Waals gas is not necessarily intrinsically unstable at all temperatures *via* $\kappa_T < 0$, the instability must be in the behavior of C_v. Differentiate $1/T$ implicitly at constant v.

$$-dT/T^2 = Rc^2\,du\,\text{sech}^2\,c\,(u + a/v)\,,$$

or

$$C_v = \left(\frac{\partial u}{\partial T}\right)_v = \frac{-\cosh^2 c(u + a/v)}{Rc^2T^2} < 0 \text{ for all } T.$$

12. From the Clapeyron equation,

$$\frac{dP}{dT} = \frac{\ell}{T\delta v} \simeq \frac{\ell}{Tv_g} = \frac{\ell}{RT^2/P}, \text{ or}$$

$$\frac{dP}{P} = \frac{\ell dT}{RT^2} = \frac{a(T_c - T)^2\,dT}{RT^2} = \frac{a}{R}\left(\frac{T_c^2}{T^2} - \frac{2T_c}{T} + 1\right)dT$$

$$\ln P = -\frac{aT_c^2}{RT} - \frac{2aT_c}{R}\ln T + \frac{aT}{R} + \ln P_0, \text{ by integration,}$$

so that

$$A = \ln P_0; \quad B = 2aT_c/R; \quad C = aT_c^2/R; \quad D = a/R.$$

13.

$$\kappa_s = -\frac{1}{v}\left(\frac{\partial v}{\partial P}\right)_s.$$

$$\left(\frac{\partial v}{\partial P}\right)_s = \frac{\partial(v,s)}{\partial(P,s)} = \left(\frac{\partial(v,s)}{\partial(P,T)}\right)\left(\frac{\partial(P,s)}{\partial(P,T)}\right)^{-1} = \begin{vmatrix} \left(\frac{\partial v}{\partial P}\right)_T & \left(\frac{\partial v}{\partial T}\right)_P \\ \left(\frac{\partial s}{\partial P}\right)_T & \left(\frac{\partial s}{\partial T}\right)_P \end{vmatrix} \times \begin{vmatrix} 1 & 0 \\ \left(\frac{\partial s}{\partial P}\right)_T & \left(\frac{\partial s}{\partial T}\right)_P \end{vmatrix}^{-1}$$

$$= \left[\left(\frac{\partial v}{\partial P}\right)_T\left(\frac{\partial s}{\partial T}\right)_P - \left(\frac{\partial v}{\partial T}\right)_P^2\right]\left(\frac{\partial s}{\partial T}\right)_P^{-1} = \left[(-v\kappa_T)(C_P/T) + (\alpha v)^2\right]\left(\frac{T}{C_P}\right),$$

or

$$\kappa_s = \kappa_T - Tv\alpha^2/C_P.$$

14.a. As in the solution to problem 6,

$$(\partial C_v/\partial v)_T = T(\partial^2 P/\partial T^2)_v = 0$$

for this gas law. Therefore $C_v(v,T) = C_v(T)$. b. From

$$\kappa_T = (v - b)/vP = \left[RT/(v - b)^2 + f'(v)\right]^{-1}/v$$

it follows that

$$f(v) = [(v_0 - b)/(v - b)]\,f_0,$$

where v_0 and f_0 are constants, and this correction seems to be the equivalent of replacing T by $T - T_c$, where T_c is a constant.

15.a.

$$P_1V/2 = N_1RT = RT; \quad P_2V/2 = N_2RT = 2RT.$$

$$U = (3/2)N_1 RT + (5/2)N_2 RT; \quad T = (2U/R)(1/(3N_1 + 5N_2)) = 2U/13R.$$
$$P_1 = 4U/13V; \quad P_2 = 8U/13V.$$

b. At new equilibrium,

$$P_1' = P_2' = P;$$

T is unchanged since U is constant. Then

$$V_1 = N_1 RT/P; \quad V_2 = N_2 RT/P; \quad V_1 + V_2 = V;$$

and

$$(N_1 + N_2)RT/P = V, \text{ or } P = (N_1 + N_2)RT/V.$$

From the ideal gas fundamental relation, with unchanged energies,

$$\Delta S = N_1 R \ln (V_1/(V/2)) + N_2 R \ln (V_2/(V/2))$$
$$= N_1 R \ln [2N_1/(N_1 + N_2)] + N_2 R \ln [2N_2/(N_1 + N_2)]$$
$$= R \ln(2/3) + 2R \ln(4/3) = R \ln(32/27) \approx 0.17R.$$

16. From α,

$$(\partial v/\partial T)_p = (Rv^2/P) + (Av^2/T^2), \text{ or}$$
$$v = [(A/T) - (RT/P) + g(P)]^{-1}.$$

Then for κ_T,

$$\kappa_T = \frac{(RT/P^2) + g'(P)}{(A/T) - (RT/P) + g(P)} = \frac{Tf(P)}{(A/T) - (RT/P) + g(P)},$$

or

$$f(P) = R/P^2; \quad g'(P) = 0; \quad g(P) = C.$$
$$v(P,T) = [(A/T) - (RT/P) + C]^{-1} = PT/(AP - RT^2 + CPT).$$

17. The equations of state reduce to $1/T = S/3U$ and $P/T = S/3V$, from which it follows that $S = NR^3 T^2/9A_0 P$ and $U = (RT/3)^{3/2}(VN/A_0)^{1/2} = PV$, where $A_0 = U_0 V_0 N_0$. From these, one obtains $C_P = 2R^3 T^2/9A_0 P$, and $C_v = (3/2)(U/NT) = S/2N = R^3 T^2/18A_0 P$. Since $V = U/P = (N/A_0 P^2)(RT/3)^3$, one obtains $\alpha = 3/T$ and $\kappa_T = 2/P$. Then

$$C_v + \frac{TV\alpha^2}{N\kappa_T} = \frac{R^3 T^2}{18A_0 P} + \left(\frac{T}{A_0 P^2}\right)\left(\frac{RT}{3}\right)^3 \left(\frac{3}{T}\right)^2 \left(\frac{P}{2}\right) = \frac{R^3 T^2}{18A_0 P} + \frac{R^3 T^2}{6A_0 P} = \frac{2R^3 T^2}{9A_0 P} = C_P.$$

18.

$$dS = dS_1 + dS_2 = (N_1 C_{v1} dT_1'/T_1') + (N_2 C_{v2} dT_2'/T_2') = (R/\theta)(N_1 dT_1' + N_2 dT_2').$$

For maximum work, $\Delta S = 0$, or

$$0 = \Delta S = (R/\theta)\left[N_1 \int_{T_1}^{T_f} dT_1' + N_2 \int_{T_2}^{T_f} dT_2'\right] = (R/\theta)[(N_1 + N_2)T_f - N_1 T_1 - N_2 T_2]$$

giving

$$T_f = (N_1 T_1 + N_2 T_2)/(N_1 + N_2).$$

The energy added to the source is

$$\Delta U_1 = (RN_1/\theta) \int_{T_1}^{T_f} T_1' dT_1' = (RN_1/2\theta)(T_f^2 - T_1^2) < 0.$$

The energy added to the sink is

$$\Delta U_2 = (RN_2/\theta) \int_{T_2}^{T_f} T_1' dT_1' = (RN_2/2\theta)(T_f 1^2 - T_2^2) > 0.$$

The energy available for work is $-\Delta U_1 - \Delta U_2$, or with T_f for maximum work,

$$\text{work} = RN_1 N_2 (T_1 - T_2)^2 / 2\theta(N_1 + N_2),$$

as required.

19.
$$1/T = (\partial S/\partial U)_{V,N} = nNR/U, \text{ or } U = nNRT.$$

Then
$$C_v = (1/N)(\partial U/\partial T)_N = nR.$$

Since
$$P/T = (\partial S/\partial V)_{U,N} = NR/V$$

the ideal gas law is satisfied.

20. In general,
$$C_P = C_v + Tv\alpha^2/\kappa_T.$$

For an ideal gas,
$$\alpha = 1/T, \quad \kappa_T = 1/P.$$

Therefore
$$C_P = C_v + TvP/T^2 = C_v + Pv/T = C_v + R.$$

21. The initial entropy of each is given by

$$S_i = (5Nk/2) + Nk \ln \left[(V_i/V_0)(U_i/U_0)^{C_v/R}(N/N_0)^{-(1+C_v/R)} \right].$$

The final entropy is

$$S = 5Nk + 2Nk \ln \left[\left[\frac{V_1 + V_2}{V_0}\right] \left[\frac{U_1 + U_2}{U_0}\right]^{C_v/R} \left[\frac{N_1 + N_2}{N_0}\right]^{-(1+C_v/R)} \right].$$

Then

$$\Delta S = Nk \ln \left[\left[\frac{(V_1 + V_2)^2}{V_1 V_2}\right] \left[\frac{(T_1 + T_2)^2}{T_1 T_2}\right]^{C_v/R} \left(\frac{4N^2}{N^2}\right)^{-(1+C_v/R)} \right]$$

$$= 2Nk \ln \left[\left[\frac{T_1 + T_2}{2\sqrt{T_1 T_2}}\right]^{1+C_v/R} \right].$$

Note the geometric construction showing that

$$(T_1 + T_2)/2 > \sqrt{T_1 T_2}$$

unless $T_1 = T_2$. Therefore $\Delta S > 0$.

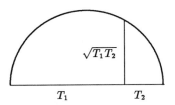

22.a. Since $P/T = (\partial s/\partial v)_u = R/v$, it follows from prob. 20, that $C_P - C_v = R$. b. From $\tanh(au/u_0) = 3aRT/2u_0$, it follows that

$$C_v = \frac{du}{dT} = \frac{(3R/2)}{\mathrm{sech}^2(au/u_0)} = \frac{3RT_0^2}{2T_0^2 - T^2},$$

where

$$u_0 = 3aRT_0/\sqrt{2}.$$

23. Treat all Q's and W's as intrinsically positive. The net heat supplied to the house is $Q_H = Q_H' + 0.1W_p$, where Q_H' comes from the heat pump and $0.1W_p$ is the 10% energy loss by the motor (which is in the house). Also $Q_H = Q_S + W'$, where Q_s is the heat pumped from the surroundings and $W' = .9W_p$ is the work done by the motor to drive the pump. If the pump (which is an ideal reversible heat engine) were reversed, the engine efficiency would be

$$\eta = W'/Q_H' = \Delta T/T_H = 30°/300° = 0.1.$$

Then $W' = 0.1Q_H' = .9W_p$, or $Q_H' = 9W_p$, where W_p is the total electrical energy supplied to the motor. Then $Q_H = 9.1W_p$, or $W_p = Q_H/9.1$, the total energy paid for. When the cost of Q_H directly is \$30, the cost of the same energy *via* the heat pump is \$30/9.1 = \$3.30.

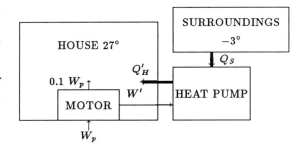

24.

$$đq \equiv C_\Gamma dT = du + Pdv.$$

But from the general ideal gas property that u is a function of T alone, it follows that $du = C_v dT$. From the equation for the polytropic, $\Gamma P v^{\Gamma-1} dv + v^\Gamma dP = 0$, and from the ideal gas law, $Pdv + vdP = RdT$. Then $vdP = -\Gamma Pdv$, from which it follows that $Pdv - \Gamma Pdv = RdT = (1 - \Gamma)Pdv$, or $Pdv = RdT/(1 - \Gamma)$. These results permit the C_Γ equation to be written

$$C_\Gamma dT = C_v dT + \frac{RdT}{1 - \Gamma},$$

or

$$C_\Gamma = C_v + \frac{C_P - C_v}{1 - \Gamma} = C_v \left(\frac{\gamma - \Gamma}{1 - \Gamma}\right).$$

For $\Gamma = 0$, the process is one of constant pressure, and the formula yields, for $\Gamma = 0$, $C_0 = \gamma C_v (= C_P)$, as expected. For $\Gamma = 1$, the process is evidently an isotherm for the ideal gas, for which the heat capacity is formally infinite. For $\Gamma = \gamma$, the process is an adiabatic, for which there is a change of temperature for no added heat, or zero heat capacity. For $\Gamma = \infty$ the Γth root of the defining equation, in the limit of $\Gamma \to \infty$, becomes a constant-volume process, in agreement with the same limiting result for C_Γ.

25. From Eq. (2.9.15), the entropy of the system before mixing is seen to be

$$S_0 = N_1 k \ln(V/N_1) + M_1 k \ln(V/M_1) + N_1 f_N(T) + M_1 f_M(T) +$$
$$+ N_2 k \ln(V/N_2) + M_2 k \ln(V/M_2) + N_2 f_N(T) + M_2 f_M(T).$$

The entropy after mixing is found from the same formula to be, with $N_1 + N_2 = N$, and $M_1 + M_2 = M$,

$$S_f = N k \ln(2V/N) + M k \ln(2V/M) + N f_N(T) + M f_M(T).$$

The entropy of mixing is

$$\Delta S = S_f - S_0 = N_1 k \ln(2N_1/N) + N_2 k \ln(2N_2/N) + M_1 k \ln(2M_1/M) + M_2 k \ln(2M_2/M).$$

Evidently, when $N_1 = N_2$ and $M_1 = M_2$, $\Delta S = 0$, and when $N_1 = M_2 = N$ and $N_2 = M_1 = 0$, $\Delta S = 2Nk \ln 2$. Thus there is no discontinuous change in the entropy of mixing.

3. EQUILIBRIUM STATISTICAL THERMOPHYSICS

3.1 INTRODUCTION

The primary objective of equilibrium statistical thermophysics is the derivation from microphysical principles of the fundamental relations necessary for the calculation of macroscopic thermodynamic properties. The source of fundamental relations is postulated to be the information-theoretic uncertainty appropriately chosen to represent the underlying microphysics, maximized subject to the existing constraints and boundary conditions, and expressed in suitable units. Thermodynamics has been introduced in Chapter 2 in a form compatible with the statistical entropy-maximum postulate, and the macroscopic intensive and extensive thermal parameters have been defined and shown to behave as required. In the present chapter, the principal postulate is that the Jaynes information-theoretic maximized uncertainty derived in Chapter 1 gives the required fundamental relation of Chapter 2 when correctly based in microphysics. Once this postulate is accepted, the consequent identifications follow the pattern established in Chapter 2. (Jaynes [1957a]; Jaynes [1957b])

Although entropy plays the leading role in the conceptual development of the subject, the partition function is usually the most convenient quantity for purposes of calculation, and the entropy is often derived from the partition function.

Gibbs considered three kinds of entities, called *ensembles*, in his pioneering treatment of statistical mechanics. (Gibbs [1960]) His ensembles were infinite sets of macroscopically identical systems, each represented by a point in his phase space. His reason for developing ensemble theory apparently was the belief that probabilities should be given objective representation, even if the representation was entirely fictitious. It is much more realistic to accept probabilities as measures of expectation based on our knowledge, rather than as objective measures in fiction space. The mathematical operations that go into a calculation are basically identical, whether expressed in terms of information-theoretic probabilities or densities in phase space. Therefore the names given by Gibbs to his various ensemble calculations will be preserved, in so far as practical, for their information-theoretic equivalents.

Gibbs's *microcanonical* ensemble was made up of identical systems, all at the same fixed energy; the microcanonical system considered in the next section consists of one closed system. Gibbs's *canonical* ensemble permitted otherwise identical systems to exchange energy thermally, so the system energy was not constant, though its expectation value was regarded as a known constant; the canonical system of Section 3.3 is a single system in thermal contact with a heat bath. Finally, Gibbs's *grand canonical* ensemble permitted matter flow as well as thermal exchange of energy; the grand canonical system of Section 3.4 is visualized as a definite volume without walls, contained in a large canonical system.

Evidently other ensembles and their equivalents could be devised, corresponding to work transfers *via* movable system walls, restricted matter flow with semipermeable membranes, and any other available form of direct, reciprocal energy transfer. The advantage of exploring several of these is that ease of calculation varies considerably from one to another, and quite intractable problems in one treatment sometimes become almost trivial in another.

3.2 MICROCANONICAL SYSTEMS

In the microcanonical system, the energy is regarded as known, and all allowed states correspond to the same energy. The basic equations are

$$S = -k\Sigma p_i \ln p_i, \tag{3.2.1}$$

and

$$1 = \Sigma p_i, \tag{3.2.2}$$

with the usual third one

$$\langle E \rangle = \Sigma p_i E_i \tag{3.2.3}$$

reducing to Eq. (3.2.2), since all the E_i are equal to $\langle E \rangle$. Maximization of S subject to Eq. (3.2.2) gives

$$p_i = 1/\Omega, \tag{3.2.4}$$

where Ω is the total number of distinguishable states corresponding to the energy $\langle E \rangle$, and all the probabilities turn out to be equal, as expected. With the $\{p_i\}$ given by Eq. (3.2.4), the entropy of Eq. (3.2.1) becomes

$$S = k \ln \Omega. \tag{3.2.5}$$

The entire thermodynamic structure follows from Eq. (3.2.5) and the postulate that this is the thermodynamic entropy. The only work involved is the calculation of Ω, which will be done in three illustrative examples. It may come as a surprise that calculation of Ω is usually more difficult than calculation of S by other methods. Appendix C provides some guidance in counting numbers of states.

3.2.1 Quantum Oscillators. In this section we consider a collection of N identical quantum harmonic oscillators, supposed distinguishable because of their positions. They share M indistinguishable quanta, which may be thought of as photons or phonons that may be exchanged, but not lost. It will be convenient to regard M and N as very large numbers for later stages of the calculation.

The energy of an oscillator having n quanta is $\hbar\omega(n+1/2)$ where $\hbar\omega/2$ is the zero point energy. The total energy is

$$U = \hbar\omega(M + N/2), \tag{3.2.6}$$

and the problem is to determine Ω, the number of distinguishable ways in which N distinguishable oscillators can share M indistinguishable quanta. The result, as explained in Appendix C, is

$$\Omega = \frac{N(M + N - 1)!}{M!\, N!} = \frac{(M + N - 1)!}{(N - 1)!M!}. \tag{3.2.7}$$

From Eq. (3.2.5) and the simplified form of Stirling's approximation as given in Appendix E, with the assumption that $N \gg 1$, it follows that

$$S/k = (M + N)\ln(M + N) - M \ln M - N \ln N. \tag{3.2.8}$$

Since $\langle E \rangle = U$, and U is specified by M, use of Eq. (3.2.6) and Eq. (2.3.5) gives

$$\frac{1}{kT} = \frac{\partial(S/k)}{\partial U} = \frac{1}{\hbar\omega}\big[\ln(M + N) - \ln M\big], \tag{3.2.9}$$

or

$$M = \frac{N}{e^{\beta \hbar \omega} - 1},$$ (3.2.10)

where $\beta = 1/kT$. The energy now becomes

$$U = \frac{N\hbar\omega}{e^{\beta \hbar \omega} - 1} + \frac{N\hbar\omega}{2} = \frac{N\hbar\omega}{2} \coth(\beta\hbar\omega/2).$$ (3.2.11)

This is a well-known result, closely related to Planck's early considerations of black-body equilibria. The entropy, from Eq. (3.2.8) and modest algebra, is

$$S/k = -N\ln(e^{\beta \hbar \omega} - 1) + \frac{N\beta\hbar\omega}{1 - e^{-\beta\hbar\omega}},$$ (3.2.12)

which reduces, when $\beta\hbar\omega \ll 1$, to

$$S = Nk(1 - \ln\beta\hbar\omega),$$ (3.2.13)

in agreement with the classical equilibrium result of Eq. (1.7.70). Under these conditions also, the energy of Eq. (3.2.11) reduces to $U = NkT$, the classical result corresponding to Eq. (1.7.65). The derivation of Eqs. (3.2.11) and (3.2.12) by the canonical formalism will be shown for comparison in Section 3.3.1.

3.2.2 Schottky Defects.
A simple crystalline solid is considered as composed of a single species of atom (or molecule) with two possible kinds of energy states. Each atom could be in its correct lattice position, in which case it is assigned the energy zero, or it could be in an interstitial position, with energy ϵ. The atoms are assumed to be noninteracting and other kinds of energy such as vibrations are ignored. It is further assumed that the atoms are indistinguishable, the lattice sites and interstitial positions are distinguishable because of their geometry, and the total number of atoms M in interstitial positions is large (for Stirling approximation purposes) but much less than the total number N of atoms. The energy is given by

$$U = M\epsilon,$$ (3.2.14)

and the entropy is given by

$$S = k\ln\Omega,$$ (3.2.15)

where the problem again is the evaluation of Ω, the number of ways of taking M atoms from N lattice sites and distributing them over the available interstitial sites, here assumed to be N'. The first of the M atoms may be selected in N ways, the second in $N - 1$, *etc.*, giving $N!/(N - M)!$ ways of removing M *distinguishable* atoms in numerical order from N lattice sites. But the M atoms are indistinguishable, and the order of removal is of no consequence. Therefore the number of ways of removing M indistinguishable atoms from N lattice sites is $N!/(N - M)!\, M!$, the binomial coefficient. The number of ways of placing the M atoms in N' interstitial sites is calculated in the same way to be $N'!/(N' - M)!\, M!$, giving Ω as the product of these independent numbers, or

$$\Omega = \binom{N}{M}\binom{N'}{M}.$$ (3.2.16)

When $N' = N$, as is typical, Eq. (3.2.15) and the simplified Stirling approximation give

$$S/k = 2N \ln N - 2M \ln M - 2(N - M)\ln(N - M), \tag{3.2.17}$$

from which it follows that

$$\beta = \frac{\partial(S/k)}{\partial U} = \frac{2}{\epsilon}[-\ln M + \ln(N - M)]$$

$$= \frac{2}{\epsilon} \ln\left(\frac{N}{M} - 1\right), \tag{3.2.18}$$

or

$$M = \frac{N}{e^{\beta\epsilon/2} + 1}. \tag{3.2.19}$$

The energy now becomes

$$U = \frac{N\epsilon}{e^{\beta\epsilon/2} + 1}, \tag{3.2.20}$$

and the entropy is

$$S = 2Nk\left[\ln\left(1 + e^{\beta\epsilon/2}\right) - \frac{(\beta\epsilon/2)}{1 + e^{-\beta\epsilon/2}}\right]. \tag{3.2.21}$$

At very low temperature ($\beta \to \infty$) the energy and entropy vanish; at very high temperature ($\beta\epsilon \ll 1$), the energy becomes $N\epsilon/2$ and the entropy becomes $2Nk \ln 2$, indicating that half the lattice sites are vacant and half the interstitial sites are filled. (Presumably the solid would melt before this happened, but the physical assumptions become rather strained when an appreciable fraction of the interstitial sites no longer are surrounded by their full quota of occupied lattice sites.)

This particular calculation is a rarity; it is easiest in the microcanonical formalism.

3.2.3 The Monatomic Ideal Gas.
The calculation of the fundamental relation for a monatomic ideal gas by the microcanonical formalism is included as motivation for the study of other formalisms, since it is a rather awkward calculation for an almost trivial problem.

The model for an ideal monatomic gas is a collection of indistinguishable, structureless, noninteracting point particles confined to a volume V. In order to simplify the calculation as much as possible, the volume is taken as a cubic box of side a, with $V = a^3$. The single-particle energy levels are easily obtained from the Schrödinger equation to be

$$\epsilon_{qrs} = \frac{\pi^2 \hbar^2}{2ma^2}\left(q^2 + r^2 + s^2\right), \tag{3.2.22}$$

where q, r, and s are positive integers that arise from the standing-wave modes in the box. The total energy of N particles is given by

$$U = \frac{h^2}{8ma^2}\sum_{i=1}^{N}\left(q_i^2 + r_i^2 + s_i^2\right). \tag{3.2.23}$$

The constancy of U requires that the point in $3N$-dimensional number space that represents the integers in Eq. (3.2.23) lie on the surface of a sphere given by

$$\sum_{i=1}^{N} \left(q_i^2 + r_i^2 + s_i^2 \right) = 8ma^2 U/h^2, \tag{3.2.24}$$

where the right-hand side can be regarded as the squared radius R^2 of a sphere in the $3N$-dimensional space. The representative point must also lie in the sector of space for which all the integers are positive.

If the points were allowed to be chosen from anywhere in the volume, instead of just on the constant energy surface, there would then be one possible point per unit $3N$-volume. But since the volume of the $3N$-dimensional sphere can be written, with the help of Appendix F, as

$$\mathcal{V}_{3N} = \mathcal{A}_{3N} R^{3N} \tag{3.2.25}$$

where $\mathcal{A}_{3N}$ is a constant, it follows that the fractional volume in the shell of thickness dR is given by

$$\frac{d\mathcal{V}_{3N}}{\mathcal{V}_{3N}} = \frac{3N\,dR}{R}, \tag{3.2.26}$$

and a shell of thickness $dR = R/3N$ approximately accounts for the entire volume, *i.e.*, for spheres in a space of a large number of dimensions, nearly the entire volume of the sphere lies in a thin surface shell! Therefore the error will be negligible, for large N, if we count the number of possible states by relating it to the entire volume of the $3N$-sphere, rather than to the surface shell. As shown in Appendix F, $\mathcal{A}_{3N}$ of Eq. (3.2.25) is given by

$$\mathcal{A}_{3N} = \frac{\pi^{3N/2}}{(3N/2)\Gamma(3N/2)}, \tag{3.2.27}$$

and the number of distinguishable energy states Ω is given by

$$\Omega = \frac{\mathcal{V}_{3N}}{2^{3N} N!}, \tag{3.2.28}$$

where $\mathcal{V}_{3N}$ counts one set of integers in Eq. (3.2.24) per unit volume in the 3N-dimensional number space, 2^{3N} reduces the counted volume to the sector in which all the integers are positive (corresponding to the first octant in 3-space) and the $N!$ reduces Ω because of the nonindividuality of the point particles. (We can describe an N-particle state as made up of N single-particle states, each prescribed by its set of numbers q, r, s, but we are forbidden by quantum mechanics to say which particle is in which state, or even that there is an individual particle occupying any state. This point is discussed further in this chapter, in Appendix C, and in Chapter 4. Provided there is negligible multiple occupancy of single-particle states, there are then too many states by a factor of $N!$ in the sector volume $\mathcal{V}_{3N}/2^{3N}$, since $N!$ is the factor by which classical counting overweights the quantum N-body states that can be formed from the N single-particle states. This feature will be re-examined in the next section.)

The volume $\mathcal{V}_{3N}$ in Eq. (3.2.19) can be replaced by means of Eqs.(3.2.25), (3.2.27), and R from Eq. (3.2.24), to give

$$\Omega = \frac{(2\pi mU/h^2)^{3N/2}V^N}{(3N/2)\Gamma(3N/2)N!}.$$ (3.2.29)

The entropy follows from Eq. (3.2.5) as

$$S = \frac{5Nk}{2} + Nk\ln\left[\left(\frac{4\pi m}{3h^2}\right)^{3/2}\left(\frac{VU^{3/2}}{N^{5/2}}\right)\right],$$ (3.2.30)

which is the famous Sackur-Tetrode equation, the fundamental relation for a monatomic ideal gas. The derivation of ideal-gas properties from a fundamental relation of this form has already been carried out in Chapter 2, and need not be repeated.

3.3 CANONICAL SYSTEMS

The canonical formalism has already been developed in Section 3.2.2 without being identified as such. The energy expectation value is U, given by

$$U = \sum p_i E_i,$$ (3.3.1)

with the usual definition of information-theoretic entropy and the condition on the probabilities, as in Eqs. (3.2.1) and (3.2.2). The probabilities are calculated for entropy maximum to be

$$p_i = e^{-\beta E_i}/Z,$$ (3.3.2)

with

$$Z = \sum_i e^{-\beta E_i} = \sum_\epsilon \Omega(\epsilon)e^{-\beta\epsilon},$$ (3.3.3)

where the first sum is over distinguishable states and the second over energy levels with $\Omega(\epsilon)$ as the number of states with energy ϵ. The expectation value of the energy is seen to be

$$U = -\left(\frac{\partial \ln Z}{\partial \beta}\right)_{\{E_i\}},$$ (3.3.4)

and the entropy is

$$S = k\beta U + k\ln Z.$$ (3.3.5)

The postulate that S is indeed the thermodynamic entropy and the use of Eq. (2.3.9) give the results

$$\beta = 1/kT$$ (3.3.6)

and

$$kT\ln Z = PV - \sum \mu_j N_j = -F,$$ (3.3.7)

by Eqs. (2.5.13) and (2.4.6), where $F = F(V, T, \{N\})$ is the Helmholtz potential.

Since F contains the same amount of information as any other fundamental relation, it is often more convenient to derive the thermodynamic properties from it (or from Z) instead of returning to the entropic fundamental relation. The definition $F = U - TS$ gives

$$S = -k\beta\frac{\partial \ln Z}{\partial \beta} + k\ln Z,$$ (3.3.8)

and the pressure is obtained from

$$P = -\left(\frac{\partial F}{\partial V}\right)_{T,\{N\}} = kT\left(\frac{\partial \ln Z}{\partial V}\right)_{\beta,\{N\}} \tag{3.3.9}$$

The canonical formalism often offers calculational advantages when the composite energy levels are sums of levels of non-interactive components. Since these component systems are regarded as being in equilibrium with the heat bath rather than as having reached equilibrium as a consequence of direct interaction with each other, there is no conceptual difficulty in treating them as independent. Suppose each of two such component systems comes to equilibrium with the heat bath and establishes its own probability set, say $\{p_i\}$ for one and $\{w_j\}$ for the other, where

$$p_i = e^{-\alpha_1 - \beta E_i}, \tag{3.3.10}$$

as in Eq. (1.4.5), and

$$w_j = e^{-\alpha_2 - \beta \epsilon_j}, \tag{3.3.11}$$

with the same β because of equilibrium with the heat bath. The individual partition functions are

$$Z_1 = e^{\alpha_1} = \sum_i e^{-\beta E_i}, \tag{3.3.12}$$

and

$$Z_2 = e^{\alpha_2} = \sum_j e^{-\beta \epsilon_j}. \tag{3.3.13}$$

The composite probabilities are

$$W_{ij} = p_i w_j = e^{-(\alpha_1+\alpha_2)-\beta(E_i+\epsilon_j)}, \tag{3.3.14}$$

and the composite partition function is

$$Z_{12} = e^{\alpha_1+\alpha_2} = \sum_{i,j} e^{-\beta(E_i+\epsilon_j)}$$
$$= Z_1 Z_2. \tag{3.3.15}$$

The analysis can easily be iterated to any number of component subsystems.

For an ideal gas system, for example, with the separate sets of energy levels for translation, rotation, and vibration, the partition function can be factored, with appreciable simplification of calculations. In the treatment of the Schottky defects of Section 3.2.2 by the canonical procedure, the lattice vibrations could be included in a separate factor of the partition function, instead of being ignored.

There is one point at which caution is necessary however. The composite probabilities Eq. (3.3.14) for two noninteracting systems are written with clearly distinguishable energies E_i and ϵ_j, implying that it is meaningful to regard the state of the composite system as a superposition of the states with identifiable energies. The partition function for a composite system is properly of the form

$$Z_c = \sum_n e^{-\beta E_{cn}},$$

where the index n enumerates the states of the composite system and E_{cn} is the energy when the system is in state n. Any state counted in Z_c must be an identifiable combination state of the composite system and not an alternative or incomplete representation of an improperly defined composite state.

As an illustration, suppose Z_1 and Z_2 in Eqs.(3.3.12) and (3.3.13) have just three states each; let

$$e^{-\beta E_i} = a_i \text{ and } e^{-\beta \epsilon_j} = b_j.$$

Then the composite Z_{12} of Eq. (3.3.15) becomes

$$Z_{12} = (a_1 + a_2 + a_3)(b_1 + b_2 + b_3),$$

or

$$Z_{12} = a_1 b_1 + a_1 b_2 + a_1 b_3 + a_2 b_1 + a_2 b_2 + a_2 b_3 + a_3 b_1 + a_3 b_2 + a_3 b_3,$$

for nine composite states. But if the subsystems are identical and indistinguishable, then $a_i = b_i$ and the nine terms can be written as $a_1^2 + 2a_1 a_2 + a_2^2 + 2a_1 a_3 + a_3^2 + 2a_2 a_3$. A term such as $2a_1 a_2$ arises as $a_1 b_2 + a_2 b_1$, with $a_i = b_i$. In quantum theory, however, the state of the system, is properly described as a two-particle state with states 1 and 2 occupied, *but by particles with no individuality*, so it cannot be said that one particle is in state 1 and the other in state 2. The formalism says that each particle is in both single-particle states, and that therefore only one such state exists. (This point is discussed further in the commentary of this chapter, in Chapter 4, and in Appendix C.) Thus the proper partition function should be

$$Z_c = a_1^2 + a_1 a_2 + a_2^2 + a_1 a_3 + a_3^2 + a_2 a_3,$$

with only six (single) terms, rather than nine (counting the double ones).

If, instead of two such systems, there are N, then the function corresponding to Z_1^N is

$$(a_1 + a_2 + a_3)^N = \sum_{m_1,m_2,m_3} (N; m_1, m_2, m_3) a_1^{m_1} a_2^{m_2} a_3^{m_3},$$

which generates the number of permutations of N elements into three groups, where $(N; m_1, m_2, m_3)$ is the multinomial coefficient and $\sum m_i = N$. The proper partition function should be simply

$$\sum_{m_1,m_2,m_3} a_1^{m_1} a_2^{m_2} a_3^{m_3}, \quad \text{again with } \sum m_i = N,$$

but such a thing is difficult to write for a general partition function with an infinite number of energy levels. However, when the number of levels becomes infinite, (or much larger than N) the single-system partition function to the Nth power will still be of the form

$$\left(\sum_{i=1}^{\infty} a_i\right)^N = \sum_{\{m_i\}} (N; \{m_i\}) \prod_{i=1}^{\infty} a_i^{m_i}, \tag{3.3.16}$$

with $\sum_{i=1}^{\infty} m_i = N$. Evidently, then, most of the m_i will be zero, and the largest terms in Eq. (3.3.9) are those for which all the m_i are 1 or 0. This is just the regime for which the system behaves classically, since quantum symmetry requirements become restrictive only when multiple occupancy of classical single-particle states is highly probable. Since the number of single-particle levels far exceeds any finite N, it is evident that most of the terms in Eq. (3.3.9) will be those for which the m_i are all 1 or 0, and these all have $N!$ as their multinomial coefficient. Therefore an excellent approximation to the correct composite partition function for N nonindividualizable subsystems, each with $Z = Z_1$, is

$$Z_N = Z_1^N / N!. \tag{3.3.17}$$

The $N!$ is a necessity introduced in a natural way by quantum mechanics; it arises entirely because of our need to count quantum N-body states, rather than classical permutations, since we describe a composite state merely as having $\{m_i\}$ particles sharing the single-system states $\{i\}$. Our desire to use a classical canonical formalism, rather than to impose the full quantum symmetry requirements used in Chapter 4, can be fulfilled if we restrict the approximation to regimes that have orders of magnitude more single-particle states than particles to occupy them, and we correct the canonical N-body partition function as shown in Eq. (3.3.17). The need to introduce the $N!$ factor in order to obtain an extensive entropy function is known in the literature as the *Gibbs paradox*. It is discussed further in the commentary section of this chapter. For systems of distinguishable, individual particles, including those that are so only because of position in a lattice, there is no $N!$ divisor. The examples in the next three subsections illustrate the use of factored partition functions.

3.3.1 Quantum Oscillators. The problem of N distinguishable oscillators and M quanta discussed in Section 3.2.1 is treated here from the point of view of the canonical system. The states of a single oscillator correspond to the energy levels

$$\epsilon_n = \hbar\omega \left(n + \tfrac{1}{2}\right), \tag{3.3.18}$$

and the partition function Z_1 for a single oscillator is

$$Z_1 = \sum_{n=0}^{\infty} e^{-\beta\epsilon_n} = e^{-\beta\hbar\omega/2} \Big/ \left(1 - e^{-\beta\hbar\omega}\right)$$

$$= 1/2\sinh\left(\beta\hbar\omega/2\right). \tag{3.3.19}$$

From Z_1, the average energy of the single oscillator can be found as

$$\langle\epsilon\rangle = -\frac{\partial \ln Z_1}{\partial \beta} = \frac{\hbar\omega/2}{\tanh\left(\beta\hbar\omega/2\right)}, \tag{3.3.20}$$

and the expectation that $U = N\langle\epsilon\rangle$ immediately gives Eq. (3.2.11). Since the oscillators are distinguishable, however, the N-oscillator partition function Z_N is given by

$$Z_N = Z_1^N, \tag{3.3.21}$$

and $\ln Z_N = N \ln Z_1$, from which Eq. (3.2.11) follows directly. The entropy, from Eq. (3.3.8), is found to be

$$S = Nk\beta\hbar\omega/2\tanh\left(\beta\hbar\omega/2\right) - Nk\ln\left[2\sinh\left(\beta\hbar\omega/2\right)\right], \tag{3.3.22}$$

a result that can be shown to agree exactly with Eq. (3.2.12). From Eqs. (3.3.18) and (3.3.20), the expectation value $\langle n \rangle$, the average number of quanta per oscillator, may be written

$$\langle n \rangle = 1/\left(e^{\beta \hbar \omega} - 1\right), \tag{3.3.23}$$

and in terms of $\langle n \rangle$ the entropy is

$$S/Nk = (\langle n \rangle + 1)\ln\left(\langle n \rangle + 1\right) - \langle n \rangle \ln \langle n \rangle, \tag{3.3.24}$$

a result that often arises in field-theoretic treatments of radiation.

The identification of the microcanonical and canonical expressions for S (Eqs. (3.2.5), (3.2.8) or (3.2.12), and (3.3.5) or (3.3.22)) with thermodynamic entropy when the starting expressions are evidently different is often puzzling, especially since the final results are found to be identical. A more formal analysis of this seemingly coincidental identity is left as an exercise based on postulate (c) of information theory; Section 1.3.

3.3.2 Monatomic Ideal Gas. The single-particle energy levels for a monatomic ideal gas atom in a box of dimensions a, b, c are, similar to Eq. (3.2.22),

$$\epsilon_{qrs} = \frac{h^2}{8m}\left(\frac{q^2}{a^2} + \frac{r^2}{b^2} + \frac{s^2}{c^2}\right), \tag{3.3.25}$$

and the single-atom partition function is

$$\begin{aligned}
Z_1 &= \sum_{q,r,s=1}^{\infty} e^{-\beta \epsilon_{qrs}} \\
&= \left[\sum_{q=1}^{\infty} \exp\left(\frac{-\pi \lambda^2 q^2}{4a^2}\right)\right]\left[\sum_{r=1}^{\infty} \exp\left(\frac{-\pi \lambda^2 r^2}{4b^2}\right)\right]\left[\sum_{s=1}^{\infty} \exp\left(\frac{-\pi \lambda^2 s^2}{4c^2}\right)\right],
\end{aligned} \tag{3.3.26}$$

where λ is the so-called *thermal wavelength*, given by

$$\lambda \equiv \frac{h}{\sqrt{2\pi m k T}}. \tag{3.3.27}$$

Here even the single-particle partition function can be factored. Each sum can be evaluated by the Euler-Maclaurin technique, shown in Appendix E, when $\lambda/2 \ll a$, b, or c as the case may be. The mathematical summation procedure becomes faulty just when, for physical reasons, the particles begin not to fit into the box very well. The Boltzmann ideal gas, for which symmetry of the eigenstates is not considered, must be abandoned well before the failure of the Euler-Maclaurin technique, as shown in Chapter 4, so for present purposes the thermal wavelength is taken to be much smaller than the dimensions of the box. Equation (3.3.26) then becomes

$$Z_1 \approx \left(\frac{a}{\lambda} - \frac{1}{2}\right)\left(\frac{b}{\lambda} - \frac{1}{2}\right)\left(\frac{c}{\lambda} - \frac{1}{2}\right) \approx V/\lambda^3, \tag{3.3.28}$$

since $abc = V$, and the $1/2$ factors can be ignored. Since the particles are indistinguishable, interchangeable, and without individuality, the N-particle partition function is written, as in Eq. (3.3.17)

$$Z_N = Z_1^N/N! = (V/\lambda^3)^N/N!, \tag{3.3.29}$$

and the energy is given by

$$U = -\left(\frac{\partial \ln Z_N}{\partial \beta}\right)_{V,N} = 3N\frac{\partial \ln \lambda}{\partial \beta} = \frac{3N}{2\beta} = \frac{3}{2}NkT. \tag{3.3.30}$$

The pressure is calculated from the Helmholtz potential

$$F = -kT \ln Z_N$$

as

$$P = -\left(\frac{\partial F}{\partial V}\right)_{T,N} = kT\left(\frac{\partial \ln Z_N}{\partial V}\right)_{\beta,N} = \frac{NkT}{V}. \tag{3.3.31}$$

Similarly the chemical potential is given by

$$\mu = \left(\frac{\partial F}{\partial N}\right)_{V,T} = -kT\left(\frac{\partial \ln Z_N}{\partial N}\right)_{V,\beta} = kT \ln\left(N\lambda^3/V\right), \tag{3.3.32}$$

showing an interesting relationship between the chemical potential and the number of thermal volumes λ^3 available per particle. The chemical potential is negative in the range of parameters for which the ideal gas can be treated as a Boltzmann gas, but just when μ changes sign in Eq. (3.3.32) quantum effects become important and the treatment of this section is no longer valid. At this point, the volume per particle V/N just equals the thermal volume. A slightly different aspect of the limitation at large thermal wavelengths is the statement that the number of N-particle states with energy levels less than $3NkT/2$ goes from order $\gg N$ to a relatively small number, of order N or smaller, at which point the problem of defining and properly counting N-body eigenstates of energy must be faced, as shown in Chapter 4.

The monatomic ideal gas may also be treated entirely classically in terms of the canonical Liouville function, discussed for continuous probability distributions in Section 1.6. The partition function of Eq. (3.3.3) can be replaced by an integral, with the number of states of energy ϵ, $\Omega(\epsilon)$, replaced by a density-of-states function. The classical calculation is usually done in terms of integrals over phase space, however, since there the density of states is regarded as one state per volume of size h^{3N}, where h is a small but arbitrary constant with units of action. The single-atom partition function written in this way is

$$Z_1 = \frac{1}{h^3}\int d^3r \int d^3p\, e^{-\beta p^2/2m}, \tag{3.3.33}$$

where the momentum integrals extend to infinity and the coordinate integrals cover the volume of the container. Since the integrand does not depend upon V in this case, the spatial integral simply yields V; the momentum space integral is well-known, and the result is

$$Z_1 = \frac{V}{h^3}\left(\frac{2\pi m}{\beta}\right)^{3/2} = \frac{V}{\lambda^3}, \tag{3.3.34}$$

if the same formal definition (Eq. (3.3.27)) is used for λ. This is the same result as that of (3.3.28), and the derived results follow in the same way. When the quantum energy levels are spaced much more closely than kT, the classical calculation is valid.

The momentum integral can be transformed to energy space, with

$$\epsilon = \frac{p^2}{2m}; \quad p = \sqrt{2m\epsilon}; \quad dp = \sqrt{\frac{2m}{\epsilon}}\, d\epsilon/2,$$

and the volume element in momentum space becomes

$$4\pi p^2\, dp = 2\pi (2m)^{3/2}\epsilon^{1/2}\, d\epsilon.$$

Equation (3.3.33) then transforms to energy space, after the volume integral is done, as

$$Z_1 = \left(\frac{V}{h^3}\right)(2\pi)(2m)^{3/2}\int_0^\infty \epsilon^{1/2}e^{-\beta\epsilon}d\epsilon$$
$$= \int_0^\infty g(\epsilon)e^{-\beta\epsilon}d\epsilon = \frac{V}{\lambda^3}, \tag{3.3.35}$$

as before. The factor $2\pi V(2m)^{3/2}\epsilon^{1/2}/h^3$ is called the density of states, $g(\epsilon)$, and the number of states in the interval $d\epsilon$ is $g(\epsilon)d\epsilon$.

The relationship

$$Z(\beta) = \int_0^\infty g(\epsilon)e^{-\beta\epsilon}d\epsilon \tag{3.3.36}$$

may be regarded as saying that $Z(\beta)$ is the Laplace transform of $g(\epsilon)$. If $Z(\beta)$ is already known, the density of states can be found by the inverse Laplace transform

$$g(\epsilon) = \int_{c-i\infty}^{c+i\infty} Z(\beta)e^{\beta\epsilon}d\beta, \tag{3.3.37}$$

where the path of integration is to the right of all singularities of $Z(\beta)$. In simple cases $g(\epsilon)$ may also be obtained by use of tables of Laplace transforms.

The classical probability of finding a gas molecule with energy in the interval $d\epsilon$ around ϵ is given by

$$\mathcal{F}(\epsilon)d\epsilon = \frac{g(\epsilon)e^{-\beta\epsilon}d\epsilon}{Z_1(\beta)}, \tag{3.3.38}$$

from which the Maxwell probability density for finding a gas molecule in the momentum interval dp around p is found to be

$$f(p)dp = \frac{4\pi}{(2\pi mkT)^{3/2}}p^2 e^{-\beta p^2/2m}dp, \tag{3.3.39}$$

where the density of states $g(\epsilon)$ was taken from Eq. (3.3.35), transferred back to momentum space, and the normalizing factor Z_1^{-1} was taken from Eq. (3.3.34).

Much of the classical work on kinetic theory of gases and even a substantial fraction of current work on plasma physics is based on Eq. (3.3.39). For this reason, the user should be aware that the Maxwellian probability density is strictly valid for noninteracting molecules in a temperature range such that the spacing of energy levels is very small compared to kT. The most frequent usage in violation of these conditions is the ignoring of interactions between pairs of particles. The reason for ignoring evident Coulomb and multipole interactions is one of convenience, rather than of ignorance, since the statistical physics of interacting particles is complicated, and the results are seldom in a form that is convenient for further analytical treatment or physical description. The deliberate misuse, then, of the Maxwell density often may be justified pragmatically, but the derivations so obtained should be viewed with some skepticism. When interactions are disregarded, the clustering of some particles and the mutual avoidance of others (the collective effects so important in plasma physics) disappear, and the entropy attributed to the system is larger than it should be, based on the information available. Correlations and anticorrelations, which reduce entropy by modifying the N-particle probability density in phase space, are ignored in the simplified treatment. Perturbation methods, however, do attempt to include correlations, density waves, and other effects of interaction, often with considerable success.

3.3.3 Langevin Paramagnetism. The example of this section is the simplest statistical treatment of magnetic phenomena, that of a system of non-interacting molecules with fixed permanent magnetic dipole moments, in a uniform magnetic field. The classical energy of each such molecule may be written as

$$\epsilon = \epsilon_t + \epsilon_r + \epsilon_v + \epsilon_m, \tag{3.3.40}$$

where the subscripts stand for translational, rotational, vibrational, and magnetic, respectively. The first three terms are entirely mechanical, and each constituent of the energy may be regarded as being in equilibrium with the heat bath. (It is assumed that these mechanical energies are not altered by the magnetic field). The single-particle partition function may therefore be written as

$$Z_1 = Z_{1t} \, Z_{1r} \, Z_{1v} \, Z_{1m}, \tag{3.3.41}$$

with all the magnetic information in Z_{1m}. Since the classical energy of a magnetic dipole of moment m in a uniform magnetic field B is

$$\epsilon_m = -\mathbf{m} \cdot \mathbf{B} = -mB \cos \theta, \tag{3.3.42}$$

the magnetic factor in Eq. (3.3.41) may be written

$$Z_{1m} = \int_0^\pi e^{\beta mB \cos \theta} (1/2) \sin \theta d\theta, \tag{3.3.43}$$

where the weighting factor $(1/2) \sin \theta$ is required to convert the integrand in the polar angle θ to the proper solid-angle expression. Direct integration yields

$$Z_{1m} = \sinh y/y, \tag{3.3.44}$$

where

$$y = \beta mB. \tag{3.3.45}$$

The mean magnetic energy per particle is

$$\langle \epsilon_m \rangle = -\frac{\partial \ln Z_{1m}}{\partial \beta} = -mB(\coth y - \frac{1}{y}), \tag{3.3.46}$$

or, since $\langle \epsilon_m \rangle$ may be written as $-\langle m \rangle B$, where $\langle m \rangle$ is the average component of **m** along **B**,

$$\langle m \rangle = m \left(\coth y - \frac{1}{y} \right) \equiv mL(y), \qquad (3.3.47)$$

where $L(y)$ is the Langevin function. For $|y| < \pi$, the Langevin function is readily expanded as

$$L(y) = \frac{y}{3} - \frac{y^3}{45} + \frac{2y^5}{945} + \cdots + \frac{2^{2n} B_{2n}}{(2n)!} y^{2n-1} + \cdots, \qquad (3.3.48)$$

where the B_n's are the Bernoulli numbers (Appendix A). Thus for small y (high temperature), $L(y)$ starts out as linear in y and the average moment is

$$\langle m \rangle \approx \frac{m^2 B}{3kT}, \qquad (3.3.49)$$

or the molecular susceptibility is

$$\chi = \left(\frac{\partial \langle m \rangle}{\partial B} \right)_T = \frac{m^2}{3kT}. \qquad (3.3.50)$$

At low temperature, the Langevin function becomes

$$L(y) \approx 1 - \frac{1}{y} + 2e^{-2y} + \mathcal{O}\left(e^{-4y}\right), \quad y \gg 1, \qquad (3.3.51)$$

indicating that as $T \longrightarrow 0$, the permanent dipoles all approach complete alignment with the applied field.

Paramagnetism of this kind is always associated with mechanical angular momentum (orbital or spin) of electrons or nuclei. It should properly be treated with the angular momenta and their associated magnetic moments quantized. Since the quantized treatment is easier than the classical one [because of the need for the solid angle factor in Eq. (3.3.43)], it has been relegated to the exercises.

3.3.4 Debye Heat Capacity.
The classical theory of the molar heat capacity at constant volume, C_v, of a homogeneous, isotropic solid gives the result

$$C_v = 3R = 3N_0 k$$

on the argument that each atom of the solid has three degrees of freedom, each of which can store $kT/2$ of kinetic energy and a like amount of potential energy. The value of the molar heat capacity, $C_v = 3R$, is independent of material, a fact known as the law of Dulong and Petit. Its validity is remarkably good at high temperatures, where "high" must be defined later in terms of parameters of the solid. At low temperatures, however, C_v for electrically insulating solids is found experimentally to behave as aT^3, where a is a constant. Einstein first obtained qualitative agreement with the general shape of C_v *vs* T by treating the solid as if it were composed of $3N_0$ identical quantum oscillators, each with average energy as given by Eq. (3.3.20). The details of Einstein's calculation are left as an exercise. Good quantitative agreement was not to be found in Einstein's results; especially notable was the failure of the calculated C_v to go as T^3 at low temperatures.

Debye proposed a simple but remarkably effective improvement that not only gave the correct T^3 dependence at low temperatures and the Dulong-Petit result at high, but also gave better quantitative agreement over the entire temperature range than should be expected from such a simple model. (Debye [1912]) Debye chose to regard the solid as a homogeneous, isotropic ideal elastic body. Such a body is well-known to vibrate in one or more of a set of independent normal modes, corresponding mathematically to the eigenfunctions of the free particle in a box, already used in writing Eq. (3.3.25), and physically to the independent standing wave patterns characteristic of the solid. Since there should be $3N_0$ degrees of freedom per mole, as in the classical calculation, Debye decided arbitrarily to consider energy storage as possible in only the first $3N_0$ normal modes, and to regard each of these as an independent harmonic oscillator. The partition function may therefore be written

$$Z = \prod_{i=1}^{3N_0} Z_i, \tag{3.3.52}$$

or

$$\ln Z = \sum_{i=1}^{3N_0} \ln Z_i, \tag{3.3.53}$$

where each Z_i is the harmonic oscillator partition function (3.3.19), appropriate to the normal-mode frequency ω_i. When (3.3.53) is written in terms of frequencies instead of modes it becomes

$$\ln Z = \sum_{\omega=0}^{\omega_D} \Omega(\omega) \ln Z(\omega) \tag{3.3.54}$$

where ω_D (D for Debye) is taken as the maximum value of ω, such that

$$3N_0 = \sum_{\omega=0}^{\omega_D} \Omega(\omega), \tag{3.3.55}$$

and $\Omega(\omega)$ is evidently the number of normal modes corresponding to a particular ω.

The solid is taken as a rectangular parallelopiped of dimensions a, b, c, containing one mole of molecules. (Debye chose a sphere!) The standing-wave modes of the solid contain terms like $\exp(i\mathbf{k} \cdot \mathbf{r})$, where $\mathbf{k}$ is the wave vector. To satisfy the wave equation, $\mathbf{k}$ must satisfy

$$k_x^2 + k_y^2 + k_z^2 = \omega^2/c_a^2, \tag{3.3.56}$$

where c_a is the appropriate acoustic velocity. To satisfy the boundary conditions (here taken to be periodic) the components of $\mathbf{k}$ must be

$$k_x = 2q\pi/a; \ k_y = 2r\pi/b; \ k_z = 2s\pi/c, \tag{3.3.57}$$

where q, r, and s are integers. The use of Eqs. (3.3.57) in Eq. (3.3.56) gives the ellipsoid in number space

$$\frac{q^2}{a^2} + \frac{r^2}{b^2} + \frac{s^2}{c^2} = \frac{\omega^2}{4\pi^2 c_a^2}, \tag{3.3.58}$$

with volume

$$\mathcal{V} = v\omega^3/6\pi^2 c_a^3, \tag{3.3.59}$$

where v is the volume per mole. Since there is one normal mode per unit volume of number space, the number of normal modes in the interval $d\omega$ is

$$d\mathcal{V} \equiv g_a(\omega)d\omega = \omega^2 v d\omega/2\pi^2 c_a^3. \tag{3.3.60}$$

Since there are two independent transverse modes, with $c_a = c_t$, and one longitudinal mode, with $c_a = c_\ell$, for each ω the total density of states function becomes

$$g(\omega) = 2g_t + g_\ell = \frac{v\omega^2}{2\pi^2}\left(\frac{2}{c_t^3} + \frac{1}{c_\ell^3}\right). \tag{3.3.61}$$

The two acoustic velocities may be measured directly or calculated from the elastic constants. For present purposes, however, the rather unwieldy expression of Eq. (3.3.61) will be replaced *via* the use of Eq. (3.3.55) with integration instead of summation; thus

$$3N_0 = \int_0^{\omega_D} g(\omega)d\omega = \frac{v\omega_D^3}{6\pi^2}\left(\frac{2}{c_t^3} + \frac{1}{c_\ell^3}\right), \tag{3.3.62}$$

or

$$\frac{v}{2\pi^2}\left(\frac{2}{c_t^3} + \frac{1}{c_\ell^3}\right) = \frac{9N_0}{\omega_D^3}. \tag{3.3.63}$$

Equation (3.3.54) in integral form, with Eq. (3.3.63) used in Eq. (3.3.61) and Eq. (3.3.19) used for $Z(\omega)$, becomes

$$\ln Z = -\frac{9N_0}{\omega_D^3}\int_0^{\omega_D} d\omega\, \omega^2 \ln\left[2\sinh\beta\hbar\omega/2\right]. \tag{3.3.64}$$

The molar energy u is obtained as

$$\begin{aligned} u = -\frac{\partial \ln Z}{\partial \beta} &= \frac{9N_0\hbar}{2\omega_D^3}\int_0^{\omega_D}\frac{\omega^3 d\omega}{\tanh\beta\hbar\omega/2}, \\ &= \frac{9N_0\hbar\omega_D}{8} + \frac{9N_0\hbar}{\omega_D^3}\int_0^{\omega_D}\frac{\omega^3 d\omega}{e^{\beta\hbar\omega}-1}, \end{aligned} \tag{3.3.65}$$

where the first term, merely the total zero-point energy of the $3N_0$ oscillators, contributes nothing to the heat capacity. The substitutions $x = \beta\hbar\omega$ and $\hbar\omega_D \equiv k\Theta_D$, where Θ_D is called the *Debye temperature*, permit Eq. (3.3.65) to be rewritten as

$$u = \frac{9R\Theta_D}{8} + \frac{9RT^4}{\Theta_D^3}\int_0^{\Theta_D/T}\frac{x^3 dx}{e^x-1}. \tag{3.3.66}$$

The integral itself is usually written in terms of the Debye function $D(y)$, defined to be

$$D(y) \equiv \frac{3}{y^3}\int_0^y \frac{x^3 dx}{e^x-1}. \tag{3.3.67}$$

When $y < 2\pi$, or $T > \Theta_D/2\pi$, the integral can be evaluated in terms of the generating function of the Bernoulli numbers,

$$\frac{x}{e^x - 1} = \sum_{n=0}^{\infty} B_n x^n / n! \tag{3.3.68}$$

as in Appendix A. In this limit, $D(y)$ becomes

$$D(y) = 3 \sum_{n=0}^{\infty} \frac{B_n}{n!} \frac{y^n}{n+3}$$

$$= 1 - \frac{3}{8}y + \frac{y^2}{20} - \frac{3y^4}{7!} + \frac{4y^6}{9!} - \frac{9y^8}{11!} + \cdots, \tag{3.3.69}$$

where the first few Bernoulli numbers have been used in the expanded expression.

In terms of $D(y)$, Eq. (3.3.66) can be written

$$u = 9R\Theta_D/8 + 3RTD(\Theta_D/T), \tag{3.3.70}$$

and the molar heat capacity at constant volume is

$$C_v = \left(\frac{\partial u}{\partial T}\right)_v = 3RD\left(\Theta_D/T\right) + 3RT\left(\partial D\left(\Theta_D/T\right)/\partial T\right)_{\Theta_D}. \tag{3.3.71}$$

At $T \gg \Theta_D$, it is evident from Eq. (3.3.69) that $D\left(\Theta_D/T\right) \to 1$, $u = 3RT$ plus the zero-point energy, and $C_v \to 3R$ in agreement with Dulong and Petit. An expression equivalent to Eqs. (3.3.69) and (3.3.71) is

$$C_v/3R = 3 \sum_{n=0}^{\infty} \left(\frac{B_n}{n!}\right)\left(\frac{1-n}{3+n}\right)\left(\frac{\Theta_D}{T}\right)^n,$$

$$= 1 - \frac{(\Theta_D/T)^2}{20} + \frac{9(\Theta_D/T)^4}{7!} - \frac{20(\Theta_D/T)^6}{9!} + \frac{56(\Theta_D/T)^8}{11!} + \cdots \tag{3.3.72}$$

valid for $T > \Theta_D/2\pi$.

For $T \ll \Theta_D$, the y in $D(y)$ becomes very large; the upper limit of the integral in Eq. (3.3.67) can be replaced by ∞ with acceptably small error, and the definite integral becomes

$$\int_0^{\infty} \frac{x^3 \, dx}{e^x - 1} = \int_0^{\infty} dx \sum_{n=1}^{\infty} x^3 e^{-nx} = \Gamma(4)\zeta(4) = \pi^4/15. \tag{3.3.73}$$

In this limit, then, $D(y)$ becomes

$$D(y) = \pi^4/5y^3, \tag{3.3.74}$$

and the molar energy is

$$u = 9R\Theta_D/8 + 3\pi^4 RT^4/5\Theta_D^3. \tag{3.3.75}$$

The low temperature expression for C_v is then

$$C_v = \left(\frac{\partial u}{\partial T}\right)_v = 12\pi^4 RT^3/5\Theta_D^3, \tag{3.3.76}$$

in agreement with the T^3 experimental results.

The basic calculation is a good one, with the method clearly showing the points at which physical assumptions are introduced and could be modified. The two points at which the outcome of the calculation should be responsive to improved contact with the real world are (1) the assumption that the solid is an elastic continuum with the density of states $g(\omega)$ given by Eq. (3.3.61), and (2) the assumption that the modes are those of harmonic oscillators, with partition functions given by Eq. (3.3.19). Studies of lattice dynamics lead to a density of states with much more structure than the one used by Debye, even for harmonic lattices, and the introduction of anharmonic couplings in the calculations can modify both the single-oscillator partition functions and the density of states. An excellent reference for the information of those who want to gain further acquaintance with an area in which theory and experiment are in close contact is the book by Maradudin, Montroll and Weiss [1963].

3.4 GRAND CANONICAL SYSTEMS

In the grand canonical formalism, the entropy is given as usual as

$$S = -k \sum p_i \ln p_i,$$

and maximized subject to the conditions

$$U = \sum p_i \, E_i,$$
$$1 = \sum p_i,$$

and

$$\langle N \rangle = \sum p_i N_i,$$

the final condition corresponding to an indeterminacy in the number of particles, and generalizable to more than one species of particle. As used here, N_i represents the number of particles in the system when its state is designated by the subscript i. When S is maximized, the p_i's are found to be

$$p_i = e^{-\beta E_i + \beta \mu N_i}/Z_G, \tag{3.4.1}$$

where $\beta\mu$ is the undetermined multiplier for the $\langle N \rangle$ equation, and Z_G is given by

$$Z_G = \sum_i e^{-\beta E_i + \beta \mu N_i}. \tag{3.4.2}$$

It is convenient to write the energy levels as $E_i \to E_j(N)$, where $E_j(N)$ is the energy of the jth state of an N-particle system. With this notation, Eq. (3.4.2) can be rewritten as

$$Z_G = \sum_{N=0}^{\infty} e^{\beta \mu N} \sum_j e^{-\beta E_j(N)} = \sum_{N=0}^{\infty} e^{\beta \mu N} Z_N, \tag{3.4.3}$$

where the second line follows from the observation that the N-particle canonical partition function Z_N is

$$Z_N = \sum_j e^{-\beta E_j(N)}.$$

It is often useful to write $z \equiv e^{\beta\mu}$, where z is called *the absolute activity*, a quantity of considerable interest in chemical applications. (Guggenheim [1967]) Thus the grand partition function is written

$$Z_G = \sum_{n=0}^{\infty} z^N Z_N; \tag{3.4.4}$$

the expectation value of the energy is seen to be

$$U = -\left(\partial \ln Z_G / \partial \beta\right)_z; \tag{3.4.5}$$

and the average value of N is

$$\langle N \rangle = z \left(\partial \ln Z_G / \partial z\right)_\beta. \tag{3.4.6}$$

The probability of the N-particle state $E_j(N)$ is given by

$$p_j(N) = z^N e^{-\beta E_j(N)} / Z_G, \tag{3.4.7}$$

and the entropy is

$$\begin{aligned}
S &= -k \sum_{N=0}^{\infty} \sum_j p_j(N) \ln p_j(N) \\
&= -k \sum_{N=0}^{\infty} \sum_j p_j(N) \left[N \ln z - \beta E_j(N) - \ln Z_G\right] \\
&= -\langle N \rangle k\beta\mu + k\beta U + k \ln Z_G. \tag{3.4.8}
\end{aligned}$$

The postulate that S is the thermodynamic entropy given by Eq. (2.3.9) leads to the identification that μ is the chemical potential and

$$PV = kT \ln Z_G, \tag{3.4.9}$$

where $-PV \equiv \Phi$ is often called *the grand potential*, and

$$Z_G = e^{-\beta\Phi}. \tag{3.4.10}$$

No particular advantage recommends the use of Z_G instead of Z_N if the latter must be calculated in order that Eq. (3.4.4) can be used to calculate Z_G. Therefore only a single example will be given in this chapter of a calculation by means of Z_G. The best uses of Z_G are those for which Z_N need not be calculated first; two of these are presented in Chapter 4, where it is possible to obtain the generating functions for counting states most easily in terms of Z_G. The relation of Z_G to Z_N is left as an exercise.

3.4.1 Monatomic Ideal Gas. From Eq. (3.3.29), Z_N for a monatomic ideal gas is known to be

$$Z_N = Z_1^N/N!,$$

where $Z_1 = V/\lambda^3$. From Eq. (3.4.4) it follows that

$$Z_G = \sum_{N=0}^{\infty} (zZ_1)^N /N! = e^{zZ_1}. \tag{3.4.11}$$

The average value of N is found from Eq. (3.4.6) to be

$$\langle N \rangle = zZ_1, \tag{3.4.12}$$

and Eqs. (3.4.9), (3.4.11), and (3.4.12) combine to give

$$PV = kTzZ_1 = \langle N \rangle kT, \tag{3.4.13}$$

all without any specification of the structure of Z_1. It is therefore evident that the ideal gas law, Eq. (3.4.13), follows equally well for gases more complicated than monatomic, provided the molecules can be regarded as noninteracting. Only in the calculation of the energy by means of Eq. (3.4.5) does the internal structure become important, and for the monatomic ideal gas the result is

$$U = \frac{3}{2} \langle N \rangle kT,$$

in agreement with Eq. (3.3.30). Since further details of the monatomic ideal gas have been treated in Section 3.2, they need not be repeated here.

3.5 STATISTICAL OPERATOR METHODS

The statistical operator ρ, sometimes called the density operator or density matrix, is given in Eq. (1.8.12) as

$$\rho = e^{-\beta \mathcal{H}}/Z, \tag{3.5.1}$$

where $\mathcal{H}$ is the Hamiltonian operator and the partition function is

$$Z = \operatorname{Tr} e^{-\beta \mathcal{H}}. \tag{3.5.2}$$

The statistical expectation value of some observable f, corresponding to the operator $\mathcal{F}$, is given by

$$\langle \bar{f} \rangle = \operatorname{Tr} \rho \mathcal{F}, \tag{3.5.3}$$

as shown in Eq. (1.8.5). Three examples of techniques for the use of ρ are exhibited in the following subsections. The first is trivial and offers no computational advantage as presented; the second provides a useful operator technique; and the third demonstrates a means of calculating partition functions without first knowing the energy eigenvalues.

For notational and computational convenience, the not-yet-normalized operator ρ^* will be introduced as

$$\rho^* = e^{-\beta \mathcal{H}}, \tag{3.5.4}$$

with Eq. (3.5.2) then written as

$$Z = \text{Tr } \rho^*. \tag{3.5.5}$$

The operator ρ^* evidently satisfies the Bloch equation, which is

$$\frac{\partial \rho^*}{\partial \beta} = -\mathcal{H}\rho^*. \tag{3.5.6}$$

Its solution is subject to the boundary condition

$$\rho^* \left(\beta = 0^+ \right) = I, \tag{3.5.7}$$

where I is the appropriate unit operator. The boundary condition Eq. (3.5.7) follows from Eq. (1.8.2), rewritten without normalization of the p's as

$$\rho^* (\beta) = \Sigma_\phi \mid \phi \rangle \, e^{-\beta E_\phi} \langle \phi \mid , \tag{3.5.8}$$

since $\rho^*(\beta = 0)$ is the unit operator. The Bloch equation is just the Schrödinger equation with $\beta = it/\hbar$; its use is exhibited in Sections 5.5.3 and 5.5.4.

3.5.1 The Simple Harmonic Oscillator.   The eigenkets of the simple harmonic oscillator are denoted by $\mid n \rangle$ given by

$$\mathcal{H} \mid n \rangle = \hbar\omega \left(n + \frac{1}{2} \right) \mid n \rangle = E_n \mid n \rangle . \tag{3.5.9}$$

The statistical operator ρ^* has matrix elements in terms of the $\mid n \rangle$ given by

$$\langle m \mid \rho^* \mid n \rangle = e^{-\beta E_n} \langle m \mid n \rangle = e^{-\beta E_n} \delta_{mn}, \tag{3.5.10}$$

and the partition function is

$$Z = \text{Tr} \, \rho^* = \sum_{n=0}^{\infty} \langle n \mid \rho^* \mid n \rangle = \frac{1}{2 \sinh \left(\beta\hbar\omega/2 \right)}, \tag{3.5.11}$$

in agreement with Eq. (3.3.19). Other results follow as in Section 3.1.

3.5.2 Spin Paramagnetism and Negative Temperatures.   The magnetic behavior of a substance composed of well-separated molecules possessing permanent dipole moments can be modeled simply in terms of isolated spin-1/2 particles, each with a dipole moment $\mu\sigma$, where σ is the spin-1/2 operator, and μ is the magnitude of the dipole moment. Since the molecules are taken to be noninteracting, it is sufficient to study a single one. The magnetic part of the Hamiltonian operator is

$$\mathcal{H}_m = -\mu\mathbf{B} \cdot \sigma, \tag{3.5.12}$$

and for convenience the coordinate axes are chosen such that $\mathbf{B} \cdot \sigma = B\sigma_z$. The Pauli spin operators are

$$\sigma_x = \begin{pmatrix} 0 & 1 \\ 1 & 0 \end{pmatrix}, \ \sigma_y = \begin{pmatrix} 0 & -i \\ i & 0 \end{pmatrix}, \ \text{and } \sigma_z = \begin{pmatrix} 1 & 0 \\ 0 & -1 \end{pmatrix}. \tag{3.5.13}$$

In terms of σ_z, the statistical operator ρ^* becomes

$$\rho^* = e^{\beta\mu B \sigma_z} \equiv e^{x\sigma_z} = I \cosh x + \sigma_z \sinh x, \tag{3.5.14}$$

where I is the 2×2 unit matrix, and the last step follows from the expansion

$$
\begin{aligned}
e^{x\sigma_z} &= I + x\sigma_z + (x\sigma_z)^2/2! + (x\sigma_z)^3/3! + \cdots \\
&= I\left[1 + x^2/2! + x^4/4! + \cdots\right] + \sigma_z\left[x + x^3/3! + x^5/5! + \cdots\right],
\end{aligned} \tag{3.5.15}
$$

and the fact that $\sigma_z^2 = I$. From Eq. (3.5.14) it is seen that

$$Z = \operatorname{Tr}\rho^* = 2\cosh x, \tag{3.5.16}$$

and the expectation value of σ_z is

$$\langle \sigma_z \rangle = \left[\operatorname{Tr}\rho^*\sigma_z\right]/Z = 2\sinh x/2\cosh x = \tanh x. \tag{3.5.17}$$

The energy expectation value is then

$$\langle E \rangle = -\mu B \langle \sigma_z \rangle = -\mu B \tanh x, \tag{3.5.18}$$

which ranges from $-\mu B$ to $+\mu B$ for $-\infty < x < \infty$.

One would normally expect a reversal of B to change two signs in Eq. (3.5.18) and result in no net change. This expectation is usually fulfilled for molecules, where the mechanical terms in the total Hamiltonian are sufficiently well coupled to the magnetic terms to permit rapid exchanges of energy. Thus a magnetic moment that suddenly finds the B-field reversed and its energy $+\mu B$ instead of $-\mu B$ can dispose of $2\mu B$ of excess energy by giving it to lattice vibrations, for example. Nuclear magnetic moments, on the other hand, may be very weakly coupled to the molecular motions and vibrational modes of the system, even though they are well enough coupled to each other to be regarded as in equilibrium. Reversal of the B-field for such nuclei results in the sign of $\langle E \rangle$ being changed to $+$ while the sign of B and x both appear to change in Eq. (3.5.18). But since the equation properly must read

$$\langle E \rangle = \mu B \tanh x, \tag{3.5.19}$$

the sign of x cannot change. If β becomes $-\beta$ when B becomes $-B$, then Eq. (3.5.19) follows from Eq. (3.5.18), and the system can be regarded as having the negative temperature $T = 1/k\beta$, where β is negative.

Two physical conditions must be met in order for negative temperatures to arise; the subsystem possessing the negative temperature must be well insulated thermally from the other modes of energy storage of the complete system, and the subsystem energy levels must be bounded from above and below. Thus a normally populated state with probabilities proportional to $e^{-\beta\epsilon_i}$ can be inverted by reversal of the order of the energy levels while populations remain intact because there is no convenient energy sink available.

Examination of the entropy gives further insight to the idea of negative temperature. The probabilities p_+ and p_- for spin parallel or antiparallel to B are just the diagonal elements of ρ, or

$$p_+ = \frac{1}{2}\left(1 + \tanh x\right) \text{ and } p_- = \frac{1}{2}\left(1 - \tanh x\right). \tag{3.5.20}$$

For an N-element system, the entropy is given by

$$S_N(x)/Nk = -p_+ \ln p_+ - p_- \ln p_-.$$
$$(3.5.21)$$

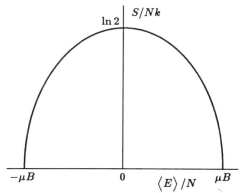

Figure 3.1 Entropy of spin-1/2 system *vs.* its energy.

A plot of Eq. (3.5.21) is shown in Fig. 3.1; the abscissa is chosen as $\langle E \rangle$ from Eq. (3.5.19). For $-\mu B < \langle E \rangle /N < 0$, it may be seen that $\partial S/\partial \langle E \rangle = 1/T$ is positive, corresponding to a normal system. But for $0 < (\langle E \rangle /N) < \mu B$, the derivative is negative. This behavior violates fundamental statement 2, Section 2.3.2, where the entropy was supposed to be a monotone increasing function of the energy. When the energy levels are bounded from above as well as from below, however, zero entropy can occur for both minimum and maximum energy.

Any negative temperature is hotter than the highest possible positive temperature. As T rises through positive values, $\tanh x \to 0$, and the probabilities of Eq. (3.5.20) approach equality. But if a negative temperature system is placed in thermal contact with a system at any positive temperature, there will be energy transfer from the negative to the positive temperature system as the inverted population loses energy and approaches a normal set of probabilities. Since the energy transfer is into the microscopic modes of the positive-temperature system, it is regarded as heat, and the hotter body must be the one at negative temperature. Thus the temperature in Fig. 3.1 changes from $+\infty$ to $-\infty$ as the energy increases from 0_- to 0_+, and the highest temperature is at $\langle E \rangle /N = \mu B$, where the temperature is 0_-. There is no direct access to $T = 0_+$ *via* negative temperatures.

When $\beta = 1/kT$ is used as the temperature parameter, then the changes are well behaved. When T jumps from $+\infty$ to $-\infty$, β merely passes through zero, and the end points of the curve where $S = 0$ correspond to $\beta = \pm\infty$. For this reason and also because the unattainability of absolute zero seems more palatable when stated as the unattainability of infinite β, it is sometimes advocated that β be used to replace T as the temperature parameter. The evident utility of both T and β precludes the disappearance of either from the literature, however.

No system really has energy levels bounded from above when the entire composite energy level structure is considered. Therefore negative temperatures evidently must correspond to nonequilibrium states of a system. But the nuclear spin systems can be so well insulated from their surroundings that the spin temperature is quite as valid a concept as is the temperature of, say, liquid helium in a well-insulated Dewar flask. No particular difficulties arise in the logical structure of statistical mechanics or of thermodynamics as a consequence of these negative-temperature systems, although it is usually more convenient to recognize them as special cases than to attempt to state every theorem or postulate in such a way that these cases are properly included.

3.5.3 Free Particle. Because of the similarity of the Bloch equation, Eq. (3.5.6), and the Schrödinger equation, it is often expedient to carry out the calculation of Z as the solution of a differential equation. For this purpose, the matrix elements of the operator ρ^* are used.

We have, for the eigenfunctions of the Schrödinger equation ϕ_i,

$$\rho^* \phi_i = \sum_{n=0}^{\infty} \frac{1}{n!} \left(-\beta \mathcal{H}\right)^n \phi_i = \sum_{n=0}^{\infty} \frac{1}{n!} \left(-\beta E_i\right)^n \phi_i = e^{-\beta E_i} \phi_i, \tag{3.5.22}$$

from which it follows that

$$\sum_i \int \phi_i^* \rho^* \phi_i \, d^3 r = \sum_i e^{-\beta E_i} = Z. \tag{3.5.23}$$

The quantity

$$R\left(\mathbf{r}', \mathbf{r}, \beta\right) = \sum_i \phi_i^* \left(\mathbf{r}'\right) \rho^* \phi_i \left(\mathbf{r}\right) \tag{3.5.24}$$

is sometimes called the *statistical matrix*, as is the normalized version using ρ in place of ρ^*. Since $\mathcal{H}$ in ρ^* acts only on $\phi_i(\mathbf{r})$, an equation of the form of (3.5.6) is satisfied by R, or

$$\frac{\partial R}{\partial \beta} = -\mathcal{H}(\mathbf{r}) R. \tag{3.5.25}$$

The solution to Eq. (3.5.25), with the condition

$$R\left(\mathbf{r}', \mathbf{r}, 0\right) = \delta\left(\mathbf{r} - \mathbf{r}'\right) \tag{3.5.26}$$

corresponding to Eq. (3.5.7) can then be used to calculate Z as

$$Z = \int R\left(\mathbf{r}, \mathbf{r}, \beta\right) d^3 r. \tag{3.5.27}$$

For a free particle, Eq. (3.5.25) becomes

$$\frac{\partial R}{\partial \beta} = \frac{\hbar^2}{2m} \nabla^2 R, \tag{3.5.28}$$

which is just the diffusion equation, and Eq. (3.5.26) makes the solution that of an initial point source in an infinite medium, given by

$$R\left(\mathbf{r}', \mathbf{r}, \beta\right) = \frac{1}{\lambda^3} \exp\left[-\frac{\pi}{\lambda^2} \left\{ (x - x')^2 + (y - y')^2 + (z - z')^2 \right\} \right], \tag{3.5.29}$$

where λ is the thermal wavelength of Eq. (3.3.27). When $\mathbf{r}' = \mathbf{r}$, then $R(\mathbf{r}, \mathbf{r}, \beta) = 1/\lambda^3$, and Eq. (3.5.27) gives

$$Z = (1/\lambda^3) \int d^3 r, \tag{3.5.30}$$

which would give V/λ^3 for a finite volume. For $V/\lambda^3 \gg 1$, the free-particle approximation is valid in the calculation of R, but a proper calculation should impose the further condition that $R = 0$ at the bounding surfaces of the container. In the so-called thermodynamic limit, for which the volume of the system becomes infinite in such a way that the volume per particle remains constant, boundary effects vanish and the partition function for one particle in a finite volume V of the infinite system is found from Eq. (3.5.30) to be

$$Z_1 = V/\lambda^3. \tag{3.5.31}$$

The procedure of going to the thermodynamic limit is closely related to the procedure of evaluating the sums in Eq. (3.3.26), in the limit of the box dimensions greatly exceeding the thermal wavelength, by integration.

3.5.4 Feynman Path-Integral Methods.   The function $R(\mathbf{r}', \mathbf{r}, \beta)$ of Eq. (3.5.24) is sufficiently useful to merit further exploration, and its relationship to Feynman path-integral techniques will be developed in this section. For notational convenience, the development will be carried out with only one spatial dimension. Generalization to three dimensions is easy and direct.

The relationship of R to propagators or Green functions becomes evident when it is expressed in its original quantum-mechanical context. Suppose $\{\phi_n(x)\}$ is a complete orthonormal set of eigenfunctions of the Hamiltonian $\mathcal{H}$, with eigenvalues E_n. Then any function $f(x)$ may be written as

$$f(x) = \sum_1^\infty a_n \phi_n(x), \tag{3.5.32}$$

with

$$a_r = \int \phi_r^*(x) f(x) dx. \tag{3.5.33}$$

From this standard result it follows that

$$f(x) = \int dx' \left[\sum_1^\infty \phi_n(x)\phi_n^*(x') \right] f(x') = \sum_1^\infty \phi_n(x) \int dx' \phi_n^*(x') f(x'),$$

giving a familiar expression for the delta function,

$$\delta(x - x') = \sum_1^\infty \phi_n(x)\phi_n^*(x'),$$

but also in agreement with Eq. (3.5.26). Any solution of the Schrödinger equation may be written

$$\psi(x,t) = \sum_1^\infty c_n e^{-iE_n t/\hbar} \phi_n(x).$$

Suppose that at $t = t_1$, a solution is known to be $\psi(x, t_1) = f(x)$, or

$$f(x) = \sum_1^\infty c_n e^{-iE_n t_1/\hbar} \phi_n(x) = \sum_1^\infty a_n \phi_n(x),$$

with $f(x)$ presumed known. It follows that $c_n = a_n \exp(iE_n t_1/\hbar)$, so that for the later time t_2, the wave function may be written

$$\psi(x, t_2) = \sum_1^\infty a_n e^{iE_n(t_1 - t_2)/\hbar} \phi_n(x). \tag{3.5.34}$$

But from Eq. (3.5.33), the a's are known, and Eq. (3.5.34) may be written

$$\psi(x, t_2) = \int dx' \sum_n \phi_n(x)\phi_n^*(x')\psi(x', t_1)e^{-iE_n(t_2-t_1)/\hbar} = \int K(x, t_2; x', t_1)\psi(x', t_1)\, dx',$$

$$(3.5.35)$$

where

$$K(x_2, t_2; x_1, t_1) = \sum_n \phi_n(x_2)\phi_n^*(x_1)e^{-iE_n(t_2-t_1)/\hbar}; \quad t_2 > t_1,$$

$$= 0; \quad t_2 < t_1. \tag{3.5.36}$$

The kernel $K(x_2, t_2; x_1, t_1)$ propagates the wave function from x_1, t_1 to x_2, t_2. It has the basic property that it can be iterated; thus

$$\psi(x_3, t_3) = \int dx_2 \int dx_1 K(x_3, t_3; x_2, t_2)K(x_2, t_2; x_1, t_1)\psi(x_1, t_1), \tag{3.5.37}$$

and continuations are evident.

If the initial time value is chosen to be zero, then, since only the time difference appears in K, the notation can be changed to replace $K(x_2, t_2; x_1, 0)$ by $K(x_2, x_1; t)$, where $t = t_2 - t_1 = t_2$, and the variable t will continue to represent the time interval between the initial and final coordinates in K. It is now evident, in this notation, that K, here written

$$K(x, x'; t) = \sum_n \phi_n(x)\phi_n^*(x')e^{-iE_n t/\hbar}, \tag{3.5.38}$$

resembles a one-dimensional version of $R(\mathbf{r}, \mathbf{r}'; \beta)$ of Eq. (3.5.24), with the replacement $\beta = it/\hbar$. (The $\beta > 0$ regime corresponds to negative imaginary times.) It is convenient at this point to define a variable with the units of time, here called the *pseudotime*, as $w = \beta\hbar$. When t in Eq. (3.5.38) is replaced by $-i\beta\hbar = -iw$, the K first becomes identical with the R of Eq. (3.5.24), except that it is a one-dimensional version, and then it emerges as

$$R(x, x_0; w) = \sum_n \phi_n(x)\phi_n^*(x_0)e^{-wE_n/\hbar}. \tag{3.5.39}$$

Thus R is seen as a propagator of the state function from its initial value at some position x_0 to a final position x in the pseudotime interval w. Details of this propagation lie at the heart of the Feynman path-integral development. (Feynman [1972])

If $\mathcal{H}$ is regarded as acting on x but not on x', then the propagator R is seen to satisfy the equation

$$\mathcal{H}R(x, x'; w) = -\hbar \frac{\partial R}{\partial w}, \tag{3.5.40}$$

which is equivalent to the Bloch equation, Eq. (3.5.25), and has the formal solution

$$R(w) = e^{-\mathcal{H}w/\hbar}, \tag{3.5.41}$$

where the position dependencies are not expressed.

The essence of the Feynman path-integral idea is that the interval w can be divided into segments, the propagator can be iterated to carry the state function forward in pseudotime, segment by segment, over all possible paths, with the contributions by the various paths automatically weighted in proportion to their importance. Suppose the interval w is divided into N segments, each of length $\delta = w/N$. The propagator can be written

$$R(w) = e^{-\mathcal{H}\delta/\hbar}e^{-\mathcal{H}\delta/\hbar}\cdots e^{-\mathcal{H}\delta/\hbar} = R(\delta)R(\delta)\cdots R(\delta), \quad N\text{factors},$$

which can be represented as

$$R(x, x_0; w) = \int \cdots \int R(x, x_{N-1}; \delta)R(x_{N-1}, x_{N-2}; \delta)\cdots R(x_1, x_0; \delta)dx_1 dx_2 \cdots dx_{N-1}.$$

Feynman considered this to symbolize the propagation of a particle from position x_0 at $w = 0$ to position x at pseudotime w, *via* intermediate steps that define a path. The propagator $R(x, x_0; w)$ represents the total amplitude for all paths with the specified beginning and ending. As the pseudotime increment $\delta = w/N$ is allowed to go to zero, the number of integrations becomes infinite, and the equation for R can be written

$$R(x, x_0; W) = \int\!\!\!\int \Phi[x(w)]\mathcal{D}x(w), \tag{3.5.42}$$

with

$$\Phi[x(w)] = \lim_{\delta \to 0} R(x, x_{N-1}; \delta)R(x_{N-1}, R_{N-2}; \delta)\cdots R(x_1, x_0; \delta), \tag{3.5.43}$$

and

$$\mathcal{D}x(w) = \lim_{N \to \infty} dx_1 dx_2 \cdots dx_{N-1}. \tag{3.5.44}$$

The techniques by which R is obtained without the need to perform an infinite number of integrations will be illustrated first for the free particle, then extended to the case of a particle in a conservative field, and later particularized to the harmonic oscillator. Since R for a free particle has already been obtained in Eq. (3.5.29) as the solution of the diffusion equation, a one-dimensional version of that solution, written in terms of w instead of β, will be used as a starting point, and the procedure will lead to a reconstruction of the input equation along a path of analysis that introduces some of the techniques. From Eq. (3.5.29), the free-particle propagator is known to be

$$R(x, x_0; w) = \sqrt{\frac{m}{hw}} \exp \frac{-\pi m}{hw}(x - x_0)^2, \tag{3.5.45}$$

with the same form when w is replaced by δ. When δ is small, the principal contributions to the propagator come from paths such that $|x - x_0| \ll h\delta/\pi m$. In a sequence of such paths, all in the same pseudotime interval $w = N\delta$, it is evident that paths deviating appreciably from the straight line that joins the two end points in (x, w)-space will contribute little to the propagator. For the free particle, the propagator is

$$R(x, x_0; W) = \lim_{\delta \to 0} \int \cdots \int \exp\left\{\frac{-m\delta}{2\hbar}\left[\left(\frac{x - x_{N-1}}{\delta}\right)^2 + \left(\frac{x_{N-1} - x_{N-2}}{\delta}\right)^2 + \cdots\right.\right.$$
$$\left.\left. + \left(\frac{x_1 - x_0}{\delta}\right)^2\right]\right\}\left(\frac{dx_1}{\sqrt{h\delta/m}}\right)\left(\frac{dx_2}{\sqrt{h\delta/m}}\right)\cdots\left(\frac{dx_{N-1}}{\sqrt{h\delta/m}}\right), \tag{3.5.46}$$

and the functional $\Phi[x(w)]$ is seen to be the exponential factor in this expression, in the limit as $\delta \to 0$. But the definition of the derivative gives

$$\lim_{\delta \to 0} \left(\frac{x_k - x_{k-1}}{\delta} \right) \to \frac{dx(w)}{dw}\bigg|_{w=k\delta} = \dot{x}(w)\big|_{w=k\delta},$$

which permits $\Phi[x(w)]$ to be written

$$\Phi[x(W)] = \exp\left[\frac{-1}{\hbar} \int_0^W \frac{m}{2} [\dot{x}(w)]^2 dw \right]. \tag{3.5.47}$$

This expression looks appropriate as an enhanced version of Eq. (3.5.41), since the integrand, written in this way, looks like the kinetic energy.

Before completion of the free-particle calculation, it is instructive to examine the case of a single particle in a conservative field, so that there exists a potential energy. In this case, it seems plausible that the propagator for an infinitesimal pseudotime interval, δ, can be written

$$R(x_k, x_{k-1}; \delta) = \sqrt{\frac{m}{h\delta}} \exp\left[\frac{-\pi m}{h\delta} (x_k - x_{k-1})^2 \right] e^{-V(x_k)\delta/\hbar},$$

where the first factor is the one already written for a free particle, and the second is the obvious factor to come from the potential energy. The problem that arises from making this believable step is that the kinetic and potential energy operators in the hamiltonian do not commute, and the usual rules for the behavior of exponential functions require modification. But for infinitesimal intervals, it is easy to show that the expression for the propagator is valid to first order in δ. It is then possible, in general, to write the path-integral expression for R as

$$R(x, x_0; W) = \int\!\!\!\int \exp\left[\frac{-1}{\hbar} \int_0^W \left[\frac{m}{2} [\dot{x}(w)]^2 + V[x(w)] \right] dw \right] \mathcal{D}x(w),$$

$$= \int\!\!\!\int \Phi[x(w)] \mathcal{D}x(w). \tag{3.5.48}$$

Evaluation of the path integrals turns out to be nontrivial, in general, but nevertheless usually not as imposing as it may appear at first sight. The first step is the expansion of the integrand in $\Phi[x(w)]$ about the principal path, which is the path for which the integral is minimum. In general, this path can be found as the solution to the Euler equation. If $G(x, \dot{x})$ represents the integrand of $\Phi[x(w)]$, the Euler equation is

$$\frac{\partial}{\partial w} \frac{\partial G}{\partial \dot{x}} - \frac{\partial G}{\partial x} = 0.$$

For the free particle, $G = m\dot{x}^2/2$, and the Euler equation gives $m\ddot{x} = 0$. The solution to this equation is denoted as $X(w)$; it can be written as

$$X(w) = Aw + B,$$

where A and B are constants to be determined by the boundary conditions. At $w = 0$, $X(0) = x_0$, and at $w = W$, $X(W) = x$, giving $B = x_0$, and $A = (x - x_0)/W$. In this case, $\dot{X} = A$, with known value, and the expansion for $x(w)$ is easy:

$$x(w) = X(w) + y(w);$$
$$\dot{x}(w) = \dot{X}(w) + \dot{y}(w) = A + \dot{y}(w).$$

The free-particle integrand now becomes

$$m[\dot{x}(w)]^2/2 = m(A^2 + 2A\dot{y} + \dot{y}^2)/2,$$

and $R(x, x_0; W)$ becomes

$$R(x, x_0; W) = \int\!\!\!\int \exp\left[\frac{-1}{\hbar}\int_0^W \frac{m}{2}\left(A^2 + 2A\dot{y} + \dot{y}^2\right) dw\right] \mathcal{D}x(w),$$

$$= e^{-mA^2W/2\hbar}\int\!\!\!\int \exp\left[\frac{-1}{\hbar}\int_0^W \frac{m}{2}\dot{y}^2 dw\right] \mathcal{D}y(w), \qquad (3.5.49)$$

since the term linear in $\dot{y}$ integrates to zero because y vanishes at the end points. The remaining path integral, which at first may have seemed to present a serious problem, need not be evaluated directly. Since it does not depend on the end points, it may be written as a function of W alone, $F(W)$, leading to

$$R(x, x_0; W) = F(W)e^{-m(x-x_0)^2/2\hbar W}, \qquad (3.5.50)$$

where

$$F(W) = \int\!\!\!\int \exp\left[\frac{-m}{2\hbar}\int_0^W \dot{y}^2 dw\right] \mathcal{D}y.$$

Since R is already known, $F(W)$ can be determined, but there is a useful general technique for calculating F in cases where it is necessary to do so. The function R can be iterated, in accordance with Eq. (3.5.47), permitting the relationship

$$R(x, x_0; w_1 + w_2) = \int R(x, x_a; w_2)R(x_a, x_0; w_1)dx_a.$$

Substitution of the expression for R from Eq. (3.5.50) in the iteration formula gives

$$F(w_1 + w_2)e^{-m(x-x_0)^2/2\hbar(w_1+w_2)} = F(w_1)F(w_2)\int_{-\infty}^{\infty} dx_a \left[e^{-m(x_a-x_0)^2/2\hbar w_1}\right]\left[e^{-m(x-x_a)^2/2\hbar w_2}\right],$$

$$= F(w_1)F(w_2)\sqrt{\frac{2\pi\hbar w_1 w_2}{m(w_1 + w_2)}}e^{-m(x-x_0)^2/2\hbar(w_1+w_2)}. \qquad (3.5.51)$$

It follows directly that

$$\sqrt{2\pi\hbar(w_1 + w_2)/m}\, F(w_1 + w_2) = \left(\sqrt{2\pi\hbar w_1/m}\, F(w_1)\right)\left(\sqrt{2\pi\hbar w_2/m}\, F(w_2)\right).$$

Now note that this equation is of the form $g(w_1 + w_2) = g(w_1)g(w_2)$, which has the general solution $g(w) = e^{\alpha w}$. This gives the result that

$$R(x, x_0; W) = \sqrt{\frac{m}{2\pi\hbar W}} e^{\alpha W} e^{-m(x-x_0)^2/2\hbar W},$$

with α arbitrary. (It is already known from Eq. (3.5.45) that the correct value is $\alpha = 0$, but the consequences of a more general solution should be examined.) The partition function is, with the length L corresponding to V in Eq. (3.5.30),

$$Z(W) = \int_0^L R(x, x; W)dx = L\sqrt{\frac{m}{2\pi\hbar W}} e^{\alpha W},$$

which becomes, with $\beta\hbar = W$,

$$Z(\beta) = L\sqrt{\frac{2\pi m}{h^2\beta}} e^{\beta\alpha\hbar}.$$

The expectation value for the energy in this case is

$$\langle \bar{\epsilon} \rangle = -\frac{\partial \ln Z}{\partial \beta} = \frac{kT}{2} + \alpha\hbar.$$

Thus the only effect of α is to shift the zero of energy. Since this is usually an unwanted freedom, there is no loss in generality if α is set to be zero.

For the harmonic oscillator, Eq. (3.5.48) becomes

$$R(x, x_0; W) = \int\!\!\int \exp\left[\frac{-1}{\hbar}\int_0^W \left(\frac{m\dot{x}^2}{2} + \frac{m\omega^2 x^2}{2}\right)dw\right]\mathcal{D}x(w), \qquad (3.5.52)$$

and the exponential function is seen to be

$$\Phi[x(w)] = \exp\left[\frac{-1}{\hbar}\int_0^W \left(\frac{m\dot{x}^2}{2} + \frac{m\omega^2 x^2}{2}\right)dw\right]. \qquad (3.5.53)$$

The Euler equation is used to minimize the argument of the exponential and thus find the principal path, yielding $m\ddot{x} - m\omega^2 x = 0$, and

$$X(w) = Ae^{\omega w} + Be^{-\omega w} \qquad (3.5.54)$$

for the principal path, with $X(0) = x_0$, and $X(W) = x$. As before, the path $x(w)$ is written

$$x(w) = X(w) + y(w),$$

where $X(w)$ is the solution to the Euler equation and the given boundary conditions. With this substitution, Eq. (3.5.53) may be written

$$\ln \Phi = \frac{-m}{2\hbar}\left[\int_0^W (\dot{X}^2 + \omega^2 X^2)dw + 2\int_0^W (\dot{X}\dot{y} + \omega^2 Xy)dw + \int_0^W (\dot{y}^2 + \omega^2 y^2)dw\right].$$

The middle term vanishes, as may be seen by integration of $\dot{X}\dot{y}$ by parts, recognition that $y(0) = y(W) = 0$, and use of the equation of motion $\ddot{X} = \omega^2 X$. The first term can be simplified by writing

$$\int_0^W \dot{X}^2 \, dw = X\dot{X} \Big|_0^W - \int_0^W \ddot{X}X \, dw,$$

which results in cancellation of everything in the first term except the already integrated part. With these steps taken, Eq. (3.5.52) becomes

$$R(x, x_0; W) = \exp\left[\frac{-m}{2\hbar}\left(X(W)\dot{X}(W) - X(0)\dot{X}(0)\right)\right] F(W), \qquad (3.5.55)$$

where

$$F(W) = \int\!\!\!\int \exp\left[\frac{-m}{2\hbar}\left(\dot{y}^2 + \omega^2 y^2\right) dw\right] \mathcal{D}y(w). \qquad (3.5.56)$$

By use of the iterative property as in Eq. (3.5.51), together with the solution of Eq. (3.5.54), the final expression for the propagator is found to be

$$R(x_1, x_2; W) = \sqrt{\frac{m\omega}{2\pi\hbar \sinh \omega W}} \exp\left[\frac{-m\omega}{2\hbar}\frac{(x_2^2 + x_1^2)\cosh \omega W - 2x_1 x_2}{\sinh \omega W}\right],$$

$$= \sqrt{\frac{m\omega}{2\hbar \sinh \omega W}} \exp\left[\frac{-m\omega}{2\hbar}\frac{(x_2 - x_1)^2 \cosh^2(\omega W/2) + (x_2 + x_1)^2 \sinh^2(\omega W/2)}{\sinh(\omega W/2)\cosh(\omega W/2)}\right]. \quad (3.5.57)$$

The diagonal elements of the propagator, or statistical matrix, are

$$R(x, x; W) = \sqrt{\frac{m\omega}{2\pi\hbar \sinh \omega W}} \exp\left[\frac{-m\omega}{\hbar}(x^2 \tanh(\omega W/2)\right],$$

and the partition function is found to be

$$Z = \int_{-\infty}^{\infty} R(x, x; W) dx,$$

$$= \sqrt{\frac{m\omega}{4\pi\hbar \sinh(\omega W/2)\cosh(\omega W/2)}} \sqrt{\frac{\pi\hbar}{m\omega \tanh(\omega W/2)}},$$

$$= \frac{1}{2\sinh(\omega W/2)} = \frac{1}{2\sinh(\beta\hbar\omega/2)}. \qquad (3.5.58)$$

This is the correct single-particle partition function for the harmonic oscillator, including the zero-point energy, in agreement with Eq. (3.5.11) and earlier computations of the same result.

The power and usefulness of the path-integral method are to be found in treatments of many-particle systems, and in systems for which perturbation approximations are necessary. A further example of this kind of statistical-matrix calculation is to be found in the treatment of Landau diamagnetism of free electrons, in Chapter 5.

3.6 THE GENERAL IDEAL GAS

Ideal gases are composed of noninteracting molecules, each of which may have internal motions – rotations and vibrations – as well as states of electronic excitation. These motions are regarded as independent, so that the single-molecule partition function can be written as

$$Z_1 = Z_{1t} Z_{1r} Z_{1v} Z_{1e} Z_{1n}, \tag{3.6.1}$$

where the subscripts denote *translation, rotation, vibration, electronic,* and *nuclear,* respectively. The N-molecule partition function is given by

$$Z_N = Z_1^N / N!,$$

in agreement with Eq. (3.3.17), when all of the molecules are considered identical. If there are several identifiable species, each with a number of molecules given by N_i, such that $N = \sum N_i$, then the canonical partition function becomes

$$Z_N = \prod \left(\frac{Z_{1i}^{N_i}}{N_i!} \right). \tag{3.6.2}$$

The translational partition function has the same form for all ideal gases, *e.g.*, $Z_{1t} = V/\lambda^3$, where λ is the thermal wave length, given by Eq. (3.3.27). The nuclear energy, for most purposes in this book, may be taken as zero, since other nuclear energy levels are inaccessible thermally, so the nuclear partition function becomes $Z_{1n} = g_{ns}$, where g_{ns} is the nuclear spin statistical weighting, which will be included as a factor in the rotational partition function. (The nuclear factor is omitted from the treatment of the monatomic ideal gas because it contributes nothing more than a constant multiplier to the partition function, which has no effect on most of the macroscopic properties of interest. When chemical reactions are studied, the nuclear factor may be important.)

3.6.1 The Rotational Partition Function. The basic expression to be summed is

$$Z_{1r} = \sum_i g_{ir} e^{-\beta \epsilon_{ir}}, \tag{3.6.3}$$

where g_{ir} is the statistical weighting of the ith rotational level, including the already mentioned nuclear weighting factor, and ϵ_{ir} is its energy. It is appropriate to include the nuclear weighting factor in g_{ir}, rather than as a part of another factor in the total partition function because only the nuclear and rotational terms are linked by parity considerations, to be explained in what follows. For diatomic molecules, the rotational energy levels are

$$\epsilon_{rJ} = \frac{\hbar^2}{2\mathcal{I}} J(J+1), \tag{3.6.4}$$

where J is the rotational quantum number, and $\mathcal{I}$ is the moment of inertia of the molecule about an axis through its center of mass and perpendicular to the molecular axis. (There are two rotational degrees of freedom, about independent perpendicular axes of rotation.) Each level is $(2J+1)$-fold degenerate, so the statistical weighting factor for the Jth rotational level of a diatomic molecule with nonidentical nuclei is $g_{ns}(2J+1)$, and the rotational partition function is

$$Z_{1r} = g_{ns} \sum_{J=0}^{\infty} (2J+1) e^{-\beta \hbar^2 J(J+1)/2\mathcal{I}} = g_{ns} \sum_{J=0}^{\infty} (2J+1) e^{-\Theta_r J(J+1)/T}, \tag{3.6.5}$$

where g_{ns} is the nuclear spin weighting factor, given by $g_{ns} = (2s_1+1)(2s_2+1)$, s_i is the spin of the ith nucleus, and $\Theta_r = \hbar^2/2k\mathcal{I}$, the rotational characteristic temperature. In general, there is no simple expression for the sum. The difference in energy between adjacent levels is $\Delta\epsilon_J = \epsilon_J - \epsilon_{J-1} = 2Jk\Theta_r$, and the quantity $k\Theta_r$ is typically in the range 10^{-4}–10^{-3} eV. Since room temperature is near 0.03 eV, it is evident that room-temperature collisions are sufficiently energetic to excite the rotational energy levels. The usual approximations for Z_{1r} are: (1) at temperatures much less than Θ_r, only the first two terms in the sum are kept, giving the low-temperature approximation as $Z_{1r} = g_{ns}(1 + 3e^{-\Theta_r/T})$, and (2) at high temperatures, the sum can be approximated by an integral, as

$$Z_{1r} \approx g_{ns} \int_0^\infty e^{-J(J+1)\Theta_r/T}(2J+1)\,dJ = g_{ns}T/\Theta_r. \tag{3.6.6}$$

For diatomic molecules with identical nuclei, the Pauli Principle requires that the overall wave function be antisymmetric in the exchange of any two identical fermions, or symmetric in the exchange of two identical bosons. If the nuclei are identical and their spins are parallel, the nuclear factor in the total wave function is symmetric, no matter whether the nuclei are fermions or bosons. If the spins are antiparallel, the nuclear wave function is symmetric if it is the normalized sum of the two possible product wave functions, and it is antisymmetric if it is the normalized difference in these product functions. Molecules with symmetric nuclear wave functions are called *ortho* and those with antisymmetric nuclear wave functions are called *para*. For such a molecule with spin-1/2 nuclei, so that the spins can be designated as $+$ or $-$, the three symmetric nuclear wave functions are $\psi_1(+)\psi_2(+)$, $2^{-1/2}[\psi_1(+)\psi_2(-) + \psi_1(-)\psi_2(+)]$, and $\psi_1(-)\psi_2(-)$. The antisymmetric nuclear wave function is $2^{-1/2}[\psi_1(+)\psi_2(-) - \psi_1(-)\psi_2(+)]$. Evidently, in general, there are $g_{ns} = (2s+1)^2$ arrangements of the two nuclear spins, and in $2s+1$ of these, the nuclear spins are parallel, as in the first and third of the spin-1/2 symmetric wave functions. The remaining $2s(2s+1)$ configurations are equally divided into symmetric-antisymmetric pairs, like the second symmetric spin-1/2 wave function and the antisymmetric one. There are, then, $s(2s+1)$ antisymmetric, or *para* nuclear states, and $s(2s+1)+(2s+1) = (s+1)(2s+1)$ *ortho* states. For the para states, the rotational wave function must be symmetric (of even parity) in order for the overall wave function to be of odd parity, if the nuclei are fermions, and similar arguments apply to the ortho states. This restricts the rotational quantum number, J, to even integers for the para states, since the parity of the rotational wave function is even for even J and odd for odd J, and to odd integers for the ortho states. Thus, for spin-1/2 nuclei, the rotational partition functions are

$$Z_{1r}^{ortho} = 3\sum_{K=0}^\infty (4K+3)\,e^{-2\Theta_r(K+1)(2K+1)/T} = 3\left(3e^{-2\Theta_r/T} + 7e^{-12\Theta_r/T} + \cdots\right),$$

where $J = 2K+1$, and

$$Z_{1r}^{para} = \sum_{K=0}^\infty (4K+1)\,e^{-2\Theta_r K(2K+1)/T} = 1 + 5e^{-6\Theta_r/T} + \cdots,$$

where $J = 2K$. In the high-temperature approximation, since three-fourths of the states are ortho, but each series contains only half as many terms as in Eq. (3.6.5), the approximation for a diatomic molecule with identical spin-1/2 nuclei is, since $g_{ns} = 2j+1 = 2$,

$$Z_{1r}^{ortho} \approx 3T/2\Theta_r \quad \text{and} \quad Z_{1r}^{para} \approx T/2\Theta_r.$$

Under most experimental conditions, the transition between ortho and para molecules is extremely slow, with typical relaxation times of the order of days, since the nuclei are not strongly coupled to each other or to external influences. Therefore, the two types of molecule behave as if they are distinct species, each with its own thermodynamic properties. At high temperatures, the thermodynamic properties are essentially identical, since the partition functions are seen to differ only by a constant factor. At low temperatures, the partition functions differ significantly, and there are major thermodynamic consequences for temperatures such that $T < \Theta_r$. In practice, the only material approximating an ideal gas that is not yet condensed into a liquid or solid at usual laboratory conditions in this temperature range is hydrogen, which has the smallest value of $\mathcal{I}$, and its isotopes. Both the ortho and para forms of hydrogen can be prepared in the laboratory, and their thermodynamic properties agree well with the theoretical predictions implicit in the partition functions. In particular, C_v for parahydrogen is as much as 50% above the normal value for an equilibrium mixture, and for orthohydrogen it is up to 12% below normal at temperatures near 150K. Departures from normal are observed from 50K to above 300K.

Of more significance than the thermodynamic properties resulting from quantum-mechanical symmetry requirements are the spectroscopic data that support the predictions of the theory. Since ortho and para molecules behave as distinct species, at least for short times, the spectral lines produced by rotational transitions are distinct in the two species, and, in particular, their intensities are different because of the different sizes of the two populations. Since the population ratio of ortho to para is $(s + 1)/s$, the intensities of the alternating rotational-transition lines in the band spectrum are approximately in this same ratio. This kind of evidence is regarded as establishing without doubt the relationship of spin to statistics, at least for the particles that can be so observed. (Herzberg [1939]; Herzberg [1945])

For most diatomic molecules, the ortho-para distinction is either unimportant or nonexistent. The nuclear-spin degeneracy factor is $g_{ns} = (2s_1 + 1)(2s_2 + 1)$, and the high-temperature rotational single-molecule partition function is given by

$$Z_{1r} = (2s_1 + 1)(2s_2 + 1)T/\sigma\Theta_r, \tag{3.6.7}$$

with $\Theta_r = \hbar^2/2\mathcal{I}$, and σ is the symmetry factor, which measures the number of indistinguishable ways of orienting the molecule in space. (σ equals 1 plus the number of possible transpositions of identical atoms in the molecule. For example, $\sigma = 1$ for heteronuclear diatomic molecules and $\sigma = 2$ for homonuclear ones.) For linear polyatomic molecules at high temperatures, the rotational term is different only because there are more nuclear spins. For these linear molecules, Eq. (3.6.7) is generalized to give

$$Z_{1r} = \prod(2s_i + 1)T/\sigma\Theta_r. \tag{3.6.8}$$

For nonlinear polyatomic molecules, there are three principal moments of inertia, and three independent rotational degrees of freedom, rather than the two of diatomic molecules. The approximate treatment is similar to that of diatomic molecules, and the exact treatment, seldom of critical importance in statistical thermophysics, can be found in references on molecular spectroscopy. (Herzberg [1939]; Herzberg [1945]) The high-temperature single-molecule rotational partition function for nonlinear polyatomic molecules is

$$Z_{1r} = g_{ns}\left(\frac{\sqrt{\pi}}{\sigma}\right)\prod_{i=1}^{3}\left(\frac{T}{\Theta_{ri}}\right)^{1/2} = g_{ns}\left(\frac{\sqrt{\pi}}{\sigma}\right)\left(\frac{T}{\Theta_r}\right)^{3/2}, \tag{3.6.9}$$

where $\Theta_r = (\Theta_{r1}\Theta_{r2}\Theta_{r3})^{1/3}$, and σ is again the symmetry factor.

The contribution of molecular rotation to heat capacities and other thermodynamic properties will be evaluated after the other factors in the partition function are obtained.

3.6.2 The Vibrational Partition Function. A gas molecule of n atoms has $3n$ total degrees of freedom; three of these are translation of the center of mass; two, for linear, or three, for nonlinear molecules, are given to rotations about axes through the center of mass; and the rest are assigned to the normal-mode vibrations or approximations thereto. If the interatomic forces were harmonic, the normal-mode treatment would be exact, but since real molecules are not harmonically bound, the treatment is only approximate. If each normal mode is treated as a single harmonic oscillator, the vibrational partition function can be written, by use of the canonical partition function for such an oscillator in Eq. (3.3.19), as

$$Z_{1v} = \prod_i \left(1/2\sinh(\beta\hbar\omega_i/2)\right), \qquad (3.6.10)$$

where the product is over the normal modes, each with its characteristic ω_i. In accord with the treatment of rotations, it is common to define vibrational characteristic temperatures, given by $\Theta_i = \hbar\omega_i/k$. The vibrational single-molecule partition function is then written as

$$Z_{1v} = \prod_i \left(1/2\sinh(\Theta_i/2T)\right). \qquad (3.6.11)$$

Vibrational characteristic temperatures are typically in the thousands of Kelvins for diatomic molecules. For example, Θ_1 for H_2 is approximately 6300K; for O_2, 2250K; for N_2, 3350K; and for NO, 2700K. (These numbers are not of high precision, because the vibrations are not precisely harmonic, and the characteristic temperatures are dependent on the actual temperature, since the degree of anharmonicity is amplitude dependent.) The lowest characteristic vibrational temperatures for polyatomic gases are usually lower. For example, the two orthogonal bending vibrational modes of CO_2, which is linear and symmetric, have $\Theta_{v1} = \Theta_{v2} = 960$K; the asymmetric and symmetric stretching modes have $\Theta_{v3} = 1990$K, and $\Theta_{v4} = 3510$K, respectively. It is evident that there is a range of temperatures over which the vibrational modes contribute to the thermodynamic properties of molecules, and the contributions of the various modes can occur in quite different temperature regions. Usually there is only a small range of temperatures for which the vibrational modes are fully excited before dissociation and ionization occur.

3.6.3 The Electronic Partition Function. There is little that can be done with the contribution to the partition function arising from the energy levels of the electrons in the molecule. The electronic partition function remains in its basic form, *e.g.*,

$$Z_{1e} = \sum_n e^{-\beta\epsilon_n}, \qquad (3.6.12)$$

where ϵ_n is the nth electronic energy level of the molecule. There is usually no simplifying pattern of energy levels that permits the partition function to be expressed more compactly. The lowest excited electronic levels are typically several eV above the ground state, corresponding to temperatures of tens of thousands of K. For temperatures much less than ϵ_1/k, the electronic partition function can be approximated by its ground-state term, since the higher-energy terms are negligible. Even when only the ground state contribution is important, there can be the further point of possible degeneracy or fine structure, which is usually expressed as a ground-state degeneracy factor, $g_e(0)$. The electronic single-molecule partition function is thus written

$$Z_{1e} = \sum_\epsilon g_e(\epsilon)e^{-\beta\epsilon} \to g_e(0),\qquad(3.6.13)$$

where the approximation is valid when higher terms can be neglected, and the fine-structure levels of the ground state, if any, are more closely spaced than kT.

3.6.4 Calculation of Thermodynamic Properties. For a single-component system, the N-molecule ideal-gas partition function is given by Eq. (3.3.17) as

$$Z_N = Z_1^N/N!,$$

no matter how much internal structure the molecules may have. The Helmholtz potential is, by Eq. (3.3.7),

$$F = -kT\ln Z_N = -NkT[\ln(Z_1/N) + 1].\qquad(3.6.14)$$

The single-molecule partition function for linear molecules can now be approximated, from Eqs. (3.5.31), (3.6.8), (3.6.11), and (3.6.14), as

$$Z_1 = \left(\frac{g_{ns}g_e(0)}{\sigma}\right)\left(\frac{V}{\lambda^3}\right)\left(\frac{T}{\Theta_r}\right)\prod_i\left(\frac{1}{2\sinh(\Theta_i/2T)}\right),\qquad(3.6.15)$$

and for nonlinear molecules, from Eq. (3.6.9),

$$Z_1 = \left(\frac{\sqrt{\pi}g_{ns}g_e(0)}{\sigma}\right)\left(\frac{V}{\lambda^3}\right)\left(\frac{T}{\Theta_r}\right)^{3/2}\prod_i\left(\frac{1}{2\sinh(\Theta_i/2T)}\right).\qquad(3.6.16)$$

Since $dF = -SdT - PdV + \mu dN$, it follows that

$$P = -\left(\frac{\partial F}{\partial V}\right)_{T,N} = \frac{NkT}{V},$$

since only Z_{1t} contains V. The internal energy is found as

$$U = -\left(\frac{\partial \ln Z_N}{\partial \beta}\right)_{V,N}$$
$$= \frac{3}{2}NkT + \frac{2}{2}NkT + N\sum_i\left(\frac{\hbar\omega_i/2}{\tanh(\beta\hbar\omega_i/2)}\right),\qquad(3.6.17)$$

for linear molecules, and

$$U = -\left(\frac{\partial \ln Z_N}{\partial \beta}\right)_{V,N}$$
$$= \frac{3}{2}NkT + \frac{3}{2}NkT + N\sum_i\left(\frac{\hbar\omega_i/2}{\tanh(\beta\hbar\omega_i/2)}\right),\qquad(3.6.18)$$

for nonlinear ones. The first term is from translation, the second from rotation, and the third (summed) term is from the vibrational modes, of which there are $3n - 3 - r$, where n is the number of atoms in the molecule, and r is 2 for linear molecules or 3 for nonlinear ones. The vibrational energies evidently go from 0 at $T = 0$ to kT for $T \gg \Theta_i$. The molar heat capacity at constant volume is found from the energy equations to be

$$C_v = \frac{5}{2}R + R \sum_i \left(\frac{\Theta_i/2T}{\sinh(\Theta_i/2T)} \right)^2, \qquad (3.6.19)$$

for linear molecules, or

$$C_v = 3R + R \sum_i \left(\frac{\Theta_i/2T}{\sinh(\Theta_i/2T)} \right)^2, \qquad (3.6.20)$$

for nonlinear ones.

The chemical potential is found from Eq. (3.6.14) to be

$$\mu = \left(\frac{\partial F}{\partial N} \right)_{T,V} = -kT \ln(Z_1/N), \qquad (3.6.21)$$

from which it is evident that all factors of the partition function contribute to the chemical potential.

For a mixture of ideal gases, Eq. (3.6.2) permits the Helmholtz potential to be written

$$F = -kT \sum_i N_i \left(\ln \frac{Z_{1i}}{N_i} + 1 \right). \qquad (3.6.22)$$

This equation shows that the total entropy and pressure of the mixture are the sums of the separate contributions of the several gases. The pressure equation, for example, is

$$P = -\left(\frac{\partial F}{\partial V} \right)_{T,N_i} = \sum_i \left(\frac{N_i kT}{V} \right) = \sum P_i, \qquad (3.6.23)$$

where the P_i's are called the partial pressures of the component ideal gases, and each is independent of the presence of the other gases. The chemical potential of any one component gas is also independent of the presence of the others, or

$$\mu_j = \left(\frac{\partial F}{\partial N_j} \right)_{T,V,N_i \neq N_j} = -kT \ln \frac{Z_{1j}}{N_j}. \qquad (3.6.24)$$

These results will be used in the next section.

3.7 CHEMICAL EQUILIBRIUM

For an equilibrium mixture of the diatomic molecules X_2, Y_2, and XY, as indicated in Section 2.10, at constant temperature and volume, a Helmholtz-potential minimum requires that

$$\mu_1 + \mu_2 - 2\mu_3 = 0, \qquad (3.7.1)$$

where the subscripts refer to the molecules in the order listed. The corresponding chemical reaction is

$$X_2 + Y_2 \Leftrightarrow 2XY,$$

as shown, where the stoichiometric coefficients are read directly from Eq. (3.7.1). Substitution of the chemical potentials from Eq. (3.6.24) gives

$$-kT \left[\ln \left(\frac{Z_{11}}{N_1} \right) + \ln \left(\frac{Z_{12}}{N_2} \right) - 2 \ln \left(\frac{Z_{13}}{N_3} \right) \right] = 0, \qquad (3.7.2)$$

or, as a direct consequence,

$$\frac{Z_{13}^2 N_1 N_2}{Z_{11} Z_{12} N_3^2} = 1,$$

which says that at equilibrium,

$$\frac{N_1 N_2}{N_3^2} = \frac{Z_{11} Z_{12}}{Z_{13}^2}.$$

Since the partial pressures of Eq. (3.6.23) are proportional to the N_i's at fixed volume and temperature, the equilibrium condition can be written in terms of the partial pressures as

$$\frac{P_1 P_2}{P_3^2} = \frac{Z_{11} Z_{12}}{Z_{13}^2} = K(T), \qquad (3.7.3)$$

where the *equilibrium constant*, $K(T)$, is a function of T alone, since the only other dependence is on V, and the volume factors in the partition functions quite generally can be combined with the N_i's on the other side of the equation to give the partial pressures. (For this example, they cancel anyway, but not in general.) The relationship stated in Eq. (3.7.3) is an example of the *law of mass action*, which is used to determine the chemical composition of a system with several possible chemical species, in terms of the pressure and temperature. For a gas mixture in which several reactions can take place simultaneously, as in Eq. (2.10.3), the law of mass action can be derived for each reaction by a procedure similar to the one used to obtain Eq. (3.7.3). The procedure for reactions of polyatomic molecules is identical to that of diatomic molecules, except for details of the partition functions. (Tribus [1961a])

The fact that equilibrium concentrations of the various chemical species present in a gas mixture can be determined uniquely and unambiguously does not mean that the actual composition of such a mixture is the one for equilibrium. For example, a mixture of H_2 and O_2 can exist quietly and uneventfully, but not necessarily safely, for many years at room temperature and atmospheric pressure, although such a mixture is not at all the one specified by the law of mass action. Such mixtures are called metastable. The H_2 molecule is in a bound state with energy much below that of the dissociated hydrogen atoms. As a diatomic molecule, it is not chemically very active, nor is the O_2 molecule. On the other hand, atomic hydrogen and oxygen *are* chemically active; they find it energetically favorable to combine into diatomic or larger molecules, with the release of energy that shows up as radiation or kinetic and potential energy of the molecules. The details of equilibration in chemical reactions are sufficiently complicated to preclude a comprehensive treatment here, and they are often covered over by the remark that an activation energy is usually required to stimulate the original molecules out of their bound states and into a condition in which reactions toward equilibrium can proceed at an easily observable rate. When the activation energy ϵ is such that $\beta\epsilon \gg 1$, the reaction rate, which is proportional to the Boltzmann factor, $e^{-\beta\epsilon}$, is quite slow, but as the temperature rises, the rate can increase to the point that energy is released by the reacting molecules fast enough to provide further heating of the gases, and the reaction rate can become so high that a chemical explosion can occur.

3.7.1 Dissociation, Ionization, and the Saha Equation. As the temperature rises, molecular collisions become more energetic, and at temperatures of the order of a few thousand K, diatomic molecules dissociate into atoms. At even lower temperatures, polyatomic molecules begin to come apart, and at temperatures of the order of a few tens of thousands K, most atoms are at least singly ionized. The resulting dissociation and ionization equilibria can be studied in a way that is not different in principle from the procedure used for chemical equilibrium.

The binding energy of a typical diatomic molecule is of the order of 5 eV, which means that the ground-state electronic energy level of the diatomic molecule is about 5 eV below the sum of the ground-state electronic energy levels of the separated atoms. For a dissociation reaction of the type

$$X_2 \Leftrightarrow 2X,$$

conservation of atoms requires that $2N_{X_2} + N_X = N$, where N is the (constant) total number of atoms of X, either bound into molecules or single. At constant volume and temperature, a Helmholtz-potential minimum requires that $\mu_{X_2} dN_{X_2} + \mu_X dN_X = 0$, or $(\mu_{X_2} - 2\mu_X)dN_{X_2} = 0$. By use of Eq. (3.6.24) and the equilibrium condition on the chemical potentials, it follows that

$$\frac{N_X^2}{N_{X_2}} = \frac{Z_{1X}^2}{Z_{1X_2}},$$

or, by multiplication of each N_i by kT/V, the resulting equation for the partial pressures is

$$\frac{P_X^2}{P_{X_2}} = \frac{kT}{V}\frac{Z_{1X}^2}{Z_{1X_2}}.$$

Under the conditions that the temperature is high enough to cause some dissociation, but not high enough to produce a significant population of excited electronic states, the single-atom partition function can be written

$$Z_{1X} = \frac{V g_{0X}}{\lambda_X^3}(2s+1)e^{-\beta\epsilon_{0X}},$$

where s is the nuclear spin, $2s+1$ is the nuclear-spin degeneracy, and λ_X is the thermal wavelength of the atoms, and ϵ_{0X} is the ground-state electronic term. With the further assumptions that the temperature is much greater than the characteristic rotational and vibrational temperatures of the molecule, but not high enough to produce a significant population of excited molecular electronic states, the single-molecule partition function can be approximated as

$$Z_{1X_2} = \frac{V g_{0X_2}}{\lambda_{X_2}^3}\frac{(2s+1)^2}{2}\left(\frac{T}{\Theta_r}\right)\left(\frac{T}{\Theta_v}\right)e^{-\beta\epsilon_{0X_2}},$$

where ϵ_{0X_2} is the ground-state electronic energy of the molecule, and the g's are the ground-state degeneracies. With these partition functions, the equilibrium relation for the pressures becomes

$$\frac{P_X^2}{P_{X_2}} = \left(\frac{kT}{V}\right)\left(\frac{V}{\lambda_X^3}\right)^2\left(\frac{\lambda_{X_2}^3}{V}\right)\left(\frac{(2s+1)^2 g_{0X}^2 e^{-\beta\epsilon_{0X}}}{(2s+1)^2 g_{0X_2} e^{-\beta\epsilon_{X_2}}}\right)\left(\frac{\Theta_r}{T}\right)\left(\frac{\Theta_v}{T}\right).$$

After substitution of the expressions for the λ's, the recognition that $m_{X_2} = 2m_X$, and the use of the dissociation energy ϵ_d, where $2\epsilon_d = 2\epsilon_{0X} - \epsilon_{X_2}$, the pressure equation may be rewritten as

$$\frac{P_X^2}{P_{X_2}} = kT\left[\left(\frac{\Theta_r\Theta_v}{T^2}\right)\left(\frac{g_{0X}^2}{g_{0X_2}}\right)\left(\frac{\pi m_X kT}{h^2}\right)^{3/2} e^{-2\beta\epsilon_d}\right]. \tag{3.7.4}$$

The partial pressures may be expressed in terms of the degree of dissociation, ζ, which is defined by

$$\zeta = \frac{N_{X_2}^0 - N_{X_2}}{N_{X_2}^0} = \frac{N_X}{2N_{X_2}^0},$$

so that $P_{X_2} = (1 - \zeta)P_0$, and $P_X = 2\zeta P_0$, where P_0 is the pressure of the undissociated gas. With this substitution, Eq. (3.7.4) becomes

$$\begin{aligned}\frac{\zeta^2}{1-\zeta} &= \frac{kT}{4P_0}\left[\left(\frac{\Theta_r\Theta_v}{T^2}\right)\left(\frac{g_{0X}^2}{g_{0X_2}}\right)\left(\frac{\pi m_X kT}{h^2}\right)^{3/2} e^{-2\beta\epsilon_d}\right]\\ &= \frac{1}{4n_{X_2}^0}\left[\left(\frac{\Theta_r\Theta_v}{T^2}\right)\left(\frac{g_{0X}^2}{g_{0X_2}}\right)\left(\frac{\pi m_X kT}{h^2}\right)^{3/2} e^{-2\beta\epsilon_d}\right],\end{aligned} \tag{3.7.5}$$

where $n_{X_2}^0$ is the number of molecules per unit volume before any were dissociated. The only dependencies on the right-hand side of the equation are on temperature and initial density. When $\zeta \ll 1$, the approximate dependence is $\zeta \approx A(n_{X_2}^0)^{-1/2}T^{-1/4}e^{-\beta\epsilon_d}$, where A is a constant. Thus the fractional dissociation increases rapidly with temperature at the onset, and decreases slowly with density. As $\zeta \to 1$, the fractional dissociation becomes much less strongly dependent on temperature. Equation (3.7.4) is another example of the law of mass action.

At still higher temperatures, further electronic excitation and ionization occur. These processes can be treated in much the same manner. It should be pointed out that in many laboratory situations for which these developments are valid, the electrons are not in thermal equilibrium with the gas atoms and molecules. In the familiar plasmas of glow discharges (neon signs), for example, the container is warm to the touch, the atoms, ions, and molecules are in a thermodynamic steady state that approximates equilibrium, at a temperature near that of the container walls. The electrons, both free and bound, are also in a state of near equilibrium, with their own temperature, which is determined by the rate at which they acquire energy from the applied electric field, and the rate at which they lose energy to the more massive particles. The energy lost by an electron in an elastic collision with an atom is a small fraction of its initial energy, but electrons redistribute energy quite well among themselves. In an inelastic collision with an atom or molecule, energy is transferred from the free electron to a bound electron, resulting in an electronic-state excitation, or an ionization. It is typical to find laboratory plasmas of this sort with electron temperatures of 10 eV (116,000K), in glass tubes that would melt if the gas temperature approached 1000K.

For the process of ionizing an already n-times ionized atom, X_n, to the state of being $(n + 1)$-times ionized, the reaction of interest is

$$X_n \Leftrightarrow X_{n+1} + e.$$

For each such reaction, there is a separate and independent condition on the chemical potentials for a Helmholtz- or Gibbs-potential minimum. For this reaction, since $N_n + N_{n+1} = const.$, and since $N_{n+1} = N_e$, the equilibrium condition $\sum \mu_i dN_i = 0$ gives

$$(\mu_n - \mu_{n+1} - \mu_e)\, dN_n = 0.$$

From this it follows in the usual way that

$$\frac{N_n}{N_{n+1} N_e} = \frac{Z_{1n}}{Z_{1,n+1} Z_{1e}},$$

and only the appropriate partition functions are required in order to obtain the Saha equation for ionization equilibrium. The translational factors of Z_{1n} and $Z_{1,n+1}$ cancel, since the masses are almost the same, and the nuclear degeneracy factors also cancel. The free electron partition function is taken to be that of an ideal gas with the electron mass and a twofold spin degeneracy. The only remaining problem is that of the electronic factors in the two ionic partition functions.

The electronic partition function for a single ion that has lost n electrons may be written

$$Z_{1n}^{el} = \sum g_i e^{-\beta \epsilon_i},$$

where the ϵ_i are the electronic energy levels of the n-times ionized atom. If the ion's ground-state energy term is factored out, the partition function may be restated as

$$Z_{1n}^{el} = e^{-\beta \epsilon_0} \sum g_i e^{-\beta (\epsilon_i - \epsilon_0)} = e^{-\beta \epsilon_0} \xi_n(T),$$

where

$$\xi_n(T) = g_{n0} + g_{n1} e^{-\beta \eta_1} + g_{n2} e^{-\beta \eta_2} + \cdots,$$

with $\eta_i = \epsilon_i - \epsilon_0$. The ionization potential of the nth ion is $\phi_n = \epsilon_{0\,n+1} - \epsilon_{0\,n}$. With these notational conventions, the Saha equation follows, as

$$\frac{n_{n+1} n_e}{n_n} = 2\frac{\xi_{n+1}}{\xi_n} \left(\frac{2\pi m_e kT}{h^2} \right)^{3/2} e^{-\beta \phi_n} = K_{n+1}(T). \qquad (3.7.6)$$

The evaluation of the ξ-functions is not always easy, but because of the wide spacings typical of electronic levels, it is often unnecessary to sum more than the first few terms in these functions for a satisfactory approximation. The Saha equation is important in plasma physics, astrophysics, spectroscopy, and the study of flames and shock waves. A satisfactory treatment of the applications would take up too much space in this book. The equation was presented because it is widely used but often seemingly plucked from the air, rather than derived in the texts that use it. Since it is usual to assume that the derivation is a part of the reader's background, it seems appropriate to make it so.

3.8 COMMENTARY

The postulational approach to thermostatics, presented in detail by Tisza [1966] and Callen [1985], is widely accepted, and Callen's excellent book on the subject is a standard reference work, as well as a textbook. As both authors have emphasized, the accepted results of equilibrium macroscopic thermodynamics follow plausibly and efficiently from the axioms given in Chapter 2, and the calculations pertinent to a particular physical system are found to give the correct results, once the fundamental relation of each of its subsystems is known. There was originally a body of resistance to their presentation, with claims that it is unphysical or too formalistic. In time, however, the clarity and power of their approach has led to general acceptance of the axiomatic formulation of, as Tisza expresses it, the *macroscopic thermodynamics of equilibrium.*

Thermodynamics has always been a theory in close contact with measurement. In a real and important sense, the successes of the axiomatic formulation of thermostatics can be attributed to the clearly evident way in which the results of measurement are incorporated into the theoretical analysis, leading to revision of the fundamental relation, when prediction and experiment diverge, without damage to the underlying theoretical structure. This feature is an esthetically satisfying attribute of a mature science.

The Jaynes information-theoretic approach to the microscopic thermodynamics of equilibrium, in my opinion, offers exactly the same advantages that are found in the Gibbs-Tisza-Callen treatment of macroscopic thermodynamics of equilibrium. (Jaynes [1957a]; Jaynes [1957b]) There is one additional axiom needed to supplement those of Chapter 2:

6. *The equilibrium information-theoretic entropy, maximized subject to the constraints imposed by our knowledge of the system, is exactly the entropic fundamental relation of the Tisza-Callen formulation.*

Here, then, is offered a single axiom to be appended to the already successful set of Tisza and Callen, that provides a unified axiomatic basis for the macroscopic and microscopic thermodynamics of equilibrium. Both structures are based on a variational principle, which is a characteristic feature of theoretical physics. Both permit revision of the fundamental relation (the maximized entropy), based on macroscopic experimental evidence. And both demonstrably lead to the same macroscopic results in every case that follows from an agreed-upon description of the system. Moreover, this added axiom leads to the standard results of equilibrium statistical mechanics and offers a direct formalism for the treatment of nonequilibrium systems. There are other advantages, such as the avoidance of ergodic theory as an attempt to provide a sound basis for statistical mechanics, the automatically provided extra terms that are needed to account for surface and edge effects when they are present, and the treatment of fluctuation phenomena, that further recommend this theoretical structure. Furthermore, there are no calculational or analytical disadvantages in its use. I therefore fail to see why there should be any pragmatic objection to it, although I recognize the residual abhorrence caused by misinterpretation of *subjective* as used by Jaynes. Essentially the

same concepts were used by Tribus in his demonstrations that the laws of classical thermodynamics could be defined in terms of information-theoretic entropy, and his presentation
was found objectionable by most of the community of thermodynamicists, including those
that have come to accept Callen's axiomatization (Costa de Beauregard and Tribus [1974];
Tribus [1961a]; Tribus [1961b]; Tribus, Shannon and Evans [1966]). By the overt and
explicit assertion that the information-theoretic maximum entropy is exactly that of the
Tisza-Callen formulation, with all of its consequences, I hope to contribute to the diminu-
tion of the communication gap that persists within the field of thermal physics. A careful
treatment of the foundations of the subject, with emphasis on the interplay of statistical
entropy maximization and measurement of macroscopic thermodynamic parameters is given
by Grandy [1987].

3.8.1 The Thermodynamic and Continuum Limits.

It is often useful to examine the
consequences of a theory or a calculation in the limit that certain parameters vanish or
become infinite. As examples, the classical limit of a quantum-mechanical result is found by
setting h equal to zero, and the nonrelativistic limit is found by setting c to be infinite. Two
such limits are of particular use in statistical thermophysics: the so-called *thermodynamic
limit* and the *continuum limit*. These limits are discussed briefly in this subsection.

The thermodynamic limit is the best known and most widely used of the two. Applied
to a homogeneous system, it may be characterized as the limit of an infinite system with
the same densities and intensive parameters as the idealized finite system being studied.
The volume becomes infinite, while the volume per particle remains constant, as do the
energy density, mass density, temperature, pressure, and chemical potential. These are not
all independent, and it is not, in general, necessary to set the limits and restrictions as
if they were so. For example, in the canonical system, it is usually necessary to set only
the volume per particle and the temperature constant as the volume (or the number of
particles) becomes infinite. The thermodynamic limit eliminates edge and surface effects
(for three-dimensional systems) and often permits the rigorous evaluation of the partition
function and the thermodynamic properties if the system. Another advantage of going to
the thermodynamic limit may be seen in the treatment of noninteracting free particles, for
which the momentum and energy values are quantized by the geometry of the container, and
there are problems with low-energy particles that have thermal wavelengths that may exceed
the dimensions of the container. In the thermodynamic limit, the momentum values become
a continuum, as do the energy levels, and sums can be replaced rigorously by integrals. In
Chapter 7, evaluation of some of the integrals in the theory of phase transitions requires
that the system be infinite. A comparison of finite and infinite systems is to be found in
Problems 9 and 10 of this chapter.

The thermodynamic limit leads to rigorous mathematical results, at the expense of
physical reality. It seems unsatisfactory to require the universe in order to calculate the
pressure in a thimblefull of gas, no matter how noble and noninteracting. It must be re-
membered that the objective of mathematical rigor, or even of calculability, often provides
results that only approximately correspond to the reality of the laboratory. The approximation is usually quite good, because Avogadro's number is suitably large, and for this
reason the thermodynamic limit provides a useful tool. In at least one class of problems,
however, the thermodynamic limit is not useful: systems in which the potential energy
of interaction increases other than linearly with the volume of the system. Examples of
such systems are those that have a uniform, nonzero electrical charge density, or systems

where the gravitational interaction of the particles cannot be ignored. These are not *normal* systems; the technique of seeking rigor in the thermodynamic limit works only for systems that are classified as normal—*i.e.* for systems that do not have energy terms that vary in any way other than linearly with volume. (Landsberg [1984])

It is sometimes more attractive to examine systems in the continuum limit, in which the quantities of macroscopic thermodynamics remain invariant, but the underlying microstructure loses its graininess and attains the continuous, smooth, homogeneous texture of the hypothesized medium in classical continuum mechanics. The density, pressure, temperature, energy density, local flow velocity, and other such macroscopically accessible quantities are unchanged, but the underlying per-particle quantities vanish in a prescribed way. Use of the continuum limit often clarifies the interrelationship of thermodynamics and statistical mechanics. (Thermodynamics was developed originally with no reference to the microstructure of matter, and many of its practitioners resisted, on principle, any attempt to dilute the power and generality of their discipline by resorting to what they regarded as fictitious models of unseen and unnecessary atomicity.)

The formalities of approaching the continuum limit are somewhat dependent upon the details of the system being studied, but some general procedures are evident. Typically, the volume is kept constant, as is the density, so the number of particles must become infinite as the particle mass vanishes, in such a way that mN remains constant. The energy per particle is typically proportional to kT, and the energy density to NkT/V. This implies that $k \to 0$ but Nk remains constant. Similar considerations can be invoked to characterize the limiting process for whatever system is being considered.

An example given by Compagner considers a monatomic fluid subject to an external field. Here the treatment is for three dimensions, whereas Compagner presents a more general version in d dimensions. The external potential energy of a particle is given by the expression $w_1(\mathbf{r}_1) = \epsilon_1 \phi_1(\mathbf{r}_1)$, where ϕ_1 is a dimensionless shape factor, and ϵ_1 is the microscopic coupling constant between the particle and the external field. In addition, a pair potential is assumed, of the form $w_2(r_{ij}) = \epsilon_2 \phi_2(r_{ij}/\ell)$, where r_{ij} is the particle separation, ϵ_2 is the coupling strength, and ℓ is a scaling length such that $N\ell^3$ is constant. The passage to the continuum limit is then characterized by the requirement that $k \to 0$, subject to the conditions that these quantities are kept constant: k/ℓ^3, m/ℓ^3, $N\ell^3$, $N_0\ell^3$, ϵ_1/ℓ^3, ϵ_2/ℓ^3, h/ℓ^4, λ/ℓ, v, c, ϕ_1, ϕ_2, V, and T, where N is number of particles in the constant volume V (and N becomes infinite), N_0 is Avogadro's number, λ is the thermal wave length, v is the flow velocity, and c is the speed of light. The condition on Planck's constant may be obtained from the requirement that the number of particles per cubic thermal wavelength remains constant. This vanishing of h is not the equivalent of the classical limit, for which h vanishes without constraint.

In most cases of comparison, the continuum and thermodynamic limits are in agreement on the calculated macroscopic parameters. The continuum limit offers the advantage that it can be applied to nonequilibrium systems and abnormal systems, such as those with matter flows and thermal energy flows, or with energy terms that do not vary linearly with volume, but which exhibit locally uniform conditions that are characterized as local thermodynamic equilibrium. This means, among other things, that local temperatures, pressures, (or the tensor generalizations), energy densities, entropy densities, mass densities, and flow velocities can be defined, even though these quantities are not constant throughout the system. The thermodynamic limit cannot treat such systems.

As a simple example, consider the monatomic ideal gas in a canonical system. The partition function is already known to be

$$Z_N = \frac{1}{N!} \left(\frac{V}{\lambda^3} \right)^N, \tag{3.8.1}$$

from which the thermodynamics emerges. The internal energy, $U = 3NkT/2$, a typical extensive property, keeps the energy per particle, U/N, constant in the thermodynamic limit, and keeps U constant in the continuum limit, since in that limit, Nk is constant. The Helmholtz potential is only slightly more complicated:

$$F = -kT \ln Z_N = -NkT \left[1 + \ln(V/N\lambda^3) \right]. \tag{3.8.2}$$

In the thermodynamic limit, it is clear that F/N remains constant as $N \to \infty$ with N/V constant. In the continuum limit, the only new feature is that $N\lambda^3/V$, the number of particles per cubic thermal wavelength, remains constant. The pressure, $P = NkT/V$, behaves in the thermodynamic limit, with N/V constant as N becomes infinite, or in the continuum limit, with Nk constant as k vanishes. The chemical potential (per particle), $\mu = kT \ln(N\lambda^3/V)$, remains constant in the thermodynamic limit, where the particle density is invariant, and the quantity $N\mu$ stays constant in the continuum limit. Other examples are easy to examine with this guidance.

The excellent paper by Compagner [1989] provides much more detail, and several carefully developed examples of calculations using limiting techniques.

3.8.2 The Gibbs Paradox.   When Willard Gibbs attempted to calculate the partition function for an ideal gas (Gibbs [1960]), he obtained an expression equivalent to Eq. (3.2.29) without the $N!$ in the denominator. His result leads to the correct ideal gas law and the correct energy, but his entropy was not equivalent to a homogeneous first order function of the extensive parameters. From consideration of the entropy of mixing, as in Section 2.9, he recognized that he had to introduce the $N!$ into the denominator of his version of Ω, as was done in Eq. (3.2.28). A corresponding factor was introduced into Eq. (3.3.17), after some discussion, but with no real demonstration that the underlying states of a system of N identical particles necessitate this factor. There are those who believe that the $N!$ can be placed on a sound theoretical basis without invoking quantum mechanics (van Kampen [1984]), just as the entropy-of-mixing analysis led to the correct N-dependence of the thermodynamic entropy. I find these analyses quite difficult to follow, and I always leave them with a nagging feeling that I may be the victim of some sort of chicanery. I therefore do not categorically deny the possibility of a classical explanation, but I claim that if the result were not known *a priori*, the classical explanation would still be lacking. The quantum-theory-based appearance of the $N!$ is both natural and easy to understand. It is presented in the commentary section of Chapter 4, and it is examined from a different point of view in Appendix C.

The paper by van Kampen offers the following criticism of the way Gibbs contrived to introduce the required correction factor: "Gibbs argued that, since the observer cannot distinguish between different molecules, 'it seems in accordance with the spirit of the statistical method' to count all microscopic states that differ only by a permutation as a single one. Actually it is exactly opposite to the basic idea of statistical mechanics, namely that the probability of a macrostate is given by the measure of the Z-star, i.e., the number of corresponding, macroscopically indistinguishable microstates....Unfortunately, Planck revived the Gibbs mystique and this view seems to dominate the literature." (Planck [1923]; Planck [1932])

As an illustration of the usual weighting of macrostates in statistical theories, consider the dog-flea model in Chapter 1. With a total of N indistinguishable fleas, the number of ways dog A can have N fleas is 1. It can have $N-1$ fleas in N ways, $N-2$ fleas in $N(N-1)/2$ ways, *etc.*, the binomial coefficients. One does not say, since the fleas are indistinguishable, that there is only one state for which dog A has $N-1$ fleas and dog B has 1 flea, because identical, indistinguishable fleas are *individuals*, that can be localized and isolated. The fleas exist on the dogs as individuals, each with its own separate identity. They are indistinguishable in the sense that if I remove one of them from a dog, examine it, return it to its home, and sometime later repeat the process, I cannot tell whether I have caught the same flea or another. Perhaps I merely choose not to distinguish them. The classical view of ideal gas molecules is no different from that of fleas: individual but identical and indistinguishable members of an N-body system that is called a gas.

Gibbs and Planck arrived at their *ad hoc*, and obviously needed, correction factor before the advent of quantum theory. They offered what seemed to be plausible words to justify what had to be done mathematically. It is surprising that their words are still used to describe the many-body states that emerge in the quantum-mechanical treatment of the problem. As is shown specifically in the commentary section of Chapter 4 and discussed in Appendix C, the symmetry properties of the Hamiltonian, and the fact that it commutes with the particle-interchange operator for identical particles, lead to the description of the many-body system in terms of many-body states that are properly symmetrized composites of single-particle states, in which every particle is associated with every occupied single-particle state. It is as if there are N fleas on M dogs, with a macrostate being described as having each classical flea replaced by a group essence, a superposition of all the fleas! It seems to me that the formalism describes the N-body state as an entity into which the particles have melded, like drops of water falling into a pool, and do not exist as individuals. Upon extraction from the N-body state, a single particle may exist as an entity, an individual, but it was not so before extraction, and it will not be so upon return to the system.

A familiar example of this sort of behavior is that of an electron in a beta-decay process. The electron is emitted by the nucleus, but it was not ever *in* the nucleus. The nucleus as an entity undergoes a transformation to a different nuclear entity, and the beta particle emerges, itself as an entity. Electrons in an atom or molecule could be similarly regarded, as could systems in general consisting of assemblages of identical particles, subject to the appropriate symmetry conditions.

The van Kampen [1984] paper discusses the entropy-of-mixing problem in the situation that the gases are so similar that the experimenter has no convenient way to distinguish them. He then concludes that there is no entropy change upon mixing, and that the entropy is correctly given by the revised entropy formula of Section 2.9, or by the Sackur-Tetrode equation, with $N = N_a + N_b$. According to van Kampen, and with which I agree, "The point is, *this is perfectly justified* and that he will not be led to any wrong results.... The expression for the entropy (which he constructs by one or the other processes mentioned above) depends on whether or not he is willing or able to distinguish between molecules A and B. This is a paradox only for those who attach more physical reality to the entropy than is implied by its definition." In other words, he seems to say that the entropy of a system is not an unambiguously measured or intrinsically existing quantity, as is its mass, but rather it depends upon the information used by the experimenter in defining it. This is exactly my point of view, and that of information theory.

It is therefore surprising that he makes the following statement, two paragraphs later, in discussing the probability distribution in phase space, which determines the entropy: "It is *not* determined by the information or lack of information of the observer, or by a fair representation of his ignorance and other anthropomorphic considerations. No-one seriously believes that statistical mechanics ceases to be valid when he is not looking." He sets up a straw man and ridicules his position as an argument against the use of information theory, after having expressed the real and valid viewpoint of this formulation in his previous discussion. Surely no sincere and thoughtful advocate of the information-theoretic approach to statistical mechanics has ever even hinted that his subject ceases to exist when he is not looking. Such arguments do nothing to resolve the disparate points of view of the two camps, and they stand as a barrier to serious discussion.

3.8.3 Fugacity and Activity.

A term frequently seen in texts on statistical mechanics and chemical thermodynamics is *fugacity*. It was originally proposed by G. N. Lewis (Guggenheim [1967]; Lewis [1907]) as a generalization of the property of pressure as the tendency to escape. (Fugacity is derived from the Latin *fugere*–to flee.) Lewis considered other possibilities, such as the chemical potential, but finally chose the fugacity P^* to be defined by the relation

$$g = RT \ln P^* + b(T), \tag{3.8.3}$$

where g is the molar Gibbs potential. (This is not standard notation for the fugacity, but f is already pre-empted, and the notation P^* emphasizes that the fugacity has the units of pressure.) If $b(T)$ is written as $b(T) = -RT \ln P_0(T)$, where P_0 is a function of T alone, having the units of pressure, then the defining equation may be rewritten

$$g = RT \ln(P^*/P_0). \tag{3.8.4}$$

Lewis further required that at constant T, $P^* \to P$ as $v \to \infty$. This requirement can be utilized to obtain expressions for P^* by use of Eq. (2.5.19) to write

$$v = \left(\frac{\partial g}{\partial P} \right)_T = RT \left(\frac{\partial \ln P^*}{\partial P} \right)_T,$$

or, along an isotherm,

$$\frac{dP^*}{P^*} = \frac{v}{RT} dP. \tag{3.8.5}$$

For a perfect gas, evidently $dP^*/P^* = dP/P$, so the requirement that $P^* \to P$ as $P \to 0$ leads to the unambiguous identification for this case that $P^* = P$.

In general, the fugacity may be calculated by integration of Eq. (3.8.5). As an illustration, for the van der Waals gas,

$$dP = \left[-\frac{RT}{(v-b)^2} + \frac{2a}{v^3} \right] dv,$$

so that integration along an isotherm gives

$$\ln[P_2^*(v_2 - b)] - \frac{b}{v_2 - b} + \frac{2a}{RTv_2} = \ln[P_1^*(v_1 - b)] - \frac{b}{v_1 - b} + \frac{2a}{RTv_1} = C(T).$$

As $v_2 \to \infty$, it is required that $P_2^* \to RT/v_2$, so $P_2^*(v_2 - b) \to RT$, and $C(T) = \ln RT$. Therefore, since the other end point can be any point on the isotherm, the subscript will be dropped, giving the van der Waals fugacity as

$$P^* = \frac{RT}{v - b} \exp\left(\frac{b}{v - b} - \frac{2a}{RTv}\right). \tag{3.8.6}$$

There is sometimes defined a *fugacity coefficient*, ν, defined as

$$\nu = P^*/P,$$

the ratio of the fugacity to the pressure. It is a measure of the nonideality of the substance. The fugacity coefficient is sometimes confused with the *relative fugacity*, also known as the *activity*, defined by

$$a = P^*/P_s^*, \tag{3.8.7}$$

where P_s^* is some arbitrarily defined standard, or reference, value of the fugacity. This definition is equivalent to choosing some particular reference point for the zero of the molar Gibbs potential, or

$$g_s = 0 = RT\ln(P_s^*/P_0),$$

so that

$$g - g_s = RT\ln(P^*/P_s^*) = RT\ln a. \tag{3.8.8}$$

For a pure substance, the molar Gibbs potential is equivalent to the molar chemical potential. The use of a relative value, referred to some arbitrary reference level, for the chemical potential results in a relative activity a, given by

$$a = e^{(\mu - \mu_s)/RT}, \tag{3.8.9}$$

which begins to show its relationship to the absolute activity, z, defined following Eq. (3.4.3) to be

$$z = e^{\beta\mu}.$$

There are two differences between z and a: The chemical potential in z is for one molecule, whereas that in a is for one mole, so the latter is Avogadro's number times the former, and a is expressed in terms of a chemical potential referred to an arbitrary zero, where z is expressed in terms of an absolute value of μ, derived from the underlying microphysics. Examples of this absolute value are to be found in Eq. (3.4.13), for the ideal gas, and in Eqs. (4.5.4) and (4.5.5) for an ideal fermion gas. For most purposes in chemical thermodynamics, there is no advantage in choosing between a and z.

The frequently seen statement that z is the fugacity is wrong in two senses. First, z is dimensionless, whereas P^* has the units of pressure. Second, from Eq. (3.8.7) the activity a can be set numerically equal to P^* if P_s^* is chosen as unity. This implies that the reference fugacity is set equal to one something (usually one atmosphere), so that there is numerical equivalence. But clearly, even if a is numerically equal to P^*, such an arbitrary choice of reference point cannot result in the general equivalence of a and z. These distinctions are wasted on those who regard density and specific gravity as equals, conceptually and numerically.

3.9 REFERENCES

A few of the principal references are listed here; all references are listed in the Bibliography at the end of the book.

CALLEN, H. B. [1985], *Thermodynamics and an Introduction to Thermostatistics.* John Wiley & Sons, New York, NY, 1985.

FEYNMAN, R. P. [1972], *Statistical Mechanics.* W. A. Benjamin, Inc., New York, Amsterdam, 1972.

GIBBS, J. W. [1960], *Elementary Principles in Statistical Mechanics.* Dover, Mineola, NY, 1960, Reprint of the original 1902 edition.

GRANDY, JR., W. T. [1987], *Foundations of Statistical Mechanics V.I: Equilibrium Theory.* D. Reidel, Dordrecht–Boston–London, 1987.

JAYNES, E. T. [1957b], Information Theory and Statistical Mechanics. *Physical Review* **106** (1957), 620–630.

JAYNES, E. T. [1957a], Information Theory and Statistical Mechanics II. *Physical Review* **108** (1957), 171–190.

TISZA, L. [1966], *Generalized Thermodynamics.* MIT Press, Cambridge, MA, 1966.

TRIBUS, M. [1961a], *Thermostatics and Thermodynamics.* D. van Nostrand Co., Inc., New York, 1961.

3.10 PROBLEMS

1. The partition function for a system of some kind of particles is reported to be

$$Z_N = [(V - Nb)/\lambda^3]^N \quad \exp(\beta a N^2/V),$$

where λ is the thermal wave length $h/(2\pi m k T)^{\frac{1}{2}}$, and a and b are constants.

a. Find $U(N, T, V)$. b. Find $P(N, T, V)$. c. Find $S(U, V, N)$. d. Is this S a valid fundamental relation except perhaps at $T = 0$? If not, what is wrong, and how can Z_N be corrected?

2. In a simple quantized model of rubber elasticity, a polymer chain of N links, each of length a, is subject to a force F in the $+x$ direction. Each link can point independently along any one of the x, y, or z axes, in either the $+$ or $-$ direction. The energy ϵ is only x dependent; $\epsilon = aF$ for the link pointing along $-x$, $\epsilon = -aF$ for the link along $+x$, and $\epsilon = 0$ for the link along $\pm y, \pm z$.

a. Calculate the partition function for the N-link chain.
b. Show that the linear thermal expansivity is negative, as it is for rubber.

3. The partition function for a diatomic van der Waals gas (at temperature such that $kT \gg$ rotational energy level spacing) is given by the expression

$$Z_N = \frac{1}{N!} \left(\frac{V - Nb}{\lambda^3} \right)^N \left(\frac{16\pi^2 I}{h^2 \beta \sinh(\beta h\nu/2)} \right)^N e^{\beta a N^2/V},$$

where b and a are the usual constants, λ is the thermal wavelength, I is the moment of inertia of the molecule, and ν is its vibration frequency. Calculate

a. its energy, $U = -\partial \ln Z_N / \partial \beta$,
b. its Helmholtz potential, $F = U - TS = -kT \ln Z_N$,
c. its entropy, $S = U/T + k \ln Z_N$,
d. $C_v = (\partial u / \partial T)_v$, and
e. sketch C_v vs. T.

4. A monatomic Boltzmann ideal gas of spin 1/2 atoms in a uniform magnetic field has, in addition to its usual kinetic energy, a magnetic energy of $\pm mB$ per atom, where m is the magnetic moment. (It is assumed that the gas is so dilute that the interaction of magnetic moments may be neglected.)

a. What is the partition function for a canonical ensemble of N such atoms?
b. Calculate C_v from the partition function.
c. Sketch C_v as a function of T.

5. The classical energy of a simple rigid rotator is $L^2/2I$, where L is the angular momentum and I is the moment of inertia. In quantum mechanics, L^2 is quantized to values $\hbar^2 J(J+1)/2I$, where J is an integer. Each energy level $\epsilon_J = \hbar^2 J(J+1)/2I$ is $2J+1$-fold degenerate.

a. Write out the partition function for a diatomic ideal gas (ignoring vibration). (Each molecule has translational kinetic energy, and independently, rotational kinetic energy.)
b. Assume that the moment of inertia and the temperature are such that the rotational energy levels are small compared to kT, so these levels can be approximated as a continuum. Write $x = \beta \hbar^2 J(J+1)/2I$, replace the sum in the rotational part of the partition function by an integral, and calculate the average rotational energy of a molecule.
c. Calculate C_v for this gas.

6. A two dimensional model of a simple solid with impurities is shown below:

The solid at temperature T contains N negatively charged impurity ions ($\ominus$) per cm^3, these ions replacing some of the ordinary atoms ($\odot$) of the solid. Since the solid is electrically neutral, each negative ion (with charge $-e$) has near it, in one of the four interstitial lattice sites ($\oplus$), an ion of charge $+e$. The positive ion is small and free to move from one adjacent interstitial site to another, but not (for this model) to more distant sites. Calculate the electric polarization, *i.e.*, the mean electric dipole moment per unit volume, when an electric field $\mathbf{E}$ is applied in the x-direction.

7. The energy of an anharmonic oscillator is given by $\epsilon = p^2/2m + bx^{2n}$, where $2n > 2$. Consider a thermodynamic system composed of a large number of these identical, non-interacting oscillators. (n is a positive integer.)

a. Write out the single oscillator partition function.
b. Calculate the average kinetic energy of an oscillator. (Do this without evaluating an integral, by the technique of writing, say, $\beta p^2/2m = r$, a dimensionless variable of integration, so that the integral appears merely as a number, and the pertinent variables appear outside.)
c. Calculate the average potential energy of an oscillator. (Again, do not evaluate an integral.)
d. Show that the heat capacity is $C = N_0 \left(\frac{1}{2} + \frac{1}{2n} \right) k$.

8. Consider a classical ideal gas consisting of N particles. Suppose the energy E of one particle is given by the extreme relativistic expression $E = cp$, where c is the speed of light, and p is the magnitude of the momentum. Find the thermodynamic functions of this ideal gas without considering the internal structure of the particles (U, F, H, P, C_v, C_p).

9. A small subsystem of a very large container filled with an ideal gas of identical particles is regarded as a grand canonical ensemble. Show that the probability of finding the subsystem with N_1 atoms is given by the Poisson distribution

$$P(N_1) = \left\langle N_1 \right\rangle^{N_1} \Big/ N_1! \, e^{\langle N_1 \rangle}$$

where $\left\langle N_1 \right\rangle$ is the mean number of atoms present in the subsystem.

10. A subsystem of volume V_1 of a larger container of volume V, filled with an ideal gas of N indistinguishable, noninteracting particles, is regarded as a classical system. Denote the probability of finding a particular gas molecule in the subsystem volume as $p = V_1/V$. Find a classical expression for the probability $P(N_1)$ that there are N_1 molecules in the subsystem. From this expression, find the expected number of particles, $\left\langle N_1 \right\rangle$, the value for $\left\langle N_1^2 \right\rangle$, and the relative variance $\left\langle \left(N_1 - \left\langle N_1 \right\rangle \right)^2 \right\rangle / \left\langle N_1 \right\rangle^2$. Show that these results agree with those of the previous problem in the thermodynamic limit, *i.e.*, when $V \to \infty$, with N/V constant.

11. An attempt to model an unusual elastic solid is made as follows: Assume that the normal mode frequency spectrum is just as usual; *i.e.*, $g(\omega)d\omega = A\omega^2 d\omega$, where A is a constant derived from the elastic properties of the solid; but the normal mode energy levels are restricted to the ground state and the first excited state of each normal mode oscillation. The total number of normal modes is $3N$. Derive expressions for the total energy and C_v. Sketch U and C_v *vs.* T, and indicate particularly the form of the high- and low-temperature dependencies.

12. Show that if another quantity, the gufacity, is defined in terms of the Helmholtz potential instead of the Gibbs potential, then the gufacity for a van der Waals gas satisfies the Dieterici equation of state, or

$$P^{*'} = \frac{RT}{v - b} e^{-a/RTv}.$$

13.a. According to Eq. (2.7.10), a system in equilibrium with a heat bath at constant temperature T_b minimizes the Helmholtz potential, subject to the constraint that $T_b = $ constant. Derive an expression for the partition function of such a system, in terms of F, by the variational technique used to maximize entropy at constant energy.

b. Show that the Helmholtz potential so obtained is a minimum.

c. Repeat part a., this time to minimize the Gibbs potential for a system constrained to be at constant temperature T_b and constant pressure P_b, in accord with Eq. (2.7.12).

3.11 SOLUTIONS

1.a.
$$U = -\frac{\partial \ln Z_N}{\partial \beta} = 3N\frac{\partial \ln \lambda}{\partial \beta} - \frac{aN^2}{V} = \frac{3}{2}NkT - \frac{aN^2}{V}$$

b.
$$F = -kT \ln Z_N; \quad dF = -SdT - PdV + \mu dN$$
$$P = -\left(\frac{\partial F}{\partial V}\right)_{T,N} = \frac{NkT}{V-Nb} - \frac{aN^2}{V^2}$$
$$\left(P + \frac{aN^2}{V^2}\right)(V - Nb) = NkT; \quad \text{van der Waals eq'n.}$$

c.
$$S = -\left(\frac{\partial F}{\partial T}\right)_{V,N} = S(T,V,N) = k \ln Z_N + \frac{3}{2}Nk - \frac{aN^2}{VT}$$
$$= Nk \ln\left(\frac{V-Nb}{\lambda^3}\right) + \frac{3}{2}Nk.$$

But
$$T = \frac{2}{3Nk}\left(U + \frac{aN^2}{V}\right),$$

or
$$\frac{1}{\lambda^3} = \left(\frac{4\pi m(U + aN^2/V)}{3Nh^2}\right)^{3/2},$$

and
$$S(U,V,N) = Nk \ln[(V - Nb)/\lambda^3(U,V,N)] + \frac{3}{2}Nk.$$

d. This S is not homogeneous, first-order in U, V, N, because the term $\ln(V - Nb)$. If, however, the partition function were
$$Z_N = \frac{1}{N!}\left(\frac{V-Nb}{\lambda^3}\right)^N e^{\beta aN^2/V},$$

then there would be added to S the terms $-Nk \ln N + Nk$, giving
$$S(U,V,N) = \frac{5}{2}Nk + Nk \ln\left(\frac{V-Nb}{N\lambda^3(U,V,N)}\right).$$

Note that when $a \to 0$ and $b \to 0$, then $S \to S(\text{ideal gas})$.

2.a.
$$Z_1 = \sum_i e^{-\beta\epsilon_1} = e^{-\beta aF} + e^{\beta aF} + 4 = 2\cosh\beta aF + 4$$
$$Z_N = Z_1^N = 2^N(2 + \cosh\beta aF)^N.$$

b. The average projection of a link in the +x direction is $\langle \ell \rangle = ap_+ - ap_-$, where the probabilities are
$$p_+ = e^{-\beta\epsilon_+}/Z_1 = e^{\beta aF}/Z_1; \quad p_- = e^{-\beta\epsilon_-}/Z_1 = e^{-\beta aF}/Z_1.$$

Then

$$\langle \ell \rangle = \frac{2a \sinh \beta a F}{Z_1} = a \frac{\sinh \beta a F}{\cosh \beta a F + 2},$$

and

$$L = N \langle \ell \rangle .$$

The linear thermal expansivity is

$$\alpha_\ell = \left(\frac{\partial \ln L}{\partial T} \right)_F = \left(\frac{\partial \ln \langle \ell \rangle}{\partial T} \right)_F = -\frac{a^2 F}{kT^2} \left(\frac{2 \cosh \beta a F + 1}{(\cosh \beta a F + 2)^2} \right) < 0.$$

3.

$$U = -\frac{\partial \ln Z_n}{\partial \beta} = \frac{5}{2} NkT + \frac{Nh\nu}{2} \coth \frac{\beta h\nu}{2} - \frac{aN^2}{V}. \tag{a}$$

$$F = -kT \ln Z_N = -NkT \ln \left[\left[\frac{V - Nb}{N\lambda^3} \right] \left[\frac{16\pi^2 I}{h^2 \beta \sinh \beta h\nu/2} \right] \right] - NkT - \frac{aN^2}{V}. \tag{b}$$

$$S = \frac{U - F}{T} = \frac{7}{2} Nk + \frac{Nh\nu}{2T} \coth \beta h\nu/2 + Nk \ln \left[\left(\frac{V - Nb}{N\lambda^3} \right) \left(\frac{16\pi^2 I}{h^2 \beta \sinh \beta h\nu/2} \right) \right]. \tag{c}$$

$$C_v = \left(\frac{\partial u}{\partial T} \right)_v = \frac{5}{2} N_0 k + N_0 k \left(\frac{h\nu}{2kT} \right)^2 \operatorname{csch}^2 \frac{\beta h\nu}{2} = \frac{5}{2} R + \frac{R(h\nu/2kT)^2}{\sinh^2(h\nu/2kT)} \tag{d}$$

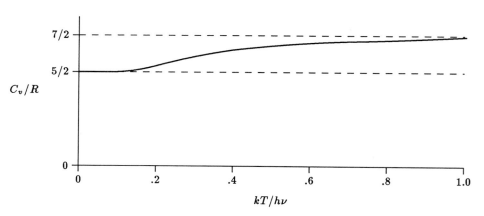

4.

$$Z_1 = Z_{1k} Z_{1m} = (V/\lambda^3)(e^{\beta m B} + e^{-\beta m B}) = (2V/\lambda^3) \cosh \beta m B.$$

a. $Z_N = Z_1^N/N! = \dfrac{1}{N!} \left(\dfrac{2V}{\lambda^3} \right)^N \cosh^N \beta m B$

b. $U = -\dfrac{\partial \ln Z_N}{\partial \beta} = \dfrac{3}{2} NkT - NmB \tanh \beta m B$

$$C_v = \left(\frac{\partial u}{\partial T} \right)_V = \frac{3}{2} N_0 k + N_0 k (\beta m B)^2 \operatorname{sech}^2 \beta m B = R \left[\frac{3}{2} + \frac{(\beta m B)^2}{\cosh^2 \beta m B} \right].$$

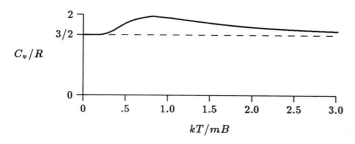

5.

    a. $Z_1 = \sum e^{-\beta \epsilon_i} = \sum e^{-\beta(\epsilon_{itrans} + \epsilon_{irot})} = Z_{1trans} Z_{1rot}.$

$$Z_{1trans} = \frac{V}{\lambda^3}; \quad Z_{1rot} = \sum_{J=0}^{\infty} (2J+1) e^{-\beta \hbar^2 J(J+1)/2I}.$$

$$Z_N = \frac{1}{N!} Z_1^N = \frac{1}{N!} \left(\frac{V}{\lambda^3} \right)^N (Z_{1rot})^N.$$

b. Let $x = \beta \hbar^2 J(J+1)/2I; \quad dx = (\beta \hbar^2/2I)(2J+1)dJ.$ Then

$$Z_{1rot} \to \frac{2I}{\beta \hbar^2} \int_0^{\infty} e^{-x} dx = \frac{2I}{\beta \hbar^2};$$

$$\left\langle \epsilon_{rot} \right\rangle = -\frac{\partial \ln Z_{1rot}}{\partial \beta} = \frac{1}{\beta} = kT.$$

c. Since

$$\left\langle \epsilon \right\rangle = \left\langle \epsilon_{trans} \right\rangle + \left\langle \epsilon_{rot} \right\rangle = \frac{3}{2} kT + kT = \frac{5}{2} kT,$$

then it follows that

$$U = \frac{5}{2} NkT, \quad \text{and} \quad C_v = \frac{5}{2} R.$$

6. The energies are $\pm aeE/2$. The single-site partition function is

$$Z_1 = 2(e^{-\beta aeE/2} + e^{\beta aeE/2}) = 4\cosh(\beta aeE/2) \equiv 4\cosh x.$$

The dipole moment for energy $+aeE/2$ is $-aE/2$, and for $-aeE/2$ is $+aE/2$. Therefore the average dipole moment is

$$\left\langle m_e \right\rangle = \frac{(-ae/2)(2e^{-x}) + (ae/2)(2e^{x})}{4\cosh x} = \frac{1}{\beta} \frac{\partial \ln Z_1}{\partial E} = (ae/2)\tanh x,$$

and

$$P_e = (Nae/2)\tanh x.$$

7.a.

$$Z_1 = \sum e^{-\beta \epsilon} = \frac{1}{h} \int_{x=-\infty}^{\infty} \int_{p=-\infty}^{\infty} e^{-\beta(p^2/2m + bx^{2n})} dx dp$$

$$= \frac{1}{h} \left[\int_{-\infty}^{\infty} dp\, e^{-\beta p^2/2m} \right] \left[\int_{-\infty}^{\infty} dx\, e^{-\beta bx^{2n}} \right].$$

b. Let $\beta p^2/2m = r$; $\beta b x^{2n} = s$. Then, since $n > 0$,

$$Z_1 = \left(\frac{1}{h}\right)\left(\frac{2m}{\beta}\right)^{1/2}\left(\frac{1}{n}\right)\left(\frac{1}{\beta b}\right)^{1/2n}\left[\int_0^\infty \frac{e^{-r}\,dr}{\sqrt{r}}\right]\left[\int_0^\infty e^{-s}s^{\frac{1}{2n}-1}\,ds\right] = Z_{1\,kin}Z_{1\,pot}.$$

$$\langle \epsilon_{1\,kin}\rangle = -\frac{\partial \ln Z_{1\,kin}}{\partial\beta} = \frac{1}{2\beta} = \frac{1}{2}kT.$$

c.

$$\langle \epsilon_{1\,pot}\rangle = -\frac{\partial \ln Z_{1\,pot}}{\partial\beta} = kT/2n$$

d.

$$\langle \epsilon \rangle = (kT/2) + (kT/2n) = (kT/2)\left(1 + \frac{1}{n}\right);$$

$$U = N\langle \epsilon \rangle;\quad C_v = (R/2)\left(1 + \frac{1}{n}\right).$$

8.

$$Z_1 = \frac{V}{h^3}\int_0^\infty e^{-\beta cp}4\pi p^2\,dp = \left[\frac{4\pi V}{h^3}\right]\left[\frac{2}{\beta^3 c^3}\right] = 8\pi V(kT/hc)^3 \equiv \xi.$$

$$Z_N = \xi^N/N!.$$

$$U = -\frac{\partial \ln Z_N}{\partial\beta} = \frac{3N}{\beta} = 3NkT.$$

$$F = -kT\ln Z_N = NkT[3\ln(hc/kT) - \ln(8\pi V/N) - 1].$$

$$P = -\left(\frac{\partial F}{\partial V}\right)_{T,N} = NkT/V;\quad \text{(ideal gas law)}.$$

$$H = U + PV = 4NkT.$$

$$S = -\left(\frac{\partial F}{\partial T}\right)_{V,N} = -Nk[3\ln(hc/kT) - \ln(8\pi V/N) - 1] + 3Nk$$

$$= 4Nk - Nk[3\ln(hc/kT) - \ln(8\pi V/N)].$$

$$C_v = \left(\frac{\partial U}{\partial T}\right)_{V,N_0} = 3R.$$

$$C_p = (N_0/N)\left(\frac{\partial H}{\partial T}\right)_{P,N} = 4R.$$

9. Note that, in general, $Z_G = \sum_{N=0}^\infty z^N Z_N$; for indistinguishable, non-interacting particles, $Z_N = Z_1^N/N!$, so that $Z_G = e^{zZ_1}$. Also, in general $\langle N \rangle = z(\partial \ln Z_G/\partial z)_\beta$, giving $\langle N \rangle = zZ_1$ in this case. By definition, the average number of particles is $\langle N \rangle \equiv \sum_{N=0}^\infty Np(N)$; by direct differentiation of $\ln Z_G$, one obtains $\langle N \rangle = \sum_{N=0}^\infty N\left[z^N Z_1^N/N!\right]/Z_G$, with the coefficient of N identified as $p(N)$. Therefore, when this is written for the subsystem population, N_1, it becomes

$$P(N_1) = \frac{(zZ_1)^{N_1}}{N_1!Z_G} = \frac{\langle N_1 \rangle^{N_1}}{N_1!}e^{-\langle N_1\rangle},$$

as required. Note that this result is valid only in the thermodynamic limit, with no finite upper bound to the population N_1.

10. Since each classical molecule is a localizable individual, the probability $P(N_1)$ of finding N_1 molecules in the subsystem is given by the binomial expression

$$P(N_1) = \binom{N}{N_1}p^{N_1}(1-p)^{N-N_1},$$

with $p = V_1/V$. The expected number of particles is

$$\langle N_1 \rangle = \sum_{N_1=0}^{N} N_1 P(N_1) = Np,$$

and the mean squared population is

$$\langle N_1^2 \rangle = \sum_{N_1=0}^{N} N_1^2 P(N_1) = N^2 p^2 - Np^2 + Np.$$

The variance is then found to be

$$\langle N_1^2 \rangle - \langle N_1 \rangle^2 = Np(1-p) = \langle N_1 \rangle (1 - V_1/V),$$

and the relative variance is

$$\frac{\langle N_1^2 \rangle - \langle N_1 \rangle^2}{\langle N_1 \rangle^2} = \frac{1}{\langle N_1 \rangle} - \frac{1}{N}.$$

In the thermodynamic limit, when $N \to \infty$ and $V \to \infty$, with N/V constant, these values agree exactly with those that can be obtained from the Poisson distribution of the previous problem.

11.

$$\int_0^{\omega_m} g(\omega)d\omega \equiv 3N = A\omega_m^3/3; \quad A = 9N/\omega_m^3.$$

$$Z_1 = e^{-\beta\hbar\omega/2} + e^{-3\beta\hbar\omega/2} = 2e^{-\beta\hbar\omega}\cosh(\beta\hbar\omega/2).$$

$$\langle \epsilon \rangle = -\frac{\partial \ln Z_1}{\partial \beta} = \hbar\omega - (\hbar\omega/2)\tanh(\beta\hbar\omega/2).$$

$$U = (9N/\omega_m^3)\int_0^{\omega_m} [\hbar\omega - (\hbar\omega/2)\tanh(\beta\hbar\omega/2)]\omega^2 d\omega$$

$$= 9N\hbar\omega_m/4 - (9N\hbar/2\omega_m^3)\int_0^{\omega_m} \omega^3 \tanh(\beta\hbar\omega/2)d\omega.$$

$$= 9Nk\Theta/4 - (9NkT^4/\Theta^3)\int_0^{\Theta/T} x^3 \tanh(x/2)dx$$

$$= 9Nk\Theta/8 + 9Nk\left(\frac{T^4}{\Theta^3}\right)\int_0^{\Theta/T} \frac{x^3 dx}{e^x + 1},$$

where $\hbar\omega_m \equiv k\Theta$ and $x = \beta\hbar\omega$.
For $T \gg \Theta$, $u \approx 9R\Theta/8 + 9R\Theta/8 - 9R\Theta^2/20T$, or $u \approx (9R\Theta/4)(1 - \Theta/5T)$, and $C_v \approx (9R/20)(\Theta/T)^2$. For $T \ll \Theta$,

$$u \approx \frac{9R\Theta}{8} + 9R\left(\frac{T^4}{\Theta^3}\right)\int_0^{\infty} \frac{x^3 dx}{e^x + 1}.$$

The integral may be written

$$\int_0^{\infty} \frac{x^3 dx}{e^x + 1} = -\sum_{n=1}^{\infty}(-)^n \int_0^{\infty} x^3 e^{-nx}dx = \frac{21}{4}\zeta(4) = \frac{7\pi^4}{120},$$

and the energy per mole then becomes

$$u \approx \frac{9R\Theta}{8} + \left(\frac{21\pi^4}{40}\right)\left(\frac{T^4}{\Theta^3}\right).$$

From this approximation, the low-temperature expression for C_v is found to be

$$C_v = \left(\frac{21\pi^4}{10}\right)\left(\frac{T}{\Theta}\right)^3 = 204.6\left(\frac{T}{\Theta}\right)^3.$$

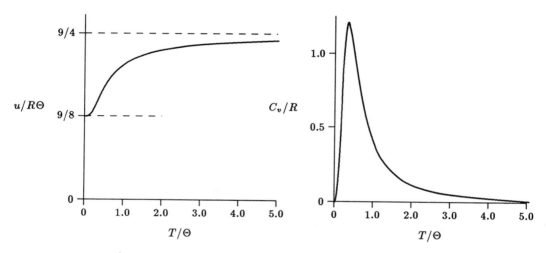

12. The gufacity, $P^{*'}$, is defined, in analogy to the fugacity, by the expression

$$f = RT \ln[P^{*'}/P_0(T)],$$

where f is the molar Helmholtz potential. Since $P = -(\partial f/\partial v)_T$, the defining equation can be differentiated at constant T to give

$$P = -\left(\frac{\partial f}{\partial v}\right)_T = -RT\left(\frac{\partial \ln P^{*'}}{\partial v}\right)_T,$$

so that along an isotherm the result is

$$\frac{dP^{*'}}{P^{*'}} = -\frac{P\,dv}{RT}.$$

For a van der Waals gas, this becomes

$$\frac{dP^{*'}}{P^{*'}} = -\frac{dv}{v-b} - \frac{a\,dv}{RTv},$$

which integrates to

$$\ln(P^{*'}/C) = -\ln(v-b) - a/RTv,$$

or $P^{*'}/C = (v - b)^{-1}e^{-a/RTv}$, where C is a to-be-determined function of T. In the limit of very dilute gas, require that $P^{*'} = P_{ideal} = RT/v$, and find that $C = RT$, or

$$P^{*'} = \frac{RT}{v - b}e^{-a/RTv},$$

which is in the form of the Dieterici equation of state.

13.a. Write $F = U - T_b S = \sum_n (p_n \epsilon_n + kT_b p_n \ln p_n)$, now subject to the single constraint $1 = \sum_n p_n$. Vary the p_n to minimize $F + \alpha \sum p_n$ to get $\sum_n \delta p_n [\epsilon_n + kT_b (\ln p_n + 1) + \alpha] = 0$, or

$$p_n = e^{-\beta_b \epsilon_n} e^{-\alpha \beta_b - 1} = Z^{-1} e^{-\beta_b \epsilon_n},$$

where $\beta_b = 1/kT_b$, and

$$Z = e^{\alpha \beta_b + 1} = \sum_n e^{-\beta_b \epsilon_n}.$$

The Helmholtz potential is found to be

$$F = \sum_n p_n [\epsilon_n + kT_b(-\ln Z - \beta_b \epsilon_n)] = -kT_b \ln Z,$$

$$Z = e^{-\beta_b F}.$$

b. To show that the probability set found in a. minimizes F, assume the existence of some other probability set, $\{w_n\}$, that leads to the same macroscopic energy U, but a different expression for the Helmholtz potential; $F' = \sum_n w_n [\epsilon_n + kT_b \ln w_n]$. Their difference is

$$F - F' = \sum_n [\epsilon_n(p_n - w_n) - kT_b(w_n \ln w_n - p_n \ln p_n)]$$

$$= kT_b \sum_n [p_n \ln p_n + w_n \ln p_n - w_n \ln p_n - w_n \ln w_n]$$

$$= kT_b \sum_n [w_n \ln(p_n/w_n) - (w_n - p_n)(-\ln Z - \beta_b \epsilon_n)]$$

$$= kT_b \sum_n w_n \ln(p_n/w_n) \le kT_b \sum_n w_n [(p_n/w_n) - 1] = 0.$$

Therefore $F \le F'$ for any F' generated by an arbitrary probability set; the equilibrium Helmholtz potential and its defining probability set are those found in part a. (For some reason, this derivation is regarded as acceptable even by some anti-information-theorists, because it does not require that the energy or the Helmholtz potential be regarded as known, and the condition that equilibrium at constant temperature implies a Helmholtz-potential minimum is a result of thermostatics. The expression $S = -k \sum p_n \ln p_n$, used in the derivation, is not usually the point of dispute; it is rather the idea that the $\{p_n\}$ are determined in a way that depends on one's knowledge of the energy that is regarded as objectionable.)

c. Similarly, $G = U - T_b S + P_b V = \sum_n p_n [\epsilon_n + kT_b \ln p_n + P_b V_n]$. Minimization of G subject to $\sum_n p_n = 1$ gives

$$p_n = Z^{-1} e^{-\beta_b(\epsilon_n + P_b V_n)},$$

with

$$Z = \sum_n e^{-\beta_b(\epsilon_n + P_b V_n)}.$$

Substitution of the $\{p_n\}$ in the expression for G gives

$$Z = e^{-\beta_b G}.$$

4. *NONINTERACTING FERMIONS AND BOSONS*

4.1 INTRODUCTION

The essential difference in classical and quantum descriptions of a system of N *identical particles* is in their individuality, rather than in their indistinguishability. The classical attitude, even when we speak of quantized states of noninteracting particles, is that individual particles occupy particular states, even though we may not be able to tell which particle is where. The quantum-mechanical description of such a system insists that if certain single-particle states are occupied, then a mixed N-body state exists with each particle in all occupied states. There is no individual-particle occupancy of single-particle states in the conventional description of a quantum system of N identical particles.

The quantum-mechanical foundation for the statistical treatment of identical particles is discussed in the commentary found in the final section of this chapter. It should be emphasized here that the derivations of this chapter follow from two simple statements about how the single-particle states may be occupied. The underlying basis for these statements may be taken as experimental, as hypothetical, or as a profound revelation of quantum field theory; probably a combination of the three is best.

For indistinguishable, noninteracting particles, the states of the system are taken as composites of single-particle states, and the microstate of the system is specified by the set of occupation numbers of the single-particle states. This is the first simple statement, and it is self-evident. The second statement, entirely generally, is that no single-particle state is allowed to have an occupation number larger than p, where p is some positive integer that cannot exceed the total number of particles. For $p = 1$, the particles are called *fermions*, and they are said to obey *Fermi-Dirac statistics*. Apparently all particles with half-integral spin are fermions; certainly electrons, protons, and neutrons are observed to be fermions, as are nuclei with half-integral total nuclear spins.

For $p \neq 1$, there seems to be no natural limit to the single-particle-state occupation number, other than N, the total number of particles. For large numbers of particles, it is usually convenient to set $p = \infty$. The particles that exhibit this behavior are called *bosons*, and they are said to obey *Bose-Einstein statistics*. Apparently all particles with integral spin (including 0) are bosons; certainly photons and nuclei with integral spins are observed to be bosons, even though the nuclei may be regarded as composite structures of nucleons, which are fermions.

There may exist particles that obey other kinds of statistics, usually called *parastatistics*. Whenever a new particle is discovered, it is almost certain (1) that someone will attribute its behavior to the property that it obeys some form of parastatistics, where $1 < p < \infty$, but (2) that further evidence subsequently will reinforce the hypothesis that all particles are either fermions or bosons. If there is a fundamental reason that all particles must be either fermions or bosons, it is so subtle that it may remain undiscovered for some time to come. Further discussion of this point may be found in the commentary.

4.2 THE GRAND PARTITION FUNCTION FOR FERMIONS

In this section and the next, the power and convenience of the *grand partition function* become evident, and I choose to use it rather than base the derivation directly on information theory. (For such a derivation, see Kapur [1989].) For fermions, the N-particle canonical partition function, Z_N, can be written as the sum of two terms,

$$Z_N = Z'_N + Z'_{N-1}e^{-\beta\epsilon_1}, \qquad (4.2.1)$$

where ϵ_1 is the energy level of the state labeled 1, and the primes denote terms in the partition functions for which there is no particle in state 1. Since fermion systems can have, at most, one particle in any one state, there is need for only the two terms given. The grand partition function is thus written as

$$Z_G = \sum_{N=0}^{\infty} z^N Z_N = \sum_N z^N Z'_N + ze^{-\beta\epsilon_1}\sum_N z^{N-1}Z'_{N-1}$$
$$= (1 + ze^{-\beta\epsilon_1})Z'_G = \prod_i (1 + ze^{-\beta\epsilon_i}). \qquad (4.2.2)$$

In this derivation, it is assumed that the canonical partition function for 0 particles is $Z_0 = 1$. The second line shows that the term in Z_G ascribable to state 1 is just the factor $(1 + ze^{-\beta\epsilon_1})$, and there is no reason to select state 1 as differing from any other state in the way it appears as a factor of the grand partition function; hence the final expression, where the product is over all single-particle states. Note that there was no need, in this derivation, to find the canonical partition function Z_N.

4.3 THE GRAND PARTITION FUNCTION FOR BOSONS

This derivation proceeds as in the previous section, except that bosons are allowed unlimited occupancy of a state. Therefore the equations analogous to those of Eq. (4.2.2) become

$$Z_G = \sum_{N=0}^{\infty} z^N Z_N = Z'_G + ze^{-\beta\epsilon_1}\sum z^{N-1}Z'_{N-1}$$
$$+ z^2 e^{-2\beta\epsilon_1}\sum z^{N-2}Z'_{N-2} + z^3 e^{-3\beta\epsilon_1}\sum z^{N-3}Z'_{N-3} + \cdots$$
$$= Z'_G \sum_{n=0}^{\infty}(ze^{-\beta\epsilon_1})^n = \frac{Z'_G}{(1 - ze^{-\beta\epsilon_1})} = \prod_i (1 - ze^{-\beta\epsilon_i})^{-1}. \qquad (4.3.1)$$

The boson Z_G turns out to be no more complicated than that for fermions; for any other maximum-occupancy number p (*i.e.*, for parastatistics), the expression for Z_G is not quite as simple, nor can the derivation be done quite as elegantly. For these reasons, those who set simplicity and elegance as criteria for the validity of a basic theory argue against the existence of particles obeying parastatistics. As shown in Section 4.10.8, however, the sacrifices in simplicity and elegance are not as severe as they may appear at first glance.

4.4 PRELIMINARY DEVELOPMENTS

It is convenient to treat fermions and bosons by a common formalism, at least in the initial stages. By using σ to represent ± 1, we may write Z_G, for both fermions and bosons, as

$$Z_G = \prod_{i=1}^{\infty}(1 + \sigma ze^{-\beta\epsilon_i})^{\sigma}, \qquad (4.4.1)$$

where $\sigma = +1$ corresponds to fermions and $\sigma = -1$ to bosons. From this, we write

$$\ln Z_G = \sigma \sum_{i=1}^{\infty} \ln(1 + \sigma z e^{-\beta \epsilon_i}). \tag{4.4.2}$$

The expectation value for the number of particles is given by

$$\langle N \rangle = z \frac{\partial \ln Z_G}{\partial z} = \sigma^2 \sum_{i=1}^{\infty} \frac{z e^{-\beta \epsilon_i}}{1 + \sigma z e^{-\beta \epsilon_i}},$$

$$= \sum_i (z^{-1} e^{\beta \epsilon_i} + \sigma)^{-1} = \sum_i \langle n_i \rangle, \tag{4.4.3}$$

where $\sigma^2 = 1$, and, since each term in the sum involves only one state, it follows that such a term must represent $\langle n_i \rangle$, the expectation value for the population of the ith state. For fermions, the result is

$$\langle n_i \rangle = \frac{1}{z^{-1} e^{\beta \epsilon_i} + 1}, \tag{4.4.4}$$

which is seen to be a number such that $0 \le \langle n_i \rangle \le 1$, as it should be. (Recall that $z = e^{\beta \mu} > 0$.) For bosons, the corresponding expression is

$$\langle n_i \rangle = \frac{1}{z^{-1} e^{\beta \epsilon_i} - 1}, \tag{4.4.5}$$

which has no evident upper bound. In both cases, the value of z must be set such that the calculated value for $\langle N \rangle$ agrees with the input data.

The expectation value for the energy is obtained, as usual, from (4.4.2), to be

$$U = -\frac{\partial \ln Z_G}{\partial \beta} = \sum_{i=1}^{\infty} \frac{\epsilon_i}{z^{-1} e^{\beta \epsilon_i} + \sigma} = \sum_i \epsilon_i \langle n_i \rangle, \tag{4.4.6}$$

as is evident for a system of noninteracting particles.

For further development of these systems in general terms, before the analyses diverge to the point that fermions and bosons must be treated separately, it is useful to replace the sums by integrals. For this purpose, recall that the density of states for a single-particle system is given by Eq. (3.3.35) to be

$$g(\epsilon) = 2\pi g_s V (2m)^{3/2} \epsilon^{1/2} / h^3, \tag{4.4.7}$$

where $g_s = 2s + 1$, the spin-degeneracy factor. Use of this expression in Eq. (4.4.2) gives

$$\ln Z_G = \sigma \int_0^{\infty} d\epsilon \, g(\epsilon) \ln(1 + \sigma z e^{-\beta \epsilon}),$$

$$= \frac{2\pi g_s V (2m)^{3/2} \sigma}{h^3} \int_0^{\infty} d\epsilon \, \epsilon^{1/2} \ln(1 + \sigma z e^{-\beta \epsilon}), \tag{4.4.8}$$

from which the expectation values for number of particles and energy can be derived as

$$\langle N \rangle = \left(\frac{2\pi g_s V (2m)^{3/2}}{h^3} \right) \int_0^\infty \frac{d\epsilon \, \epsilon^{1/2}}{z^{-1} e^{\beta \epsilon} + \sigma}, \tag{4.4.9}$$

and

$$U = \left(\frac{2\pi g_s V (2m)^{3/2}}{h^3} \right) \int_0^\infty \frac{d\epsilon \, \epsilon^{3/2}}{z^{-1} e^{\beta \epsilon} + \sigma}. \tag{4.4.10}$$

From the standard relationship $PV = kT \ln Z_G$, Eq. (3.4.9), and from Eq. (4.4.8), the pressure is found to be

$$\begin{aligned}
PV &= \sigma kT \left(\frac{2\pi g_s V (2m)^{3/2}}{h^3} \right) \int_0^\infty d\epsilon \, \epsilon^{1/2} \ln(1 + \sigma z e^{-\beta \epsilon}) \\
&= \frac{2}{3} \sigma kT \left(\frac{2\pi g_s V (2m)^{3/2}}{h^3} \right) \left[\epsilon^{3/2} \ln(1 + \sigma z e^{-\beta \epsilon}) \big|_0^\infty \right. \\
&\quad \left. + \beta \sigma \int_0^\infty \frac{d\epsilon \, \epsilon^{3/2}}{(z^{-1} e^{\beta \epsilon} + \sigma)} \right] = \tfrac{2}{3} U, \tag{4.4.11}
\end{aligned}$$

where $\ln Z_G$ was integrated by parts, the integrated term was observed to vanish at both limits, and the remaining integral was found to be identical to that of Eq. (4.4.10). The relationship of the pressure to the energy density found in Chapter 3 for Boltzmann ideal gases is seen to be valid as well for gases of noninteracting fermions or bosons, at least when the sums can be replaced by integrals.

The expressions found in Eqs. (4.4.9) and (4.4.10) are valid for Boltzmann particles, with $\sigma = 0$. It is not acceptable to set $\sigma = 0$ at an earlier point in the development, so any assertion about Boltzmann particles being encompassed in the general derivation must be made cautiously.

Further results require evaluation of the integrals. Since the fermion and boson integrals behave quite differently, and the techniques for integrating the expressions are not equivalent, it becomes necessary at this point to treat the two systems separately.

4.5 FERMION SYSTEMS

The ideal fermion systems to be considered are almost always at temperatures such that $kT \ll \mu$. The reasons for this are that (1) the effects of the exclusion principle can be studied most easily and most graphically in this regime, and (2) many physical systems can be approximated with some success as ideal Fermi gases with $kT \ll \mu$. Some of the properties of conduction *electrons* in metals can be understood in terms of the behavior of ideal fermion systems, as will be shown, but more detailed and accurate calculations require that the electrons interact with each other, with the lattice vibrations (which can be represented as a phonon gas), and with the Coulomb potential of the lattice. These topics are usually treated in courses in condensed-matter physics. Also, approximate theories of the stars known as *white dwarfs* and *neutron stars* provide other examples of the use of ideal fermion systems.

The evaluation of the integrals for fermions is carried out in Appendix B, and the results are used in this section as needed. From Eq. (4.4.9), with $\sigma = 1$, the expectation value for the number of particles in the system is, for spin-1/2 systems, such that $g_s = 2$,

$$
\langle N \rangle = \left(\frac{4\pi V (2m)^{3/2}}{h^3} \right) \int_0^\infty \frac{d\epsilon\, \epsilon^{1/2}}{z^{-1} e^{\beta \epsilon} + 1} = \left(\frac{4\pi V (2m)^{3/2}}{h^3} \right) F_{3/2}(\beta, \mu),
$$
$$
= \left(\frac{8\pi V (2m\mu)^{3/2}}{3h^3} \right) \left[1 + \left(\frac{1}{8} \right) \left(\frac{\pi}{\beta\mu} \right)^2 + \left(\frac{7}{640} \right) \left(\frac{\pi}{\beta\mu} \right)^4 + \cdots \right], \quad (4.5.1)
$$

and

$$
U = \left(\frac{4\pi V (2m)^{3/2}}{h^3} \right) \int_0^\infty \frac{d\epsilon\, \epsilon^{3/2}}{z^{-1} e^{\beta \epsilon} + 1} = \left(\frac{4\pi V (2m)^{3/2}}{h^3} \right) F_{5/2}(\beta, \mu),
$$
$$
= \left(\frac{8\pi V (2m)^{3/2} \mu^{5/2}}{5h^3} \right) \left[1 + \left(\frac{5}{8} \right) \left(\frac{\pi}{\beta\mu} \right)^2 - \left(\frac{7}{384} \right) \left(\frac{\pi}{\beta\mu} \right)^4 + \cdots \right]. \quad (4.5.2)
$$

From Eq. (4.5.1), when $T = 0$, the expression simplifies to

$$
\langle N \rangle = \frac{8\pi V}{3h^3} (2m\mu_0)^{3/2}, \quad (4.5.3)
$$

or,

$$
\mu_0 = \left(\frac{3 \langle N \rangle}{\pi V} \right)^{2/3} \frac{h^2}{8m}. \quad (4.5.4)
$$

This limiting value of the chemical potential, μ_0, is usually called the *Fermi energy* and written as ϵ_F. The use of this expression for μ_0 in Eq. (4.5.1) leads to a series expressing the temperature dependence of μ:

$$
\mu = \epsilon_F \left[1 + \left(\frac{1}{8} \right) \left(\frac{\pi}{\beta\mu} \right)^2 + \left(\frac{7}{640} \right) \left(\frac{\pi}{\beta\mu} \right)^4 + \cdots \right]^{-2/3},
$$
$$
= \epsilon_F \left[1 - \left(\frac{1}{12} \right) \left(\frac{\pi}{\beta\mu} \right)^2 + \left(\frac{139}{2880} \right) \left(\frac{\pi}{\beta\mu} \right)^4 + \cdots \right]. \quad (4.5.5)
$$

With these expressions, the energy U can be rewritten as

$$
U = \frac{3}{5} \langle N \rangle \epsilon_F \left[1 + \left(\frac{5}{12} \right) \left(\frac{\pi}{\beta\epsilon_F} \right)^2 + \cdots \right], \quad (4.5.6)
$$

where it is correct, to this order, to replace μ in the right-hand side of this expression by ϵ_F. There is seldom any point in carrying the approximation to higher orders, since the second-order term provides a key to the interesting behavior, and there is no experimental basis for seeking a more accurate expression. The energy per mole is obtained from Eq. (4.5.6) by writing N_0 in place of $\langle N \rangle$, as

$$
u = \frac{3}{5} N_0 \epsilon_F \left[1 + \left(\frac{5}{12} \right) \left(\frac{\pi}{\beta\epsilon_F} \right)^2 + \cdots \right]. \quad (4.5.7)
$$

The molar heat capacity at constant volume is then seen to be

$$C_v = \frac{\pi}{2} R \left(\frac{\pi k T}{\epsilon_F} \right), \tag{4.5.8}$$

which consists of a factor that is almost equal to that of a monatomic gas, $3R/2$, multiplied by a factor linear in the temperature that is typically much less than unity. This result provides a solution to one of the problems that perplexed classical physicists who attempted to explain the thermal and electrical properties of metals by invoking a sea of free electrons as the carriers of thermal and electrical currents. How, they asked, could the expected N_0 free electrons per mole fail to add $3R/2$ to the molar heat capacity? Electrons were clearly needed as charge carriers, and for this purpose they must be free to move through the material. Thermal conductivities of metals are generally much greater than those of nonmetals, and it seemed evident that the higher conductivity should be attributable to the transport of energy by the free electrons. But if they are capable of thermal energy transport, then why should they not cause metals to exhibit molar heat capacities significantly greater than the classical $3R$-value of the Dulong-Petit law? Part of the puzzle has been solved by Eq. (4.5.8), which shows that the electronic contribution to C_v is typically quite small. The solution to the remainder of the puzzle requires a demonstration that, despite the small contribution to C_v, the electrons exhibit the transport properties necessary to explain the characteristic enhanced thermal and electrical conductivities of metals. A treatment of these transport properties appears in Chapter 9.

For approximate calculations with degenerate fermion systems, it is often useful to treat them as if the temperature were zero, since the fermi level, ϵ_F, is typically so much higher than kT that the latter is negligible. The energy levels are regarded as filled, up to the Fermi level, and empty above that. From Eq. (4.4.7), in this approximation, it is seen that

$$N = \int_0^{\epsilon_F} g(\epsilon) d\epsilon = \left(\frac{4\pi V g_s}{3} \right) \left(\frac{2m}{h^2} \right)^{3/2} \epsilon_F^{3/2}, \tag{4.5.9}$$

from which it follows directly that

$$\epsilon_F = \left(\frac{h^2}{2m} \right) \left(\frac{3N}{4\pi V g_s} \right)^{2/3},$$

as already found in Eq. (4.5.4).

This result may be understood in another way. From Eq. (4.5.4), the Fermi energy may be written

$$\epsilon_F = \left(\frac{3N_0}{\pi v} \right)^{2/3} \frac{h^2}{8m}, \tag{4.5.10}$$

where N_0 is the Avogadro number, and v is the molar volume. For $\epsilon_F \gg kT$, it follows from Eqs. (4.4.11) and (4.5.7) that

$$Pv = \frac{2}{3} u \approx \frac{2}{5} N_0 \left(\frac{3N_0}{\pi v} \right)^{2/3} \frac{h^2}{8m},$$

from which it follows that the pressure-volume relationship for these degenerate fermions is

$$Pv^{5/3} = \frac{2}{5}N_0^{5/3}\left(\frac{3}{\pi}\right)^{2/3}\frac{h^2}{8m} = K_{5/3},\qquad(4.5.11)$$

where $K_{5/3}$ is an obvious constant.

The adiabatic pressure-volume relationship for any gas satisfying Eq. (4.4.11) is found from $dq = du + Pdv = 0$ to be $Pv^{5/3} = K_{5/3}$, the same as that of Eq. (4.5.11). Evidently, then, the degenerate fermion gas behaves as if it were insulated from its surroundings, and all processes are approximately adiabatic, for which the heat capacity is zero. The higher-order expression of Eq. (4.5.8) is a refinement of this result. The principle, once expressed, is seen to be generally valid for degenerate fermion systems, whether or not they obey Eq. (4.4.11).

From another point of view, one can define a *Fermi momentum*, p_F, as the momentum corresponding to the Fermi energy, and regard the $T = 0$ state as one in which the phase space is filled as tightly as possible, up to the level of p_F, and empty above that. This means that there are g_s particles per volume h^3 of phase space, or

$$N = \left(\frac{g_s}{h^3}\right)\int_0^{p_F}d^3p\int_0^V d^3x = \left(\frac{4\pi V g_s}{3h^3}\right)p_F^3,\qquad(4.5.12)$$

from which it follows that

$$p_F = \left(\frac{3Nh^3}{4\pi V g_s}\right)^{1/3} = \left(\frac{3N_0h^3}{4\pi v g_s}\right)^{1/3}.\qquad(4.5.13)$$

This result is identical to that of Eq. (4.5.9) when $\epsilon_F = p_F^2/2m$, as is the case for an ideal, monatomic, nonrelativistic gas. But the result of Eq. (4.5.13) is more general, and more convenient for calculating the Fermi level of a relativistic gas, such that

$$\epsilon_F = \sqrt{(cp_F)^2 + (mc^2)^2} - mc^2.\qquad(4.5.14)$$

In the extreme relativistic case, when $cp_F \gg mc^2$, Eq. (4.5.14) gives $\epsilon_F \approx cp_F$. The principle that a degenerate collection of fermions is effectively an insulated, adiabatic system has already been established in the nonrelativistic case. The same kind of calculation may be made for these extremely relativistic systems, giving the pressure-volume relationship for the adiabat as

$$Pv^{4/3} = K_{4/3} = hc\left(\frac{N_0}{4}\right)^{4/3}\left(\frac{3}{\pi g_s}\right)^{1/3}.\qquad(4.5.15)$$

Equations of the form $Pv^\Gamma = K_\Gamma$, such as Eq. (4.5.11) and Eq. (4.5.15), are called *polytropes*, and they are much easier to work with than the adiabatics that may be derived from the intermediate ranges of the relativistic expression relating momentum and kinetic energy, given by Eq. (4.5.14). (See Problem 24, Chapter 2.)

In order to treat the full range of relativistic fermion energies in a unified manner, it is useful to derive the density of states by the procedure used to obtain Eq. (3.3.35), which converts the phase-space density of states, $1/h^3$, to energy space. For this purpose, write $p = mc \sinh\theta$, and $dp = mc \cosh\theta d\theta$. Then, from Eq. (4.5.14), it follows that $\epsilon = mc^2(\cosh\theta - 1)$, and

$$\frac{Vg_s}{h^3}(4\pi p^2 dp) = g(\epsilon)d\epsilon = \left(\frac{4\pi g_s V(mc^2)^3}{(ch)^3}\right)\sinh^2\theta\cosh\theta d\theta. \qquad (4.5.16)$$

For a degenerate Fermi gas, effectively at $T = 0$, the Fermi momentum defines the upper limit for θ, or,

$$\sinh\theta_F = p_F/mc, \qquad (4.5.17)$$

where p_F is given by Eq. (4.5.13). The Pv equation of state for any temperature is given by

$$Pv = kT\ln Z_G = kT\int_0^\infty d\epsilon g(\epsilon)\ln\left(1 + ze^{-\beta\epsilon}\right),$$

$$= kTG(\epsilon)\ln\left(1 + ze^{-\beta\epsilon}\right)\Big|_0^\infty + \beta kT\int_0^\infty \frac{d\epsilon G(\epsilon)}{z^{-1}e^{\beta\epsilon} + 1},$$

$$= \int_0^\infty \frac{d\epsilon G(\epsilon)}{z^{-1}e^{\beta\epsilon} + 1}, \qquad (4.5.18)$$

where

$$G(\epsilon) = \int d\epsilon' g(\epsilon').$$

For $T = 0$, Eq. (4.5.18) becomes

$$Pv = \int_0^{\epsilon_F} d\epsilon G(\epsilon),$$

or, by use of Eq. (4.5.16) and related definitions, including $d\epsilon = mc^2\sinh\theta d\theta$,

$$Pv = \frac{4g_s\pi v(mc^2)^4}{3(hc)^3}\int_0^{\theta_F} d\theta\sinh^4\theta = \frac{\pi v g_s m^4 c^5}{6h^3}A(x),$$

where x, the relativity parameter, is defined as $x = \sinh\theta_F = p_F/mc$, and

$$A(x) = 8\int_0^{\theta_F} d\theta\sinh^4\theta.$$

Integration yields

$$A(x) = x\sqrt{1 + x^2}\left(2x^2 - 3\right) + 3\sinh^{-1}x.$$

The Pv equation of state for noninteracting, spin 1/2 fermions of any energy can now be written, since $g_s = 2$,

$$Pv = \left(\frac{\pi v}{3}\right)\left(\frac{mc}{h}\right)^3(mc^2)A(x). \qquad (4.5.19)$$

The function $A(x)$ can be evaluated numerically as required, but it will be approximated for extreme cases after the next derivation.

Similarly, the molar energy can be written as

$$u = \int_0^{\epsilon_F} \epsilon g(\epsilon) d\epsilon,$$

which becomes

$$u = \left[\frac{8\pi v m^4 c^5}{3h^3}\right]\left[3\int_0^{\theta_F} d\theta \sinh^2\theta \cosh\theta(\cosh\theta - 1)\right] = \frac{1}{8}\left[\frac{8\pi v m^4 c^5}{3h^3}\right]B(x), \quad (4.5.20)$$

where

$$B(x) = 8x^3\left(\sqrt{1-x^2}-1\right) - A(x).$$

For $x \ll 1$, corresponding to the nonrelativistic limit $p_F \ll mc$, and for $x \gg 1$, corresponding to the extreme relativistic limit $p_F \gg mc$, approximations for $A(x)$ and $B(x)$ are

$$A(x) = \frac{8x^5}{5} - \frac{4x^7}{7} + \frac{x^9}{3} - \frac{5x^{11}}{22} + \cdots \quad \text{for } x \ll 1,$$
$$= 2x^4 - 2x^2 + 3\ln 2x - 7/4 + 5/4x^2 + \cdots \quad \text{for } x \gg 1 \quad (4.5.21)$$

and

$$B(x) = \frac{12x^5}{5} - \frac{3x^7}{7} + \frac{x^9}{6} - \frac{15x^{11}}{176} + \cdots \quad \text{for } x \ll 1,$$
$$= 6x^4 - 8x^3 + 6x^2 - 3\ln 2x + 3/4 - 3/4x^2 + \cdots \quad \text{for } x \gg 1. \quad (4.5.22)$$

These results will be used in the following section.

4.6 ELEMENTARY THEORY OF WHITE DWARFS AND NEUTRON STARS

As an application of the methods of this chapter to a real physical system, an elementary and oversimplified study of white-dwarf and neutron stars is presented in this section. The nature of white dwarfs was puzzling to astronomers before 1925 because of the observed color-brightness (or temperature-luminosity) relationship, the Hertsprung-Russell plot, which correlates red color with low brightness and white color with high brightness, much like the behavior of an incandescent bulb with a brightness control. The white dwarf's color implied a high temperature, and its low brightness therefore implied small size. (Both white dwarfs and red giants are off the main sequence in the H-R diagram, and neutron stars do not appear at all, unless the luminosity measure is extended to include microwave radiation.)

White dwarfs came to be regarded as approximately Earth-size stars with masses and central temperatures of the order of those of the Sun. They seem to be stars that have burned all of their hydrogen in thermonuclear processes, and are made up mostly of helium. At temperatures of the order of 10^7 K, or 10^3 eV, the helium is fully ionized, and the star consists primarily of electrons and alpha particles. At the apparent density, the Fermi level for electrons is $\approx 10^6$ eV, making $\mu/kT \approx 10^3$, which allows the electron gas to be treated as fully degenerate. The α-gas is not degenerate; furthermore it is a boson gas, with its contribution to the pressure being somewhat less than that of a corresponding Boltzmann gas. Therefore the electron gas must support the star against gravitational collapse.

The star can be approximated as a nonuniform system in local thermodynamic equilibrium, which means that throughout the star there exist well-defined temperatures, pressures, and densities, but these can vary with radius. Spherical symmetry is assumed, and the density is expected to vary with position in accordance with the requirement of hydrostatic equilibrium:

$$\rho \nabla \phi = -\nabla P,$$

or,

$$\nabla \phi = -\frac{1}{\rho} \nabla P,$$

where ϕ is the gravitational potential, which satisfies the Poisson equation

$$\nabla^2 \phi = 4\pi G \rho.$$

In spherical coordinates with no angular dependence, these equations can be combined to give

$$\frac{1}{r^2} \frac{d}{dr} \left(\frac{r^2}{\rho} \frac{dP}{dr} \right) = -4\pi G \rho. \tag{4.6.1}$$

In the approximate treatment to be presented here, the pressure will be attributed entirely to the fully degenerate electron population, but the density will be attributed to the total collection of particles, and must therefore arise principally from the nucleons. Charge neutrality leads to the electron density matching the mass density in its radial dependence. Even with charge neutrality, there is a Coulomb contribution to the pressure, but in this treatment it will be ignored, as will the effects of rotation and possible high magnetic fields. At the electron densities that are soon to be calculated, it will be seen that the Fermi energies can correspond to relativistic electrons. Therefore it will be expedient to treat both the nonrelativistic and relativistic cases together, although it would be more forceful to show that the nonrelativistic treatment leads to results that compel examination of the relativistic situation.

A complete theory of even this simplified model of white dwarfs is complicated by the need to solve Eq. (4.6.1), and to present its solution and consequences in suitable textbook form. Instead, two limiting regimes of white-dwarf existence will be examined: (a) the nonrelativistic and (b) the extreme relativistic. These provide bounds for intermediate behavior. The reason for these choices is that Eq. (4.6.1) can be solved easily, and the results presented conveniently, when the pressure-density relation is a polytrope; Eqs. (4.5.11) and (4.5.15) already provide the Emden polytropic relationships in these cases. (Chandrasekhar [1939])

In order to convert $Pv^\Gamma = \text{constant}$ to $P = K\rho^\Gamma$, it is necessary to write the molar volume as

$$v = M_m / \rho,$$

where M_m is the mass associated with Avogadro's number of electrons. Then write M_m as

$$M_m = N_0 \mu_e m_u,$$

where μ_e is the effective nuclear mass number per electron ($\mu_e = 2$ for helium), and m_u is the mass of one atomic mass unit ($m_u = 1.660 \times 10^{-27}$ kg). With these substitutions, Eq. (4.5.11) for nonrelativistic, completely degenerate fermions becomes

$$P = K_{5/3}\rho^{5/3},$$

where

$$K_{5/3} = \left(\frac{h^2}{20m_e}\right)\left(\frac{3}{\pi}\right)^{2/3}(\mu_e m_u)^{-5/3}. \qquad (4.6.2)$$

Similarly, Eq. (4.5.15) for extreme relativistic electrons becomes

$$P = K_{4/3}\rho^{4/3},$$

where

$$K_{4/3} = \left(\frac{hc}{8}\right)\left(\frac{3}{\pi}\right)^{1/3}(\mu_e m_u)^{-4/3}. \qquad (4.6.3)$$

The substitution in Eq. (4.6.1) of $P = K\rho^{\Gamma}$ gives

$$\frac{\Gamma}{r^2}\frac{d}{dr}\left(\frac{r^2 P}{\rho^2}\frac{d\rho}{dr}\right) = -4\pi G\rho.$$

The Emden trick is to write

$$\rho = y^n \rho_c, \quad \text{and} \quad r = aw,$$

where ρ_c is the central density, a is a constant to be determined, and n was called by Emden the *index* of the polytrope. The indicated manipulations lead to the choices $n = (\Gamma - 1)^{-1}$, or $\Gamma = 1 + 1/n$,

$$a = \left[\frac{(n+1)K\rho_c^{\frac{1-n}{n}}}{4\pi G}\right]^{1/2}, \qquad (4.6.4)$$

and, for the differential equation to be solved,

$$\frac{1}{w^2}\frac{d}{dw}\left(w^2\frac{dy}{dw}\right) = -y^n. \qquad (4.6.5)$$

Even with this simplification, the equation is not conveniently solvable analytically, but a numerical solution is an easy exercise. Evidently one boundary condition is that $y(0) = 1$. The other follows from the requirement that the pressure gradient at the center must be zero, leading to $y'(0) = 0$. For the two polytropes of direct interest, the plot of $y(w)$ is a monotone decreasing function that reaches zero at a finite value, w_1, with a finite negative slope, y_1'. The radius of the star is

$$R = aw_1 = \left[\frac{(n+1)K}{4\pi G}\right]^{1/2}\rho_c^{\frac{1-n}{2n}}w_1, \qquad (4.6.6)$$

and the mass is obtained by integration to be

$$M = \int_0^R 4\pi r^2 \rho(r) dr = 4\pi a^3 \rho_c \int_0^{w_1} w^2 y^n dw$$

$$= -4\pi a^3 \rho_c \int_0^{w_1} w^2 \left[\frac{1}{w^2} \frac{d}{dw} \left(w^2 \frac{dy}{dw} \right) \right] dw$$

$$= -4\pi a^3 \rho_c \int_0^{w_1} d \left(w^2 \frac{dy}{dw} \right) = -4\pi a^3 \rho_c w_1^2 y_1'$$

$$= 4\pi \left[\frac{(n+1)K}{4\pi G} \right]^{3/2} \rho_c^{\frac{3-n}{2n}} w_1^2 |y_1'|. \tag{4.6.7}$$

The central density can be eliminated to yield a general mass-radius relationship:

$$M = 4\pi \left[\frac{(n+1)K}{4\pi G} \right]^{\frac{n}{n-1}} R^{\frac{3-n}{1-n}} w_1^{\frac{3-n}{n-1}} w_1^2 |y_1'|. \tag{4.6.8}$$

In the nonrelativistic case, with $\Gamma = 5/3$, or $n = 3/2$, the numerical calculation gives $w_1 = 3.65375$, $|y_1'| = 0.20330$, $w_1^2 |y_1'| = 2.71406$, and $K_{5/3} = 1.00436 \times 10^7$, in SI units. These values give

$$R = \left[\frac{5K_{5/3}}{8\pi G} \right]^{1/2} w_1 \rho_c^{-1/6} = 6.3298 \times 10^8 (\rho_c \mu_e^5)^{-1/6} \text{m}, \tag{4.6.9}$$

$$M = 4\pi \left[\frac{5K_{5/3}}{8\pi G} \right]^{3/2} w_1^2 |y_1'| \rho_c^{1/2} = 1.7733 \times 10^{26} \left(\rho_c \mu_e^{-5} \right)^{1/2} \text{kg}, \tag{4.6.10}$$

or, upon elimination of the central density,

$$M = 4\pi \left[\frac{5K_{5/3}}{8\pi G} \right]^3 R^{-3} w_1^3 w_1^2 |y_1'| = 4.497 \times 10^{52} R^{-3} \mu_e^{-5} \text{kg}, \tag{4.6.11}$$

with all values expressed in SI units. A more common measure for astrophysical purposes is

$$M = 22.6094 \mu_e^{-5} R^{-3} \quad \text{in} \quad M_\odot,$$

with R in units of 10^4 km.

A well-measured white dwarf, the companion of 40 Eridani, designated as 40 Eri B, has an observed mass of 0.48 $M_\odot$, and a radius of 0.0124 $R_\odot$, both reported to be good to within about 4%. If the mass is used as input, Eq. (4.6.10) gives the central density to be $\rho_c = 9.28 \times 10^9$ kg/m^3, from which the radius, calculated from Eq. (4.6.9), turns out to be $R = 1.14 \times 10^7$ m or 0.0163 $R_\odot$, both calculations based on the assumption that $\mu_e = 2$. The agreement is not excellent, but the order of magnitude is clearly right. The relativity parameter x for this case is 1.68, for which the Fermi velocity is 0.86c. It is therefore not unreasonable to find some disagreement with the nonrelativistic theory.

At the central densities of white dwarfs, the possibility exists of inverse beta decay, *i.e.*, a reaction of the form

$$e^- + p \rightarrow n + \nu,$$

where the reaction may take place with free particles or with those bound in nuclei. Much like the chemical reactions of Chapter 3, this reaction proceeds in the direction determined by the respective chemical potentials, but with additional considerations. For charge neutrality, the electron and proton number densities must be equal, and there is the further feature that in the fully degenerate fermion sea of electrons, the beta decay of the neutron may be blocked by the absence of a vacancy at the energy level of the to-be-emitted electron. For inverse beta decay, there must be electrons present with enough energy to balance the neutron-proton mass difference (1.29 MeV for free particles), which sets the minimum density level at which such reactions can occur near 10^{10} kg/m^3. For further details of the internal nuclear physics of white dwarfs, it is appropriate to consult the specialized references (Chandrasekhar [1939]; Chandrasekhar [1978]; Shapiro and Teukolsky [1983]).

For the extreme relativistic case, with $\Gamma = 4/3$, $n = 3$, the numerical calculations give $w_1 = 6.89685$, $|y_1'| = 0.042430$, and $w_1^2|y_1'| = 2.01824$. For this case,

$$\left[\frac{K_{4/3}}{\pi G}\right] = 5.93695 \times 10^{19} \mu_e^{-5/3},$$

$$R = \left[\frac{K_{4/3}}{\pi G}\right]^{1/2} w_1 \rho_c^{-1/3} = 5.3154 \times 10^{10} \rho_c^{-1/3} \mu_e^{-2/3} = 3.3477 \times 10^{10} \rho_c^{-1/3} \text{ m for } \mu_e = 2,$$
$$(4.6.12)$$

and

$$M = 4\pi \left[\frac{K_{4/3}}{\pi G}\right]^{3/2} w_1^2 |y_1'| = 1.1602 \times 10^{31} \mu_e^{-2} = 2.9005 \times 10^{30} \text{ kg for } \mu_e = 2. \quad (4.6.13)$$

In terms more often used in astronomy,

$$M = \frac{5.83}{\mu_e^2} M_\odot = 1.458 M_\odot \text{ for } \mu_e = 2.$$

The surprising result in this case is that the mass is fixed and independent of radius or central density. This is an odd result that requires further exploration. The mass specified by Eq. (4.6.13) is called the Chandrasekhar limit, M_{Ch}; it represents the upper limit to possible white-dwarf masses, and indeed no white dwarf has been observed to have a mass in excess of this limit.

In order to examine the phenomenon further, it is useful to recall that the pressure is given by two different expressions, the polytrope of Eq. (4.6.3), and the general relationship developed for relativistic fermions, Eq. (4.5.19). These may be equated, to give, for the central pressure and density,

$$P_c = K_{4/3} \rho_c^{4/3} = \left[\left(\frac{\pi}{3}\right)\left(\frac{mc}{h}\right)^3 (mc^2)\right] A(x_c).$$

Since the polytrope was derived to represent extreme relativistic conditions, it follows that the relativity parameter x should be large, and the expansion of Eq. (4.5.21) is appropriate.

In the approximation to be used here, $A(x)$ may be written as $A(x) \approx 2x^4(1 - x^{-2})$. With the appropriate values for x and the constants, the two expressions for P_c become

$$\left(\frac{hc}{8}\right)\left(\frac{3}{\pi}\right)^{1/3}(m_u\mu_e)^{-4/3} \approx 2\left[\left(\frac{\pi}{3}\right)\left(\frac{mc}{h}\right)^3(mc^2)\right]\left[\left(\frac{3h^3\rho_c}{8\pi\mu_e m_u}\right)^{4/3}\left(\frac{1}{mc}\right)^4\right] \times$$

$$\times \left[1 - (mc)^2\left(\frac{8\pi\mu_e m_u}{3h^3\rho_c}\right)^{2/3}\right]$$

$$\approx \left[\left(\frac{3}{\pi}\right)^{1/3}\left(\frac{hc}{8}\right)(\mu_e m_u)^{-4/3}\right]\left[1 - 4\left(\frac{mc}{h}\right)^2\left(\frac{\pi}{3}\right)^{2/3}\left(\frac{\mu_e m_u}{\rho_c}\right)^{2/3}\right], \qquad (4.6.14)$$

or, when like factors are cancelled,

$$1 \approx 1 - 4\left(\frac{mc}{h}\right)^2\left(\frac{\pi}{3}\right)^{2/3}\left(\frac{\mu_e m_u}{\rho_c}\right)^{2/3},$$

$$\approx 1 - 4\left(\frac{\pi}{3}\right)^{2/3}\left[\frac{(\mu_e m_u)/(h/mc)^3}{\rho_c}\right]^{2/3}. \qquad (4.6.15)$$

Thus the polytropic approximation is seen to be good only when the central density is very much larger than the mass per Compton volume (the volume of a cubic electron Compton wave length), which turns out to be 2.3×10^8 kg/m^3. As already seen, central densities of white dwarfs are calculated to exceed this value by an order of magnitude, but in order for the equality to hold, the central density must become infinite. This result, apparently, is the meaning of the Chandrasekhar limit, because infinite central density implies zero radius. (Chandrasekhar's original calculation, which assumed constant density throughout the star, arrived at the limiting mass as that for which the radius goes to zero. His result differed from that of Eq. (4.6.13), giving about $0.8M_\odot$, instead of the result there of $1.458M_\odot$.

The significance of the mass/Compton volume as a unit of density measure is as follows: When the electrons are packed so densely that the volume per electron equals the cube of the Compton wavelength, then by the Heisenberg uncertainty principle the electron momentum is $\approx h/(h/mc) = mc$, giving for the relativity parameter $x = p/mc \approx 1$, which is a useful boundary between somewhat relativistic and quite relativistic. But the mass associated with each electron is $\mu_e m_u$, and at this electron density, the mass density must therefore be $\rho = \mu_e m_u/(h/mc)^3$.

It is interesting to examine the interrelationship of the fundamental constants of quantum theory, gravitation, and relativity in the theory of white dwarfs. As an aid to this endeavor, it is useful to recall Eq. (4.6.8), relating the mass to the radius for any polytrope. The factor K_Γ may be written

$$K_\Gamma = a_\Gamma\left(\frac{mc^2}{\mu_e m_u}\right)\left[\left(\frac{3}{\pi}\right)\left(\frac{h}{mc}\right)^3\left(\frac{1}{\mu_e m_u}\right)\right]^{1/n},$$

where the powers of the fundamental constants are obtained unambiguously from dimensional analysis, and a_Γ is an unspecified dimensionless constant. Note that $a_{5/3} = 1/20$, and $a_{4/3} = 1/8$. The use of this expression in Eq. (4.6.8), followed by algebraic sorting into terms that are, or are not, n-dependent, gives

$$\left[\left(\frac{G\mu_e m_u M/R}{mc^2}\right)\left(\frac{w_{1n}/(n+1)a_\Gamma}{w_{1n}^2|y_{1n}'|}\right)\right]^n = \left(\frac{M/(4\pi R^3/3)}{\mu_e m_u/(h/mc)^3}\right)\left(\frac{w_{1n}^3}{\pi w_{1n}^2|y_{1n}'|}\right). \quad (4.6.16)$$

Aside from constants, the left-hand side contains the ratio of the gravitational binding energy of a nucleon at the surface of the star to the electron rest energy. The right-hand side has the ratio of the average density of the star to the already examined Compton density. When $n = 3$, the radial dependence vanishes, giving the mass M_{Ch} in terms of fundamental constants and determinate numerical factors. For any n, there is a unique relationship between the average density and the gravitational binding energy of a surface nucleus. Although this relationship is convenient for exhibiting the groupings of fundamental constants in the white-dwarf problem, it seems to be more a consequence of polytropes than of any basic properties of these stars.

It is evident from the shape of the $\Gamma = 4/3$ polytrope (with $y_1' = -.042430$ at the boundary) that the density drops off quite slowly with radius in the outer regions of the star, disregarding the central-density catastrophe that is necessary for the strict validity of the polytropic approximation. Therefore the argument that the Fermi energy is high enough to demand a relativistic treatment throughout the star is clearly invalid, and it is evident that a more careful analysis, based on Eq. (4.5.19) and the hydrostatic equations, is needed. Such a treatment is inappropriate in the present context, as are the inclusions of other complicating factors such as spinning stars, intense magnetic fields, pulsations, and the pycnonuclear reactions that can occur at white-dwarf densities (Shapiro and Teukolsky [1983]).

At sufficiently high central densities, the electron pressure is no longer able to prevent gravitational collapse. Stellar radii shrink until the densities reach those of the nuclei themselves, and the stellar matter becomes a mixture of neutrons, protons, and electrons that can be treated approximately as if it were all neutrons. (The principal evidence that these neutron stars are not composed entirely of neutrons is that they typically exhibit extremely high magnetic fields, which require current carriers.) A single simple approximate theory of neutron stars can be presented without invoking either the complications mentioned in the previous paragraph or those required by general relativity. In this model, the star is regarded as composed entirely of cold, degenerate, noninteracting, nonrelativistic fermions with the mass of neutrons. The pressure of the degenerate ideal gas supports the mass against the gravitational force. Under these conditions, the molar volume is related to the density by $v = N_0 m_n/\rho$, and the pressure-density relation is again a polytrope with $\Gamma = 5/3$, given by

$$P = \left(\frac{h^2}{20m_n^{8/3}}\right)\left(\frac{3}{\pi}\right)^{2/3}\rho^{5/3} = K_{5/3}\rho^{5/3}. \quad (4.6.17)$$

From Eqs. (4.6.9), (4.6.10), and (4.6.11), with the value of $K_{5/3}$ from the neutron polytrope, the radius and mass are given as

$$R = 1.459 \times 10^7 \rho_c^{-1/6}\text{m}, \quad (4.6.18)$$

$$M = 2.199 \times 10^{21}\rho_c^{1/2}\text{kg} = 6.829 \times 10^{42}R^{-3}. \quad (4.6.19)$$

For $M = M_\odot$, these give $\rho_c = 8.18 \times 10^{17}$ kg/m³, which is higher than nuclear densities, and $R = 1.5065 \times 10^4$m, which is less than ten miles. (It is interesting to observe that one cubic millimeter of matter at this density has a mass of about 800,000 tons.) For neutron stars, the relativity parameter, is

$$x = \frac{p_F}{m_n c} = \left(\frac{h}{2c}\right)\left(\frac{3\rho}{\pi}\right)^{1/3} m_n^{-4/3} = 5.46 \times 10^{-7} \rho^{1/3}. \tag{4.6.20}$$

For the given numerical example, $x_c = 0.509$, based on the density calculated from the model, indicating that even at the calculated supernuclear density, the neutrons are only mildly relativistic.

A similar calculation for the extreme relativistic model of neutron stars gives

$$P \approx \frac{1}{8}\left(\frac{3}{\pi}\right)^{1/3}\left(\frac{hc}{m_n^{4/3}}\right)\rho^{4/3}, \tag{4.6.21}$$

so that

$$\begin{aligned} R &= \left[\frac{K_{4/3}}{\pi G}\right]^{1/2} \rho_c^{-1/3} w_1 \\ &= 5.282 \times 10^{10} \rho_c^{-1/3}, \end{aligned} \tag{4.6.22}$$

and

$$\begin{aligned} M &= 4\pi \left[\frac{K_{4/3}}{\pi G}\right]^{3/2} w_1^2 |y_1'|, \\ &= 1.139 \times 10^{31} \text{kg} = 5.73 M_\odot. \end{aligned} \tag{4.6.23}$$

Further examination of the structure of neutron stars becomes too much involved with equations of state of nuclear matter at high densities, general relativity, and other astrophysical considerations to be appropriate for inclusion in a text on statistical thermophysics. Current theories estimate that neutron stars have a limiting mass in the range of one to six solar masses, with the results being sensitive to the details of the model.

A question that often occurs to the nonspecialist in this area who first learns of the existence of limiting masses for white dwarfs and neutron stars is "What happens to stars with masses too great to collapse into these known compact objects?" An early answer seemed to be that the radiation pressure in these more massive stars was always sufficient to preclude collapse. It has become apparent that as these massive stars cool, they must ultimately become unstable, and either collapse to black holes or expand violently, perhaps as the first stage of a recurrent fluctuation in temperature, radius, and residual mass (Shapiro and Teukolsky [1983]).

4.7 BOSON SYSTEMS

The most interesting properties of boson systems, like those of fermion systems, are manifest when quantum degeneracy prevails. For bosons, however, there is no exclusion-principle-based mandate against multiple occupancy of the single-particle states, and therefore no forcing of the chemical potential to a high positive value in order to provide enough states to accommodate all of the particles. Examination of Eq. (4.4.5) for the number of particles in the ground state (at $\epsilon_0 = 0$) shows that $\langle n_0 \rangle = (z^{-1} - 1)^{-1} \geq 0$, so that $z \leq 1$, and $\mu < 0$.

Use of the density of states to replace sums by integrals, as was done for fermions, leads from Eq. (4.4.8) and Eq. (B.2.2) to the result

$$\langle N \rangle = \left(\frac{2\pi g_s V (2m)^{3/2}}{h^3} \right) \int_0^\infty \frac{\epsilon^{1/2} d\epsilon}{z^{-1} e^{\beta \epsilon} - 1},$$

$$= \left(\frac{2\pi g_s V (2m)^{3/2}}{h^3} \right) \frac{\Gamma(3/2)}{\beta^{3/2}} g_{3/2}(z),$$

$$= \frac{g_s V}{\lambda^3} g_{3/2}(z). \tag{4.7.1}$$

A problem becomes apparent at once. The first factor, proportional to $T^{3/2}$, can become vanishingly small as $T \to 0$, and the second factor, $g_{3/2}(z)$, has as its maximum permissible value, corresponding to $z = 1$, the zeta function $\zeta(3/2) = 2.612$. But with real particles of non-zero mass, not even the magic of quantum mechanics can make them disappear with cooling or reappear when warmed. The source of the difficulty is that the density-of-states approximation assigns zero weight to the ground state at $\epsilon = 0$, whereas the ground-state term in Eq. (4.4.5) evidently can become infinite as $z \to 1$. The obvious and usual procedure for correcting the disappearing-particle problem is to add the slightly rewritten ground-state term to the result of Eq. (4.7.1), to get

$$\langle N \rangle = \frac{g_s V}{\lambda^3} g_{3/2}(z) + \frac{z}{z - 1}. \tag{4.7.2}$$

This equation is clearly the condition that determines z. If the system is not highly degenerate, the first term is adequate to accommodate all but a few of the particles, and z is not close to 1. As the temperature is lowered, a crisis occurs near the temperature for which z is extremely close to its maximum value and the first term is required to contain all of the particles. (Note that if $z = 1 - 10^{-6}$, the ground-state term can contain only 999,999 particles, which is negligible when compared to the Avogadro number.) Therefore the excellent approximation is made that T_c is the lowest temperature for which all of the particles can be contained in the nonground-state terms, with $z = 1$, and $g_{3/2}(1) = \zeta(3/2)$. At this temperature, $\lambda = \lambda_c$, and Eq. (4.7.2) gives

$$\langle N \rangle = \frac{g_s V}{\lambda_c^3} \zeta(3/2),$$

or

$$kT_c = \left(\frac{h^2}{2\pi m} \right) \left(\frac{\langle N \rangle}{V g_s \zeta(3/2)} \right)^{2/3}. \tag{4.7.3}$$

This value of kT_c may be compared with the Fermi energy μ_0 of Eq. (4.5.4) or (4.5.8) to show that they depend in like manner on the particle concentration and mass, and that for particles of equal mass, their ratio is $kT_c/\mu_0 \approx 0.436$, showing that the interesting effects of degeneracy occur in both cases when the appropriate wavelength, either thermal or Fermi, is approximately equal to the particle separation.

At temperatures below T_c, the boson population may be separated, in accordance with Eq. (4.7.2), into two distinct segments—the so-called *normal* population, N_e, contained in the first term and consisting of a gas of particles in states excited above the ground state, and the *ground-state* population, $N_g = \langle n_0 \rangle$, contained in the second term. Evidently, from Eqs. (4.7.2) and (4.7.3), the normal population may be written

$$N_e = \langle N \rangle \left(\frac{T}{T_c}\right)^{3/2}, \tag{4.7.4}$$

and, since $N_g = \langle N \rangle - N_e$, it follows that

$$N_g = \langle N \rangle \left[1 - \left(\frac{T}{T_c}\right)^{3/2}\right]. \tag{4.7.5}$$

The accumulation of a large portion of the particle population into one single-particle state is usually called *Bose-Einstein condensation*. The ground-state wave function for a particle in a rectangular parallelopiped, for example, shows that the particle density is proportional to $\sin^2(\pi x/a)\sin^2 \pi y/b)\sin^2(\pi z/c)$, where a, b, and c are the dimensions of the box. This evidently represents a concentration of particles near the center of the box, but it is nothing like the formation of droplets in a vapor-to-liquid phase transition, which results from intermolecular forces. The momentum vector has each of its components condense into one of two possible values: $p_x = \pm\pi\hbar/a$; $p_y = \pm\pi\hbar/b$; and $p_z = \pm\pi\hbar/c$. Thus the condensation is much more pronounced in momentum space. Above $T = T_c$, the population is considered to be entirely normal.

The energy is found from Eq. (4.4.10) and Eq. (B.2.2) to be

$$U = \left(\frac{2\pi g_s V(2m)^{3/2}}{h^3}\right) \int_0^\infty \frac{d\epsilon\,\epsilon^{3/2}}{z^{-1}e^{\beta\mu} - 1} = \left(\frac{2\pi g_s V(2m)^{3/2}}{h^3}\right) \frac{\Gamma(5/2)}{\beta^{5/2}} g_{5/2}(z),$$

$$= (kT)\left(\frac{3g_s V}{2\lambda^3}\right) g_{5/2}(z). \tag{4.7.6}$$

The ground-state term need not be considered, since its particles have zero energy. The quantity $V g_s/\lambda^3$ may be replaced in terms of $\langle N \rangle$. For the appropriate temperature ranges, the energy per mole is

$$u = \frac{3}{2} RT \frac{g_{5/2}(z)}{g_{3/2}(z)}; \quad T \geq T_c,$$

$$= \frac{3}{2} RT \left(\frac{\zeta(5/2)}{\zeta(3/2)}\right)\left(\frac{T}{T_c}\right)^{3/2}; \quad T \leq T_c. \tag{4.7.7}$$

The pressure, found from $PV = kT \ln Z_G$, agrees with that derived in Eq. (4.4.11) for $T > T_c$, but it must be re-examined for $T < T_c$. The pressure contribution from the ground state, P_g, is found directly from the ground-state term in the partition function:

$$P_g V = -kT \ln(1 - z),$$

which seems to become infinite as $z \to 1$. But the population of the ground state is

$$\langle n_0 \rangle = z/(1-z),$$

which permits z to be expressed as

$$z = \langle n_0 \rangle /(1 + \langle n_0 \rangle).$$

Therefore the ground-state contribution to the pressure becomes

$$P_g V = kT \ln \left(\langle n_0 \rangle + 1 \right). \tag{4.7.8}$$

For $T > T_c$, this partial pressure is clearly zero, as expected. For $T < T_c$, even Avogadro's number of molecules in the ground state will contribute only about $55kT$ to the PV term, which is completely negligible for most purposes. On this basis, then, the ground-state partial pressure will be regarded as zero, and the total pressure for all temperatures will be taken as

$$P = \tfrac{2}{3}(U/V).$$

(Note (a) that the ground-state partial pressure is not intensive, and (b) that it vanishes in the so-called thermodynamic limit, where $V \to \infty$ at constant density. It is therefore an artifact of the approximations inherent in the calculation.)

The entropy can be expressed in two ways: (1) in terms of the occupation numbers of the single-particle states, and (2) in terms of the already calculated thermodynamic quantities. The first derivation, done generally for bosons and fermions, begins with the expression for $\langle n_i \rangle$ given in Eq. (4.4.3), solved for ϵ_i:

$$\epsilon_i = kT \left(\ln z + \ln(1 - \sigma \langle n_i \rangle) - \ln \langle n_i \rangle \right). \tag{4.7.9}$$

From the same expression for $\langle n_i \rangle$, the grand partition function can be written as

$$\ln Z_G = -\sigma \sum_i \ln(1 - \sigma \langle n_i \rangle).$$

The internal energy and the Helmholtz potential appear as

$$U = \sum_i \langle n_i \rangle \, \epsilon_i = kT \sum_i \langle n_i \rangle \left(\ln z + \ln(1 - \sigma \langle n_i \rangle) - \ln \langle n_i \rangle \right), \quad \text{and}$$

$$F = -PV + \mu \langle N \rangle = -kT \ln Z_G + \langle N \rangle \, kT \ln z = \sigma kT \sum_i \ln(1 - \sigma \langle n_i \rangle) + \langle N \rangle \, kT \ln z.$$

From these, the entropy is found by $S = (U - F)/T$ to be

$$S = -k \sum_i \left[\langle n_i \rangle \ln \langle n_i \rangle + \sigma(1 - \sigma \langle n_i \rangle) \ln(1 - \sigma \langle n_i \rangle) \right]. \tag{4.7.10}$$

This expression for the entropy clearly vanishes for fermions at $T = 0$, since every $\langle n_i \rangle$ is either 1, for states below the Fermi level, or 0, for states above it. For bosons, at

$T = 0$, $\langle n_0 \rangle = \langle N \rangle$, and all other occupation numbers vanish. The entropy for completely degenerate bosons is given by Eq. (4.7.10) to be the nonextensive expression

$$S_0 \approx k[\ln \langle N \rangle + 1 + 1/\langle N \rangle], \qquad (4.7.11)$$

which is another artifact of the approximations used in the calculation. It is a negligible magnitude in comparison with normal molar entropies, which are of the order of RT, and the entropy per particle given by this expression vanishes in the thermodynamic limit. On this basis, and from the original information-theoretic expression in terms of a probability set, the entropy for a system of bosons at $T = 0$ is found to be zero. This does not say, as is sometimes asserted, that particles in the ground state have zero entropy, and Bose-Einstein condensation is the process of depositing the particles into this zero-entropy state. The entropy becomes zero because the ground-state probability becomes unity, while all other probabilities vanish. The same zero entropy would occur if any state j could have $p_j = 1$. If the ground state were r-fold degenerate, its contribution to the entropy would be, since each of the ground-state probabilities is $1/r$ times the total ground-state probability,

$$\Delta S_0/\langle N \rangle = -kr[(p_0/r)\ln(p_0/r)] = -kp_0[\ln p_0 - \ln r],$$

so that when $p_0 \to 1$, the total entropy becomes

$$S_0 = \langle N \rangle k \ln r,$$

which agrees with the expected result when $r = 1$, and extends it for degenerate ground states. (The often-stated Third Law of Thermodynamics assigns the value zero to the entropy of any system at $T = 0$. Here it is evident that the validity of the third law rests upon the assumption that the ground state of every system is nondegenerate, which is clearly not true when spin degeneracy exists.)

The other calculation of the entropy comes from $U = TS - PV + \mu \langle N \rangle$. Since $PV = 2U/3$, it follows that

$$S = \frac{5U}{3T} - \frac{\mu \langle N \rangle}{T} = \frac{5U}{3T} - \langle N \rangle k \ln z. \qquad (4.7.12)$$

When $\langle N \rangle = N_0$, the entropy per mole is found to be

$$s = \frac{5}{2}R\frac{g_{5/2}(z)}{g_{3/2}(z)} - R\ln z, \quad T > T_c$$

$$= \frac{5}{2}R\frac{\zeta(5/2)}{\zeta(3/2)}\left(\frac{T}{T_c}\right)^{3/2}, \quad T < T_c. \qquad (4.7.13)$$

Note that there is no discontinuity in energy, pressure, or entropy at $T = T_c$.

In order to calculate C_v, it is necessary first to examine the derivative of z with respect to T, which is obtained from Eq. (4.7.1), rewritten for one mole, or

$$N_0 = \frac{g_s v}{\lambda^3}g_{3/2}(z).$$

Differentiation with respect to T at constant volume gives

$$\left(\frac{\partial z}{\partial T}\right)_v = -\frac{3z}{2T}\frac{g_{3/2}(z)}{g_{1/2}(z)}. \tag{4.7.14}$$

From this result and that of Eq. (4.7.4), C_v is found to be

$$C_v = \left(\frac{\partial u}{\partial T}\right)_v = \frac{15R}{4}\frac{g_{5/2}(z)}{g_{3/2}(z)} - \frac{9R}{4}\frac{g_{3/2}(z)}{g_{1/2}(z)}, \quad T > T_c,$$

$$= \frac{15R}{4}\frac{\zeta(5/2)}{\zeta(3/2)}\left(\frac{T}{T_c}\right)^{3/2}, \quad T < T_c. \tag{4.7.15}$$

Since $g_{1/2}(1) = \zeta(1/2)$ is infinite, C_v is continuous at $T = T_c$, but since z diminishes toward zero as the temperature rises, the ratios of the g-functions approach unity, and $C_v \to 3R/2$ as T becomes large, in agreement with the expectation that quantum ideal gases become increasingly like Boltzmann gases at temperatures well above the region of degeneracy. The temperature derivative of C_v is

$$\frac{\partial C_v}{\partial T} = \frac{45R}{8T}\frac{g_{5/2}(z)}{g_{3/2}(z)} - \frac{9R}{4T}\frac{g_{3/2}(z)}{g_{1/2}(z)} - \frac{27R}{8T}\frac{g_{3/2}^2(z)g_{-1/2}(z)}{g_{1/2}^3(z)}, \quad T > T_c,$$

$$= \frac{45R}{8T}\frac{\zeta(5/2)}{\zeta(3/2)}\left(\frac{T}{T_c}\right)^{3/2}, \quad T < T_c. \tag{4.7.16}$$

There is a discontinuity in the slope of C_v at T_c. The slope for $T = T_c^-$ is positive, and that for $T = T_c^+$ is negative. The difference is found to be

$$\left(\frac{\partial C_v}{\partial T}\right)_{T_c^-} - \left(\frac{\partial C_v}{\partial T}\right)_{T_c^+} = -\frac{27R}{16\pi T_c}[\zeta(3/2)]^2 = -3.665\frac{R}{T_c}. \tag{4.7.17}$$

Electrons have small enough mass and occur in metals at high enough densities to offer a physical system that is a reasonable approximation to a degenerate fermion gas. No such system is readily available for bosons. The least-massive boson gas that is ordinarily available is ^{4}He, with zero spin. Bose-Einstein condensation is sometimes presented as the introductory theory of liquid helium, which exhibits the phenomenon of superfluidity, and has a phase transition at the temperature of superfluidity onset, 2.19K at atmospheric pressure. Remarkably enough, substitution of the data for helium into Eq. (4.7.3), *i.e.* $v = 27.6 \times 10^{-3}\text{m}^3/\text{kmole}$, $m = 6.65 \times 10^{-27}$ kg, the result for T_c is 3.13K, a value encouragingly close to the so-called λ-temperature of liquid helium, at which the transition to superfluidity onsets. But liquid helium is not even approximately an ideal gas, since interatomic binding energies at the required low temperatures are not negligible with respect to kT. (It should be noted, however, that the behavior of liquid ^{3}He, which is a fermion, is quite unlike that of ^{4}He. ^{3}He becomes superfluid only at millikelvin temperatures, and by a different mechanism, indicating that boson properties are important in understanding of superfluidity of ^{4}He.)

Possibly the best naturally occurring example of Bose-Einstein condensation might be found in the α-particle populations of some white dwarfs. At a typical white-dwarf central density of $\rho_c = 10^{10}$ kg/m^3, the condensation temperature for particles of α-particle mass is $T_c = 2.5 \times 10^6$ K, which may be above the existing core temperature. The actual temperature may be high enough to dominate interparticle potential energies, and there may be an appreciable population of ground-state particles. It seems unlikely that this speculation will ever be examined experimentally.

4.8 PHOTONS AND BLACK-BODY RADIATION

Photons are massless, spin-1 particles, and therefore bosons. The zero-mass property not only requires re-examination of such results as Eq. (4.7.3) for T_c, but also the general question of particle conservation. When the grand-canonical formalism was introduced at the beginning of Chapter 3, it was stated that a grand-canonical system could be regarded as a subsystem of a canonical system that is open to matter flow. The expectation value for the number of particles is simply the particle density of the canonical system—the total number of particles divided by its volume—multiplied by the volume of the grand-canonical system. This is, at least in principle, objective information, valid as an input expectation value for the determination of μ. No such situation exists for photons in a cavity. The number is not conserved, even in a closed cavity that might be regarded as a canonical system. It will turn out that the number of photons is subject to major fluctuations. Therefore a reformulation is necessary.

In general, even without good input data for $\langle N \rangle$, it is possible to use the grand-canonical formalism as if $\langle N \rangle$ were a valid number. But after the formal results, the required step is the minimization of the Helmholtz potential at constant volume and temperature, in order to find the equilibrium value of z (or μ), since, in the absence of a valid input constraint for $\langle N \rangle$, the correct equilibrium state is the one that minimizes F. From Eqs. (4.7.6) and (4.7.12), F is found to be

$$F = U - TS = -PV + \mu \langle N \rangle = -kT[\ln Z_G - \langle N \rangle \ln z] = kT[\langle N \rangle \ln z - \ln Z_G].$$

From this expression, the derivative gives

$$\left(\frac{\partial F}{\partial \langle N \rangle} \right)_{T,V} = kT \left[\ln z + \left(\frac{\langle N \rangle}{z} - \left(\frac{\partial Z_G}{\partial z} \right)_{T,V} \right) \bigg/ \left(\frac{\partial \langle N \rangle}{\partial z} \right)_{T,V} \right].$$

The last two terms cancel, in accordance with Eq. (3.4.6), leaving as the requirement for minimum F,

$$\left(\frac{\partial F}{\partial \langle N \rangle} \right)_{T,V} = kT \ln z = 0. \tag{4.8.1}$$

As a consequence, the choice $z = 1$, or $\mu = 0$, is necessary in order to minimize the Helmholtz potential at constant volume and temperature, in the absence of a valid physical constraint on $\langle N \rangle$. When this choice is made, the Helmholtz potential and the grand potential become equal, as do the canonical and grand partition functions.

In order to derive an expression for the photon density of states, a procedure much like the one used for the Debye modes of an elastic solid, Sec. 3.3.4, is used. There, each normal mode of the solid is treated as an independent quantum oscillator, and the density of states is found by counting the distribution of normal modes. For a cavity filled with electromagnetic radiation in thermal equilibrium with the fixed-temperature walls, two equivalent treatments are available. Either the electromagnetic waves that fit in a region of space with periodic boundary conditions may be regarded as independent quantized oscillators, as in the Debye treatment of the elastic solid, or the photons may be regarded as independent particles of energy $\hbar\omega$, and treated by the already developed grand-canonical formalism. In either case, the density of states is found from the normal-mode (or equivalent) distribution for the cavity. There are two independent transverse waves for each normal mode of a cavity, and no longitudinal wave for empty space. Again, periodic boundary conditions are imposed on a rectangular-parallelopiped cavity of dimensions a_1, a_2, and a_3, with volume $V = a_1 a_2 a_3$. The components of the wave vector satisfy

$$k_1^2 + k_2^2 + k_3^2 = \omega^2/c^2, \tag{4.8.2}$$

where c is the velocity of light. The periodic boundary conditions require that the k_i satisfy

$$k_i = 2q_i\pi/a_i,$$

leading to the number-space ellipsoid

$$\sum_{i=1}^{3}\left(\frac{q_i}{a_i}\right)^2 = \left(\frac{\omega}{2\pi c}\right)^2,$$

with a volume in number space of

$$\mathcal{V} = \frac{V\omega^3}{6\pi^2 c^3}.$$

Since there is one normal mode per unit volume of number space, the number of normal modes in the interval $d\omega$ is

$$d\mathcal{V} \equiv g_t(\omega)d\omega = V\omega^2 d\omega/2\pi^2 c^3.$$

When the two transverse modes are taken into account, the density-of-states function becomes

$$g(\omega) = 2g_t = V\omega^2/\pi^2 c^3. \tag{4.8.3}$$

Unlike the Debye model, there is not an upper limit to the allowed frequency of electromagnetic waves in a cavity. In the absence of an upper limit to the number of modes, the zero-point energy of the cavity becomes infinite, since each normal-mode oscillator is regarded as having its own zero-point energy. For most thermodynamic processes, this energy does nothing. It is usually no more to be considered than is the rest energy of any object having mass. It is usual, therefore, to disregard the zero-point energy, even though it is formally infinite. (Its effects are not always negligible, and its existence should be kept in mind (Boyer [1974a]; Boyer [1974b]).) The energy per photon is $\hbar\omega$, whether or not the normal mode has zero-point energy, and from the grand-canonical point of view that the photons constitute a gas of noninteracting bosons, the grand partition function can now be written

$$\ln Z_G = -\left(\frac{V}{\pi^2 c^3}\right)\int_0^\infty d\omega\omega^2 \ln(1 - e^{-\beta\hbar\omega}). \tag{4.8.4}$$

As already mentioned, this result is exactly the same as would have been derived by a Debye-like procedure that leads to the canonical partition function, with the zero-point energy disregarded. Integration by parts gives

$$
\begin{aligned}
\ln Z_G &= -\left(\frac{V\omega^3}{3\pi^2 c^3}\right)\ln(1-e^{-\beta\hbar\omega})\bigg|_0^\infty + \left(\frac{V\beta\hbar}{3\pi^2 c^3}\right)\int_0^\infty \frac{\omega^3\, d\omega}{e^{\beta\hbar\omega}-1}, \\
&= \left(\frac{V\beta\hbar}{3\pi^2 c^3}\right)\left(\frac{kT}{\hbar}\right)^4 \int_0^\infty dx\, x^3 e^{-x}\left(1+e^{-x}+e^{-2x}+\cdots\right), \\
&= \frac{V}{3\pi^2 c^3}\left(\frac{kT}{\hbar}\right)^3 \Gamma(4)\zeta(4) = \left(\frac{\pi^2 V}{45 c^3}\right)\left(\frac{kT}{\hbar}\right)^3.
\end{aligned}
\tag{4.8.5}
$$

The pressure is found, from $PV = kT\ln Z_G$, to be

$$
P = \frac{\pi^2}{45(c\hbar)^3}(kT)^4 = \frac{1}{3}\sigma T^4,
\tag{4.8.6}
$$

and the energy density, $u_v = U/V$ is

$$
u_v = -\frac{1}{V}\left(\frac{\partial Z_G}{\partial\beta}\right)_V = \frac{\pi^2 k^4}{15(\hbar c)^3}T^4 = \sigma T^4 = 3P,
\tag{4.8.7}
$$

where

$$
\sigma = \frac{\pi^2 k^4}{15(\hbar c)^3}.
\tag{4.8.8}
$$

This result is the *Stefan-Boltzmann Law*, and σ is called the *Stefan-Boltzmann Constant*.

The energy density per unit frequency interval is obtained by differentiating the first line of Eq. (4.8.5) with respect to β. The result is

$$
du_v = \frac{8\pi\nu^2}{c^3}\frac{h\nu\, d\nu}{e^{\beta h\nu}-1},
\tag{4.8.9}
$$

a result known as the *Planck Radiation Law*. When $kT \gg h\nu$, the Planck law is approximated by the *Rayleigh-Jeans Law*,

$$
du_v = \frac{8\pi\nu^2}{c^3}kT\, d\nu,
\tag{4.8.10}
$$

which is a purely classical result that, in its disagreement with experiment, led Planck to his well-known theoretical explorations.

The heat capacity per unit volume of the cavity in equilibrium with its walls is

$$
C_V = \left(\frac{\partial u_v}{\partial T}\right)_V = 4\sigma T^3.
\tag{4.8.11}
$$

The entropy per unit volume, s_v, is found to be

$$
s_v = \frac{u_v + P}{T} = \frac{4}{3}\sigma T^3,
\tag{4.8.12}
$$

in agreement with Eq. (4.8.11). Since the entropy of the radiation in a cavity of volume V is $S = V s_v = 4\sigma T^3 V/3$, it follows that a reversible adiabat is given by

$$VT^3 = 3S/4\sigma = \text{constant.}$$

When this expression is solved for T and the result substituted into Eq. (4.8.6), the P-V expression for the adiabatic is

$$PV^{4/3} = \frac{1}{4}\left(\frac{3}{4\sigma}\right)^{1/3} S^{4/3} = \text{constant.} \tag{4.8.13}$$

It might be expected that adiabats are always of the form $PV^\gamma = \text{const.}$, and that $\gamma = 4/3 = C_P/C_v$ would automatically give C_P. But since P is a function of T alone, the isotherms in PV-space are parallel to the V axis, and C_P is therefore infinite. (No finite amount of heat can change the temperature at constant pressure.)

The expected number of photons can now be calculated. From Eq. (4.4.3) with $z = 1$, it follows that

$$\langle N \rangle = \sum_i (e^{\beta\hbar\omega_i} - 1)^{-1}.$$

When the density-of-states expression is used to replace the sum by an integral, the result is

$$\langle N \rangle = \frac{V}{\pi^2 c^3} \int_0^\infty \frac{\omega^2 \, d\omega}{e^{\beta\hbar\omega} - 1} = \left(\frac{2V}{\pi^2}\right)\left(\frac{kT}{\hbar c}\right)^3 \zeta(3). \tag{4.8.14}$$

The integrand may be considered to represent the expected number of photons per frequency interval, as was written for the energy in Eq. (4.8.9). As will be shown in the next section, these expressions for the expected photon population are not to be viewed with great confidence, because of the uncommonly large fluctuations in photon numbers (and boson numbers in general).

4.9 FLUCTUATIONS IN IDEAL GASES AT EQUILIBRIUM

The extraction of values for the macroscopic thermal properties from the statistical averages over the underlying microstructure of a system, shifting and grainy though it may be, is a major triumph for theoretical physics. A further feature of the statistical treatment, and one that reveals the existence of the unobserved microscopic activities that in sum manifest themselves as the macroscopic properties of the system, is the study of fluctuations. There are various levels of such a study. In Chapter 1, several calculations gave the time-dependent behavior of the probabilities that describe the microstates of systems as they evolve to the states that may be regarded as equilibrium, characterized by probabilities that become independent of time. But even in these analyses, the final probability sets contain nonvanishing probabilities that the system can be in states that would not be regarded as equilibrium states, *if it were observed to be in one of those states.* When a distribution of probabilities exists, not all of which lead to the same value of a macroscopic system parameter, it is usual to say that the value of the parameter fluctuates, and to quantify the fluctuation by the variance, or mean squared deviation from the mean. This quantification is a passive one, in that it says nothing of the time scale or power spectrum of the fluctuation. (These aspects are studied in Chapter 10.) The development of this section is concerned only with the passive quantification of fluctuations in ideal gases, and in the related occupation numbers of the single-particle states (Pathria [1972]; Rahal [1989]).

It is instructive at this point to examine these equilibrium fluctuations of the macroscopic thermal variables and, when readily accessible, the underlying fluctuations in the microscopic quantities, for ideal gases, both with and without the quantum constraints of this chapter. What is allowed to fluctuate depends upon the system under consideration. A microcanonical system has a fixed number of particles in a fixed volume, with the energy fixed, in that no energy exchange with the surroundings is permitted. There is a presumed small uncertainty δE in the known value of the energy, so systems that are prepared to be identical may vary in energy over the range δE. (Similar practical ranges of variation exist for volume and particle number.) Because of this intrinsic uncertainty, the system energy in a microcanonical ensemble is said to fluctuate over the range δE, although the energy of a single such system is regarded as constant, but imperfectly known. Thus there are no proper fluctuations in U, V, or N in the microcanonical system. The only other extensive variable is the entropy S, here regarded as given in terms of our information about the system and therefore not subject to fluctuation under normal circumstances for any equilibrium system described by information theory.

The canonical system has fixed volume and particle numbers, but it exchanges energy with a reservoir at constant temperature. Therefore its only variables that can fluctuate are the energy and any other variables that are directly related to the energy, such as pressure and the occupation numbers of the individual single-particle states. (The pressure fluctuation requires some discussion, which is forthcoming.)

The grand-canonical system is open to matter flow. As a consequence, the total population can fluctuate, as can the occupation numbers of the single-particle quantum states. The volume is considered fixed. Thus there can be fluctuations in the energy, particle numbers, and variables dependent upon these quantities. These fluctuations are quite different for fermions, bosons, and Boltzmann particles, and a comparison is instructive.

4.9.1 General Formulation. In order to calculate these fluctuations, and other related ones, it is expedient to review the structure of the grand partition function in a generalized development that is not restricted to ideal gases. Recall that

$$Z_G = \sum_{N=0}^{\infty} z^N Z_N,$$

where the N-particle canonical partition function Z_N may be written

$$Z_N = \sum_i \exp\left(-\beta E_i(N)\right),$$

where $E_i(N)$ is the energy of the ith N-particle state. In this treatment, the states are regarded as different if the particle quantum numbers, including spin, are not identical. The probability of there being N particles in the system is

$$W_N = z^N Z_N / Z_G, \tag{4.9.1}$$

and the expectation value for N is

$$\langle N \rangle = \sum_{N=0}^{\infty} N W_N = z \left(\frac{\partial \ln Z_G}{\partial z} \right)_{V,T}. \tag{4.9.2}$$

Another application of the same operator gives

$$\left(\frac{z\partial}{\partial z}\right)^2_{V,T} \ln Z_G = \sum_N N^2 W_N - \left(\sum_N N W_N\right)^2$$
$$= \langle N^2 \rangle - \langle N \rangle^2$$
$$= \langle (\Delta N)^2 \rangle_G , \tag{4.9.3}$$

which is the same as $\langle (N - \langle N \rangle)^2 \rangle_G$, the variance, where the subscript on the average refers to the grand canonical system.

Energy variances are found similarly. The energy expectation value is already known to be

$$U = \langle E \rangle_N = -\left(\frac{\partial \ln Z_N}{\partial \beta}\right)_{V,N} = \langle E \rangle_G = -\left(\frac{\partial \ln Z_G}{\partial \beta}\right)_{V,z},$$

where the subscript N denotes the N-particle canonical system. A second application of the β-derivative gives the appropriate variance, but this time different for canonical and grand systems.

$$\langle (\Delta E)^2 \rangle_N = -\left(\frac{\partial U}{\partial \beta}\right)_{V,N}, \tag{4.9.4}$$

and

$$\langle (\Delta E)^2 \rangle_G = -\left(\frac{\partial U}{\partial \beta}\right)_{V,z}. \tag{4.9.5}$$

One other useful expectation value for the grand canonical system is a measure of the correlation between particle number and energy, given by the E-N covariance

$$\langle \Delta E N \rangle_G = \langle E N \rangle_G - U \langle N \rangle = -\left(\frac{\partial \langle N \rangle}{\partial \beta}\right)_{V,z} = z\left(\frac{\partial U}{\partial z}\right)_{V,\beta}. \tag{4.9.6}$$

These measures of expected fluctuation can be related to macroscopic thermodynamic parameters. For example,

$$\langle (\Delta E)^2 \rangle_N = kT^2 \left(\frac{\partial U}{\partial T}\right)_{V,N} = kT^2 \frac{N}{N_0} C_v, \tag{4.9.7}$$

where N_0 is the Avogadro number. (Note that the canonical energy variance can be found without first obtaining the canonical partition function, because equivalent canonical and grand-canonical systems are regarded as having the same energy expectation value.) This result says, first of all, that $C_v > 0$, since the variance and all other quantities are non-negative. This agrees with one of the stability criteria established in Chapter 3. But beyond that, it says that at a given temperature and population, the larger the molar heat capacity, the larger the energy variance. This sometimes-puzzling result can be understood in terms of the Le Chatelier principle: As the energy of the system fluctuates, say positively, the kinetic temperature of the particles increases, resulting in a stabilizing flow of energy back to the reservoir. The larger the molar heat capacity, the smaller the kinetic temperature response to that energy fluctuation, or the more the energy can fluctuate for a given change in the kinetic temperature. Other aspects of the energy variance will be explored in later sections.

From Eq. (4.9.3), the variance in particle number can be written

$$\langle(\Delta N)^2\rangle_G = z\left(\frac{\partial\langle N\rangle}{\partial z}\right)_{V,\beta} = z\left(\frac{\partial\langle N\rangle}{\partial\mu}\right)_{V,\beta}\left(\frac{\partial\mu}{\partial z}\right)_\beta$$

$$= kT\left(\frac{\partial\langle N\rangle}{\partial\mu}\right)_{V,\beta} = \frac{\langle N\rangle^2 kT\kappa_T}{V}, \tag{4.9.8}$$

where the final step requires further explanation. From Eq. (2.4.7), rewritten as

$$d\mu = \frac{V}{N}dP - \frac{S}{N}dT,$$

and from Eq. (2.5.19), it follows that

$$\left(\frac{\partial\mu}{\partial N}\right)_{T,V} = \frac{V}{N}\left(\frac{\partial P}{\partial N}\right)_{T,V}$$

$$= \left(\frac{-V}{N}\right)\left(\frac{\partial P}{\partial V}\right)_{T,N}\left(\frac{\partial V}{\partial N}\right)_{T,P}$$

$$= \left(\frac{V}{N}\right)\left(\frac{1}{V\kappa_T}\right)\left(\frac{\partial\mu}{\partial P}\right)_{T,N},$$

which is the relationship needed in Eq. (4.9.8). Again one of the stability criteria of Chapter 3, $\kappa_T \geq 0$, is established from consideration of a variance. This result also says, among other things, that the more compressible the material, the more its number of particles can fluctuate, for a given volume. In Eq. (4.9.8), the density can be written everywhere in place of N, to give

$$\langle(\Delta\rho)^2\rangle = \langle\rho\rangle_G^2 kT\kappa_T/V,$$

relating the density variance to the compressibility. The explanation, in accordance with the Le Chatelier principle, is that a material that has high compressibility has a relatively small increase in pressure per unit increase in density, and therefore is relatively tolerant of density fluctuations.

The canonical and grand-canonical energy variances can be related, as follows:

$$\langle(\Delta E)^2\rangle_G = -\left(\frac{\partial U}{\partial\beta}\right)_{V,z}$$

$$= -\left(\frac{\partial U}{\partial\beta}\right)_{V,N} - \left(\frac{\partial N}{\partial\beta}\right)_{V,z}\left(\frac{\partial U}{\partial N}\right)_{V,\beta}$$

$$= \langle(\Delta E)^2\rangle_N + \langle(\Delta NE)\rangle_G\left(\frac{\partial U}{\partial N}\right)_{V,\beta}$$

$$= \langle(\Delta E)^2\rangle_N + \langle(\Delta N)^2\rangle_G\left(\frac{\partial U}{\partial N}\right)_{V,T}^2. \tag{4.9.9}$$

The terms in this result are clear. The first is the energy variance attributable to heat exchange with the surroundings, but no matter flow, and the second is the additional contribution resulting from the exchange of particles between the system and its surroundings.

Other measures of system fluctuation are the relative quantities:

$$\langle (\Delta E)^2 \rangle / U^2, \quad \langle (\Delta N)^2 \rangle / N^2, \quad \text{and} \langle \Delta N E \rangle / \langle N \rangle U.$$

The first two are called *relative variances*, and the third seems not to have a common name, but it could be called the *relative covariance*. A more universally used measure of the relative correlation is the correlation coefficient, $\eta_{E,N}$, given by

$$\eta_{E,N} = \frac{\langle \Delta E N \rangle}{\sqrt{\langle (\Delta E)^2 \rangle} \sqrt{\langle (\Delta N)^2 \rangle}}, \tag{4.9.10}$$

which takes values between -1 and $+1$.

These quantities will be compared for Boltzmann, Fermi, and Bose gases in subsequent sections.

4.9.2 Pressure Fluctuations. Pressure is an intensive variable, whereas energy and particle number are extensive. Furthermore, it is sometimes narrowly regarded only as the force per unit area *on the walls of the container*, rather than as a well-defined local parameter that can exist throughout the system. As a consequence, some authors have considered it off limits to discuss fluctuations in pressure, or have considered it possible only in the context of the detailed power spectrum of wall-particle interactions. In reality, pressure fluctuations present no difficulty that has not already been encountered in the study of fluctuations of energy and particle number.

The pressure tensor is defined quite satisfactorily in classical kinetic theory as

$$P_{ij} = \frac{1}{\Delta V} \sum_k m_k \langle (v_{ik} - \langle v_i \rangle)(v_{jk} - \langle v_j \rangle) \rangle,$$

where i and j can stand for x, y, or z, m_k is the mass of the kth particle, the sum on k is over the particles in the volume ΔV being considered, the v_i are components of the individual particle velocities, and the $\langle v_i \rangle$ are the averages of these components over the chosen volume. Since this part of the discussion is primarily for the purpose of background information, it is notationally more convenient to choose a reference frame in which the average velocity is zero, in agreement with the usual thermostatic treatment. The pressure tensor then becomes

$$P_{ij} = \frac{1}{\Delta V} \sum_k m_k \langle (v_{ik} v_{jk}) \rangle,$$

and further simplification is possible. In an ideal isotropic gas at equilibrium, the orthogonal components of a molecule's velocity are independent, so the off-diagonal elements of the pressure tensor vanish, and the diagonal elements are found to be equal. In this case, the pressure is defined to be

$$P = P_{xx} = P_{yy} = P_{zz} = \frac{1}{\Delta V} \sum_k m_k v_{ik}^2 = \frac{1}{3\Delta V} \sum_k m_k v_k^2.$$

Since the translational kinetic energy density is given by

$$\frac{\Delta E_K}{\Delta V} = \frac{1}{2\Delta V} \sum_k m_k v_k^2,$$

it follows that

$$P = \frac{2}{3} \frac{\Delta E_K}{\Delta V}, \qquad (4.9.11)$$

and the pressure must fluctuate exactly as does the translational kinetic energy density, since exactly the same microscopic fluctuations are responsible. It should be noted that the pressure, defined in this manner, is a local property of the chosen region of space, in no way dependent upon the existence of walls or material boundaries. For monatomic, classical ideal gases, the pressure fluctuation matches that of the internal energy density.

Pressure fluctuations can be discussed more generally, without any restriction to the kinetic theory of classical ideal gases. An N-particle state for an ideal gas of noninteracting particles has the energy $E_i(N) = \sum_j n_{ij}(N)\epsilon_j$, where the ϵ_j are single-particle energies, the $n_{ij}(N)$ are the occupation numbers of the single-particle states for the ith N-particle state, and $\sum_j n_{ij}(N) = N$. For a microcanonical system, all permitted N-particle states have essentially the same energy. For each such state of energy E_i, the pressure is $P_i = -(\partial E_i/\partial V)_N$.

The canonical system is a subsystem of the microcanonical system, at fixed T and N. The Helmholtz potential is $F = -kT \ln Z_N$, where $Z_N = \sum_i \exp[-\beta E_i(N)]$. The pressure is given by

$$P = -\left(\frac{\partial F}{\partial V}\right)_{T,N} = kT \left(\frac{\partial \ln Z_N}{\partial V}\right)_{T,N},$$

which says that

$$P = -\sum_i \left(\frac{\partial E_i(N)}{\partial V}\right)_N \frac{e^{-\beta E_i(N)}}{Z_N}$$

$$= \sum_i P_i W_i(N), \qquad (4.9.12)$$

where P_i is the pressure of the ith N-particle state, and $W_i(N)$ is the canonical probability of that state. As may be seen in Eq. (3.2.22), the translational kinetic energy for every single-particle ideal-gas state is proportional to $V^{-2/3}$, so for every $E_i(N)$, it follows that

$$P_i = -\left(\frac{\partial E_i(N)}{\partial V}\right)_N = \frac{2E_{iK}}{3V},$$

where $E_{iK}(N)$ is the kinetic part of the ith N-particle state. Substitution of this result in Eq. (4.9.12) gives

$$P = \frac{2}{3V} \sum_i E_{iK}(N)W_i = \frac{2}{3} \frac{\langle E_K(N)\rangle}{V},$$

in agreement with Eq. (4.9.11). For monatomic ideal gases, $E_K(N) = U$, since the internal energy is entirely kinetic. In this case, the pressure variance is given by

$$\langle (\Delta P)^2 \rangle = \left(\frac{2}{3V} \right)^2 \langle (\Delta E)^2 \rangle,$$

and the relative variance is seen to be exactly that of the energy. The energy-pressure correlation coefficient for a monatomic gas is equal to one, showing perfectly correlated fluctuations. For other ideal gases, where energy can be stored in rotation and vibration that do not affect the pressure, the correlation coefficient is accordingly reduced.

4.9.3 Relationship of Canonical and Grand-Canonical Quantities. Before proceeding further, it is useful to clarify, in more detail than is usual, the relationships of equivalent canonical and grand-canonical systems. For the treatment of equivalent systems, the grand-canonical system may be regarded as contained in the volume originally occupied by the equivalent canonical system, but now open to matter flow from a myriad of canonical systems identical to the original one. (Equivalently, the g-c system is in equilibrium with a thermal reservoir at temperature T, as is the canonical system, and with a matter reservoir at chemical potential μ, equal to the chemical potential of the equivalent canonical system.) The canonical energy is

$$U_N = \sum_i W_i(N) E_i(N) = \sum_j \langle n_j \rangle_N \epsilon_j, \qquad (4.9.13)$$

where $W_i(N) = e^{-\beta E_i(N)}/Z_N$, $E_i(N) = \sum_j n_{ij}(N)\epsilon_j$, and $\langle n_j \rangle_N = \sum_i W_i(N) n_{ij}(N)$. The $n_{ij}(N)$ are the occupation numbers of the single-particle states with energies ϵ_j in the ith N-particle state. As already noted, these numbers define the N-particle states, and are therefore not subject to fluctuation.

The grand-canonical energy is

$$U_G = \sum_j \langle n_j \rangle_G \epsilon_j. \qquad (4.9.14)$$

When the two systems are equivalent, evidently $U_N = U_G$, the canonical total particle number must equal the grand-canonical expectation value, $\langle N \rangle$, and the canonical energy can be written as $U_{\langle N \rangle}$ to specify that the canonical system is a particular one, and not an arbitrarily selected one of the set that is summed over to define Z_G. From Eqs. (4.9.13) and (4.9.14), it follows that

$$\langle n_i \rangle_G = \langle n_i \rangle_{\langle N \rangle}, \qquad (4.9.15)$$

for all i. It is therefore unnecessary to use any subscript to differentiate between canonical and grand-canonical occupation numbers for equivalent systems.

The probability of N particles in the grand canonical system is seen from the definition of Z_G to be

$$W_N = z^N Z_N / Z_G,$$

and the energy is given by

$$U_G = -\left(\frac{\partial \ln Z_G}{\partial \beta} \right)_{V,z} = -\frac{\sum_N z^N (\partial Z_N/\partial \beta)_V}{Z_G} = \frac{\sum_N z^N Z_N U_N}{Z_G} = \sum_N W_N U_N. \qquad (4.9.16)$$

Therefore, since $U_G = U_{\langle N \rangle}$, it follows that $U_{\langle N \rangle} = \sum_N W_N U_N$.

The entropies are compared by means of the thermodynamic relation $S = (U - F)/T$. Since $U_G = U_{\langle N \rangle}$, it is necessary only to compare the expressions for F, from $F = U - TS = -PV + \mu N$. The PV term is clearly the same for both systems, as is μ, and for both systems, the N becomes $\langle N \rangle$. Therefore $F_G = F_{\langle N \rangle}$, and the entropies are found to satisfy $S_G = S_{\langle N \rangle}$. The partition functions can be related through expressions for F:

$$F_{\langle N \rangle} = -kT \ln Z_{\langle N \rangle} = F_G = -PV + \mu \langle N \rangle$$

$$= -kT \ln Z_G + \langle N \rangle kT \ln z = -kT \ln \left(z^{-\langle N \rangle} Z_G \right), \qquad (4.9.17)$$

from which it follows that

$$Z_G = z^{\langle N \rangle} Z_{\langle N \rangle}. \qquad (4.9.18)$$

Related expressions are to be found in the solution to Problem 1.

4.9.4 Fluctuations in a Boltzmann Ideal Gas. In this and the next two subsections, the formalism of Section 4.9.1 will be applied to Boltzmann, Fermi, and Bose ideal gases. The Boltzmann gas is a useful reference system with which the others can be compared. The canonical partition function for the Boltzmann gas is $Z_N = Z_1^N/N!$, where Z_1 is the single-particle partition function. The grand partition function is $Z_G = e^{zZ_1}$, as shown in Chapter 3. From this, it follows that $\langle N \rangle = z Z_1$, and the probability that there are N particles in the system is given by Eq. (4.9.1) to be

$$W_N = \frac{z^N Z_N}{Z_G} = \frac{(zZ_1)^N}{N! Z_G} = \frac{\langle N \rangle^N}{N!} e^{-\langle N \rangle}, \qquad (4.9.19)$$

which is the Poisson distribution. The variance is

$$\langle (\Delta N)^2 \rangle_G = \langle N \rangle, \qquad (4.9.20)$$

which is characteristic of the Poisson distribution. The relative variance is seen to be

$$\langle (\Delta N)^2 \rangle / \langle N \rangle^2 = 1/ \langle N \rangle,$$

which becomes negligibly small, for most purposes, in macroscopic systems. This is a typical result, and one that accounts for the excellent agreement between the macroscopic and statistical theories.

The energy is found to be $U = \langle N \rangle U_1$, where U_1 is the expectation value of the single-particle energy, as should be evident for a Boltzmann ideal gas. For these gases in general, little can be done to improve the result of Eq. (4.9.7), since C_v can be temperature dependent. For the monatomic gas, however, $C_v = 3N_0 k/2$, and the canonical energy variance is

$$\langle (\Delta E)^2 \rangle_N = (3/2) \langle N \rangle (kT)^2. \qquad (4.9.21)$$

Since $U = C_v T$ for the monatomic gas, the relative variance of the canonical energy is

$$\langle (\Delta E)^2 \rangle_N / U^2 = 2/(3 \langle N \rangle). \qquad (4.9.22)$$

The grand canonical energy variance is found from Eq. (4.9.9) and the result of Eq. (4.9.20) to be

$$\langle(\Delta E)^2\rangle_G = \langle(\Delta E)^2\rangle_N + \langle N\rangle U_1^2$$
$$= (3/2)\langle N\rangle(kT)^2 + (9/4)\langle N\rangle(kT)^2 = (15/4)\langle N\rangle(kT)^2. \quad (4.9.23)$$

Evidently sixty percent of the energy variance for a monatomic Boltzmann gas is attributable to the fluctuation in particle density. But even in this case, the relative variance for a micromicromole of the gas is about 10^{-12}.

The *E-N* covariance is found from Eq. (4.9.6) to be

$$\langle\Delta EN\rangle_G = (3/2)\langle N\rangle kT, \quad (4.9.24)$$

and the *E-N* correlation coefficient is

$$\eta_{E,N} = 3/\sqrt{15} = 0.7746, \quad (4.9.25)$$

which is $\sqrt{0.6}$, as expected from the particle-number fluctuation.

Nondegenerate boson and fermion gases exhibit the same fluctuations as Boltzmann gases.

4.9.5 Fluctuations in Degenerate Fermion Gases. For a monatomic spin-1/2 fermion gas, the energy and particle number were derived to be

$$U = \left(\frac{4\pi V(2m)^{3/2}}{h^3}\right)\mathcal{F}_{5/2}(\beta,\mu) = \frac{3}{5}\langle N\rangle\mu_0\left[1 + \frac{5}{12}\left(\frac{\pi}{\beta\mu_0}\right)^2 + \cdots\right],$$

and

$$\langle N\rangle = \left(\frac{4\pi V(2m)^{3/2}}{h^3}\right)\mathcal{F}_{3/2}(\beta,\mu).$$

From these, the variances are found as they were in the previous subsection, to be

$$\langle(\Delta N)^2\rangle = \frac{kT}{2}\left(\frac{4\pi V(2m)^{3/2}}{h^3}\right)\mathcal{F}_{1/2}(\beta,\mu)$$
$$= \left(\frac{3\langle N\rangle kT}{2\mu_0}\right)\left[1 - \frac{1}{6}\left(\frac{\pi}{\beta\mu_0}\right)^2 + \cdots\right], \quad (4.9.26)$$

$$\langle(\Delta E)^2\rangle_G = \langle(\Delta E)^2\rangle_N + \langle(\Delta N)^2\rangle_G\left(\frac{\partial U}{\partial N}\right)^2_{\beta,V} = \frac{3}{2}\langle N\rangle(kT)^2\left[\frac{1}{3}\left(\frac{\pi^2}{\beta\mu_0}\right)\right]$$
$$+ \left(\frac{3\langle N\rangle kT}{2\mu_0}\right)\left[1 - \frac{1}{6}\left(\frac{\pi}{\beta\mu_0}\right)^2 + \cdots\right][\mu_0^2]\left[1 + \frac{1}{6}\left(\frac{\pi}{\beta\mu_0}\right)^2 + \cdots\right]$$
$$= \frac{3}{2}\langle N\rangle kT\mu_0\left[1 + \frac{1}{3}\left(\frac{\pi^2}{\beta\mu_0}\right)^2 + \cdots\right], \quad (4.9.27)$$

where the major contribution is seen to come from fluctuations in particle number, and

$$\langle \Delta N E \rangle = kT \left(\frac{\partial U}{\partial \mu} \right)_{\beta, V} = \left(\frac{3kT}{2} \right) \left(\frac{4\pi V (2m)^{3/2}}{h^3} \right) \mathcal{F}_{3/2}(\beta, \mu) = \frac{3}{2} \langle N \rangle kT, \quad (4.9.28)$$

where the last result is exact and identical to the Boltzmann value. The relative energy and particle number variances for the fermion gas are

$$\frac{\langle (\Delta E)^2 \rangle}{U^2} = \left(\frac{25}{6 \langle N \rangle} \right) \left(\frac{kT}{\mu_0} \right) \left[1 - \frac{1}{2} \left(\frac{\pi}{\beta \mu_0} \right)^2 + \cdots \right], \quad (4.9.29)$$

which is smaller than the Boltzmann equivalent by a factor of order kT/μ_0, and similarly

$$\frac{\langle (\Delta N)^2 \rangle}{\langle N \rangle^2} = \left(\frac{3}{2 \langle N \rangle} \right) \left(\frac{kT}{\mu_0} \right) \left[1 - \frac{1}{6} \left(\frac{\pi}{\beta \mu_0} \right)^2 + \cdots \right]. \quad (4.9.30)$$

The E-N correlation coefficient is

$$\eta_{E,N} \approx \frac{(3/2) \langle N \rangle kT}{\left[(3/2) \langle N \rangle kT \mu_0 \right]^{1/2} \left[(3/2) \langle N \rangle (kT/\mu_0) \right]^{1/2}} = 1, \quad (4.9.31)$$

which reinforces the result of Eq. (4.9.27) that virtually all of the energy fluctuation in the degenerate fermion gas is associated with particles entering or leaving the system at the Fermi level.

Degenerate fermions are seen to have much less freedom to fluctuate than the nondegenerate Boltzmann particles, a result that is a clear consequence of the exclusion principle. Boltzmann relative fluctuations go as $\langle N \rangle^{-1}$, and corresponding degenerate fermion quantities go as kT/μ_0 times the Boltzmann equivalents.

4.9.6 Fluctuations in Degenerate Boson Gases. For a monatomic boson ideal gas with spin degeneracy g_s, the particle number and energy for $T \geq T_c$ were derived to be

$$\langle N \rangle = \frac{g_s V}{\lambda^3} g_{3/2}(z),$$

and

$$U = \frac{3}{2} \langle N \rangle kT \frac{g_{5/2}(z)}{g_{3/2}(z)}.$$

From these, the variances are found to be

$$\langle (\Delta N)^2 \rangle_G = \frac{g_s V}{\lambda^3} g_{1/2}(z) = \langle N \rangle \frac{g_{1/2}(z)}{g_{3/2}(z)}, \quad (4.9.32)$$

and

$$\langle (\Delta E)^2 \rangle_G = -\left(\frac{\partial U}{\partial \beta} \right)_{V,z} = \frac{15 g_s V}{4 \beta^2 \lambda^3} g_{5/2}(z) = \frac{5}{2} U kT. \quad (4.9.33)$$

By use of Eq. (4.9.9) and the result

$$\left(\frac{\partial U}{\partial N}\right)_{T,V} = \frac{3}{2}kT\left(\frac{g_{3/2}(z)}{g_{1/2}(z)}\right),$$

it follows that the canonical energy variance is

$$\langle(\Delta E)^2\rangle_N = \frac{5}{2}UkT\left[1 - \frac{3g_{3/2}^2(z)}{5g_{1/2}(z)g_{5/2}(z)}\right]. \tag{4.9.34}$$

The second term in the brackets vanishes as $z \to 1$. The relative variance is

$$\frac{\langle(\Delta E)^2\rangle_G}{U^2} = \frac{5}{2}\frac{kT}{U} = \frac{5g_{3/2}(z)}{3\langle N\rangle g_{5/2}(z)} \leq \frac{3.246}{\langle N\rangle}, \tag{4.9.35}$$

where the number in the numerator comes from substitution of the zeta-functions as the limiting values of the g-functions. (The comparable Boltzmann value is $5/3 = 1.67$.) The boson expression approaches the Boltzmann limit as $z \to 0$.

The relative variance of the particle number is

$$\frac{\langle(\Delta N)^2\rangle}{\langle N\rangle^2} = \left(\frac{1}{\langle N\rangle}\right)\left(\frac{g_{1/2}(z)}{g_{3/2}(z)}\right), \tag{4.9.36}$$

which greatly exceeds $1/\langle N\rangle$ when $z \to 1$.

The E-N covariance is

$$\langle\Delta EN\rangle_G = -\left(\frac{\partial\langle N\rangle}{\partial\beta}\right)_{V,z} = \frac{3}{2}\langle N\rangle kT, \tag{4.9.37}$$

exactly, as for fermions and Boltzmann particles. The E-N correlation coefficient is easily found to be

$$\eta_{E,N} = \frac{3}{\sqrt{15}}\frac{g_{3/2}(z)}{g_{1/2}(z)g_{5/2}(z)}, \tag{4.9.38}$$

which approaches the Boltzmann value as $z \to 0$, and vanishes as T decreases to T_c and $z \to 1$.

For a degenerate boson gas, with $T \leq T_c$, the number density and energy have been found to be

$$\langle N\rangle = \frac{g_s V}{\lambda^3}\zeta(3/2) + \frac{z}{1-z} = N_e + N_g,$$

and

$$U = \frac{3}{2}kT\frac{g_s V}{\lambda^3}\zeta(5/2).$$

For most purposes, z can be set equal to unity, and the ground-state population determined by the expression given in Eq. (4.7.5). For examining fluctuations, however, it must be realized that z is slightly less than 1, and not independent of temperature. The particle-number variance is

$$\langle (\Delta N)^2 \rangle_G = \frac{g_s V}{\lambda^3} \frac{g_{1/2}(z)}{z} + \frac{z}{1-z} + \left(\frac{z}{1-z} \right)^2, \tag{4.9.39}$$

with z very close to its maximum value of 1 ($z = N_g/(N_g + 1)$, with N_g presumably large). From the last two terms in Eq. (4.9.39), the variance of the ground state population is seen to be

$$\langle (\Delta N_g)^2 \rangle_G = N_g(N_g + 1), \tag{4.9.40}$$

and the relative variance of the ground state is

$$\frac{\langle (\Delta N_g)^2 \rangle_G}{N_g^2} = 1 + 1/N_g = \frac{1}{z}. \tag{4.9.41}$$

With such a large relative variance, it is clear that statistical estimates of the ground-state and other populations are not to be taken as good approximations to their instantaneous values.

The energy variance for the degenerate boson gas is

$$\langle (\Delta E)^2 \rangle_G = -\left(\frac{\partial U}{\partial \beta} \right)_{V,z} = \frac{15}{4}(kT)^2 \frac{g_s V}{\lambda^3} \zeta(5/2) = \frac{5}{2}kTU, \tag{4.9.42}$$

which agrees with that for bosons at $T \geq T_c$. The relative variance of the energy is

$$\frac{\langle (\Delta E)^2 \rangle_G}{U^2} = \frac{5kT}{2U} = \frac{5}{3 \langle N \rangle} \frac{\zeta(3/2)}{\zeta(5/2)} \left(\frac{T_c}{T} \right)^{3/2}, \tag{4.9.43}$$

which is seen to diverge as $T \to 0$, but agrees with the result of Eq. (4.9.35) at the condensation temperature.

The N-E covariance in this region is

$$\langle \Delta N E \rangle = \frac{3}{2} N_e kT, \tag{4.9.44}$$

which is a departure from previous results in that only the population of the excited-states appears in this correlation. But since the ground-state population in the other cases is negligible, Eq. (4.9.44) can be regarded merely as exhibiting a feature of the E-N covariance that was not explicitly evident in the previous calculations. Ground-state particles, at zero energy, cannot correlate with the energy, because their comings and goings do not affect the energy. Because of the huge fluctuations of population, the E-N correlation coefficient effectively vanishes for $T \leq T_c$, as may be seen in the following:

$$\begin{aligned}
\eta_{E,N} &= \frac{\langle \Delta E N \rangle_G}{\sqrt{\langle (\Delta E)^2 \rangle_G} \sqrt{\langle (\Delta N)^2 \rangle_G}} \\
&= \frac{(3 \langle N \rangle kT/2)(T/T_c)^{3/2}}{\sqrt{(15/4)(kT)^2(T/T_c)^{3/2}[\zeta(5/2)/\zeta(3/2)]} \sqrt{\langle (\Delta N)^2 \rangle_G}} \\
&= \frac{3}{\sqrt{15}} \left(\frac{\zeta(5/2)}{\zeta(3/2)} \right)^{1/2} \frac{(T/T_c)^{3/4}}{\sqrt{\langle (\Delta N)^2 \rangle}}. \tag{4.9.45}
\end{aligned}$$

Since all other factors in the final expression are finite (and their overall numerical value is less than the Boltzmann 0.7746), the correlation coefficient is determined finally by the population variance (or its square root, the standard deviation) in the denominator. Since this term is very large (formally infinite if z is permitted to equal 1), the correlation coefficient may be regarded as vanishingly small.

4.9.7 Fluctuations in Photons. The expectation values for photon number and total energy have been found in Section 4.8 to be

$$U = \sigma V T^4,$$

and

$$\langle N \rangle = \left(\frac{2V}{\pi^2} \right) \left(\frac{kT}{\hbar c} \right)^3 \zeta(3).$$

From these expressions, the variance in E can be found by the usual method as

$$\langle (\Delta E)^2 \rangle = - \left(\frac{\partial U}{\partial \beta} \right)_{V,z} = 4kTU, \qquad (4.9.46)$$

and the relative variance is

$$\frac{\langle (\Delta E)^2 \rangle}{U^2} = \frac{4kT}{U} = \frac{120\zeta(3)}{\pi^4 \langle N \rangle} = \frac{1.481}{\langle N \rangle}. \qquad (4.9.47)$$

The energy variance is in agreement with Eq. (4.9.7), when the value of C_V from Eq. (4.8.11) is used and recognized to be defined per unit volume instead of per mole. The relative variance is somewhat larger than that of a Boltzmann gas, but less than that of a moderately degenerate boson gas, and is satisfactorily small for macroscopic systems.

The E-N covariance is likewise well behaved. It is found from Eq. (4.8.14) to be

$$\langle \Delta E N \rangle = - \left(\frac{\partial \langle N \rangle}{\beta} \right)_{V,z} = 3 \langle N \rangle kT, \qquad (4.9.48)$$

which agrees with the other E-N covariances except for a factor of 2 that arises because photons are fully relativistic particles. Thus it would seem that the photon number should be well behaved, since the energy and E-N covariance are exemplary.

Two ways can be found to calculate the variance in the photon number. The formal problem is that since $z = 1$ for a photon gas, the usual derivatives with respect to z are not available. The first technique is the obvious one of restoring z to its parameter status, carrying out the formal manipulations, and then setting $z = 1$. As a check, $\langle N \rangle$ should be calculated by this procedure. With z restored, and the photon density of states of Eq. (4.8.3), the photon grand partition function of Eq. (4.8.4) becomes

$$\ln Z_G = - \left(\frac{V}{\pi^2 c^3} \right) \int_0^\infty d\omega \, \omega^2 \ln(1 - z e^{\beta \hbar \omega}). \qquad (4.9.49)$$

In this formulation, the expected photon number is found to be

$$\langle N \rangle = \lim_{z \to 1} z \left(\frac{\partial \ln Z_G}{\partial z} \right)_{V,\beta} = \lim_{z \to 1} \frac{2V}{\pi^2} \left(\frac{kT}{c\hbar} \right)^3 g_3(z) = \frac{2V}{\pi^2} \left(\frac{kT}{c\hbar} \right)^3 \zeta(3),$$

in agreement with Eq. (4.8.14). By this procedure, then, the photon-number variance should be

$$
\begin{aligned}
\langle (\Delta N)^2 \rangle &= \lim_{z \to 1} z \left(\frac{\partial \langle N \rangle}{\partial z} \right)_{V,\beta} \\
&= \lim_{z \to 1} \frac{2V}{\pi^2} \left(\frac{kT}{c\hbar} \right)^3 \frac{d g_3(z)}{dz} \\
&= \frac{2V}{\pi^2} \left(\frac{kT}{c\hbar} \right)^3 \zeta(2) \\
&= \langle N \rangle \frac{\zeta(2)}{\zeta(3)} = 1.368 \langle N \rangle.
\end{aligned}
\tag{4.9.50}
$$

This result is somewhat larger than that for a Boltzmann gas, but it is by no means extraordinary. The relative variance in photon number is seen to be $1.368/\langle N \rangle$, which is suitably small for macroscopic systems.

The other evident way to calculate the photon-number variance is by use of Eq. (4.9.8), in terms of the isothermal compressibility. Since the photon pressure is a function only of temperature, as shown in Eq. (4.8.6) it follows that $(\partial P / \partial V)_T = 0$, so the compressibility is infinite, as is also evident physically. The resulting photon-number variance is therefore infinite! Which result is to be believed?

The first calculation is dubious in that z was allowed to vary before it was set equal to its limiting value. The second seems equally dubious in that μ was allowed to do the same thing. Neither calculation is totally free of objection, and neither seems more rigorous than the other. This puzzle, which requires further study, is offered as one of the several challenges that seem to be a persistent goad to deeper understanding.

4.9.8 Fluctuations in Microstate Populations. The expected population of a general monatomic ideal gas can be written, in accordance with Eq. (4.4.3) as

$$\langle N \rangle = \sum_i (z^{-1} e^{\beta \epsilon_i} + \sigma)^{-1}$$

where $\sigma = +1$ for fermions and $\sigma = -1$ for bosons. It turns out that at this point it is acceptable to set $\sigma = 0$ to include Boltzmann particles in the general formulation. The expectation value for the population of the ith single-particle state is

$$\langle n_i \rangle = \frac{1}{z^{-1} e^{\beta \epsilon_i} + \sigma}, \tag{4.9.51}$$

and these populations may fluctuate independently. In this section, some of the aspects of these fluctuations will be examined.

The energy of the ith N-particle state is written

$$E_i(N) = \sum_j n_{ij}(N)\epsilon_j, \tag{4.9.52}$$

where the ϵ_j are the energies of the single-particle states, and $n_{ij}(N)$ is the occupation number of the jth single-particle state in the ith N-particle state. The canonical N-particle partition function is

$$Z_N = \sum_i e^{-\beta E_i(N)} = \sum_i \exp\left(-\beta \sum_j n_{ij}(N)\epsilon_j\right). \tag{4.9.53}$$

The expectation value $\langle n_k \rangle_N$ for the occupation number of the kth single-particle state in a system of N particles is

$$\langle n_k \rangle_N = \frac{\sum_i n_{ik}(N)e^{-\beta E_i(N)}}{Z_N} = \sum_i W_N(i)n_{ik}(N) = -kT\left(\frac{\partial \ln Z_N}{\partial \epsilon_k}\right)_{V,\beta,\epsilon_j \neq \epsilon_k}. \tag{4.9.54}$$

In agreement with an already established pattern, the variance of this expectation value is seen to be

$$\langle (\Delta n_k)^2 \rangle_N = -kT\left(\frac{\partial \langle n_k \rangle_N}{\partial \epsilon_k}\right), \tag{4.9.55}$$

where the derivative is subject to the same conditions as in the previous equation. The covariances are similarly determined:

$$\langle \Delta n_k n_l \rangle_N = -kT\left(\frac{\partial \langle n_k \rangle_N}{\partial \epsilon_l}\right) = -kT\left(\frac{\partial \langle n_l \rangle_N}{\partial \epsilon_k}\right). \tag{4.9.56}$$

For the grand-canonical system, the expected occupation numbers for the single-particle states are

$$\langle n_k \rangle_G = \sum_{N=0}^{\infty} W_N \langle n_k \rangle_N = \langle n_k \rangle_{(N)}, \tag{4.9.57}$$

as shown in Eq. (4.9.15), where W_N is the probability of there being N particles in the system, as defined in Eq. (4.9.1). The variance is then seen to be

$$\langle (\Delta n_k)^2 \rangle_G = -kT\left(\frac{\partial \langle n_k \rangle}{\partial \epsilon_k}\right)_{V,z,\epsilon_j \neq \epsilon_k}. \tag{4.9.58}$$

The covariance is similarly found to be

$$\langle \Delta n_k n_l \rangle_G = -kT\left(\frac{\partial \langle n_k \rangle}{\partial \epsilon_l}\right) = -kT\left(\frac{\partial \langle n_l \rangle}{\partial \epsilon_k}\right). \tag{4.9.59}$$

By use of Eqs. (4.9.58) and (4.9.59), the variances and covariances may be found from the general expressions of Eq. (4.9.51):

$$\left\langle (\Delta n_i)^2 \right\rangle = \langle n_i \rangle_G - \sigma \langle n_i \rangle^2 , \qquad (4.9.60)$$

and

$$\left\langle \Delta n_i n_j \right\rangle_G = 0. \qquad (4.9.61)$$

The first of these shows that fermion gases have smaller fluctuations of the single-particle-state occupation numbers than do Boltzmann gases, and boson gases have correspondingly larger fluctuations. When the gas is nondegenerate, so that $\langle n_i \rangle \ll 1$, all the variances are approximately equal. The result that the Boltzmann variance equals the expectation value of the population suggests that the individual populations will have Poisson probability distributions. This conjecture will be confirmed subsequently.

The result that the covariance is zero says that the separate populations fluctuate independently. In order for this to occur, the sets of probabilities of single-particle-state occupation numbers must be independent of the properties of the other states. For fermion gases, the probability set is found as follows:

$$\langle n_i \rangle = (1 + z^{-1} e^{\beta \epsilon_i})^{-1} = \sum_{n=0}^{1} n w_i(n) = w_i(1), \qquad (4.9.62)$$

from which $w_i(0)$ is seen to be $w_i(0) = 1 - w_i(1)$, or

$$w_i(0) = \frac{1}{1 + z e^{-\beta \epsilon_i}} = 1 - \langle n_i \rangle ,$$

and

$$w_i(1) = \frac{z e^{-\beta \epsilon_i}}{1 + z e^{-\beta \epsilon_i}} = \langle n_i \rangle . \qquad (4.9.63)$$

These two probabilities could have been read off directly from the terms in Eq. (4.2.2), normalized by dividing by Z_G. (All the terms that involve other states cancel.)

The boson probability set may be read from Eq. (4.3.1) to be

$$w_i(n) = \left(z e^{-\beta \epsilon_i} \right)^n \left(1 - z e^{-\beta \epsilon_i} \right). \qquad (4.9.64)$$

These probability sets are seen to be independent of properties specific to any state other than the ith.

A similar result can be obtained for a Boltzmann gas. The Boltzmann canonical partition function is

$$Z_N = \frac{Z_1^N}{N!} = \frac{1}{N!} \left(\sum_i e^{-\beta \epsilon_i} \right)^N ,$$

or, as explained in Eq. (3.3.16),

$$Z_N = \frac{1}{N!} \sum_i (N; n_1, n_2, \cdots)_i \left(e^{-\beta \epsilon_1} \right)^{n_{1i}} \left(e^{-\beta \epsilon_1} \right)^{n_{2i}} \cdots .$$

From this expression, the grand partition function is written

$$
Z_G = \sum_{N=0}^{\infty} \sum_i \prod_j \left(\frac{e^{-\beta n_{ji}\epsilon_j}}{n_{ji}!} \right)
$$

$$
= \left(1 + ze^{-\beta\epsilon_1} + \frac{\left(ze^{-\beta\epsilon_1}\right)^2}{2!} + \frac{\left(ze^{-\beta\epsilon_1}\right)^3}{3!} + \cdots \right) Z_G'
$$

$$
= \exp\left(ze^{-\beta\epsilon_1}\right) Z_G' = \prod_i \exp\left(ze^{-\beta\epsilon_i}\right)
$$

$$
= \exp\left(z \sum_i e^{-\beta\epsilon_i} \right) = e^{zZ_1}, \tag{4.9.65}
$$

where the terms for state 1 have been factored out explicitly, as in Eq. (4.3.1) for the boson gas, and Z_G' again denotes that part of the grand partition function that has no terms from state 1. The final expression confirms the standard result for a Boltzmann gas, and it could have been used directly to write the factored versions in the previous steps. The single-particle-state probability set can now be obtained from Z_G, as was done for the bosons and fermions:

$$
w_i(n) = \frac{\left(ze^{-\beta\epsilon_i}\right)^n /n!}{\exp(ze^{-\beta\epsilon_i})}.
$$

Since the probability set can be used to calculate $\langle n_i \rangle$, giving $\langle n_i \rangle = ze^{-\beta\epsilon_i}$, in agreement with Eq. (4.9.51) for $\sigma = 0$, the Boltzmann probability set can be rewritten to give

$$
w_i(n) = \frac{\langle n_i \rangle^n}{n!} e^{-\langle n_i \rangle}, \tag{4.9.66}
$$

the promised Poisson distribution. It is interesting to note that for the Boltzmann gas the total population has a Poisson distribution, as shown in Eq. (4.9.19), and the uncorrelated single-particle-state populations are similarly distributed.

Another point of interest is the comparison of single-particle factors of the three ideal-gas partition functions. The terms for zero particles and for one particle in a particular state are identical in all three types of ideal gas, as may be seen from Eqs. (4.2.2), (4.3.1), and (4.9.65). There are only two terms for fermions; for bosons, the term for n particles in the ith state is $(ze^{-\beta\epsilon_i})^n$ and for Boltzmann particles the corresponding term is $(ze^{-\beta\epsilon_i})^n/n!$. The Boltzmann $n!$ came from the multinomial coefficient of Eq. (3.3.16), which came from writing $Z_N = Z_1^N/N!$. As noted in the discussion of this procedure in Chapter 3, its validity is dependent on there being many more accessible states than particles, so that the occupation numbers are almost always 0 or 1. When this is the case, the pertinent multinomial coefficients in Eq. (3.3.16) are all equal to $N!$, which cancels out of the expression for Z_G. The correct expression for Z_G has the Boltzmann factors $\left(\prod_i n_i!\right)^{-1}$ replaced by 1. As seen in Eq. (4.9.66) and above, the expected population of the ith state is $ze^{-\beta\epsilon_i}$. When conditions for the validity of the Boltzmann partition function prevail, the expected populations are much less than 1. Therefore the dominant terms in the partition function are those for $n_i = 0$ or $n_i = 1$, which are identical in all three statistics.

For bosons, the probability that there are n particles in the ith single-particle state is given by Eq. (4.9.64). But Eq. (4.9.51) says that for bosons,

$$ze^{-\beta\epsilon_i} = \frac{\langle n_i \rangle}{\langle n_i \rangle + 1},$$

so that $w_i(n)$ may be written

$$w_i(n) = \frac{\langle n_i \rangle^n}{(\langle n_i \rangle + 1)^{n+1}}, \tag{4.9.67}$$

a result that differs in form and behavior from a Poisson distribution. It says, among other things, that the probability ratio $w_i(n+1)/w_i(n)$ is independent of n. This means that the probability of a boson state with n occupants acquiring another occupant is independent of the number of occupants. There is no prejudice among bosons against overcrowding.

The relative variance of the single-particle-state occupancy number is, from Eq. (4.9.60),

$$\frac{\langle (\Delta n_j)^2 \rangle_G}{\langle n_j \rangle^2} = \frac{1}{\langle n_j \rangle} - \sigma. \tag{4.9.68}$$

This result is valid for all three statistics, including photons and bosons with $T \leq T_c$, as is the equation for the variances, Eq. (4.9.60).

Further aspects of fluctuation phenomena are examined in Chapters 9 and 10, where subjects such as Brownian motion, the fluctuation-dissipation theorem, the power spectra of fluctuations, and the principle of detailed balance (and its relatives) are treated. The underlying microscopic activity that is responsible for fluctuations is the theme of Chapter 10, whereas the mere passive existence of probability sets that characterize our knowledge of particular systems has been the basis for the treatment of this section. Both points of view are required for a general understanding of the success of the statistical description of thermophysical phenomena.

4.10 COMMENTARY ON QUANTUM THEORY AND STATISTICS

The developments of this chapter began with the statement that for indistinguishable, noninteracting particles, the states of the system are to be taken as composites of single-particle states, and the state of the system is to be specified by the set of occupation numbers of the single-particle states. No single-particle state is allowed, for reasons to be explored in this commentary, to have an occupation number larger than p, some positive integer. It was noted that for $p = 1$, the particles are called *fermions*, for $p = \infty$, the particles are called *bosons*, and at least all common particles are one or the other. Furthermore, there is a relationship between the spin of the particle and the statistics it obeys. Why should nature choose only the two extreme values for p? Does the spin-statistics connection permit other values of p, and why or why not? What are the consequences of other kinds of statistics? These are topics that I regard as fundamental to my understanding of physics, even if they seem to be a distraction from the generally accepted main task of statistical mechanics—that of calculating partition functions and their consequences. Because they are a distraction, I have included them in an isolated and appendix-like section, rather than in the main body of the text. But because they are of fundamental importance, I could not consider omitting them.

It is not within my ability, in the space I consider to be appropriate, to treat the subject matter of spin, statistics, and quantization definitively. To do so would require another volume, entire unto itself. As a substitute, I have provided some illustrative examples of the nature of N-body states and how the permutation operator leads to such states. This is followed by a discussion of a generalization of quantization called *paraquantization*, with illustrations of its consequences in statistical physics, with its *parabosons* and *parafermions*. The introduction to this topic is presented in enough detail to provide the flavor of the subject. A few results and theorems from the advanced developments are quoted without proof, but with references. Then some of the observational consequences of parafermi statistics are examined, as an indication of what kinds of phenomena could be expected if an example of parafermion statistics were ever encountered. A brief examination of the experimental evidence for the choices of statistics provides a number of certainties, but also offers the possibility that there may exist paraparticles among those already known. Finally, the number operators for ordinary fermions and bosons are used to illustrate in more detail the nature of the N-body state, and its essential feature in quantum theory— the nonindividual, or nonlocal, nature of the particles in such a state.

4.10.1 Permutation Operators and Two-Particle States. Instead of a general treatment of the permutation operator and its effect on the quantum states of a system of many identical particles, I have chosen to examine in some detail a system of two identical particles, for which the permutation operator reduces to the particle-interchange operator. For noninteracting particles, it is convenient to describe the system in terms of a complete orthonormal set of single-particle states, $\{|i\rangle\}$, such that the usual wave functions are

$$\phi_i(r) = \langle r|i\rangle, \tag{4.10.1}$$

with

$$\langle i|j\rangle = \int \langle i|r\rangle \, dr \, \langle r|j\rangle = \int dr \phi_i^*(r)\phi_j(r) = \delta_{ij}, \tag{4.10.2}$$

where r represents the coordinate variables of the physical space, and, if appropriate, spin; dr represents the volume element in that space. The integrations are over the entire available space, and they imply sums over spin variables. For this treatment, the wave-function representation of the states will be used, since some of the details are most-conveniently displayed in these terms, but for much of the analysis, state vectors could be used equally well.

The two-particle Hamiltonian is, in the absence of interactions, the sum of the two single-particle Hamiltonians, or $\mathcal{H}_{1,2} = \mathcal{H}(r_1)+\mathcal{H}(r_2)$. A suitable solution to the Schrödinger equation is $\phi_{ab} = \phi_a(r_1)\phi_b(r_2)$, for which particle 1 is in state ϕ_a and particle 2 in state ϕ_b. The interchange operator, $\mathcal{P}_{12}$, interchanges the two particles. It clearly commutes with $\mathcal{H}_{1,2}$, since both terms in $\mathcal{H}$ have the same functional form, and their order is immaterial. The eigenvalue of $\mathcal{H}_{1,2}$ for the eigenstate ϕ_{ab} is $E_{ab} = \epsilon_a + \epsilon_b$, the sum of the energies for the two states. It is then possible to write

$$\mathcal{H}_{1,2}\mathcal{P}_{12}\phi_{ab} = \mathcal{P}_{12}\mathcal{H}_{1,2}\phi_{ab}$$
$$= \mathcal{P}_{12}E_{ab}\phi_{ab}$$
$$= E_{ab}\mathcal{P}_{12}\phi_{ab} = E_{ab}\phi_{ba}. \tag{4.10.3}$$

Thus the function $\mathcal{P}_{12}\phi_{ab} = \phi_{ba}$ is another eigenfunction of $\mathcal{H}_{1,2}$ with the same eigenvalue as ϕ_{ab}. It is customary to construct a pair of orthonormal simultaneous eigenfunctions of $\mathcal{H}_{1,2}$ and $\mathcal{P}_{12}$ as linear combinations of ϕ_{ab} and ϕ_{ba}, but not always as follows:

$$\Phi_{ab}(r_1, r_2) = (1 + \alpha\alpha^*)^{-1/2} \left[\phi_a(r_1)\phi_b(r_2) + \alpha\phi_a(r_2)\phi_b(r_1)\right]$$
$$\Psi_{ab}(r_1, r_2) = (1 + \beta\beta^*)^{-1/2} \left[\phi_a(r_2)\phi_b(r_1) + \beta\phi_a(r_1)\phi_b(r_2)\right], \tag{4.10.4}$$

where α and β are c-numbers that need not be real. The two functions are already normalized (except when $a = b$). The requirement that they be orthogonal says, for $a \neq b$,

$$\begin{aligned}
0 &= \int dr_1 dr_2 \, \Phi_{ab}^* \Psi_{ab} \\
&= \frac{1}{\sqrt{1 + \alpha\alpha^*}} \frac{1}{\sqrt{1 + \beta\beta^*}} \int dr_1 dr_2 \left[\phi_a^*(r_2)\phi_b^*(r_1) + \beta^* \phi_a^*(r_1)\phi_b^*(r_2)\right] \\
&\quad \times \left[\phi_a(r_1)\phi_b(r_2) + \alpha\phi_a(r_2)\phi_b(r_1)\right] \\
&= \frac{1}{\sqrt{1 + \alpha\alpha^*}} \frac{1}{\sqrt{1 + \beta\beta^*}} (\alpha + \beta^*), \tag{4.10.5}
\end{aligned}$$

which requires that $\beta^* = -\alpha$, or $\beta = -\alpha^*$.

It is often stated that since $\mathcal{P}_{12}^2 = 1$, it must follow that $\mathcal{P}_{12} = \pm 1$, which in turn requires that $\alpha = \pm 1$, or that the particles be either fermions or bosons. A somewhat more general result may be possible, and should be explored. The action of $\mathcal{P}_{12}$ on either of the new states must produce a state that is a linear combination of the two new states, and a second application of the operator must return the state to the original one, since an even number of interchanges should be equivalent to no interchange. If nothing more is required, the action of the interchange operator can be written out directly, to give

$$\begin{aligned}
\mathcal{P}_{12}\Phi_{ab} &= (1 + \alpha\alpha^*)^{-1/2} \left[\phi_a(r_2)\phi_b(r_1) + \alpha\phi_a(r_1)\phi_b(r_2)\right] \\
&= c_1 \Phi_{ab} + c_2 \Psi_{ab}, \tag{4.10.6}
\end{aligned}$$

where c_1 and c_2 are numbers. The two equations produce the pair of relations $\alpha = c_1 - \alpha^* c_2$, and $1 = \alpha c_1 + c_2$, leading to the results

$$\begin{aligned}
c_1 &= (\alpha + \alpha^*)/(1 + \alpha\alpha^*), \\
c_2 &= (1 - \alpha^2)/(1 + \alpha\alpha^*). \tag{4.10.7}
\end{aligned}$$

A similar calculation for Ψ_{ab} gives

$$\begin{aligned}
\mathcal{P}_{12}\Psi_{ab} &= (1 + \alpha\alpha^*)^{-1/2} \left[\phi_a(r_1)\phi_b(r_2) - \alpha^* \phi_a(r_2)\phi_b(r_1)\right] \\
&= d_1 \Phi_{ab} + d_2 \Psi_{ab}, \tag{4.10.8}
\end{aligned}$$

with equations for the constant d's: $1 = d_1 - \alpha^* d_2$, and $-\alpha^* = \alpha d_1 + d_2$, leading to

$$\begin{aligned}
d_1 &= (1 - \alpha^{*2})/(1 + \alpha\alpha^*) = c_2^*, \\
d_2 &= -(\alpha + \alpha^*)/(1 + \alpha\alpha^*) = -c_1. \tag{4.10.9}
\end{aligned}$$

One possible choice for α is $\alpha = 0$, which leads to the two original two-particle states $\phi_{ab} = \phi_a(r_1)\phi_b(r_2)$, and its counterpart $\phi_{ba} = \phi_a(r_2)\phi_b(r_1)$. This choice leads to Boltzmann statistics. Otherwise, when the properly symmetrized states are combinations of the two Boltzmann states, it is fitting, but not necessary, that the two appear with equal amplitudes. In this case, all choices of α are of the form $\alpha = e^{i\theta}$. The two states are then written

$$\Phi_{ab} = 2^{-1/2}\left[\phi_a(r_1)\phi_b(r_2) + e^{i\theta}\phi_a(r_2)\phi_b(r_1)\right],$$
$$\Psi_{ab} = 2^{-1/2}\left[\phi_a(r_2)\phi_b(r_1) - e^{-i\theta}\phi_a(r_1)\phi_b(r_2)\right]. \quad (4.10.10)$$

When $\mathcal{P}_{12}$ is written in matrix form, it becomes

$$\mathcal{P}_{12} = \begin{pmatrix} \cos\theta & (1 - e^{2i\theta})/2 \\ (1 - e^{-2i\theta})/2 & -\cos\theta \end{pmatrix}. \quad (4.10.11)$$

The usual choices are $\theta = 0$ and $\theta = \pi$. For the former, Φ_{ab} represents the symmetric, or boson, state, and Ψ_{ab} represents the antisymmetric, or fermion, state. There is, however, no requirement imposed by the interchange operator that restricts the choice of θ. (There is another requirement, to be introduced just in time to prevent disaster, but the consequences of its omission are amusing, and will be pushed a bit further.)

In the state Φ_{ab}, the probability that a particle is in the interval dr_1 and a particle is also in dr_2 is

$$P_{ab}\,dr_1\,dr_2 = \Phi_{ab}^*\Phi_{ab}\,dr_1\,dr_2$$
$$= \frac{1}{2}\Big[\phi_a^*(r_1)\phi_a(r_1)\phi_b^*(r_2)\phi_b(r_2) + \phi_a^*(r_2)\phi_a(r_2)\phi_b^*(r_1)\phi_b(r_1) +$$
$$+ e^{i\theta}\phi_a^*(r_1)\phi_b(r_1)\phi_b^*(r_2)\phi_a(r_2) + e^{-i\theta}\phi_a^*(r_2)\phi_b(r_2)\phi_b^*(r_1)\phi_a(r_1)\Big]. \quad (4.10.12)$$

When $a \neq b$, $\int P_{ab}\,dr_1\,dr_2 = 1$, with all θ-dependent terms vanishing because of orthogonality, as expected from the imposed normalization of the wave function. But when $a = b$, the behavior seems to be quite different. Coherence effects in the θ terms reduce Eq. (4.10.12) to

$$P_{aa}\,dr_1\,dr_2 = \left[\phi_a^*(r_1)\phi_a(r_1)\phi_a^*(r_2)\phi_a(r_2)(1 + \cos\theta)\right]dr_1\,dr_2, \quad (4.10.13)$$

with the result that

$$\int P_{aa}\,dr_1\,dr_2 = 1 + \cos\theta. \quad (4.10.14)$$

For bosons ($\theta = 0$), the result is 2; for fermions ($\theta = \pi$), it is zero; and the result varies continuously with θ from one extreme to the other.

A calculation of the expected energy of the two-particle system provides additional interest. When $a \neq b$, the expected energy is

$$\overline{E}_{ab} = \int \Phi_{ab}^*\mathcal{H}_{1,2}\Phi_{ab} = \epsilon_a + \epsilon_b, \quad (4.10.15)$$

which is the eigenvalue E_{ab} of $\mathcal{H}_{1,2}$ corresponding to Φ_{ab}. But when $a = b$, the result is

$$E_{aa} = 2\epsilon_a(1 + \cos\theta), \quad (4.10.16)$$

even though the energy eigenvalue is $E_{aa} = 2\epsilon_a$. Thus the energy for a pair of like fermions in the same state is zero, as might be expected, since the exclusion principle forbids the existence of such a pair. But the result that $\overline{E}_{aa} = 4\epsilon_a$ for a boson pair is startling, and indicative that something must be wrong. How do two particles, each with energy ϵ_a, form a state with energy $4\epsilon_a$? The apparent answer is that this is a manifestation of the wavelike aspects of the quantum system. It is a typical interference phenomenon. Bosons in the same state interfere constructively; fermions destructively; and quantum particles in general interfere in a way dependent upon the phase angle θ. Particles with $\theta = \pi/2$ have the two parts of the wave function in quadrature, and the particles behave independently. Coherence phenomena in optical systems exhibit related behavior. But something is wrong; the creation of energy by coherence effects could have unphysical consequences in boson systems.

A resolution of this difficulty is easy, when it is recognized that the integral of P_{ab} in Eq. (4.10.12) over the entire available space, *i.e.*, the probability that there are two particles in the space, cannot exceed unity. Therefore the normalization of the two-particle wave functions must be different when both particles are in the same state. For complete orthonormality, the last line of Eq. (4.10.5) should have the term $(\alpha + \beta^*)$ replaced by $(\alpha + \beta^*) + \delta_{ab}^2(1 + \alpha\beta^*)$. For $a \neq b$, again it follows that $\beta^* = -\alpha$; assume that this result is always valid. For $a = b$, $1 + \alpha\beta^* = 0$, which then gives $\alpha^2 = 1$, or $\alpha = \pm 1 = e^{i\theta}$. Therefore the only allowed values of θ are integer multiples of π, and only fermions or bosons are permitted by the formalism to exist.

Under these apparently inviolable restrictions, the only unobjectionable pair of two-particle functions is that equivalent to $\theta = 0$:

$$\Phi_{ab} = [2(1 + \delta_{ab})]^{-1/2} \left[\phi_a(r_1)\phi_b(r_2) + \phi_a(r_2)\phi_b(r_1) \right],$$
$$\Psi_{ab} = [2(1 + \delta_{ab})]^{-1/2} \left[\phi_a(r_2)\phi_b(r_1) - \phi_a(r_1)\phi_b(r_2) \right]. \tag{4.10.17}$$

The first of these is a boson function, normalized to unity; the second a fermion function, also normalized to unity except when $a = b$, in which case the function vanishes, as does its inner product. The pair is orthogonal, as is required. The interchange operator leaves the boson function unchanged and reverses the sign of the fermion function, or $\mathcal{P}_{12}\Phi_{ab} = \Phi_{ab}$, and $\mathcal{P}_{12}\Psi_{ab} = -\Psi_{ab}$.

Other choices for the effect of $\mathcal{P}_{12}$ appear in some theories of high-temperature super-conductivity, fractional quantum Hall effect, and certain quark models. In two dimensions, where the distinction between bosons and fermions virtually disappears (See problem 10 of this chapter), and wherein the properties of high-temperature superconductors have been posited to reside, the consequences of $\mathcal{P}_{12}\Phi = e^{i\alpha}\Phi$ have been explored extensively for values of α other than 0 or π. For such values, the resulting states are called *anyons*; when $\alpha = \pi/2$, they are called *semions*. Without exploring these theories further, since they are in fact theories of interacting particles, and without invoking the spin-statistics theorem, which will be considered later, it is nevertheless interesting to note the existence of these states (Chen et al. [1989]; Forte [1992]; Leinaas and Myrheim [1977]; Wilczek [1982]; Wilczek [1990]; Wu [1984]).

Much of this discussion can be extended to particles with the Hamiltonian

$$\mathcal{H}_{1,2} = \frac{\hbar^2}{2m} \left(\nabla_1^2 + \nabla_2^2 \right) + V(r_1) + V(r_2) + V_{12}(|r_2 - r_1|), \tag{4.10.18}$$

which again commutes with the interchange operator $\mathcal{P}_{12}$. The orthonormal two-particle wave functions become

$$\Phi_{ab} = \left[2(1 + \delta_{ab}\cos\theta)\right]^{-1/2}\left[\phi_{ab}(r_1, r_2) + e^{i\theta}\phi_{ab}(r_2, r_1)\right]$$

$$\Psi_{ab} = \left[2(1 + \delta_{ab}\cos\theta)\right]^{-1/2}\left[\phi_{ab}(r_2, r_1) - e^{-i\theta}\phi_{ab}(r_1, r_2)\right], \qquad (4.10.19)$$

where the index ab designates a particular two-particle state. Note that when the functional dependencies of ϕ_{ab} on the first and second variables are different, the two particles can be considered as being in different states, even though ϕ_{ab} is strictly a two-particle state that is not the product of single-particle states.

If a two-particle system is initially in one of the combination states, say Φ_{ab}, then it will remain so as time goes on, provided $\mathcal{H}_{1,2}$ is not time dependent. Suppose $U_{ab}(1,2,t)$ is a solution of the time-dependent Schrödinger equation

$$\mathcal{H}_{1,2}U_{ab} = i\hbar\frac{\partial U_{ab}}{\partial t}.$$

Since Φ_{ab} is an eigenfunction of $\mathcal{H}_{1,2}$ corresponding to the energy E_{ab}, a solution for U_{ab} is

$$U_{ab} = \Phi_{ab}e^{-iE_{ab}t/\hbar}. \qquad (4.10.20)$$

(The other solution, W_{ab}, replaces Φ_{ab} by Ψ_{ab}.) The time evolution of U_{ab} is seen to be

$$\begin{aligned}
U_{ab}(1,2,t+\Delta t) &= U_{ab}(1,2,t) + \frac{\partial U_{ab}}{\partial t}\Delta t \\
&= U_{ab}(1,2,t) - (i\Delta t/\hbar)\mathcal{H}_{1,2}U_{ab} \\
&= U_{ab}(1,2,t)\left(1 - iE_{ab}\Delta t/\hbar\right) \\
&= \Phi_{ab}e^{-iE_{ab}t/\hbar}\left(1 - iE_{ab}\Delta t/\hbar\right), \qquad (4.10.21)
\end{aligned}$$

a result that can be iterated. It follows, therefore, that

$$\left(U_{ab}(1,2,t_1), W_{ab}(1,2,t_2)\right) = \int dr_1 dr_2 U_{ab}^*(t_1)W_{ab}(t_2) = 0 \qquad (4.10.22)$$

for any t_1, t_2. Thus a fermion state remains as a fermion state, a boson state remains as a boson state, and intermediate states, if such can exist, remain so, with constant phase.

For more than two particles, there are additional possibilities for the construction of composite wave functions that are simultaneously eigenfunctions of the Hamiltonian and of the set of permutation operators, expressed as matrices. (A general permutation of the particles can be expressed, not necessarily uniquely, as a product of interchange operators, which usually do not commute.) The only N-particle wave functions, typically, that can be devised easily from the set of single-particle functions are the fermion function, which is totally antisymmetric in all two-particle interchanges, and the boson function, which is totally symmetric. It is an interesting exercise to try to construct, for a three-particle system, the other four orthonormal wave functions necessary to form a complete set.

4.10.2 Standard Fermion and Boson Many-Particle States.
Fermion systems of indistinguishable particles are described by state functions that are antisymmetric in the exchange

of any two particles. For noninteracting particles, the usual way to write the completely antisymmetric fermion state is as a Slater determinant:

$$\Psi_j(r_1, r_2, \cdots, r_n) = C \begin{vmatrix} \phi_1(r_1) & \phi_1(r_2) & \cdots & \phi_1(r_n) \\ \phi_2(r_1) & \phi_2(r_2) & \cdots & \phi_2(r_n) \\ \vdots & \vdots & \ddots & \vdots \\ \phi_n(r_1) & \phi_n(r_2) & \cdots & \phi_n(r_n) \end{vmatrix}. \tag{4.10.23}$$

In the Slater determinant, the interchange of two particles is represented by the exchange of the two columns that are functions of the coordinates of the exchanged particles. As a consequence of the properties of determinants, this operation results in a change in the sign of Ψ_j. Clearly any number p of such interchanges results in the multiplication of the sign of Ψ_j by the factor $(-)^p$, and if any two columns represent the same single-particle state, then $\Psi_j = 0$, as demanded by the Pauli principle. The factor C in Eq. (4.10.23) is a constant that normalizes Ψ_j; it is usually given by $C = 1/\sqrt{N!}$, where N is the number of particles represented in the antisymmetrized state Ψ_j. This is the quantum-mechanical origin of the $1/N!$ factor in the N-particle partition function, and the resolution of the Gibbs paradox.

A similar representation for bosons is called a *permanent*. Written out in full detail, a permanent corresponds term for term with the expanded Slater determinant, but all terms are positive. Thus the formalisms for writing the completely antisymmetric, or fermion, states, and the completely symmetric, or boson, states are easy to present and to understand (Minc [1978]).

Within these formalisms, however, students should wonder about the apparent necessity to include, for example, every electron in the universe in the composite wave function employed to represent any electron system, since these composite wave functions seem to say that every electron is in all single-particle states. A concrete, but easily generalized, example is offered as a means of clarifying this point. Consider a universe of three electrons; two are in a box, always, and the third is permanently outside the box. (Disregard the composition of the box, or replace the electrons by trions, defined to be massive, stable, noninteracting fermions, only three of which exist in the universe. Two of them are in our box, and the other is still at large.) The single-particle wave functions for a particle in the box are denoted by ϕ_a and ϕ_b; for a particle outside the box by ϕ_c. The particles in the box have coordinates r_1 and r_2; the outside particle's coordinates are r_3. Therefore $\phi_a(r_i)$ can differ from zero only for $i = 1, 2$, and only for values of these coordinates that represent the inside of the box. Similarly for $\phi_b(r_i)$. But $\phi_c(r_i)$ can differ from zero only when $i = 3$, and only for values of r_3 that lie outside the box. Under these conditions, the Slater determinant becomes

$$\Phi_{abc}(r_1, r_2, r_3) = C \begin{vmatrix} \phi_a(r_1) & \phi_a(r_2) & \phi_a(r_3) \\ \phi_b(r_1) & \phi_b(r_2) & \phi_b(r_3) \\ \phi_c(r_1) & \phi_c(r_2) & \phi_c(r_3) \end{vmatrix} = C \begin{vmatrix} \phi_a(r_1) & \phi_a(r_2) & 0 \\ \phi_b(r_1) & \phi_b(r_2) & 0 \\ 0 & 0 & \phi_c(r_3) \end{vmatrix},$$

where the second Slater determinant is written to reflect conditions known to apply to the single-particle wave functions. As a result, the state function becomes

$$\Phi_{abc}(r_1, r_2, r_3) = C \begin{vmatrix} \phi_a(r_1) & \phi_a(r_2) \\ \phi_b(r_1) & \phi_b(r_2) \end{vmatrix} \times \phi_c(r_3). \tag{4.10.24}$$

This state function is factored into the product of inside and outside terms. Evidently the same two kinds of factors would occur for any number of inside and outside particles. For problems that concern only inside particles, a reduced state function can be written, either informally by simply dropping the outside factor and any reference to it, or formally by multiplying by the Hermitian conjugate of the outside factor and integrating over the outside variables: in this example

$$\Phi_{ab}(r_1, r_2) = \int \phi_c^*(r_3)\Phi_{abc}(r_1, r_2, r_3)dr_3 = C \begin{vmatrix} \phi_a(r_1) & \phi_a(r_2) \\ \phi_b(r_1) & \phi_b(r_2) \end{vmatrix}. \tag{4.10.25}$$

This is the kind of state function that is usually written for a set of particles that are regarded as separated from the rest of the universe. Note that in this case, the normalizing factor C is properly $1/\sqrt{2}!$ and not the $1/\sqrt{3}!$ value that is usual for 3×3 Slater determinants.

4.10.3 Paraquantization and Parastatistics. Instead of choosing the operators for position and momentum in some standard picture (Schrödinger or Heisenberg, for example) and allowing the choice to determine the commutation relations for the chosen operators, a more basic approach is to proceed from one of the most fundamental equations of quantum mechanics, the Heisenberg equation of motion,

$$i\hbar \frac{dA}{dt} = [A, \mathcal{H}], \tag{4.10.26}$$

where A is an operator that is not explicitly time-dependent, and $[A, B] = AB - BA$, the commutator, with $\mathcal{H}$ the Hamiltonian. For a system with a classical analog, a generally valid guiding principle is that the Heisenberg equations of motion should agree with their classical counterparts. For example, consider a one-dimensional harmonic oscillator, with Hamiltonian

$$\mathcal{H} = \frac{p^2}{2m} + \frac{1}{2}kq^2,$$

with Heisenberg equations of motion

$$i\hbar\dot{q} = [q, \mathcal{H}] = \frac{1}{2m}(qp^2 - p^2q).$$

When the classical Euler-Lagrange equation $\dot{q} = p/m$ is invoked, there results the relation

$$i\hbar p = \frac{1}{2}(qp^2 - p^2q) = \frac{1}{2}[q, p^2] = \frac{1}{2}(p[q, p] + [q, p]p), \tag{4.10.27}$$

which is consistent with the canonical commutation relation $[q, p] = i\hbar$, but more general. (A useful relation in commutator algebra is $[A, BC] = B[A, C] + [A, B]C$.) A similar treatment of the equation for $\dot{p}$ gives

$$-i\hbar q = \frac{1}{2}(pq^2 - q^2p) = \frac{1}{2}[p, q^2] = \frac{1}{2}(q[p, q] + [p, q]q). \tag{4.10.28}$$

Thus the canonical commutation relations are seen to be sufficient, but not necessary, in this first example of what is called *paraquantization* (Ohnuki and Kamefuchi [1982]). The consequences of this procedure are to be examined, with more convenient notation, in the next subsection.

The attitude displayed in this section is that unless there exists some *a priori* reason to choose the canonical quantization procedure, the problems of quantum physics should be studied from the most general standpoint possible. If canonical quantization turns out to be the only possibility consistent with the experimental evidence, then there still remains the fundamental problem of understanding why nature has made such a choice.

4.10.4 A Parabose Oscillator. With properly chosen units, the Lagrangian for the harmonic oscillator of the last subsection can be written

$$\mathcal{L} = \frac{1}{2}(\dot{q}^2 - q^2),$$

so that $p = \dot{q}$. The Hamiltonian is

$$\mathcal{H} = \frac{1}{2}(q^2 + p^2),$$

and the combined canonical and Heisenberg equations of motion are

$$i\dot{q} = [q, \mathcal{H}] = ip,$$

and

$$i\dot{p} = [p, \mathcal{H}] = -iq.$$

It is convenient to define the operators

$$a = \frac{1}{\sqrt{2}}(q + ip), \quad a^\dagger = \frac{1}{\sqrt{2}}(q - ip). \tag{4.10.29}$$

These have the useful property

$$[a, a^\dagger] = i[p, q]. \tag{4.10.30}$$

The Hamiltonian can be written as

$$\mathcal{H} = \frac{1}{2}(a^\dagger a + aa^\dagger) = N, \tag{4.10.31}$$

where N is the number operator, representing the number of quanta that account for the total energy. The Heisenberg equations for a and $\dot{a}$ are found from their definitions and the equations of motion for q and p to be

$$[a, N] = i\dot{a} = a; \quad [a^\dagger, N] = i\dot{a}^\dagger = -a^\dagger. \tag{4.10.32}$$

These equations may be expanded to give, in analogy to Eqs. (4.10.27) and (4.10.28),

$$2a = [aa, a^\dagger] = \{a, [a, a^\dagger]\}; \quad 2a^\dagger = [\{a^\dagger, a\}, a^\dagger], \tag{4.10.33}$$

the generalized trilinear commutation relations, where the anticommutator is $\{A, B\} = AB + BA,$.

All eigenvalues of N, a positive-definite self-adjoint operator, must be nonnegative, so a smallest eigenvalue, N_0, must exist, with eigenket $|0\rangle$, or $N|0\rangle = N_0|0\rangle$. The operators a and $a^\dagger$ may be used to generate new kets. If use is made of Eqs. (4.10.32), it follows that

$$Na^\dagger|0\rangle = a^\dagger(N+1)|0\rangle = (N_0+1)a^\dagger|0\rangle,$$

which says that $a^\dagger|0\rangle$ is an eigenket of N with eigenvalue N_0+1. For any $|n\rangle$ such that $N|n\rangle = (N_0+n)|n\rangle$, the same procedure gives

$$Na^\dagger|n\rangle = (N_0+n+1)a^\dagger|n\rangle, \qquad (4.10.34)$$

which represents a new state with $n+1$ quanta, generated from the n-quanta state. For this reason, $a^\dagger$ is called a *creation operator*. Similarly, the operator a acts on $|n\rangle$ to create a state with $n-1$ quanta:

$$Na|n\rangle = a(N-1)|n\rangle = (N_0+n-1)a|n\rangle, \qquad (4.10.35)$$

for which it is called an *annihilation operator*. Since N_0 is the least eigenvalue of N, it is necessary that $a|0\rangle = 0$; there cannot be a state with fewer quanta than N_0. The operator $a^\dagger$ acting on $|n\rangle$ does not produce the exact base vector $|n+1\rangle$, but rather a vector in the same direction, but of some to-be-determined length. It may be represented formally as

$$a^\dagger|n\rangle = |n+1\rangle\left\langle n+1|a^\dagger|n\right\rangle, \qquad (4.10.36)$$

where the matrix element is the required coefficient, since $\langle n'|n\rangle = \delta_{n'n}$. A similar procedure for a yields

$$a|n\rangle = |n-1\rangle\langle n-1|a|n\rangle. \qquad (4.10.37)$$

The adjoints of these equations are, since $a^{\dagger\dagger} = a$:

$$\langle n|a = \langle n|a|n+1\rangle\langle n+1|,$$

and

$$\langle n|a^\dagger = \left\langle n|a^\dagger|n-1\right\rangle\langle n-1|.$$

With these relationships, the matrix elements of N can be found directly.

$$\langle n|N|n'\rangle = \frac{1}{2}\left\langle n|(aa^\dagger + a^\dagger a)|n'\right\rangle$$
$$= \frac{1}{2}\left(\langle n|a|n+1\rangle\langle n+1|n'+1\rangle\left\langle n'+1|a^\dagger|n'\right\rangle\right) +$$
$$+ \left(\left\langle n|a^\dagger|n-1\right\rangle\langle n-1|n'-1\rangle\langle n'-1|a|n'\rangle\right)$$
$$= \frac{1}{2}\delta_{nn'}\left(\langle n|a|n+1\rangle^2 + \langle n-1|a|n\rangle^2\right), \qquad (4.10.38)$$

since the matrix elements can always be chosen real. Evidently the only nonzero elements are those on the diagonal. Since the matrix elements of N are known, those of a and $a^\dagger$ can be determined, beginning at $n = 0$:

$$\langle 0|N|0\rangle = N_0 = \frac{1}{2}\langle 0|a|1\rangle^2,$$

or

$$\langle 0|a|1\rangle^2 = 2N_0.$$

Systematic progression through successive values of n leads to the general results:

$$\langle n|a|n+1\rangle^2 = 2N_0 + n; \quad n \text{ even},$$
$$= n + 1; \quad n \text{ odd}. \tag{4.10.39}$$

From these values, the matrix elements of the commutator can be obtained:

$$\left\langle n|[a, a^\dagger]|n'\right\rangle = \delta_{nn'}\left(\langle n|a|n+1\rangle^2 - \langle n-1|a|n\rangle^2\right).$$

Again only the diagonal elements differ from 0, and they are found to be

$$\left\langle n|[a, a^\dagger]|n\right\rangle = 2N_0, \quad n \text{ even},$$
$$= 2(1 - N_0), \quad n \text{ odd}. \tag{4.10.40}$$

An esthetic argument for ordinary boson quantization is that $N_0 = 1/2$ is the only value for which the same formula holds for odd and even n. In this case, $\left\langle n|[a, a^\dagger]|n'\right\rangle = \delta_{nn'}$, so $[a, a^\dagger] = 1$, the canonical boson commutation relation. Otherwise, the more general relation $2a = [a, \{a^\dagger, a\}]$ must be used. In any case, the Hamiltonian in ordinary units is $\mathcal{H} = h\nu N$, with eigenvalues $E_n = h\nu(N_0 + n)$. The zero-point energy is $E_0 = N_0 h\nu \geq 0$.

Some useful commutation relations, derivable from the basic ones of Eqs. (4.10.33) are the following:

$$[aa^\dagger, a^\dagger a] = 0,$$
$$[aa, aa^\dagger] = [aa, a^\dagger a] = 2aa,$$
$$[a^\dagger a^\dagger, a^\dagger a] = [a^\dagger a^\dagger, aa^\dagger] = 2a^\dagger a^\dagger,$$
$$[aa, a^\dagger a^\dagger] = 2\{a, a^\dagger\}. \tag{4.10.41}$$

An interesting and useful set of operators can be defined in terms of a and $a^\dagger$:

$$J_1 = \frac{1}{4}(aa + a^\dagger a^\dagger); \quad J_2 = \frac{i}{4}(aa - a^\dagger a^\dagger); \quad J_3 = \frac{1}{4}(a^\dagger a + aa^\dagger) = \frac{N}{2}. \tag{4.10.42}$$

These satisfy the commutation relations

$$[J_1, J_2] = -iJ_3; \quad [J_2, J_3] = iJ_1; \quad [J_3, J_1] = iJ_2, \tag{4.10.43}$$

which are the Lie commutation relations for SO(2,1), the three-dimensional Lorentz group. This group is noncompact, and has an infinite number of eigenvalues, in agreement with the result that there is no finite upper bound for n in Eq. (4.10.34).

4.10.5 A Parafermi Oscillator. In this excursion, which treats a system with no classical analog, the development of Section 4.10.4 is paraphrased for an interchange of the expressions for the Hamiltonian and the commutator (or the commutator and the anticommutator). Since these commute, formally either could be regarded as the Hamiltonian and the other as a constant of the motion. In this section, then, the Hamiltonian is taken to be

$$\mathcal{H} = \frac{1}{2}(a^\dagger a - a a^\dagger) = N = \frac{1}{2}[a^\dagger, a], \qquad (4.10.44)$$

and the anticommutator, with as yet unspecified properties, is

$$A = a^\dagger a + a a^\dagger = \{a^\dagger, a\}. \qquad (4.10.45)$$

The Heisenberg equations of motion are still given by Eq. (4.10.32), since they depend only on the definitions of a and $a^\dagger$. The generalized trilinear commutation relations that arise from expansion of these equations now become

$$2a = [a, [a^\dagger, a]]; \quad 2a^\dagger = [[a^\dagger, a], a^\dagger]. \qquad (4.10.46)$$

The creation and annihilation operators continue to work as they did in Eqs. (4.10.34) and (4.10.35); the formal effects of these operators on the ket $|n\rangle$ are given in Eqs. (4.10.36) and (4.10.37); and the same procedure as before is used to evaluate the coefficients in these equations. The principal difference arises from the structure of N, which is no longer positive definite. Suppose, again, that there is a least eigenvalue of N, written before as N_0, and now written as $-p/2$, so that $N|0\rangle = (-p/2)|0\rangle$. The procedures of the last subsection give, instead of Eq. (4.10.38),

$$\langle n|N|n'\rangle = \frac{1}{2}\delta_{nn'}\left(\langle n|a^\dagger|n-1\rangle\langle n-1|a|n\rangle - \langle n|a|n+1\rangle\langle n+1|a^\dagger|n\rangle\right), \qquad (4.10.47)$$

or

$$-p + 2n = \langle n-1|a|n\rangle^2 - \langle n|a|n+1\rangle^2.$$

The matrix elements of a are found to be

$$\langle n|a|n+1\rangle^2 = (n+1)p - n(n+1) = (n+1)(p-n). \qquad (4.10.48)$$

These quantities are then used to find the matrix elements of A, given by

$$\langle n|A|n\rangle = \langle n-1|a|n\rangle^2 + \langle n|a|n+1\rangle^2 = 2n(p-n) + p. \qquad (4.10.49)$$

There are thus $p+1$ nonzero elements of the matrix $\langle n|A|n'\rangle$, all along the diagonal, with $\langle n|A|n\rangle = \langle p-n|A|p-n\rangle$. When $p = 1$, the anticommutator has the canonical value $A = 1$, and the quantization is that of ordinary fermions.

A set of operators can be formed for a parafermion system, corresponding to those of Eq. (4.10.42), and also satisfying the commutation relations for a group. The operators are

$$J_1 = \frac{1}{2}(a^\dagger + a); \quad J_2 = \frac{1}{2i}(a^\dagger - a); \quad J_3 = \frac{1}{2}[a^\dagger, a] = N. \tag{4.10.50}$$

The commutation relations satisfied by these operators are:

$$[J_1, J_2] = iJ_3; \quad [J_2, J_3] = iJ_1; \quad [J_3, J_1] = iJ_2, \tag{4.10.51}$$

which are those of the generators of the group $SO(3)$, the three-dimensional rotation group. This group is compact, and possesses an irreducible unitary representation of finite dimension, $D_{p/2}$, for $p = 0, 1, 2, \cdots$. The operator J_3 has $p + 1$ eigenvalues, $-p/2, -p/2 + 1, \cdots, p/2 - 1, p/2$. As already evident, p must be a nonnegative integer.

4.10.6 Many-Body Paraquantization Relations. The paraquantization procedures can be extended to a system of many identical, indistinguishable parasystems, each of which obeys the appropriate paraquantization rules already developed in the previous two subsections. The results will be stated here without proof, but with references (Green [1972]; Ohnuki and Kamefuchi [1982]). The general trilinear parafermi and parabose paracommutation relations for a system consisting of f oscillators, the kth one of which is represented by the operators a_k and $a_k^\dagger$, which obey the rules of Eq. (4.10.33) or of Eq. (4.10.46), are

$$[a_k, [a_l^\dagger, a_m]_\mp] = 2\delta_{kl} a_m,$$
$$[a_k, [a_l^\dagger, a_m^\dagger]_\mp] = 2\delta_{kl} a_m^\dagger \mp 2\delta_{km} a_l^\dagger,$$
$$[a_k, [a_l, a_m]_\mp] = 0, \tag{4.10.52}$$

where $[A, B]_+ = \{A, B\} = AB + BA$. The upper signs apply to parafermions; the lower to parabosons. A general number operator can be written as

$$\mathcal{N} = N - N_0 = N \pm p/2 = \frac{1}{2}[a^\dagger, a]_\mp \pm p/2. \tag{4.10.53}$$

For all k, $a_k |0\rangle = 0$, and for the vacuum state, $|0\rangle$, which is regarded as unique, a general rule for systems of order p is

$$a_k a_l^\dagger |0\rangle = p\delta_{kl} |0\rangle. \tag{4.10.54}$$

It is possible to develop p-specific paracommutation rules, as a consequence of results such as those in Eq. (4.10.49). These rules are often easier to work with than the general ones of Eq. (4.10.52), at least for the order of the paraquantization, p, no greater than 4. Some of these, for the parafermi case, are exhibited in the next few lines, as points of information.

For $p = 0$, the result is the trivial one $a_k = a_k^\dagger = 0$. For $p = 1$, the procedure for obtaining the p-specific commutation rules is shown, as an example of the required technique. Any state of a $p = 1$ system can be written as $|g\rangle = \alpha |0\rangle_k + \beta |1\rangle_k$, where α and β are c-numbers, and only the two possible occupation numbers of the kth single-particle state are considered. Note that $a|0\rangle = 0$, since $|0\rangle$ is the ground level, and $a^\dagger|1\rangle = 0$ because $|1\rangle$ is the highest level. It follows that $aa|g\rangle = 0$, and $a^\dagger a^\dagger|g\rangle = 0$, because of the annihilation and creation properties of the operators. This result is equivalent to

$$aa = a^\dagger a^\dagger = 0. \tag{4.10.55}$$

Also, since $N_g|0\rangle = -\frac{1}{2}|0\rangle$, and $N_g|1\rangle = \frac{1}{2}|1\rangle$, it follows that $N_g N_g|g\rangle = \frac{1}{4}|g\rangle$, for any general state $|g\rangle$, or, equivalently,

$$N_g^2 = 1/4. \tag{4.10.56}$$

From Eq. (4.10.46), after use of Eqs. (4.10.55), it is seen that

$$aa^\dagger a = a, \quad \text{and} \quad a^\dagger aa^\dagger = a^\dagger. \tag{4.10.57}$$

Expansion of Eq. (4.10.56) and use of Eqs. (4.10.55) leads to $\{a, a^\dagger aa^\dagger\} = \{a^\dagger, aa^\dagger a\} = 1$, which combines with Eq. (4.10.57) to give

$$\{a, a^\dagger\} = \{a^\dagger, a\} = 1, \tag{4.10.58}$$

confirming the result that parafermion quantization of order 1 is canonical fermion quantization. From the first of Eqs. (4.10.52), with $l = k$, the expanded expression is

$$[a_k, [a_k^\dagger, a_m]] = a_k a_k^\dagger a_m - a_k a_m a_k^\dagger - a_k^\dagger a_m a_k + a_m a_k^\dagger a_k = 2a_m.$$

The anticommutator of this equation with a_k yields, with the help of Eqs. (4.10.55),

$$\{a_k, a_m\} = 0, \tag{4.10.59}$$

the basic equation, along with its complex conjugate, for the Pauli exclusion principle. As will be seen in a subsequent subsection, this equation leads to the requirement, for canonical fermions, that the wave function change sign upon exchange of any two identical particles.

For $p = 2$, the relations become trilinear. With a general state vector written as $|g\rangle = \alpha|0\rangle_k + \beta|1\rangle_k + \gamma|2\rangle_k$, it may be seen that $a_k a_k a_k|g\rangle = 0$, $a_k^\dagger a_k^\dagger a_k^\dagger|g\rangle = 0$, and $N_k^3|g\rangle = N_k|g\rangle$. These results may then be written

$$a_k a_k a_k = 0; \quad a_k^\dagger a_k^\dagger a_k^\dagger = 0; \quad \text{and} \quad N_k^3 = N_k. \tag{4.10.60}$$

These results, specific for $p = 2$, together with the general relations of Eq. (4.10.52), lead to the self-contained set of commutation relations of order 2 for parafermions:

$$a_k a_l^\dagger a_m + a_m a_l^\dagger a_k = 2\delta_{kl} a_m + 2\delta_{lm} a_k,$$
$$a_k^\dagger a_l a_m + a_m a_l a_k^\dagger = 2\delta_{kl} a_m,$$
$$a_k a_l a_m + a_m a_l a_k = 0. \tag{4.10.61}$$

These and their hermitian conjugates provide the rules for constructing properly symmetrized states for parafermions of order 2.

Higher order parafermion rules can be developed by use of the procedures exhibited in these examples. The existence of a ket of highest occupation number, $|p\rangle_k$, for each parafermion single-particle state leads to $(a_k^\dagger)^p = 0$, and the existence of an equation of degree p for the operator N_k provides the p-specific relations that can be combined with the general ones of Eq. (4.10.52) to produce the set of rules for any parafermion system of order p. The parabose systems are more difficult, because they lack the convenient p-specific relations of parafermions. But as might be expected from the general expressions of Eqs. (4.10.52), which differ for parafermions and parabosons only in a few signs, it turns out to be possible to express the p-specific rules for the two types of noninteracting paraparticles in a unified formulation. With the notation $\langle A, B, C \rangle_\pm = ABC \pm CBA$, the p-specific rules for paraparticles are here summarized, with the upper sign referring to parafermions, as before:

For $p = 0$,

$$a_k = a_k^\dagger = 0;$$

for $p = 1$,

$$[a_k, a_l^\dagger]_\pm = \delta_{kl}, \quad [a_k, a_l]_\pm = 0;$$

for $p = 2$,

$$\left\langle a_k, a_l^\dagger, a_m \right\rangle_\pm = 2\delta_{kl}a_m \pm +2\delta_{lm}a_k,$$

$$\left\langle a_k, a_l, a_m^\dagger \right\rangle_\pm = 2\delta_{lm}a_k,$$

$$\left\langle a_k, a_l, a_m \right\rangle = 0. \tag{4.10.62}$$

4.10.7 Further Developments. The next steps involve rewriting the commutation rules in terms of the equal-time field operators, showing that the technique of paraquantization meets the basic requirements of quantum field theory for field quantization, and then developing the theorems on n-particle states and the statistics they obey. I regret that the inclusion of this material would fill more space than can be made available. The most important theorem, as related to statistical mechanics, is the spin-statistics theorem (Belinfante [1939a]; Belinfante [1939b]; Belinfante [1939c]; Pauli [1940]; Pauli and Belinfante [1940]). It was obtained by Belinfante, in a series of three papers in 1939 (in English, with abstracts in German and Esperanto), based on charge-conjugation invariance of his version of the field operators, which he called *undors*. At about the same time, Pauli derived the theorem using postulates based on relativity. His paper was to have been presented first at the Solvay Conference of 1939, but Hitler's mischief disrupted all normal scientific activities, and the paper appeared in *Physical Review* in 1940. That same year, Pauli and Belinfante wrote a joint paper to clear up the differences in the two approaches. (According to N. Kemmer, who was Pauli's student and colleague, the spin-statistics theorem was Pauli's proudest accomplishment.) Pauli explained and expanded his original paper, in more modern notation, ten years later (Pauli [1950]). In this paper, he repeats his original conditions leading to the spin-statistics theorem for noninteracting particles:

(1.) The vacuum is the state of lowest energy.

(2.) Physical quantities (observables) commute with each other in two space-time points with space-like separation. (Measurements at two such points cannot disturb each other.)

(3.) The metric in the Hilbert space of quantum-mechanical states is positive definite. (This guarantees that physical probabilities are non-negative.)

He also comments on papers by Schwinger and Feynman, some of which gave results that seemed to violate the theorem. Pauli shows the necessity for treating *self-adjoint* and *Hermitian* as noninterchangeable properties, since negative probabilities can result from regarding the terms as equivalent. Every proposed violation of the theorem violates at least one of Pauli's conditions.

The original spin-statistics theorem for fields of noninteracting particles, says:

Fields of half-integer and integer spin must satisfy Fermi and bose commutation relations, respectively; or, particles of half-integer and integer spin must obey Fermi and Bose statistics, respectively.

The theorem has been generalized to replace *fermi* and *bose* in the original statement by *parafermi* and *parabose*, and the derivation has been generalized to include interacting particles. For a system containing $n > p$ particles, at most p of them can form the completely symmetric (antisymmetric) configuration in the parafermi (parabose) case. In the parafermi (parabose) case, any number of particles can form the completely antisymmetric (symmetric) configuration. That is, in the parafermi (parabose) case of order p, at most p (any number of) particles can occupy any single-particle state. The parafermi situation represents a natural generalization of the Pauli principle. In cases for which $p \geq 2$, state vectors of more general types of symmetry, or of mixed symmetry, are allowed. Specific results for parafermion statistics of noninteracting particles are given in the following subsection. A more general discussion is given by Forte [1992].

4.10.8 Ideal Parafermion Gases. The derivation of the grand partition function for parafermions proceeds as for fermions and bosons, in Eqs. (4.2.2) and (4.3.1), except that any single-particle state can have at most p occupants. The equations analogous to those for fermions and bosons become

$$Z_G = \sum_{N=0}^{\infty} z^N Z_N = Z'_G + z e^{-\beta\epsilon_1} \sum z^{N-1} Z'_{N-1}$$
$$+ z^2 e^{-2\beta\epsilon_1} \sum z^{N-2} Z'_{N-2} + \cdots + z^p e^{-p\beta\epsilon_1} \sum z^{N-p} Z'_{N-p}$$
$$= Z'_G \sum_{n=0}^{p} (z e^{-\beta\epsilon_1})^n = Z'_G \frac{1-(z e^{-\beta\epsilon_1})^{p+1}}{1-z e^{-\beta\epsilon_1}} = \prod_i \frac{1-(z e^{-\beta\epsilon_i})^{p+1}}{1-z e^{-\beta\epsilon_i}}. \quad (4.10.63)$$

From this result, the development parallels that of ordinary fermions. First write $\ln Z_G$ as a sum, and replace the sum by an integral, using the appropriate single-particle density of states, to get

$$\ln Z_G = \int_0^\infty d\epsilon\, g(\epsilon) \left[\ln \left(1 - (ze^{-\beta\epsilon})^{(p+1)} \right) - \ln \left(1 - ze^{-\beta\epsilon} \right) \right]. \tag{4.10.64}$$

The expected particle number and total energy are then found to be

$$\langle N \rangle = \int_0^\infty d\epsilon\, g(\epsilon) \left[\frac{1}{z^{-1}e^{\beta\epsilon} - 1} - \frac{p+1}{(z^{-1}e^{\beta\epsilon})^{(p+1)} - 1} \right]; \tag{4.10.65}$$

and

$$U = \int_0^\infty d\epsilon\, \epsilon g(\epsilon) \left[\frac{1}{z^{-1}e^{\beta\epsilon} - 1} - \frac{p+1}{(z^{-1}e^{\beta\epsilon})^{(p+1)} - 1} \right]. \tag{4.10.66}$$

The density-of-states integrations can be generalized easily to embrace spaces of d dimensions, and an energy-momentum relationship of the form $\epsilon = ap^\alpha$, where a is a constant. The single-particle partition function is given by the standard integral

$$\begin{aligned} Z_1 &= \left(\frac{V_d A_d d}{h^d} \right) \int_0^\infty dp\, p^{d-1} e^{-\beta a p^\alpha} \\ &= \left(\frac{V_d A_d d}{\alpha h^d a^{(d/\alpha)-1}} \right) \int_0^\infty d\epsilon\, \epsilon^{(d/\alpha)-1} e^{-\beta\epsilon}, \end{aligned} \tag{4.10.67}$$

where the d-dimensional sphere in momentum space has volume $A_d p^d$, and Appendix F gives $A_d = \pi^{d/2}/(d/2)\Gamma(d/2)$. The single-particle density of states in this generalization is seen from Eq. (4.10.67) to be

$$g_d(\epsilon; \alpha) = \frac{2\pi^{d/2} V_d a^{1-d/\alpha}}{h\Gamma(d/2)} \epsilon^{(d/\alpha)-1} = C(d, \alpha)\epsilon^{(d/\alpha)-1}, \tag{4.10.68}$$

where V_d is the d-dimensional volume of the box containing the particles, and $C(d, \alpha)$ is used for notational convenience. The parafermi integrals given in Appendix B, Eq. (B.3.4), together with the density of states $g_d(\epsilon, \alpha)$, give

$$\langle N \rangle = \frac{C(d, \alpha)\alpha p\mu^{d/\alpha}}{d} \left[1 + \frac{\Gamma(d/\alpha + 1)}{3(p+1)\Gamma(d/\alpha - 1)} \left(\frac{\pi}{\beta\mu} \right)^2 + \cdots \right] = \frac{C(d, \alpha)\alpha p\mu_0^{d/\alpha}}{d}, \tag{4.10.69}$$

and

$$\begin{aligned} U &= \frac{C(d, \alpha)\alpha p\mu^{d/\alpha+1}}{\alpha + d} \left[1 + \frac{\Gamma(d/\alpha + 2)}{3(p+1)\Gamma(d/\alpha - 1)} \left(\frac{\pi}{\beta\mu} \right)^2 + \cdots \right] \\ &= \frac{\langle N \rangle \mu_0 d}{d + \alpha} \left[1 + \frac{d + \alpha}{3(p+1)\alpha} \left(\frac{\pi}{\beta\mu_0} \right)^2 + \cdots \right]. \end{aligned} \tag{4.10.70}$$

The molar heat capacity at constant volume, to lowest order, is found from Eq. (4.10.70) to be

$$C_v = \frac{2\pi^2 dRkT}{3(p+1)\alpha\mu_0}, \tag{4.10.71}$$

where

$$\mu_0 = \left[\frac{Nd}{\alpha p C(d,\,\alpha)} \right]^{\alpha/d}. \tag{4.10.72}$$

Thus for any finite maximum occupation number p, the parafermion gas behaves qualitatively like a fermion gas. The Fermi energy decreases, according to Eq. (4.10.72), as p increases, as expected, and the approximations used to evaluate the fermion integrals require for acceptable accuracy either lower temperature or more terms. The linear variation of C_v with temperature persists for $T \ll \mu_0$ at all finite values of p, and bosonic behavior appears, for $d/\alpha > 1$, only as a nonuniform convergence in the limit $p \to \infty$ (Vieira and Tsallis [1987]).

4.10.9 The Statistics of Elementary Particles. It is appropriate to examine the evidence for assigning statistical types to the so-called elementary particles. It should first be noted that the techniques used to decide on the proper statistics for a particle are not dependent upon that particle's being elementary, whatever that may mean. Thus it is valid, for example, to determine that the α-particle is an ordinary boson by the same methods that are used for determining that its constituent protons and neutrons are ordinary fermions.

The basic method of ascertaining the statistics of a particle is, in principle, to assemble a collection of them and find the maximum number that can exist in completely symmetric or antisymmetric states. If it is found that no more than p particles can exist in any completely symmetric (antisymmetric) state, the particles are certifiable as parafermions (parabosons) of order p, provided they are not adherents to some totally alien kind of statistics. This method works well for the more common particles — electrons, protons, photons, and neutrons — as evinced in studies of electrons in metals, atoms, and molecules, photons in cavity radiation, protons in ortho and para hydrogen, and nucleons in nuclear shell structure and in scattering experiments. For other particles, less direct methods must be used.

The superselection rules of paraquantum field theory, based on the same locality condition as Pauli's spin-statistics theorem, provide further guidance in determining the statistics of particles that can be observed in interaction, when all the other particles are known. Examples of such theorems are:

Interactions such that one of the initial (or final) particles is a paraparticle and all others are ordinary bosons or fermions are forbidden.

For a family of paraparticles of even p, the total number of paraparticles of this family that are present in both the initial and final states of a transition must be even.

For a family of paraparticles of odd p, the total number of paraparticles of this family that are present in both the initial and final states of a transition can take any value except an odd number less than p.

The first of these forbids the decay of a paraparticle into ordinary particles; any particle that is observed to decay into ordinary particles cannot be a paraparticle. As an example, consider the determination of the nature of the lambda hyperon (Λ^0), an unstable particle with a lifetime of about 10^{-10}s, observed to decay in the transition

$$\Lambda^0 \to p + \pi^-.$$

The proton is an ordinary fermion, and the π^- is observed in the transition

$$\pi^- + d \to n + n,$$

where the deuteron is known to be a boson, and the neutrons are fermions. Therefore the pion cannot be a paraparticle (it is seen to be a boson), and the Λ^0 is determined to be a fermion. All such particles that ultimately decay into the well-known particles, electrons, protons, neutrons, pions, gammas, and electron neutrinos, can thus be certified as ordinary bosons or fermions and added to the list for use in establishing the nature of other particles.

Apparently every particle that can be observed in a transition to a state in which it either appears or disappears can be classified by the general procedure of the previous paragraph, and all of them have been found to be ordinary fermions or bosons, not paraparticles. Since quarks apparently have not been so observed, it is conceivable that they are parafermions of order $p = 3$ (Ohnuki and Kamefuchi [1982]). Otherwise there seems to be no compelling reason to assign a nonzero probability to the existence of paraparticles, but one is still beset with the question, "Why not?"

4.11 REFERENCES

A few of the principal references are listed here; all references are listed in the Bibliography at the end of the book.

CHANDRASEKHAR, S. [1939], *An Introduction to the Study of Stellar Structure.* Dover, Mineola, NY, 1939.

CHANDRASEKHAR, S. [1978], *Why are the stars as they are?* In *Physics and Astrophysics of Neutron Stars and Black Holes*, R. Giaconni and R. Ruffini, Eds. International School of Physics "Enrico Fermi," North-Holland, Amsterdam, 1978.

OHNUKI, Y. AND KAMEFUCHI, S. [1982] *Quantum Field Theory and Parastatistics.* Springer-Verlag, New York–Heidelberg–Berlin, 1982.

PAULI, W. [1940], Connection Between Spin and Statistics. *Physical Review* **58** (1940), 716–722.

SHAPIRO, S. L. AND TEUKOLSKY, S. A. [1983], *Black Holes, White Dwarfs, and Neutron Stars.* John Wiley & Sons, New York, NY, 1983.

VIEIRA, M. C. de S. AND TSALLIS, C. [1987], D-dimensional Ideal Gas in Parastatistics: Thermodynamic Properties. *J. Stat. Phys.* **48** (1987), 97–120.

4.12 PROBLEMS

1.a. Use the expressions for, and relations among, the various thermodynamic potentials (U, F, *etc.*), together with the relationships for Z_N and Z_G to thermodynamic quantities, to solve for $Z_{\langle N \rangle}$ in terms of Z_G, in general.

b. For an ideal fermion or boson gas ($\sigma = \pm 1$), derive an expression for the entropy S entirely in terms of the mean occupation numbers $\langle n_i \rangle$ of the single particle states, and σ. Note the necessity for writing Z_G in terms of the $\langle n_i \rangle$, as well as for so expressing the E_i. The results of a. should be useful in this calculation.

2. Let the density of states of the electrons in some solid be a constant, D, for all energies greater than zero, and zero for all energies less than zero. The total number of electrons is N.

a. Calculate the Fermi energy μ_0 at 0 K.
b. What is the condition on the temperature in order that the system be non-degenerate?

3. For a two-dimensional cavity (say a box) filled with quanta (photons), in thermal equilibrium with the walls of the cavity at temperature T, derive

a. the number of normal modes of the cavity in the interval $d\omega$.
b. the two-dimensional Planck radiation law. (Find $(1/A)dU/d\nu$, where A is the area.)
c. the Stefan-Boltzmann law.
d. the entropy. (Your result should yield $2TS = 3U$.)

4. For temperature low enough to give high degeneracy, calculate C_v for an ideal Fermi gas having a density of states $g(\epsilon) = a\epsilon^n$, where a and n are constants.

5. Consider an ideal Bose gas composed of particles that have internal degrees of freedom, and assume for simplicity that only the first excited level, ϵ_1, of these internal levels need be considered, besides the ground level, $\epsilon_0 = 0$. Determine the Bose-Einstein condensation temperature as a function of ϵ_1.

6. Find expressions for $\langle N \rangle$, U, and P for a system of relativistic, spin 1/2, non-interacting fermions, where $\epsilon = c[p^2 + (mc)^2]^{1/2}$. It is helpful to write $p = mc \sinh \theta$, and to include the rest energy mc^2 in the chemical potential.

7. Calculate C_v and μ for the extreme relativistic ideal Fermi gas that is highly degenerate. In this case $\epsilon = cp$.

8. In the Debye theory of the specific heats of solids, both transverse and longitudinal normal modes act as degrees of freedom and contribute to the heat capacity.
a. Derive a Debye-like theory for the heat capacity of liquid helium, which should not exhibit transverse waves.
b. Compare the temperature dependence of C_v so obtained with that derived from the Bose-Einstein condensed boson fluid.
c. The observed density (0.145 gm/cm^3) and longitudinal sound velocity $(2.39 (10)^4 \text{ cm/sec})$ of ^{4}He give what value for the Debye temperature?
d. Compare the numerical results of both theories with the observed values:

$$C_v(0.1 \text{ K}) = 2.35(10)^{-5} \text{joules/gmK},$$

and

$$C_v(0.5 \text{ K}) = 2.94(10)^{-3} \text{joules/gmK}.$$

9.a. Find the density of states $g(\epsilon)$ of a system with an energy spectrum given by

$$\epsilon_{\ell m n} = \gamma \left(\frac{\lambda_1 \ell}{L_1} + \frac{\lambda_2 m}{L_2} + \frac{\lambda_3 n}{L_3} \right)$$

where ℓ, m, and n are non-negative integers, the L's are the dimensions of the container, γ and the λ's are constants, not necessarily integers. (Hint: find the volume in integer space between an energy surface and the coordinate planes.)
b. Given a Bose-Einstein gas of N particles having the energy spectrum of part a., find the transition

temperature T.
c. Calculate the heat capacity for $T < T_c$.
d. Calculate the discontinuity in C_v at $T = T_c$.

10. Show that the entropy vanishes at $T = 0$, and the quantum mechanical expression for the entropy evolves to the Sackur-Tetrode equation at high temperatures for
a. An ideal boson gas.
b. An ideal fermion gas.

11. For a spinless monatomic ideal gas of noninteracting particles in n dimensions:
a. Calculate the single-particle density of states $g_n(\epsilon)$.
b. Calculate C_v for fermions, bosons, and Boltzmann particles.
c. Compare these expressions for C_v when $n = 2$, especially at temperatures for which quantum effects are important.

4.13 SOLUTIONS

1.a.

$$Z_N = \sum_i e^{-\beta E_{Ni}}; \quad F_N = U_N - T S_N = -P_N V + \mu N = -kT \ln Z_N$$

$$Z_G = \sum_{N=0}^{\infty} z^N Z_N; \quad \Phi = U - TS - \mu \langle N \rangle = -PV = -kT \ln Z_G$$

When

$$N = \langle N \rangle, P_N = P, \text{ and } PV = kT \ln Z_G = kT \ln Z_{\langle N \rangle} + \mu \langle N \rangle.$$

Since $\mu = kT \ln z$, it follows that

$$\ln Z_G = \ln Z_{\langle N \rangle} + \langle N \rangle \ln z.$$

This relationship is only approximate, since it is the equivalent of setting Z_G equal to the largest term in $\sum z^N Z_N$.
b. For fermions or bosons, $\ln Z_G = \sigma \sum_i \ln \left(1 + \sigma z e^{-\beta \epsilon_i}\right)$, and

$$\langle N \rangle = \sum_i \frac{1}{z^{-1} e^{\beta \epsilon_i} + \sigma} = \sum \langle n_i \rangle = \sum (z^{-1} e^{\beta \epsilon_i} + \sigma)^{-1}, \text{ and}$$

$$\epsilon_i = kT \ln z + kT \ln \left[(1 - \sigma \langle n_i \rangle)/ \langle n_i \rangle\right]$$

Since $S \equiv S_{\langle N \rangle} = \left(U_{\langle N \rangle} - F_{\langle N \rangle}\right) \Big/ T$, with

$$F_{\langle N \rangle} = -kT \ln Z_{\langle N \rangle} = -kT \ln Z_G + \langle N \rangle kT \ln z, \text{ and}$$

$$U_{\langle N \rangle} = \sum \langle n_i \rangle \epsilon_i = \langle N \rangle kT \ln z + kT \sum_i \langle n_i \rangle \ln(1 - \sigma \langle n_i \rangle) - kT \sum_i \langle n_i \rangle \ln \langle n_i \rangle,$$

it follows that

$$S = -k \sum_i \left[\langle n_i \rangle \ln \langle n_i \rangle + \sigma(1 - \sigma \langle n_i \rangle) \ln(1 - \sigma \langle n_i \rangle) \right].$$

2.a. For a constant density of states, D,

$$\ln Z_G = D \int_0^\infty d\epsilon \ln(1 + z e^{-\beta\epsilon}); \quad \langle N \rangle = D \int_0^\infty \frac{d\epsilon}{z^{-1} e^{\beta\epsilon} + 1}.$$

At $T = 0$, $\langle N \rangle = D \int_0^{\mu_0} d\epsilon = D\mu_0$, or $\mu_0 = \langle N \rangle / D$.

b. When non-degenerate, $z^{-1} e^{\beta\epsilon} \gg 1$ for $0 \le \epsilon \le \infty$, so that

$$\langle N \rangle \approx D z \int_0^\infty e^{-\beta\epsilon} d\epsilon = D z k T,$$

or, since $z^{-1} \gg 1$, $\langle N \rangle \ll DkT =$ number of energy levels in the interval kT. From $\mu_0 = \langle N \rangle / D$, it follows that $kT \gg \mu_0$ is another statement of the criterion.

c. From Eq. (B-14), the Fermi integral, when $\nu = 1$ and the system is highly degenerate, is $\mathcal{F}_1(\beta, \epsilon) = \mu$, which says $\langle N \rangle = \mu D$, or $\mu = \mu_0$. This approximation gives

$$U = (D/2)[\mu^2 + (\pi kT)^2/3], \text{ or with } \langle N \rangle = N_0,$$
$$C_v = (D/2)(\pi k)^2 (2T/3) = R\pi^2 kT/3\mu_0.$$

3.a. Photons satisfy the wave equation

$$\nabla^2 \psi = \psi_{tt}/c^2; \quad \psi = \psi_0 \exp i(\mathbf{k} \cdot \mathbf{r} - \omega t); \quad k^2 = \omega^2/c^2.$$

For a two-dimensional rectangular cavity of sides a, b, periodic boundary conditions require $a k_x = 2\pi p$; $b k_y = 2\pi q$; p, q integers. Then $k^2 = \omega^2/c^2$ yields

$$\left(\frac{2\pi p}{a}\right)^2 + \left(\frac{2\pi q}{b}\right)^2 = \frac{\omega^2}{c^2}, \text{ or } \frac{p^2}{(a\omega/2\pi c)^2} + \frac{q^2}{(b\omega/2\pi c)^2} = 1,$$

an ellipse in number space. The area is $\mathcal{A} = \pi(a\omega/2\pi c)(b\omega/2\pi c) = ab\omega^2/4\pi c^2$. The number of normal modes in the interval $d\omega$ is $ab\omega \, d\omega/2\pi c^2$.

b. The average energy of a mode of frequency ω is $\hbar\omega/(e^{\beta\hbar\omega} - 1)$. The total energy is then

$$U = \frac{ab\hbar}{2\pi c^2} \int_0^\infty \frac{\omega^2 \, d\omega}{e^{\beta\hbar\omega} - 1} = \frac{ab(kT)^3 \zeta(3)}{\pi c^2 \hbar^2};$$
$$\frac{1}{ab} \frac{dU}{d\nu} = \frac{2\pi}{ab} \frac{dU}{d\omega} = \frac{2\pi h \nu^2}{c^2 (e^{\beta h\nu} - 1)}.$$

c. The total energy per unit area (energy density) is $U/ab = \zeta(3) k^3 T^3 / \pi c^2 \hbar^2$. The energy radiated per unit length of radiator in equilibrium with a two-dimensional black-body field is

$$(U/ab)(c/\pi) = \zeta(3) k^3 T^3 / \pi^2 c \hbar^2 = \sigma_2 T^3,$$

where σ_2 is the two-dimensional Stefan-Boltzmann constant.

d. Use $TS = U + kT \ln Z_G$. The last term is given by

$$kT \ln Z_G = \frac{-kTab}{2\pi c^2} \int_0^\infty d\omega \, \omega \ln(1 - e^{-\beta\hbar\omega}) = U/2,$$

after integration by parts. Then $TS = (3/2)U$, or $S = 3U/2T$.

4. For $g(\epsilon) = a\epsilon^n$,

$$\ln Z_G = a \int_0^\infty d\epsilon\, \epsilon^n \ln(1 + ze^{-\beta\epsilon})$$

$$U = a \int_0^\infty \frac{\epsilon^{n+1}\, d\epsilon}{z^{-1}e^{\beta\epsilon} + 1} = a\frac{\mu^{n+2}}{n+2}\left[1 + \left(\frac{(n+2)(n+1)}{3\times 2!}\right)\left(\frac{\pi}{\beta\mu}\right)^2 + \cdots\right];$$

$$\langle N \rangle = a \int_0^\infty \frac{\epsilon^n\, d\epsilon}{z^{-1}e^{\beta\epsilon} + 1} = a\frac{\mu^{n+1}}{n+1}\left[1 + \left(\frac{(n+1)n}{3\times 2!}\right)\left(\frac{\pi}{\beta\mu}\right)^2 + \cdots\right].$$

At $T = 0$, it follows that $\langle N \rangle = a\mu_0^{n+1}/(n+1)$, as in Eq. (4.5.3), and

$$\mu = \mu_0\left[1 + \frac{n(n+1)}{6}\left(\frac{\pi}{\beta\mu}\right)^2 + \cdots\right]^{-1/n+1} \approx \mu_0\left[1 - \frac{n}{6}\left(\frac{\pi}{\beta\mu_0}\right)^2\right],$$

and

$$U \approx \left(\frac{a}{n+2}\right)\mu_0^{n+2}\left[1 + \frac{n+2}{6}\left(\frac{\pi}{\beta\mu_0}\right)^2\right] = \frac{(n+1)\langle N \rangle \mu_0}{n+2}\left[1 + \frac{n+2}{6}\left(\frac{\pi}{\beta\mu_0}\right)^2\right].$$

With $\langle N \rangle = N_0$, so that $U = u$, it follows that

$$C_v = (n+1)\pi^2 R/3\beta\mu_0.$$

5. For an ideal boson gas with internal degrees of freedom, and with only the first excited internal state and the ground state of importance,

$$\ln Z_G = -\sum_i \left[\ln(1 - ze^{-\beta\epsilon_i}) + \ln(1 - z'e^{-\beta\epsilon_i})\right],$$

where $z' = ze^{-\beta\epsilon_1}$, the result of including the first excited internal level. The population divides into states: those with ϵ_1 excited and those without. Then

$$\langle N \rangle = (V/\lambda^3)g_{3/2}(z) - \ln(1 - z) + (V/\lambda^3)g_{3/2}(z') - \ln(1 - z').$$

When $\epsilon_1 \to \infty$,

$$T_c = T_c(\epsilon_1 = \infty) \equiv T_c(\infty) = (h^2/2\pi mk)\left(\langle N \rangle /V\zeta(3/2)\right)^{2/3}.$$

With this notation, for $\epsilon_1 \neq \infty$,

$$T_c(\epsilon_1) = \left(\frac{h^2}{2\pi mk}\right)\left(\frac{\langle N \rangle}{V}\right)^{2/3}\left(\frac{1}{\zeta(3/2) + g_{3/2}(e^{-\beta\epsilon_1})}\right)$$

$$= T_c(\infty)\left[1 + \frac{g_{3/2}(e^{-\beta\epsilon_1})}{\zeta(3/2)}\right]^{-2/3}$$

$$\approx T_c(\infty)\left[1 - \frac{2}{3}\frac{g_{3/2}(e^{-\beta\epsilon_1})}{\zeta(3/2)} + \cdots\right].$$

6. The number of states in the interval dp is $(2V/h^3)(4\pi p^2 \, dp) \equiv dg$. Write $p = mc \sinh \theta$, so that $\epsilon = [(cp)^2 + (mc^2)^2]^{1/2} = mc^2 \cosh \theta$. Then $dg = (8\pi V/h^3)(mc)^3 \sinh^2 \theta \cosh \theta d\theta$, and

$$\langle N \rangle = \frac{8\pi V(mc)^3}{h^3} \int_0^\infty \frac{d\theta \, \sinh^2 \theta \, \cosh \theta}{e^{\beta(mc^2 \cosh \theta - \mu)} + 1},$$

$$U = \frac{8\pi V m^4 c^5}{h^3} \int_0^\infty \frac{d\theta \, \sinh^2 \theta \, \cosh \theta}{e^{\beta(mc^2 \cosh \theta - \mu)} + 1}.$$

$$P = \frac{kT}{V} \ln Z_G = \left(\frac{8\pi}{\beta}\right)\left(\frac{mc}{h}\right)^3 \int_0^\infty d\theta \, \sinh^2 \theta \, \cosh \theta \ln(1 + ze^{-\beta mc^2 \cosh \theta})$$

$$= \frac{8\pi m^4 c^5}{3h^3} \int_0^\infty \frac{d\theta \, \sinh^4 \theta}{e^{\beta(mc^2 \cosh \theta - \mu)} + 1}$$

after integration by parts.

7. When $\epsilon = cp$, $dg = (8\pi V/c^3 h^3)\epsilon^2 \, d\epsilon$, and

$$\ln Z_G = \frac{8\pi V}{c^3 h^3} \int_0^\infty d\epsilon \, \epsilon^2 \ln(1 + ze^{-\beta \epsilon}).$$

$$U = \frac{8\pi V}{c^3 h^3} \int_0^\infty \frac{\epsilon^3 \, d\epsilon}{e^{\beta(\epsilon - \mu)} + 1};$$

$$N = \frac{8\pi V}{c^3 h^3} \int_0^\infty \frac{\epsilon^2 \, d\epsilon}{e^{\beta(\epsilon - \mu)} + 1}$$

These integrals have been evaluated in problem 4 for the highly degenerate case. With n = 2, the results of 4 give

$$C_v = \frac{R\pi^2 kT}{\mu_0}$$

$$\mu_0 = (hc/2)(3 \langle N \rangle /V)^{1/3},$$

$$\mu = \mu_0 \left[1 - \frac{1}{3}(\frac{\pi kT}{\mu_0})^2 + \cdots \right].$$

8.a. $dW(\omega) = V\omega^2 \, d\omega/2\pi^2 c_a^3$, the number of modes in the interval $d\omega$.

$$\int_0^{\omega_D} dW(\omega) \equiv 3N = V\omega_D^3/6\pi^2 c_a^3.$$

$$\ln Z_N = -\int_0^{\omega_D} \ln(1 - e^{-\beta \hbar \omega}) \, dW(\omega).$$

$$U = -\frac{\partial \ln Z_N}{\partial \beta} = \frac{9N\hbar}{\omega_D^3} \int_0^{\omega_D} \frac{d\omega \, \omega^3 e^{-\beta \hbar \omega}}{1 - e^{-\beta \hbar \omega}} = \frac{9N\hbar(kT)^4}{\hbar^4 \omega_D^3} \int_0^{\theta/T} dx \, x^3 \sum_{n=1}^\infty e^{-nx},$$

with $k\theta = \hbar \omega_D$. With $\theta/T \gg 1$,

$$U = \frac{9NkT^4}{\theta^3} \Gamma(4)\zeta(4) = \frac{3Nk\pi^4 T^4}{5\theta^3},$$

and

$$C_v = \frac{12R\pi^4 T^3}{5\theta^3} = \frac{16R\pi^5 V k^3 T^3}{15 h^3 N c_a^3}.$$

The given numbers yield $\theta = 28.7$ K, with $N/V = 2.18(10)^{22}$ atoms/cm³.

c. Then

$$C_v = 8.24(10)^5 T^3 \text{ ergs/mole K} = 2.060(10)^{-2} T^3 \text{J/gm K};$$

$$C_v(0.1\text{K}) = 2.060(10)^{-5} \text{J/gm K};$$

$$C_v(0.5\text{K}) = 2.575(10)^{-3} \text{J/gm K}.$$

b. For B-E fluid,

$$C_v/R = (1514)(T/T_c)^{3/2}(\zeta(5/2))/\zeta(3/2) = 1.93(T/T_c)^{3/2}.$$

$$T_c = \left(N/V\zeta(3/2)\right)^{2/3} \left(h^2/2\pi m k\right) = 3.126\text{K}.$$

Then

$$C_v = 2.904(10)^7 T^{3/2} \text{ergs/mole K} = 0.726 T^{3/2} \text{J/gm K}.$$

$$C_v(0.1\text{K}) = 2.29(10)^{-3} \text{J/gm K}$$

$$C_v(0.5\text{K}) = 0.257 \text{J/gm K}.$$

Thus the Debye theory fits the experimental results much better.

9.a.

$$\epsilon_{\ell m n} = \gamma \left[\frac{\ell\lambda_1}{L_1} + \frac{m\lambda_2}{L_2} + \frac{n\lambda_3}{L_3}\right] = \frac{\ell}{(L_1/\gamma\lambda_1)} + \frac{m}{(L_2/\gamma\lambda_2)} + \frac{n}{(L_3/\gamma\lambda_3)};$$

the volume in integer space between the plane $\epsilon = $ const. and the coordinate planes is the volume of a pyramid, given by

$$N(\epsilon) = (\tfrac{1}{3})(\tfrac{1}{2}) \left[\frac{\epsilon L_1}{\gamma\lambda_1}\right] \left[\frac{\epsilon L_2}{\gamma\lambda_2}\right] \left[\frac{\epsilon L_3}{\gamma\lambda_3}\right] = \frac{\epsilon^3 V}{6\gamma^3 \lambda_1\lambda_2\lambda_3}.$$

$$dN(\epsilon) = V\epsilon^2 d\epsilon / 2\lambda_0^3 \gamma^3,$$

where

$$\lambda_0^3 \equiv \lambda_1\lambda_2\lambda_3.$$

b.

$$\ln Z_G = -\sum_i \ln(1 - ze^{-\beta\epsilon i}) \to -\frac{V}{2\lambda_0^3\gamma^3} \int_0^\infty d\epsilon\, \epsilon^2 \ln(1 - ze^{-\beta\epsilon})$$

$$= \frac{V}{(\lambda_0\gamma\beta)^3} g_4(z) - \ln(1 - z).$$

$$\langle N \rangle = z\left(\frac{\partial \ln Z_G}{\partial z}\right)_\beta = \frac{V}{(\lambda_0\gamma\beta)^3} g_3(z) + \frac{z}{1 - z}.$$

At $T = T_c$, $\dfrac{\langle N \rangle}{V} = \left(\dfrac{kT_c}{\lambda_0\gamma}\right)^3 \zeta(3)$, or $kT_c = \lambda_0\gamma/\left(v_c\zeta(3)\right)^{1/3}$.

c.

$$U = -\left(\frac{\partial \ln Z_G}{\partial \beta}\right)_{z,v} = \frac{3V g_4(z)}{(\lambda_0\gamma)^3 \beta^4} = \frac{3\langle N \rangle kT^4 g_4(z)}{T_c^3 \zeta(3)}.$$

For $T < T_c$, $U = \dfrac{3\langle N\rangle kT^4\zeta(4)}{T_c^3\zeta(3)}$, and

$$C_v = \frac{12RT^3\zeta(4)}{T_c^3\zeta(3)}$$

d. For $T > T_c$,

$$C_v = \frac{12RT^3 g_4(z)}{T_c^3\zeta(3)} + \frac{3RT^4 g_3(z)}{zT_c^3\zeta(3)}\left(\frac{\partial z}{\partial T}\right)_{N,V},$$

or, since

$$\left(\frac{\partial z}{\partial T}\right)_{N,V} = -\frac{3zg_3(z)}{Tg_2(z)}, \quad C_v = \frac{12RT^3}{T_c^3\zeta(3)}\left(g_4(z) - \frac{3g_3^2(z)}{4g_2(z)}\right)$$

Then

$$C_v(T_c^+) - C_v(T_c^-) = -9R\zeta(3)/\zeta(2).$$

10.a. From the results of problem 1.b.,

$$S/k = -\sum_i \langle n_i\rangle \ln\langle n_i\rangle - \sigma\sum_i (1 - \sigma\langle n_i\rangle)\ln(1 - \sigma\langle n_i\rangle)$$

$$= \sigma\sum_i \frac{(z^{-1}e^{\beta\epsilon_i} + \sigma)\ln(z^{-1}e^{\beta\epsilon_i} + \sigma) - z^{-1}e^{\beta\epsilon_i}\ln(z^{-1}e^{\beta\epsilon_i})}{(z^{-1}e^{\beta\epsilon_i} + \sigma)}$$

For bosons, $\sigma = -1$; for $T < T_c$, $z = 1$, and

$$S/k = \sum_i \frac{\beta\epsilon_i - (1 - e^{\beta\epsilon_i})\left[\beta\epsilon_i + \ln(1 - e^{-\beta\epsilon_i})\right]}{1 - e^{-\beta\epsilon_i}}$$

$$= \sum_i \left(\frac{\beta\epsilon_i e^{-\beta\epsilon_i}}{1 - e^{-\beta\epsilon_i}} - \ln(1 - e^{-\beta\epsilon_i})\right).$$

$$\lim_{\beta\to\infty}(S/k) = 0.$$

At $T \gg T_c$, $z^{-1} \gg 1$, and σ can be neglected in comparison. Thus,

$$S/k \simeq \sigma\sum_i \left[\beta(\epsilon_i - \mu) + \sigma z e^{-\beta\epsilon_i} - \beta(\epsilon_i - \mu)(1 - \sigma z e^{-\beta\epsilon_i})\right]$$

$$= \sum_i z e^{-\beta\epsilon_i}\left[1 + \beta(\epsilon_i - \mu)\right]$$

$$= zZ_1 + \beta zZ_1\langle\epsilon\rangle - \beta\mu zZ_1,$$

where Z_1 is the one-particle partition function, and $\langle\epsilon\rangle$ is the average single-particle energy. From $Z_G = \sum(zZ_1)^N/N! = e^{zZ_1}$, it follows that $zZ_1 = \langle N\rangle$. Also, since $z = \langle N\rangle/Z_1 = \langle N\rangle\lambda^3/V$, it follows that $\mu = kT\ln(\langle N\rangle\lambda^3/V)$, and

$$S/k = \langle N\rangle\left[1 + \beta\langle\epsilon\rangle - \beta\mu\right]$$

$$= \frac{5}{2}\langle N\rangle + \langle N\rangle\ln\left[\left(\frac{2\pi m}{z}\right)^{3/2}\left(\frac{V}{\langle N\rangle}\right)\left(\frac{2U}{3\langle N\rangle}\right)^{3/2}\right].$$

for both bosons and fermions.

b. For fermions at low T, write

$$S/k = - \sum_i \langle n_i \rangle \ln \langle n_i \rangle - \sum_i (1 - \langle n_i \rangle) \ln(1 - \langle n_i \rangle).$$

The exclusion principle requires that $0 \leq \langle n_i \rangle \leq 1$ for all T. At $T = 0$, $\langle n_i \rangle = 1$ for $\epsilon_i \leq \mu_0$; $\langle n_i \rangle = 0$ for $\epsilon_i > \mu_0$. Therefore every term in S/k vanishes at $T = 0$.

11.a. The single-particle partition function may be written

$$Z_1 = \frac{V_n}{h^n} \int_0^\infty \left(\frac{\pi^{n/2}}{\Gamma(1 + n/2)} \right) n \, p^{n-1} e^{-\beta p^2/2m} \, dp.$$

With $\epsilon = p^2/2m$, the transfer from momentum space to energy space gives the density of states to be

$$g_n(\epsilon) = \frac{V_n}{\Gamma(n/2)} \left(\frac{2\pi m}{\hbar^2} \right)^{n/2} \epsilon^{n/2-1}$$

b. For fermions and bosons, the standard formalism gives

$$\ln Z_G = \sigma \int_0^\infty d\epsilon \, g_n(\epsilon) \ln(1 + \sigma z e^{-\beta \epsilon}).$$

The energy and particle number are then found to be

$$U = -\frac{\partial \ln Z_G}{\partial \beta} = \int_0^\infty \frac{d\epsilon \, \epsilon \, g_n(\epsilon)}{z_{-1} e^{\beta \epsilon} + \sigma},$$

$$\langle N \rangle = z \frac{\partial \ln Z_G}{\partial z} = \int_0^\infty \frac{d\epsilon \, g_n(\epsilon)}{z_{-1} e^{\beta \epsilon} + \sigma}.$$

At this point, the results are even valid for Boltzmann particles, for which $\sigma = 0$. For degenerate fermions, the integrals are written out by means of the formalism in Appendix B to be

$$U = \left[\left(\frac{V_n}{\Gamma(n/2)} \right) \left(\frac{2\pi m}{h^2} \right)^{n/2} \right] \left(\frac{\mu^{n/2+1}}{n/2+1} \right) \left[1 + \frac{(n/2+1)(n/2)}{3 \times 2!} \left(\frac{\pi}{\beta \mu} \right)^2 + \cdots \right],$$

$$\langle N \rangle = \left[\left(\frac{V_n}{\Gamma(n/2)} \right) \left(\frac{2\pi m}{h^2} \right)^{n/2} \right] \left(\frac{\mu^{n/2}}{n/2} \right) \left[1 + \frac{(n/2)(n/2-1)}{3 \times 2!} \left(\frac{\pi}{\beta \mu} \right)^2 + \cdots \right].$$

At $T = 0$, the $\langle N \rangle$-equation gives μ_0 to be

$$\mu_0 = \frac{h^2}{2\pi m} \left(\frac{\langle N \rangle \Gamma(n/2+1)}{V_n} \right)^{2/n}.$$

When $T \neq 0$, μ is found to be

$$\mu = \mu_0 \left[1 - \frac{(n/2-1)}{6} \left(\frac{\pi}{\beta \mu} \right)^2 + \cdots \right] \approx \mu_0 \left[1 - \frac{(n/2-1)}{6} \left(\frac{\pi}{\beta \mu_0} \right)^2 \right].$$

To this approximation, U can be calculated as

$$U \approx \frac{n\langle N\rangle\mu_0}{n+2}\left[1+\frac{n+2}{12}\left(\frac{\pi}{\beta\mu_0}\right)^2\right],$$

and C_v is found to be

$$C_v = \frac{N_0}{\langle N\rangle}\left(\frac{\partial U}{\partial T}\right)_{\langle N\rangle,V} = \frac{nR\pi^2 kT}{6\mu_0}.$$

Similarly for bosons, the energy and particle number are, by use of Appendix B,

$$U = \left[\left(\frac{V_n}{\Gamma(n/2)}\right)\left(\frac{2\pi m}{h^2}\right)^{n/2}\right]\left(\frac{\Gamma(n/2+1)}{\beta^{(n/2+1)}}\right)g_{n/2+1}(z);$$

$$\langle N\rangle = \left[\left(\frac{V_n}{\Gamma(n/2)}\right)\left(\frac{2\pi m}{h^2}\right)^{n/2}\right]\left(\frac{\Gamma(n/2)}{\beta^{(n/2)}}\right)g_{n/2}(z).$$

These can be combined to give, for $T > T_c$,

$$\langle N\rangle = V_n g_{n/2}/\lambda^n;$$

$$U = \frac{n}{2}\langle N\rangle kT\frac{g_{n/2+1}(z)}{g_{n/2}(z)}.$$

Since the equation for $\langle N\rangle$ gives

$$\left(\frac{\partial z}{\partial T}\right)_v = -\frac{zng_{n/2}(z)}{2Tg_{n/2-1}(z)},$$

the molar heat capacity can be found from U to be

$$C_v = \frac{nR}{2}\left[\left(1+\frac{n}{2}\right)\frac{g_{n/2+1}(z)}{g_{n/2}(z)} - \frac{n}{2}\frac{g_{n/2}(z)}{g_{n/2-1}(z)}\right]; \quad T > T_c.$$

For $T \gg T_c$, since the g's become unity, the molar heat capacity becomes $C_v = nR/2$. As $T \to T_c$ from above,

$$C_v(T_c^+) = \frac{nR}{2}\left[\left(1+\frac{n}{2}\right)\frac{\zeta(n/2+1)}{\zeta(n/2)}\right].$$

For $T \leq T_c$, the results become

$$\langle N\rangle = \frac{V_n\zeta(n/2)}{\lambda^n} + \frac{z}{1-z} = N_e + N_g;$$

$$U = \frac{nV_n kT}{2\lambda^n}\zeta(n/2+1) = \frac{n\langle N\rangle kT}{2\zeta(n/2)}\zeta(n/2+1)\left(\frac{T}{T_c}\right)^{n/2};$$

$$C_v = \frac{n(n+2)R\zeta(n/2+1)}{4\zeta(n/2)}\left(\frac{T}{T_c}\right)^{n/2}.$$

For the Boltzmann gas, $U = (n/2)\langle N\rangle kT$, and $C_v = nR/2$.

(c) For $n = 2$, the fermion calculation gives $C_v = \pi^2 RkT/3\mu_0 = \pi^2 Rv_2/3\lambda^2$. For bosons with $n = 2$, $T_c = 0$ because $\zeta(1)$ is infinite, and there is no need for condensation into the ground state. At low T, $z < 1$, but $z \approx 1$, so $g_n(z) \approx \zeta(n)$ for all n. Then $U \approx (V_2 kT/\lambda^2)\zeta(2)$, and $C_v \approx \pi^2 Rv_2/3\lambda^2$, as for fermions.

5. DIELECTRIC AND MAGNETIC SYSTEMS

5.1 INTRODUCTION

Confusion characterizes the treatment of the thermodynamics of systems in which electric or magnetic work terms are involved. The problem resides in there being several suitable expressions for the electric or magnetic energy, depending upon choices of which parts of the energy to include in the system, which to include in the surroundings, and what are the sources of the fields. In the treatment required by the formalism of statistical thermophysics, however, the appropriate energy terms must be compatible with the corresponding microscopic ones in the Hamiltonian. It is therefore appropriate to begin with the system Hamiltonian, see where the microscopic energies must be placed, and then express these energies in terms of the macroscopic variables that are used in thermodynamics. From the underlying microphysics, the energy of any particle in the system is determined entirely by a local Hamiltonian, not by fields in distant parts of space that have their sources outside the system. The macroscopic formulations presented in this chapter are therefore given in terms of local fields, and the geometry is chosen to be compatible with this objective.

Most of this chapter is concerned with the analysis of simple magnetic systems, principally those of noninteracting electrons. Systems of interacting particles are examined in Chapter 7. But since the statistical and macroscopic theories of magnetic systems, even simple ones, are more complicated than the corresponding theories of purely electric systems, the latter will be examined first, as a guide to the more complex magnetic ones.

It is appropriate for equilibrium systems to be treated as being in states that are independent of time, even though microscopic fluctuations are recognized as essential to their statistical behavior. In keeping with the customary time-independence of the macroscopic variables, only constant (zero-frequency) macroscopic fields will be considered. Electric fields arise from macroscopically fixed charge distributions or potentials, and magnetic fields from time-independent currents or fluxes. Appropriate references in condensed-matter theory treat a far wider range of topics than can be included here (van Vleck [1932]). The purposes of this chapter are to exhibit the basic statistical and macroscopic theories, together with their interrelationships, and to provide the access to more specialized topics. References to some of these are given throughout the chapter.

5.2 SIMPLE ELECTRIC SYSTEMS

The canonical dielectric system to be examined is a perfect electric insulator of finite size, in equilibrium with a heat bath at constant temperature, and subject at first to the electric field produced by fixed external charges. There is no free charge within such a system, but surface charges can reside on its boundary. The system is regarded as made up of atoms, which themselves are composed of, for present purposes, point electrons and point nuclei. The Hamiltonian of the system contains a sum of Coulomb potential energy terms, representing the potential energy of each charged particle in the system in the fields of all other charged particles, both within the system and in the surroundings. The mutual potential energies of the system particles are regarded macroscopically as part of the internal energy of the system. Similarly the mutual potential energies of the external particles, which are not included in the system Hamiltonian, are considered to be outside energy, of no more importance than the energy of the heat bath. The important energies to understand are the mutual potential energy terms that involve one system particle and one outside particle. Where do these energies belong? The surprising answer is that the system Hamiltonian places the entire mutual energy within the system! Therefore the macroscopic treatment must do likewise to be compatible. (This procedure agrees with our usual practice of assigning all of the mutual gravitational potential energy of a mass near the surface of the earth, for example, to the mass itself, instead of dividing it equally with the earth. There is no physical reality in the localization of field energies.)

5.2.1 Notation and Review of Electrostatics. In this section, a brief review of pertinent parts of electrostatics is included, both as a reminder of the principles and as a means of introducing the notation, which varies in detail from author to author.

Specifically, the pertinent Coulomb energy terms in the Hamiltonian of the system are

$$
\begin{aligned}
PE &= \frac{1}{4\pi\epsilon_0} \sum_i \sum_{j>i} \frac{q_i q_j}{r_{ij}} \\
&= \frac{1}{8\pi\epsilon_0} \sum_{i \neq j} \frac{q_i q_j}{r_{ij}} \\
&= (1/2) \sum_i q_i \phi(\mathbf{r}_i),
\end{aligned}
\tag{5.2.1}
$$

where the sums include all terms in which at least one of the indices refers to a charge within the system. Both indices must cover the external charges, but terms for which q_i and q_j are both external charges are excluded. (The second form of the potential energy term counts each interaction twice, and then takes half of the sum.) The potential term is obviously

$$
\phi(\mathbf{r}_i) = \frac{1}{4\pi\epsilon_0} \sum_{j \neq i} \frac{q_j}{r_{ij}}.
\tag{5.2.2}
$$

In macroscopic terms, the charges are replaced by a charge density $\rho(\mathbf{r})$, the potential by an integral over the charge density, and the potential energy sum by an integral. Thus, the first step in the macroscopic-variable description of the system is to write the total potential energy, W, as

$$
W = \frac{1}{2} \int \rho(\mathbf{r})\phi(\mathbf{r})dV,
\tag{5.2.3}
$$

where the potential can be written

$$
\phi(\mathbf{r}) = \frac{1}{4\pi\epsilon_0} \int \frac{\rho(\mathbf{r}')dV'}{|\mathbf{r} - \mathbf{r}'|}.
\tag{5.2.4}
$$

The integrations extend over both the system and the external charges that can interact with it; therefore W includes the internal potential energy of the system, the interaction energy, which also must be regarded as belonging to the system, and the energy of the outside charges. These terms will be separated after further development.

Since the field equations are linear, it is convenient to write the total electric field as $\mathbf{E} = \mathbf{E}_o + \mathbf{E}_s$, where o and s denote the fields with sources *outside* and within the *system*, respectively. Each of these fields can be derived as the gradient of a potential with its source taken as the appropriate charge density, ρ_o or ρ_s. But since the system is a dielectric with no internal free charges, it is useful to express its potential in terms of its polarization. As may be easily derived by taking the limit as the charge separation $\mathbf{a}$ goes to zero while the dipole moment $\mathbf{p} = q\mathbf{a}$ of a charge pair $\pm q$ remains constant, the potential at $\mathbf{r}_0$ of a point dipole at $\mathbf{r}_1$ is

$$\phi_d(\mathbf{r}_0) = \frac{1}{4\pi\epsilon_0} \frac{\mathbf{p} \cdot \hat{\mathbf{r}}}{(|\mathbf{r}_1 - \mathbf{r}_0|)^2},\tag{5.2.5}$$

where $\mathbf{r} = \mathbf{r}_0 - \mathbf{r}_1$, and $\hat{\mathbf{r}} = \mathbf{r}/r$. Then for the potential from a macroscopically continuous distribution of polarized matter, with polarization $\mathbf{P}(\mathbf{r}')$ (Polarization is defined to be the dipole moment per unit volume, in the continuum limit as the volume goes to zero.), the potential at point $\mathbf{r}_0$ is given by

$$\phi_s(\mathbf{r}_0) = \frac{1}{4\pi\epsilon_0} \int_V \frac{\hat{\mathbf{r}} \cdot \mathbf{P}(\mathbf{r}')dV'}{(\mathbf{r}' - \mathbf{r}_0)^2},\tag{5.2.6}$$

where $\mathbf{r} = \mathbf{r}_0 - \mathbf{r}'$, and $\hat{\mathbf{r}} = \mathbf{r}/r$. This expression may be rewritten by noting an operator identity:

$$\nabla'\left(\frac{1}{|\mathbf{r}' - \mathbf{r}_0|}\right) = \frac{\hat{\mathbf{r}}}{(\mathbf{r}' - \mathbf{r}_0)^2}.\tag{5.2.7}$$

By use of this identity, Eq. (5.2.6) becomes

$$\begin{aligned}
\phi_s(\mathbf{r}_0) &= \frac{1}{4\pi\epsilon_0} \int_V \mathbf{P}(\mathbf{r}') \cdot \nabla'\left(\frac{1}{|\mathbf{r}' - \mathbf{r}_0|}\right) dV'\\
&= \frac{1}{4\pi\epsilon_0} \left[\int_V \nabla' \cdot \left(\frac{\mathbf{P}(\mathbf{r}')}{|\mathbf{r}' - \mathbf{r}_0|}\right) dV' - \int_V \frac{\nabla' \cdot \mathbf{P}(\mathbf{r}')}{|\mathbf{r}' - \mathbf{r}_0|}dV'\right]\\
&= \frac{1}{4\pi\epsilon_0} \left[\int_{surface} \frac{\mathbf{P}(\mathbf{r}') \cdot d\mathbf{A}'}{|\mathbf{r}' - \mathbf{r}_0|} - \int_V \frac{\nabla' \cdot \mathbf{P}(\mathbf{r}')}{|\mathbf{r}' - \mathbf{r}_0|}dV'\right].
\end{aligned}\tag{5.2.8}$$

The first integral is over the surface of the system and is equivalent to the potential produced by a bound surface charge of charge density $\sigma_b(\mathbf{r}') = \mathbf{P}(\mathbf{r}') \cdot \hat{\mathbf{n}}'$, where $\hat{\mathbf{n}}'$ is the outward normal unit vector at the surface point $\mathbf{r}'$. The second integral is from the bound charge density within the volume, $\rho_b(\mathbf{r}') = -\nabla' \cdot \mathbf{P}(\mathbf{r}')$.

From the basic Maxwell equations for electrostatics,

$$\epsilon_0 \nabla \cdot \mathbf{E} = \rho = \rho_f + \rho_b,\tag{5.2.9}$$

and

$$\nabla \times \mathbf{E} = 0,\tag{5.2.10}$$

it follows that

$$\nabla \cdot (\epsilon_0 \mathbf{E} + \mathbf{P}) \equiv \nabla \cdot \mathbf{D} = \rho_f,\tag{5.2.11}$$

where the so-called displacement $\mathbf{D}$ is defined as $\mathbf{D} = \epsilon_0 \mathbf{E} + \mathbf{P}$. A linear dielectric is one for which the polarization is proportional to the local electric field, or

$$\mathbf{P} = \epsilon_0 \chi_e \mathbf{E}, \tag{5.2.12}$$

where χ_e is called the electric susceptibility, a function of temperature, density, and perhaps other parameters, but not of $\mathbf{E}$. In this case, $\mathbf{D}$ may be written as $\mathbf{D} = \epsilon \mathbf{E}$, where $\epsilon = \epsilon_0(1 + \chi_e)$. (Both ϵ and χ may be symmetric tensors of rank 2, but this generalization will not be expressed explicitly, since it clutters the notation.) For linear dielectrics, the bound and free charge densities are seen to be proportional:

$$\rho_f = \nabla \cdot \mathbf{D} = \epsilon_0(1 + \chi_e)\nabla \cdot \mathbf{E} = -\rho_b(1 + \chi_e)/\chi_e, \tag{5.2.13}$$

thus demonstrating that for ideal linear dielectrics with no internal free charge density, there is also no internal bound charge. Because of the complete absence of any internal charge density, free or bound, it becomes possible to model the dielectric, at zero frequency, as a volume of empty space bounded by a surface charge layer of density $\sigma_b = \mathbf{P} \cdot \hat{\mathbf{n}}$.

A convenient system to study is a volume of dielectric that is a subsection of a homogeneous dielectric that fills the volume of a parallel-plate capacitor in equilibrium with a heat bath at temperature T. The system is chosen to be well away from the fringing fields at the edges of the capacitor, but extending from one plate to the other. (The dielectric buffer within the capacitor but outside the system can be regarded as part of the heat bath.) The external charge produces the field $\mathbf{E}_o$, with $\epsilon_0 \mathbf{E}_o = \mathbf{D}$, and $D = \sigma_f$. Since the source of $\mathbf{D}$ is the free charge, which is here taken to be fixed, the displacement $\mathbf{D}$ is independent of the presence of the dielectric. The field produced by the outside charge, $\mathbf{E}_o$, is also independent of the presence of the dielectric. The field produced by the system charges is found from the equivalent bound-charge distribution. Since $\nabla \cdot \mathbf{D}_s = 0 = \nabla \cdot (\epsilon_0 \mathbf{E}_s + \mathbf{P})$, it follows from the Gauss theorem that $\mathbf{D}_s = 0$, and $\mathbf{E}_s = -\mathbf{P}/\epsilon_0$. From Eq. (5.2.3), the total potential energy of the system and its surroundings becomes

$$W = \frac{1}{2} \int_{allspace} \phi(\mathbf{r})\rho(\mathbf{r})dV = \frac{\epsilon_0}{2} \int \phi(\mathbf{r})\nabla \cdot \mathbf{E}(\mathbf{r})dV. \tag{5.2.14}$$

But because of the vector identity

$$\nabla \cdot (\phi \mathbf{E}) = \nabla\phi \cdot \mathbf{E} + \phi\nabla \cdot \mathbf{E}, \tag{5.2.15}$$

the potential energy may be written

$$\begin{aligned}
W &= \frac{\epsilon_0}{2} \int_{allspace} \nabla \cdot (\phi(\mathbf{r})\mathbf{E}(\mathbf{r}))\, dV - \frac{\epsilon_0}{2} \int_{allspace} \nabla\phi(\mathbf{r}) \cdot \mathbf{E}(\mathbf{r})dV \\
&= \frac{\epsilon_0}{2} \int_{surface} \phi(\mathbf{r})\mathbf{E}(\mathbf{r}) \cdot d\mathbf{A} + \frac{\epsilon_0}{2} \int_{allspace} E^2(\mathbf{r})dV \\
&= \frac{\epsilon_0}{2} \int_{allspace} E^2\, dV,
\end{aligned} \tag{5.2.16}$$

since the surface integral vanishes for finite, bounded charge distributions.

It may be useful to remind the reader who may have forgotten some of the details of this development that the potential energy is clearly non-negative in Eq. (5.2.16), whereas it could have either sign in Eq. (5.2.1). It is usually pointed out that the macroscopic potential includes the self potential of each charge, now considered as a charge distribution, since the potential in the continuum approximation no longer excludes the self-potential of each particle (Griffiths [1987]; Jackson [1976]). It is demonstrated in the references that despite the positive-definite nature of the new energy expression, the interaction terms are the correct ones. While it is true that the energy in terms of the macroscopic ϕ and ρ, or E^2, is positive definite because of the self energies of the particles, a question remains: Why does this energy vanish when any of the macroscopic variables become zero, since the particle self energies are still present, and the absence of polarization charge should not affect them? The answer requires an examination of the process of constructing macroscopic fields from microscopic ones. The local macroscopic charge density ρ, for example, is a time-independent quantum-mechanical average of the charge in a macroscopically small volume (say in a cube with edges of a few tens of atoms in length). The microscopic charge density, ρ_t, is a true point function of position, and it must fluctuate enormously within the small volume that is used to define ρ. Write ρ_t as

$$\rho_t = \rho + \tilde{\rho}, \tag{5.2.17}$$

where $\tilde{\rho}$ is the microscopic quantity that measures the difference between ρ and ρ_t. A similar expression can be written for ϕ. Note that $\langle \rho_t \rangle = \rho$, and $\langle \tilde{\rho} \rangle = 0$, where the average is over the small volume used to define the macroscopic quantities. The actual energy, properly including all of the self-energy is

$$W_t = \frac{1}{2} \int dV \, \phi_t \rho_t = \frac{1}{2} \int dV (\phi\rho + \phi\tilde{\rho} + \tilde{\phi}\rho + \tilde{\phi}\tilde{\rho}) = \frac{1}{2} \int dV (\phi\rho + \tilde{\phi}\tilde{\rho}), \tag{5.2.18}$$

since the two middle terms vanish because the microscopic factors average to zero over every one of the small volumes used to define the macroscopic quantities. But the final term does not vanish; it gives the missing self-energy that is not accounted for by the macroscopic field. The first term is the usual one given in Eq. (5.2.16), and it really should vanish in the absence of macroscopic fields. The microscopic term, which is independent of the presence of macroscopic fields, is properly regarded as part of the internal energy of the system.

The total macroscopic field $\mathbf{E}$ is given by

$$\mathbf{E} = \mathbf{E}_o + \mathbf{E}_s = \mathbf{E}_o - \mathbf{P}/\epsilon_0, \tag{5.2.19}$$

so the energy density in Eq. (5.2.16) is

$$\begin{aligned} w = W/V &= \left(\frac{\epsilon_0}{2}\right)(\mathbf{E}_o + \mathbf{E}_s)^2 \\ &= \left(\frac{\epsilon_0}{2}\right)\left(E_o^2 + 2\mathbf{E}_o \cdot \mathbf{E}_s + E_s^2\right) \\ &= \frac{\epsilon_0 E_o^2}{2} - \mathbf{E}_o \cdot \mathbf{P} + \frac{P^2}{2\epsilon_0}. \end{aligned} \tag{5.2.20}$$

Evidently the energy density given by the first term in the last line of Eq. (5.2.20) is that of the outside charges interacting with each other, independent of the presence of the dielectric. Similarly, the final term is the self-energy of the dielectric in its state of polarization, independent of the external charges. It is usually taken to be part of the internal energy of the system. The middle term represents the interaction energy. One is normally inclined to partition this energy equally between the system and the surroundings, but the earlier discussion has already indicated that it must be regarded as part of the energy of the system.

5.2.2 A Dielectric Thermal System with Fixed External Charges. For convenience of notation, the capacitor system of the previous subsection is further simplified. The dielectric filling the parallel-plate capacitor was taken to be homogeneous, so that the polarization is uniform, and the interaction energy term can be written

$$W_{int} = -\int \mathbf{E}_o \cdot \mathbf{P} \, dV = -\mathbf{E}_o \cdot \boldsymbol{\mathcal{P}}, \tag{5.2.21}$$

where $\boldsymbol{\mathcal{P}}$ is the total electric dipole moment of the system. (Henceforth, terms containing $(\cdots)\boldsymbol{\mathcal{P}}$ are intended to be replaced by $\int(\cdots)P\,dV$ when appropriate.) The change in interaction energy is given by

$$\delta W_{int} = -\mathbf{E}_o \cdot \delta \boldsymbol{\mathcal{P}} - \boldsymbol{\mathcal{P}} \cdot \delta \mathbf{E}_o. \tag{5.2.22}$$

The first of these terms is clearly not a consequence of anything done by the external charges; it therefore must be regarded as merely a shifting of the system energy from one form to another, as will be reconsidered subsequently. The other term, $-\boldsymbol{\mathcal{P}} \cdot \delta \mathbf{E}_o$, represents work done on the system by a change in the field produced by the external charges; it is therefore the proper work term to appear in the differential of the Helmholtz potential for electric systems with fixed external charge configurations.

In review, then, the procedure is as follows:
(1) Write a Hamiltonian for the system: $\mathcal{H} = \mathcal{H}_0 + \mathcal{H}_D$, where $\mathcal{H}_0$ is the Hamiltonian in the absence of externally produced electric fields, and $\mathcal{H}_D$ contains the potential-energy terms that are present as a consequence of fields produced by a fixed configuration of external charges. Note that it must include all potential energy terms that affect particles (charges) in the system.
(2) Obtain the partition function Z_D from the Hamiltonian; methods for this procedure are presented in Chapter 3.
(3) Write the Helmholtz potential as $F_D = -kT \ln Z_D$.
(4) Write dF_D in terms of the proper macroscopic variables:

$$dF_D = -SdT - \boldsymbol{\mathcal{P}} \cdot d\mathbf{E}_o, \tag{5.2.23}$$

but note that other work terms, such as the usual $-PdV$, might also be included. A more general expression for the work term is, of course, $-\int \mathbf{P} \cdot d\mathbf{E}_o$. (The subscripts on F and $\mathcal{H}$ are used in anticipation of there being other choices for the Helmholtz potential and D was chosen for this one because constant external charge configuration implies constant $\mathbf{D}$.)

This formulation leads to the result that equilibrium of a dielectric system occurs at a Helmholtz potential minimum for a system in equilibrium with a heat bath at constant T and a fixed configuration of external charges. For further simplicity of notation, regard the dielectric as isotropic, so that the polarization is parallel to the internal and applied fields. Then dF_D can be written

$$dF_D = -SdT - \mathcal{P}dE_o = \left(\frac{\partial F_D}{\partial T}\right)_{E_o} dT + \left(\frac{\partial F_D}{\partial E_o}\right)_T dE_o. \tag{5.2.24}$$

For thermodynamic stability, it is necessary that F_D be a concave function of its intensive parameters and a convex function of its extensive ones, as pointed out on Section 2.7.3. This means that, since $\mathbf{D} = \epsilon_0 \mathbf{E}_o$,

$$\left(\frac{\partial^2 F_D}{\partial T^2} \right)_D = - \left(\frac{\partial S}{\partial T} \right)_D \equiv - \left(\frac{N}{N_0} \right) \left(\frac{C_D}{T} \right) \leq 0, \qquad (5.2.25)$$

or $C_D \geq 0$, where C_D is the molar heat capacity at fixed external charge configuration; and

$$\left(\frac{\partial^2 F_D}{\partial E_o^2} \right)_T = - \left(\frac{\partial \mathcal{P}}{\partial E_o} \right)_T \leq 0. \qquad (5.2.26)$$

By Eq. (5.2.12), the total electric dipole moment may be written

$$\mathcal{P} = VP = V\epsilon_0 \chi_e E = V\epsilon_0 \chi_e E_o / (1 + \chi_e). \qquad (5.2.27)$$

The last two equations combine to give the result

$$\chi_e / (1 + \chi_e) \geq 0, \qquad (5.2.28)$$

as a stability condition on the electric susceptibility. This result is not quite equivalent to a requirement that the susceptibility must be non-negative, but it clearly precludes the possibility of there being stable dielectrics with small ($|\chi_e| < 1$) negative susceptibilities. Recall, in anticipation of forthcoming topics, that magnetic systems with small negative magnetic susceptibilities exist, and are common states of condensed matter.

5.2.3 A Dielectric System with Fixed External Potentials. The dielectric system examined so far is one with a fixed configuration of external charges. The other common type of system is one with fixed external potentials. A model of such a system is the parallel-plate capacitor of the previous subsection, but now connected to a fixed-voltage power supply, so that changes in polarization of the medium result in a transfer of energy from the power supply to maintain a constant potential difference between the plates. The appropriate thermodynamic potential for this situation is obtained by a Legendre transformation of the previously obtained Helmholtz potential. Since there seem to be no agreed-upon names or notations for these potentials, and since both can be obtained as $F = -kT \ln Z$, where Z is calculated from the proper Hamiltonian for the system, I have designated the one already used (with fixed external charge configuration) as F_D, and the one with fixed external potentials as F_E, since fixed external potentials imply constant internal electric fields. The Legendre transformation, which corresponds to adding a charge equal in magnitude to the polarization charge at the potential difference of the plates, is given formally by

$$F_E = F_D + \mathcal{P} E_o. \qquad (5.2.29)$$

The differential of this equation is

$$dF_E = dF_D + \mathcal{P} dE_o + E_o d\mathcal{P} = -S dT + E_o d\mathcal{P}, \qquad (5.2.30)$$

and, as in Eq. (5.2.24), it may contain other work terms.

The internal self-energy density of the polarized dielectric is, as before, $P^2/2\epsilon_0$, where this time the polarization is assumed to be established already in the dielectric at a value corresponding to the internal electric field, $E = E_o$. The external charge is then enhanced as necessary to produce this internal field. The self-energy associated with the external charges is of no direct interest. The mutual energy is the work done by the power supply on the external charges in order to keep the internal electric field equal to its original value, E_o. This energy is $\int E_o P \, dV = \mathcal{P} E_o$, which is exactly the term added to F_D to produce the Legendre transformation to F_E. But because the internal electric field is constant, the potential energies of all charges within the system remain constant, since these charges are regarded as bound and fixed in position. Therefore there can be no mutual interaction term in the system Hamiltonian, and only the terms that comprise the internal self energy are present.

There can be a mutual energy transfer, as a consequence of a change in $\mathcal{P}$, but this transfer is between the power supply and the mutual energy, not between the system and the mutual energy. This is the source of the term $E_o d\mathcal{P}$ in the expression for dF_E.

In terms of F_E, the condition for stable equilibrium is a Helmholtz-potential minimum at constant temperature and constant external potentials. For stability, F_E must be a concave function of T and a convex function of the extensive parameter $\mathcal{P}$. From Eq. (5.2.30), the first of these conditions is

$$-\left(\frac{\partial S}{\partial T}\right)_{\mathcal{P}} \leq 0,$$

or

$$C_{\mathcal{P}} \equiv \frac{N_0 T}{N}\left(\frac{\partial S}{\partial T}\right)_{\mathcal{P}} \geq 0, \tag{5.2.31}$$

where $C_{\mathcal{P}}$ is the molar heat capacity at constant dipole moment (or polarization and volume). The second stability condition is

$$\left(\frac{\partial E}{\partial \mathcal{P}}\right)_T \geq 0, \tag{5.2.32}$$

which may be rewritten as its reciprocal, with $\mathcal{P}$ expressed as $\mathcal{P} = V\epsilon_0\chi_e E$:

$$\left(\frac{\partial \mathcal{P}}{\partial E}\right)_T = V\epsilon_0\chi_e \geq 0. \tag{5.2.33}$$

This is a stronger condition on the electric susceptibility than that from Eq. (5.2.28), and its counterpart for magnetic systems seems destined to be surprising.

5.2.4 Summary of Thermostatic Formulas for Dielectric Systems.
The two usual types of thermostatic systems for dielectrics are those with constant external charge configurations, which therefore have constant $\mathbf{D}$, and those with constant external potentials, which must then have constant $\mathbf{E}$. The Helmholtz potential F_D is obtained from

$$F_D = -kT \ln Z_D, \tag{5.2.34}$$

where Z_D is the partition function for which the Hamiltonian contains potential energy terms that represent the self energy of the system and the mutual energy of the system charges in the potential that has as its source the external charges.

The Helmholtz potential F_E is similarly obtained from

$$F_E = -kT \ln Z_E, \qquad (5.2.35)$$

where Z_E is the partition function for which the Hamiltonian of the system contains no terms that represent mutual interaction of the system with external charges, since the field originating with these charges is kept constant. The differentials for these two functions are

$$dF_D = -S dT - \mathcal{P} dE_o, \qquad (5.2.36)$$

and

$$dF_E = -S dT + E d\mathcal{P}. \qquad (5.2.37)$$

The energy functions associated with the F's can be obtained by a Legendre transformation of the temperature-entropy term:

$$dU_E = T dS + E d\mathcal{P}, \quad \text{and} \quad dU_D = T dS - \mathcal{P} dE_o. \qquad (5.2.38)$$

The first of these is in the more usual form for an energy function in the style of Chapter 2, because U_E is a function of the extensive variables. Since U_D is obtainable from U_E by a Legendre transformation in the polarization work term, it is sometimes called the dielectric enthalpy and denoted by H_D. Similarly, F_D is called the dielectric Gibbs potential and denoted by G_D. Both of the F's, however, are derived from the canonical partition function by equations of the form Eq. (5.2.36), and there seems to be no advantage in renaming them.

Maxwell relations are easily derivable from the F's and U's, as are comparisons of C_D and $C_{\mathcal{P}}$. Several particular examples of simple dielectric systems are offered in the problems at the end of this chapter. I recommend that these problems be worked out and understood as preparation for further developments in this and later chapters.

In these derivations it has been assumed that the volume of the system remains constant, or at least that the volume is not affected by the electric field. Electrostriction and piezoelectricity are common phenomena that should be analyzed in a comprehensive treatise on the subject (Kubo and Nagamiya [1969]). But I am not aware of any treatments of these areas that are suitable for inclusion in a book that emphasizes the underlying statistical microphysics; therefore I have not attempted to discuss them. Modification of the basic thermostatic functions to include electric-field-induced volume changes can be done in some cases, but I know of no general treatment (Smith-White [1949]; Smith-White [1950]).

5.3 MICROSCOPIC MODELS OF DIELECTRIC SYSTEMS

In Chapter 3, a weakly paramagnetic system was modeled as a collection of noninteracting permanent magnetic dipoles that could be aligned by an external magnetic field, with alignment opposed by thermal stirring. In this section, a similar dielectric system will be examined, but with somewhat more care. The system here consists of identical, noninteracting permanent electric dipoles, a typical one having moment $\mathbf{p}$. The energy of a dipole is written, as in Eq. (3.3.42), as

$$\epsilon_p = -\mathbf{p} \cdot \mathbf{E} = -pE \cos \theta, \qquad (5.3.1)$$

and the mean energy per dipole follows from Eq. (3.3.46) to be

$$\langle \epsilon_p \rangle = -pEL(y) = -pE(\coth \beta pE - 1/\beta pE) = -\langle p \rangle E, \qquad (5.3.2)$$

where $\langle p \rangle$ is the average component of **p** along **E**, the Langevin function $L(y)$ was defined before in Eq. (3.3.47), and $y = \beta pE$. If the external electric field is produced by fixed external charges, then, in accordance with Eq. (5.2.18), the electric field that aligns the dipole is $E_o - P/\epsilon_0$. But P is the polarization produced by the aligned dipoles, or

$$P = n \langle p \rangle, \qquad (5.3.3)$$

where n is the number of dipoles per unit volume. In order to obtain the correct expression for the average dipole energy per unit volume, it is useful to think of the individual dipoles as aligning themselves from an initially unaligned state, under the influence of both the external field and the polarization field that they produce as they become aligned. Therefore $n \langle p \rangle$ must be replaced by dP, and the energy per unit volume becomes

$$w = n \langle \epsilon_p \rangle = -\int_0^P dP'(E_o - P'/\epsilon_0) = -E_o P + P^2/2\epsilon_0. \qquad (5.3.4)$$

This result is in agreement with that of Eq. (5.2.19) and the discussion that follows it. Not only is the internal self-energy correctly given, but also the entire interaction energy is assigned to the system.

In this calculation, even though the dipoles were taken as noninteracting, so that a single-dipole partition function could be used, the internal self-energy appeared, but not as a consequence of direct dipole-dipole interactions, since such terms did not occur in the equation for the energy of a single dipole, Eq. (5.3.1). As is evident from the calculation, the self-energy came from the interactions of the individual dipoles with the average polarization field of the medium. It is of interest to compare this result with those of so-called mean-field theories of systems of interacting particles, examined in Chapter 7. In those theories, a representative particle interacts with its neighbors, represented as dipoles or spin systems, but each interacting neighbor is then replaced by one that is the required system average. The average energy of the representative particle is then calculated in terms of the already-assumed average for the system, and a self-consistent solution yields the sought-for average in terms of other system parameters, including coupling constants and temperature. Because of the interaction of the particles with each other, it is possible for self-polarization to occur, resulting in a phase transition.

In the present calculation, if the objective is to express the polarization in terms of the field of the external charges, a similar kind of calculation is necessary. From Eq. (5.3.2), the average dipole moment can be written

$$\langle p \rangle = pL(y), \qquad (5.3.5)$$

with $y = \beta pE = \beta p(E_o - P/\epsilon_0)$, and $P = n \langle p \rangle$, according to Eq. (5.3.3). When Eq. (5.3.5) is multiplied by n, it becomes an equation for P in terms of P, among other parameters.

A general analytic solution for P is at least improbable, but for $y \ll 1$, the approximation $L(y) = y/3$ permits the solution to be expressed as

$$P = n \langle p \rangle = \frac{np^2 E_o}{3kT + (np^2/\epsilon_0)} \approx \frac{np^2 E_o}{3kT}, \qquad (5.3.6)$$

when $np^2 \ll 3\epsilon_0 kT$. Thus, for small enough values of n the polarization obeys the Langevin formula, with $E_o \approx E$. But for systems that are sufficiently polarized to cause the internal field to differ appreciably from the applied field, the modification offered in Eq. (5.3.6) is appropriate. The susceptibility is usually defined in terms of the local field E, rather than E_o, and it is found in this approximation to be

$$\chi_e = np^2/3\epsilon_0 kT, \qquad (5.3.7)$$

the Langevin formula.

The dielectric contribution to the single-particle partition function, as found in Chapter 3, is

$$Z_1 = \sinh y/y, \qquad (5.3.8)$$

where $y = \beta E p$. For a dielectric system of $\mathcal{N}$ dipoles, the dielectric part of F_D is

$$F_D = -kT\mathcal{N} \ln Z_1, \qquad (5.3.9)$$

and the term in dF_D with T constant is

$$(dF_D)_T = -\mathcal{N}kT(\coth y - 1/y)(\beta p dE) = -\mathcal{N}pL(y)dE = -\mathcal{P}(dE_o - dP/\epsilon_0), \qquad (5.3.10)$$

including the contribution of the polarization field to the internal energy, as advertised in Eq. (5.2.24). The energy term contributed by the dielectric is

$$U_D = -\mathcal{N} \left(\frac{\partial \ln Z_1}{\partial \beta} \right)_{E,V} = -\mathcal{P}E = -\mathcal{P}E_o + \frac{\mathcal{P}P}{2\epsilon_0}, \qquad (5.3.11)$$

where the 2 in the denominator of the final term is the one introduced in Eq. (5.3.4). As already noted, that term is usually assigned to the proper internal energy of the system, and only the interaction term is normally exhibited explicitly.

The Legendre transformation from F_D to F_E is tantamount, on a microscopic scale, to shifting the zero of energy of each dipole to its mean value, since adding the energy $\mathcal{P}E$ adds $\langle p \rangle E$ to the energy of each dipole. The dipole energy is now written

$$\epsilon_p = E(\langle p \rangle - p \cos \theta). \qquad (5.3.12)$$

The partition function corresponding to this energy assignment is

$$Z_1 = e^{-\beta \langle p \rangle E} \frac{\sinh y}{y}, \qquad (5.3.13)$$

which leads to zero as the expectation value for the (polarization) energy. This partition function gives F_E as

$$F_E = E\mathcal{P} - \mathcal{N}kT(\ln\sinh y - \ln y), \tag{5.3.14}$$

and the contribution at constant temperature to dF_E is

$$(dF_E)_T = Ed\mathcal{P} + \mathcal{P}dE - \mathcal{N}kTL(y)(\beta pdE) = Ed\mathcal{P}, \tag{5.3.15}$$

again as advertised.

An equivalent treatment taken from macroscopic thermodynamics considers, in analogy to the definition of a perfect paramagnet, a perfect dielectric to be one such that the polarization is a function only of E/T, or $P = P(E/T)$. For such a dielectric, the energy U_E is independent of E. (Recall that for the system with energy function U_E, the internal and external fields are equal, because the potential difference between the capacitor plates is fixed by an external power supply.) The energy differential is

$$dU_E = TdS + Ed\mathcal{P}, \tag{5.3.16}$$

which leads to

$$\left(\frac{\partial U_E}{\partial E}\right)_T = T\left(\frac{\partial S}{\partial E}\right)_T + E\left(\frac{\partial \mathcal{P}}{\partial E}\right)_T. \tag{5.3.17}$$

But a Maxwell relation from Eq. (5.2.24) says

$$\left(\frac{\partial S}{\partial E}\right)_T = \left(\frac{\partial \mathcal{P}}{\partial T}\right)_E, \tag{5.3.18}$$

which leads directly from the given functional structure of $\mathbf{P}$ to the result that for a perfect dielectric, U_E is independent of E, in agreement with the sentence that includes Eq. (5.3.13), which followed from Eq. (5.3.5). (As might be expected, a similar result is obtained for a perfect paramagnet.)

5.3.1 Induced Polarization.   The dielectric systems treated so far have been made up of ideal electric dipoles that are being aligned by the local electric field in competition with thermal scrambling of the alignment. When these dipoles are regarded as noninteracting, the Langevin equation Eq. (5.3.5) gives the average dipole moment as an odd function of E/T. Another polarization mechanism exists for all atoms and molecules, polar or not, but it is usually evident only for nonpolar systems, because otherwise it is obscured by the much more easily accomplished alignment of permanent dipoles. This mechanism, called *induced polarization*, results from the separation of the positive and negative charge centers of the atom or molecule by action of the macroscopic local field. In the absence of an applied field, the positive and negative charge centers of a nonpolar particle (atom or molecule) are at the same point, by definition. The applied field moves the positive charge center slightly in the direction of the field, and the negative charge center slightly in the opposite direction, producing a net separation of charge centers, x, and a net dipole moment $p = Qx$, where Q is the total positive charge, and both charge distributions are assumed to be approximately spherically symmetric. Because applied fields are typically orders of magnitude smaller than the internal fields of the atom, the charge separation x is small in terms of atomic dimensions. The restoring force that opposes the charge separation can be expanded in a power series in x, with only the first power term retained for present purposes. Thus the restoring force is written $-Kx$, as if Hooke's law pertained. The energy stored in the pseudospring is $\epsilon_s = Kx^2/2$, and the change in electric potential energy is $\epsilon_e = -QEx = -pE$. At equilibrium, the restoring force balances the applied force, or $QE - Kx = 0$. The separation is therefore $x = QE/K$, and the induced dipole moment is $p = Q^2E/K = \alpha E$, where α is called the *polarizability*. Since the polarization is given by $P = np$, where n is the number of polarizable particles per unit volume, and P is also given by $P = \epsilon_0\chi_e E$, it follows that the polarizability is related to the susceptibility by

$$\alpha = \epsilon_0\chi_e/n, \tag{5.3.19}$$

and the pseudospring constant is related to measurable parameters by

$$K = nQ^2/\epsilon_0\chi_e = Q^2/\alpha. \tag{5.3.20}$$

The total energy per particle is given by

$$\epsilon = \epsilon_s + \epsilon_e = -\frac{1}{2}\alpha E^2, \tag{5.3.21}$$

and the single-particle partition function is

$$Z_1 = e^{\beta\alpha E^2/2}. \tag{5.3.22}$$

The expected value of the energy is

$$\langle\epsilon\rangle = -\left(\frac{\partial\ln Z_1}{\partial\beta}\right)_E = -\frac{1}{2}\alpha E^2 = -\frac{1}{2}pE, \tag{5.3.23}$$

which is independent of temperature, as might have been anticipated, since there is no opportunity for thermal agitation in such a model. For this reason, it is usually omitted from statistical treatments of polarization, and students may miss part of the phenomenology.

5.4 SIMPLE MAGNETIC SYSTEMS

The canonical magnetic system to be examined in this chapter is a weakly diamagnetic or paramagnetic material of finite size, in a region of space for which the geometry allows all pertinent fields to be confined to the volume occupied by the system. It will become evident that magnetic systems, much more so than electrical ones, must be examined first from a microscopic point of view, and the Hamiltonian will show the necessity for quantum mechanics. The canonical system is in equilibrium with a heat bath at constant temperature, and in analogy to the treatment of dielectrics, the system will be established to have fixed external **H** or fixed external **B**. At first, magnetization will be regarded as having its source in microscopic ampèrean current loops, each of which produces a magnetic dipole moment. The consequences of this model will be developed in analogy to the treatment of dielectrics. But divergences are shown to be necessary; quantum theory is invoked, and then applied to the magnetic behavior of an ideal fermion gas of noninteracting electrons.

5.4.1 Notation and Review of Magnetostatics. As in the section on electrostatics, this review is intended to remind the reader of material that is presumed to be familiar, and to exhibit the details of notation, which can vary considerably with choice of units. It is not intended to be a comprehensive presentation of the subject to readers who are seeing it for the first time. The notation used here is essentially that of Griffiths [1987], an excellent source of the full coverage that is presumed to be already a part of the reader's body of knowledge.

The absence of isolated magnetic poles and the use of Ampère's law allow the equations for the static magnetic field to be written

$$\nabla \cdot \mathbf{B} = 0, \tag{5.4.1}$$

and

$$\nabla \times \mathbf{B} = \mu_0 \mathbf{J}, \tag{5.4.2}$$

where $\mathbf{J}$ is the total local current density (Griffiths [1987]). The B-field can then be obtained as the curl of a vector potential,

$$\mathbf{B} = \nabla \times \mathbf{A}, \tag{5.4.3}$$

and it is always possible, by adding the gradient of a properly chosen scalar, to construct $\mathbf{A}$ such that

$$\nabla \cdot \mathbf{A} = 0. \tag{5.4.4}$$

With this choice of $\mathbf{A}$, Eq. (5.4.2) can be written

$$\mu_0 \mathbf{J} = \nabla \times (\nabla \times \mathbf{A}) = \nabla(\nabla \cdot \mathbf{A}) - \nabla^2 \mathbf{A} = -\nabla^2 \mathbf{A}, \tag{5.4.5}$$

because of Eq. (5.4.4). The equation for the vector potential in terms of its source has been reduced to a three-dimensional version of the Poisson equation, with the evident solution

$$\mathbf{A}(\mathbf{r}_0) = \frac{\mu_0}{4\pi} \int \frac{\mathbf{J}(\mathbf{r}')dV'}{|\mathbf{r}_0 - \mathbf{r}'|}. \tag{5.4.6}$$

The Biot-Savart law states

$$\mathbf{B}(\mathbf{r}_0) = \frac{\mu_0}{4\pi} \int \frac{\mathbf{J}(\mathbf{r}') \times \hat{\mathbf{r}}\,dV'}{|(\mathbf{r}_0 - \mathbf{r}')|^2}, \tag{5.4.7}$$

where $\mathbf{r} = \mathbf{r}_0 - \mathbf{r}'$ and $\hat{\mathbf{r}} = \mathbf{r}/r$. One further relationship follows from the parallel forms of the $\mathbf{A}$ equations, Eq. (5.4.4) and Eq. (5.4.3), and the $\mathbf{B}$ equations, Eqs. (5.4.1) and (5.4.2); the parallel of Eq. (5.4.7) that relates $\mathbf{A}$ to $\mathbf{B}$:

$$\mathbf{A}(\mathbf{r}_0) = \frac{1}{4\pi} \int \frac{\mathbf{B}(\mathbf{r}') \times \hat{\mathbf{r}}\,dV'}{|(\mathbf{r}_0 - \mathbf{r}')|^2}. \tag{5.4.8}$$

From Eq. (5.4.6), the vector potential of an ampèrian loop carrying the current I can be written

$$\mathbf{A}(\mathbf{r}_0) = \frac{\mu_0 I}{4\pi} \oint \frac{d\mathbf{r}'}{|\mathbf{r}_0 - \mathbf{r}'|} = \frac{\mu_0 I}{4\pi} \sum_{n=0}^{\infty} \frac{1}{r_0^{n+1}} \oint (r')^n P_n(\cos\theta)d\mathbf{r}', \tag{5.4.9}$$

where P_n is a Legendre function, and θ is the angle between $\mathbf{r}_0$ and $\mathbf{r}'$. The zeroth order term in the multipole expansion is identically zero, as expected from the absence of magnetic monopoles. The $n = 1$ dipole term dominates the others, particularly in the limit of an infinitesimal spatial extent for the current loop. When all but the dipole term are neglected, the vector potential of an infinitesimal ampèrian loop becomes

$$\mathbf{A}(\mathbf{r}_0) = \frac{\mu_0 I}{4\pi r_0^2} \oint r' \cos\theta \, dr' = \frac{\mu_0 I}{4\pi r_0^2} \oint (\mathbf{r}' \cdot \hat{\mathbf{r}}_0) dr'. \tag{5.4.10}$$

The identities $d[(\hat{\mathbf{r}}_0 \cdot \mathbf{r}')\mathbf{r}'] = (\hat{\mathbf{r}}_0 \cdot d\mathbf{r}')\mathbf{r}' + (\hat{\mathbf{r}}_0 \cdot \mathbf{r}')d\mathbf{r}'$ and $\oint d[\cdots] = 0$ lead to the result

$$\hat{\mathbf{r}}_0 \times \oint \mathbf{r}' \times d\mathbf{r}' = -2 \oint (\hat{\mathbf{r}}_0 \cdot \mathbf{r}')d\mathbf{r}'. \tag{5.4.11}$$

For a plane area $\mathbf{a}$, it is evident that

$$\mathbf{a} = \frac{1}{2} \oint \mathbf{r}' \times d\mathbf{r}', \tag{5.4.12}$$

where the line integral is along the boundary. These geometric results allow the vector potential to be written

$$\mathbf{A}(\mathbf{r}_0) = \frac{\mu_0}{4\pi} \frac{\mathbf{m} \times \hat{\mathbf{r}}_0}{r_0^2}, \tag{5.4.13}$$

where the magnetic dipole moment of the current loop is

$$\mathbf{m} = \frac{I}{2} \oint \mathbf{r}' \times d\mathbf{r}'. \tag{5.4.14}$$

When the current loop is planar, the result can be written, as is well known from elementary physics, $\mathbf{m} = I\mathbf{a}$, where $\mathbf{a}$ is the area vector of the current loop.

Inside a magnetized system, the vector potential is obtained by integration, as was the scalar potential in Eq. (5.2.8), where here the magnetic dipole moment is replaced by $\mathbf{M}dV$, or

$$\mathbf{A}(\mathbf{r}_0) = \frac{\mu_0}{4\pi} \int \frac{\mathbf{M}(\mathbf{r}') \times \hat{\mathbf{r}}_0 dV'}{(|\mathbf{r}_0 - \mathbf{r}'|^2)}. \tag{5.4.15}$$

By use of Eq. (5.2.7), this equation may be rewritten

$$\mathbf{A}(\mathbf{r}_0) = \frac{\mu_0}{4\pi} \int \mathbf{M}(\mathbf{r}') \times \nabla' \frac{1}{|\mathbf{r}_0 - \mathbf{r}'|} dV', \tag{5.4.16}$$

and further polished, in analogy with Eq. (5.2.8), by means of a vector identity

$$\nabla' \times \frac{\mathbf{M}(\mathbf{r}')}{|\mathbf{r}_0 - \mathbf{r}'|} = \frac{\nabla' \times \mathbf{M}(\mathbf{r}')}{|\mathbf{r}_0 - \mathbf{r}'|} - \mathbf{M}(\mathbf{r}') \times \nabla' \frac{1}{|\mathbf{r}_0 - \mathbf{r}'|}, \tag{5.4.17}$$

to give

$$\mathbf{A}(\mathbf{r}_0) = \frac{\mu_0}{4\pi} \left[\int \frac{\nabla' \times \mathbf{M}(\mathbf{r}')}{|\mathbf{r}_0 - \mathbf{r}'|} dV' - \int \nabla' \times \left(\frac{\mathbf{M}(\mathbf{r}')}{|\mathbf{r}_0 - \mathbf{r}'|} \right) dV' \right]. \tag{5.4.18}$$

Use of the divergence theorem, with the vector quantity written as $\mathbf{M} \times \mathbf{c}$, where $\mathbf{c}$ is an arbitrary constant vector that ultimately disappears from the resulting identity, permits the second term to be written as a surface integral, giving

$$\mathbf{A}(\mathbf{r}_0) = \frac{\mu_0}{4\pi} \left[\int \frac{\nabla' \times \mathbf{M}(\mathbf{r}')}{|\mathbf{r}_0 - \mathbf{r}'|} dV' + \int \frac{\mathbf{M}(\mathbf{r}') \times d\mathbf{a}'}{|\mathbf{r}_0 - \mathbf{r}'|} \right]. \tag{5.4.19}$$

The first term resembles that of Eq. (5.4.7), with the source term, here the bound current density, given by

$$\mathbf{J}_b = \nabla \times \mathbf{M}. \tag{5.4.20}$$

The second term must therefore represent the part of the vector potential attributable to bound surface current, with the surface current linear density $\mathbf{K}_b$, (the current per unit length perpendicular to the direction of $\mathbf{K}_b$) given by

$$\mathbf{K}_b = \mathbf{M} \times \hat{\mathbf{n}}', \tag{5.4.21}$$

where $\hat{\mathbf{n}}'$ is the outward normal unit vector at the surface point.

For magnetizable systems, Ampère's law, Eq. (5.4.2), can be written

$$\nabla \times \mathbf{B} = \mu_0(\mathbf{J}_f + \mathbf{J}_b) = \mu_0(\mathbf{J}_f + \nabla \times \mathbf{M}). \tag{5.4.22}$$

The two curl terms can be combined, as were the two divergence terms in Eq. (5.2.11), to give

$$\nabla \times (\mathbf{B}/\mu_0 - \mathbf{M}) = \mathbf{J}_f \equiv \nabla \times \mathbf{H}, \tag{5.4.23}$$

where $\mathbf{H}$ apparently is determined by the free currents, just as $\mathbf{D}$ is apparently determined by the free charges. (But only apparently, since there is always an additional equation to be satisfied.) The magnetizable systems chosen for analysis in this chapter will allow $\mathbf{H}$ to be determined by the free currents, just as the chosen dielectric systems allowed $\mathbf{D}$ to be determined by the free charges. Evidently $\mathbf{H}$ is given by

$$\mathbf{H} = \mathbf{B}/\mu_0 - \mathbf{M}. \tag{5.4.24}$$

The divergence of this equation gives $\nabla \cdot \mathbf{M} = -\nabla \cdot \mathbf{H}$, the aforementioned additional equation that shows the second possible source of $\mathbf{H}$. For linear magnetizable materials, the magnetization is found to be proportional to $\mathbf{H}$, written as

$$\mathbf{M} = \chi_m \mathbf{H}, \tag{5.4.25}$$

where χ_m is usually called the *magnetic susceptibility*. This notation is not quite the analog of the dielectric case, but it is rooted in tradition. With this expression for $\mathbf{M}$, Eq. (5.4.24) can be written

$$\mathbf{B} = \mu_0(\mathbf{H} + \chi_m \mathbf{H}) = \mu_0(1 + \chi_m)\mathbf{H} = \mu\mathbf{H}, \tag{5.4.26}$$

where the so-called *permeability* of the material is $\mu = \mu_0(1 + \chi_m)$.

In analogy with Eq. (5.2.13) relating the bound and free charge densities, the bound current density can be written

$$\mathbf{J}_b = \nabla \times \mathbf{M} = \nabla \times (\chi_m \mathbf{H}) = \chi_m \mathbf{J}_f, \tag{5.4.27}$$

showing the absence of bound currents where there are no free currents. This is one of the results that permits the analysis of magnetizable systems to develop in close parallel to the dielectric ones.

The derivation of the appropriate magnetic energy in terms of the macroscopic fields is not as direct or as intuitively satisfying as that for the dielectric energy that culminated in Eq. (5.2.16), because the vector potential is not as easy to relate to energy as is the scalar potential. For this reason, I have chosen to present a Poynting-like derivation of both the electric and the magnetic energy terms. The fact that the thus-derived electric term is the correct one of Eq. (5.2.16), together with the use of the Maxwell equations in a form that includes both free and bound charges, should offer assurance that the magnetic term is also the correct one.

From the Maxwell expression of Faraday's law,

$$\nabla \times \mathbf{E} = -\frac{\partial \mathbf{B}}{\partial t}, \tag{5.4.28}$$

the continuity equation,

$$\nabla \cdot \mathbf{J} + \frac{\partial \rho}{\partial t} = 0, \tag{5.4.29}$$

and the Maxwell expression of the Gauss law, Eq. (5.2.9), which includes all charges, the correct form of the displacement current term in the corrected Maxwell expression of Ampère's law is

$$\nabla \times \mathbf{B} = \mu_0 \mathbf{J} + \mu_0 \epsilon_0 \frac{\partial \mathbf{E}}{\partial t}. \tag{5.4.30}$$

The Lorentz force equation for a particle of charge q and velocity $\mathbf{v}$,

$$\mathbf{F} = q(\mathbf{E} + \mathbf{v} \times \mathbf{B}), \tag{5.4.31}$$

leads to an expression for the rate at which work is done by the fields on the particle:

$$\dot{w} = \mathbf{v} \cdot \mathbf{F} = q\mathbf{v} \cdot \mathbf{E}. \tag{5.4.32}$$

For a collection of particles, the total rate that work is done on them by the fields is

$$\dot{W} = \sum q_i \mathbf{v}_i \cdot \mathbf{E}(\mathbf{r}_i) \longrightarrow \int \mathbf{J} \cdot \mathbf{E} dV, \tag{5.4.33}$$

where the transition from sum to integral has been indicated. The current density is replaced by means of Eq. (5.4.30) to give

$$\dot{W} = \int \left(\frac{\nabla \times \mathbf{B}}{\mu_0} - \epsilon_0 \frac{\partial \mathbf{E}}{\partial t} \right) \cdot \mathbf{E} dV. \tag{5.4.34}$$

By use of a vector identity and Eq. (5.4.28), the first term in the integral can be rewritten:

$$(\nabla \times \mathbf{B}) \cdot \mathbf{E} = -\nabla \cdot (\mathbf{E} \times \mathbf{B}) + \mathbf{B} \cdot (\nabla \times \mathbf{E}) = -\nabla \cdot (\mathbf{E} \times \mathbf{B}) - \mathbf{B} \cdot \frac{\partial \mathbf{B}}{\partial t}. \tag{5.4.35}$$

With this result, the work-rate equation becomes

$$\dot{W} = -\frac{1}{\mu_0} \int \nabla \cdot (\mathbf{E} \times \mathbf{B}) dV - \frac{\partial}{\partial t} \left[\frac{\epsilon_0}{2} \int E^2 dV + \frac{1}{2\mu_0} \int B^2 dV \right]$$

$$= -\frac{1}{\mu_0} \int (\mathbf{E} \times \mathbf{B}) \cdot d\mathbf{a} - \frac{\partial}{\partial t} \left[\frac{\epsilon_0}{2} \int E^2 dV + \frac{1}{2\mu_0} \int B^2 dV \right], \tag{5.4.36}$$

where the Gauss theorem has been used to convert the first term to a surface integral. For present purposes, systems will be chosen such that, at equilibrium, either $\mathbf{E}$ or $\mathbf{B}$ is zero; therefore the Poynting term that represents the rate at which energy is transported across the boundaries of the system by the fields is always zero for these systems. Furthermore, the boundaries of the systems are taken to be fixed, or at worst, slowly moving, so there is not a convective contribution to the total time derivative, $\dot{W}$. Therefore, in this case there is no difference between the total and the partial time derivative. With these considerations in mind, Eq. (5.4.36) can be written

$$\frac{\partial}{\partial t} \left[W + \frac{\epsilon_0}{2} \int E^2 dV + \frac{1}{2\mu_0} \int B^2 dV \right] = 0, \tag{5.4.37}$$

leading to the result that

$$W + \frac{\epsilon_0}{2} \left[\int E^2 dV + \frac{1}{2\mu_0} \int B^2 dV \right] = \text{constant.} \tag{5.4.38}$$

This expression, with the proper choice of constant, can represent the total energy of the system, including electric and magnetic terms. In any case, the term in E^2 agrees with that of Eq. (5.2.16). Therefore, the term in B^2 will be taken as the total magnetic energy, including that of the external fields, with the appropriate system energy to be extracted from the total, as in the treatment of dielectric systems.

The B-field can be written as the sum of the field produced by external currents, $\mathbf{B}_o$, and that produced by magnetization of the system material, $\mathbf{B}_s$. The total magnetic energy density is thus

$$W/V = (1/2\mu_0)(B_o^2 + 2\mathbf{B}_o \cdot \mathbf{B}_s + B_s^2). \tag{5.4.39}$$

Clearly the first term can be assigned to the outside world, and the third term is the self-energy of the magnetized system. The problem, as with dielectric systems, is the assignment of the interaction term.

5.5 MACROSCOPIC MODELS OF MAGNETIZABLE SYSTEMS

The system is chosen to be a section of linear magnetizable material near the center section of a long solenoid that is completely filled with the material, all in equilibrium with a heat bath at temperature T. The system extends radially to the full size of the solenoid, but it is restricted longitudinally to ensure that the field produced by the solenoid is uniform throughout the system. The magnetizable material carries no free currents, and therefore no bound currents; these can exist only as a current sheet on the surface of the medium, almost coincident with the solenoid current. The system can be thus modeled as empty space bounded by a current sheet of density $\mathbf{K}_b$ as given by Eq. (5.4.21), and all electromagnetic interactions of the system with its surroundings are modeled as interactions of this current sheet with the solenoid.

As with dielectrics, two convenient ways exist to establish the outside field: (a) with constant current in the solenoid, which results in constant $\mathbf{H}$, or (b) with constant flux inside the solenoid, which implies constant $\mathbf{B}$. (The latter can be attained by means of a superconducting solenoid winding.) The system with constant $\mathbf{H} = \mathbf{H}_o$ has magnetization $\mathbf{M} = \chi_m \mathbf{H}$, and $\mathbf{B}_o = \mu_0 \mathbf{H}$. The total B-field is thus found to be

$$\mathbf{B} = \mu_0(\mathbf{H}_o + \mathbf{M}) = \mathbf{B}_o + \mu_0 \mathbf{M}, \tag{5.5.1}$$

and the combined energy density is

$$W/V = (1/2\mu_0)(B_o^2 + 2\mu_0 \mathbf{B}_o \cdot \mathbf{M} + \mu_0{}^2 M^2). \tag{5.5.2}$$

The question to be resolved is that of correct placement of the various energy terms.

Fortunately, the magnetizable system chosen for study offers the possibility of clear and unambiguous gedanken experiments to determine where the various energies should be placed and what the proper interaction terms should be. As before, the external field is produced in a long solenoid, at first by current from a current-regulated power supply. The system is now modeled, not as a sheet of surface current, but as another solenoid, closely fitting inside the external one, carrying a current in its coils that is equivalent to that in the former current sheet, and also driven by a current-regulated power supply. Each power supply is conceived to be equipped with an integrating wattmeter, which can provide data on the energy that it has supplied. (The steady joule-heating power can be eliminated either by subtraction or by regarding the coils as having zero resistance, without the other properties of superconductors.) One way to simulate the system in operation is to start with the magnetization current already established in the inner (system-simulating) solenoid, and bring up the external solenoid current to its operating value. The increase in this solenoid current, while the magnetization current is held constant, produces an emf, $\mathcal{E}$, in each coil of the inner solenoid equal to $a\, dB_o/dt$; this emf increases the system wattmeter reading by the amount $I_s a\, dB_o/dt = m_s dB_o/dt$, where m_s is the magnetic moment of that coil. The total increase in the wattmeter reading is $\mathcal{M} dB_o/dt$, where $\mathcal{M}$ is the total magnetic moment of the system. Integration of this power over the time interval necessary to reach the full external field gives the mutual energy $\mathcal{M}B_o$. (Each coil has also developed its own self-energy, but these energies are always attributed to their proper owners.) The sequence of establishing the two fields is unimportant; the state of the system, with its two self-energies and the mutual energy, is to be regarded as independent of history. The critical question is: What is the proper work term for such a configuration, $\mathcal{M}dB_o$ or $B_o d\mathcal{M}$? A related question is: Where should the mutual energy be placed, within the system, or as part of the surroundings? Recall that the two dielectric systems assigned the mutual energy differently. The wattmeters show that the term $B_o d\mathcal{M}$ can result in no energy transfer to the system from the outside, because the constant external field produces no emf in the system coils. (There can be no work done on moving charges by a fixed B-field.) The energy transfer, like that of the constant-E_o dielectric system, is between the mutual energy and the external power supply. Therefore the mutual energy corresponding to ampèrean currents must be placed outside the system. The term $\mathcal{M}dB_o$ represents work done *by* the system *on* the mutual energy; therefore the proper term for the work done *on* the system by the mutual energy must be $-\mathcal{M}dB_o$, and the energy equation should be

$$dU_H = TdS - \mathcal{M}dB_o. \tag{5.5.3}$$

The corresponding equation for the Helmholtz potential is

$$dF_H = -SdT - \mathcal{M}dB_o. \tag{5.5.4}$$

The total magnetic moment is

$$\mathcal{M} = VM = -\left(\frac{\partial F_H}{\partial B_o}\right)_T, \tag{5.5.5}$$

The isothermal susceptibility is given by

$$\chi_{mT} = \left(\frac{\partial M}{\partial H_o}\right)_T = \mu_0 \left(\frac{\partial M}{\partial B_o}\right)_T, \tag{5.5.6}$$

and the thermodynamic stability criterion that F_H be a concave function of its intensive parameters says that

$$\left(\frac{\partial^2 F_H}{\partial B_o^2}\right)_T = -V\left(\frac{\partial M}{\partial B_o}\right)_T = -\frac{V\chi_{mT}}{\mu_0} \leq 0. \tag{5.5.7}$$

So now the analysis, aided by wattmeters that tell unambiguously the form of the energy transfer term, comes up with the result that for stability the susceptibility must be positive! Since diamagnetic materials exist, the result must be incorrect. The subject is complicated and controversial; further discussion is deferred to the Commentary section.

Legendre transformations convert the formulation of Eq. (5.5.3) and Eq. (5.5.4) to one describable as having constant total flux within the system, generated in the gedanken experiment by an external solenoid with a superconducting winding. Since the flux is constant, the field produced by the external solenoid is $B_o = B - \mu_0 M$. The mutual energy is, as before, $\mathcal{M}B_o$, but its location must be determined. The term $\mathcal{M}dB_o = \mathcal{M}dB - \mu_0\mathcal{M}dM = -\mu_0\mathcal{M}dM$ is seen to be nothing more than a shift of system energy, since B is constant. But this behavior shows that the mutual energy is now placed within the system, as might be expected from the analysis of dielectric systems. The term B_odM therefore represents the work term in dU_B and dF_B. The transformed equations are

$$dU_B = d(U_H + \mathcal{M}B_o) = TdS + B_odM, \tag{5.5.8}$$

and

$$dF_B = d(F_H + \mathcal{M}B_o) = -SdT + B_odM. \tag{5.5.9}$$

The second of these equations yields essentially the same stability criterion for the susceptibility as that given in Eq. (5.5.7). Maxwell relations and other response functions can be obtained from these two equations and the corresponding ones with subscript H;

e.g., isentropic susceptibility, and molar heat capacities at constant H and at constant magnetization.

5.6 MICROSCOPIC MODELS OF MAGNETISM

The next task is to compare the microscopic physics derived from quantum mechanics with the results of thermostatics.

5.6.1 The Single-Particle Hamiltonian.
Ampèrian currents are usually thought of as attributable to the motions of single charged particles, rather than as microscopic wires carrying continuous currents. The Hamiltonian for such a particle must reproduce the Lorentz force as a consequence of the canonical equations of motion. First, the Lagrangian for a particle of charge q, in the absence of spin and its related intrinsic magnetic dipole moment, is well known to be

$$\mathcal{L} = (1/2)mv^2 - q\phi + q\mathbf{v} \cdot \mathbf{A}, \tag{5.6.1}$$

which defines the canonical momentum components; for example

$$p_x = \frac{\partial \mathcal{L}}{\partial v_x} = mv_x + qA_x. \tag{5.6.2}$$

The Hamiltonian is

$$\mathcal{H} = \mathbf{p} \cdot \mathbf{v} - \mathcal{L} = \frac{1}{2m}\left(\mathbf{p} - q\mathbf{A}\right)^2 + q\phi. \tag{5.6.3}$$

Note that

$$\dot{p}_x = -\frac{\partial \mathcal{H}}{\partial x} = q\left[-\frac{\partial \phi}{\partial x} + \frac{(\mathbf{p} - q\mathbf{A})}{m} \cdot \left(\frac{\partial \mathbf{A}}{\partial x}\right)\right], \tag{5.6.4}$$

which reduces to the x component of the Lorentz equation,

$$m\dot{v}_x = q[E_x + v_y B_z - v_z B_y] = q[E_x + (\mathbf{v} \times \mathbf{B})_x]. \tag{5.6.5}$$

All of weak classical magnetism, and all of weak quantum magnetism that depends upon the existence of ampèrean currents, can be derived from Eq. (5.6.3). Effects that include the interactions of magnetic dipoles with each other, rather than with average fields, require more discussion. Such effects are examined to some extent in Chapter 7; further details, especially those that are not properly a part of statistical thermophysics, are discussed in texts on condensed matter and also in specialized treatises on particular aspects of magnetic behavior (Callaway [1974]; Kubo and Nagamiya [1969]; Mattis [1965]; van Vleck [1932]). The only topics on magnetism to be analyzed in this chapter are those that are attributable to single-particle behavior, and thus are derivable from Eq. (5.6.3). For much of this work, the potential energy term, $q\phi$, will be set equal to zero, although it is used, in effect, when particles having spin, with their associated magnetic dipole moments, are discussed.

5.6.2 Classical Diamagnetism.
The classical single-particle partition function is

$$Z_1 = \frac{1}{h^3}\int e^{-\beta\mathcal{H}}d^3p\,d^3r = \frac{1}{h^3}\int \exp\left[-\beta\left((\mathbf{p} - q\mathbf{A})^2/2m + q\phi\right)\right]d^3p\,d^3r$$
$$= \frac{m^3}{h^3}\int \exp\left[-\beta(mv^2/2 + q\phi)\right]d^3v\,d^3r. \tag{5.6.6}$$

As Niels Bohr explained in his doctoral dissertation in 1911 (or even earlier, in his M.Sc. examination paper in 1909), and as Johanna H. van Leeuwen independently rediscovered in 1919, there can be no classical magnetism! The Boltzmann factor in Eq. (5.6.6) is clearly independent of the magnetic field in the absence of permanent magnetic dipoles, and the Jacobean of the phase-space volume transformation is equal to unity, when the m^3 factor is taken into account. The B-field therefore cannot influence the equilibrium distribution of the charged particles or the orientation of their ampèrean magnetic moments, on the average (Bohr [1909]; Bohr [1911]; van Leeuwen [1919]; van Leeuwen [1921]; van Vleck [1932]).

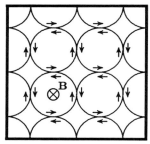

Figure 5.1 Equilibrium cancellation of ampèrean currents in a B-field.

Figure 5.1 is a diagram that indicates schematically the reason that there is no classical magnetism attributable to the ampèrean currents that result from the motions of charged particles in a magnetic field. Motion along the field axis, which is here taken to be the z axis, and into the plane of the paper, produces no net current in an equilibrium assemblage of particles, because of symmetry. Therefore only the motions perpendicular to the field are indicated. For the subset of particles that all have the same speed in the x-y plane, the centers of the gyration orbits are distributed uniformly over the cross section of the

container. The classical magnetic moment attributable to one of the current loops is clearly in the direction opposite to $\mathbf{B}$, and the observed diamagnetism was explained as a consequence of these ampèrean currents. But as the diagram indicates, the current that produces the diamagnetism in any one loop is cancelled at every point (on the average) by opposite currents from orbits with different centers. As van Vleck has remarked, the less said about classical magnetism, the better. My only other comment is that the diamagnetism that seems to be denied by thermodynamic stability in the absence of intrinsic (spin-related) magnetic dipoles cannot have its origin in the classical models; yet there need be no explicit or implicit use of quantum mechanics in the formulation of the laws of thermostatics as presented in Chapter 2. So if the $\geq$ in $\chi_m \geq 0$ had been strictly $>$, the magnetic stability condition might be said to have foreshadowed the need to supersede classical theory. The actual existence of diamagnetic materials, not to mention paramagnetic and ferromagnetic ones, demands that the microscopic theory be other than classical, and that the macroscopic theory be revised accordingly.

The efforts to confine hot, dense plasmas by complicated magnetic fields is not really as hopeless as the Bohr-van Leeuwen theorem might make it seem, because the attempt is to delay the escape of nonequilibrium plasmas, rather than to produce equilibrium confinement. The Bohr-van Leeuwen theorem implies that the plasmas cannot be confined magnetically for the length of time required for them to reach equilibrium, but it has nothing to say about the relative escape rates of plasmas from various magnetic confinement configurations.

5.6.3 Quantum Single-Particle Partition Function.
The quantum partition function of a single point-like charged particle in a uniform magnetic field should illustrate first that diamagnetism is a quantum phenomenon, and second that the density of states for such a system is wonderfully more complicated than that for either the quantum particle in

the absence of the B-field, or the classical particle in its presence. As a consequence, the phenomenology associated with magnetism is much richer and more varied than anything previously treated in this book. For this reason, I have presented the full three-dimensional calculation, rather than the simpler and more skeletal two- dimensional one that may serve to introduce the topic, but which, in my opinion, thereby loses much of its character. A semiclassical derivation that yields the same single-particle partition function is given beginning on p. 269.

The uniform B-field is chosen to be in the $+z$ direction, and its vector potential can be written

$$\mathbf{A} = -(B/2)(\hat{\mathbf{x}}y - \hat{\mathbf{y}}x).\tag{5.6.7}$$

For a free electron, with $q = -e$, Eq. (5.6.3) becomes

$$\begin{aligned}\mathcal{H} &= \frac{p^2}{2m} + \frac{e}{2m}\left(\mathbf{p}\cdot\mathbf{A} + \mathbf{A}\cdot\mathbf{p}\right) + \frac{e^2}{2m}A^2\\ &= -\frac{\hbar^2}{2m}\nabla^2 + \frac{ie\hbar B}{2m}\left(y\frac{\partial}{\partial x} - x\frac{\partial}{\partial y}\right) + \frac{e^2 B^2}{8m}\left(x^2 + y^2\right),\\ &= -\frac{\hbar^2}{2m}\nabla^2 - iB\mu_B\left(x\frac{\partial}{\partial y} - y\frac{\partial}{\partial x}\right) + \frac{e^2 B^2}{8m}(x^2 + y^2).\end{aligned}\tag{5.6.8}$$

Note that in quantum mechanics, where $\mathbf{p} \to -i\hbar\nabla$, there is no problem because of $\mathbf{p}$ and $\mathbf{A}$ possibly not commuting, since the requirement that $\nabla\cdot\mathbf{A} = 0$ ensures that $[\mathbf{p}, \mathbf{A}] = 0$. The middle term is $(e/2m)\mathbf{B}\cdot\mathbf{l}$, where the angular momentum vector is $\mathbf{l} = \mathbf{r}\times\mathbf{p} \to -i\hbar\mathbf{r}\times\nabla$. The Bohr magneton is defined as $\mu_B = e\hbar/2m$.

The single-particle partition function is derived from the Hamiltonian here by means of the Bloch equation, Eq. (3.5.25), here written

$$\frac{\partial R(\mathbf{r}', \mathbf{r}, \beta)}{\partial\beta} = -\mathcal{H}R,\tag{5.6.9}$$

where $\mathcal{H}$ operates only on $\mathbf{r}$. Instead of solving the Bloch equation formally, or exhibiting the path-integral solution technique used in Chapter 3, I shall demonstrate still another procedure: guess the form of the solution and require that the guessed form satisfy the Bloch equation as an identity. The use of Eq. (3.5.29) for the free particle as a starting point suggests the guess

$$R(\mathbf{r}', \mathbf{r}, \beta) = f(\beta)\exp\left[-i\alpha(x'y - y'x) - g(\beta)\left((x - x')^2 + (y - y')^2\right) - \frac{\pi}{\lambda^2}(z - z')^2\right],\tag{5.6.10}$$

where α is a constant.

The equations needed for substitution are:

$$\frac{\partial R}{\partial\beta} = \left[(f'/f) - g'\left((x - x')^2 + (y - y')^2\right) - (\pi/\beta\lambda^2)(z - z')^2\right]R;$$

$$\frac{\partial R}{\partial x} = \left[i\alpha y' - 2g(x - x')\right]; \quad \frac{\partial^2 R}{\partial x^2} = \left[-2g + \left(i\alpha y' - 2g(x - x')\right)^2\right]R;$$

$$\frac{\partial R}{\partial y} = \left[-i\alpha x' - 2g(y - y') \right]; \quad \frac{\partial^2 R}{\partial y^2} = \left[-2g + \left(-i\alpha x' - 2g(y - y') \right)^2 \right] R;$$

$$\text{and} \quad \frac{\partial R}{\partial z} = -\frac{2\pi}{\lambda^2}(z - z')R; \quad \frac{\partial^2 R}{\partial z^2} = -\frac{2\pi}{\lambda^2} \left[1 - \frac{2\pi}{\lambda^2}(z - z')^2 \right] R.$$

Substitution of these results into the Bloch equation, Eq. (5.6.9), with the Hamiltonian of Eq. (5.6.8) yields identities from the coefficients of particular terms. From the coefficients of the xx' terms it follows that

$$g' = -(2\hbar^2/m)g^2 + \alpha B \mu_B/2, \tag{5.6.11}$$

and from the $xy' - x'y$ terms,

$$\alpha = eB/2\hbar. \tag{5.6.12}$$

The solution for g is evidently of the form $g = a \coth \beta b$, where a and b are to be determined. Substitution of this form into Eq. (5.6.11) gives

$$g(\beta) = \frac{m\omega_L}{2\hbar} \coth \beta \hbar \omega_L, \tag{5.6.13}$$

where the Larmor frequency is $\omega_L = eB/2m$ and it is noted that $\hbar\omega_L = B\mu_B$. (Formally, tanh could have been used instead of coth, but the requirement that $\lim_{B \to 0} g(\beta) = -\pi/\lambda^2$ identifies coth as the correct choice.)

The coordinate-independent terms can now be solved for f to give

$$f(\beta) = \frac{\beta \hbar \omega_L}{\lambda^3 \sinh \beta \hbar \omega_L}, \tag{5.6.14}$$

where use has been made of the $B = 0$ limit to evaluate the constant of integration.

The single-particle partition function for a spinless electron in a magnetic field is found by means of Eq. (3.5.27) to be

$$Z_1 = \frac{V}{\lambda^3} \left(\frac{\beta \hbar \omega_l}{\sinh \beta \hbar \omega_l} \right). \tag{5.6.15}$$

If electron spin-dependent dipole magnetic moment is included, the partition function has an additional factor, as given by Eq. (3.5.16). The magnitude of the spin dipole moment is μ_B, the Bohr magneton. The single-particle partition function, including spin, is

$$Z_1 = \frac{V}{\lambda^3} \left(\frac{\beta \hbar \omega_L}{\sinh \beta \hbar \omega_L} \right) (2 \cosh \beta \mu_B B). \tag{5.6.16}$$

A system of N of these independent particles, obeying Boltzmann statistics, has the partition function $Z_N = Z_1^N/N!$, and the resulting Helmholtz potential is

$$F = -kT \ln Z_N$$
$$= -kT[N \ln(V/\lambda_3) - \ln N! + N \ln(\beta \hbar \omega_L / \sinh \beta \hbar \omega_L) +$$
$$+ N \ln(2 \sinh \beta \mu_B B)]. \tag{5.6.17}$$

The first two terms are the usual ones for an ideal Boltzmann gas, and they will be ignored in the subsequent development. The portion of dF with V and T constant is

$$(dF)_{VT} = -NkT \left[-\beta \hbar \omega_L L(\beta \hbar \omega_L) + \beta \mu_B B \tanh \beta \mu_B B \right] (dB/B)$$
$$= N \left[\left(\frac{e\hbar}{2m} \right) L(\beta \hbar \omega_L) - \mu_B \tanh \beta \mu_B B \right] dB. \tag{5.6.18}$$

If this is of the form $-\mathcal{M}dB$, then, with $n = N/V$, the magnetization is given by

$$M = n\left[-\mu_B L(\beta\hbar\omega_L) + \mu_B \tanh\beta\mu_B B\right] \approx n\left[-\frac{\mu_B^2 B}{3kT} + \frac{\mu_B^2 B}{kT}\right], \quad (5.6.19)$$

where the approximations are for magnetic energy terms much smaller than kT, and the relation $\hbar\omega_L = \mu_B B$ has been used. The ampèrean and spin-dipole magnetic moments are seen to be of opposite sign, and both vanish, in accordance with the Bohr-van Leeuwen theorem, if $\hbar = 0$. The isothermal susceptibility attributable to the spin dipoles is seen to be $\chi_{Td} = n\mu_0\mu_B^2/kT$, and that attributable to ampèrean currents is $\chi_{Ta} = -n\mu_0\mu_B^2/3kT$. The ratio is

$$\frac{\chi_{Ta}}{\chi_{Td}} = -\frac{1}{3}, \quad (5.6.20)$$

a ratio that persists in the quantum calculations, although the quantum values are generally much reduced. The quantum-statistical calculations of the susceptibilities of noninteracting electrons are presented in the next several subsections.

5.7 PAULI PARAMAGNETISM

The easiest calculation of quantum susceptibilities for systems of noninteracting electrons is that given originally by Pauli (Pauli [1927]). In the absence of spin-orbit interactions, which are negligible to a good approximation in a free-electron gas, the energy of each electron is simply altered by the dipole energy due to its presence in the magnetic field. There are thus two populations of electrons: those aligned with the B-field, with energies $\epsilon - \mu_B B$, and those aligned opposite to the B-field, with energies $\epsilon + \mu_B B$, where ϵ is the energy the electron would have had in the absence of B, and $\mu_B = e\hbar/2m$ is the Bohr magneton. For each population, the grand partition function can be determined from Eq. (4.4.8) as

$$\ln Z_{G\pm} = \left(\frac{2\pi V(2m)^{3/2}}{h^3}\right)\int_0^\infty d\epsilon\,\epsilon^{1/2}\ln\left(1 + ze^{-\beta(\epsilon\pm B\mu_B)}\right)$$
$$= \left(\frac{2\pi V(2m)^{3/2}}{h^3}\right)\int_0^\infty d\epsilon\,\epsilon^{1/2}\ln(1 + z_\pm e^{-\beta\epsilon}), \quad (5.7.1)$$

where $z_\pm = ze^{\pm\beta B\mu_B}$. (Note that the spin degeneracy factor, g_s, has disappeared because the two spin populations are treated separately.) From Eq. (4.5.1), the populations are found to be, with $\eta_\pm = \epsilon_F \pm B\mu_B$,

$$\langle N\rangle_\pm = \left(\frac{2\pi V(2m)^{3/2}}{h^3}\right)\mathcal{F}(\beta,\eta_\pm)$$
$$= \left(\frac{4\pi V\beta^{3/2}\eta_\pm^{3/2}}{3\sqrt{\pi}\lambda^3}\right)\left[1 + \frac{1}{8}\left(\frac{\pi kT}{\eta_\pm}\right)^2 + \frac{7}{640}\left(\frac{\pi kT}{\eta_\pm}\right)^4 + \cdots\right]$$
$$= \left(\frac{4\pi V(\beta\epsilon_F)^{3/2}}{3\sqrt{\pi}\lambda^3}\right)\left(1 \pm \frac{B\mu_B}{\epsilon_F}\right)^{3/2}\left[1 + \frac{1}{8}\left(\frac{\pi kT}{\epsilon_F}\right)^2\left(1 \pm \frac{B\mu_B}{\epsilon_F}\right)^{-2} + \cdots\right]$$
$$= \left(\frac{4\pi V(\beta\epsilon_F)^{3/2}}{3\sqrt{\pi}\lambda^3}\right)\left[1 + \frac{1}{8}\left(\frac{\pi kT}{\epsilon_F}\right)^2 \pm \frac{3}{2}\left(\frac{B\mu_B}{\epsilon_F}\right) + \right.$$
$$\left. \mp\frac{1}{16}\left(\frac{B\mu_B}{\epsilon_F}\right)\left(\frac{\pi kT}{\epsilon_F}\right)^2 + \frac{3}{8}\left(\frac{B\mu_B}{\epsilon_F}\right)^2 + \cdots\right], \quad (5.7.2)$$

where some of the routine algebraic steps have been omitted. The total population is

$$\langle N \rangle = \langle N \rangle_+ + \langle N \rangle_- = \left(\frac{8\pi V (\beta \epsilon_F)^{3/2}}{3\sqrt{\pi}\lambda^3} \right) \left[1 + \frac{1}{8} \left(\frac{\pi kT}{\epsilon_F} \right)^2 + \frac{3}{8} \left(\frac{B\mu_B}{\epsilon_F} \right)^2 + \cdots \right]$$

$$= \left(\frac{8V}{3\sqrt{\pi}} \right) \left(\frac{2\pi m}{h^2} \right)^{3/2} \epsilon_{F0}^{3/2}, \tag{5.7.3}$$

where ϵ_{F0}, the chemical potential at $T = 0$ and $B = 0$, may be obtained from Eq. (5.7.3) to write

$$\epsilon_F = \epsilon_{F0} \left[1 - \frac{1}{12} \left(\frac{\pi kT}{\epsilon_F} \right)^2 - \frac{1}{4} \left(\frac{B\mu_B}{\epsilon_F} \right)^2 + \cdots \right]. \tag{5.7.4}$$

The total magnetic moment is $\mathcal{M} = \mu_B(\langle N \rangle_+ - \langle N \rangle_-)$, and the magnetization is, to first order in B,

$$M = \frac{\mathcal{M}}{V} = \left(\frac{4\pi \sqrt{\epsilon_F} B\mu_B^2}{\sqrt{\pi}} \right) \left(\frac{2\pi m}{h^2} \right)^{3/2} \left[1 - \frac{1}{24} \left(\frac{\pi kT}{\epsilon_F} \right)^2 + \cdots \right]$$

$$= (4\sqrt{\pi \epsilon_{F0}} B\mu_B^2) \left(\frac{2\pi m}{h^2} \right)^{3/2} \left[1 - \frac{1}{12} \left(\frac{\pi kT}{\epsilon_{F0}} \right)^2 + \cdots \right]$$

$$= \frac{3}{2} \frac{n\pi B\mu_B^2}{\epsilon_{F0}} \left[1 - \frac{1}{12} \left(\frac{\pi kT}{\epsilon_{F0}} \right)^2 + \cdots \right]. \tag{5.7.5}$$

The Pauli susceptibility, which is the isothermal paramagnetic susceptibility obtained from Eq. (5.7.5), is

$$\chi_P = \frac{\mu_0 M}{B} = \frac{3}{2} \left(\frac{n\mu_0 \mu_B^2}{\epsilon_{F0}} \right) \left[1 - \frac{1}{12} \left(\frac{\pi kT}{\epsilon_{F0}} \right)^2 + \cdots \right]. \tag{5.7.6}$$

Except for the 3/2 factor, the essential difference in χ_P and χ_{Td} of the Boltzmann-statistics calculation is that ϵ_{F0} has replaced kT, as might be expected. Therefore the behavior is quite different, since χ_P is virtually independent of T for degenerate fermion gases for which $\epsilon_{F0} \gg kT$. The reason for this temperature independence is that the two spin populations are shifted along the energy axis, each by an amount equal to the energy of the dipole in the magnetic field. But since equality of the chemical potentials is necessary for equilibrium with respect to matter flow, both populations fill up their respective energy distributions to the same Fermi level, with the result that there are more electrons with spins parallel to the field than opposite to it. The populations of the two spin components are only weakly dependent on temperature, so the difference in populations is also nearly temperature independent.

The calculation of Pauli paramagnetism given here has ignored the possible effects of the diamagnetic distortion of the partition function. It turns out that for the degenerate fermion gas, paramagnetic and diamagnetic effects are almost completely independent; confirmation of this statement is the subject of one of the problems at the end of this chapter.

5.8 LANDAU DIAMAGNETISM

The Landau calculation of the diamagnetism of a degenerate system of noninteracting electrons is much more complicated than that for Pauli paramagnetism (Landau [1930]).

There are three major steps, one of which is already completed: (1) derive the single-particle partition function, (2) use the partition function to derive the density of states, $g(\epsilon)$, and (3) use this density of states to obtain the grand partition function, from which the required results follow with little additional effort. The next task, therefore, is the derivation of the density of states from Eq. (5.6.15).

5.8.1 The Density of States. It was noted in Eq. (3.3.37) that the density of states can be derived from the single-particle partition function by the Laplace-transform inversion formula, which in this case becomes

$$g(\epsilon) = \int_{c-i\infty}^{c+i\infty} \left(\frac{V}{\lambda^3}\right) \left(\frac{\beta\hbar\omega_L}{\sinh \beta\hbar\omega_L}\right) e^{\beta\epsilon} d\beta. \tag{5.8.1}$$

The usual procedure for evaluating the inversion formula is by means of a contour integration, where the contour is completed by a circular arc that encloses the origin and the entire imaginary axis, with cuts and other contours chosen to keep the integrand analytic within the contour. A convergence problem arises with the integrand of Eq. (5.8.1), suggesting that a modification could be helpful. Note that two integrations by parts permit Eq. (5.6.15) to be written

$$Z_1(\beta) = \int_0^\infty g(\epsilon)e^{-\beta\epsilon}d\epsilon = \beta^2 \int_0^\infty d\epsilon \, G(\epsilon)e^{-\beta\epsilon}, \tag{5.8.2}$$

where $d^2 G(\epsilon)/d\epsilon^2 = g(\epsilon)$. The inversion formula for $G(\epsilon)$ is

$$\begin{aligned}
G(\epsilon) &= \left(\frac{V}{2\pi i}\right) \int_{c-i\infty}^{c+i\infty} \frac{d\beta e^{\beta\epsilon} \beta\hbar\omega_L}{\lambda^3 \beta^2 \sinh \beta\hbar\omega_L} \\
&= V\hbar\omega_L \left(\frac{2\pi m}{h^2}\right)^{3/2} \left(\frac{1}{2\pi i}\right) \int_{c-i\infty}^{c+i\infty} \frac{d\beta e^{\beta\epsilon}}{\beta^{5/2} \sinh \beta\hbar\omega_L}.
\end{aligned} \tag{5.8.3}$$

Let $x = \beta\hbar\omega_L$, and rewrite Eq. (5.8.3) as

$$G(\epsilon) = \left[V(\hbar\omega_L)^{5/2} \left(\frac{2\pi m}{h^2}\right)^{3/2} \right] \mathcal{I}, \tag{5.8.4}$$

where

$$\mathcal{I} = \frac{1}{2\pi i} \int_{\gamma-i\infty}^{\gamma+i\infty} \frac{dx \, e^{\epsilon x/\hbar\omega_L}}{x^{5/2} \sinh x}. \tag{5.8.5}$$

The integral is to be evaluated by closing the path with a counterclockwise circular arc that encloses the entire imaginary axis. The integral vanishes along this arc of infinite radius, but there are poles on the imaginary axis. Furthermore, a cut is required, to exclude the origin and the negative real axis. These are standard procedures in contour integration, and they need not be expounded here. The required integral is thus converted into the expression

$$\mathcal{I} = \mathcal{I}_{cut} + \sum_{n=\pm 1}^{\pm\infty} \mathcal{I}_n, \tag{5.8.6}$$

where the integral around the cut, $\mathcal{I}_{cut}$, starts on the infinite semicircular arc just below the negative real axis, proceeds along this axis while remaining in the third quadrant, circles the origin counterclockwise, and returns to its starting point along a path in the second quadrant. The $\mathcal{I}_n$ integrals are around the poles on the imaginary axis, which occur at $x = in\pi$, with $n \neq 0$, since the origin has already been removed from the contour. To evaluate $\mathcal{I}_n$, let $x = in\pi + \zeta$, and note that $\sinh x = (-)^n \sinh \zeta \approx (-)^n \zeta$, for small ζ. It follows, for $n > 0$, that

$$
\mathcal{I}_n = \frac{1}{2\pi i} \oint_n \frac{dx\, e^{\epsilon x/\hbar\omega_L}}{x^{5/2} \sinh x} = -\frac{(-)^n}{2\pi i} \oint \frac{d\zeta \exp i\,(n\pi\epsilon/\hbar\omega_L - \pi/4)}{(n\pi)^{5/2}\zeta}
$$
$$
= -\frac{(-)^n}{(n\pi)^{5/2}} \exp i\left(\frac{n\pi\epsilon}{\hbar\omega_L} - \frac{\pi}{4}\right), \quad \text{for } n > 0. \tag{5.8.7}
$$

The terms for $n < 0$ can be treated similarly, and then summed to give

$$
\sum_{i=1}^{\infty} (\mathcal{I}_n + \mathcal{I}_{-n}) = -\frac{2}{\pi^{5/2}} \sum_{n=1}^{\infty} \frac{(-)^n}{n^{5/2}} \cos\left(\frac{n\pi\epsilon}{\hbar\omega_L} - \frac{\pi}{4}\right). \tag{5.8.8}
$$

Each cosine term is evidently an oscillatory function, periodic in $1/B$. The significance of these terms will be discussed subsequently, but their contribution to $G(\epsilon)$ is labeled $G_{osc}(\epsilon)$, and given as

$$
G_{osc}(\epsilon) = -2V \left(\frac{\hbar\omega_L}{\pi}\right)^{5/2} \left(\frac{2\pi m}{h^2}\right)^{3/2} \sum_{n=1}^{\infty} \frac{(-)^n}{n^{5/2}} \cos\left(\frac{n\pi\epsilon}{\hbar\omega_L} - \frac{\pi}{4}\right). \tag{5.8.9}
$$

The oscillatory terms vanish for $B = 0$, as expected. Also to be noted: this result is for each spin state, since it was derived from the single-particle partition function of a spinless particle.

The integral along the cut requires more care. With $\alpha = \epsilon/\hbar\omega_L$, the integral to be evaluated is

$$
\mathcal{I}_{cut} = \frac{1}{2\pi i} \int_{cut} \frac{dx\, e^{\alpha x}}{x^{5/2} \sinh x}, \quad \alpha = \epsilon/\hbar\omega_L. \tag{5.8.10}
$$

For convergence purposes, it is helpful to write the denominator of the integral as

$$
\frac{1}{x^{5/2} \sinh x} \equiv \frac{1}{x^{7/2}} - \frac{1}{3!x^{3/2}} - \left(\frac{1}{x^{7/2}} - \frac{1}{3!\,x^{3/2}} - \frac{1}{x^{5/2} \sinh x}\right)
$$
$$
= \frac{1}{x^{7/2}} - \frac{1}{3!\,x^{3/2}} - f(x), \tag{5.8.11}
$$

where $f(x)$ is defined as the expression in parentheses. For small x, the function $f(x)$ becomes

$$
f(x) = \frac{6\sinh x - x^2 \sinh x - 6x}{3!\,x^{7/2} \sinh x}
$$
$$
\approx \frac{(6 - x^2)(x + x^3/3! + x^5/5!) - 6x}{3!\,x^{9/2}} = \frac{-7x^{1/2}}{360} + \mathcal{O}(x^{5/2}), \tag{5.8.12}
$$

so $f(0) = 0$. The integral of the $f(x)$ term around the cut is

$$\frac{1}{2\pi i} \int_{cut} f(x) e^{\alpha x} dx = \frac{1}{2\pi i} \int_{\infty}^{\eta} dr e^{-i\pi} e^{-\alpha r} f\left(re^{-i\pi}\right) + \frac{\eta}{2\pi i} \int_{-\pi}^{\pi} id\theta e^{i\theta} e^{-\alpha\eta e^{i\theta}} f\left(\eta e^{i\theta}\right)$$
$$+ \frac{1}{2\pi i} \int_{\eta}^{\infty} dr e^{i\pi} e^{-\alpha r} f\left(re^{i\pi}\right)$$
$$= \frac{i}{\pi} \int_0^{\infty} dr e^{-\alpha r} f(-r), \tag{5.8.13}$$

since the middle term vanishes as $\eta \to 0$, and the other two terms add. Numerical analysis shows that, even with $\alpha = 0$, the value of the integral is less than 0.06, which turns out to be of negligible consequence, and it will henceforth be ignored. (It should be noted that this approximate treatment violates one of the assumptions that led to the use of $G(\epsilon)$, the second integral of the density of states—that $dG/d\epsilon = 0$ for $\epsilon = 0$, as required by the integrations by parts in Eq. (5.8.2). I have not been able to provide a treatment that resolves this difficulty. I mention it as a challenging problem in contour integration.)

The other two integrals are of the form

$$\frac{1}{\Gamma(s)} = \frac{1}{2\pi i} \int_{cut} dx \; x^{-s} e^x, \tag{5.8.14}$$

although tables usually write this result in a more complicated form with the cut along the positive real axis. The integrals needed are thus

$$\frac{1}{2\pi i} \int_{cut} dx \; x^{-s} e^{\epsilon x/\hbar\omega_L} = \frac{(\epsilon/\hbar\omega_L)^{s-1}}{\Gamma(s)}. \tag{5.8.15}$$

The required function $G(\epsilon)$ can now be written

$$G(\epsilon) = \left(\frac{2\pi m}{h^2}\right)^{3/2} V(\hbar\omega_L)^{5/2} \left[\frac{(\epsilon/\hbar\omega_L)^{5/2}}{\Gamma(7/2)} - \frac{(\epsilon/\hbar\omega_L)^{1/2}}{6\Gamma(3/2)} - \frac{2}{\pi^{5/2}} \sum_{r=1}^{\infty} \frac{(-)^r}{r^{5/2}} \cos\left(\frac{r\pi\epsilon}{\hbar\omega_L} - \frac{\pi}{4}\right)\right]$$
$$= \left(\frac{2\pi m}{h^2}\right)^{3/2} V\left[\left(\frac{8}{15\sqrt{\pi}}\right)\epsilon^{5/2} - \left(\frac{(\hbar\omega_L)^2}{3\sqrt{\pi}}\right)\epsilon^{1/2} + \right.$$
$$\left. - 2\left(\frac{\hbar\omega_L}{\pi}\right)^{5/2} \sum_{r=1}^{\infty} \frac{(-)^r}{r^{5/2}} \cos\left(\frac{r\pi\epsilon}{\hbar\omega_L} - \frac{\pi}{4}\right)\right]. \tag{5.8.16}$$

5.8.2 The Grand Partition Function and the Grand Potential.

The grand partition function for any gas of noninteracting fermions is given by

$$\ln Z_G = \int_0^{\infty} d\epsilon \, g(\epsilon) \ln(1 + ze^{-\beta\epsilon}) = \beta \int_0^{\infty} \frac{d\epsilon \, G(\epsilon)}{z^{-1}e^{\beta\epsilon} + 1}$$
$$= \beta^2 \int_0^{\infty} \frac{d\epsilon \, G(\epsilon)}{(z^{-1}e^{\beta\epsilon} + 1)(ze^{\beta\epsilon} + 1)}, \tag{5.8.17}$$

where there have been two integrations by parts, the integrated terms have vanished at the limits in both cases, $\mathcal{G}' = g$, and $G'' = g$, as in the previous subsection. Use of Eq. (5.8.16) and the results of Appendix B allow the grand partition function to be evaluated easily. Note that $\phi(\epsilon)$ in Appendix B corresponds to $\beta\mathcal{G}(\epsilon)$ here, and $\Phi(\epsilon)$ similarly corresponds to $\beta^2 G(\epsilon)$. Also note that the absence of Pauli energies in this calculation requires that the spin degeneracy factor, 2, be redisplayed. In order to avoid the confusion of too many μ's, the chemical potential is written as ϵ_F; its value at $T = 0$ and $B = 0$ is written as ϵ_{F0}; the quantity s is used as $s = \beta\epsilon_F$; and $s_0 = \beta\epsilon_{F0}$. The non-oscillatory terms in Eq. (5.8.17) can be written out immediately by use of Eq. (B.1.8) as

$$
\begin{aligned}
\ln Z_G &= 2V\beta^2 \left(\frac{2\pi m}{h^2}\right)^{3/2} \int_0^\infty \frac{d\epsilon}{\left(e^{\beta(\epsilon-\epsilon_F)}+1\right)\left(e^{\beta(\epsilon_F-\epsilon)}+1\right)} \left[\frac{8\epsilon^{5/2}}{15\sqrt{\pi}} - \frac{(\hbar\omega_L)^2\epsilon^{1/2}}{3\sqrt{\pi}}\right] \\
&= 2\left(\frac{2\pi m}{\hbar^2}\right)^{3/2} V\beta \left[\left(\frac{8\epsilon_F^{5/2}}{15\sqrt{\pi}}\right)\left[1 + \frac{5}{8}\left(\frac{\pi kT}{\epsilon_F}\right)^2 + \cdots\right] + \right.\\
&\qquad \left. - \left(\frac{(\hbar\omega_L)^2\epsilon_F^{1/2}}{3\sqrt{\pi}}\right)\left[1 - \frac{1}{24}\left(\frac{\pi kT}{\epsilon_F}\right)^2 + \cdots\right]\right] \\
&= \frac{16Vs^{5/2}}{15\lambda^3\sqrt{\pi}}\left[1 + \frac{5}{8}\left(\frac{\pi}{s}\right)^2 + \cdots\right] - \frac{2(\beta\hbar\omega_L)^2 s^{1/2} V}{3\sqrt{\pi}\lambda^3}\left[1 - \frac{1}{24}\left(\frac{\pi}{s}\right)^2 + \cdots\right], \quad (5.8.18)
\end{aligned}
$$

where the first group of terms is the nonmagnetic one that agrees with the results of Chapter 4, and the second group, which vanishes when $B = 0$, gives Landau diamagnetism.

The oscillatory terms, which have been omitted from Eq. (5.8.18), are usually negligible. If a magnetic temperature is defined by $k\Theta_m = B\mu_B = \hbar\omega_L$, a useful referrent is that at $B = 1$ Tesla (10,000 G), the magnetic temperature is $\Theta_m = 0.67$ K. Therefore, except at appropriate combinations of low temperature and high B-field, the effects attributable to the oscillatory terms are not usually observable. The consequences of these terms will be explored subsequently.

The total population can be obtained from Eq. (5.8.18) as

$$
\begin{aligned}
\langle N \rangle &= z\left(\frac{\partial \ln Z_G}{\partial z}\right)_{V,\beta,B} = \left(\frac{\partial \ln Z_G}{\partial s}\right)_{V,T,B} \\
&= \frac{8Vs^{3/2}}{3\lambda^3\sqrt{\pi}}\left[1 + \frac{1}{8}\left(\frac{\pi}{s}\right)^2 - \frac{1}{8}\left(\frac{\beta\hbar\omega_L}{s}\right)^2 + \cdots\right] + \text{osc. terms} \\
&= \frac{8V\epsilon_{F0}^{3/2}}{3\sqrt{\pi}}\left(\frac{2\pi m}{h^2}\right)^{3/2}. \quad (5.8.19)
\end{aligned}
$$

This result gives

$$
\epsilon_{F0} = \left(\frac{3\langle N\rangle}{\pi V}\right)^{2/3}\left(\frac{h^2}{8m}\right) = \left(\frac{3n}{\pi}\right)^{2/3}\left(\frac{h^2}{8m}\right), \quad (5.8.20)
$$

and, after the usual algebraic steps,

$$
\epsilon_F = \epsilon_{F0}\left[1 - \frac{1}{12}\left(\frac{\pi}{s}\right)^2 + \frac{1}{12}\left(\frac{\beta\hbar\omega_L}{s}\right)^2 + \cdots\right], \quad (5.8.21)
$$

in agreement with Eq. (4.5.4) and Eq. (4.5.5).

The Landau diamagnetic susceptibility will be obtained in the next subsection, by way of the grand potential. Recall that $\Phi = -kT \ln Z_G = F - \mu \langle N \rangle$, leading to

$$d\Phi = dF - \mu d \langle N \rangle - \langle N \rangle d\mu. \tag{5.8.22}$$

From Eq. (3.4.8) and Eq. (2.5.16), together with the identification that $F = F_H$, the explicit differential of Φ is

$$d\Phi_H = -SdT - PdV - \mathcal{M}dB_o - \langle N \rangle d\epsilon_F, \tag{5.8.23}$$

so the correct expression for the magnetization is

$$M = -\frac{1}{V}\left(\frac{\partial \Phi}{\partial B_o}\right)_{TV\epsilon_F} = \frac{kT}{V}\left(\frac{\partial \ln Z_G}{\partial B_o}\right)_{TV\epsilon_F}. \tag{5.8.24}$$

The oscillatory terms in the grand partition function can be treated in a similar fashion, but a better way to do it comes from the use of Eq. (B.1.11). The terms needed for $\ln Z_G$ are of the form

$$A_r = \int_0^\infty \frac{dx \cos(\alpha_r x - \pi/4)}{(e^{x-s}+1)(e^{s-x}+1)}, \tag{5.8.25}$$

where $x = \beta\epsilon$, $s = \beta\epsilon_F$, and $\alpha_r = r\pi/\beta\hbar\omega_L$. By the procedure used in Appendix B for the evaluation of Fermi integrals, $x - s$ is replaced by y, the lower limit of integration, which has become $-s$, is extended to $-\infty$, and the cos term becomes $\cos(\alpha_n x - \pi/4) = \cos(\alpha_r y + \alpha_r s - \pi/4) = \cos\alpha_r y \cos(\alpha_r s - \pi/4) - \sin\alpha_r y \sin(\alpha_r s - \pi/4)$. The integral has become, except for constant factors that can be moved outside the integral sign, two integrals, one of which is zero because its integrand is an odd function. Only the cosine integral remains, and it can be evaluated by means of Eq. (B.1.11) and Eq. (B.1.13):

$$A_r = \cos(\alpha_r - \pi/4)\int_{-\infty}^\infty \frac{\cos\alpha_r y}{(e^y+1)(e^{-y}+1)} = \cos(\alpha_r s - \pi/4)\frac{\alpha_r \pi}{\sinh \alpha_r \pi}. \tag{5.8.26}$$

From this result, the oscillatory terms can be written

$$(\ln Z_G)_{osc} = -\frac{4V(\beta\hbar\omega_L)^{3/2}}{\sqrt{\pi}\lambda^3}\sum_{r=1}^\infty \frac{(-)^r}{r^{3/2}}\frac{\cos\left(\frac{r\pi\epsilon_F}{\hbar\omega_L} - \frac{\pi}{4}\right)}{\sinh\left(\frac{r\pi^2 kT}{\hbar\omega_L}\right)}. \tag{5.8.27}$$

The complete expression for $\ln Z_G$ is the sum of Eq. (5.8.18) and Eq. (5.8.27).

5.8.3 Landau Diamagnetism. The diamagnetism of a degenerate gas of noninteracting electrons is obtained from the nonoscillatory terms of the grand potential, Φ_H, which are given by

$$(\Phi_H)_{nonosc} = -kT \ln (Z_G)_{nonosc}. \tag{5.8.28}$$

The Landau magnetization is found as

$$M_L = -\frac{1}{V}\left(\frac{\partial \Phi_H}{\partial B_o}\right)_{TVs} = \frac{kT}{V}\left(\frac{\partial \ln Z_G}{\partial B_o}\right)_{TVs}, \tag{5.8.29}$$

which gives from Eq. (5.8.18)

$$M_L = -\left(\frac{16s^{5/2}}{15\lambda^3\sqrt{\pi}}\right)\left(\frac{5}{4}\right)\left(\frac{\beta B_o\mu_B^2}{s^2}\right) = -\frac{4s_0^{3/2}B_o\mu_B^2}{3\lambda^3\sqrt{\pi}\epsilon_{F0}} = -\frac{n\mu_B^2 B_o}{2\epsilon_{F0}}. \tag{5.8.30}$$

The Landau isothermal diamagnetic susceptibility is seen to be

$$\chi_L = \frac{M_L}{H_o} = \frac{\mu_0 M_L}{B_o} = -\frac{n\mu_0\mu_B^2}{2\epsilon_{F0}}, \tag{5.8.31}$$

with the recognition that higher order terms could be included, but that a better approximation is usually not warranted.

The ratio $\chi_L/\chi_P = -1/3$ is the same as that for the nondegenerate fermion gas of the preceding section, where χ_{Ta}/χ_{Td} was given in Eq. (5.6.20). The degenerate fermion results are not dependent on temperature to first order, whereas the nondegenerate fermion gas shows classical $1/T$ dependence of its susceptibilities.

5.8.4 The Oscillatory Terms. A similar calculation gives the oscillatory terms, this time obtained from Eq. (5.8.27), with the substitutions $V = 3\sqrt{\pi}\lambda_F^3\langle N\rangle/8$ and $\gamma_r = (r\pi s/\beta\hbar\omega_L) - \pi/4$:

$$M_{osc} = -\frac{kT}{V}\left(\frac{\partial\ln(Z_G)_{osc}}{\partial B_o}\right)_{TVs}$$
$$= -\left(\frac{3nkT\mu_B^{3/2}B_o^{1/2}}{2\epsilon_F^{3/2}}\right)\sum_{r=0}^{\infty}\frac{(-)^r}{r^{3/2}}\left[\frac{3\cos\gamma_r}{2\sinh(r\pi^2/\beta\hbar\omega_L)}+\right.$$
$$\left.\frac{r\pi\epsilon_{F0}\sin\gamma_r}{\hbar\omega_L\sinh(r\pi^2/\beta\hbar\omega_L)} + \frac{r\pi^2\cos\gamma_r\cosh(r\pi^2/\beta\hbar\omega_L)}{\beta\hbar\omega_L\sinh^2(r\pi^2/\beta\hbar\omega_L)}\right] \tag{5.8.32}$$

The oscillatory terms in the susceptibility, here defined as $\chi_{osc} = M_{osc}/H_o = M_{osc}\mu_0/B_o$, rather than by means of the derivative, are

$$\chi_{osc} = -\left(\frac{3nkT\mu_B^{3/2}}{2\epsilon_F^{3/2}B_o^{1/2}}\right)\sum_{r=0}^{\infty}\frac{(-)^r}{r^{3/2}}\left[\frac{3\cos\gamma_r}{2\sinh(r\pi^2/\beta\hbar\omega_L)}+\right.$$
$$\left.\frac{r\pi\epsilon_{F0}\sin\gamma_r}{\hbar\omega_L\sinh(r\pi^2/\beta\hbar\omega_L)} + \frac{r\pi^2\cos\gamma_r\cosh(r\pi^2/\beta\hbar\omega_L)}{\beta\hbar\omega_L\sinh^2(r\pi^2/\beta\hbar\omega_L)}\right] \tag{5.8.33}$$

These terms are of little importance unless $\beta\hbar\omega_L \geq 1$. In this case, the magnetization and the susceptibility derived from it exhibit oscillations, from the trigonometric terms, that are periodic in $1/B_o$. In this regime, the amplitude of the oscillations in the susceptibility is roughly proportional to $\sqrt{B_o}$.

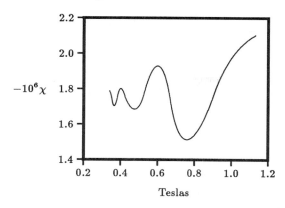

Figure 5.2 de Haas-van Alphen plot of χ for bismuth at 4 K.

Although this derivation applies to a degenerate ideal Fermi gas, it provides the start of an explanation for an observation of behavior of this general description, first reported by de Haas and van Alphen in 1930 (de Haas and van Alphen [1930a]; de Haas and van Alphen [1930b]). They described measurements of the susceptibility of bismuth at 4K as a function of B_o. Their results were nearly those shown in Fig. 5.2.

An easier and somewhat more physical derivation of the single-particle partition function, Eq. (5.6.15), makes use of its property that independent components of the energy contribute independent factors to the partition function. An electron is considered to be in a volume $V = L_x L_y L_z$, moving under the influence of a magnetic field $\mathbf{B} = \hat{z}B$. The motion along the z direction is unaffected by B and is also independent of motion in the x-y plane. The classical motion of the electron in the x-y plane is uniform circular motion at the cyclotron frequency $\omega_c = eB/m$. In quantum mechanics, this perpendicular energy is quantized as

$$\epsilon_\perp = \hbar\omega_c(j + 1/2), \tag{5.8.34}$$

where j is an integer that takes on values from 0 to infinity. The total energy can be written

$$\epsilon = \epsilon_\perp + \epsilon_z = \hbar\omega_c(j + 1/2) + p_z^2/2m. \tag{5.8.35}$$

The z-factor of the partition function is

$$Z_z = \frac{L_z}{\lambda}, \tag{5.8.36}$$

and the perpendicular factor is

$$Z_\perp = \sum_{j=0}^{\infty} g_j \exp[-\beta\hbar(j + 1/2)], \tag{5.8.37}$$

where g_j is the number of states with quantum number j. Note that a continuum (or nearly so in a large volume) of states has coalesced into a countable set. Every state with energy in the interval $\hbar\omega_c j \leq \epsilon_\perp \leq \hbar\omega_c(j + 1)$ condenses to the jth quantum level. The number of these states is

$$g_j = \frac{L_x L_y}{h^2} \int dp_x dp_y = \frac{L_x L_y}{h^2} \int 2\pi p_\perp dp_\perp$$

$$= \frac{L_x L_y}{h^2} \int_{\hbar\omega_c j}^{\hbar\omega_c(j+1)} 2\pi m \, d\epsilon_\perp = \frac{L_x L_y}{h^2}(2\pi m\hbar\omega_c) = \frac{L_x L_y eB}{h}, \tag{5.8.38}$$

where the integration limits were finally specified in the last integral, and the result turns out to be independent of j. The sum in Eq. (5.8.37) can now be completed to give

$$Z_\perp = \frac{L_x L_y eB}{h} \sum_{j=0}^{\infty} \exp[-\beta\hbar\omega_c(j+1/2)] = \frac{L_x L_y}{2h \sinh \beta\hbar\omega_c/2}. \qquad (5.8.39)$$

The combined single-particle partition function is, with the Larmor-cyclotron frequency relationship, $\omega_L = \omega_c/2$, used to improve notation,

$$Z_1 = Z_z Z_\perp = \frac{VeB}{2\lambda h \sinh \beta\hbar\omega_L} = \frac{V}{\lambda^3} \frac{\beta\hbar\omega_L}{\sinh \beta\hbar\omega_L}, \qquad (5.8.40)$$

in agreement with Eq. (5.6.15). The $\epsilon_\perp$ energies are known as Landau levels. It is evident from Eq. (5.8.38) that the degeneracy of the Landau levels increases with B, as does the level spacing. The degeneracy of a Landau level represents the fact that the centers of the cyclotron orbits, projected on the x-y plane can be located at many independent sites without violation of the Pauli principle.

A degenerate fermion gas occupies the lowest set of energy levels available to it, subject to the Pauli principle. As B increases, the allowed occupation number of each Landau level increases, and the levels move further apart. It is clear that an increase in B must result in transitions from higher to lower Landau levels in order that the newly created vacancies at the lower levels can be kept filled. At $T = 0$, the Fermi level is the highest occupied energy level. Suppose that at some particular value of B the energy of the highest occupied level is the same as that for $B = 0$, but only very few electrons occupy the highest Landau level. If B is increased slightly, the Landau energies change only slightly, but the allowed occupation numbers of the lower levels can change enough to permit all electrons to vacate the highest level that was occupied at the field value B. Thus there can be a sudden and dramatic change in the Fermi level at some critical value of the field, say B_t, for *transition*. There are many such transition field values; at each one, as B is increased, the Fermi level drops by the amount $\beta\hbar\omega_{Lt}$. The system energy and the other thermodynamic potentials also change discontinuously at the transition fields, at $T = 0$. At higher temperatures, thermal smoothing of the upper energy boundary softens the transitions, with the result that Eq. (5.8.33) yields a continuous function. As the field is increased above any particular value of B_t, the Fermi level rises, and the population of the highest occupied state decreases, until the next value of B_t is reached, at which value the Fermi level again decreases abruptly.

The total magnetic moment is given by $\mathcal{M} = (\partial F_H/\partial B_o)_T$; it is expected, therefore, to show variation in response to the undulations in Fermi level caused by shifts in population of the Landau levels. Other low-temperature properties of metals, such as heat capacity, electrical and thermal conductivity, and susceptibility all have an oscillatory component for the same reason. It is therefore useful to mention two more aspects of this basic phenomenon.

The flux through a cyclotron orbit is quantized. The radius of a cyclotron orbit is $r = mv_\perp/eB$, and the flux through this orbit is $\Phi = (\pi r^2)B = (\pi/B)(mv_\perp/e)^2$. But the kinetic energy is known, from Eq. (5.8.34) to be quantized: $mv_\perp^2 = 2\hbar\omega_c(j+1/2)$. This substitution into the flux equation gives

$$\Phi_j = (h/e)(j+1/2), \qquad (5.8.41)$$

as the only permitted values for the flux enclosed by one of the cyclotron orbits that are the ampèrean currents. It is noteworthy that these flux quanta are independent of B and of the mass of the gyrating particle. (Flux quanta appear throughout quantum physics. For example, the flux enclosed by a superconducting ring is an integral multiple of the basic flux quantum $\Phi_0 = h/q$, where q is the charge of the carriers of current in the superconductor. For currents carried by Cooper pairs, the value of the charge is $q = -2e$, and the basic unit of flux is then $\Phi_0 = h/2e$.)

Note that the phase of the de Haas-van Alphen oscillations depends on the Fermi energy ϵ_F, as well as on known quantities or measured parameters (such as B_o). In the study of metals, the experimental data on the low temperature periodicity of susceptibility, or other response function, at different orientations of the magnetic field with respect to the crystal axes is used as a means of mapping the structure of the Fermi surface (Blakemore [1988]; Callaway [1974]; Kittel [1986]; Kubo and Nagamiya [1969]).

5.9 ATOMIC PARAMAGNETISM

As is well known from quantum mechanics, the magnetic moment of a free atom is given by

$$\mathbf{m} = -g_J\mu_B\mathbf{J}, \tag{5.9.1}$$

where the total angular momentum is $\hbar\mathbf{J} = \hbar(\mathbf{L}+\mathbf{S})$, the sum of the orbital and spin angular momenta. The factor g_J, the Landé g-factor, is given by

$$g_J = 1 + \frac{J(J+1) + S(S+1) - L(L+1)}{2J(J+1)}. \tag{5.9.2}$$

(There is a quantum-electrodynamic correction to this expression for g_J, but for present purposes it is negligible. It is most important for free electrons, where the proper g-value should be 2.0023, instead of the value 2 given by Eq. (5.9.2).) The energy is given by $\epsilon = -\mathbf{m}\cdot\mathbf{B}$. The possible values for ϵ are

$$\epsilon_j = -m_jB = g_J\mu_BBj; \quad -J \leq j \leq J. \tag{5.9.3}$$

The magnetic part of the single-atom partition function is

$$Z_{1m} = \sum_{j=-J}^{J} \exp[-\beta\epsilon_j] = \frac{\sinh(2J+1)x}{\sinh x}, \tag{5.9.4}$$

where $x = \beta g_J\mu_BB/2$. The expectation value for ϵ is

$$\langle\epsilon\rangle = -\left(\frac{\partial\ln Z_{1m}}{\partial\beta}\right)_B = -\langle m\rangle B, \tag{5.9.5}$$

where the expectation value of m along B is

$$\langle m\rangle = g_J\mu_BJB_J(x). \tag{5.9.6}$$

The Brillouin function of order J is defined as

$$B_J(x) = \frac{2J+1}{2J}\coth(2J+1)x - \frac{1}{2J}\coth x. \qquad (5.9.7)$$

As $x \to \infty$, the value of B_J approaches 1, and for $x \ll 1$, $B_J(x) \approx 2(J+1)x/3$. For a single electron, $J = S = 1/2$, and the result agrees with that of Eq. (5.6.19).

A basic assumption in this calculation is that the atom is in its $(2J+1)$-fold degenerate ground state, and that there is no other nearby state that is likely to be occupied. When this assumption is valid, as in salts of most of the rare-earth and iron-group elements at low temperatures, there is good experimental agreement with the simple theory. But if excited levels with energies of the order of kT above the ground state are present, the theory must be modified to include these states in the partition function.

Similar calculations can be made for nuclear magnetic moments. In this case, it is usual to replace μ_B by the nuclear magneton, defined similarly, but with the electron mass replaced by that of the proton.

5.10 COMMENTARY

The thermodynamic behavior of polarizable or magnetizable matter in the presence of externally imposed fields is a confusing and relatively difficult subject, with a remarkable divergence of opinions among those who have written about it (Brown [1953]; Guggenheim [1936]; Guggenheim [1945]; Guggenheim [1967]; Heine [1956]; Landau and Lifshitz [1960]; Leupold [1969]; Livens [1945]; Livens [1947]; Livens [1948]; Privorotskii [1976]). The student should be warned that not everything written in the various textbooks can possibly be correct, since they often disagree. Every assertion should be viewed with skepticism, and each student of the subject should accept the challenge to straighten up this mess, once and for all. Good luck!

The statistical thermodynamics of magnetism is a rich and complex subject, involving techniques and details from condensed matter physics that are not necessarily a part of the background of every student who will eventually know about these matters. Furthermore, many facets of magnetism do not require thermodynamics or statistical physics as a primary concern. It is therefore appropriate to apprise the student that this is, and will probably continue to be, an interesting field of study, but also to warn that there is much more to be learned than can be covered in any one text book, especially one that does not have magnetism as its central concern. Beyond that, I comment that aspects of magnetism are treated in the next three chapters, which deal with the thermodynamics of phase transitions, the statistical mechanics of interacting systems, and the renormalization-group approach to critical phenomena.

The principal topic of this commentary is the magnetic stability puzzle. The proponents of macroscopic thermodynamics assert, and I agree, that their discipline should be independent of the validity of microscopic models of the structure of matter. Furthermore, there is no separation of quantum and classical thermodynamics. The stability criteria, properly derived from macroscopic principles, should be correct, and if the consequences of a microscopic model are nonsense, then the model must be wrong. A careful and correct examination of the macroscopic and microscopic details of the systems chosen for the study of weak magnetism should be able to resolve all discrepancies. But if a small cloud persists on the horizon, perhaps it, too, should be followed, as were those that Rayleigh made famous.

If the fundamental assumption is adopted that the thermodynamic behavior of a body in an electromagnetic field depends on only the field that is present within that body, then the microscopic physics should agree with thermodynamics. A further criterion, and one that I have used in this chapter, is that the system energy should not include field energy that would be present even if the medium were not there. Thus, in the dielectric system, the external field energy density, $\epsilon_0 E_o^2/2$, is not regarded as part of the system energy. It is not ordinarily included in the Hamiltonian that describes the microphysics of the system, and its presence with or without the system seems to me sufficient reason to exclude it. But Landau and Lifshitz disagree.

They argue that the free energy density must exist outside the body, as well as inside, and that the energy density at any point within the body can depend only on the field actually present in the body, and not on the field that would be there if the body were not present. Although $dF_D = -SdT - \mathcal{P}dE_o$, which they imply is derivable from the Hamiltonian, they deny that the function F_D is a proper Helmholtz potential for the system. Instead, they offer a function $\mathcal{F}$, that satisfies

$$\delta\mathcal{F} = -SdT + \int \mathbf{E} \cdot \delta\mathbf{D}dV, \qquad (5.10.1)$$

which agrees with Eq. (5.3.16) when E_o is fixed and $\delta\mathbf{D} = \delta\mathbf{P}$. Thus their Helmholtz potential seems to be the equivalent of F_E. In this case, with fixed potentials on the capacitor plates, the self energy density of the external field cannot change in response to changes in the dielectric. But they conclude from the stability requirements that $(\partial E/\partial D)_T \geq 0$, which is not a strong result, since apparently $\epsilon \geq \epsilon_0$ from the intrinsic properties of dielectrics. Their Legendre transformation is with ED rather than with PE_o, and their dielectric stability criteria are all in terms of ϵ.

For magnetic systems, similarly, they write the differential of the Helmholtz potential as

$$\delta\mathcal{F} = -SdT + \int \mathbf{H} \cdot \delta\mathbf{B}dV, \qquad (5.10.2)$$

with the work term equivalent to $B_o d\mathcal{M}$ at constant H_o. They obtain the stability criterion

$$\left(\frac{\partial^2 \mathcal{F}}{\partial B^2}\right)_T = \left(\frac{\partial H}{\partial B}\right)_T = \frac{1}{\mu} \geq 0.$$

There is no puzzle, and no conflict with experiment, if the strongest thermodynamic stability requirement is that $\mu \geq 0$. They point out that this is the only theoretical limitation on μ; its translation in terms of the susceptibility merely says that $\chi_{mT} \geq -1$. They may ultimately be right. Their criterion certainly does not contradict the facts of magnetic susceptibility. But their formulation includes fields that are present even when the material of the system is missing, and their energy term is not derivable, except in a contrived way, from a Hamiltonian that describes the microphysics. So either we are forced to include extraneous fields in the system, and concede defeat in the goal of deriving the thermodynamics from microphysics, or we must choose to accept the derivations based on microphysics but renounce the stability criterion because the partition function does not yield the true expressions for the energy and the Helmholtz potential.

The one remaining hope, not yet discussed, is that the magnetization was attributed entirely to ampèrean currents, and quantum-theoretical analysis confirms that such currents always give rise to diamagnetism. Paramagnetism, as was seen, comes from the intrinsic magnetic dipole moments associated with spin. For these, the energy term is in the exact form of that for intrinsic electric dipoles in an electric field: $\epsilon_d = -\mathbf{m}_d \cdot \mathbf{B}$. The intrinsic dipoles are aligned by the local B-field, just as are electric ones in an E-field. The medium is thus polarized by these aligned dipoles, and their total energy may be written as $-V\mathbf{M}_d \cdot \mathbf{B}$, where $\mathbf{M}_d$ is the magnetization attributable to the intrinsic dipoles. The model must be supplemented to include this energy, and the notation is modified to distinguish M_a, the magnetization attributable to ampèrean currents, and M_d, that attributable to intrinsic dipoles. The mutual energy term now becomes $V(M_a - M_d)B_o$, with the sign of the ampèrean term changed to permit exploration of the consequences of the sign change. *The two principal objections to this conjectured thermodynamic resolution of the stability paradox are: (1) there was no error in the sign of the work term in Eq. (5.5.4), and (2) it disagrees with the work term derived from the partition function.* Thus there seems to be no thermodynamic resolution to the puzzle, and a cloud remains on the horizon. Even though the quantum-mechanical treatment of ampèrean currents clearly identifies these as the source of diamagnetism, thermodynamics continues to insist that stable systems require positive susceptibility, unless the Landau and Lifshitz point of view prevails.

Heine [1956] offers one of the most detailed and careful thermodynamic analyses that I have seen of bodies in static electromagnetic fields. He obtains the electrostatic work term $dW = -\int dV \mathbf{P} \cdot d\mathbf{E}_o$ for a system with constant external charges, in agreement with that of dF_D. His magnetic work term is $dW = \int dV \mathbf{B}_o \cdot d\mathbf{M}$ for a system with fixed external currents, in agreement with that of dF_H. He later objects to the expression that is the usual macroscopic translation of the magnetic energy term derived by statistical mechanics,

$$-\int_V \int_0^{B_o} \mathbf{M} \cdot d\mathbf{B}_o dV,$$

carried out at constant temperature, which is equivalent to the work term $dW = -\mathcal{M}dB_o$ for a system with constant flux, that of dF_B. He objects, apparently, to calling the expression by the name "free energy," because "this expression" has no unique meaning, among other things. I agree with him if he is saying that the term "free energy" has no unique meaning, and I do not use the term to name anything. On the other hand, if he means that the expression itself has no unique meaning, or that the work term is ambiguously defined for a constant-flux system of the type discussed in this chapter, then I disagree.

In Appendix B of Callen [1985], the magnetic energy term that should be used in macroscopic thermostatics is given as $B_o d\mathcal{M}$, so that he can write

$$dU = TdS - PdV + B_o d\mathcal{M} + \sum \mu_j dN_j,$$

in keeping with his postulate that U is an extensive function of the extensive parameters of the system. This result is in agreement with that of Eq. (5.5.3) for dU_H. His book has a careful and detailed treatment of thermodynamic stability, but no mention of the stability of magnetic systems. Perhaps his prudence in avoiding the subject should be regarded implicitly as good advice.

Interesting details about the historic development of the theory of susceptibilities and of magnetism *per se* are to be found in Mattis [1965] and van Vleck [1932]. Also, the lecture given by van Vleck when he received the Nobel Prize in 1978 provides a briefer and more up-to-date historical survey of the subject. (van Vleck [1978])

5.11 REFERENCES

A few of the principal references are listed here; all references are listed in the Bibliography at the end of the book.

GRIFFITHS, D. J. [1987], *Introduction to Electrodynamics*. Prentice-Hall, Englewood Cliffs, NJ, 1987.

GUGGENHEIM, E. A. [1945], Notes on Magnetic Energy. *Physical Review* **68** (1945), 273–276.

HEINE, V. [1956], The Thermodynamics of Bodies in Static Electromagnetic Fields. *Proceedings of the Cambridge Philosophical Society* **52** (1956), 546–552.

JACKSON, J. D. [1976], *Classical Electrodynamics, 2nd ed.*. John Wiley & Sons, New York, NY, 1976.

LANDAU, L. D. AND LIFSHITZ, E. M. [1960], *Electrodynamics of Continuous Media*. Addison Wesley, Reading, MA, 1960.

LEUPOLD, H. A. [1969], Notes on the Thermodynamics of Paramagnets. *American Journal of Physics* **37** (1969), 1047–1054.

MATTIS, D. C. [1965], *The Theory of Magnetism*. Harper and Row, N.Y., 1965.

VAN VLECK, J. H. [1932], *The Theory of Electric and Magnetic Susceptibilities*. Oxford University Press, London, 1932.

VAN VLECK, J. H. [1978], Quantum Mechanics: The Key to Understanding Magnetism. *Science* **201** (1978), 113–120.

5.12 PROBLEMS

1. A dielectric system in thermostatics is chosen to be the material that completely fills a parallel-plate capacitor, held at constant temperature in equilibrium with a heat bath. One plate is at ground potential ($\phi = 0$), and the other holds the fixed charge Q. The plate area is A, and the separation is a. The dielectric is homogeneous and isotropic, with permitivity ϵ. Ignore edge effects.

a. Find the capacitance, the potential difference of the plates, the electric field E_o produced by the free charge Q, the displacement D, the free surface charge density σ_f, the bound surface charge density σ_b, the internal field in the dielectric E, the polarization P, and the relationship of E to E_o and P.

b. Evaluate the total energy of the capacitor (dielectric plus free charge) in terms of E_o, ϵ_0, and P for the following three expressions that are often given as the energy density for such a system: $\epsilon_0 E^2/2$, $\mathbf{D} \cdot \mathbf{E}/2$, and $D^2/2\epsilon_0$.

c. The energy of a capacitor is usually calculated to be $Q\phi/2$, where ϕ is the potential of the charged capacitor, presumably found in (a), and Q is the free charge. Express the energy given by this formula in terms of E_o, ϵ_0, and P, and identify it with one of the energies found in part (b).

d. The net charge at the potential ϕ is $Q_{net} = Q + Q_b$, where Q_b is the bound charge on the surface of the dielectric. Find an expression for Q_b in terms of P. Calculate the energy required to bring the charge Q_{net} from infinity to the ungrounded plate of a vacuum capacitor of the same dimensions as C. Express this result in terms of E_o, ϵ_0, and P, and identify it with one of the energies found in part (b). Note that the dielectric system of Sec. 1 is modeled as a vacuum bounded by the appropriate polarization surface charge.

e. Similarly, calculate the energy required to bring the free charge Q from infinity to the same vacuum capacitor, and identify this energy with one of those found in part b.

2.a. Derive the Maxwell relations for a dielectric system with variables T, S, E_o, and $\mathcal{P}$.

b. Draw a mnemonic diagram in the style of Fig. 2.2 that corresponds to these relations.

3. A quantity analogous to the volume expansivity α is the response function $\alpha_E = (\partial P/\partial T)_E$. Use α_E and χ_{eT} to relate C_P and C_E in a way that corresponds to Eq. (2.5.23).

4. Similarly, relate the isentropic susceptibility χ_{eS} to the isothermal susceptibility χ_{eT}.

5. Use the response function $\alpha_P = (\partial E_o/\partial T)_P$ instead of α_E to relate the two heat capacities of problem (3).

6. A magnetic system is considered to be the medium that completely fills a long solenoid, held at constant temperature in equilibrium with a heat bath. The coil carries a fixed current I. It is of length l, and cross-sectional area A, with N closely and uniformly spaced turns. Ignore fringing fields and end effects.

a. Find the self-inductance L, the H-field produced by the outside current, the field B_o attributable to the external current, the total system field B, the magnetization M, the bound surface current density, and the relationship of B, B_o, and M.

b. Evaluate the total energy of the solenoid in terms of B_o, M, and $V = lA$ for the following three expressions that are often given as the energy density for such a system: $B^2/2\mu_0$, $\mathbf{B}\cdot\mathbf{H}/2$, and $\mu_0 H^2/2$.

c. The energy of an inductor is usually given as $W_L = LI^2/2$, where I is the coil current. Express the energy given by this formula in terms of B_o, M, and V, and identify it with one of the energies found in part (b).

d. The surface magnetization current could be regarded as actually sharing the coil with the external current. Find the magnitude of the coil current I_m that is equivalent to the surface-current density of the magnetization current, and write the net effective current as $I_{net} = I + I_m$. Calculate the energy required to bring the current of an empty solenoid of the same geometry from zero to the value I_{net}, and identify it with one of the energies of part (b).

e. Similarly, calculate the energy required to bring the current up from zero to the value I, when the solenoid is regarded as empty of all matter.

7.a. Derive the Maxwell relations for a magnetic system with variables T, S, B_o, and $\mathcal{M}$.

b. Draw a mnemonic diagram in the style of Fig. 2.2 that corresponds to these relations.

8. A quantity analogous to the volume expansivity α is the response function $\alpha_H = (\partial M/\partial T)_H$. Use α_H and χ_{mT} to relate $C_{\mathcal{M}}$ and C_H in a way that corresponds to Eq. (2.5.23).

9. Similarly, relate the isentropic susceptibility χ_{mS} to the isothermal susceptibility χ_{mT}.

10. Use the response function $\alpha_{\mathcal{M}} = (\partial H_o/\partial T)_{\mathcal{M}}$ instead of α_H to relate the two heat capacities of problem (8).

11. A perfect paramagnet, in analogy to a perfect gas, is defined as a material for which the energy

is a function only of temperature. Show that if U_B is taken as the energy, assumed to be a function of T and B_o, then any material such that $M = M(B_o/T)$ is a perfect paramagnet, according to the laws of thermostatics.

12. Consider a Debye solid as being also a nonconductor with permanent, noninteracting, alignable, quantized magnetic moments m such that the z-component can take the values $m_j = jg\mu_B$, with $-J \leq j \leq J$, and g is the gyromagnetic ratio. The lattice vibrations are regarded as the thermalizing environment that competes with the magnetic field in determining the equilibrium magnetization of the medium, but the canonical partition functions of the Debye solid and the magnetic moments are considered independent. The system comes to equilibrium with a thermal reservoir at temperature $T \ll \Theta_D$ and a magnetic field B_o produced by fixed external solenoid current. Assume that $\beta g\mu_B B_o \gg 1$. It is then thermally isolated, and the current is reduced to zero. Derive an approximate expression for the change in temperature. Why is this not a possible process for reaching $T = 0$?

13. Combine the Landau and Pauli calculations for an electron gas, regarded as an ideal fermion gas of electrons with spin. Modify each half of the Landau partition function to include the effect of the spin-associated dipole moment on the chemical potential. Obtain the combined susceptibility to second order in the usual expansion parameters, and compare the result with the summed susceptibilities of the separate calculations.

5.13 SOLUTIONS

1. The parts of this problem are all elementary physics, included as reminders of things already known. Therefore only the answers are given.

a. $C = \epsilon A/a$; $\Delta V = Q/C = aE$; $E_o = Q/\epsilon_0 A$; $D_o = \epsilon_0 E_o = \sigma_f = Q/A$; $\sigma_b = -P = -(D_o - \epsilon_0 E)$; $E = D_o/\epsilon = \epsilon_0 E_o/\epsilon = \Delta V/a$; $P = \epsilon_0 \chi_e E = D_o - \epsilon_0 E$.

b.

$$W_1 = aA\epsilon_0 E^2/2 = aA(\epsilon_0/2)(E_o - P/\epsilon_0)^2 = aA[\epsilon_0 E_o^2/2 - E_o P + P^2/2\epsilon_0];$$

$$W_2 = aA\mathbf{D} \cdot \mathbf{E}/2 = (aA/2)[\epsilon_0 E_o \cdot (\mathbf{E}_o - \mathbf{P})] = aA[\epsilon_0 E_o^2/2 - \mathbf{E}_o \cdot \mathbf{P}/2];$$

$$W_3 = aAD^2/2\epsilon_0 = aA\epsilon_0 E_o^2/2.$$

c. $W_c = Q\phi(Q)/2 = (AD_o)(aE)/2 = aA(\epsilon_0 E_o)(E_o - P/\epsilon_0)/2 = W_2.$

d. $Q_b = A\sigma_b = -AP$; $Q_{net} = Q - AP = A(\epsilon_0 E_o - P)$.

$$W_d = \int_0^{Q_{net}} (q/C)dq = (1/2C)Q_{net}^2 = (a/2A\epsilon_0)(A^2)(\epsilon_0 E_o - P)^2 = W_1.$$

e.

$$W_e = (1/2C)Q^2 = (a/2A\epsilon_0)(A\epsilon_0 E_o)^2 = W_3.$$

It is useful to understand the differences in the various work and energy expressions.

2. The differentials of the potentials are: $dU_D = TdS - \mathcal{P}dE_o$; $dU_E = TdS + E_odP$; $dF_D = -SdT - \mathcal{P}dE_o$; and $dF_E = -SdT + E_odP$. From these, the Maxwell relations can be read off:

$$\left(\frac{\partial T}{\partial E_o}\right)_S = -\left(\frac{\partial \mathcal{P}}{\partial S}\right)_{E_o},$$

$$\left(\frac{\partial T}{\partial P}\right)_S = \left(\frac{\partial E_o}{\partial S}\right)_{\mathcal{P}},$$

$$\left(\frac{\partial S}{\partial E_o}\right)_T = \left(\frac{\partial \mathcal{P}}{\partial T}\right)_{E_o}, \text{ and}$$

$$\left(\frac{\partial S}{\partial P}\right)_T = -\left(\frac{\partial E_o}{\partial T}\right)_{\mathcal{P}}.$$

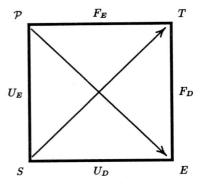

Figure 5.3 Mnemonic diagram for obtaining the dielectric Maxwell relations.

The mnemonic diagram for the Maxwell relations for dielectric systems, Fig. 5.3, may be used to summarize the Maxwell relations that were derived from U_E, F_E, U_D, and F_D in exactly the same way as a similar diagram, Fig. 2.2, was used in Chapter 2.

3. This problem is a straightforward application of the techniques presented in Chapter 2. Here molar quantities are used, so $\mathcal{P} = NP$, where P is the molar polarization, $\mathcal{P}$ is the total dipole moment, and N is the mole number. Also, E has been written for E_o.

$$C_E = \frac{T}{N}\left(\frac{\partial S}{\partial T}\right)_E = \frac{T}{N}\frac{\partial(S,E)}{\partial(T,E)} = \frac{T}{N}\frac{\partial(S,E)}{\partial(T,\mathcal{P})}\frac{\partial(T,\mathcal{P})}{\partial(T,E)}$$

$$= C_{\mathcal{P}} + T\left(\frac{\partial E}{\partial T}\right)_{\mathcal{P}}^2 \epsilon_0\chi_{eT} = C_{\mathcal{P}} + \frac{T\alpha_E^2}{\epsilon_0\chi_{eT}},$$

since

$$\left(\frac{\partial E}{\partial T}\right)_{\mathcal{P}} = -\left(\frac{\partial E}{\partial \mathcal{P}}\right)_T\left(\frac{\partial \mathcal{P}}{\partial T}\right)_E = \frac{\alpha_E}{\epsilon_0\chi_{eT}}.$$

4. In the same kind of calculation,

$$\chi_{eS} = \frac{1}{\epsilon_0}\left(\frac{\partial P}{\partial E}\right)_S = \frac{1}{\epsilon_0}\frac{\partial(P,S)}{\partial(E,S)}$$

$$= \chi_{eT} - \alpha_E^2 T/\epsilon_0 C_E.$$

From the results of Problems 3 and 4, it follows that

$$\chi_{eS}/\chi_{eT} = C_{\mathcal{P}}/C_E.$$

5. The use of $\alpha_{\mathcal{P}}$ leads to

$$C_E = C_{\mathcal{P}} + T\alpha_{\mathcal{P}}^2\epsilon_0\chi_{eT}.$$

6. The parts of this problem are all elementary physics, included as reminders of things already known. Therefore only the answers are given. a. $L = \mu N^2 A/l$; $H_o = IN/l$; $B = \mu H_o = \mu_0(H_o+M)$;

$K_s = M.$

b.

$$W_1 = \frac{1}{2\mu_0} \int B^2 \, dV = \frac{V}{2\mu_0}(B_o + \mu_0 M)^2 = V\left(\frac{B_o^2}{2\mu_0} + B_o M + \frac{\mu_0 M^2}{2}\right).$$

$$W_2 = \frac{1}{2} \int \mathbf{B} \cdot \mathbf{H} \, dV = \frac{V}{2} H_o(B_o + \mu_0 M) = V\left(\frac{B_o^2}{2\mu_0} + \frac{B_o M}{2}\right).$$

$$W_3 = \frac{\mu_0}{2} \int H^2 \, dV = \frac{V}{2\mu_0} B_o^2.$$

c.

$$W_c = LI^2/2 = V\left(\frac{B_o^2}{2\mu_0} + \frac{B_o M}{2}\right) = W_2.$$

d. Since $I_m = lK_s/N$, and $I_{net} = I + I_m$, the empty solenoid energy with this net current is

$$W_d = L_0 I_{net}^2/2 = \frac{\mu_0 V}{2}(H_o + M)^2 = W_1.$$

e.

$$W_e = L_0 I^2/2 = W_3.$$

7. The differentials of the potentials are: $dU_H = T dS - \mathcal{M} dB_o$; $dU_B = T dS + B_o d\mathcal{M}$; $dF_H = -S dT - \mathcal{M} dB_o$; and $dF_B = -S dT + B_o d\mathcal{M}$. From these, the Maxwell relations can be read off:

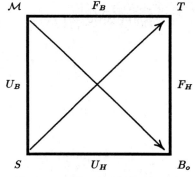

Figure 5.4 Mnemonic diagram for obtaining the magnetic Maxwell relations.

$$\left(\frac{\partial T}{\partial B_o}\right)_S = -\left(\frac{\partial \mathcal{M}}{\partial S}\right)_{B_o},$$

$$\left(\frac{\partial S}{\partial B_o}\right)_T = \left(\frac{\partial \mathcal{M}}{\partial T}\right)_{B_o},$$

$$\left(\frac{\partial T}{\partial \mathcal{M}}\right)_S = \left(\frac{\partial B_o}{\partial S}\right)_{\mathcal{M}}, \text{ and}$$

$$\left(\frac{\partial S}{\partial \mathcal{M}}\right)_T = -\left(\frac{\partial B_o}{\partial T}\right)_{\mathcal{M}}.$$

The mnemonic diagram for the Maxwell relations for magnetic systems, Fig. 5.4, may be used to summarize the Maxwell relations that were derived from U_B, F_B, U_H, and F_H in exactly the same way as a similar diagram, Fig. 2.2, was used in Chapter 2.

8. This problem is a straightforward application of the techniques presented in Chapter 2. Here molar quantities are used, so $\mathcal{M} = NM$, where M is the molar magnetization, $\mathcal{M}$ is the total magnetic moment, and N is the mole number. Also, H has been written for H_o.

$$C_H = \frac{T}{N}\left(\frac{\partial S}{\partial T}\right)_H = \frac{T}{N}\frac{\partial(S, H)}{\partial(T, H)} = \frac{T}{N}\frac{\partial(S, H)}{\partial(T, \mathcal{M})}\frac{\partial(T, \mathcal{M})}{\partial(T, H)}$$

$$= C_\mathcal{M} + T\mu_0\left(\frac{\partial H}{\partial T}\right)_\mathcal{M}^2 \chi_{mT} = C_\mathcal{M} + \frac{T\mu_0\alpha_H^2}{\chi_{mT}},$$

since

$$\left(\frac{\partial H}{\partial T}\right)_{\mathcal{M}} = -\left(\frac{\partial H}{\partial \mathcal{M}}\right)_T \left(\frac{\partial \mathcal{M}}{\partial T}\right)_H = -\frac{\alpha_H}{\chi_{mT}}.$$

9. In the same kind of calculation,

$$\chi_{mS} = \mu_0 \left(\frac{\partial M}{\partial B}\right)_S = \frac{\mu_0}{N} \frac{\partial(\mathcal{M}, S)}{\partial(B, S)} = \chi_{mT} - \mu_0 \alpha_H^2 T/C_H.$$

From the results of Problems 8 and 9, it follows that

$$\chi_{mS}/\chi_{mT} = C_{\mathcal{M}}/C_H.$$

10. The use of $\alpha_{\mathcal{M}}$ leads to

$$C_H = C_{\mathcal{M}} + T\mu_0\alpha_{\mathcal{M}}^2\chi_{mT}.$$

11. Since $dU_B = TdS + B_o d\mathcal{M}$, it follows for $M = M(B_o/T)$ that

$$\left(\frac{\partial U_B}{\partial B_o}\right)_T = T\left(\frac{\partial S}{\partial B_o}\right)_T + B_o\left(\frac{\partial M}{\partial B_o}\right)_T = 0,$$

since $(\partial S/\partial B_o)_T = (\partial \mathcal{M}/\partial T)_{B_o} = -(B_o/T^2)\mathcal{M}'$, and $(\partial \mathcal{M}/\partial B_o)_T = (1/T)\mathcal{M}'$, where $\mathcal{M}'$ is the derivative of $\mathcal{M}$ with respect to its argument.

12. For $T \ll \Theta_D$, the molar entropy of a Debye solid is $s_D = 4\pi^4 RT^3/5\Theta_D^3$. The entropy of the magnetic system is found from the single-magnetic-moment partition function, $Z_{1H} = \sinh[(2J+1)x]/\sinh x$ to be, with $x = \beta g_J \mu_B B_o/2$,

$$s_m = (u_H/T) + R\ln Z_{1H}.$$

At large x, $s_m \approx 0$, since all of the magnetic moments are fully aligned with the B-field. (This statement should be verified by means of the above equation for s_m.) At $B_o = 0$, the magnetic moments are assumed to be thermalized, such that $u_H = 0$, and the limiting value of the partition function gives $s_m = R\ln(2J+1)$. The system starts at temperature T_1, and after adiabatic (isentropic) demagnetization, the temperature is T_2. Equating the entropies gives

$$s(T_1) = \frac{4\pi^4 RT_1^3}{5\Theta_D^3} = s(T_2) = \frac{4\pi^4 RT_2^3}{5\Theta_D^3} + R\ln(2J+1).$$

Finally, T_2 is given by

$$T_2^3 = T_1^3 - (5\Theta_D^3/4\pi^4)\ln(2J+1).$$

Apparently this process permits attainment of $T_2 = 0$, but only apparently. Note that reaching $T_2 = 0$ is usually stated to correspond to a state of zero entropy, in disagreement with the above result that the entropy at $T_2 = 0$ is $s(T_2 = 0) = R\ln(2J+1)$. If the ground state were truly $(2J+1)$-fold degenerate, then the attainment of $T_2 = 0$ by adiabatic demagnetization is conceivably possible. In this case, however, interactions between the magnetic dipoles, which have been disregarded in this analysis, result in alignment of the dipoles and a nondegenerate ground state, for which the entropy is indeed zero. Thus, if the final entropy is zero at $T_2 = 0$, the initial entropy must also be zero for an isentropic process, and clearly no such process can exist.

13. The Landau result gives

$$\frac{\ln Z_G}{2} = \frac{8Vs^{5/2}}{15\lambda^3\sqrt{\pi}}\left[1 + \frac{5}{8}\left(\frac{\pi}{s}\right)^2 + \cdots\right] + \frac{(\beta\hbar\omega_L)^2 s^{1/2}V}{3\sqrt{\pi}\lambda^3}\left[1 - \frac{1}{24}\left(\frac{\pi}{s}\right)^2 + \cdots\right] +$$

$$- \frac{2V(\beta B_o\mu_B)^{3/2}}{\lambda^4\sqrt{\pi}}\sum_{r=1}^{\infty}\frac{(-)^r\cos\left(r\pi s/\beta B_o\mu_B - \pi/4\right)}{r^{3/2}\sinh(r\pi^2/\beta B_o\mu_B)}.$$

Replace $s = \beta\epsilon_F$ by $s_{\pm} = \beta(\epsilon_F \pm B_o\mu_B) = s(1 \pm \beta B\mu_B/s)$. The results of the calculation, which is routine but tedious, are:

$$\epsilon_F = \epsilon_{F_0}\left[1 - \frac{1}{12}\left(\frac{\pi}{s}\right)^2 - \frac{1}{6}\left(\frac{\beta B\mu_B}{s}\right)^2 + \cdots\right] + \text{ oscillatory terms};$$

which is the arithmetic average of the Landau and Pauli values,

$$\chi_{comb} = \frac{n\mu_0\mu_B^2}{\epsilon_{F_0}}\left[1 - \frac{1}{24}\left(\frac{\pi}{s}\right)^2 - \frac{1}{12}\left(\frac{\beta B\mu_B}{s}\right)^2 + \cdots\right] + \text{ oscillatory terms};$$

the sum of the separate susceptibilities, to this order. The combined oscillatory terms for the magnetization are

$$M_{osc} = \frac{-4\mu_B}{\lambda^3\sqrt{\pi}}\left[\frac{3\sqrt{\beta B_o\mu_B}}{2}\sum_{r=1}^{\infty}\frac{\cos\left(r\pi s/\beta B_o\mu_B - \pi/4\right)}{r^{3/2}\sinh(r\pi^2/\beta B_o\mu_B)} + \right.$$

$$\left. - \frac{\pi}{\sqrt{\beta B_o\mu_B}}\sum\frac{\sin(\ )}{r^{1/2}\sinh(\ )} - \frac{\pi^2}{\sqrt{\beta B_o\mu_B}}\sum\frac{\cos(\ )\cosh(\ )}{r^{1/2}\sinh^2(\ )}\right],$$

where the trigonometric functions all have the same argument, and the hyperbolic functions also have their own argument. This is a complicated result, leading to an even more complicated expression for the susceptibility. It is hardly worth developing or displaying explicitly, but its qualitative periodicity in $1/B$ is evident. It is not expected to agree in detail with experiments.

6. *THERMODYNAMICS OF PHASE TRANSITIONS*

6.1 INTRODUCTION

Phase transitions have been studied formally since the emergence of the science of thermodynamics. Well-known examples of the phenomenon are the condensation-evaporation, or liquid-vapor transition, and the freezing-melting, or solid-liquid transition. As preliminary definition, a phase transition is a process in which a thermodynamic system changes some of its macroscopic properties discontinuously (or apparently so) at fixed values of its intensive parameters. Such behavior results in discontinuities or singularities in the thermodynamic potentials and derived properties. The microphysical processes that underlie the phenomenon are: (1) the direct interactions of the components of the system (atoms, spins, molecules) with each other, and (2) thermal stirring of the interacting components. The interactions offer the opportunity for cooperation that can result in ordering of the interacting components to form a lattice, or clumping to form droplets, for example, while the thermal stirring acts to oppose any such cooperation. Ordinary states of matter are those for which one or the other of these competing processes dominates; the system is ordered or it is not (or perhaps it is ordered in one crystal structure instead of another; the concept of order is not trivial, and it must be examined in more detail). A phase transition occurs at the boundary between the domination of thermal stirring and that of interactive cooperation. It is evidently temperature dependent.

The study of phase transitions from the microscopic point of view requires consideration of systems of interacting particles. (The Bose-Einstein condensation is the only exception to this statement of which I am aware.) Up to this point, all microsystems have been treated as noninteracting, so that single-particle partition functions could be compounded to yield the partition function for a macroscopic system. For interacting systems, such simplification is not possible. It becomes necessary, in order to derive a proper N-body partition function, to write the N-body Hamiltonian and, in effect, sum $\exp(-\beta\mathcal{H})$ over the N-body states of the system. As might be imagined, such an undertaking is usually impossible, and substitute procedures are invoked. Exact solutions of simplified systems are examined in Chapter 7, and a remarkable analysis that transcends the details of specific interactions and reaches quite general conclusions about phase transitions is considered in Chapter 8. In the present chapter, the thermodynamics of phase transitions is reconsidered, after the brief treatment in Chapter 3, with some interesting results that provide precursors to the generalizations of Chapter 8.

6.2 PHASE TRANSITIONS AND STABILITY

The thermodynamic approach to phase transitions is in terms of stability. As an example, the Maxwell construction for the van der Waals gas in Section 3.8 was based on the Gibbs-potential-minimum criterion for stability. The generalized Maxwell-construction procedure is made necessary by the implicit assumption of homogeneity in the writing of an equation of state. The underlying partition function, if indeed one has been found, also assumes a homogeneous system, and equations of state that are derived from it will produce the same kind of unstable segments as are seen in the van der Waals equation. (Brout [1965]).

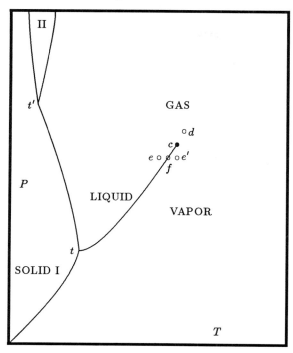

Figure 6.1 Representative phase boundaries in the T-P plane, showing the triple point, t, the critical point, c, the usual solid-vapor and solid-liquid phase boundaries and a possible bifurcation, t', on the solid-gas line. The points d, e, e' and f are for use in Fig. 6.2.

Stability conditions were derived in Chapter 2, and the Maxwell construction was obtained as a consequence of the equilibrium requirement of a Gibbs-potential minimum at constant T and P. Figure 2.6 shows the transfer of equilibrium from one branch of the molar Gibbs function to another along an isothermal process. A set of such plots for different isotherms would generate the coordinates in P-T space of the phase boundary, which is the curve that marks the transition from a branch of the Gibbs potential that represents a particular phase to another branch that represents a different phase. Such a phase diagram is shown in Fig. 6.1, in more detail than that of Fig. 2.1. Notable features of such a diagram are the boundary curves that separate these phases, the termination of the liquid-vapor boundary at the critical point, and the existence of one or more triple points, where three phases can coexist simultaneously. It is evident that no junctions more complicated than triple points can exist for a pure substance, as shown by the Gibbs phase rule, Eq. (2.4.11). The slopes of the boundary curves are given by Eq. (2.4.8), the Clapeyron equation. There are phase-to-phase discontinuities in such properties as molar entropy, density, and the various response functions. Most of these features were understood by the end of the nineteenth century and are part of the subject matter of undergraduate thermodynamics courses. They are not the concern of this chapter. The negative slope of the solid-liquid line is typical of water, for which the density of the solid is less than that of the liquid, but it is unusual. (The property of ice that it floats on water contributes significantly to the existence of terrestrial life forms.) Other such phase-boundary plots differ in detail from this one, but it will turn out that there are deeply rooted common features.

The principal topic of this chapter is the behavior of systems in the neighborhood of the critical point. This behavior, it turns out, is not always as marked as that described in the first paragraph of this chapter. Phase transitions exist for which there is not a discontinuity in most macroscopic physical characteristics or in the response functions. There can be discontinuities in the derivatives of the response functions (such as that in C_v for an ideal boson gas at the Bose-Einstein condensation temperature, but not in the response functions *per se*. These are sometimes called *second-order* phase transitions. It is apparent that at the critical point of a liquid-vapor system, the densities of the two phases become equal, as do a number of other properties. In this sense, the behavior of a system near the critical point resembles that of a second-order phase transition. The behavior of the liquid-vapor system in the neighborhood of the critical point can be understood in terms of the Gibbs potential for the system, here examined from a point of view that differs from that of Fig. 2.6. The shape of G as a function of volume at constant temperature is indicated in Fig. 6.2 for the points designated c, d, e, e', and f on Fig. 6.1.

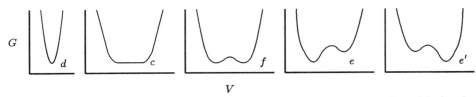

V

Figure 6.2 Behavior of G as a function of volume, parametric in T, near the critical point, c. The labels correspond to points of the same label in Fig. 6.1. Point d is along an approximate extrapolation of the liquid-vapor phase-boundary line, at $T > T_c$. At c, there is a broad Gibbs-potential minimum; at f there are two symmetrically placed minima, corresponding to the now-separated liquid and vapor phases with different densities; and the two side points, e and e' show the asymmetry in the depth of the minima for points that do not lie on the phase boundary.

The broad, flat minimum at the critical point shows the absence of distinct separation of the phases. It corresponds to the zero-slope critical inflection point in the isotherm that led to evaluation of the van der Waals critical constants in Eq. (2.8.5). The phenomenon of critical opalescence, first reported by Andrews in 1869, is a consequence of light being scattered by transient clusters of molecules; these clusters are the beginnings of droplet formation, and many have the dimensions of the order of the wavelengths of the scattered light, which may be of the order of a hundred times a molecular dimension. The compressibility at point c is infinite, and large-scale fluctuations are to be found in the other response functions. The behavior of G at points f, e, and e' can be followed to lower temperatures, with the result that the minima become deeper and more separated, as expected. (Andrews [1869])

Order-disorder transitions in alloys and magnetic systems are typical of those that never exhibit so-called *first-order* phase transitions, in which macroscopic properties change. In magnetic systems, for example, in which macroscopic magnetization can exist in the absence of an external magnetizing field, as in a permanent magnet, the microscopic magnetic moments interact to cause localized, and often transient, alignments of nearest neighbors in a lattice. At temperatures far above the critical temperature, alignments are short-lived because of thermal agitation; as T_c is approached from above, the local patches of ordered moments increase in typical size and persistence, exhibiting considerable short-range correlation. The ordered domains interact, each competing to impose its local orientation on neighboring domains, while trying to maintain order despite the thermal agitation of the lattice vibrations. There is no enduring long-range order, but there are transient waves of alignment, like gusts of wind in a wheat field, sweeping through the system. When the temperature reaches T_c, thermal stirring and dipole-dipole interactions have reached a balance, such that the more important struggle is between neighboring domains, which have been randomly oriented, in so far as is permitted by quantum theory and lattice structure. The struggle is statistical, and the outcome is heavily weighted in favor of the orientation that happens to be dominant as T_c is reached. Long-range order is established, in the weak sense that maximally distant magnetic moments within a single crystal have alignment correlation coefficients that are persistently greater than zero. Individual spins are still subject to thermal disalignment, and then to realignment by interaction with neighbors, but the dipole moment of the system has acquired a (barely) nonzero value, which maintains its direction and increases in magnitude as the temperature is lowered to values below T_c.

The magnetic moment of the system, or its magnetization M, may be taken as a measure of the long-range order. When $M = 0$, above or at T_c, there is no long-range order. As T is lowered below T_c, the magnitude M increases as ferromagnetism spontaneously establishes itself. In this system, M is called the *order parameter*, and the system behavior can be described in terms of M. The concept of an order parameter is readily generalized to include a wide range of seemingly unrelated systems, and their characteristics near the critical point can be treated on a common basis when expressed in terms of the order parameter. It should be noted that all of the microscopic activity just described is completely invisible to the classical thermodynamicist. Nevertheless, the existence of an order parameter could be conjectured, without defining it in terms of the underlying microphysics, and theoretical consequences of its existence can be explored.

As another example, the particular brass alloy composed of equal numbers of zinc and copper atoms has two different crystal structures: The high-temperature structure is a body-centered cubic lattice with each site occupied by either an atom of copper or one of zinc, with no evident ordering. As the temperature is lowered, localized ordering occurs, in which copper atoms preferentially occupy one of the interlocking simple cubic lattices that form the original lattice, and zinc atoms occupy the other. Energetically, the interlocking simple cubic lattice structure is preferred, but weakly, and at high temperatures thermal agitation precludes segregation of the two populations into separate but equal sublattices. As the temperature approaches T_c from above, the locally ordered domains increase in size and duration, while competing with neighboring domains over the choice of which sublattice belongs to copper and which to zinc. As T_c is reached, long-range order begins. Throughout the lattice, the configuration with copper atoms on A sites and zinc atoms on B sites (or the opposite) begins to prevail. As the temperature is lowered below T_c, the alloy becomes increasingly ordered. A measure of the order, the order parameter, can be taken as $(\langle N_{Zn}^A \rangle - \langle N_{Cu}^A \rangle)/N^A$. This parameter is zero for $T \geq T_c$, and it can be either positive or negative, depending on which sublattice is labelled A, as T decreases below T_c.

It turns out to be useful, in the development of theories of critical phenomena, to choose the order parameter such that it is zero at or above T_c, but nonzero below. (Landau and Lifshitz [1969]) It is also convenient to choose the parameter to be symmetric about zero, as was done for the brass system. For a simple magnetic system, when the aligned entities are electron spins, the order parameter can be taken as M_z, the component of **M** along the z-axis. For the liquid-vapor transition of the nonideal gas, a suitable order parameter is $\rho - \rho_c$, or $v - v_c$, where ρ is the density and v is the molar volume. Order parameters can be devised for all systems that exhibit critical points. For the purposes of further analysis, the order parameter is written as η. (Note that η can be a vector, such as **M**.)

6.3 THE LANDAU THEORY

In 1937, Landau offered a general theory of the behavior of systems in the immediate neighborhood of the critical point. (Landau and Lifshitz [1969]; Tolédano and Tolédano [1987]) He proposed that the Gibbs potential be written as $G(P, T, \eta)$, with T and P specified as usual, but η determined by the equilibrium condition–a Gibbs-potential minimum at constant T and P. He then expanded G as a power series in η, even though c is often a singular point, and there is no real expectation that such an expansion is possible, or that it can be carried to arbitrarily high powers of η. Nevertheless, the procedure is worth exploring. He wrote a series expansion equivalent to

$$G(P, T, \eta) = G_0(P, T) + G_1\eta + \frac{G_2}{2}\eta^2 + \frac{G_3}{3}\eta^3 + \frac{G_4}{4}\eta^4 + \cdots, \qquad (6.3.1)$$

where each G_n is a function of T and P. If it is assumed that the expansion is carried out around points on the P-T phase-boundary curve (the coexistence curve), or a reasonable extrapolation of it, for small values of η it seems valid to assume that G is symmetric in η, which is chosen to be zero at the critical point. In such a case, the odd coefficients are zero, and Eq. (6.3.1) becomes

$$G(P, T, \eta) = G_0 + \frac{G_2}{2}\eta^2 + \frac{G_4}{4}\eta^4 + \cdots. \qquad (6.3.2)$$

Along the coexistence curve, P and T are not independent, and the coefficients then become functions only of T and η. The shape of G near the critical point, insofar as the expansion is valid, is determined by the signs of the expansion coefficients. If terms no higher than fourth order in η are included, the general appearance of G at constant T is indicated in Fig. 6.3.

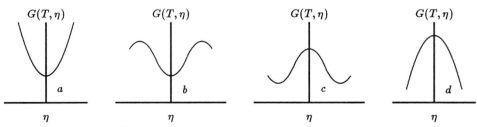

Figure 6.3 Behavior of $G(T, \eta)$ near the critical point as determined by the signs of the coefficients G_2 and G_4. (a) $G_2 > 0$, $G_4 > 0$; (b) $G_2 > 0$, $G_4 < 0$; (c) $G_2 < 0$, $G_4 > 0$; (d) $G_2 < 0$, $G_4 < 0$.

The curves (a) and (c) resemble the behavior of G when plotted against the volume in Fig. 6.2 d and 6.2 f, respectively. Indeed Fig. 6.2 could have been plotted with the proper order parameter $v - v_c$ as the abscissa. Evidently the difference between the curves 6.2d and 6.2f, which are on the coexistence curve or its extrapolation, is the sign of G_2. The critical point, therefore, must correspond to the condition $G_2 = 0$.

In this analysis, in which potentials are expanded in power series of a small parameter, and the expansion coefficients themselves are then expanded, it is convenient to choose the expansion parameters as unitless quantities, when possible. Therefore, for the fluid system, the order parameter is taken as $\eta = (\rho - \rho_c)/\rho_c = (v_c - v)/v$. (In the magnetic system, since $M_c = 0$, it is customary to use M as the order parameter.) The coefficients G_n in Eq. (6.3.2), which are functions of T, are written instead as functions of the dimensionless temperature $\tau = (T - T_c)/T_c$. The advantage, in addition to dimensionlessness, is that many systems exhibit essentially identical behavior in terms of these parameters that are standardized to their critical values. This behavior, to be examined subsequently, is called *the Law of Corresponding States.*

The coefficient G_2 of Eq. (6.3.2), which changes sign at T_c, is now written in terms of τ, with all coefficients G_{ij} chosen to be non-negative constants, as

$$G_2(\tau) = \tau^{2n+1} \left(G_{21} + \tau^2 G_{23} + \cdots \right), \qquad (6.3.3)$$

where n is an integer, and G_2 is represented as a series in odd powers of τ so that its sign will change at $\tau = 0$. Apparently stable systems have $G_4 > 0$ at temperatures on both sides of T_c; therefore the series for G_4 is represented as one of even powers of τ, or

$$G_4(\tau) = \tau^{2m} \left(G_{40} + \tau^2 G_{42} + \cdots \right). \qquad (6.3.4)$$

The condition for stability along the coexistence curve C is that $G(T, \ P(T), \ \eta)$ must be a minimum as a function of η. From Eq. (6.3.2), this condition may be expressed as

$$\left(\frac{\partial G}{\partial \eta} \right)_C = 0 = G_2 \eta + G_4 \eta^3, \qquad (6.3.5)$$

where terms above η^4 are presumed negligible near T_c. For $T > T_c$, the only real root is $\eta = 0$; for $T < T_c$, the root $\eta = 0$ is seen from Fig. 6.3c to correspond to a local maximum, and therefore not to equilibrium. The other two roots are found to be

$$\langle \eta \rangle = \pm \left(\frac{-G_2}{G_4} \right)^{1/2} = \pm \sqrt{\frac{G_{21}}{G_{40}}} (-\tau)^{n-m+1/2}, \qquad (6.3.6)$$

where only the first terms in the expansions for the G_n's were used. Since the equilibrium roots $\langle \eta \rangle$ merge at zero for $\tau = 0$, it follows that $n \geq m$. It is usual to take $m = n = 0$, but since there is disagreement between the consequences of that choice and experimental results, other choices should be examined. Thus the analysis predicts that the equilibrium order parameter near the critical point should depend on the temperature as

$$\langle \eta \rangle = A(-\tau)^\beta, \qquad (6.3.7)$$

where $\beta = 1/2, \ 3/2, \ 5/2$, and higher values.

Guggenheim, in 1945, plotted the data for eight gases in the form of Eq. (6.3.7). The data points were found to lie on essentially the same curve for all gases, and Guggenheim reported that they seemed to fit Eq. (6.3.7) with $\beta \approx 1/3$. (Guggenheim [1945]; Guggenheim [1967]) Although this value of β was observed to fit the data to an accuracy of 0.5% or better, there were criticisms on two accounts. First, it was pointed out that data near the critical point are quite difficult to obtain because of the large fluctuations. The extremely high compressibility makes even the Bernoulli-equation variation in pressure within the dimensions of the container so important that it is necessary to consider data from separate horizontal layers as being at different values of η, and the huge values of the molar heat capacities slow the equilibration process, requiring hours or even days, at very carefully controlled temperatures, to be sure that the points are those at equilibrium. For these reasons, Guggenheim's data did not include reliable points that were as close to the critical point as his critics thought were necessary in order to fit the $\beta = 1/2$ prediction of the classical theory. Second, a number of very careful experiments were performed to yield the quality of data needed to determine the critical exponent β (and others, to be defined). Some of these, such as a 1952 measurement on xenon (Weinberger and Schneider [1952]), which gave the value $\beta = 0.345 \pm 0.015$, were not inconsistent with Guggenheim's value of $1/3$, while other measurements seemed to give slightly higher values, such as $\beta = 0.342$ or 0.346 ± 0.01. (Kadanoff et al. [1967]) There is no theoretical basis for requiring that β be a rational fraction, other than the clearly erroneous results of the Landau theory. But since the smallest non-negative Landau value from Eq. (6.3.6), $\beta = 1/2$, is closest to that found by Guggenheim, it is usual to carry out the details of the Landau theory with n in Eq. (6.3.3) and m in Eq. (6.3.4) set equal to 0.

Despite disagreements in numerical detail, insight developing from these attempts to understand critical phenomena in quite general terms has led to a remarkable resurgence of interest and understanding in the field of thermodynamics. It is useful, before examining the modern theories, to complete the presentation of Landau's results, in order to show most simply how critical exponents can arise in theory. A critical exponent is defined, in general, by means of some property $f(\tau)$, presumed representable near the critical point as proportional to a power of τ. The exponent, say β, is defined as

$$\beta = \lim_{\tau \to 0} \frac{\ln |f(|\tau|)|}{\ln |\tau|}. \tag{6.3.8}$$

The existence of a critical exponent does not imply that $f(\tau) = a\tau^\beta$, because there can be terms in $f(\tau)$ of higher order in τ, or even a logarithmic dependence.

There are usually four basic critical exponents in thermodynamics. (Others appear in statistical theories, and these will be introduced in later chapters.) These are:

1. For the order parameter $\eta = (\rho - \rho_c)/\rho_c$, or M, since $M_c = 0$ when $B_o = 0$: $\eta \sim (-\tau)^\beta$.

2. For the heat capacities C_P or C_{B_o} : $C \sim \tau^{-\alpha}$ for $T > T_c$, or $C \sim |\tau|^{-\alpha'}$ for $T < T_c$. The heat capacities diverge at the critical point for $\alpha > 0$. They may also exhibit a discontinuity or a logarithmic divergence.

3. For the generalized susceptibilities κ_T or χ_T: κ_T or $\chi_T \sim \tau^{-\gamma}$ for $T > T_c$, or $\sim (-\tau)^{-\gamma'}$ for $T < T_c$.

4. On the critical isotherm, the order parameter and its intensive conjugate, h, if there is one, satisfy $\eta \sim h^{1/\delta}$. ($M \sim B_o^{1/\delta}$ or $(\rho - \rho_c)/\rho_c \sim [(P - P_c)/P_c]^{1/\delta}$).

According to the Landau theory, with G written as in Eq. (6.3.2), the molar Gibbs potential becomes

$$g(T, \eta) = g_0 - h\eta + \frac{g_2}{2}\eta^2 + \frac{g_4}{4}\eta^4 + \cdots, \qquad (6.3.9)$$

where the $h\eta$ term represents the contribution of the intensive parameter h for points off the $h = 0$ coexistence curve. The connection with thermodynamics comes from the expression for dg from Eq. (2.5.19), rewritten for molar quantities, and identified with dg from Eq. (6.3.9) in the region of the critical point. The equilibrium equation of state is $(\partial g/\partial \eta)_{\tau,h} = 0$, or

$$g_{21}\tau\eta + g_{40}\eta^3 = h, \qquad (6.3.10)$$

a generalization of Eq. (6.3.5) for any small h; it reduces to its former representation for $h = 0$. Thus, with the $h\eta$ term chosen to have the proper units,

$$\begin{aligned} dg &= dg_0 + [g_{21}\tau\eta + g_{40}\eta^3 - h]d\eta + (g_{21}/2)\eta^2 d\tau - \eta dh \\ &= (-s_0 + g_{21}\eta^2/2T_c)dT + (\eta_0 - \eta)dh, \end{aligned} \qquad (6.3.11)$$

since the term in square brackets vanishes along the coexistence curve, by use of Eq. (6.3.5). The entropy per mole is

$$s = -\left(\frac{\partial g}{\partial T}\right)_h = s_0 - \frac{g_{21}}{2T_c}\eta^2, \qquad (6.3.12)$$

where the subscript h implies that the intensive variable conjugate to η is equal to its value at the critical point. Note that s_0 comes from g_0, which is independent of the order parameter. For $T > T_c$, at equilibrium the order parameter is zero, and $s = s_0$. For $T < T_c$, the order parameter is given by Eq. (6.3.6), and the entropy becomes

$$s = s_0 + \frac{g_{21}^2\tau}{2T_c g_{40}} \; ; \quad T < T_c. \qquad (6.3.13)$$

For T just above T_c, the heat capacity is

$$C_h^+ = T_c\left(\frac{\partial s_0}{\partial T}\right)_h = C_{h0}. \qquad (6.3.14)$$

Below T_c, the result from Eq. (6.3.13) is

$$C_h^- = C_{h0} + \frac{g_{21}^2}{2g_{40}T_c} > C_{h0}. \qquad (6.3.15)$$

These results show no dependence of C_h on τ, either above or below T_c; therefore the Landau values for the exponents α and α' are both zero. The theory shows a discontinuity in the heat capacity, but no divergence. The discontinuity is

$$\Delta C_h = C_h^+ - C_h^- = -g_{21}^2/2g_{40}T_c. \qquad (6.3.16)$$

At the critical point, there is no discontinuity in s or η; therefore $\Delta s = 0$, and $\Delta \eta = 0$. For a more specific example, choose $\eta = v$, and differentiate along the coexistence curve through the critical point to show that the Landau theory predicts discontinuities in the other response functions.

$$0 = \frac{d\Delta v}{dT} = \left(\frac{\partial \Delta v}{\partial T}\right)_P + \left(\frac{\partial \Delta v}{\partial P}\right)_T \frac{dP}{dT}$$

$$= \Delta\left(\frac{\partial v}{\partial T}\right)_P + \Delta\left(\frac{\partial v}{\partial P}\right)_T \frac{dP}{dT}$$

$$= v_c\Delta\alpha - v_c\Delta\kappa_T \frac{dP}{dT} = \Delta\alpha - \Delta\kappa_T \frac{dP}{dT}, \tag{6.3.17}$$

where α here is the isothermal volume expansivity, not the critical exponent. Similarly,

$$0 = \frac{d\Delta s}{dT} = \Delta\left(\frac{\partial s}{\partial T}\right)_P + \Delta\left(\frac{\partial s}{\partial P}\right)_T \frac{dP}{dT}$$

$$= \frac{\Delta C_P}{T_c} - v_c\Delta\alpha \frac{dP}{dT}, \tag{6.3.18}$$

since $(\partial v/\partial T)_P = -(\partial s/\partial P)_T$. The discontinuity in C_P was found in Eq. (6.3.16), and the slope of the coexistence curve is expected to be finite but not zero. Thus Eq. (6.3.18) establishes that the discontinuity in α is nonvanishing, as does Eq. (6.3.17) for κ_T. The discontinuity in C_P arises from the second term in Eq. (6.3.13); therefore the same discontinuity occurs for C_v, and the same critical exponent, $\alpha = \alpha' = 0$, is found. This result is used in the next section.

The more general versions of these equations are

$$0 = \Delta\alpha - \Delta\chi_T \frac{dh}{d\tau}, \quad \text{and} \quad 0 = \Delta C_h - \Delta\alpha \frac{dh}{d\tau}, \tag{6.3.19}$$

where χ_T is the generalized notation for the response function.

The susceptibility may be found from Eq. (6.3.10) as

$$\chi_T^{-1} = \left(\frac{\partial h}{\partial \eta}\right)_\tau = g_{21}\tau + 3g_{40}\eta^2. \tag{6.3.20}$$

For $\tau > 0$, $\eta = 0$, and $\chi_T^{-1} = g_{21}\tau$, which gives the critical exponent $\gamma = 1$. For $\tau < 0$, with $h = 0$, Eq. (6.3.10) gives $\eta^2 = (g_{21}/g_{40})(-\tau)$, and therefore $\chi_T^{-1} = 2g_{21}(-\tau)$, or $\gamma' = 1$. Thus the Landau values are $\gamma = \gamma' = 1$.

Along the critical isotherm, $\tau = 0$, and $h = g_{40}\eta^3$, giving the Landau value of the critical exponent as $\delta = 3$.

For magnetic systems with M as the order parameter, the Landau critical exponents are identical to those of fluid systems. The analysis seems simple and direct; it predicts specific values for the critical exponents and discontinuities in response functions at the critical point. It predicts even more detail, such as the ratio of the coefficients, above and below T_c, in the expressions for χ_T. The fact that the predicted behavior does not agree with experiment has spoiled many a good theory, including this one. But the idea that there could be universal values for the critical exponents that transcend the microscopic details of interaction and apply to widely disparate kinds of systems has made a lasting and fruitful contribution to the development of the subject. The insights gained from the sequellae of the Landau study have deepened our level of understanding and pointed the way to remarkable discoveries of universal behavioral properties of systems in the neighborhood of the critical point, reported in this chapter and in Chapter 8. It should also be noted that Landau himself was well aware that his analysis of critical phenomena might not be valid, since it predicted singularities or discontinuities in the response functions, which are second derivatives of the thermodynamic potentials that the theory attempts to expand as power series around the critical point. It is evident that if the derivatives needed for a Taylor-series expansion are discontinuous at the expansion point, the expansion itself cannot be valid, and consequences of the expansion cannot be expected to be correct.

6.4 CRITICAL EXPONENTS

The Landau theory, together with Guggenheim's plot that seemed to show the value $\beta = 1/3$, inspired a substantial body of work, both experimental and theoretical, on the nature, values, and interrelations of critical exponents. Some of these results are examined in this section.

6.4.1 A Basic Lemma. An almost trivial lemma should be exhibited before invoking it to establish relationships among the critical exponents. For $0 < f(x) \sim x^p$ and $0 < g(x) \sim x^q$, with $0 < x < 1$, and $f(x) \leq g(x)$ for sufficiently small x, it follows that $p \geq q$. Proof: The condition $f(x) \leq g(x)$ implies $\ln f(x) \leq \ln g(x)$. Since $0 < x < 1$, $\ln x < 0$, and the ratios that define the critical exponents in the limit become

$$p = \lim_{x \to 0} \frac{\ln f(x)}{\ln x} \geq \lim_{x \to 0} \frac{\ln g(x)}{\ln x} = q, \tag{6.4.1}$$

thus proving the lemma.

6.4.2 Critical-Exponent Inequalities. One more critical exponent is required in order that a basic inequality may be established among the already defined exponents. It will be used only in the present subsection, because it is not an independent exponent, and because its use introduces some notational ambiguity. The required exponent is, in the most familiar context, that for the volume expansivity, also denoted by α. The temporary critical exponent θ (or θ') is defined by $\alpha \sim \tau^{-\theta}$, where the volume expansivity is given near the critical point by

$$\alpha = \frac{1}{v_c} \left(\frac{\partial v}{\partial T} \right)_P = \frac{1}{T_c} \frac{d\eta}{d\tau}$$
$$= \frac{1}{2T_c} \sqrt{\frac{g_{21}}{g_{40}}} (-\tau)^{-1/2} \sim (-\tau)^{-\theta'}; \quad T = T_c^-, \tag{6.4.2}$$

where Eq. (6.3.10) with $h = 0$ has been used in the second line. Thus the Landau value for the critical exponent for thermal expansivity shown to be $\theta' = 1/2$. But in general, since $\eta \sim (-\tau)^\beta$, the first line of Eq. (6.4.2) shows that $\theta' = 1 - \beta$. Since θ is so simply related to β, there is no need to continue its usage.

As a consequence of the observation that $C_P \geq C_v \geq 0$, and that the two heat capacities exhibit identical discontinuities at the critical point, it follows from Eq. (2.5.23) that a relationship must exist among the exponents α, γ, and θ. Without imposing any particular value on α', note that $C_P \sim (-\tau)^{-\alpha'}$, and therefore, along the coexistence curve for $T = T_c^-$, the inequality

$$C_p \geq T_c v_c \alpha^2 / \kappa_T \sim (-\tau)^{(-2\theta' + \gamma')}$$

leads by way of Eq. (6.4.1) and Eq. (6.4.2) to the inequality

$$-2\theta' + \gamma' = 2(\beta - 1) + \gamma' \leq -\alpha'. \tag{6.4.3}$$

This result is well known as the Rushbrooke inequality, usually written (Rushbrooke [1963])

$$\alpha' + 2\beta + \gamma' \geq 2. \tag{6.4.4}$$

Several other such inequalities have been found, mostly in the period from 1963 to 1970, to be joined by some strange equalities, discussed in the following section. They provided stimulus for the explosion of new results that burst forth in the next two decades. (Stanley [1971]) These results are discussed in the rest of this chapter and in Chapter 8, after a background is developed in Chapter 7. (Griffiths [1965a], [1965b])

6.5 GAP EXPONENTS

Essam and Fisher [1963] proposed that the singular parts of the Gibbs and Helmholtz potentials, in the neighborhood of the critical point, could have other interesting derivatives, of higher order than those already used, and that the singularities might be represented in terms of new critical exponents. These so-called gap exponents are written as Δ_r, and defined by the relations, here given in magnetic terms, but as easily expressed in terms of the generalized intensive parameter h:

$$\left(\frac{\partial g_s}{\partial B}\right)_T = g^{(1)} \sim (-\tau)^{-\Delta_1'} g_s^{(0)},$$

$$\left(\frac{\partial^2 g_s}{\partial B^2}\right)_T = g^{(2)} \sim (-\tau)^{-\Delta_2'} g_s^{(1)},$$

$$\vdots$$

$$\left(\frac{\partial^r g_s}{\partial B^r}\right)_T = g^{(r)} \sim (-\tau)^{-\Delta_r'} g_s^{(r-1)}, \tag{6.5.1}$$

where g_s is the singular part of the molar Gibbs potential. The divergent part of C_H, for $T < T_c$ is expressed in terms of its critical exponent as

$$C_H \sim (-\tau)^{-\alpha'} \sim \left(\partial^2 g_s / \partial (-\tau)^2\right)_B .$$

It follows that $g_s \sim (-\tau)^{2-\alpha'}$, with a similar result for $g_s^{(0)}$. Also, since $M \sim g^{(1)}$, and $M \sim (-\tau)^\beta$, the first definition of Eq. (6.5.1) gives $-\Delta_1' + 2 - \alpha' = \beta$, or

$$\alpha' + \beta + \Delta_1' = 2. \tag{6.5.2}$$

In the same manner, since $\chi_T \sim (-\tau)^{-\gamma'}$, the second definition leads to the result

$$\beta + \gamma' + \Delta_2' = 0. \tag{6.5.3}$$

Since Landau theory gives $\alpha' = 0$, $\beta = 1/2$, and $\gamma' = 1$, these equations say that $\Delta_1' = \Delta_2' = 3/2$, leading to the conjecture that all gap exponents for any particular system are equal. Numerical calculations for various model systems were not in disagreement with the conjecture, and the one exact calculation from statistical mechanics, the two-dimensional Ising model, supported it rigorously, giving $\Delta_1' = \Delta_2' = \cdots = 15/8$.

The pattern that emerged from numerical approximations was that for temperatures below T_c, gap exponents of all orders are equal; for temperatures above T_c, the odd field derivatives vanish, but the even gap exponents, $\Delta_2 = \Delta_4 = \cdots = \Delta_{2n}$, are all equal, and are equal to those for $T < T_c$. There was no experimental evidence in support of the higher gap exponents, but the numerical results indicated that the gap exponents may be somewhat dependent upon details of the interaction that leads to the phase transition.

6.6 CRITICAL EXPONENTS AND THE SCALING HYPOTHESIS

Puzzle upon puzzle developed, proliferating in the early 60's, but with roots in the Landau theory of the late 30's, the Guggenheim plots in the 40's, Onsager's exact solution of the two-dimensional Ising problem (Onsager [1944]), and a growing body of experimental measurement and theoretical approximation of critical exponents. It had become evident that the Landau theory, in all its attractiveness, was wrong in detail, but probably right in the concept that critical exponents seemed to exhibit a universality that transcends the details of the particular microscopic interactions that give rise to phase transitions. Therefore it seemed apparent that there must exist an underlying order parameter, just as Landau had hypothesized, and that a good candidate for this parameter could be the correlation length, defined in various ways, but intuitively conceived as a measure of the distance over which correlation in the alignment (or some other indicator of the influence of the system particles on each other) is significant. In large enough systems, the correlation length should be the only significant distance. The problem with the Landau theory, perhaps, was that the wrong order parameter was chosen, and therefore the wrong critical exponents emerged. A modification of the Landau theory was needed that would allow for the various thermodynamic variables to scale with the correlation length, but in a way that is sufficiently general to permit critical exponents that differ from those of the Landau theory. Such a modification is the subject of this section.

6.6.1 Generalized Homogeneous Functions.
The postulates of thermostatics, in Chapter 2, included the scaling law that the entropy is a first-order homogeneous function of the extensive parameters of the system. This postulate was seen to mean that the entropy of a uniform system is proportional to its size, or that such variables as U, S, and N, are proportional to the volume V. The Gibbs and Helmholtz potentials, which were used to determine the Landau critical exponents, are not entirely functions of the extensive parameters, and the further problem is to determine how, if at all, they scale with an unspecified order parameter, presumed to be a correlation length. The evident choice for a scaling function is a homogeneous function, or a generalization thereof.

A homogeneous function of one variable may be written

$$f(\lambda x) = g(\lambda)f(x), \qquad (6.6.1)$$

where g is as yet an unspecified function. Note that by the choice $x = 1$, Eq. (6.6.1) becomes $f(\lambda) = g(\lambda)f(1)$. Thus the function f for any argument is known if $f(1)$ and the scaling function g are known. But since this kind of scaling can be iterated, the scaling function cannot be arbitrary. Note that $f(\lambda\mu x) = g(\lambda\mu)f(x) = g(\lambda)f(\mu x) = g(\lambda)g(\mu)f(x)$, or

$$g(\lambda)g(\mu) = g(\lambda\mu) \ (= g(\nu)\ ). \tag{6.6.2}$$

Differentiate partially with respect to λ, multiply the resulting equation by λ, and divide both sides by $g(\lambda)g(\mu)$ to get, if g is differentiable,

$$\frac{\lambda g'(\lambda)}{g(\lambda)} = \frac{\nu g'(\nu)}{g(\nu)} = c,$$

where each side must be constant, since they are clearly independent. The only possible solution that satisfies Eq. (6.6.2) is

$$g(\lambda) = \lambda^c. \tag{6.6.3}$$

Thus the only possible differentiable scaling function is the scaling factor to some power.

For functions of more than one variable, the scaling law becomes

$$f(\lambda^p x, \lambda^q y, \ldots) = \lambda f(x, y, \ldots). \tag{6.6.4}$$

It is clearly not more general to write λ^n as the factor on the right-hand side of Eq. (6.6.4), since the rewritten factor could then be defined as a new λ.

6.6.2 Critical Scaling by Generalized Homogeneous Functions. The singular part of the Gibbs potential is assumed to be a generalized homogeneous function of $\tau = (T - T_c)/T_c$ and h, the intensive parameter conjugate to η. It can then be written

$$g_s(\lambda^p \tau, \lambda^q h) = \lambda g_s(\tau, h). \tag{6.6.5}$$

This assumption can be used to express all thermodynamic critical exponents in terms of the scaling exponents p and q. The singular part of the entropy is $s_s = -(\partial g_s/\partial T)_h \sim (\partial g_s/\partial \tau)_h$. Differentiation of Eq. (6.6.5) with respect to τ gives

$$\lambda^p s_s(\lambda^p \tau, \lambda^q h) = \lambda s_s(\tau, h). \tag{6.6.6}$$

Repetition of the differentiation yields, for the singular part of C_h,

$$\lambda^{2p} C_h(\lambda^p \tau, \lambda^q h) = \lambda C_h(\tau, h). \tag{6.6.7}$$

For $h = 0$, Eq. (6.6.7) becomes

$$\lambda^{2p} C_h(\lambda^p \tau, 0) = \lambda C_h(\tau, 0),$$

and for $\tau = 1$, this becomes

$$\lambda^{2p} C_h(\lambda^p, 0) = \lambda C_h(1, 0). \tag{6.6.8}$$

The scaling behavior of the generalized homogeneous function is now evident. For $T > T_c$, set $\lambda^p = \tau$, to obtain

$$C_h(\tau, 0) = \tau^{-2+1/p} C_h(1, 0). \tag{6.6.9}$$

The critical exponent is then read off from the relation $C_h \sim \tau^{-\alpha}$ to be

$$\alpha = 2 - 1/p. \tag{6.6.10}$$

For $T < T_c$, set $\lambda^p = -\tau$, and carry out the same analysis, to obtain

$$\alpha' = 2 - 1/p \ (= \alpha). \tag{6.6.11}$$

Now that the scaling behavior has been displayed explicitly in Eq. (6.6.8), a somewhat shorter procedure will be used for the other critical exponents. The order parameter, given by $\eta = (\partial g_s / \partial h)_\tau$, scales from Eq. (6.6.5) to give

$$\lambda^q \eta(\lambda^p \tau, \lambda^q h) = \lambda \eta(\tau, h). \tag{6.6.12}$$

For $h = 0$ and $T < T_c$, let $\lambda^p(-\tau) = 1$, so Eq. (6.6.12) becomes

$$(-\tau)^{-q/p} \eta(-1, 0) = (-\tau)^{-1/p} \eta(\tau, 0),$$

or

$$\eta(\tau, 0) = (-\tau)^{(1-q)/p} \eta(-1, 0) \sim (-\tau)^{(1-q)/p}. \tag{6.6.13}$$

But since the critical exponent for the order parameter η is denoted by β, it follows that

$$\beta = (1 - q)/p. \tag{6.6.14}$$

The susceptibility, $\chi_T \sim (\partial \eta / \partial h)_T$, similarly leads from Eq. (6.6.13) by the same steps to the critical exponents γ and γ':

$$\begin{aligned} \chi_T(\tau, 0) &= \tau^{(1-2q)/p} \chi_T(1, 0), \ T > T_c \\ &= (-\tau)^{(1-2q)/p} \chi_T(-1, 0), \ T > T_c. \end{aligned} \tag{6.6.15}$$

From these, it follows that

$$\gamma = \gamma' = (1 - 2q)/p. \tag{6.6.16}$$

Along the critical isotherm, $\tau = 0$, and Eq. (6.6.12) gives

$$\lambda^q \eta(0, \lambda^q h) = \lambda \eta(0, h). \tag{6.6.17}$$

The choice $\lambda = h^{-1/q}$ in Eq. (6.6.17) leads to

$$h^{-1} \eta(0, 1) = h^{-1/q} \eta(0, h),$$

or

$$\eta(0, h) = h^{(1-q)/q} \eta(0, 1) \sim h^{1/\delta}. \tag{6.6.18}$$

Thus the critical exponent δ is seen to be

$$\delta = q/(1 - q). \tag{6.6.19}$$

A quite general calculation can be made in this vein for the gap exponents, defined in Eq. (6.5.1), and repeated, in part, here:

$$g_s^{(r)} = \left(\frac{\partial^r g_s(\tau, h)}{\partial h^r} \right)_T \sim (-\tau)^{-\Delta_r} g_s^{(r-1)}. \tag{6.6.20}$$

Repeated differentiation of Eq. (6.6.5) with respect to h produces

$$\lambda^{qr} g_s^{(r)}(\lambda^p \tau, \lambda^q h) = \lambda g_s^{(r)}(\tau, h). \tag{6.6.21}$$

The expression for the gap exponent is valid at, or near, $h = 0$. When the ratio of Eq. (6.6.21) and an equivalent equation for $r \to r - 1$ is taken, the result at $h = 0$ is

$$\frac{\lambda^{qr} g_s^{(r)}(\lambda^p \tau, 0)}{\lambda^{q(r-1)} g_s^{(r-1)}(\lambda^p \tau, 0)} = \frac{g_s^{(r)}(\tau, 0)}{g_s^{(r-1)}(\tau, 0)} \sim (-\tau)^{-\Delta'_r}. \tag{6.6.22}$$

When $\lambda^p \tau = -1$, or $\lambda = (-\tau)^{-1/p}$, Eq. (6.6.22) gives

$$\Delta'_r = q/p = \Delta', \tag{6.6.23}$$

independent of r, and therefore valid for all orders. (A similar calculation could be made for Δ, with the same value as that of Δ', again to all orders.)

6.6.3 Derived Relations Among Critical Exponents.
Every critical exponent that has been introduced so far has been expressed, in the previous subsection, in terms of just two independent parameters, the exponents p and q that appear in the generalized homogeneous function of Eq. (6.6.5). Therefore any two critical exponents can be used to solve for p and q, and then to generate further relationships among the remaining exponents. For example, the use of Eq. (6.6.11) for α and Eq. (6.6.14) for β gives

$$p = 1/(2 - \alpha) \text{ and } q = 1 - \beta/(2 - \alpha). \tag{6.6.24}$$

When these expressions are substituted in Eq. (6.6.16) the result is

$$\alpha + 2\beta + \gamma = 2, \tag{6.6.25}$$

the Rushbrooke inequality of Eq. (6.4.4), now presented as an equality, and not in violation of Rushbrooke's proof. Similarly, the use of Eq. (6.6.19) gives

$$\alpha + \beta(1 + \delta) = 2, \tag{6.6.26}$$

and the use of Eq. (6.6.23) gives

$$\alpha + \beta + \Delta = 2. \tag{6.6.27}$$

By systematic expression of p and q in terms of all possible pairs of critical exponents, a family of critical-exponent relationships can be displayed. Clearly they are not all independent, but that they can be derived from only the two scaling exponents is both surprising and profound, especially since all derived, measured, and numerically estimated sets of critical exponents satisfy these equalities, either exactly for theoretically derived exponents, or within experimental or computational accuracy. For the five critical exponents defined so far, a total of ten different-looking equations are obtained from this procedure. All are either the ones already exhibited as Eqs. (6.6.25) – (6.6.27), or they can be derived from these three.

Numerical values of the parameters p and q for exactly solvable models are presented here, together with their calculated critical exponents. For the Landau theory, and also for the mean-field theory, to be presented in the next chapter, the values are

$$p = 1/2; \ q = 3/4; \ \alpha = 0; \ \beta = 1/2;$$
$$\gamma = 1; \ \delta = 3; \ \Delta = 3/2, \quad \text{(Landau)} \tag{6.6.28}$$

with the same values for the primed exponents, when applicable. The two-dimensional Ising model, also solved exactly in the next chapter, gives

$$p = 1/2; \ q = 15/16; \ \alpha = 0; \ \beta = 1/8;$$
$$\gamma = 7/4; \ \delta = 15; \ \Delta = 15/8. \quad \text{(2d Ising)} \tag{6.6.29}$$

The three-dimensional spherical model (Berlin and Kac [1952]), not solved in this book, gives

$$p = 1/3; \ q = 5/6; \ \alpha = -1; \ \beta = 1/2;$$
$$\gamma = 2; \ \delta = 5; \ \Delta = 5/2. \quad \text{(3d spherical)} \tag{6.6.30}$$

Numerical and experimental results are not listed here, since they are subject to revision and improvement.

6.7 CRITICAL FLUCTUATIONS

The usual Landau theory of critical phenomena begins with a series expansion of the Gibbs potential as a power series in the order parameter, in the neighborhood of the critical point. The existence of fluctuations and inhomogeneities has been ignored, although it has long been evident, in such phenomena as critical opalescence, for example, that fluctuations can become manifest on a macroscopic scale. Theoretical work on these critical fluctuations preceded the Landau theory by several decades, but in the absence of good experimental data the incentive for complicating the Landau theory by trying to include the effects of fluctuations was lacking. L. S. Ornstein and Frits Zernike presented a detailed treatment of density variations, molecular clustering, and light scattering at the critical point. (Ornstein and Zernike [1914]; Ornstein and Zernike [1917]; Ornstein and Zernike [1918]; Ornstein and Zernike [1926]; Zernike [1916]) Their work was improved and extended by Laszlo Tisza. (Klein and Tisza [1949]) The treatment of fluctuations in this section is based on the synthesis by Vitaly Ginzburg of the Ornstein-Zernike-Tisza theory with that of Landau. (Ginzburg [1961]) (The theory of Landau is a masterful reduction to its essence of the Gibbs description of the behavior of thermodynamic functions near T_c. (Gibbs [1878]))

6.7.1 The Ginzburg Criterion. In order to examine the effects of fluctuations on critical phenomena, Ginzburg proposed a continuum-limit modification of the Landau theory, and thus a reduction of the Ornstein-Zernike-Tisza theory to its bare essentials. He wrote the Gibbs potential per unit volume as

$$g(\mathbf{r}) = g_0 - h(\mathbf{r})\eta(\mathbf{r}) + \frac{1}{2}g_2\eta^2 + \frac{1}{4}g_4\eta^4 + \frac{c}{2}(\nabla\eta)^2, \tag{6.7.1}$$

which is a modification of Eq. (6.3.9) to include an energy term that models the effect of nonuniformity. Spatial variations in the order parameter are expected to cause an increase in energy and to be independent of the sign of the gradient. Therefore the constant c is expected to be positive, and the lowest-order gradient term allowed is the one given. Higher-order terms in the gradient, which should be to even powers to satisfy the symmetry requirement, are omitted from this treatment. The total Gibbs potential for the system is

$$G = \int dV \left[g_0 - h(\mathbf{r})\eta(\mathbf{r}) + \frac{1}{2}g_2\eta^2 + \frac{1}{4}g_4\eta^4 + \frac{c}{2}(\nabla\eta)^2 \right]. \qquad (6.7.2)$$

For equilibrium, the order parameter η should be chosen to minimize the Gibbs potential, so that its variation is zero for small, arbitrary variations of η, or

$$\delta G = 0 = \int dV \delta\eta \left[-h + g_2\eta + g_4\eta^3 - c\nabla^2\eta \right]. \qquad (6.7.3)$$

Since $\delta\eta$ is arbitrary, the terms in brackets must vanish; therefore η must satisfy

$$-c\nabla^2\eta + g_2\eta + g_4\eta^3 = h. \qquad (6.7.4)$$

Suppose the order parameter consists of a uniform, spatially independent term, $\langle\eta\rangle$, and a presumably small spatially varying term, η_1, and similarly for h, or

$$\eta(\mathbf{r}) = \langle\eta\rangle + \eta_1(\mathbf{r}). \qquad (6.7.5)$$

The terms of zeroth order in η_1, when Eq. (6.7.5) is used in Eq. (6.7.4), and the same expressions for the g_n's are used as in Eq. (6.3.10), is

$$g_{21}\tau\langle\eta\rangle + g_{40}\langle\eta\rangle^3 = h_0, \qquad (6.7.6)$$

which is equivalent to Eq. (6.3.10), and with the same consequences. The first-order terms in η_1, the fluctuation under study, must satisfy

$$-c\nabla^2\eta_1 + g_{21}\tau\eta_1 + 3g_{40}\langle\eta\rangle^2\eta_1 = h_1(\mathbf{r}). \qquad (6.7.7)$$

This equation may be rewritten as

$$\left(\nabla^2 - \kappa^2\right)\eta_1 = -h_1/c, \qquad (6.7.8)$$

where

$$\kappa^2 = (g_{21}\tau + 3g_{40}\langle\eta\rangle^2)/c = 1/\xi^2, \qquad (6.7.9)$$

and the quantity ξ, with the evident units of length, defines the length scale in this analysis. In order to determine the range of correlations in this system, it is useful to choose the perturbation h_1 to be a highly localized thermal fluctuation at the origin, such that it represents an energy of thermal magnitude, kT, or

$$\int \eta_0 h_1 dV = kT, \qquad (6.7.10)$$

where $h_1 = (kT/\eta_0)\delta(\mathbf{r})$, and η_0 is the local value of the order parameter conjugate to h_1, such that the perturbation is of the required magnitude. In three dimensions, the solution to Eq. (6.7.8) in an isotropic medium is

$$\eta_1(r) = \frac{kT}{4\pi\eta_0 c r} e^{-\kappa r}. \qquad (6.7.11)$$

A measure of correlation in the system is the covariance, given here by

$$g(\mathbf{r}, \mathbf{r}') = \langle \eta(\mathbf{r})\eta(\mathbf{r}') \rangle - \langle \eta(\mathbf{r}) \rangle \langle \eta(\mathbf{r}') \rangle = g(\mathbf{r}', \mathbf{r}). \qquad (6.7.12)$$

In this case, choose $\mathbf{r}' = 0$, note Eq. (6.7.5), and observe that $\langle \eta_0 \rangle = 0 = \langle \eta_1 \rangle$, so that Eq. (6.7.12) becomes

$$g(0, \mathbf{r}) = \langle \eta_0 \eta_1(\mathbf{r}) \rangle = \frac{kT}{4\pi c r} e^{-\kappa r}, \qquad (6.7.13)$$

which is independent of the sign of the fluctuation η_0. The range of influence of a localized perturbation is thus determined by the correlation length $\xi = 1/\kappa$.

A new critical exponent defines the behavior of the correlation length in the neighborhood of the critical point: $\xi \sim \tau^{-\nu}$ for $\tau > 0$, and $\xi \sim (-\tau)^{-\nu'}$ for $\tau < 0$. The Landau values for ν and ν' are found from Eq. (6.7.9). If $h_0 = 0$ in Eq. (6.7.6), then $\langle \eta \rangle = 0$ for $T > T_c$, and $\xi^2 = c/g_{21}\tau$, or $\langle \eta \rangle^2 = (g_{21}/g_{40})(-\tau)$ for $T < T_c$, and $\xi^2 = c/2g_{21}(-\tau)$. Thus the Landau values are $\nu = \nu' = 1/2$.

Ginzburg proposed as a reasonable criterion for the validity of the Landau theory that the fluctuation in the order parameter should be much smaller than its average value, so that fluctuations can be ignored. He used as his measure of fluctuation the covariance of Eq. (6.7.13), summed over a volume of linear dimension equal to the correlation length, or

$$\langle \eta_0 \eta_1 \rangle_\xi = \frac{kT}{4\pi c} \int_0^\xi 4\pi r^2 dr \frac{e^{-\kappa r}}{r} = \frac{kT\xi^2}{c}[1 - 2e^{-1}]. \qquad (6.7.14)$$

The Ginzburg requirement for validity then becomes

$$0.264 kT\xi^2/c \ll (4\pi/3)\xi^3 \langle \eta \rangle^2, \qquad (6.7.15)$$

where a numerical value has been used for the quantity in square brackets in Eq. (6.7.14). Substitutions of the known Landau exponents for ξ and η, and evaluation of the numerical coefficient, lead to the result near the critical point that

$$8 \times 10^{-3} \frac{k^2 T_c^2 g_{40}^2}{c^3 g_{21}} \ll |\tau| \ll 1, \qquad (6.7.16)$$

as the Ginzburg criterion for the validity of the Landau theory in three dimensions. The numerical values are not particularly informative, but one consequence is clear. Since the left-hand side of (6.7.16) is constant, independent of temperature, there exist values of τ that are small enough to violate the inequality. Thus the Ginzburg criterion says that the Landau theory of critical phenomena is not valid at the critical point of a three-dimensional system. But the actual numerical value of the left-hand side of (6.7.16) is not evident.

Therefore the theory might be valid quite near the critical point for some systems. A better formulation of the criterion is needed, with universal results.

In a d-dimensional system with radial symmetry, Eq. (6.7.8) becomes

$$\frac{1}{r^{d-1}}\frac{d}{dr}\left(r^{d-1}\frac{d\eta_1}{dr}\right) - \kappa^2\eta_1 = -\frac{h_1}{c}. \qquad (6.7.17)$$

The solution, with a point thermal perturbation at the origin, as before, is of the form

$$\eta_1 = \frac{kT(\xi r)^{(2-d)/2}}{(2\pi)^{d/2}\eta_0 c}K_{\pm(d-2)/2}(\kappa r), \qquad (6.7.18)$$

where $K_\nu(z)$ is the modified Bessel function. (See Abramowitz and Stegun, Sec. 9.6.) Only for $d = 1$ and $d = 3$ are the solutions given by elementary functions. The $d = 3$ solution is that of Eq. (6.7.11); for $d = 1$ it is

$$\eta_1(r) = \frac{kT}{2c\eta_0\kappa}\sqrt{\frac{\pi}{2}}e^{-\kappa r}. \qquad (6.7.19)$$

The solutions in other dimensions can be approximated for large and small values of their arguments. Since multiplicative constants are unimportant for the analysis of the Ginzburg criterion in terms of critical exponents, provided they do not contain powers of κ, they usually will be omitted. At first, however, the entire approximation will be displayed. For $d = 2$, the approximations of η_1 for large and small values of κr are

$$\eta_1 \approx -\frac{kT}{2\pi c\eta_0}\ln(\kappa r), \quad \kappa r \ll 1,$$

$$\approx \frac{kT}{2^{3/2}c\eta_0\sqrt{\pi\kappa r}}e^{-\kappa r}, \quad \kappa r \gg 1. \qquad (6.7.20)$$

For $d \geq 3$, the approximations are

$$\eta_1 \approx \frac{kT\Gamma(d/2-1)}{4\pi^{d/2}c\eta_0 r^{d-2}}, \quad \kappa r \ll 1$$

$$\approx \frac{kT\kappa^{(d-3)/2}}{2^{1+d/2}(\pi r)^{(d-1)/2}c\eta_0}e^{-\kappa r}, \quad \kappa r \gg 1. \qquad (6.7.21)$$

Note that the coefficients for the small-r approximations are independent of κ, and the κ-dependence of the coefficients appears only at large values of r, where the function is small. It should be remarked that the Ginzburg criterion requires integration out to $\kappa r = 1$, and it is not evident from the approximate forms of the functions that the use of either of them will correctly display the ξ dependence of the resulting inequality. Fortunately, the integration can be carried out exactly in all dimensions, and the approximations are used only in elucidating the behavior of the correlated region around the perturbation. The Ginzburg integral of the covariance is

$$\langle\eta_0\eta_1\rangle_\xi = B_d\int_0^\xi dr\, r^{d-1}\eta_0\eta_1(r)$$

$$= B_d \left(\frac{kT\xi}{c(2\pi\xi)^{d/2}} \right) \int_0^\xi dr\, r^{d/2} K_{\pm(d/2-1)}(\kappa r)$$

$$= B_d \left(\frac{kT\xi^2}{(2\pi)^{d/2}c} \right) \int_0^1 dz\, z^{d/2} K_{\pm(d/2-1)}(z)$$

$$= \left(\frac{kT\xi^2}{c} \right) \left[1 - \frac{2K_{d/2}(1)}{2^{d/2}\Gamma(d/2)} \right], \tag{6.7.22}$$

where $B_d = 2\pi^{d/2}/\Gamma(d/2)$ is the numerical coefficient in the d-dimensional volume element, as given in Appendix F. The final factor, in square brackets, is independent of ξ, and the Ginzburg covariance integral is seen to depend only on ξ^2, and thus to vary as $|\tau|^{-2\nu}$, independent of the dimensionality d. The Ginzburg condition for the validity of Landau theory becomes

$$\left(\frac{kT_c\xi^2}{c} \right) \left[1 - \frac{2K_{d/2}(1)}{2^{d/2}\Gamma(d/2)} \right] \ll \frac{B_d\xi^d}{d} \left\langle \eta \right\rangle^2 . \tag{6.7.23}$$

In terms of critical exponents, Eq. (6.7.23) can be written

$$A_1\xi^2 = A_2|\tau|^{-2\nu} \ll A_3|\tau|^{-\nu d}(-\tau)^{2\beta}, \tag{6.7.24}$$

where the A_i's are constants. Then by the lemma of Section 4.1, the critical exponents must satisfy

$$-2\nu > -\nu d + 2\beta, \tag{6.7.25}$$

or $d > d^*$, where the *marginal dimensionality* d^* is

$$d^* = 2 + 2\beta/\nu. \tag{6.7.26}$$

Landau theory gives $\beta = 1/2$ and $\nu = 1/2$, or $d^* = 4$. Thus the Ginzburg criterion leads to the expectation that Landau theory and its cognates are not valid in spaces of fewer than four dimensions, and to the extrapolated expectation that the measured or exactly calculated values of the critical exponents will depart increasingly from the Landau values as d decreases below 4.

Another formulation of the Ginzburg criterion is in terms of the generalized susceptibility, χ_T, given by

$$\chi_T = \left(\frac{\partial \eta}{\partial h} \right)_\tau = \left(\frac{\partial^2 g}{\partial h^2} \right)_\tau \sim \left\langle \eta^2 \right\rangle - \left\langle \eta \right\rangle^2 . \tag{6.7.27}$$

Here the covariance is evaluated over the entire volume of the system, rather than being restricted to a region of radius ξ. It turns out that near T_c the Ginzburg integral is a temperature-independent fraction of the susceptibility, since

$$\chi_T \sim B_d \int_0^\infty dr\, r^{d-1} \eta_0 \eta_1(r) = \frac{kT\xi^2}{c}. \tag{6.7.28}$$

(A different derivation of Eq. (6.7.28), with the spin at a lattice site as the order parameter, is given in Section 7.10.1.) It follows in Landau theory that since $\chi_T \sim \tau^{-\gamma}$, and $\xi \sim \tau^{-\nu}$, Eq. (6.7.28) requires that $\gamma = 2\nu$. This result is not derivable from the scaling considerations of Section 6.2, since the correlation length does not appear there; it is restricted to Landau theory, and it is in agreement with the already derived results that $\gamma = 1$ and $\nu = 1/2$. It is not in agreement with the exact results of the two-dimensional Ising model, for example.

The Ginzburg criterion, in terms of critical exponents, may be written

$$d^* = (2\beta + \gamma)/\nu, \tag{6.7.29}$$

which seems to establish a different expression for the marginal dimensionality. But since the analysis is based on Landau theory, only the Landau critical exponents can be used, and the same numerical value of d^* is obtained from Eq. (6.7.26) and Eq. (6.7.29).

A more-general dimension-dependent relationship connects the exponent α to ν, and therefore permits the introduction of d and ν into the critical-exponent equations of Section 6.2. This so-called *hyperscaling relation* is not a consequence of the use of generalized homogeneous functions, and its derivation is not, at this point, soundly based. It is examined further in the context of scaling theory in Chapter 8. The Gibbs potential for the entire system is an extensive function with the units of energy, proportional to its volume. Therefore, near T_c, a unitless version of the singular part of the Gibbs potential per unit volume, $G_s/kTV = \tilde{g}$, should be a nonextensive function, as is the molar Gibbs potential. The singular part of the entropy per unit volume is then

$$\tilde{s} \sim \left(\frac{\partial \tilde{g}}{\partial \tau}\right)_h,$$

and the singular part of the heat capacity per unit volume is

$$\tilde{C}_h \sim \left(\frac{\partial \tilde{s}}{\partial \tau}\right)_h \sim \tau^{-\alpha}. \tag{6.7.30}$$

Therefore $\tilde{g} \sim \tau^{2-\alpha}$, a quantity that not only does not diverge, but actually vanishes as $\tau \to 0$, since α is always less than 2. It is then argued that the quantity $\tilde{g}$ vanishes because the volume in the denominator of the defining equation for $\tilde{g}$ is proportional to the critical volume, ξ^d, which diverges at the critical point. A conceptual basis for this argument is that the only length scale of significance near the critical point is the correlation length. Since $\xi \sim \tau^{-\nu}$, it follows that

$$\tilde{g}_0 \tau^{2-\alpha} \sim \tilde{g}_0 \tau^{\nu d}, \tag{6.7.31}$$

or $\nu d = 2 - \alpha$. The argument is specious, but the result seems to be correct. A more plausible way of phrasing the same argument is to say that $\tilde{g}$ represents the energy per unit volume in units of kT, and it is proportional to the thermal energy at a lattice site, say, distributed over a coherence volume. As the coherence volume diverges, the energy density associated with each site vanishes as the energy is spread over the coherence volume.

Modifications of the Ginzburg criterion arise in anisotropic systems, where the coherence volume is not proportional to ξ^d. (Als-Nielsen and Birgeneau [1977]) For example, if the spins in a lattice are pointing in the z direction and coupled by a dipole-dipole interaction, it is found that the correlation range along the z axis goes as ξ^2, whereas in planes perpendicular to this axis, it goes as ξ. Therefore, the coherence volume for a three-dimensional system goes as $\xi^2 \times \xi \times \xi = \xi^4$, instead of as ξ^3. This suggests that both sides of Eq. (6.7.23) should be modified. The left-hand side, even after recognition that the correlation function no longer satisfies a radially symmetric differential equation, because the coherence is anisotropic, and that the integration volume also must involve angle variables, remains proportional to the susceptibility. The right-hand side is the product of the new volume element and $\langle \eta \rangle^2$. Therefore, if the new volume is proportional to ξ^{d+m}, where $m \geq 0$ represents the possibility that the dependence of the coherence volume on the correlation length may differ from that expected from spherical symmetry, the revision of Eq. (6.7.29) gives as the generalized Ginzburg criterion

$$d > d^* = \frac{2\beta + \gamma}{\nu} - m. \tag{6.7.32}$$

Thus in the example given, $m = 1$, the marginal dimensionality is $d^* = 3$, and the critical behavior is expected to be close to that predicted by Landau theory. Usually when $d = d^*$, the critical exponents are the Landau values, but a factor containing $|\ln |\tau||$ is also present. For the three-dimensional system of magnetic dipoles aligned parallel to the z axis, renormalization-group calculations of the type presented in Chapter 8 give the results

$$\chi_\tau \sim \xi^2 \sim |\tau|^{-1} |\ln |\tau||^{1/3}, \quad \text{and} \quad \eta \sim |\tau|^{1/2} |\ln |\tau||^{1/3}. \tag{6.7.33}$$

Except for the logarithmic factors, the critical behavior is classical. (Als-Nielsen and Birgeneau [1977])

6.8 MULTICRITICALITY

The critical point is reached when the stirring action of the temperature field is just balanced by the ordering action of some internal field conjugate to the order parameter, produced as a consequence of interactions over a wide range of scale lengths. Once this idea is understood, more complicated situations can be examined. There can be other fields, either ordering or nonordering, and there can be several order parameters. Examples of additional nonordering fields are the pressure in a magnetizable solid, the difference in chemical potentials in a mixture of liquid ^{3}He and ^{4}He, or the pressure in a system that undergoes a change in crystal structure as a consequence of temperature and pressure changes. Examples of additional order parameters are the staggered alignments of spins in an antiferromagnet and the competing alignments of spins parallel to, or transverse to a crystal axis. The area of multicritical phenomena promises to continue to provide new puzzles for a long time. Only a brief treatment of its simplest aspects is possible in this section. (Aharony [1983]; Lawrie and Sarbach [1984]; Plischke and Bergersen [1989]; Pynn and Skjeltorp [1983])

6.8.1 Landau Theory of Tricritical Points.
The Landau theory of Section 6.3 may be paraphrased, with slightly different notation, to provide a view of a model system with a tricritical point. Since two nonordering intensive parameters, T and an unspecified Y, are present, the coefficients in the series expansion of g in the critical region are expected to depend on both of these. The expansion is conveniently written, almost as in Eq. (6.3.9), as

$$g = g_0 - h\eta + (1/2)g_2\tau_2\eta^2 + (1/4)g_4\tau_4\eta^4 + (1/6)g_6\eta^6 + \cdots, \tag{6.8.1}$$

where

$$\tau_n = [T - T_n(Y)]/T_n(Y), \tag{6.8.2}$$

and the T_n's are the generalizations of the critical temperature, indexed to match the coefficients in the power series. The τ_n's are the reduced temperatures. It is expected that $T_2(Y)$ and $T_4(Y)$ exhibit different functional dependencies on Y. Note that now g_0 contains the explicit dependence on Y and its conjugate extensive parameter, which does not undergo an ordering transition in this analysis. As an example, Y may be thought of as the pressure on a solid, conjugate to its volume, which does not undergo a structural phase transition in response to changes in Y; but interatomic distances do change with Y, and these changes result in shifts in the internal forces that are responsible for phase changes.

Equilibrium values of η are found as those that minimize g at constant values of the intensive parameters, so that

$$h = \eta \left(g_2 \tau_2 + g_4 \tau_4 \eta^2 + g_6 \eta^4 \right), \qquad (6.8.3)$$

where terms beyond g_6 are ignored, and g_6 is taken to be independent of temperature. As before, the coefficients g_n are regarded as positive constants. For $h = 0$, solutions of Eq. (6.8.3) that minimize g are to be found, and the behavior of the order parameter is to be examined. It is convenient to examine this behavior quadrant by quadrant in τ_2-τ_4 space. Note that at a fixed value of Y, a point in this space is an isotherm, and the values of η that are solutions of Eq. (6.8.3) and also minimize g are the equilibrium values of the order parameter on that isotherm.

In the first quadrant, where $\tau_2 > 0$ and $\tau_4 > 0$, the only real root of Eq. (6.8.3) with $h = 0$, is $\langle \eta \rangle = 0$, which may be described as the disordered phase.

In the second quadrant, a search for roots with $\eta \neq 0$ uses the quadratic formula

$$\eta^2 = \frac{-g_4 \tau_4}{2 g_6} \left[1 \pm \sqrt{1 - (4 g_2 g_6 \tau_2 / (g_4 \tau_4)^2)} \right], \qquad (6.8.4)$$

where only the negative sign before the square root produces real roots, since $\tau_2 < 0$ and $\tau_4 > 0$. The square root may be expanded and the equation rewritten as

$$\eta^2 = \frac{-g_2 \tau_2}{g_4 \tau_4} \left[1 + x + x^2 + 2 x^3 + \cdots \right] = \frac{-g_2 \tau_2}{g_4 \tau_4} F(x), \qquad (6.8.5)$$

where $x = g_2 g_6 \tau_2 / (g_4 \tau_4)^2$. In the limit as $g_6 \rightarrow 0$, this result reduces to that of Eq. (6.3.10), when allowance is made for the differences in notation. For $\tau_4 > 0$, the system behaves exactly as in simple Landau theory. A critical point is found at $\tau_2 = 0$, where there is a second-order phase transition, with $\langle \eta \rangle = 0$. The only difference is that now there is a line of critical points, the positive τ_4 axis. As before, for $\tau_2 < 0$, the system behaves in the same way as that of Fig. 6.3c, exhibiting symmetrically placed equilibrium values of the order parameter.

The conditions not previously examined are those for which $\tau_4 < 0$; these would correspond to Fig. 6.3b and 6.3d, with the modification that the $g_6 \eta^6$ term makes all the curves turn upward for sufficiently large values of $|\eta|$. In the third quadrant, only the $+$ sign in Eq. (6.8.4) can lead to real values of η. These solutions are equilibrium values; they produce minima of g that are less than $g(\eta = 0)$. An isotherm corresponding to a point in either the second or third quadrant (a plot of η vs. h at constant T and Y) exhibits a discontinuity $\Delta \eta$ at $h = 0$. This is a first-order phase transition. There is little qualitative difference in the near-critical behavior of in the second and third quadrants, since Figs. 6.3c and 6.3d are qualitatively similar when a positive $g_6\eta^6$ term is included.

In the fourth quadrant, where Fig. 6.3b is modified by the added sixth-order term, as shown in Fig. 6.3a, the central minimum and the symmetric maxima exist as before, and two minima at larger values of $|\eta|$ can appear as the curve turns upward. Both signs in Eq. (6.8.4) yield values of $|\eta| > 0$ that satisfy Eq. (6.8.3) with $h = 0$; the negative sign gives the smaller values of η^2 that correspond to the already mentioned maxima, and the positive sign gives the roots of interest. In order for them to be equilibrium values, it is necessary that they give the lowest minimum of g, or $g(\eta = 0) - g(\eta) > 0$. From Eq. (6.8.1), with $h = 0$, the difference to be examined is

$$g(0) - g(\eta) = -\frac{\eta^2}{2}\left(g_2\tau_2 + {}^1\!/_2 g_4\tau_4\eta^2 + {}^1\!/_3 g_6\eta^4\right). \tag{6.8.6}$$

From Eq. (6.8.3), with $h = 0$ and $\eta^2 > 0$, it follows that

$$g_2\tau_2 = -\eta^2\left(g_4\tau_4 + g_6\eta^2\right). \tag{6.8.7}$$

The use of this expression in Eq. (6.8.6) gives

$$g(0) - g(\eta) = {}^1\!/_4 \eta^4\left[g_4\tau_4 + {}^4\!/_3 g_6\eta^2\right]. \tag{6.8.8}$$

In order that $g(\eta)$ be a minimum, the term in square brackets in Eq. (6.8.8) must be positive. Thus a boundary between the $\langle\eta\rangle = 0$ and $\langle\eta\rangle^2 > 0$ solutions is given by

$$\eta^2 = \frac{3g_4}{4g_6}(-\tau_4). \tag{6.8.9}$$

This condition, used in Eq. (6.8.7) gives

$$g_2\tau_2 = \frac{3g_4^2}{16g_6}\tau_4^2, \quad \text{or} \quad \tau_4^2 = \frac{16}{3}\frac{g_2 g_6}{g_4^2}\tau_2, \tag{6.8.10}$$

the parabola in τ_2-τ_4 space labelled 1 in Fig. 6.4, beginning at the origin and providing a useful boundary only in the fourth quadrant. For points inside the parabola (on the $+\tau_2$-axis side), $\langle\eta\rangle = 0$ is the equilibrium value; for points outside the parabola, the $\langle\eta\rangle^2 > 0$ roots of Eq. (6.8.4) are the correct ones. The parabola itself is called the *first-order line*. A process that crosses the first-order line from the disordered region results in a macroscopic jump in the order parameter. This behavior may be said to begin at the origin, which is the boundary between the line of critical points and the first-order line. The origin is then said to be the *tricritical point*. The τ_2-τ_4 plane is thus seen to be divided into two sectors: (a) a sector in which the equilibrium value of the order parameter is zero, consisting of the first quadrant and the part of the fourth quadrant above the first-order line, and (b) the rest of the plane, for which $\langle\eta\rangle^2 > 0$, with no qualitative differences in the behavior of the order parameter over the area. At the tricritical point, and everywhere along the $-\tau_2$ axis, the order parameter is given by

$$\langle\eta\rangle^4 = (g_2/g_6)(-\tau_2). \tag{6.8.11}$$

Here the critical exponent is $\beta_2 = 1/4$, rather than the Landau value, $\beta_2 = 1/2$. The subscript on the critical exponent is chosen to agree with that of its reduced temperature.

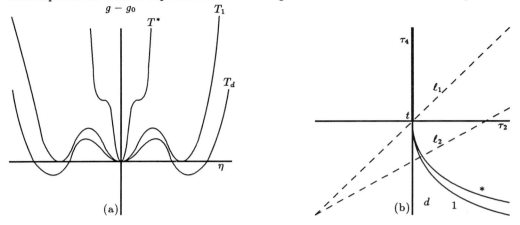

Figure 6.4 (a) Behavior of $g - g_0$ as a function of η in Eq. (6.8.6) for temperatures corresponding to points marked in (b). The T_1 curve, for which the outlying minima are just equal to that at $\eta = 0$, represents a temperature on the first-order line. The T^* curve represents a temperature on the line of points for which the slope of $g - g_0$ first becomes zero for a point other than at $\eta = 0$. The T_d curve is for a temperature in the two-phase region of the $\tau_2 - \tau_4$ diagram. (b) The $\tau_2 - \tau_4$ diagram of a tricritical system. The line of critical points is the positive τ_4 axis. The tricritical point t is at the origin. The first-order line is marked 1, and the line of unobserved singularities is marked *. The region corresponding to T_d in (a) is labelled d; it comprises the region between the negative τ_4 axis and the first-order line. The lines ℓ_1 and ℓ_2 are process lines along which T changes while the other intensive parameters remain fixed.

It is not usually possible, or even meaningful, to derive critical exponents for arbitrary points in the τ_2-τ_4 plane, since these exponents are intended to characterize the manner in which certain parameters diverge as $\tau \to 0$. If the parameter does not diverge because τ_2 and τ_4 do not simultaneously vanish, then there is no significance in its critical exponent. In certain regions, or along certain paths, critical exponents exist, and their derivations are not difficult. A revision of Eq. (6.3.11) gives

$$dg = dg_0 + [g_2\tau_2\eta + g_4\tau_4\eta^3 + g_6\eta^5 - h]d\eta + \tfrac{1}{2}g_2\eta^2 d\tau_2 + \tfrac{1}{4}g_4\eta^4 d\tau_4 - hd\eta$$
$$= \left(-s_0 + \frac{g_2\eta^2}{2T_2} + \frac{g_4\eta_4}{4T_4}\right) dT + (\eta_0 - \eta)dh, \tag{6.8.12}$$

where the terms in the square brackets vanish because of Eq. (6.8.3), and the entropy is found to be

$$s = s_0 - (g_2/2T_2)\langle\eta\rangle^2 - (g_4/4T_4)\langle\eta\rangle^4. \tag{6.8.13}$$

It should be noted that the heat capacity can no longer be identified by a single subscript, except by definition, since the behavior of the additional intensive variable Y must be specified. The present convention is that $C_h \equiv C_{h,Y}$, with all its implications. For example, at constant Y, both T_2 and T_4 are fixed, and a change in temperature must take place along a straight line in τ_2-τ_4 space that has non-negative slope and passes through the point $(-1, -1)$, the values that correspond to $T = 0$. The equation of this family of lines, implicitly parametric in Y, is

$$\tau_4 = \left(\frac{T_2}{T_4}\right)\tau_2 + \frac{T_2 - T_4}{T_4}. \qquad (6.8.14)$$

Two such temperature-changing processes are shown as dashed lines in Fig. 6.4b.

In the regions where $\langle\eta\rangle = 0$, the critical exponent for C_h is seen to be $\alpha = 0$, as in Eq. (6.3.14), including points on the critical line. Otherwise,

$$C_h = C_{h0} - T\left(\frac{g_2}{2T_2} + \frac{g_4\langle\eta\rangle^2}{2T_4}\right)\left(\frac{\partial\langle\eta\rangle^2}{\partial T}\right)_h. \qquad (6.8.15)$$

As the temperature approaches the tricritical value, T_t, from below on the only allowed path, $\tau_2 = \tau_4 \equiv \tau$, marked ℓ_1 in Fig. 6.4b, the derivative in Eq. (6.8.15) is dominated by

$$\left(\frac{\partial\langle\eta\rangle^2}{\partial T}\right)_h \simeq -\frac{1}{T_t}\sqrt{\frac{g_2}{-g_6\tau}}, \qquad (6.8.16)$$

so that

$$C_h - C_{h0} \simeq \frac{g_2}{2T_t}\sqrt{\frac{g_2}{-g_6\tau}}. \qquad (6.8.17)$$

Therefore the value of the critical exponent α' at the tricritical point is $\alpha'_t = 1/2$. This is an example of a model system for which a critical exponent takes different values when the reduced temperature τ changes from positive to negative.

From Eq. (6.8.4), the derivative is

$$\left(\frac{\partial\eta^2}{\partial T}\right)_h = \frac{-g_4}{2g_2T_4} \pm \frac{g_4^2\tau_4T_2 - 2g_2g_6T_4}{2T_2T_4g_6\sqrt{g_4^2\tau_4^2 - 4g_2g_6\tau_2}}. \qquad (6.8.18)$$

This derivative diverges as the line on which T is changing crosses the parabola in the fourth quadrant of τ_2-τ_4 space, given by

$$\tau_4^2 = \frac{4g_2g_6}{g_4^2}\tau_2. \qquad (6.8.19)$$

This parabola, closer to the τ_2 axis than the first-order line of Eq. (6.8.10), is marked * in Fig. 6.4b. It is the locus of points that mark the onset of the second minimum that develops into the first-order line, but at its onset, it does not correspond to a Gibbs-potential minimum. The curve marked T^* in Fig. 6.4a is seen to have a zero-slope inflection point, but only one minimum. For temperatures in the range $T_1 < T < T^*$, there are minima for nonzero values of η, but they are above the central minimum. Since the equilibrium value of η on * is zero, there are no physical consequences of a divergence in the derivative in Eq. (6.8.19). The generalized susceptibility is obtained from Eq. (6.8.3):

$$\chi_T^{-1} = \left(\frac{\partial h}{\partial\langle\eta\rangle}\right)_T = g_2\tau_2 + 3g_4\tau_4\langle\eta\rangle^2 + 5g_6\langle\eta\rangle^4. \qquad (6.8.20)$$

In regions for which $\langle \eta \rangle = 0$, the critical exponent for the susceptibility is $\gamma_2 = 1$. Otherwise, it is usual to find two temperature variables, each with its own exponent, and no general way to combine them to yield only one such variable. There is no divergence of the susceptibility if one τ remains finite while the other vanishes. In certain cases, again, such a calculation is possible. Along the negative τ_2 axis, where $\tau_4 = 0$, the use of Eq. (6.8.11) permits Eq. (6.8.20) to be written $\chi_T^{-1} = -4g_2\tau_2$, for which $\gamma_2' = 1$, as was the case in classical Landau theory. On the negative τ_4 axis, where $\tau_2 = 0$ and $\langle \eta \rangle^2 = -g_4\tau_4/g_6$, the result is $\chi_T^{-1} = 2g_4^2\tau_4^2/g_6$, giving the critical exponent of τ_4 as $\gamma_4' = 2$. There is no other choice on this axis, since $\tau_2 = 0$. But along the first-order line, where $\langle \eta \rangle^2 = -3g_4\tau_4/4g_6$, and $\tau_4^2 = 16g_2g_6\tau_2/3g_4^2$, it is seen that

$$\chi_T^{-1} = 4g_2\tau_2 = (3g_4^2/4g_6)\tau_4^2, \qquad (6.8.21)$$

which yields $\gamma_2 = 1$ and $\gamma_4' = 2$. These, it is to be noted, are the critical exponents that pertain as the tricritical point is approached along the first-order line. In addition, the susceptibility becomes infinite on the line of critical points, in keeping with its relationship to the correlation length, as given in (6.7.28).

The critical exponent δ may be read from Eq. (6.8.3), which is

$$h = g_2\tau_2\eta + g_4\tau_4\eta^3 + g_6\eta^5.$$

At $\tau_2 = 0$, $h \sim \eta^3$, and $\delta_2 = 3$, as in Landau theory. At $\tau_4 = 0$, $h \sim \eta$, and $\delta_4 = 1$. At the tricritical point, $\tau_2 = \tau_4 = 0$, and $h \sim \eta^5$, giving $\delta_t = 5$.

The Ginzburg criterion, Eq. (6.7.29), for the marginal dimensionality in a homogeneous system, gives $d^* = (2\beta + \gamma)/\nu$. For points in τ_2-τ_4 space such that $\beta_2 = 1/4$ and $\gamma_2 = 1$, the marginal dimensionality is $d^* = 3$, and Landau theory is marginally valid in three dimensions, rather than the usual four.

6.8.2 Other Multicritical Phenomena. Opportunities for further exploration of multicriticality are evident in several directions, even within the confines of thermodynamics. The crossover regions, where critical exponents change from one value to another have not been treated here. Terms beyond sixth order in the order parameter, and reduced critical temperatures τ_i for $i = 6, 8$, and higher can be examined. More than one order parameter may exist, and multiple values of other reduced intensive variables can be introduced into the coefficients of power-series expansions of the Gibbs function. Generalized Ginzburg criteria can be studied, and more critical exponents can be found. I have neither space nor intention to enlarge on these procedures; I merely point out that the generalizations and extensions use already developed techniques to generate ever more complexity. Once the concepts are understood, anyone who needs them to study particular systems should be able to develop the details without undue difficulty.

As an example of this kind of generalization, I write the Gibbs-potential density for a system with two order fields, η and ζ, with the notation similar to that of Eq. (6.7.2):

$$g = g_0 - \eta h - \zeta q + \sum_n \frac{1}{2n}g_{2n}\tau_{2n}\eta^{2n} + \sum_m \frac{1}{2m}g_{2m}'v_{2m}\zeta^{2m} + \frac{c}{2}(\nabla\eta)^2 + \frac{c'}{2}(\nabla\zeta)^2 + a\zeta\eta^2 \quad (6.8.22)$$

where the summation indices n and m can go as high as necessary, and the final term that couples the two fields could have some other form, if more appropriate. The reduced parameters v_{2m} are given by

$$v_{2m} = \frac{Y - Y_{2m}(T)}{Y_{2m}(T)}, \tag{6.8.23}$$

in analogy to the definitions of the τ_{2n} of Eq. (6.8.2). There will be two distinct correlation lengths, regions of competition between the order parameters for the establishment of long-range order, and complications because of the coupling term. Such systems have been studied experimentally. (Coulon, Roux and Bellocq [1991]; Pynn and Skjeltorp [1983])

6.9 COMMENTARY

The subject matter of this and the following two chapters is closely related. Phase transitions are presented in this chapter in the context of continuum thermodynamics, with no attempt to base derivations upon the underlying microphysics. Chapter 7 introduces the statistical mechanics of systems of interacting particles, with examples chosen from idealized models, and both approximate and exact solutions given. The mathematics required for exact solutions, in general, precludes treatment of problems in three dimensions or problems with realistic interactions. In Chapter 8, phase transitions are considered in terms of scaling, as in Chapter 6, but applied to the microscopic interactions of Chapter 7. The result is an understanding of the phenomenon based on scale invariance, the so-called renormalization-group theory. A more detailed commentary appears at the end of this final chapter on phase transitions.

Many aspects of the Landau theory of phase transitions and critical phenomena have been omitted, but references have been provided to assist further study. It should be mentioned here that the order parameter in the Landau-Ginzburg theory is a field, and that it can easily be generalized to other continuum field theories, and to the lattice gauge theories that are analogs of the discrete spin systems of Chapter 7. A rich and productive era has developed from the realization that the theory of critical phenomena and continuum field theory are different physical representations of essentially the same mathematical problem. Some aspects of this similarity appear at the level of Landau theory, while others emerge in the context of the renormalization group, in Chapter 8. (Baker, Jr. [1984]; David [1989]; Shenker [1984])

6.10 REFERENCES

The most important references for this chapter are listed here. The Bibliography at the end of the book contains the complete set.

AHARONY, A. [1983], *Multicritical Points*. In *Critical Phenomena*, F. J. W. Hahne, Ed. Springer-Verlag, New York–Heidelberg–Berlin, 1983, 209–258.

ALS-NIELSEN, J. AND BIRGENEAU, R. J. [1977], Mean Field Theory, the Ginzburg Criterion, and Marginal Dimensionality of Phase Transitions. *American Journal of Physics* **45** (1977), 554–560.

ESSAM, J. W. AND FISHER, M. E. [1963], Padé Approximant Studies of the Lattice Gas and Ising Ferromagnet below the Critical Point. *Journal of Chemical Physics* **38** (1963), 802–812.

GIBBS, J. W. [1876, 1878], On the Equilibrium of Heterogeneous Substances. *Transactions of the Connecticut Academy* **3** (1876, 1878), 108–248, 343-524.

GINZBURG, V. L. [1961], Some Remarks on Phase Transitions of the Second Kind and the Macroscopic Theory of Ferromagnetic Materials. *Soviet Physics-Solid State* **2** (1961), 1824–1834.

GUGGENHEIM, E. A. [1945], The Principle of Corresponding States. *Journal of Chemical Physics* **13** (1945), 253–261.

KADANOFF, L. P., GÖTZE, W., HAMBLEN, D., HECHT, R., LEWIS, E. A. S., PALCIAUSKAS, V. V., RAYL, M., SWIFT, J., ASPNES, D. AND KANE, J. [1967], Static Phenomena Near Critical Points: Theory and Experiment. *Reviews of Modern Physics* **39** (1967), 395–431.

KLEIN, M. J. AND TISZA, L. [1949], Theory of Critical Fluctuations. *Physical Review* **76** (1949), 1861–1868.

LANDAU, L. D. AND LIFSHITZ, E. M. [1969], *Statistical Physics*, 2nd. ed.. Addison Wesley, Reading, MA, 1969.

ORNSTEIN, L. S. AND ZERNIKE, F. [1918], The Linear Dimensions of Density Variations. *Physik. Zeits.* **19** (1918), 134–137.

PLISCHKE, M. AND BERGERSEN, B. [1989], *Equilibrium Statistical Mechanics*. Prentice Hall, Englewood Cliffs, NJ, 1989.

PYNN, R. AND SKJELTORP, A., Eds. [1983], *Multicritical Phenomena*. Plenum Press, New York, 1983.

RUSHBROOKE, G. S. [1963], On the Thermodynamics of the Critical Region for the Ising Model. *Journal of Chemical Physics* **39** (1963), 842–843.

STANLEY, H. E. [1971], *Introduction to Phase Transitions and Critical Phenomena*. Oxford University Press, London, 1971.

6.11 PROBLEMS

1.a. The partition function for one mole of a non-ideal gas is given by

$$Z = \frac{1}{N_0!} \left(\frac{v-b}{\lambda^3}\right)^{N_0} \exp(\beta a/v)$$

where a and b are constants. Determine the critical exponents α, β, γ, and δ, and the primed ones, when appropriate. b. Show that these satisfy the standard relations among critical exponents.

2. Suppose the molar van der Waals equation is modified to read

$$\left(P + av^{-n}\right)(v - b) = RT.$$

How do the values of the critical exponents depend upon n?

3. In the Landau theory, the molar Gibbs potential for a tricritical system is taken to be

$$g(T, h, \eta) = g_0 - h\eta + (1/2)g_2\tau_2\eta^2 + (1/4)g_4\tau_4\eta^4 + (1/6)g_6\eta^6.$$

For a process of temperature change such as that along ℓ_2 in Fig. 6.4b, calculate the latent heat of the first-order phase transition, which takes place at a temperature such as T_1 of Fig. 6.4a.

4. Show that the correlation length and the susceptibility become infinite on the line of critical points for a tricritical system.

More problems related to this chapter are to be found in the problem sets for Chapters 7 and 8. Instead of further problems, I offer several questions, some of which I have not resolved to my complete satisfaction, and others that involve more conceptual understanding than calculation, so there are no solutions given.

5. The Landau theory seems to describe a simple magnetizable system quite plausibly. For $\tau < 0$, there are two equivalent minima of the Gibbs potential, corresponding to the two possible signs of a one-dimensional order parameter, the magnetization. Since these states have the same energy and entropy when $h = 0$, there is no latent heat associated with a transition from one to the other. But for a condensable-vapor system, presumably the two Gibbs-potential minima correspond to the liquid and vapor phases, for which the Gibbs potential is indeed the same, but the entropy is not, and there is a latent heat associated with isothermal vaporization. What, then, is wrong with the entropy expression of Eq. (6.3.13), which gives the same entropy for both phases and no latent heat?

6. The correlation length ξ of Eq. (6.7.9) becomes infinite at the critical point, because $\tau = 0$ and $\eta = 0$. But for $\tau < 0$ why does the correlation length again become finite? The same question applies to the susceptibility. (The answer to this question is not difficult, but its resolution is important in the understanding of statements about range of correlations in molecular field models, in Chapter 7.)

7. Can the van der Waals system of Problem 1 be described in terms of a molar Gibbs potential that is of the same form as that of Eq. (6.3.9)? Is the answer to this question related to that of question 5, above?

6.12 SOLUTIONS

1. From the partition function, the energy is seen to be

$$u = -\frac{\partial \ln Z}{\partial \beta} = \frac{3N_0}{2\beta} - \frac{a}{b} = (\sqrt[3]{2})RT - \frac{a}{b},$$

and the pressure is

$$P = kT \left(\frac{\partial \ln Z}{\partial c} \right)_T = \frac{RT}{v - b} - \frac{a}{v^2},$$

which is the van der Waals equation. Since $C_v = (\partial u/\partial T)_v = 3R/2$, and in the Landau theory, $C_v \sim \tau^{-\alpha}$, it follows that $\alpha = \alpha' = 0$. With the critical parameters as found in Section 2.8, write $b = v_c/3$, $a = 9RT_cv_c/8$, $P_c = 3RT_c/8v_c$, $\bar{p} = (P - P_c)/P_c$, the order parameter $\eta = (\rho - \rho_c)/\rho_c$, $\tau = (T - T_c)/T_c$, and note that $v/v_c = \rho_c/\rho$. The van der Waals equation can be expressed in terms of the reduced parameters as

$$(\bar{p} + 1)(2 - \eta) = 8(\tau + 1)(\eta + 1) - 3(\eta + 1)^2(2 - \eta),$$

and without further approximation rewritten as

$$\bar{p}(2 - \eta) = 8\tau(1 + \eta) + 3\eta^3.$$

The phase boundary, or its extrapolation, may be thought of as the locus of the central Gibbs potential extremum, as shown in Fig. 6.3. This locus is given by $\eta = 0$, and the phase boundary becomes $\bar{p} = 4\tau$. Along this line, then, the rewritten van der Waals equation becomes

$$4\tau(2 - \eta) = 8\tau(1 + \eta) + 3\eta^3.$$

The solutions to this equation are $\eta = 0$ and $\eta = \pm 2\sqrt{-\tau}$. For $\tau > 0$, the only real solution is the first, but for $\tau < 0$, the two symmetrically displaced non-zero values of the order parameter are found, corresponding to the liquid and vapor phases. The critical exponent is seen to be $\beta = 1/2$. Along the critical isotherm, for which $\tau = 0$, the reduced pressure is

$$\bar{p} = \frac{3\eta^3}{2 - \eta} = \frac{3\eta^3}{2}\left(1 + \frac{\eta}{2} + \frac{\eta^2}{4} + \cdots\right),$$

from which it follows that $\delta = 3$. The critical exponent for the compressibility, γ, is found by noting that

$$\kappa_T^{-1} \sim \left(\frac{\partial \bar{p}}{\partial \eta}\right)_\tau = \frac{24\tau + 18\eta^2 - 6\eta^3}{(2 - \eta)^2}.$$

For $\tau > 0$, $\eta = 0$, and $\kappa_T^{-1} = 6\tau$, so that $\gamma = 1$. For $\tau < 0$, the known solution for η gives, to lowest order in τ, $\kappa_T^{-1} \sim -12\tau$, so $\gamma' = 1$. These are the classic Landau values.

2. The critical exponents are independent of n, as may be seen by an analysis that parallels the solution of Prob. 1. The requirement that the thermodynamic potentials be continuous at the critical point is sufficient to yield the Landau critical exponents.

3. The molar entropy, from Eq. (6.3.12), is

$$s = s_0 - (g_2/2T_2)\langle\eta\rangle^2 - (g_4/4T_4)\langle\eta\rangle^4.$$

For $T = T_1^+$, $\langle\eta\rangle = 0$, and for $T = T_1^-$, $\langle\eta\rangle^2 = -3g_4\tau_4/4g_6$. The latent heat is

$$\ell = T_1(s_+ - s_-) = \frac{3T_1 g_4 \tau_4}{8g_6}\left[-\frac{g_2}{T_2} + \frac{3g_4^2\tau_4}{8T_4 g_6}\right].$$

With τ_4 for the temperature T_1 given by $\tau_4 = (T_1 - T_4)/T_4$, the latent heat becomes

$$\ell = \frac{3T_1 g_4(T_1 - T_4)}{8g_6 T_4}\left[-\frac{g_2}{T_2} + \frac{3g_4^2}{8T_4 g_6}\left(\frac{T_1 - T_4}{T_4}\right)\right].$$

4. For a tricritical system that is the analog of that described by Eq. (6.8.1), the spatial equation corresponding to Eq. (6.7.4) is

$$-c\nabla^2\eta + g_2\tau_2\eta + g_4\tau_4\eta^3 + g_6\eta^5 = h.$$

The zeroth-order solution is that of Eq. (6.8.3), and the first-order equation becomes formally identical to Eq. (6.7.8), but with κ^2 now given by

$$\kappa^2 = (g_2\tau_2 + 3g_4\tau_4\langle\eta\rangle^2 + 5g_6\langle\eta\rangle^4)/c.$$

Along the line of critical points, $\langle\eta\rangle = 0$, and $\tau_2 = 0$, so that $0 = \kappa = 1/\xi$.

7. SYSTEMS OF INTERACTING PARTICLES

7.1 INTRODUCTION

The equilibrium statistical physics of systems of noninteracting particles is essentially a completed subject, even though the basic techniques are often carried over for approximate treatments of more complex systems. The equilibrium statistical physics of interacting systems, on the other hand, presents a vast sea of unsolved problems, based not only in physics and chemistry, but also very much in microbiology, macrobiology, ecology, sociology, economics, and many-people problems in general. Within this sea appear a few islands of already solved problems; processes of land formation are evident throughout this budding archipelago; bridges are being built; and some exploitation has begun.

While many of the solved problems are as yet too lengthy and difficult for inclusion in a book of modest size, others have become quite manageable. In particular, the class of problems generally called the "Ising problem" appears to offer an excellent balance in that the subject seems destined to remain active for a long time as new techniques, variations, and extensions are explored; yet the basic technique can be presented clearly and in a reasonable space. Furthermore, comparison of exact and approximate techniques of solution provides interesting warnings for the uninitiated.

Systems of interacting particles can exhibit correlations, so-called cooperative phenomena, and phase transitions such as spontaneous magnetization, order-disorder transitions, condensation or crystallization, and ferroelectricity. The Ising problem chosen for study in this chapter can be adapted to all of these contexts and solved exactly (in one and two dimensions) as well as approximately (in any number of dimensions). In addition to exact analytic solutions, the refinement of series expansion and Padé-approximant techniques permits highly precise numerical answers to be obtained for essentially any problem that can be formulated in terms of the Ising model. Thus the model is satisfying for physicists who want numbers to relate to experimental results and for puzzle solvers who are far less concerned about laboratory data than about exact solutions to challenging problems. Furthermore, the more general results based on similarity and the renormalization group show a surprising independence of the details of the model, as discussed in Chapters 6 and 8.

The omission of other kinds of calculations in the theory of systems of interacting problems is not a reflection on their importance, usefulness, or beauty. Excellent treatments of these theories continue to appear, and their availability to interested learners precludes the necessity for all-inclusive tomes. An abbreviated guide to the literature is included in the commentary and references at the end of this chapter.

7.2 THE ISING PROBLEM-GENERAL STATEMENT

A general Ising lattice is a regular array of elements, each of which can interact with other elements of the lattice and with an external field, such that the nonkinetic part of the

Hamiltonian is

$$\mathcal{H} = -\frac{1}{2} \sum_{i,\,j=1}^{N}{}' U_{ij}(\sigma_i,\,\sigma_j) - \sum_{i=1}^{N} V_i(\sigma_i), \tag{7.2.1}$$

where the σ_i are spin variables, capable of only the two values ± 1, $\sum'$ denotes omission of $i = j$ terms, the $1/2$ factor allows for the terms U_{ij} and U_{ji}, which both appear in the sum and represent the same interaction energy, and $V_i(\sigma_i)$ is a noninteraction term that may represent the effect of an external field or its equivalent. Somewhat more generally, the σ variables could take on a wider set of values, and indeed they often do, but the form of $\mathcal{H}$ to be treated here in most detail is

$$\mathcal{H} = -\frac{1}{2} \sum_{i,\,j=1}^{N}{}' J_{ij}\sigma_i\sigma_j - mB \sum_{i=1}^{N} \sigma_i, \tag{7.2.2}$$

with $J_{ij} = J$ if points i and j are nearest neighbors and zero otherwise; the mB term represents a magnetic dipole energy in the applied field B in this formulation.

Since the kinetic and nonkinetic parts of the partition function are separate factors, and since the interesting behavior is exhibited entirely by the nonkinetic factor, the kinetic factor will not be carried along in the development to follow, and the term *partition function* will be understood to mean the nonkinetic factor. (The partition function factors are assumed to commute.)

The partition function is written

$$Z = \sum_{\sigma_1=\pm1} \sum_{\sigma_2=\pm1} \cdots \sum_{\sigma_N=\pm1} \exp\left[(\beta/2) \sum_{i,\,j=1}^{N}{}' U_{ij}(\sigma_i,\sigma_j) + \beta \sum_{i=1}^{N} V_i(\sigma_i) \right], \tag{7.2.3}$$

where all the σ sums are written out for clarity. These sums will be indicated after this either by the notation $\sum_{\{\sigma_i=\pm1\}}$, the sum over the set of σ_i's, each of which can be ± 1 independently, or by $\sum_{\{\sigma_i\}}$, which is less explicit but more convenient and not used in any other sense.

All of the macroscopic physics is derivable from the partition function, and the problem is to extract it. Simple approximate methods of direct intuitive appeal will be examined first in several contexts, followed by exact solutions of the simplest Ising problem in one dimension. Since there is no phase transition for the usual Ising chain in this case, a somewhat more complicated one-dimensional model will be analyzed. (The absence of a phase transition in one dimension led Ising to the erroneous conclusion that the interaction term he used was unacceptable as a model for ferromagnetism, even in three dimensions. (Ising [1925]) As a consequence, the more complicated Heisenberg model was developed and studied in considerable detail before the Ising model was properly explored. The Heisenberg model involves interaction terms of the form $J_{ij}\mathbf{S}_i\cdot\mathbf{S}_j$, where the $\mathbf{S}$'s are vector spin variables, and the interaction is usually confined to nearest neighbors.) The exact solution of the two-dimensional Ising problem by the dimer-Pfaffian method is explained in the main body of this chapter, although other methods are introduced in treatment of the one-dimensional problem. The reasons for the choice of the dimer-Pfaffian technique are: (1) it is one of the shortest and clearest available; (2) the transfer matrix is readily available in brief text-book form; (3) while the transfer-matrix method has been used to solve more problems than any other exact technique so far employed, there are apparently very few unsolved problems that are amenable to its use; (4) the full power of the dimer-Pfaffian method has yet to be explored, but it may require considerable ingenuity to extend the techniques to other systems. Another method that offers great simplification is one relating closed graphs to random-walk paths on a lattice, but its inclusion would unduly lengthen this chapter. (Glasser [1970]; Vdovichenko [1965a]; Vdovichenko [1965b])

7.3 THE WEISS MOLECULAR-FIELD APPROXIMATION

The Hamiltonian given in Eq. (7.2.2) leads to the partition function

$$Z = \sum_{\{\sigma_i=\pm 1\}} \exp\left[(\beta/2) \sum_{i,\,j=1}^{N}{}' J_{ij}\sigma_i\sigma_j + \beta m B \sum_{i=1}^{N} \sigma_i\right], \qquad (7.3.1)$$

with $J_{ij} = J$ if points i and j are nearest neighbors, $J_{ij} = 0$ otherwise. The approximation proposed by Weiss, paraphrased for this context, is to regard the nearest-neighbor interactions on a particular σ_i as being the equivalent of an internal magnetic field, and then to approximate that individual field by an average field obtained by replacing σ_j in the lattice by $\langle\sigma\rangle$, for every j except i, the one being studied. Then the energy of the ith element becomes

$$\epsilon_i = -J\nu \langle\sigma\rangle \sigma_i - mB\sigma_i = -mB^\star\sigma_i, \qquad (7.3.2)$$

where ν is the number of nearest neighbors, often called the *coordination number*, $B^\star$ is called the *effective field*, or *mean field* and is the sum of the applied field B and the *molecular field* $J\nu \langle\sigma\rangle /m$. The $1/2$ factor in the interaction term is no longer needed, because now the entire lattice has fixed, unchanging $\langle\sigma\rangle$, and the only changes in energy come from σ_i. Effectively then, a single-particle partition function is to be used in place of Eq. (7.3.1), given by

$$Z_1 = \sum_{\sigma_i=\pm 1} e^{\beta m B^\star \sigma_i} = 2\cosh \beta m B^\star. \qquad (7.3.3)$$

The average energy from Eq. (7.3.2) is $-mB^\star \langle\sigma_i\rangle$, and from Eq. (7.3.3) it is found to be $-mB^\star \tanh \beta m B^\star$. For self consistency, $\langle\sigma_i\rangle$ is required to be just $\langle\sigma\rangle$, and the equating of the two expressions for energy gives

$$\langle\sigma\rangle = \tanh \beta m B^\star = \tanh(\beta J\nu \langle\sigma\rangle + \beta m B). \qquad (7.3.4)$$

The fact that Eq. (7.3.3) gives $\langle\sigma\rangle$ in terms of itself causes some analytic inconvenience, but otherwise the result is easily understood.

One of the questions to be treated in this context is the phase transition from $\langle\sigma\rangle = 0$ to $\langle\sigma\rangle \neq 0$ when $B = 0$, called *spontaneous magnetization*. When $B = 0$, Eq. (7.3.4) becomes

$$\langle\sigma\rangle = \tanh \beta J\nu \langle\sigma\rangle. \qquad (7.3.5)$$

As shown by Fig. (7.1) or by direct analysis, there is no solution to Eq. (7.3.5) except $\langle\sigma\rangle = 0$ if $\beta J\nu < 1$. For $\beta J\nu > 1$, there is an intersection of the two sides of Eq. (7.3.5) other than at $\langle\sigma\rangle = 0$, and since both signs of $\langle\sigma\rangle$ are acceptable, either sign of the average spontaneous magnetization may arise. The critical temperature is given by $\beta J\nu = 1$, or

$$T_c = J\nu/k. \tag{7.3.6}$$

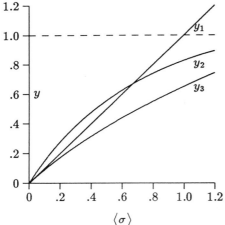

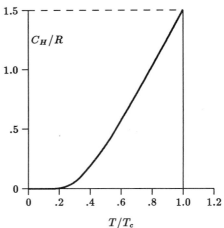

Figure 7.1 Plot of y_i vs. $\langle\sigma\rangle$, where $y_1 = \langle\sigma\rangle$; $y_2 = \tanh\alpha\langle\sigma\rangle$, with $\alpha > 1$; $y_3 = \tanh\alpha\langle\sigma\rangle$, with $\alpha < 1$.

Figure 7.2 Plot of C_H/R vs. T/T_c at $B = 0$ for the Weiss molecular-field model.

The result is entirely plausible. Large coupling constant J and large number of nearest neighbors give high onset temperature for spontaneous magnetization. The next questions concern the discontinuity in the heat capacity at T_c, when $B = 0$, and the behavior of the susceptibility at temperatures above T_c. To calculate the heat capacity at $B = 0$, we can write the energy per particle as $\langle\epsilon\rangle = -J\nu\langle\sigma\rangle \tanh\beta J\nu\langle\sigma\rangle$, where B^* has been replaced by its value when $B = 0$. The energy per mole, however is not $N_0\langle\epsilon\rangle$ but $N_0\langle\epsilon\rangle/2$, since a return to a many particle system requires that the energy be shared by nearest neighbor pairs. The molar energy at $B = 0$ is then

$$u = -N_0 J\nu\langle\sigma\rangle^2/2, \tag{7.3.7}$$

by use of Eq. (7.3.5), and the molar heat capacity at constant $B(= 0)$ is

$$C_H = \left(\frac{\partial u}{\partial T}\right)_H = N_0\nu J\langle\sigma\rangle\frac{\partial\langle\sigma\rangle}{\partial T}. \tag{7.3.8}$$

By differentiating Eq. (7.3.5), $\partial\langle\sigma\rangle/\partial T$ is found from

$$\frac{\partial\langle\sigma\rangle}{\partial T} = -\frac{1}{kT^2}\frac{\partial\langle\sigma\rangle}{\partial\beta} = -\frac{J\nu}{kT^2}\left(\langle\sigma\rangle + \beta\frac{\partial\langle\sigma\rangle}{\partial\beta}\right)(\cosh\beta J\nu)^{-2}\langle\sigma\rangle,$$

or

$$\frac{\partial\langle\sigma\rangle}{\partial T} = \frac{-\beta J\nu\langle\sigma\rangle}{T\left[\cosh^2\beta J\nu\langle\sigma\rangle - \beta J\nu\langle\sigma\rangle\right]}. \tag{7.3.9}$$

Equation (7.3.8) can now be written as

$$C_H = \frac{R(T_c \langle \sigma \rangle /T)^2 (1 - \langle \sigma \rangle^2)}{1 - (T_c/T)(1 - \langle \sigma \rangle^2)}, \tag{7.3.10}$$

where $R = N_0 k$, and Eqs. (7.3.5) and (7.3.6) have been used. Evidently for $T > T_c$, since $\langle \sigma \rangle = 0$, it follows that $C_H = 0$. For $T < T_c$ let $y = T_c \langle \sigma \rangle /T = (T_c/T) \tanh y$. In terms of y, Eq. (7.3.10) becomes

$$\frac{C_H}{R} = \frac{y^2}{\cosh^2 y - y \coth y}. \tag{7.3.11}$$

Series expansion of the denominator gives

$$\lim_{y \to 0_+} (C_H/R) = 3/2, \tag{7.3.12}$$

and for $y \gg 0$, Eq. (7.3.11) gives

$$\lim_{y \gg 0} C_H/R \to 4y^2 e^{-2y}. \tag{7.3.13}$$

The behavior of C_H/R *vs* T is shown in Fig. (7.2), with the sharp discontinuity indicating a clear phase transition.

The molar magnetization of the system is given by

$$M = N_0 m \langle \sigma \rangle. \tag{7.3.14}$$

(Since the system could be one-, two-, or three-dimensional, molar magnetization is used here instead of magnetic moment per unit volume, but N_0 can easily be replaced by N/V where appropriate.) With $B \neq 0$ and $T > T_c$, the molar susceptibility at constant T is defined as

$$\chi_T = \mu_0 \left(\frac{\partial M}{\partial B} \right)_T.$$

By a calculation similar to that for C_H, χ_T is readily found to be

$$\chi_T = \frac{\beta m^2 N_0 (1 - \langle \sigma \rangle^2)}{1 - (T_c/T)(1 - \langle \sigma \rangle^2)}. \tag{7.3.15}$$

In the limit of very small $\langle \sigma \rangle$, Eq. (7.3.15) becomes

$$\chi_T \to \frac{m^2 N_0/k}{T - T_c}, \tag{7.3.16}$$

which is in the form of the Curie-Weiss law for the susceptibility of a ferromagnet at temperatures above T_c, the Curie temperature.

Up to this point, then, the approximate treatment looks good enough to be taken seriously. Refinements are easily possible and the literature is filled with them. A few are suggested as exercises.

In order to show another approach to the approximate solution and to show how dissimilar the same approximation can look in different guises, a mathematically more formalistic treatment is given in the next section, together with paraphrases of it in other contexts.

7.4 THE BRAGG-WILLIAMS APPROXIMATION

Although the Bragg-Williams technique was first used to approximate an order-disorder transition in a binary alloy, the procedure is easily adaptable and will be presented first in terms of a Weiss system. Afterwards it will be restated to apply to the original order-disorder transition. (Bragg and Williams [1934]) The lattice gas will be mentioned only after the exact Ising problem has been treated.

The total energy of the spin system in the external field B may be written as

$$E = J\left[(N_{++} + N_{--}) - (N_{+-} + N_{-+})\right] - mB(N_- - N_+), \qquad (7.4.1)$$

where N_{++} is the number of nearest neighbor pairs with both spins $+$, *etc.*, N_+ is the number of $+$ spins and N_- the number of $-$ ones. The only difference in N_{+-} and N_{-+} is that if one proceeds through the lattice in some orderly fashion and encounters an unlike nearest neighbor pair in the order $+-$, he counts it in N_{+-}, whereas the ordering $-+$ is counted in N_{-+}. Except for possible edge effects, ignored here, $N_{+-} = N_{-+}$.

The numbers N_+ and N_- are written for convenience as

$$N_+ = \frac{N}{2}(1+x); \quad N_- = \frac{N}{2}(1-x), \qquad (7.4.2)$$

where the order-parameter x can be of either sign. The essential point of the B-W technique is the writing of the numbers of n.n. pairs as

$$N_{++} = \gamma N_+^2; \; N_{--} = \gamma N_-^2; \; N_{+-} = N_{-+} = \gamma N_+ N_-, \qquad (7.4.3)$$

where γ is a constant of proportionality chosen so that the total number of n.n. pairs is $N\nu/2$, where ν is the number of nearest neighbors of each lattice site. The constant γ is easily found to be $\nu/2N$, and the total energy becomes

$$E = -(JN\nu/2)x^2 - mBNx. \qquad (7.4.4)$$

If the system is regarded as microcanonical, with E fixed, the entropy can be written in terms of the total number of complexions with N, N_+ and N_-, according to Eq. (3.2.5), as

$$S = k\ln(N!/N_+!N_-!)$$
$$= -\frac{Nk}{2}\left[(1+x)\ln\left(\frac{1+x}{2}\right) + (1-x)\ln\left(\frac{1-x}{2}\right)\right], \qquad (7.4.5)$$

and the temperature equation is

$$\beta = \frac{\partial(S/k)}{\partial E} = \left(\frac{\partial x}{\partial E}\right)\left(\frac{\partial(S/k)}{\partial x}\right)$$
$$= \frac{1}{2(J\nu x + mB)}\ln\left(\frac{1+x}{1-x}\right). \qquad (7.4.6)$$

Solution of Eq. (7.4.6) for x gives

$$x = \tanh\beta(J\nu x + mB), \tag{7.4.7}$$

which is just Eq. (7.3.4) with x in place $\langle\sigma\rangle$. But $\langle\sigma\rangle$ can be found from the definitions Eq. (7.4.2) as

$$\langle\sigma\rangle = \frac{N_+ - N_-}{N} = x$$

so the identification is complete. All further calculations of results are exactly those of Sec. 3.

The reason for the equivalence of these apparently different approaches is that they both ignore short-range order and consider only the long-range, overall average ordering of the lattice. The Weiss theory looks at a representative spin in a lattice of average spins, and the B-W theory calculates numbers on n.n. pairs as random pairings, completely free of local influence.

This same approximation can be treated in a somewhat more transparent manner and, at no further expense, slightly more generally. Suppose the probability of a site j having the spin state σ to be $p_j(\sigma)$, and assume further that $p_i(\sigma)$ and $p_j(\sigma)$ are independent for all i and j. The interaction energy may be written $U_{ij}(\sigma_i, \sigma_j)$, usually assumed symmetric. Since the p_i's are independent, the individual site entropies are additive, and the total entropy is

$$S = -Nk \sum_{\sigma=\pm 1} p(\sigma)\ln p(\sigma). \tag{7.4.8}$$

The total energy may be similarly written

$$E = \frac{N\nu}{2}\sum_\sigma \sum_{\sigma'} u(\sigma,\ \sigma')p(\sigma)p(\sigma') - NmB \sum_\sigma \sigma p(\sigma), \tag{7.4.9}$$

where σ' is a nearest neighbor spin. Now S is maximized subject to $\sum_\sigma p(\sigma) = 1$ and $E = $ constant by the technique of Sec. (1:4) as follows:

$$\delta\left[S/k - \beta E - N\Lambda\sum_\sigma p(\sigma)\right] = -N\sum_\sigma \delta p(\sigma)\left[1 + \ln p(\sigma)\right]$$

$$-\frac{\beta N\nu}{2}\sum_{\sigma,\ \sigma'} u(\sigma,\ \sigma')\left[p(\sigma)\delta p(\sigma') + p(\sigma')\delta p(\sigma)\right]$$

$$+ \beta NmB\sum_\sigma \sigma\delta p(\sigma) - N\Lambda\sum_\sigma \delta p(\sigma) = 0. \tag{7.4.10}$$

The double sum over all spins counts each $\sigma\sigma'$ pair twice, so (7.4.10) may be written

$$0 = \sum_\sigma \delta p(\sigma)\left[1 + \Lambda + \ln p(\sigma) + \beta\nu\sum_{\sigma'} u(\sigma,\ \sigma')p(\sigma') + \beta mB\sigma\right],$$

with the $\delta p(\sigma)$ now regarded as independent. The result for $p(\sigma)$ is

$$p(\sigma) = C \exp \left[-\beta(\nu \sum_{\sigma'} u(\sigma, \, \sigma') - mB\sigma) \right], \qquad (7.4.11)$$

where C is a normalization constant. Now suppose that $u(\sigma, \, \sigma') = -J\sigma\sigma'$, with $\sigma = \pm 1$, and carry out $\sum_{\sigma'}$ in Eq. (7.4.11) to get

$$\begin{aligned} p(\sigma) &= C \exp \left[\beta\sigma\{\nu J \langle \sigma \rangle + mB\} \right], \\ &= C \left(\cosh \left[\beta \left(\nu J \langle \sigma \rangle + mB \right) \right] + \sigma \sinh \left[\beta \left(\nu J \langle \sigma \rangle + mB \right) \right] \right), \\ &= C \cosh \left[\beta \left(\nu J \langle \sigma \rangle + mB \right) \right] \left[1 + \sigma \tanh \left[\beta \left(\nu J \langle \sigma \rangle + mB \right) \right] \right]. \quad (7.4.12) \end{aligned}$$

But since $p(+1) + p(-1) = 1$, C can be found from (7.4.12), and it follows that

$$p(\sigma) = \frac{1}{2} \left[1 + \sigma \tanh \left[\beta \left(\nu J \langle \sigma \rangle + mB \right) \right] \right]. \qquad (7.4.13)$$

Then, since $\langle \sigma \rangle = \sum \sigma p(\sigma)$, we again have

$$\langle \sigma \rangle = \tanh \left[\beta \left(\nu J \langle \sigma \rangle + mB \right) \right], \qquad (7.4.14)$$

as in (7.3.4), and all of the results of Sec. 3 may be obtained.

In this treatment, the independence of probabilities on neighboring sites was made explicit, and the exclusion of short-range order from the calculation is most evident.

In its original context, the B.W. calculation considers two types of lattice sites, α and β, and two types of atoms, A and B (Bragg and Williams [1934]). Normally atoms A are at home on α sites and B on β. In this example, the numbers $N(A)$ and $N(B)$ will be taken equal and half the sites are of each type. Two types of energy are considered. First, when an atom is on its home site its energy is assigned to be zero, but if it is on the other type of site, its site energy is ϵ. Second, the interaction energy will be written so that each A prefers to be surrounded by B atoms and vice versa. In the Ising model notation, the σ variable is $+$ if A is on an α site or B on a β, $-$ if A is on a β or B on an α. The total energy for a configuration is then written

$$E = -\frac{1}{2} \sum_{i, \, j=1}^{N} J_{ij}\sigma_i\sigma_j + \frac{1}{2} \sum_{i}(1 - \sigma_i)\epsilon, \qquad (7.4.15)$$

where $J_{ij} = J$ for n.n. pairs, zero otherwise. In the B-W notation, the energy becomes

$$E = J \left[(N_{AA} + N_{BB}) - (N_{AB} + N_{BA}) \right] + \epsilon \left(N_\alpha(B) + N_\beta(A) \right). \qquad (7.4.16)$$

With the numbers of α sites and β sites both equal to N/2, the site occupancies are written

$$N_\alpha(A) = \frac{N}{4}(1 + x) = N_\beta(B)$$

$$N_\alpha(B) = \frac{N}{4}(1 - x) = N_\beta(A), \qquad (7.4.17)$$

and the numbers of n.n. pairs are written

$$N_{AA} = \gamma N_\alpha(A)N_\beta(A); \qquad N_{BB} = \gamma N_\alpha(B)N_\beta(B);$$
$$N_{AB} = \frac{\gamma}{2}\left[N_\alpha(A)N_\beta(B) + N_\alpha(B)N_\beta(A)\right] = N_{BA}.$$

From the total number of pairs, $N\nu/2$, γ is found to be $2\nu/N$, giving

$$N_{AA} = \frac{N\nu}{8}(1 - x^2) = N_{BB}, \text{ and } N_{AB} = \frac{N\nu}{8}(1 + x^2) = N_{BA} \qquad (7.4.18)$$

and

$$E = -\frac{JN\nu}{2}x^2 + \frac{N\epsilon}{2}(1 - x). \qquad (7.4.19)$$

The entropy is written

$$S/k = N\ln N - N_\alpha(A)\ln N_\alpha(A) - N_\alpha(B)\ln N_\alpha(B) - N_\beta(A)\ln N_\beta(A) - N_\beta(B)\ln N_\beta(B)$$
$$= -\frac{N}{2}\left[(1 + x)\ln\left(\frac{1 + x}{4}\right) + (1 - x)\ln\left(\frac{1 - x}{4}\right)\right], \qquad (7.4.20)$$

and the temperature equation, in analogy to Eq. (7.4.16), is

$$\beta = \frac{1}{2J\nu x + \epsilon}\ln\left(\frac{1 + x}{1 - x}\right), \text{ or } x = \tanh\beta(J\nu x + \epsilon/2). \qquad (7.4.21)$$

In a more realistic calculation, the site energy ϵ should be order-dependent, since in a highly disordered lattice there is little difference in energy between sites, but as order increases the difference becomes more marked. If one chooses to attribute the entire site energy to interaction, ϵ can be set equal to zero. A number of different physical arguments can lead to variations of the above calculation, but the basic procedure is as presented. The unrealistic treatment of short-range order accounts for the relative ease of calculation, and for the ultimate inaccuracy of the results.

7.5 EXACT ONE-DIMENSIONAL ISING MODEL

7.5.1 Direct Solution. The simple Ising problem in one dimension can be solved directly in several ways. First the chain is considered as open ended and B=0. The partition function is given by

$$Z_N = \sum_{\{\sigma_i = \pm 1\}} \exp K \sum_{i=1}^{N-1} \sigma_i\sigma_{i+1}, \qquad (7.5.1)$$

where $\beta J = K$. The exponential can be factored as a product of terms of the form $\exp K\sigma_i\sigma_{i+1}$, each of which can be written as

$$\exp K\sigma_i\sigma_{i+1} = \cosh K + \sigma_i\sigma_{i+1}\sinh K$$
$$= \cosh K(1 + \sigma_i\sigma_{i+1}y), \qquad (7.5.2)$$

where $y = \tanh K$. The partition function then becomes

$$Z_N = \cosh^{N-1} K \sum_{\{\sigma_i = \pm 1\}} \prod_{i=1}^{N-1} (1 + \sigma_i \sigma_{i+1} y)$$

$$= (\cosh K)^{N-1} \sum_{\{\sigma_i = \pm 1\}} (1 + \sigma_1 \sigma_2 y)(1 + \sigma_2 \sigma_3 y) \cdots (1 + \sigma_{N-1} \sigma_N y). \qquad (7.5.3)$$

Summation over $\sigma_1 = \pm 1$ gives $(1 + \sigma_2 y)(\cdots) + (1 - \sigma_2 y)(\cdots)$, where $(\cdots)$ represents

$$\prod_{i=2}^{N-1} (1 + \sigma_i \sigma_{i+1} y),$$

and the indicated summation yields $2(\cdots)$. Next, summation over $\sigma_2 = \pm 1$ gives another factor 2 and a product of $N - 3$ terms. Continued summations finally produce the result

$$Z_N = (2 \cosh K)^{N-1}, \qquad (7.5.4)$$

or

$$\lim_{N \to \infty} (\ln Z_N)/N = \ln 2 \cosh K,$$

$$\langle \epsilon \rangle = \lim_{N \to \infty} \left[-\frac{1}{N} \frac{\partial \ln Z_N}{\partial \beta} \right] = -J \tanh K, \qquad (7.5.5)$$

and the heat capacity C is

$$C = N_0 \frac{\partial \langle \epsilon \rangle}{\partial T} = RK^2 (1 - \tanh^2 K). \qquad (7.5.6)$$

The energy and heat capacity are smoothly varying, always finite functions of temperature, exhibiting no phase transition. Thus the molecular-field–Bragg-Williams approximation is incorrect, no matter how plausible, for a one-dimensional system, and its validity in n-dimensions then is immediately to be doubted.

Another direct technique for the open, one-dimensional chain makes use of a change in variables to $\tau_i = \sigma_i \sigma_{i+1}$, $1 \le i \le N-1$, and $\tau_N = \sigma_N$, where it may be seem from Eq. (7.5.1) or (7.5.3) that Z_N is independent of τ_N. For the open chain, the τ's are independent, and Z_N can be written as

$$Z_N = \sum_{\{\tau_i = \pm 1\}} \prod_{i=1}^{N-1} e^{K \tau_i} = (2 \cosh K)^{N-1}, \qquad (7.5.7)$$

where each sum on a τ_i yields a factor $2 \cosh K$, and the final result is the product of $N - 1$ such independent factors, in agreement with Eq. (7.5.4). For closed chains in one dimension and for Ising lattices of higher dimension, no such simple technique will work because the τ_i's are no longer independent.

The final direct calculation for the open chain is more difficult because $B \neq 0$, but more interesting as well. When $B \neq 0$, Eq. (7.5.3) becomes

$$Z_N = (\cosh K)^{N-1}(\cosh x)^N \sum_{\{\sigma_i = \pm 1\}} \prod_{i=1}^{N} (1 + \sigma_i \sigma_{i+1} y)(1 + \sigma_i z), \qquad (7.5.8)$$

where $x = \beta m B$, $z = \tanh x$, and $\sigma_{N+1} = 0$. Summation over σ_1 gives

$$Z_N = (\cosh K)^{N-1}(\cosh x)^N \sum_{\{\sigma_i = \pm 1\}} 2(1 + \sigma_2 yz) \prod_{i=2}^{N} (1 + \sigma_i \sigma_{i+1} y)(1 + \sigma_i z), \qquad (7.5.9)$$

where the term before $\prod$ comes from the σ_1 summation. After $n < N$ such summations Z_N will have the form

$$Z_N = (\cosh K)^{N-1}(\cosh x)^N \sum_{\{\sigma_i = \pm 1\}} 2^n (a_n + b_n \sigma_{n+1}) \prod_{i=n+1}^{N} (1 + \sigma_i \sigma_{i+1} y)(1 + \sigma_i z), \qquad (7.5.10)$$

where a_n and b_n are functions of y and z. The summation over σ_{n+1} gives

$$Z_N = (\cosh K)^{N-1} (\cosh x)^N \sum_{\{\sigma_i = \pm 1} 2^{n+1} \left[a_n + z b_n + (a_n yz + b_n y) \sigma_{n+2} \right] \times$$

$$\prod_{i=n+2}^{N} (1 + \sigma_i \sigma_{i+1} y)(1 + \sigma_i z), \qquad (7.5.11)$$

and identification with the form of Eq. (7.5.10) gives the matrix equation

$$\begin{pmatrix} a_{n+1} \\ b_{n+1} \end{pmatrix} = \begin{pmatrix} 1 & z \\ yz & y \end{pmatrix} \begin{pmatrix} a_n \\ b_n \end{pmatrix} = \begin{pmatrix} 1 & z \\ yz & y \end{pmatrix}^{n+1} \begin{pmatrix} 1 \\ 0 \end{pmatrix}, \qquad (7.5.12)$$

where $a_0 = 1$, $b_0 = 0$ are read off from Eq. (7.5.8). After N such summations, with $\sigma_{N+1} = 0$, the result is

$$Z_N = (\cosh K)^{N-1}(2 \cosh x)^N a_N, \qquad (7.5.13)$$

and the entire problem becomes that of calculating a_N from Eq. (7.5.12). It is convenient to denote the 2×2 matrix as

$$A = \begin{pmatrix} 1 & z \\ yz & y \end{pmatrix}, \qquad (7.5.14)$$

and to observe that $A = BC$, where

$$B = \begin{pmatrix} 1 & 0 \\ 0 & y \end{pmatrix}, \quad \text{and} \quad C = \begin{pmatrix} 1 & z \\ z & 1 \end{pmatrix}. \qquad (7.5.15)$$

The matrix $B^{1/2}$ is defined as

$$B^{1/2} = \begin{pmatrix} 1 & 0 \\ 0 & \sqrt{y} \end{pmatrix}, \qquad (7.5.16)$$

and a matrix A_1 is defined as

$$A_1 = B^{1/2} C B^{1/2} = \begin{pmatrix} 1 & z\sqrt{y} \\ z\sqrt{y} & y \end{pmatrix}, \tag{7.5.17}$$

which has the same eigenvalues as A, but is symmetric. The problem of finding a_N is just that of calculating A^N, which can be written

$$
\begin{aligned}
A^N &= BCBC \cdots BC, \quad N \text{ times} \\
&= B^{1/2} \left(B^{1/2} C B^{1/2} \right) \left(B^{1/2} C B^{1/2} \right) \cdots \left(B^{1/2} C B^{1/2} \right) \left(B^{1/2} C \right) \\
&= B^{1/2} A_1^N B^{-1/2}.
\end{aligned}
\tag{7.5.18}
$$

Since A_1 is a real symmetric matrix, it can be diagonalized by an orthogonal transformation

$$T = \begin{pmatrix} c & s \\ -s & c \end{pmatrix}, \tag{7.5.19}$$

where c and s are the cosine and sine of the rotation angle. The transformation is

$$\Lambda = T A_1 T^{-1} = \begin{pmatrix} \lambda_1 & 0 \\ 0 & \lambda_2 \end{pmatrix}, \tag{7.5.20}$$

where λ_1 and λ_2 are the eigenvalues of A_1 (and of A), with the consequence that

$$A_1^N = T^{-1} \begin{pmatrix} \lambda_1^N & 0 \\ 0 & \lambda_2^N \end{pmatrix} T. \tag{7.5.21}$$

From Eq. (7.5.18), then, the needed result is

$$
\begin{aligned}
A_N &= B^{1/2} T^{-1} \Lambda^N T B^{-1/2} \\
&= \begin{pmatrix} c^2 \lambda_1^N + s^2 \lambda_2^N & (cs/\sqrt{y})(\lambda_1^N - \lambda_2^N) \\ (cs\sqrt{y})(\lambda_1^N - \lambda_2^N) & s^2 \lambda_1^N + c^2 \lambda_2^N \end{pmatrix},
\end{aligned}
\tag{7.5.22}
$$

and

$$a_N = c^2 \lambda_1^N + s^2 \lambda_2^N = c^2 \lambda_1^N \left[1 + t^2 (\lambda_2/\lambda_1)^N \right], \tag{7.5.23}$$

where $t = \tan = s/c$. The use of this result in Eq. (7.5.13) leads to

$$\lim_{N \to \infty} (\ln Z_N / N) = \ln \cosh K + \ln(2 \cosh x) + \ln \lambda_1, \tag{7.5.24}$$

where λ_1 is the larger of the two eigenvalues. The eigenvalues of A are easily found to be

$$\lambda_{1,\, 2} = (1/2)\{1 + y \pm [(1 - y)^2 + 4yz^2]^{1/2}. \tag{7.5.25}$$

When $B = 0$, both x and z are zero, and the partition function reduces to Eq. (7.5.7). The average spin $\langle \sigma \rangle$ is given by

$$\langle \sigma \rangle = \frac{\partial}{\partial x} \left(\lim_{N \to 0} \frac{\ln Z_N}{N} \right) = z \frac{(1 + y)^2 + (1 + y)[(1 - y)^2 + 4zy^2]^{1/2}}{(1 - y)^2 + 4yz^2 + (1 + y)[(1 - y)^2 + 4yz^2]^{1/2}}, \tag{7.5.26}$$

a result that goes to zero when $B = 0$ at all finite strengths of interaction, at all positive temperatures. Thus it is again seen that there is no phase transition, this time by the absence of spontaneous magnetization.

Shorter calculations will be developed for the same problem, but the technique used in this direct evaluation of Z_N can be employed instructively in the study of double and triple Ising chains with no external field. Such problems are left as exercises.

When the chain is closed, with $\sigma_{N+1} = \sigma_1$, direct evaluation of Z_N becomes slightly more difficult, but other procedures are often simpler for the closed chain than for the open one. For the closed chain, Eq. (7.5.3) becomes

$$Z_N = \cosh^N K \sum_{\{\sigma_i = \pm 1\}} \prod_{i=1}^{N} (1 + \sigma_i \sigma_{i+1} y); \quad \sigma_{N+1} \equiv \sigma_1. \tag{7.5.27}$$

Each σ_i now appears twice, and the product of the two terms containing σ_j is $1 + (\sigma_{j-1}\sigma_j + \sigma_j\sigma_{j+1})y + \sigma_{j-1}\sigma_{j+1}y^2$. Summation over $\sigma_j = \pm 1$ gives the factor $2(1 + \sigma_{j-1}\sigma_{j+1}y^2)$, and repetition of this process for all spins gives

$$Z_N = (2\cosh K)^N (1 + y^N), \tag{7.5.28}$$

a result that differs from that of Eq. (7.5.4). In the limit of very large N, however, the y^N contribution becomes vanishingly small, since $y = \tanh K < 1$ for all finite J and β.

7.5.2 Transfer-Matrix Method. The spin part of the Hamiltonian $\mathcal{H}$ can be written quite generally for a one-dimensional chain as

$$\mathcal{H} = \sum_{i=1}^{N} [u(\sigma_i, \sigma_{i+1}) + v(\sigma_i)], \tag{7.5.29}$$

where the interaction energy $u(\sigma_i, \sigma_{i+1})$ is usually a symmetric function of the two spins, and $v(\sigma_i)$ is the energy of σ_i, perhaps in a field B. It is sometimes convenient to write the v term as $v(\sigma_i, \sigma_{i+1}) = (1/2)[v(\sigma_i) + v(\sigma_{i+1})]$, for symmetry, and to work with closed chains, so that $\sigma_1 = \sigma_{N+1}$. The partition function can then be written as

$$Z_N = \sum_{\{\sigma_i = \pm 1\}} \prod_{i=1}^{N} \exp\{-\beta u(\sigma_i, \sigma_{i+1}) - (\beta/2)[v(\sigma_1) + v(\sigma_{i+1})]\}$$

$$= \sum_{\{\sigma_i = \pm 1\}} \langle \sigma_1 | P | \sigma_2 \rangle \langle \sigma_2 | P | \sigma_3 \rangle \cdots \langle \sigma_N | P | \sigma_1 \rangle, \tag{7.5.30}$$

where the transfer matrix P is given by

$$P = \begin{bmatrix} \exp[-\beta u(+1, +1) - \beta v(+1)] & \exp[\beta u(+1, -1) - (\beta/2)v(+1, -1)] \\ \exp[-\beta u(-1, +1) - (\beta/2)v(-1, +1)] & \exp[-\beta u(-1, -1) - \beta v(-1)] \end{bmatrix}. \tag{7.5.31}$$

The summation over all $\sigma_i = \pm 1$ in Eq. (7.5.31) permits Z_N to be written as a matrix product, and closing the chain makes the final sum over the values of σ_1 just the operation of taking the trace, or

$$Z_N = \operatorname{Tr} P^N. \tag{7.5.32}$$

Since P is a real symmetric matrix, it may be diagonalized by an orthogonal transformation, which preserves its trace. Then $\operatorname{Tr} P^N$ can be written as $\sum \lambda_i^N$, and Eq. (7.5.32) becomes

$$Z_N = \lambda_1^N + \lambda_2^N = \lambda_1^N [1 + (\lambda_2/\lambda_1)^N]$$
$$\rightarrow \lambda_1^N \text{ for very large } N, \tag{7.5.33}$$

when λ_1 is the larger eigenvalue. Thus the problem can be solved generally in one dimension by the simple procedure of finding the eigenvalues of a 2×2 matrix.

As an example, let $u(\sigma_i, \sigma_{i+1}) = -J\sigma_i\sigma_{i+1}$ and $v(\sigma_i) = -mB\sigma_i$, to duplicate the problem already solved in the previous subsection. The transfer matrix then becomes, with $\beta J = K$ and $\beta m B = x$,

$$P = \begin{pmatrix} e^{K+x} & e^{-K} \\ e^{-K} & e^{K-x} \end{pmatrix}, \tag{7.5.34}$$

with eigenvalues

$$\lambda_{1,\,2} = e^K \cosh x \pm [e^{2K} \cosh^2 x - 2\sinh 2K]^{1/2},$$

or

$$\lambda_1 = \cosh x \cosh K \{1 + y + [(1 - y)^2 + 4yz^2]^{1/2}\}, \tag{7.5.35}$$

where $y = \tanh K$ and $z = \tanh x$, as before. In the limit as $N \rightarrow \infty$, Eq. (7.5.33) gives

$$\lim_{N \to \infty} [\ln Z_N/N] = \ln \lambda_1 = \ln \cosh x + \ln \cosh K +$$
$$+ \ln\{(1 + y + [(1 - y)^2 + 4yz^2]^{1/2}\}, \tag{7.5.36}$$

in agreement with Eq. (7.5.24), when Eq. (7.5.25) is noted.

More generally, the spin variables in Eq. (7.5.29) may be made to represent a greater range of choices than permitted by the restriction $\sigma_i = \pm 1$. In the next subsection, the transfer matrix method is used to solve a problem with four states of the σ's, and other examples are to be found in the exercises. For any configuration of nearest-neighbor interactions, Z_N can be written as in Eq. (7.5.30). For a full two-dimensional $M \times N$ Ising lattice, for example, each σ_i can represent the spins of an entire line of M lattice elements, each of which can take the values ± 1. Thus σ_i could be replaced by the symbol $s(\{\sigma_i\})$, where $\{\sigma_i\}$ is the set of M primitive σ's, of the ith column of the lattice, and each s therefore takes on 2^M distinct values. The transfer matrix P is of dimension $2^M \times 2^M$, and the two-dimensional Ising problem becomes that of finding the largest eigenvalue of P when $M \rightarrow \infty$. The original Onsager solution to the two-dimensional square Ising lattice proceeded in this fashion. (Kaufman [1949]; Kaufman and Onsager [1949]; Onsager [1944]) Considerable algebraic skill was required in order to produce the solution. (Marked simplification has been effected by means of techniques introduced more recently.) (Schultz, Mattis and Lieb [1964]; Thompson [1965]) For the three-dimensional cubic lattice, the matrices become $2^{M^2} \times 2^{M^2}$, and the algebra seems hopelessly forbidding. In the two-dimensional lattice calculation by the transfer-matrix method, Onsager showed that the problem could be reduced to the diagonalization of 4×4 blocks in the 2^M square matrix, and further such insight may possibly permit progress in the exact solution of the cubic lattice, but other more transparent methods may offer better opportunities.

7.5.3 Linear KDP. As an instructive example of a one-dimensional system that exhibits a phase transition, John F. Nagle devised a simplification of the Slater model of KDP (KH_2PO_4). (Nagle [1968]) KDP is typical of hydrogen-bonded solids in that each proton is shared by two negative ions (as, for example, in ice), and a proton has two possible equilibrium positions on its bond, typically about one-third the distance from one end or the other. At equilibrium, the Bernal-Fowler *ice condition*

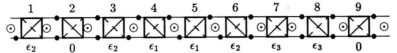

Figure 7.3 Energy assignments for linear model of KDP. The crossed squares represent the (PO_4) tetrahedrons, the ⊙ the K^+ ions, and the ● the protons.

demands that each phosphate ion have exactly two protons near it. (Bernal and Fowler, Jr. [1933]) In order to model this condition, Nagle chose the energies of all configurations that violate it to be infinite; thus in Fig. 7.3, infinite values are assigned to ϵ_2 in positions 1, 3, and 6, with one or three near protons, and to ϵ_3 in positions 7 and 8, with zero or four near protons. Zero energy is given to configurations 2 and 9, with the two near protons on the same side of the PO_4, and the finite but positive energy ϵ_1 is assigned to the configurations 4, 5, and others with the two near protons on opposite sides of the PO_4. A physical argument for setting $\epsilon_1 > 0$ is that the energy of the $K^+H^+H^+$ configuration is higher when the K^+ is directly between the protons than when it can be moved off to one side (Lieb [1967a]; Lieb [1967b]; Lieb [1967c]; Nagle [1966]).

The partition function can be written

$$Z_N = \sum_{E_N} g(E_N)e^{-\beta E_N}, \tag{7.5.37}$$

where $g(E_N)$ is the number of configurations with the energy E_N.

The linear chain may be closed to eliminate end effects. The only non-vanishing terms that contribute to Z_N are those for which all energies are finite; *i.e.* those that do not violate the ice condition. The energy $E_N = 0$ can be achieved in exactly two ways, with all unit cells like either cell 2 or cell 9; thus $g(0) = 2$. The only other possible finite energy is $N\epsilon_1$, for which $g(N\epsilon_1) = 2^N$, since there are two possible ϵ_1-positions for each proton pair between phosphates. Therefore the partition function becomes

$$Z_N = 2 + 2^N e^{-N\beta\epsilon_1}, \tag{7.5.38}$$

and the energy expectation value is

$$U_N = -\frac{\partial \ln Z_N}{\partial \beta} = N\epsilon_1 \frac{\left(2e^{-\beta\epsilon_1}\right)^N}{2 + \left(2e^{-\beta\epsilon_1}\right)^N}$$

$$= N\epsilon_1 \left[1 + 2\left(e^{\beta\epsilon_1}/2\right)^N\right]^{-1}. \tag{7.5.39}$$

The average energy per cell, as $N \to \infty$, becomes

$$\langle \epsilon \rangle = \epsilon_1, \quad \text{for} \quad e^{-\beta\epsilon_1} > 0.5$$
$$= 0, \quad \text{for} \quad e^{-\beta\epsilon_1} < 0.5. \tag{7.5.40}$$

There is a phase transition, then, at $kT_c = \epsilon_1/\ln 2$, such that for $T > T_c$ the populous ϵ_1 configurations exist throughout, and for $T < T_c$, one of the two zero-energy states must exist. When N is large enough for the transition to take place in an extremely narrow temperature range, the entropy (found from $F_N = -kT \ln Z_N = U_N - TS_N$) is given by

$$S_N = k \ln 2; \quad T < T_c,$$
$$= Nk \ln 2; \quad T > T_c, \tag{7.5.41}$$

showing the change from 2 to 2^N configurations at $T = T_c$. The latent heat, given by $\ell = T_c \Delta S_N$, is thus seen to be $(N-1)\epsilon_1$, almost as expected.

The ground-state entropy, $S_N = k \ln 2$, arises from the two configurations, corresponding to all like cell 2 or all like cell 9, being regarded as different. But if the linear chain were macroscopically turned end for end, or the corresponding closed loop were rotated 180° about any axis in its plane, one configuration would be converted into the other without any microscopic transition having taken place. Since different macroscopic orientations are not customarily regarded as distinguishable states (otherwise there may be infinities of them), the partition function Eq. (7.5.38) should properly read

$$Z_N = 1 + 2^{N-1} e^{-N\beta\epsilon_1}. \tag{7.5.42}$$

No change results from this modification except in Eq. (7.5.41), where $S_N = k \ln 1 = 0$ for $T < T_c$, but $\ell = (N-1)\epsilon_1$, as before. The coefficient $N-1$, instead of N, in ℓ arises from the ratio $g(N\epsilon_1)/g(0)$, and no apparent logic permits the ratio to be 2^N, though consideration of the top-to-bottom interchanges in the linear chain (or turning the ring inside out) could reduce the ratio to nearly 2^{N-2}. Thus there is no support in the value of ℓ for requiring the ground state to be nondegenerate.

A more realistic assignment of energies, permitting Bjerrum defects (Bjerrum [1951]), is $0 < \epsilon_1 << \epsilon_2 < \epsilon_3 = 2\epsilon_2 < \infty$. In this case many energies are possible, and the simple counting of $g(N)$ becomes tedious. The transfer-matrix method will be seen to reduce the problem to an easy solution. If we set $e^{-\beta\epsilon_1} = x$, $e^{-\beta\epsilon_2} = y$, $e^{-\beta\epsilon_3} = y^2$, since $\epsilon_3 = 2\epsilon_2$, denote the positions of the two protons in the space to the left of the ith phosphate tetrahedron by a pair of left-right symbols, and those to *its* right (or to the left of the $(i+1)^{st}$ phosphate) by a similar pair, we may construct the transfer matrix to be

$$
\begin{array}{cc}
& \begin{array}{cccc} LL & LR & RL & RR \end{array} \\
\begin{array}{c} LL \\ LR \\ RL \\ RR \end{array} &
\left[\begin{array}{cccc}
1 & y & y & y^2 \\
y & x & x & y \\
y & x & x & y \\
y^2 & y & y & 1
\end{array} \right]
\end{array}
\tag{7.5.43}
$$

The eigenvalues of this matrix are found to be $\lambda = 0$, $1 - y^2$, and $(1/2)\{1 + 2x + y^2 \pm [(1 - 2x)^2 + 18y^2 - 4xy^2 + 4y^4]^{1/2}\}$, with the largest eigenvalue corresponding to the $+$ in the quadratic pair. Therefore, for $N \to \infty$, by Eq. (7.5.33),

$$\lim_{N \to \infty} [(\ln Z_N/N)] = \ln \lambda_1$$

$$= -\ln 2 + \ln \left\{ 1 + 2x + y^2 + \left[(1 - 2x)^2 + y^2(18 - 4x + 2y^2) \right]^{1/2} \right\}. \tag{7.5.44}$$

The energy and heat capacity can be calculated from Eq. (7.5.44) by standard procedures. There is no phase transition for finite ϵ_2, but as the ratio $\gamma = \epsilon_2/\epsilon_1$ increases, the behavior approximates ever more closely the results obtained for infinite ϵ_3. A plot of normalized heat capacity *vs.* x for four values of γ, as seen in Fig. 7.4, shows incipient anomaly in C at $x = 0.5$ for large γ.

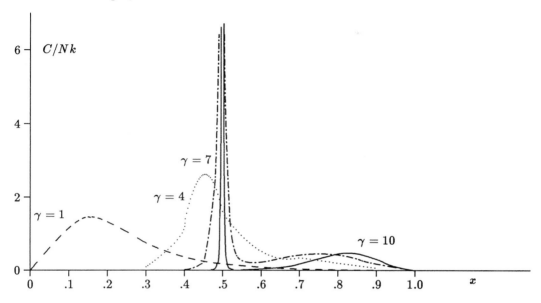

Figure 7.4 Plot of normalized heat capacity *vs.* $x = e^{-\beta\epsilon_1}$ for various values of $\gamma = \epsilon_2/\epsilon_1$ for the linear KDP model, showing the approach to a phase transition at $x = 0.5$ as γ increases (Nagle [1968]). For $\gamma = 7$, the maximum value is 8.564, at $x = .4968$; for $\gamma = 10$, the maximum value is 61.73, at $x = .49992$.

Several similar systems in one dimension are to be found in the exercises. The exact solutions to two models of KDP in two dimensions have been found by transfer-matrix methods, and other problems related to hydrogen-bonded compounds have been similarly solved (Lieb [1967a]; Lieb [1967b]; Lieb [1967c]).

7.6 COMBINATORIAL PRELIMINARIES IN TWO DIMENSIONS

The partition function for the simple Ising lattice in the absence of an applied magnetic field, and for only nearest-neighbor interactions, is

$$Z_N = \sum_{\{\sigma_{i,j}=\pm 1\}} \prod_{nn} e^{K\sigma_i\sigma_j} = (\cosh K)^{N\nu/2} \sum_{\sigma_{i,j}=\pm 1} \prod_{nn} (1 + y\sigma_i\sigma_j) \qquad (7.6.1)$$

where the product is taken over nearest-neighbor pairs and the notation of Eq. (7.5.2) was used. Expanded, Z_N becomes

$$Z_N = (\cosh K)^{N\nu/2} \sum_{\sigma_{i,j}=\pm 1} \left[1 + y\sum_{nn} \sigma_i\sigma_j + y^2 \left(\sum_{nn} \sigma_i\sigma_j\right) \left(\sum_{nn} \sigma_k\sigma_l\right) + \cdots \right]$$

$$= 2^N \left(\cosh K \right)^{N\nu/2} \sum_{r=0}^{\infty} n(r) y^r, \tag{7.6.2}$$

where the coefficient of y^r is the sum of all products of $2r$ σ's. The σ's always occur in n.n. pairs, and no *pair* may appear in any product more than once. The second line is obtained after summation over the $\sigma_i = \pm 1$. Any term containing a σ_j to an odd power vanishes upon summation on that σ_j, since $\sum_{\sigma=\pm 1} \sigma = 0$, but since $\sum_{\sigma=\pm 1} \sigma^2 = \sum_{\sigma=\pm 1} \sigma^{2n} = 2$ for all integer values of n, including $n = 0$, all nonvanishing terms contain the factor 2^N. Furthermore, every term that survives in the second line can be put in one-to-one correspondence with a closed graph on the lattice, with $\sigma_i \sigma_j$ represented by a bond connecting point i to point j (they must be nearest neighbors in this introductory presentation); the requirement that the graph be closed corresponds exactly to the requirement that each surviving σ must appear to an even power. A graph of r bonds may be made up of intersecting subgraphs, but these subgraphs cannot have bonds in common, since these would correspond to one or more n.n. pairs appearing twice in the same product. A typical graph shown in Fig. 7.5 corresponds to a thirty-σ product: (3,4) (3,9) (4,10) (5,6) (5,11) (6,12) (7,8) (7,13) (8,9) (10,16) (11,12) (13,19) (14,15) (14,20) (15,16) (16,17) (16,22) (17,23) (19,25) (20,26) (21,22) (21,27) (23,29) (25,26) (26,27) (26,32) (29,35) (32,33) (33,34) (34,35). The term $n(r)$ in the last line of Eq. (7.6.2) is the number of closed graphs of r bonds that can be drawn on the lattice. The partition function is known if all the $n(r)$ can be counted. Since $y = \tanh \beta J$, it is evident that at high temperatures, y is small, and a good high-temperature expansion

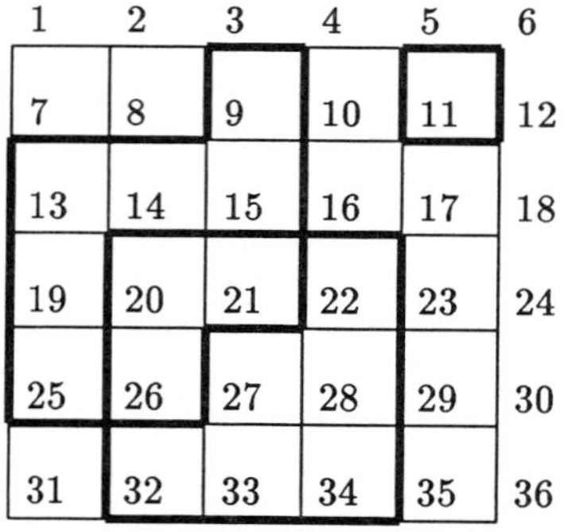

Figure 7.5 A thirty-bond graph made up of two overlapping closed graphs and an isolated four-bond one.

of the partition function may be obtained from direct counting of small-r graphs. This topic is treated in more detail in a subsequent section.

In a rectangular $N \times M$ lattice, with different coupling constants in the two directions, and $\nu = 2 + 2$, Eq. (7.6.2) becomes

$$Z_{MN} = \left(2 \cosh K_1 \cosh K_2 \right)^{MN} \sum_{r,s=0}^{\infty} n(r,s) y_1^r y_2^s, \tag{7.6.3}$$

where $n(r,s)$ is the number of closed graphs with r horizontal bonds and s vertical ones.

In the following sections, techniques are presented for counting $n(r,s)$. Interesting results may be obtained without such counting, including the correct evaluation of the critical temperature on a two-dimensional lattice. The basic procedure may be demonstrated for a square lattice with coordination number $\nu = 4$.

Let $N_1 =$ the number of σ's that are $+1$, and N_{12} the number of *unlike* n.n. pairs. Since the total number of n.n. pairs is $2N$, we may write

$$\sum_{nn} \sigma_i \sigma_j = (2N - N_{12})(+1) + N_{12}(-1), \qquad (7.6.4)$$

and

$$\sum \sigma_i = N_1(+1) + (N - N_1)(-1) = 2N_1 - N. \qquad (7.6.5)$$

With $\beta J = K$ and $\beta m B = x$, the partition function is

$$
\begin{aligned}
Z_N &= \sum_{N_1, N_{12}} g(N, N_1, N_{12}) \exp[2K(N - N_{12}) + (2N_1 - N)x] \\
&= e^{2NK} e^{-Nx} \sum_{N_1, N_{12}} g(N, N_1, N_{12}) e^{-2(KN_{12} - xN_1)},
\end{aligned}
\qquad (7.6.6)
$$

where $g(N, N_1, N_{12})$ is the number of ways of choosing N_1 and N_{12} on a lattice of N sites. Much futile effort has gone into its evaluation. When $x = 0$, Eq. (7.6.6) becomes

$$Z_N = e^{2NK} \sum_{r=0}^{\infty} m(r) e^{-2Kr}, \qquad (7.6.7)$$

where $m(r)$ is the number of ways of choosing r unlike n.n. pairs on the lattice.

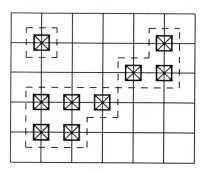

Figure 7.6 Correspondence between closed graphs on the dual lattice and unlike n.n. pairs on the original lattice.

In order to examine the problems of evaluating the $m(r)$, we select a simple square lattice and construct its *dual lattice*. The dual is constructed by connecting the midpoints of the unit cells; as shown in Fig. 7.6, the dual in this case is an identical square lattice displaced by a half unit cell vertically and horizontally. We now bisect each unlike bond on the original lattice by a line to adjacent points on the dual lattice, with the result that we have constructed closed graphs on the dual lattice with the same number of bonds as there are unlike n.n. pairs on the original lattice. Thus there is a one-to-one correspondence between graphs on the dual lattice and unlike n.n. pairs on the original lattice, and the problem of evaluating $m(r)$ in Eq. (7.6.7) is just that of finding $n(r)$ in Eq. (7.6.2), since $m(r) = n(r)$.

The problems may be made to look alike by the substitution $y^* = e^{-2K}$, so that Eq. (7.6.7) becomes, with $n(r)$ written for $m(r)$,

$$Z_N = (y^*)^{-N} \sum_{r=0}^{\infty} n(r)\,(y^*)^r = (\coth K^*)^N \sum_{r=0}^{\infty} n(r)(y^*)^r. \tag{7.6.8}$$

The series expansion in Eq. (7.6.8) should converge rapidly for $y^* \ll 1$, just as does Eq. (7.6.2) for $y \ll 1$. But small y^* means large $K = \beta J$, or low temperatures. Thus the expressions Eq. (7.6.2) and Eq. (7.6.8) can be regarded as high- and low-temperature expansions of Z_N. In general, even though $n(r)$ is the same in both expressions for all r, at any arbitrary temperature the two expressions converge at different rates, and are not identical term by term. But if a singularity exists at a temperature $T_c = J/kK$, one should also be found in the other series at $T_c = J/kK^*$, since the singularity must arise in the series itself, and not in its well-behaved outside multiplier. In that case, the two series expressions must be identical term by term, giving the two equations

$$\tanh K_c = \tanh K_c^*, \tag{7.6.9}$$

and

$$2\cosh^2 K_c = \coth K_c^*. \tag{7.6.10}$$

These combine to the result

$$\sinh 2K_c = \sinh 2K_c^* = 1, \tag{7.6.11}$$

with the numerical solution

$$K_c = \beta_c J = 0.4407, \tag{7.6.12}$$

or, if a singularity exists, it must occur at

$$T_c = 2.269\,J/k. \tag{7.6.13}$$

The molecular field model says that a singularity exists, with the critical temperature given in Eq. (7.3.6) as $T_c = 4J/k$, whereas Eq. (7.6.13) says that if a singularity exists, it must occur exactly at $T_c = 2.286\,J/k$. These results are clearly in disagreement.

Similar results can be obtained for other lattices in two dimensions, but the argument is particularly simple for the square lattice, which is self dual.

7.7 THE DIMER-PFAFFIAN METHOD

7.7.1 Foreword. The procedure to be detailed in this section is a simple and elegant method for counting the closed graphs on a lattice. The summation in the partition function, Eq. (7.6.3), may be written

$$\Phi(x,y) = \sum_{r,s} n(r,s)x^r y^s, \tag{7.7.1}$$

where $\Phi(x,y)$ is the generating function that counts the number of closed graphs on the lattice with r x-bonds and s y-bonds. The calculation of $\Phi(x,y)$ can be presented with startling brevity or with more detail than the nonspecialist cares to peruse. The exposition of this section and the next is intended to be a working compromise, short enough to show that the two-dimensional Ising problem is now rather easily solved, yet detailed enough to permit mastery of the entire procedure.

The mathematical apparatus involves little more than elementary matrix algebra. Pfaffians are assumed to be unfamiliar, and their evaluation is explained in the next subsection. Direct-product matrices may induce fear in the timid, but they should be recognized for what they are—a mere notational convenience—without any need for a separate course in their algebra. The presentation is lengthened by inclusion of specific examples, but formal proofs are omitted when these examples permit obvious and easily verified generalization. Supplementary references are given in case further details are of interest. These are listed in Section 7.13, and also in the Bibliography at the end of the book.

7.7.2 Pfaffians. A Pfaffian is an array of elements similar to a determinant in that it is evaluated to give a numerical result or its algebraic equivalent. Pfaffians were so named by Cayley, who found them well suited to the solution of a problem proposed by Pfaff. A Pfaffian is composed of the same elements a_{ij} that make up an antisymmetric determinant of even order, and it is usually written out as such a determinant with all elements deleted on and below the principal diagonal. Thus if the antisymmetric determinant of even order is

$$|D| = \begin{vmatrix} 0 & a_{12} & a_{13} & \cdots & a_{1,2N} \\ -a_{12} & 0 & a_{23} & \cdots & a_{2,2N} \\ \vdots & \cdots & \cdots & \cdots & \vdots \\ -a_{1,2N} & \cdots & & -a_{2N-1,2N} & 0 \end{vmatrix},$$

the corresponding Pfaffian, which is obtained by taking the elements above the diagonal, is seen to be

$$Pf = \begin{vmatrix} a_{12} & a_{13} & \cdots & a_{1,2N} \\ & a_{23} & \cdots & a_{2,2N} \\ & & & \vdots \\ & & & a_{2N-1,2N} \end{vmatrix}.$$

An antisymmetric determinant of even order is always a perfect square, or

$$|D| = q^2,$$

where q is made up of sums of products of the elements of D. For example, the antisymmetric determinant of order 4 is found to be $|D|_4 = [a_{12}a_{34} + a_{14}a_{23} - a_{13}a_{24}]^2$. Since the Pfaffian is roughly half of a determinant, it is fitting that the actual value of Pf is given by $Pf = q$, or $|D| = Pf^2$.

Just as $|D|$ can be evaluated as $|D| = \sum_{j=1}^{2N} a_{ij}A_{ij}$, where A_{ij} is the cofactor in $|D|$ of a_{ij}, the Pfaffian can be similarly expressed as $Pf = \sum_{j=2}^{2N} a_{ij}B_{ij}$, where the cofactor B_{ij} is the Pfaffian obtained from Pf by deleting the first row, the jth column and the other elements that correspond to those below the diagonal that would be so deleted in $|D|$ (*i.e.*, the elements in the jth row), and finally choosing the sign as $(-)^j$. This process can be iterated until the final Pfaffians consist of one element each and are defined to be equal to that element. Thus the following Pfaffian is evaluated, with $2N = 6$.

$$Pf = \begin{vmatrix} a_{12} & a_{13} & a_{14} & a_{15} & a_{16} \\ & a_{23} & a_{24} & a_{25} & a_{26} \\ & & a_{34} & a_{35} & a_{36} \\ & & & a_{45} & a_{46} \\ & & & & a_{56} \end{vmatrix} = a_{12} \begin{vmatrix} a_{34} & a_{35} & a_{36} \\ & a_{45} & a_{46} \\ & & a_{56} \end{vmatrix}$$

$$- a_{13} \begin{vmatrix} a_{24} & a_{25} & a_{26} \\ & a_{45} & a_{46} \\ & & a_{56} \end{vmatrix} + a_{14} \begin{vmatrix} a_{23} & a_{25} & a_{26} \\ & a_{35} & a_{36} \\ & & a_{56} \end{vmatrix}$$

$$- a_{15} \begin{vmatrix} a_{23} & a_{24} & a_{26} \\ & a_{34} & a_{36} \\ & & a_{46} \end{vmatrix} + a_{16} \begin{vmatrix} a_{23} & a_{24} & a_{25} \\ & a_{34} & a_{35} \\ & & a_{45} \end{vmatrix}.$$

The first of these Pfaffians is

$$\begin{vmatrix} a_{34} & a_{35} & a_{36} \\ & a_{45} & a_{46} \\ & & a_{56} \end{vmatrix} = a_{34}|a_{56}| - a_{35}|a_{46}| + a_{36}|a_{45}|$$

$$= a_{34}a_{56} - a_{35}a_{46} + a_{36}a_{45},$$

and the corresponding term of Pf is then $a_{12}a_{34}a_{56} - a_{12}a_{35}a_{46} + a_{12}a_{36}a_{45}$.

A more formal expression for Pf, and one that is quite useful in later explanations of the counting problem, is

$$Pf = \sum_P{}' (-)^P a(p_1,p_2)a(p_3,p_4)a(p_5,p_6)\cdots a(p_{2N-1},p_{2N}), \qquad (7.7.2)$$

where $\sum_P'$ is the sum over all permutations of the elements, subject to the conditions

$$p_1 < p_2; \ p_3 < p_4; \ \cdots; \ p_{2N-1} < p_{2N}, \quad \text{and}$$

$$p_1 < p_3 < p_5 < \cdots < p_{2N-1}, \qquad (7.7.3)$$

where the inequalities prevent any index p_j from appearing twice in a term, and the permutations are counted as odd or even from the normal numerical order 1, 2, 3, $\cdots$, $2N$. Thus the three terms in the previous example, $a_{12}a_{34}a_{56} - a_{12}a_{35}a_{46} + a_{12}a_{36}a_{45}$, are seen to satisfy the inequalities (7.7.3), and the negative sign of the second term is consistent with the odd permutation, the single interchange of 4 and 5.

7.7.3 Dimers. *Dimer* is a term used in place of *diatomic molecule* in treatments of the adsorption of a molecular gas on a crystalline surface. A dimer coverage of a lattice places an end of each dimer on each of two points of the lattice, with every point of the lattice covered exactly once. Dimers are usually regarded as having fixed length and capable of covering only nearest-neighbor pairs, but generalizations are easily possible. For present purposes, only n.n. coverings will be considered. Evidently the lattice must have an even number of points for complete coverage.

In order to specify a dimer coverage, the lattice points are numbered, and the points are grouped into dimer-connected pairs:

$$(p_1, p_2); \ (p_3, p_4); \ (p_5, p_6); \ \cdots; \ (p_{2N-1}, p_{2N}). \tag{7.7.4}$$

Each point must be a member of exactly one pair. A systematic listing of these dimer coverages can be made by imposing the inequalities

$$p_1 < p_2; \ p_3 < p_4; \ \cdots; \ p_{2N-1} < p_{2N}, \quad \text{and}$$

$$p_1 < p_3 < p_5 < \cdots < p_{2N-1},$$

since a coverage is unchanged by listing the pairs in arbitrary order. But these inequalities are exactly those of (7.7.3), showing that the dimer coverages of a lattice can be put in one-to-one correspondence with the terms of a Pfaffian. As an illustration of this correspondence, consider the 2×2 lattice shown in Fig. 7.7, with horizontal dimers labelled x and vertical ones y. Evidently two and only two coverages are possible–both dimers horizontal and the coverage written as x^2, since $(1,2) = x$ and $(3,4) = x$, or both dimers vertical and written y^2.

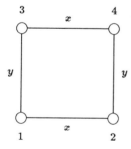
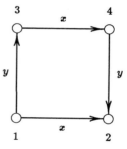

Figure 7.7 A 2×2 lattice with the x and y dimers labelled.

Figure 7.8 Directed dimers on the 2×2 lattice.

A Pfaffian can be constructed with the help of row and column labels as

$$
\begin{array}{c c c c c}
 & 1 & 2 & 3 & 4 \\
1 & 0 & x & y & 0 \\
2 &   & 0 & 0 & y \\
3 &   &   & 0 & x \\
4 &   &   &   & 0
\end{array}
\ ,
$$

where the diagonal has been included to show the absence of self bonds, and the other missing bonds are also given the value 0. The Pfaffian is then

$$
Pf = \begin{vmatrix} x & y & 0 \\ & 0 & y \\ & & x \end{vmatrix} = x^2 - y^2, \tag{7.7.5}
$$

showing that even though one-to-one correspondence is evident, the negative sign of the second term is not proper for a dimer-counting function. In more complicated examples, misplaced negative signs can lead to subtraction of terms that should be added, as would be the case in this example if $x = y$. The dimer bonds may be regarded as directed, however (they must be so regarded in order to generate the complete antisymmetric matrix), and the bond directions may then be chosen to remove unwanted negative signs. The complete determinant corresponding to the choice of bond directions in Fig. 7.8 is

$$
|D| = \begin{array}{c|cccc}
 & 1 & 2 & 3 & 4 \\
\hline
1 & 0 & x & y & 0 \\
2 & -x & 0 & 0 & -y \\
3 & -y & 0 & 0 & x \\
4 & 0 & y & -x & 0
\end{array} \, ,
$$

and the Pfaffian is evidently $Pf = x^2 + y^2$.

In Fig. 7.9, a 4×2 lattice is shown with one correct set of directed bonds indicated. The Pfaffian is easily worked out to be $Pf = x^4 + 3x^2y^2 + y^4$, corresponding to the dimer coverages shown in Fig. 7.9b.

Kasteleyn [1963] has shown that for plane lattices with no crossed bonds (such as the diagonals on a square lattice) it is always possible to choose the directions of the bonds such that the Pfaffian counts dimer coverages correctly. For the 6×4 lattice shown in Fig. 7.10a, it is evident that the corresponding term in the Pfaffian is $a(1,2) \, a(3,4) \cdots a(23,24)$, a positive term, since $P = 0$ in Eq. (7.7.2) and the inequalities (7.7.3) are satisfied. The order specified by the first of these inequalities could be violated without change in sign of the term, however, both because the term would represent the same dimer coverage if the order of the a's were changed, and because interchange of any two a's is an even permutation of the primary order, with no change of sign. Also, the order specified by the second of the inequalities (7.7.3) could be violated without change of sign because, for example, replacement of $a(3,4)$ by $a(4,3) = -a(3,4)$ is an odd permutation, and the two sign changes cancel.

If any other dimer coverage, such as that of Fig. 7.10b, is superposed on the primary coverage, as shown in Fig. 7.10c, the resulting diagram consists of closed graphs and doubly covered coincident dimers. For simplicity, consider the superposition diagram of Fig. 7.10d, in which only two dimers differ in the superposed coverages. The only difference in the Pfaffian terms of the two coverages arises from permutation of the factors $a(1,2)a(7,8)$ to $a(1,7)a(2,8)$, an odd permutation that requires $a(1,7)$ and $a(2,8)$ to have opposite signs, just as do the (1,5) and (2,6) bonds in Fig. 7.9. Since there are two choices, one is selected arbitrarily to be positive, and the other is then negative. Now proceed to a somewhat larger superposition diagram made up of the points 1, 2, 3, 4, 10, 9, 8, 7, covered in Fig. 7.10a by the standard term $a(1,2)a(3,4)a(7,8)a(9,10)$, and in the new arrangement by $a(1,7)a(2,3)a(4,10)a(8,9)$. Kasteleyn observed that the counting of permutations was onerous and proposed a systematic treatment.

The terms are arranged with indices in, say, clockwise order: $a(1,7)a(8,9)a(10,4)a(3,2)$. This term has the same sign as the correctly ordered one in the new arrangement, because

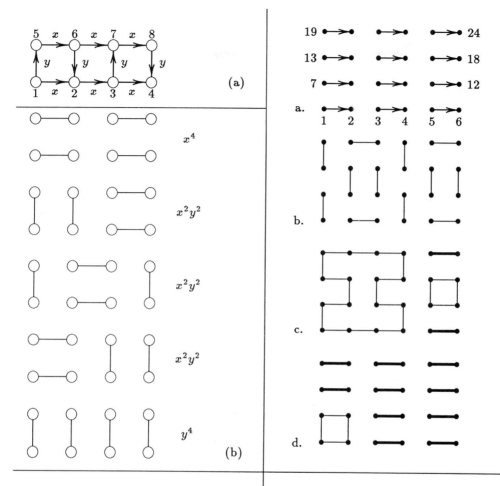

Figure 7.9 a. Bond directions on the 4 × 2 lattice. b. Terms in the Pfaffian correspond to the dimer coverages shown.

Figure 7.10 Dimer coverages and superposition diagrams. a. Primary coverage. b. Another possible coverage. c. Superposition of a. and b. d. Superposition diagram of a. and a coverage for which (1, 7) and (2, 8) replace (1, 2) and (7, 8) in a.

the order of the a's does not matter, and the internal order is backwards on exactly two of the a's. This dimer coverage is then shifted one step clockwise around the lattice, becoming $a(7,8)a(9,10)a(4,3)a(2,1)$, which has the same terms and signs as the standard term. The problem of counting permutations is thereby reduced to that of determining the number of permutations required to move index 1 from one end of the sequence to the other; *i.e.* from 1, 7, 8, 9, 10, 4, 3, 2 to 7, 8, 9, 10, 4, 3, 2, 1. Successive pairwise interchanges give $P = 7$, an odd permutation that forces at least one a in the product $a(1,7)a(2,3)a(4,10)a(8,9)$ to be negative. Since $a(1,7)$ is already chosen positive, a convenient choice is for all horizontal bonds, $a(2,3)$ and $a(8,9)$ here, to be positive and $a(4,10)$ to be negative.

Continued iteration of this procedure permits the directions of all bonds on the lattice to be assigned such that the Pfaffian terms count dimer coverages correctly, with proper sign. Kasteleyn's rule summarizing the assignment of bond directions is that the clockwise traversal of any simple closed diagram (one not containing enclosed lattice points) should encounter an odd number of clockwise-directed bonds. Such diagrams are called *clockwise odd*, and for lattices with even numbers of sides in a unit cell, clockwise odd is equivalent to counterclockwise odd. A corollary of Kasteleyn's rule is that the clockwise parity of any closed diagram (odd or even) is opposite to that of the number of enclosed points. Typical arrangements of directed bonds on planar lattices for correct Pfaffian counting of dimer coverages are shown in Fig. 7.11.

Since the bond directions for correct counting are established independent of bond magnitudes, the Pfaffian will count dimer coverages with as many kinds of dimers as necessary. Thus the generating function that counts coverages with r_1 dimers of type z_1, r_2 of type z_2, etc., is

$$\Psi(z_1, z_2, \cdots z_k) = \sum_{r_1, r_2, \cdots, r_k} n(r_1, r_2, \cdots, r_k) z_1^{r_1} z_2^{r_2} \cdots z_k^{r_k}$$
$$= Pf(z_1, z_2, \cdots, z_k), \tag{7.7.6}$$

where the signs of terms in the Pfaffian are chosen according to Kasteleyn's rule, and the magnitudes $z_1, \cdots, z_k$ express the dimer species.

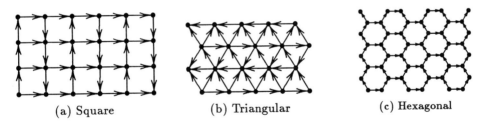

| (a) Square | (b) Triangular | (c) Hexagonal |

Figure 7.11 Typical bond directions on planar lattices for correct Pfaffian counting.

7.7.4 Dimer-Pfaffian Treatment of Ising Problem in One Dimension.
The one-dimensional Ising problem is almost too trivial for attack by the dimer-Pfaffian technique, but its simplicity makes the explanation of the mathematical steps particularly easy. It is helpful to rewrite the Hamiltonian at this point to relocate the zero of energy as

$$\mathcal{H} = -J \sum_{i=1}^{N} (\sigma_i \sigma_{i+1} - 1), \tag{7.7.7}$$

so that like nearest neighbors give zero energy. The partition function then becomes

$$Z = e^{-NK} \sum_{\{\sigma_i\}} \prod_{i=1}^{N} e^{K\sigma_i \sigma_{i+1}}$$

$$= e^{-NK} \cosh^N K \sum_{\{\sigma_i\}} (1 + x\sigma_i\sigma_{i+1})$$

$$= e^{-NK} (2\cosh K)^N \sum_r n(r)x^r, \tag{7.7.8}$$

where $n(r)$ is the number of closed graphs on the one-dimensional chain with r bonds. By inspection the only terms in the sum are those for $n(0) = 1$ and $n(N) = 1$, so the partition function is

$$Z = e^{-NK} (2\cosh K)^N (1 + x^N), \tag{7.7.9}$$

in agreement with Eq. (7.5.28) except for the shift in energy zero.

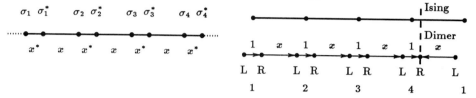

Figure 7.12 Linear chain with doubled spins, internally coupled by x^* bonds that become infinitely strong in the limit.

Figure 7.13 A four-element Ising chain and its dimer-chain equivalent.

The problem of relating $n(r)$ to dimer coverages is slightly complicated by the circumstance that the dimers do not form closed graphs. The trick for establishing correspondence is to replace each σ_i in the chain by a pair σ_i, σ_i^*, bound by a coupling constant J^* that is permitted to become infinite, thereby assuring that σ_i and σ_i^* are equal. The partition function for such a chain, shown in Fig. 7.12, is

$$Z = e^{-NK}e^{-NK^*} \sum_{\{\sigma_i,\sigma_i^*\}} \prod_{i=1}^{N} e^{K^*\sigma_i\sigma_i^* + K\sigma_i^*\sigma_{i+1}}$$

$$= e^{-N(K+K^*)} (\cosh K \cosh K^*)^N \sum_{\{\sigma_i,\sigma_i^*\}} \prod_{i=1}^{N} (1 + x^*\sigma_i\sigma_i^*)(1 + x\sigma_i\sigma_{i+1})$$

$$= e^{-N(K+K^*)} (\cosh K)^N (2\cosh K^*)^N \sum_{\{\sigma_i\}} \prod_{i=1}^{N} (1 + x^*x\sigma_i\sigma_{i+1}), \tag{7.7.10}$$

where the last step is obtained by summation over the $\{\sigma_i^* = \pm1\}$. In the limit of $J^* \to \infty$, so that $(e^{-K^*})(2\cosh K^*) \to 1$ and $x^* = \tanh K^* = 1$, the partition function becomes

$$Z = e^{-NK} (\cosh K)^N \sum_{\{\sigma_i\}} \prod_{i=1}^{N} (1 + x\sigma_i\sigma_{i+1}) = e^{-NK} (2\cosh K)^N \sum_r n(r)\, x^r,$$

as in Eq. (7.7.8). But now the chain may by regarded as one with doubled spins, as in Fig. 7.12, with each σ_i-σ_i^* bond of strength $x^* = 1$. A four-element Ising chain and its dimer equivalent are shown in Fig. 7.13, where the x-bond from $4R$ to $1L$ is reversed in keeping with the Kasteleyn rule that a simple closed diagram should be clockwise odd (no matter which direction around the loop is taken as clockwise). The Pfaffian is constructed as

		1		2		3		4	
		L	R	L	R	L	R	L	R
1	L	0	1	0	0	0	0	0	x
	R		0	x	0	0	0	0	0
2	L			0	1	0	0	0	0
	R				0	x	0	0	0
3	L					0	1	0	0
	R						0	x	0
4	L							0	1
	R								0

which yields the correct result $\sum n(r)\, x^r = 1 + x^4$. The structure of the full antisymmetric determinant related to this Pfaffian is worth further examination. It may be written as

$$|D| = \begin{vmatrix} 0 & 1 & 0 & 0 & 0 & 0 & 0 & x \\ -1 & 0 & x & 0 & 0 & 0 & 0 & 0 \\ 0 & -x & 0 & 1 & 0 & 0 & 0 & 0 \\ 0 & 0 & -1 & 0 & x & 0 & 0 & 0 \\ 0 & 0 & 0 & -x & 0 & 1 & 0 & 0 \\ 0 & 0 & 0 & 0 & -1 & 0 & x & 0 \\ 0 & 0 & 0 & 0 & 0 & -x & 0 & 1 \\ -x & 0 & 0 & 0 & 0 & 0 & -1 & 0 \end{vmatrix}. \tag{7.7.11}$$

A direct-product matrix $A \otimes B$, where A and B are matrices of arbitrary and unrelated order is defined to be the matrix

$$A \otimes B = \begin{bmatrix} a_{11}B & a_{12}B & \cdots \\ a_{21}B & a_{22}B & \cdots \\ \vdots & \vdots & \vdots \end{bmatrix}. \tag{7.7.12}$$

In terms of direct-product matrices, the matrix D can be written as

$$\begin{aligned} D &= I_4 \otimes \Gamma + E_4 \otimes X + E_4^3 \otimes X^T \\ &= \begin{bmatrix} \Gamma & X & 0 & X^T \\ -X^T & \Gamma & X & 0 \\ 0 & -X^T & \Gamma & X \\ -X & 0 & -X^T & \Gamma \end{bmatrix}, \end{aligned} \tag{7.7.13}$$

where I_4 is the unit matrix of order 4, Γ is the 2×2 internal matrix, as seen along the principal diagonal of Eq. (7.7.11), and X is the 2×2 interconnecting matrix that represents the bonds connecting the nth and $(n+1)^{st}$ spin clusters. They are given by

$$\Gamma = \begin{bmatrix} 0 & 1 \\ -1 & 0 \end{bmatrix}, \text{ and } X = \begin{bmatrix} 0 & 0 \\ x & 0 \end{bmatrix}. \tag{7.7.14}$$

X^T is the transpose of X, and $-X^T$ is the left-hand connecting matrix. The matrix E_4, the key to later calculational techniques, is

$$E_4 = \begin{bmatrix} 0 & 1 & 0 & 0 \\ 0 & 0 & 1 & 0 \\ 0 & 0 & 0 & 1 \\ -1 & 0 & 0 & 0 \end{bmatrix}. \tag{7.7.15}$$

The square and cube of E_4 are

$$E_4^2 = \begin{bmatrix} 0 & 0 & 1 & 0 \\ 0 & 0 & 0 & 1 \\ -1 & 0 & 0 & 0 \\ 0 & -1 & 0 & 0 \end{bmatrix}, \text{ and } E_4^3 = \begin{bmatrix} 0 & 0 & 0 & 1 \\ -1 & 0 & 0 & 0 \\ 0 & -1 & 0 & 0 \\ 0 & 0 & -1 & 0 \end{bmatrix}, \tag{7.7.16}$$

showing that $E_4^3 \otimes X^T$ is correctly written in Eq. (7.7.13). The most interesting property of E_4 is that $E_4^4 = -I_4$, so that E_4 is a fourth root of $-I_4$. Therefore $E_4^3 = -E_4^{-1}$, a replacement to be used in Eq. (7.7.13). The matrix E_4 can be diagonalized by a pair of matrices T_4 and T_4^{-1}, giving

$$T_4 E_4 T_4^{-1} = \Lambda_4, \tag{7.7.17}$$

such that $\Lambda_4^4 = -I_4$. Therefore the diagonal elements of Λ_4 are the fourth roots of -1, $\exp[(2r-1)i\pi/4]$, where $1 \leq r \leq 4$: these roots are written as $e^{i\phi_r}$ for convenience. If D is expressed symbolically as $D = P_4 \otimes Q$, it may be operated on by the transformation matrices $T_4 \otimes I_2$ and $T_4^{-1} \otimes I_2$, where in effect the matrix Q is unchanged and the transformation takes place only in the space of P_4. Thus we may write

$$D' = (T_4 \otimes I_2) D \left(T_4^{-1} \otimes I_2 \right) = T_4 P_4 T_4^{-1} \otimes Q$$
$$= I_4 \otimes \Gamma + \Lambda_4 \otimes X - \Lambda_4^{-1} \otimes X^T, \tag{7.7.18}$$

or, with the notation $a_r = -1 + xe^{i\phi_r}$, and a_r^* its complex conjugate,

$$D' = \begin{bmatrix} 0 & -a_1^* & 0 & 0 & 0 & 0 & 0 & 0 \\ a_1 & 0 & 0 & 0 & 0 & 0 & 0 & 0 \\ 0 & 0 & 0 & -a_2^* & 0 & 0 & 0 & 0 \\ 0 & 0 & a_2 & 0 & 0 & 0 & 0 & 0 \\ 0 & 0 & 0 & 0 & 0 & -a_3^* & 0 & 0 \\ 0 & 0 & 0 & 0 & a_3 & 0 & 0 & 0 \\ 0 & 0 & 0 & 0 & 0 & 0 & 0 & -a_4^* \\ 0 & 0 & 0 & 0 & 0 & 0 & a_4 & 0 \end{bmatrix}. \tag{7.7.19}$$

Evidently, then, since $|D| = |D'|$, the determinant $|D|$ may be written

$$|D| = \prod_{r=1}^{4} \begin{vmatrix} 0 & -a_r^* \\ a_r & 0 \end{vmatrix} = \prod_{r=1}^{4} \left(1 + x^2 - 2x \cos \phi_r \right). \tag{7.7.20}$$

Since $\cos \phi_r = \pm\sqrt{2}/2$, each value twice, Eq. (7.7.20) becomes

$$|D| = \left(1 + x^2 - x\sqrt{2} \right)^2 \left(1 + x^2 + x\sqrt{2} \right)^2 = \left(1 + x^4 \right)^2, \tag{7.7.21}$$

as expected from the relation $|D| = Pf^2$.

The point of this calculation lies in Eq. (7.7.18), where the matrices I_4, Λ_4, and Λ_4^{-1} become I_N, Λ_N, and Λ_N^{-1} for any number N of elements in the chain, and the elements of the diagonal matrix Λ_N are $e^{i\phi_r}$, where $\phi_r = (2r-1)\pi/N$; $1 \leq r \leq N$. The matrices Γ, X, and X^T are determined by the bonds of the chain, independent of N, and Eq. (7.7.18) becomes

$$D' = I_N \otimes \Gamma + \Lambda_N \otimes X - \Lambda_N^{-1} \otimes X^T. \tag{7.7.22}$$

Examples of the matrices Γ, X, and Y are given in the next section.

The generating function that counts closed graphs on the lattice in Eq. (7.7.8) is just equal to one that counts dimer coverages of the spin-doubled lattice, or

$$\sum_r n(r)\, x^r = \Phi(x) = |D|^{1/2}, \tag{7.7.23}$$

and $\ln Z$ becomes

$$\ln Z = -NK + N\ln 2 + N\ln \cosh K + (1/2)\ln |D|, \tag{7.7.24}$$

where, from Eq. (7.7.20),

$$\ln |D| = \sum_{r=1}^{N} \ln \left(1 + x^2 - 2x \cos \phi_r \right), \tag{7.7.25}$$

with $\phi_r = (2r-1)\pi/N$. For very large N, the sum in Eq. (7.7.25) can be replaced by an integral, with $dr = (N/2\pi)d\phi$, or

$$\ln |D| = \frac{N}{2\pi} \int_0^{2\pi} d\phi\, \ln \left(1 + x^2 - 2x \cos \phi \right). \tag{7.7.26}$$

With this replacement, Eq. (7.7.24) may be written

$$\lim_{N \to \infty} \left[(\ln Z)/N \right] = -K + \ln 2 + \ln \cosh K + \frac{1}{4\pi} \int_0^{2\pi} d\phi\, \ln \left(1 + x^2 - 2x \cos \phi \right). \tag{7.7.27}$$

Recall that $x = \tanh K < 1$, with the consequence that the integral is expected to be zero because it corresponds to the term $\ln \left(1 + x^N \right)$ from Eq. (7.7.9), which vanishes as $N \to \infty$.

That the integral is zero may be easily seen from the observation that the integrand may be written as $\ln\left(1 - xe^{i\phi}\right) + \ln\left(1 - xe^{-i\phi}\right)$, so that each resulting integral looks like

$$\frac{1}{2\pi i}\oint \frac{dz}{z}\frac{\ln(1-z)}{} = \ln 1 = 0, \tag{7.7.28}$$

because $|z| < 1$. In two dimensions, however, the corresponding integral is not zero; it is, in fact, the term of greatest interest, since it leads to the divergences that characterize phase transitions.

Although this development is presented as the analysis of a one-dimensional problem, it turns out that the essential mathematical apparatus for solution of the two-dimensional Ising problem has now been developed; it is shown in the next section. (This is the first of several kinds of calculations that can be explained in almost their full detail, with considerable simplification, in the context of a one-dimensional system, after which the generalization to higher dimensionality becomes straightforward and relatively easy. Further examples of this technique are to be found later in this chapter and also in Chapter 8, in the context of the renormalization group.)

7.8 EXACT SOLUTION OF THE TWO-DIMENSIONAL ISING PROBLEM

7.8.1 Preliminary Example. The structure of the dimer-counting matrix D for the $M \times N$ lattice may be seen most easily by examination of a small one such as the 4×3 example shown in Fig. 7.14. The numbered points are the actual points of the Ising lattice, but they become *cities* of points for the dimer-coverage problem. In one dimension the cities consist of two points each, considered to be bonded by infinite coupling constants, as developed in Eq. (7.7.10) *et seq.*.

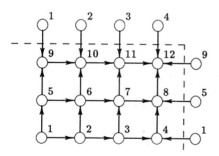

Figure 7.14 Interconnections of cities for 4×3 lattice. Horizontal arrows show directions associated with matrix X, vertical with Y.

In two dimensions the cities are more complicated, but preliminary analysis can proceed in terms of as-yet-unspecified matrices Γ, X, Y, X^T, and Y^T, corresponding to those of Eq. (7.7.13). The internal matrix, specifying the street directions within a city, is denoted by Γ. Horizontal connections from the right edge of one city to the left edge of the next, as from 1 to 2, are denoted by X, the signs of which are indicated by horizontal arrows in Fig. 7.14. Thus connection 4-1 is $-X$, in keeping with the Kasteleyn rule that closed paths should be clockwise odd.

As in Eq. (7.7.13), however, reversed horizontal bonds, connecting the left edge of one city to the right edge of the one before it are $-X^T$, not $-X$, because the RL matrix element and the LR one are in transposed positions, no matter what the structure of the X matrix. Similarly, vertical connections from the upper edge of one city to the lower edge of the next are called Y, with directions as shown in Fig. 7.14, and the opposite connections are $-Y^T$. (It is assumed here, and verified in the next two subsections, that the Γ matrix satisfies the Kasteleyn rule, without the need for alternating signs of Y as in Fig. 7.11a.) Thus the

12×12 matrix D for the lattice may be written out as

$$
\begin{array}{c}
 1 2 3 4 5 6 7 8 9 10 11 12 \\
\begin{array}{c}
1\\2\\3\\4\\ \\5\\6\\7\\8\\ \\9\\10\\11\\12
\end{array}
\left[
\begin{array}{cccccccccccc}
\Gamma & X & 0 & X^T & Y & 0 & 0 & 0 & Y^T & 0 & 0 & 0 \\
-X^T & \Gamma & X & 0 & 0 & Y & 0 & 0 & 0 & Y^T & 0 & 0 \\
0 & -X^T & \Gamma & X & 0 & 0 & Y & 0 & 0 & 0 & Y^T & 0 \\
-X & 0 & -X^T & \Gamma & 0 & 0 & 0 & Y & 0 & 0 & 0 & Y^T \\
-Y^T & 0 & 0 & 0 & \Gamma & X & 0 & X^T & Y & 0 & 0 & 0 \\
0 & -Y^T & 0 & 0 & -X^T & \Gamma & X & 0 & 0 & Y & 0 & 0 \\
0 & 0 & -Y^T & 0 & 0 & -X^T & \Gamma & X & 0 & 0 & Y & 0 \\
0 & 0 & 0 & -Y^T & -X & 0 & -X^T & \Gamma & 0 & 0 & 0 & Y \\
-Y & 0 & 0 & 0 & -Y^T & 0 & 0 & 0 & \Gamma & X & 0 & X^T \\
0 & -Y & 0 & 0 & 0 & -Y^T & 0 & 0 & -X^T & \Gamma & X & 0 \\
0 & 0 & -Y & 0 & 0 & 0 & -Y^T & 0 & 0 & -X^T & \Gamma & X \\
0 & 0 & 0 & -Y & 0 & 0 & 0 & -Y^T & -X & 0 & -X^T & \Gamma
\end{array}
\right]
\end{array}
$$

$$(7.8.1)$$

The 4×4 blocks of D are obvious direct-product matrices, so D may be written

$$
D = \begin{bmatrix}
A & I_4 \otimes Y & I_4 \otimes Y^T \\
-I_4 \otimes Y^T & A & I_4 \otimes Y \\
-I_4 \otimes Y & -I_4 \otimes Y^T & A
\end{bmatrix}, \tag{7.8.2}
$$

where $A = I_4 \otimes \Gamma + E_4 \otimes X - E_4^{-1} \otimes X^T$, and E_4 is the matrix given by Eq. (7.7.15). But Eq. (7.8.2) is evidently expressible as another direct-product matrix,

$$
D = I_3 \otimes A + E_3 \otimes (I_4 \otimes Y) - E_3^{-1} \otimes (I_4 \otimes Y^T), \tag{7.8.3}
$$

where

$$
E_3 = \begin{bmatrix}
0 & 1 & 0 \\
0 & 0 & 1 \\
-1 & 0 & 0
\end{bmatrix}, \tag{7.8.4}
$$

and the parentheses in Eq. (7.8.3) may be deleted, since their inclusion is intended only to enhance clarity on first inspection. When A is written out, and superfluous punctuation omitted, D is given by

$$
\begin{aligned}
D = &\; I_3 \otimes I_4 \otimes \Gamma + I_3 \otimes E_4 \otimes X - I_3 \otimes E_4^{-1} \otimes X^T \\
&+ E_3 \otimes I_4 \otimes Y - E_3^{-1} \otimes I_4 \otimes Y^T.
\end{aligned} \tag{7.8.5}
$$

In the $M \times N$ lattice, the indices 4 and 3 become M and N respectively.

In the same sort of notation, the transformation P_3, P_3^{-1} that diagonalizes E_3, and Q_4, Q_4^{-1} that diagonalizes E_4 may be written as a single direct-product operator on D to give

$$
D' = (P_3 \otimes Q_4 \otimes I) \; D \; (P_3^{-1} \otimes Q_4^{-1} \otimes I), \tag{7.8.6}
$$

where I is the unit matrix of the same dimensionality as Γ, X, and Y, and again the parentheses are punctuation for clarity. The P_3 and Q_4 matrices operate only in their own spaces, so if a term of D has the form $A_3 \otimes B_4 \otimes Z$ then the transformation operation gives

$$
(P_3 A_3 P_3^{-1}) \otimes (Q_4 B_4 Q_4^{-1}) \otimes Z.
$$

With this explanation of the notation, D' becomes

$$D' = I_3 \otimes I_4 \otimes \Gamma + I_3 \otimes \Lambda_4 \otimes X - I_3 \otimes \Lambda_4^{-1} \otimes X^T$$
$$+ \Lambda_3 \otimes I_4 \otimes Y - \Lambda_3^{-1} \otimes I_4 \otimes Y^T, \qquad (7.8.7)$$

where the nonzero elements of the diagonal matrix Λ_4 are fourth roots of -1 and those of Λ_3 are cube roots of -1. The structure of $I_3 \otimes \Lambda_4$ and $\Lambda_3 \otimes I_4$ assures that each of the four diagonal elements λ_4 of Λ_4 is associated exactly once with each of the three elements λ_3 of Λ_3.

Since the determinant $|D|$ is equal to $|D'|$, and since D' is diagonalized into $3 \times 4 = 12$ blocks of the size of Γ, one may write in analogy with Eq. (7.7.20),

$$|D'| = \prod_{r=1}^{4} \prod_{s=1}^{3} |D(r, s)|, \qquad (7.8.8)$$

where

$$|D(r, s)| = \left| \Gamma + X e^{i\phi_r} - X^T e^{-i\phi_r} + Y e^{i\theta_s} - Y^T e^{-i\theta_s} \right|, \qquad (7.8.9)$$

with $\phi_r = (2r - 1)\,\pi/4$ and $\theta_s = (2s - 1)\,\pi/3$. The counting problem for the 4×3 lattice is thus completely solved, and specific results corresponding to those of Eq. (7.7.23) to Eq. (7.7.25) can be obtained as soon as Γ, X, and Y are decided upon. Generalization to the $M \times N$ lattice is obvious, and the problem of the two-dimensional Ising lattice is thus solved in principle.

7.8.2 Kasteleyn Cities. Kasteleyn used a four-point city to represent each Ising lattice point in the corresponding dimer lattice. The correspondence is one-to-one except for points of the Ising lattice with no bond, as shown in Fig. 7.15. Fortunately, however, the violation

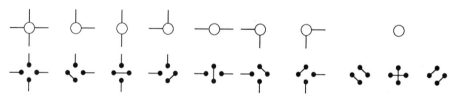

Figure 7.15 Correspondence between Ising-lattice connections and Kasteleyn cities.

(by its author) of Kasteleyn's rule that forbids crossed bonds permits choice of internal bond directions such that the three dimer coverages that represent the no-bond points give $1 + 1 - 1 = 1$ for their net contribution, and thus act as if only a single configuration were present. (Kasteleyn [1963]; Montroll [1964]) One choice of bond directions is shown in Fig. 7.16, corresponding to a Γ given by

$$\Gamma = \begin{array}{c} L \\ R \\ D \\ U \end{array} \begin{array}{cccc} L & R & D & U \\ \left[\begin{array}{cccc} 0 & -1 & 1 & 1 \\ 1 & 0 & 1 & -1 \\ -1 & -1 & 0 & 1 \\ -1 & 1 & -1 & 0 \end{array} \right] \end{array}. \qquad (7.8.10)$$

In this system, X and Y are

$$X = \begin{bmatrix} 0 & 0 & 0 & 0 \\ x & 0 & 0 & 0 \\ 0 & 0 & 0 & 0 \\ 0 & 0 & 0 & 0 \end{bmatrix}, \text{ and } Y = \begin{bmatrix} 0 & 0 & 0 & 0 \\ 0 & 0 & 0 & 0 \\ 0 & 0 & 0 & 0 \\ 0 & 0 & y & 0 \end{bmatrix}, \tag{7.8.11}$$

from which X^T and Y^T are evident.

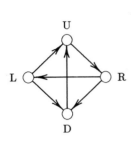

Figure 7.16 Bond directions for Γ in a Kasteleyn city.

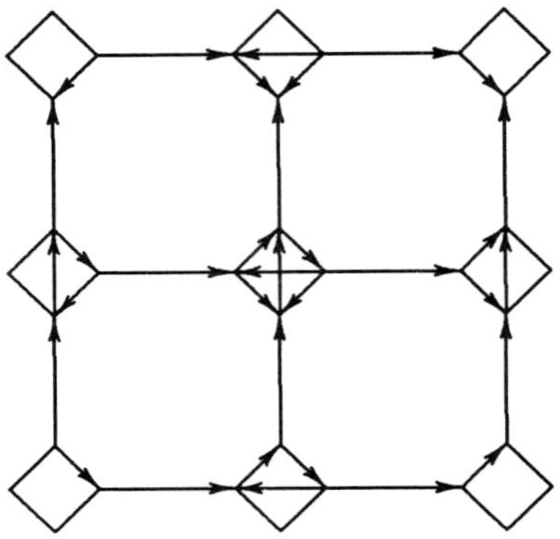

Figure 7.17 Dimer lattice corresponding to Fig. 7.16.

The dimer lattice corresponding to Fig. 7.16 is shown in Fig. 7.17. Each unit cell is both clockwise and counterclockwise odd, and paths corresponding to the closed graphs x^2y^4 or x^4y^2 on the Ising lattice may be seen to have correct clockwise parity. When diagonal paths through the city and those on its periphery are compared, it may be seen that the parity changes correctly with the number of enclosed points. Figures 7.16 and 7.17 should be regarded as clockwise-odd dimer coverages. Apparently it is not possible to set up a Kasteleyn city that satisfies all the following criteria:

(1) All diagrams have the correct clockwise parity.

(2) All x bonds are positive in the $+x$ direction.

(3) All y bonds are positive in the $+y$ direction.

(4) The city matrix elements are ordered $L\ R\ U\ D$, corresponding to $+x$, $+y$.

(5) The Pfaffian corresponding to Γ has value $+1$.

(6) The two cosine terms in $|D(\phi, \theta)|$ [see Eq. (7.8.23)] agree in sign with Onsager's original result.

All these requirements are imposed for æsthetic reasons; failure to satisfy them merely detracts from the calculation without destroying its validity. The original Kasteleyn city has

$$\Gamma_K = \begin{matrix} & \begin{matrix} R & \ \ L & \ \ U & \ \ D \end{matrix} \\ \begin{matrix} R \\ L \\ U \\ D \end{matrix} & \begin{bmatrix} 0 & -1 & -1 & -1 \\ -1 & 0 & 1 & -1 \\ 1 & -1 & 0 & 1 \\ 1 & 1 & -1 & 0 \end{bmatrix} \end{matrix}, \tag{7.8.12}$$

satisfying all criteria except Nos. 4 and 5. The choice of Fig. 7.16 violates only 6, whereas a city with every bond reversed from those of Fig. 7.16 violates both 1 and 6. The one criterion that must be satisfied for an acceptable result is that $Pf_\Gamma = \pm 1$. (Negative values are acceptable, since the determinant evaluates a squared Pfaffian, and the positive square root is always chosen.) For the Kasteleyn city specified by Eq. (7.8.10), the result of Eq. (7.8.9) is

$$|D(r,s)| = \begin{vmatrix} 0 & -1-xe^{-i\phi_r} & 1 & 1 \\ 1+xe^{i\phi_r} & 0 & 1 & 1 \\ -1 & -1 & 0 & 1-ye^{-i\theta_s} \\ -1 & 1 & -1+ye^{i\theta_s} & 0 \end{vmatrix}$$

$$= \left(1+x^2\right)\left(1+y^2\right) - 2x\left(1-y^2\right)\cos\phi_r + 2y\left(1-x^2\right)\cos\theta_s, \quad (7.8.13)$$

from which the partition function for the infinite lattice is obtained directly. (For improved symmetry, θ_s may be replaced by $\theta_s + \pi$, thus changing the sign of the final term, without, of course, any computational consequences.)

7.8.3 Fisher Cities. Fisher has pointed out that any vertex on an Ising lattice with $n > 3$ connections can be set in one-to-one correspondence with a city of $n - 2$ clusters of three points each on the dimer lattice (Fisher [1961]). Thus the square Ising lattice requires two clusters of three points each, as shown in Fig. 7.18, in order to give a one-to-one correspondence, no crossed bonds, and the easy application of all criteria listed in the previous subsection. The dimer configuration in a Fisher city corresponding to zero bonds on the Ising lattice is shown in Fig. 7.19, and the dimer lattice covering is shown in Fig. 7.20.

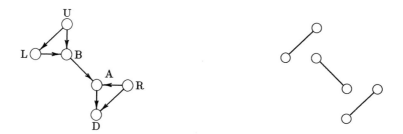

Figure 7.18 Fisher city for the square Ising lattice.

Figure 7.19 Dimers on Fisher city corresponding to no external bonds on Ising lattice.

The pertinent matrices are

$$\Gamma_F = \begin{matrix} & \begin{matrix} L & R & D & U & A & B \end{matrix} \\ \begin{matrix} L \\ R \\ D \\ U \\ A \\ B \end{matrix} & \begin{bmatrix} 0 & 0 & 0 & -1 & 0 & 1 \\ 0 & 0 & 1 & 0 & 1 & 0 \\ 0 & -1 & 0 & 0 & -1 & 0 \\ 1 & 0 & 0 & 0 & 0 & 1 \\ 0 & -1 & 1 & 0 & 0 & -1 \\ -1 & 0 & 0 & -1 & 1 & 0 \end{bmatrix} \end{matrix}, \quad (7.8.14)$$

$$X = \begin{bmatrix} 0 & 0 & 0 & 0 & 0 & 0 \\ x & 0 & 0 & 0 & 0 & 0 \\ 0 & 0 & 0 & 0 & 0 & 0 \\ 0 & 0 & 0 & 0 & 0 & 0 \\ 0 & 0 & 0 & 0 & 0 & 0 \\ 0 & 0 & 0 & 0 & 0 & 0 \end{bmatrix}, \qquad (7.8.15)$$

and

$$Y = \begin{bmatrix} 0 & 0 & 0 & 0 & 0 & 0 \\ 0 & 0 & 0 & 0 & 0 & 0 \\ 0 & 0 & 0 & 0 & 0 & 0 \\ 0 & 0 & y & 0 & 0 & 0 \\ 0 & 0 & 0 & 0 & 0 & 0 \\ 0 & 0 & 0 & 0 & 0 & 0 \end{bmatrix}, \qquad (7.8.16)$$

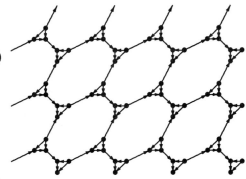

Figure 7.20 Dimer directions on Fisher-city lattice corresponding to Ising square lattice.

with $|D(r,s)|$ given by

$$|D(r,s)| = \begin{vmatrix} 0 & -xe^{-i\phi_r} & 0 & -1 & 0 & 1 \\ xe^{i\phi_r} & 0 & 1 & 0 & 1 & 0 \\ 0 & -1 & 0 & -ye^{-i\theta_s} & -1 & 0 \\ 1 & 0 & ye^{i\theta_s} & 0 & 0 & 1 \\ 0 & -1 & 1 & 0 & 0 & -1 \\ -1 & 0 & 0 & -1 & 1 & 0 \end{vmatrix}$$

$$= \left(1+x^2\right)\left(1+y^2\right) - 2x\left(1-y^2\right)\cos\phi_r - 2y\left(1-x^2\right)\cos\theta_s, \qquad (7.8.17)$$

as in Eq. (7.8.13), except that the cosine terms appear naturally with the same sign.

Although the calculation here requires 6×6 matrices instead of 4×4, the ease of satisfying æsthetic criteria and the direct applicability of already developed principles more than compensate for the slight additional work imposed by evaluation of the higher-order determinant.

7.8.4 The $M \times N$ Square Lattice.   The partition function for the $M \times N$ square Ising lattice, given in Eq. (7.6.3) and here modified to include the shift in the energy zero, as in Eq. (7.7.8), may be written

$$Z_{MN} = e^{-MN(K_1+K_2)}\left(2\cosh K_1 \cosh K_2\right)^{MN}\sum_{p,q} n(p,q)\, x^p\, y^q, \qquad (7.8.18)$$

where the generating function for the summed closed graphs is, from Eq. (7.7.6),

$$\sum_{p,q} n(p,q)\, x^p\, y^q = Pf(x,y), \qquad (7.8.19)$$

and the Pfaffian, known from Eq. (7.8.17), is given by

$$[Pf(x,y)]^2 = \prod_{r=1}^{M}\prod_{s=1}^{N} |D(r,s)|$$

$$= \prod_{r=1}^{M} \prod_{s=1}^{N} \left[(1 + x^2)(1 + y^2) - 2x(1 - y^2)\cos\phi_r - 2y(1 - x^2)\cos\theta_s \right], \quad (7.8.20)$$

with $\phi_r = (2r-1)\pi/M$ and $\theta_s = (2s-1)\pi/N$. From Eq. (7.8.20) it follows that

$$\ln Pf(x, y) = \frac{1}{2} \sum_{r=1}^{M} \sum_{s=1}^{N} \ln \left[(1 + x^2)(1 + y^2) - 2x(1 - y^2)\cos\phi_r - 2y(1 - x^2)\cos\theta_s \right]$$

$$\rightarrow \frac{MN}{8\pi^2} \int_0^{2\pi} d\phi \int_0^{2\pi} d\theta \ln \left[(1 + x^2)(1 + y^2) - 2x(1 - y^2)\cos\phi_r - 2y(1 - x^2)\cos\theta_s \right], \quad (7.8.21)$$

for very large M and N. Since $1 + x^2 = \cosh 2K_1/\cosh^2 K_1$, $1 - x^2 = 1/\cosh^2 K_1$, and similar relationships exist for y, the denominator of the ln term may be separated out in Eq. (7.8.21) and combined with the $\cosh K_1$ terms in $\ln Z$ to give

$$\ln Z_1 \equiv \lim_{M,N\to\infty} [\ln Z_{MN}/MN]$$

$$= -(K_1 + K_2) + \ln 2 + \frac{1}{8\pi^2} \int_0^{2\pi} d\phi \int_0^{2\pi} d\theta$$

$$\ln \left[\cosh 2K_1 \cosh 2K_2 - \sinh 2K_1 \cos\phi - \sinh 2K_2 \cos\theta \right], \quad (7.8.22)$$

the famous Onsager partition function (except for the first term, which comes from the zero shift of energy). (Onsager [1944]) When $K_1 = K_2$, Eq. (7.8.21) becomes

$$\ln Z_1 = -2K_1 + \ln 2 + \frac{1}{2\pi^2} \int_0^{\pi} d\phi \int_0^{\pi} d\theta \ln \left[\cosh^2 2K_1 - \sinh 2K_1(\cos\phi + \cos\theta) \right], \quad (7.8.23)$$

since symmetry permits contraction of the integration limits. The substitutions $\omega_1 = (\phi - \theta)/2$; $\omega_2 = (\phi + \theta)/2$ allow the cosine sum to become $\cos\phi + \cos\theta = 2\cos\omega_1 \cos\omega_2$, and the Jacobian is $\partial(\phi, \theta)/\partial(\omega_1, \omega_2) = 2$, permitting Eq. (7.8.23) to be rewritten

$$\ln Z_1 = -2K_1 + \ln(2\cosh 2K_1) + \frac{1}{\pi^2} \int_0^{\pi/2} d\omega_1 \int_0^{\pi} d\omega_2 \ln[1 - \gamma\cos\omega_1 \cos\omega_2], \quad (7.8.24)$$

where $\cosh^2 2K_1$ has been divided out of the ln term, and γ is given by

$$\gamma = 2\sinh 2K_1/\cosh^2 2K_1, \quad \text{or} \quad (7.8.25)$$

$$1/\gamma = (1/2)(\sinh 2K_1 + 1/\sinh 2K_1),$$

so that $-1 \leq \gamma \leq 1$.

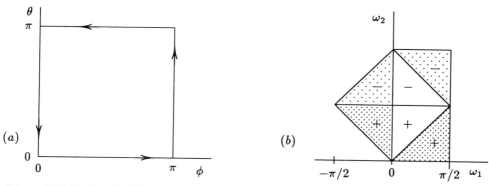

Figure 7.21 Trade-off of integrated areas to permit fixed limits in Eq. (7.8.23). The signs are those of $\cos\omega_2$; the sign of $\cos\omega_1$ is always $+$.

The existence of fixed limits in Eq. (7.8.24) is arranged by an exchange of equivalent areas in $\omega_1 - \omega_2$ space, as shown in Fig. 7.21. As ϕ and θ are followed around the borders of the integration region, ω_1 and ω_2 describe the area shown in Fig. 7.21b. An exchange of the two shaded areas corresponding to $-\pi/2 \le \omega_1 \le 0$ for equivalent ones with $\omega_1 \ge 0$ forms a rectangular area of integration. The ω_2 integral may be done easily, based on Eq. (7.7.28). Since

$$\ln\left(1 + be^{i\alpha}\right) + \ln\left(1 + be^{-i\alpha}\right) = \ln\left(1 + b^2\right) + \ln\left(1 + a\cos\alpha\right), \tag{7.8.26}$$

where $a = 2b/(1 + b^2)$ and $b = (1 - \sqrt{1 - a^2}))/a$, and since

$$\int_0^{2\pi} d\alpha \left[\ln\left(1 + be^{i\alpha}\right) + \ln\left(1 + be^{-i\alpha}\right)\right] = 0,$$

by Eq. (7.7.28), it follows that

$$\int_0^\pi d\alpha \ln(1 + \cos\alpha) = -\frac{1}{2}\int_0^{2\pi} d\alpha \ln\left(1 + b^2\right) = -\pi\ln\left(1 + b^2\right)$$

$$= \pi\ln\left[\frac{1 + \sqrt{1 - a^2}}{2}\right]. \tag{7.8.27}$$

With $a = -\gamma\cos\omega_1$, Eq. (7.8.24) becomes

$$\ln Z_1 = -2K_1 + \ln\left(2\cosh 2K_1\right) + \frac{1}{\pi}\int_0^{\pi/2} d\omega_1 \ln\left[\frac{1 + \sqrt{1 - \gamma^2\cos^2\omega_1}}{2}\right], \tag{7.8.28}$$

which is the usual form of the partition function for a square Ising lattice with equal coupling constants.

The energy per mole is given by

$$u = -N_0 \frac{\partial \ln Z_1}{\partial \beta}$$

$$= 2N_0 J\left(1 - \tanh 2K_1\right) - \frac{N_0 J\left(\sinh^2 2K_1 - 1\right)}{\sinh 2K_1 \cosh K_1}\left[\frac{2}{\pi}K(\gamma) - 1\right], \tag{7.8.29}$$

where $K(\gamma)$ is the complete elliptic integral of the first find, or

$$K(\gamma) = \int_0^{\pi/2} \frac{d\phi}{\sqrt{1 - \gamma^2\sin^2\phi}}, \tag{7.8.30}$$

in one of its standard forms. At very large or very small values of K_1 (low or high temperatures), $\gamma \to 0$ and $K(\gamma) \to \pi/2$, so that the last term then contributes nothing to u. When $\sinh 2K_1 = 1$, $\gamma = 1$ and $K(\gamma)$ exhibits logarithmic infinity which is suppressed by the external multiplier, though the slope of the curve of $u(T)$ vs. T is infinite there. A phase transition does occur at $T = T_c$, where $K_{1c} = \beta_c J = 0.4407$, or $T_c = 2.269 J/k$, as given by Eq. (7.6.13). Figure 7.22 shows a plot of Eq. (7.8.29), the exact calculation, and for comparison the first term of Eq. (7.8.29) and the energy from the molecular-field approximation of Sec. 7.3, obtained from Eq. (7.3.5), and shifted in zero to agree with Eq. (7.8.29). The cooperative term is seen to behave in a way expected in phase transitions; it lowers the energy for $T < T_c$ and raises it for $T > T_c$, in this case by reducing the short-range order.

At $T = T_c$, the energy is not that of a fully random system, as if T were infinite; it has already been reduced to about 30% of the $T = \infty$ value, in contrast to the molecular-field result, for which the system is completely random above its T_c. Thus the molecular-field model not only misplaces T_c but also misinterprets its nature.

The logarithmic singularity in $C = du/dT$ is shown in Fig. 7.22b, also in comparison with the results of molecular-field theory. The heat capacity is directly computed from Eq. (7.8.29) by differentiation with respect to T, by use of the relation

$$\frac{dK(m)}{dm} = \frac{1}{m} \left[\frac{E(m)}{1 - m^2} - K(m) \right], \tag{7.8.31}$$

where $E(m)$ is the complete elliptic integral of the second kind,

$$E(m) = \int_0^{\pi/2} d\phi \sqrt{1 - m^2 \sin^2 \phi}, \tag{7.8.32}$$

which has no singularities and goes smoothly from $\pi/2$ to 1 as m goes from 0 to 1. The expression for C is given by

$$\frac{C}{2R} = \left(\frac{K_1}{\sinh 2K_1} \right)^2 \left[\frac{2K(\gamma)}{\pi} \left(\frac{2 + \sinh^2 2K_1 + \sinh^4 2K_1}{\cosh^2 2K_1} \right) - \frac{2E(\gamma)}{\pi} \cosh^2 2K_1 - 1 \right]. \tag{7.8.33}$$

Here the singularity in $K(\gamma)$ at $\gamma = \pm 1$ is not suppressed, and C becomes infinite at T_c.

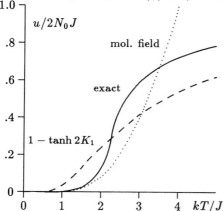
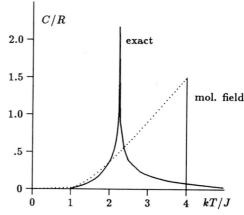

Figure 7.22

a. Energy of a square Ising lattice with equal coupling constants. The energy is calculated exactly and compared with its first term and with the molecular field model.

b. Molar heat capacity for the square Ising lattice in exact calculation and in the molecular-field model.

The elliptic integrals in this calculation are sometimes given as functions of a different parameter, usually denoted by k, to be defined. The complement of γ, written here as γ_1, is defined as

$$\gamma_1 = 2\tanh^2 2K_1 - 1, \qquad (7.8.34)$$

where $\gamma^2 + \gamma_1^2 = 1$. A standard relationship among elliptic integrals is

$$K(k) = \frac{1 + |\gamma_1|}{2} K(\gamma), \qquad (7.8.35)$$

where the definition of k is

$$k = \frac{1 - |\gamma_1|}{1 + |\gamma_1|} = \frac{1}{\sinh^2 2K_1}, \ T < T_c$$
$$= \sinh^2 K_1, \ T > T_c, \qquad (7.8.36)$$

and the alternative form of the elliptic integral is

$$|\gamma_1| K(\gamma) = \frac{2|\gamma_1|}{1 + |\gamma_1|} K(k). \qquad (7.8.37)$$

For $\gamma \to 1$, or $\gamma_1 \to 0$, approximate expressions for K, K' and E are

$$K(\gamma) \approx \ln(4/\gamma_1); \quad E(\gamma) \approx 1; \quad \text{and} \quad \frac{dK(\gamma)}{d\gamma} \approx \frac{\pi}{2}. \qquad (7.8.38)$$

The behavior of C is thus seen to differ both quantitatively and qualitatively from the results of molecular-field theory. As a consequence, since the exact, analytic calculations for the Ising model have clearly shown that the mean-field results are incorrect in one and two dimensions, four courses of action are apparent: (1) solve the Ising problem analytically in three dimensions; (2) failing that, solve it with good precision numerically; (3) improve the molecular-field model to provide a suitable approximation of the effects of short-range order; and (4) find a different approach to the problem of treating phase transitions. So far, option (1) has remained as a tempting but elusive challenge; option (2) has been done quite well, and most of the questions about Ising phase transitions in three dimensions have been answered, but without a neat formula like Eq. (7.8.29) or Eq. (7.8.33); the problems at the end of this chapter provide examples of option (3); and Chapter 8 is about option (4).

7.9 CORRELATIONS

The total spin of an N-spin system is given by

$$S \equiv NI = \left| \sum_{i=1}^{N} \sigma_i \right|, \qquad (7.9.1)$$

where $0 \leq I \leq 1$. The absolute value of the magnetic moment is then $\mathcal{M} = NIm$, and I is called the *coefficient of spontaneous magnetization*. It is evident that I is related to the x of Eq. (7.4.2) by $I = |x|$, since $N\langle\sigma\rangle = Nx$ and $N|\langle\sigma\rangle| = NI$. The probability that a particular spin, σ_j, belongs to a lattice with total spin opposite in sign to that of σ_j is

$(1/2)(1-I)$, and the probability that it belongs to a lattice with the same sign of total spin is $(1/2)(1+I)$. Thus the mean total spin of the lattice is

$$\langle S \rangle = \left\langle \sigma_j \sum_{i=1}^{N} \sigma_i \right\rangle = \frac{1}{2}(1-I)(-NI) + \frac{1}{2}(1+I)(NI) = NI^2, \qquad (7.9.2)$$

or, since most of the spins in the very large lattices considered here are distant ones, in the limit of infinite N, it follows that $\langle \sigma_i \sigma_j \rangle = I^2$. Thus, in order for long-range spin correlation to exist in the lattice, the coefficient of spontaneous magnetization must be greater than zero, no matter which direction the average spin has. Calculation of the coefficient of spontaneous magnetization is an additional challenge in the exact treatment of the Ising, or any other, spin lattice.

7.9.1 Direct Calculation in One Dimension. For a closed chain of N identical Ising spins, the correlation $\langle \sigma_i \sigma_{i+n} \rangle$ is the same as that of $\langle \sigma_1 \sigma_{1+n} \rangle$, which may be written, with $\sigma_{N+1} \equiv \sigma_1$,

$$\langle \sigma_1 \sigma_{1+n} \rangle = Z_N^{-1} \sum_{\{\sigma_i = \pm 1\}} \sigma_1 \sigma_{1+n} \prod_{i=1}^{N} e^{K \sigma_i \sigma_{i+1}}$$

$$= Z_N^{-1} \cosh^N K \sum_{\{\sigma_i\}} \sigma_1 \sigma_{1+n} \prod_{i=1}^{N} (1 + \sigma_i \sigma_{i+1} y), \qquad (7.9.3)$$

where Z_N is given by Eq. (7.5.28), and the evaluation is carried out by direct multiplication and summation over the two possible values of each σ, as was done in the steps from Eq. (7.5.27) to Eq. (7.5.28). The result is

$$\langle \sigma_1 \sigma_{1+n} \rangle = \frac{y^n + y^{N-n}}{1 + y^N}. \qquad (7.9.4)$$

In the limit as $N \to \infty$ while n remains finite, the correlation function becomes

$$\langle \sigma_1 \sigma_{1+n} \rangle = y^n. \qquad (7.9.5)$$

The coefficient of spontaneous magnetization is then found from

$$I^2 = \frac{1}{N} \sum_{n=1}^{N} \langle \sigma_1 \sigma_{1+n} \rangle$$

$$= \frac{1}{N} \sum_{n=1}^{N} y^n = \frac{1}{N} \frac{1 - y^{N+1}}{1 - y}$$

$$\to 0 \text{ as } N \to \infty. \qquad (7.9.6)$$

The nearest neighbor correlation, $\langle \sigma_i \sigma_{i+1} \rangle = \langle \sigma_1 \sigma_2 \rangle$, in the limit as $N \to \infty$, is seen from Eq. (7.9.5) to be $\langle \sigma_1 \sigma_2 \rangle = y = \tanh K$. Another way to obtain this result is by direct differentiation of the partition function with respect to K. From Eq. (7.5.1), rewritten for a closed chain,

$$\frac{\partial \ln Z_N}{\partial K} = Z_N^{-1} \sum_{\{\sigma_i = \pm 1\}} \sum_{j=1}^{N} \sigma_j \sigma_{j+1} \exp\left[K \sum_{i=1}^{N} \sigma_i \sigma_{i+1} \right]$$

$$= N Z_N^{-1} \sum_{\{\sigma_i\}} \sigma_1 \sigma_2 \exp\left[K \sum_{i=1}^{N} \sigma_i \sigma_{i+1} \right] = N \langle \sigma_1 \sigma_2 \rangle. \tag{7.9.7}$$

The actual calculation is made with the partition function of Eq. (7.5.28), giving

$$\langle \sigma_1 \sigma_2 \rangle = \frac{1}{N} \frac{\partial \ln Z_N}{\partial K} = \frac{y + y^{N-1}}{1 + y^N}, \tag{7.9.8}$$

in agreement with Eq. (7.9.4) for $n = 1$, and with Eq. (7.9.5) for $N \to \infty$.

These correlations are displayed for reference, as a verification of the results of more advanced methods of calculation.

7.9.2 Dimer-Pfaffian Calculation for the One-Dimensional Chain.   For the Pfaffian technique, the energy-shifted Hamiltonian of Eq. (7.7.7) is used to obtain the partition function of Eq. (7.7.8). In this context, the spin correlation is written, with x instead of y,

$$\langle \sigma_1 \sigma_{1+m} \rangle = Z_N^{-1} e^{-NK} \sum_{\{\sigma_i = \pm 1\}} \sigma_1 \sigma_{m+1} \prod_{i=1}^{N} e^{K \sigma_i \sigma_{i+1}}$$

$$= Z_N^{-1} e^{-NK} \cosh^N K \sum_{\{\sigma_i\}} \sigma_1 \sigma_{m+1} \prod_{i=1}^{N} (1 + x \sigma_i \sigma_{i+1})$$

$$= Z_N^{-1} e^{-NK} \cosh^N K \sum_{\{\sigma_i\}} (\sigma_1 \sigma_2 \sigma_2 \sigma_3 \sigma_3 \cdots \sigma_m \sigma_m \sigma_{m+1}) \prod_{i=1}^{N} e^{K \sigma_i \sigma_{i+1}}$$

$$= Z_N^{-1} e^{-NK} \cosh^N K \sum_{\{\sigma_i\}} (\sigma_1 \sigma_2 + x)(\sigma_2 \sigma_3 + x) \cdots (\sigma_m \sigma_{m+1} + x)$$

$$\times (1 + x \sigma_{m+1} \sigma_{m+2}) \cdots (1 + x \sigma_N \sigma_1)$$

$$= \frac{x^m}{2^N Pf(x)} \sum_{\{\sigma_i\}} \left(1 + x^{-1} \sigma_1 \sigma_2\right) \left(1 + x^{-1} \sigma_2 \sigma_3\right) \cdots \left(1 + x^{-1} \sigma_m \sigma_{m+1}\right)$$

$$\times (1 + x \sigma_{m+1} \sigma_{m+2}) \cdots (1 + x \sigma_N \sigma_1)$$

$$= \frac{x^m}{2^N Pf(x)} \sum_{\{\sigma_i\}} \prod_{i=1}^{N} (1 + x_i \sigma_i \sigma_{i+1})$$

$$= x^m Pf(\{x_i\})/Pf(x), \tag{7.9.9}$$

where $\sigma_i^2 = 1$ has been used to rewrite $\sigma_1 \sigma_{m+1}$, and $Pf(\{x_i\})$ denotes that the usual Pfaffian has been modified, with x_i replaced by x^{-1} for $1 \le i \le m - 1$, and $x_i = x$ for $i \ge m$. The square of the final equation gives

$$\langle \sigma_1 \sigma_{m+1} \rangle^2 = x^{2m} |D + \delta_m| / |D| = x^{2m} |I + D^{-1} \delta_m|, \tag{7.9.10}$$

where

$$D = I_N \otimes \Gamma + E_N \otimes X - E_N^{-1} \otimes X^T; \tag{7.9.11}$$

$$\delta_m = E_{N,m} \otimes \tilde{X} - \left(E_N^{-1}\right)_m \otimes \tilde{X}^T; \tag{7.9.12}$$

$$X = \begin{pmatrix} 0 & 0 \\ x & 0 \end{pmatrix}; \quad X^T = \begin{pmatrix} 0 & x \\ 0 & 0 \end{pmatrix}; \quad \tilde{X} = \begin{pmatrix} 0 & 0 \\ x^{-1} - x & 0 \end{pmatrix}; \tag{7.9.13}$$

$$E_{N,m} = \begin{bmatrix} 0 & 1 & 0 & 0 & 0 & \cdots \\ 0 & 0 & 1 & 0 & 0 & \cdots \\ 0 & 0 & 0 & 1 & 0 & \cdots \\ \vdots & & & \ddots & & \cdots \\ & & m & \text{of} & \text{these} & \cdots \\ & \text{all} & \text{else} & \text{zero} & & \cdots \end{bmatrix}; \tag{7.9.14}$$

and

$$\left(E_N^{-1}\right)_m = \begin{bmatrix} 0 & 0 & 0 & 0 & \cdots \\ 1 & 0 & 0 & 0 & \cdots \\ 0 & 1 & 0 & 0 & \cdots \\ 0 & 0 & \ddots & & \\ & m & \text{of} & \text{these} & \\ & \text{all} & \text{else} & \text{zero} & \end{bmatrix}. \tag{7.9.15}$$

The calculation of D^{-1} is not trivial. The procedure here is to diagonalize D, as in Eq. (7.7.19), find the inverse of D', transform $(D')^{-1}$ back to D^{-1}, and use this result to complete the calculation. From Eq. (7.7.18), D' is found to be

$$
\begin{aligned}
D' &= T_N \otimes I_2 D T_N^{-1} \otimes I_2 = I_N \otimes \Gamma + \Lambda_N \otimes X - \Lambda_N^{-1} \otimes X^T \\
&= \begin{bmatrix} d_2(\lambda_1) & 0 & 0 & \cdots \\ 0 & d_2(\lambda_2) & 0 & \cdots \\ 0 & 0 & d_2(\lambda_3) & \cdots \\ & & & \ddots \\ \cdots & 0 & 0 & d_2(\lambda_N) \end{bmatrix},
\end{aligned} \tag{7.9.16}
$$

where the 2×2 matrices along the diagonal are

$$d_2(\lambda_n) = \begin{pmatrix} 0 & 1 - x\lambda_n^{-1} \\ -1 + x\lambda_n & 0 \end{pmatrix}, \tag{7.9.17}$$

with $\lambda_n = e^{(2n-1)i\pi/N}$, for $1 \leq n \leq N$. Note that $\lambda_n^N = -1$, $\lambda_{N-n}\lambda_{n+1} = 1$, and a number of useful relations can be extracted from the definition of λ_n, such as

$$\sum_{n=1}^{N} (\lambda_r \lambda_s)^n = N\delta_{r+s}^{N+1},$$

$$\sum_{m=1}^{N} \sum_{n=1}^{N} (\lambda_m/\lambda_n)^r = 1,$$

$$\sum_{r=1}^{N} (\lambda_m/\lambda_n)^r = -1 + (N+1)\delta_n^m,$$

$$\sum_{n=r}^{r+N} (-\lambda_r)^n (-\lambda_{N-s+1})^{n+s-r} = N\delta_r^s, \tag{7.9.18}$$

where the Kronecker delta has been used. The inverse of D' is

$$(D')^{-1} = \begin{pmatrix} \ddots & & \\ & d_2^{-1}(\lambda_n) & \\ & & \ddots \end{pmatrix}, \tag{7.9.19}$$

where

$$d_2^{-1}(\lambda_n) = \begin{pmatrix} 0 & \frac{-1}{1-x\lambda_n} \\ \frac{1}{1-x\lambda_n^{-1}} & 0 \end{pmatrix}. \tag{7.9.20}$$

The transformation matrix and its inverse, needed to return from $(D')^{-1}$ to D^{-1}, are

$$T_N = \frac{1}{\sqrt{N}} \begin{bmatrix} 1 & \lambda_1^{-1} & \lambda_1^{-2} & \lambda_1^{-3} & \cdots & \lambda_1^{-(N-1)} \\ \lambda_2^1 & 1 & \lambda_2^{-1} & \lambda_2^{-2} & \cdots & \lambda_2^{-(N-2)} \\ \lambda_3^2 & \lambda_3^1 & 1 & \lambda_3^{-1} & \cdots & \lambda_3^{-(N-3)} \\ \vdots & & & & \ddots & \\ \lambda_N^{N-1} & & & \cdots & \lambda_N & 1 \end{bmatrix}; \tag{7.9.21}$$

and

$$T_N^{-1} = \frac{1}{\sqrt{N}} \begin{bmatrix} 1 & \lambda_{N-1} & \lambda_{N-2}^2 & \lambda_{N-3}^3 & \cdots & \lambda_1^{N-1} \\ \lambda_N^{-1} & 1 & \lambda_{N-2} & \lambda_{N-3}^2 & \cdots & \lambda_1^{N-2} \\ \lambda_N^{-2} & \lambda_{N-1}^{-1} & 1 & \lambda_{N-3} & \cdots & \lambda_1^{N-3} \\ \vdots & & & & \ddots & \\ \lambda_N^{-(N-1)} & \lambda_{N-1}^{-(N-2)} & \cdots & & & 1 \end{bmatrix}. \tag{7.9.22}$$

From these matrices, the inverse of D is found to be

$$D^{-1} = \left(T_N^{-1} \otimes I_2\right)(D')^{-1}\left(T_N \otimes I_2\right) = \frac{1}{N}\sum_{n=1}^{N}\sum_{r=1}^{N} E_N^r \otimes \lambda_n^{-r} d_2^{-1}(\lambda_n)$$

$$= \frac{1}{N}\sum_{n=1}^{N} \begin{bmatrix} 1 & \lambda_n^{-1} & \lambda_n^{-2} & \cdots & \lambda_n^{-(n-1)} \\ \lambda_n & 1 & \lambda_n^{-1} & \cdots & \lambda_n^{-(N-2)} \\ \lambda_n^2 & \lambda_n & 1 & 1 & \cdots \\ \vdots & & & & \vdots \\ \lambda_n^{N-1} & \lambda_n^{N-2} & \cdots & \lambda_n & 1 \end{bmatrix} \otimes d_2^{-1}(\lambda_n). \tag{7.9.23}$$

In the limit as $N \to \infty$, the sum of matrices in the last step may be replaced by an integral; $\lambda_n \to e^{i\phi} \equiv z$; and the matrix elements become

$$\frac{1}{N}\sum_{n=1}^{N} \frac{\lambda_n^{-r}}{1-x\lambda_n^{-1}} \to \frac{1}{2\pi i}\oint \frac{dz}{z^r(z-x)} = \begin{cases} 0, & r > 0 \\ 1, & r = 0 \end{cases}$$

$$\frac{1}{N}\sum_{n=1}^{N}\frac{\lambda_n^r}{1-x\lambda_n^{-1}} \to \frac{1}{2\pi i}\oint\frac{dz\,z^r}{(z-x)} = \begin{cases} x^r, & r>0 \\ 1, & r=0 \end{cases}$$

$$\frac{1}{N}\sum_{n=1}^{N}\frac{\lambda_n^{-r}}{1-x\lambda_n} \to \frac{-1}{2\pi ix}\oint\frac{dz}{z^{r+1}(z-1/x)} = \begin{cases} x^r, & r>0 \\ 1, & r=0 \end{cases}$$

$$\frac{1}{N}\sum_{n=1}^{N}\frac{\lambda_n^r}{1-x\lambda_n} \to \frac{-1}{2\pi ix}\oint\frac{dz\,z^{r-1}}{(z-1/x)} = \begin{cases} 0, & r>0 \\ 1, & r=0 \end{cases} \tag{7.9.24}$$

Note that the integration replaces the summation on n, but not the one on r in Eq. (7.9.23). With these operations, D^{-1} is found to be

$$D^{-1} = \begin{bmatrix} \begin{pmatrix} 0 & -1 \\ 1 & 0 \end{pmatrix} & \begin{pmatrix} 0 & -x \\ 0 & 0 \end{pmatrix} & \begin{pmatrix} 0 & -x^2 \\ 0 & 0 \end{pmatrix} & \begin{pmatrix} 0 & -x^3 \\ 0 & 0 \end{pmatrix} & \cdots & \begin{pmatrix} 0 & -x^n \\ 0 & 0 \end{pmatrix} & \cdots \\ \begin{pmatrix} 0 & 0 \\ x & 0 \end{pmatrix} & \begin{pmatrix} 0 & -1 \\ 1 & 0 \end{pmatrix} & \begin{pmatrix} 0 & -x \\ 0 & 0 \end{pmatrix} & \begin{pmatrix} 0 & -x^2 \\ 0 & 0 \end{pmatrix} & \cdots & & \\ \begin{pmatrix} 0 & 0 \\ x^2 & 0 \end{pmatrix} & \begin{pmatrix} 0 & 0 \\ x & 0 \end{pmatrix} & \begin{pmatrix} 0 & -1 \\ 1 & 0 \end{pmatrix} & \begin{pmatrix} 0 & -x \\ 0 & 0 \end{pmatrix} & \cdots & & \\ \vdots & \ddots & \ddots & \ddots & \ddots & & \\ \begin{pmatrix} 0 & 0 \\ x^{n-1} & 0 \end{pmatrix} & & & & & & \end{bmatrix}$$

$$\tag{7.9.25}$$

As a check, note that for $N \to \infty$, $D^{-1}D = DD^{-1} = I$.

The matrix product $D^{-1}\delta_m$ becomes

$$D^{-1}\delta_m = D^{-1}\begin{bmatrix} \begin{pmatrix} 0 & 0 \\ 0 & 0 \end{pmatrix} & \begin{pmatrix} 0 & - \\ \delta x & 0 \end{pmatrix} & \begin{pmatrix} - & - \\ 0 & - \end{pmatrix} & \cdots \\ \begin{pmatrix} 0 & -\delta x \\ 0 & 0 \end{pmatrix} & \begin{pmatrix} 0 & 0 \\ 0 & 0 \end{pmatrix} & \begin{pmatrix} 0 & - \\ \delta x & - \end{pmatrix} & \cdots \\ \begin{pmatrix} 0 & 0 \\ 0 & 0 \end{pmatrix} & \begin{pmatrix} 0 & -\delta x \\ 0 & 0 \end{pmatrix} & \begin{pmatrix} 0 & 0 \\ 0 & 0 \end{pmatrix} & \cdots \\ \vdots & (m \text{ such} & \text{terms}) & \cdots \end{bmatrix}$$

$$= \begin{bmatrix} 0 & 0 & -\delta x & 0 & -x\delta x & 0 & -x^2\delta x & 0 & \text{to} & -x^{m-1}\delta x & \cdots \\ 0 & 0 & 0 & 0 & 0 & 0 & 0 & 0 & & 0 & \cdots \\ 0 & 0 & 0 & 0 & -\delta x & 0 & -x\delta x & \cdots & & & \\ 0 & -\delta x & 0 & 0 & 0 & 0 & 0 & \cdots & & & \\ 0 & 0 & 0 & 0 & 0 & 0 & -\delta x & \cdots & & & \\ 0 & -x\delta x & 0 & -\delta x & 0 & 0 & 0 & \cdots & & & \\ \vdots & \text{etc.} & \ddots & \ddots & \ddots & & & & & & \end{bmatrix} .\tag{7.9.26}$$

The determinant of Eq. (7.9.10) consists of a $2(m+1) \times 2(m+1)$ element in the upper left-hand corner, with the rest of the terms zero, except for the 1's from I along the principal diagonal. Therefore the only factor that requires attention is the $2(m+1) \times 2(m+1)$

determinant

$$
\left|I + D^{-1}\delta_m\right| =
\begin{vmatrix}
1 & 0 & -\delta x & 0 & -x\delta x & 0 & -x^2\delta x & \cdots & \text{to} & -x^{m-1}\delta x \\
0 & 1 & 0 & 0 & 0 & 0 & 0 & \cdots & & 0 \\
0 & 0 & 1 & 0 & -\delta x & 0 & -x\delta x & \cdots & & 0 \\
0 & -\delta x & 0 & 1 & 0 & 0 & 0 & \cdots & & 0 \\
0 & 0 & 0 & 0 & 1 & 0 & -\delta x & \cdots & & 0 \\
0 & -x\delta x & 0 & -\delta x & 0 & 1 & 0 & \cdots & & 0 \\
\vdots & \vdots & & & & & & \ddots & \cdots & \vdots \\
0 & -x^{m-1}\delta x & 0 & -x^{m-2}\delta x & 0 & & & \cdots & & 1
\end{vmatrix}
$$

$$
= 1, \tag{7.9.27}
$$

from expansion by column 1, row 2, column 3, row 4, *etc.* It then follows from Eq. (7.9.10) that $\langle \sigma_1 \sigma_{m+1}\rangle^2 = x^{2m}$, or $\langle \sigma_1 \sigma_{m+1}\rangle = x^m$, as given by Eq. (7.9.5).

Here again the one-dimensional problem has been used as a vehicle for introducing a technique that can then be applied to more-complicated problems.

7.9.3 Correlations in a Two-Dimensional Ising Lattice.

The correlation $\langle \sigma_{11}\sigma_{p+1,q+1}\rangle$ in the Ising square lattice can be found by the same kind of averaging procedure as that of Eq. (7.9.9); the original entity to be averaged, $\sigma_{11}\sigma_{p+1,q+1}$ is replaced by an equivalent expression $\sigma_{11}\sigma_{21}^2 \cdots \sigma_{p1}^2 \sigma_{p+1,1}\sigma_{p+1,2}^2 \cdots \sigma_{p+1,q}^2 \sigma_{p+1,q+1}$, where intermediate σ^2's are inserted that connect the point $(1,1)$ to the point (p,q). (Any other possible path is acceptable.) The term $x^p y^q$ is then factored out of the resulting expression, and the correlation is the written

$$
\langle \sigma_{11}\sigma_{p+1,q+1}\rangle = x^p y^q P f(x_{ij}, y_{ij}), \tag{7.9.28}
$$

where $x_{ij} = x^{-1}$ for $j = 1$, $1 \le i \le p$, but $x_{ij} = x$ otherwise; $y_{ij} = y^{-1}$ for $i = p+1$, $1 \le j \le q$, but $y_{ij} = y$ otherwise. As an introduction to the procedure, the calculation will be simplified, as before, by setting $K_1 = K_2$, or $y = x$, in the final calculation. In this case, the correlation function for the lattice may be obtained in the limit of infinite p as $\langle \sigma_{1,1}\sigma_{p+1,1}\rangle$, rather than the more general expression of Eq. (7.9.28). In parallel with the 1-d procedure, the square of the correlation function becomes

$$
\langle \sigma_{1,1}\sigma_{p+1,1}\rangle^2 = x^{2p}\left|D + \delta_p\right|/\left|D\right| = x^{2p}\left|I + D^{-1}\delta_p\right|. \tag{7.9.29}
$$

As before, D is inverted by first diagonalizing it, then inverting D', and finally transforming $(D')^{-1}$ to D^{-1}.

$$
D = I_N \otimes I_M \otimes \Gamma + I_N \otimes E_M \otimes X - I_N \otimes E_M^{-1} \otimes X^T + E_N \otimes I_M \otimes Y - E_N^{-1} \otimes I_M \otimes Y^T; \tag{7.9.30}
$$

$$
D' = I_N \otimes I_M \otimes \Gamma + I_N \otimes \Lambda_M \otimes X - I_N \otimes \Lambda_M^{-1} \otimes X^T + \Lambda_N \otimes I_M \otimes Y - \Lambda_N^{-1} \otimes I_M \otimes Y^T;
$$

$$
=
\begin{bmatrix}
d(\lambda_1,\mu_1) & & & \\
& \ddots & & \\
& & d(\lambda_i,\mu_j) & \\
& & & \ddots & \\
& & & & d(\lambda_M,\mu_N)
\end{bmatrix}, \tag{7.9.31}
$$

where i and j are paired such that every possible value of i occurs exactly once with every possible value of j, μ is used for the eigenvalues in Λ_N, to avoid an extra subscript, and $d(\lambda_i, \mu_j)$ becomes $D(r, s)$, as in Eq. (7.8.20), with $D(r, s)$ given in Eq. (7.8.17), along with its determinant. The inverse of D' is

$$(D')^{-1} = \begin{bmatrix} \ddots & & \\ & D^{-1}(r, s) & \\ & & \ddots \end{bmatrix}. \tag{7.9.32}$$

If D' is written

$$D' = (P_N \otimes Q_M \otimes I) D \left(P_N^{-1} \otimes Q_M^{-1} \otimes I \right), \tag{7.9.33}$$

then $(D')^{-1}$ becomes

$$(D')^{-1} = (P_N \otimes Q_M \otimes I) D^{-1} \left(P_N^{-1} \otimes Q_M^{-1} \otimes I \right), \tag{7.9.34}$$

and this may be solved for D^{-1} to give

$$
\begin{aligned}
D^{-1} &= \left(P_N^{-1} \otimes Q_M^{-1} \otimes I \right) \left(D' \right)^{-1} (P_N \otimes Q_m \otimes I) \\
&= \frac{1}{MN} \sum_{n=1}^{N} \sum_{m=1}^{M} \sum_{r=1}^{N} \sum_{s=1}^{M} E_N^r \otimes E_M^s \otimes \lambda_n^{-r} \mu_n^{-s} D^{-1}(m, n) \\
&= \frac{1}{MN} \sum_{m=1}^{M} \sum_{n=1}^{N} \begin{bmatrix} \ddots & & \\ & \lambda_n^{-r} E_N^r & \\ & & \ddots \end{bmatrix} \otimes \begin{bmatrix} \ddots & & \\ & \mu_m^{-s} E_M^s & \\ & & \ddots \end{bmatrix} \otimes D^{-1}(m, n). \quad (7.9.35)
\end{aligned}
$$

The sums on m and n are replaced by integrals in the thermodynamic limit, as in the step from Eq. (7.9.24) to Eq. (7.9.25).

The subsequent steps are, in principle, merely an extension of the procedures already exhibited for the one-dimensional system, and indeed the calculation can be completed by following those procedures. To finish the calculation with more elegance, however, requires use of properties of Toeplitz matrices (Cheng and Wu [1967]; McCoy and Wu [1967a]; McCoy and Wu [1967b]; McCoy and Wu [1973]; Montroll, Potts and Ward [1963]; Vdovichenko [1965b]; Wu [1966]). The summed λ and μ expressions of Eq. (7.9.35) are Toeplitz matrices, as is the complete right-hand side of that equation. The final result is, for $T < T_c$,

$$
\begin{aligned}
I^2 = \lim_{p \to \infty} \left\langle \sigma_{1,1} \sigma_{1+p,1} \right\rangle &= \left[1 - \left(\frac{(1 - x^2)}{2x} \right)^4 \right]^{1/4} \\
&= \left[1 - \left(\frac{1}{\sinh^2 2K_1} \right)^4 \right]^{1/4} = \left[1 - k^2 \right]^{1/4}, \quad (7.9.36)
\end{aligned}
$$

where k has been used before, in Eq. (7.8.36). For $T > T_c$, $I = 0$. From the relation $I = \left(1 - k^2 \right)^{1/8}$, expansion about the temperature T_c leads to

$$I \sim (-\tau)^{1/8}, \tag{7.9.37}$$

leading to the exact, and clearly nonclassical, result that the critical exponent $\beta = 1/8$.

7.9.4 The Nearest-Neighbor Correlation. In one dimension, the nearest-neighbor correlation function was obtained in Eq. (7.9.8) by differentiation of the partition function with respect to K. In two dimensions, the same kind of calculation is available. In the square lattice with $K_1 = K_2$, the correlations $\sigma_{1,1}\sigma_{1,2}$ and $\sigma_{1,1}\sigma_{2,1}$ are equal to each other and to all other nearest-neighbor correlation functions. Because of the shift in the zero of energy for the Pfaffian calculation, the partition function must be slightly modified for this derivation. The correlation is

$$
\begin{aligned}
\left\langle \sigma_{1,1}\sigma_{2,1} \right\rangle &= \frac{1}{2MN} \frac{\partial \ln\left(Z_{MN}e^{2MNK_1}\right)}{\partial K_1} \\
&\to \lim_{M,N\to\infty} \frac{\partial}{\partial K_1}\left(\frac{1}{2MN}\ln Z_{MN}e^{2MNK_1}\right) \\
&= \frac{1}{2}\frac{\partial \ln Z_1^*}{\partial K_1},
\end{aligned}
\tag{7.9.38}
$$

where the modified single-site partition function is

$$
\begin{aligned}
\ln Z_1^* &= \ln 2 + \frac{1}{2\pi^2}\int_0^{2\pi}d\phi\int_0^{2\pi}d\theta \ln\left[\cosh^2 2K_1 - \sinh 2K_1(\cos\phi + \cos\theta)\right] \\
&= \ln(2\cosh 2K_1) + \frac{1}{\pi}\int_0^{\pi/2}d\omega_1 \ln\left[\frac{1 + \sqrt{1-\gamma^2\cos^2\omega_1}}{2}\right].
\end{aligned}
\tag{7.9.39}
$$

Its derivative recapitulates that of Eq. (7.8.29), giving

$$
\left\langle \sigma_{1,1}\sigma_{2,1} \right\rangle = \frac{1}{2}\frac{\partial \ln Z_1^*}{\partial K_1} = \tanh 2K_1 + \frac{1 - \sinh^2 2K_1}{\sinh 2K_1 \cosh 2K_1}\left[\frac{1}{2} - \frac{1}{\pi}K(\gamma)\right].
\tag{7.9.40}
$$

Note that, except for the shift of the energy zero, the molar energy of Eq. (7.8.29) is related to the average n-n correlation by

$$
u = -N_o J \left\langle \sigma_{1,1}\sigma_{2,1} \right\rangle,
\tag{7.9.41}
$$

as can be seen directly from the Hamiltonian. This result is direct confirmation of the importance of short-range order in the correct treatment of phase transitions.

7.10 SERIES EXPANSIONS AND PADÉ APPROXIMANTS

Exact analytic solutions are not available for Ising lattices in three or more dimensions, for even two-dimensional Ising lattices with non-zero B-fields, and for more complicated models with higher spin dimensions, such as the X-Y and Heisenberg models, to be discussed briefly in the next section.

In order to examine the critical behavior of such systems, accurate and efficient approximation procedures have been developed. As shown in Eq. (7.6.2), and extended somewhat in Eq. (7.6.3), the partition function for the square Ising lattice can be expanded as a series that counts closed graphs of all possible numbers of bonds. The expansion parameter in Eq. (7.6.2) is $y = \tanh\beta J$; at high temperatures, $y \ll 1$, and the first few terms of the series in Eq. (7.6.2) may approximate the partition function with sufficient accuracy when T is high. The actual values of $n(r)$, the number of closed graphs of r bonds, can be counted directly when r is small, and the first few terms in the series can thus be so obtained. As will be shown in the next subsection, other quantities can also be represented as series expansions, with the low-order terms accessible to direct counting. The remaining problem is that of extracting the critical behavior of the system, including values of the critical exponents, from a limited number of terms of a series expansion that is expected to be accurate, because of the paucity of terms, only at high temperatures. Two procedures for doing this, the asymptotic-ratio method and the Padé approximant method, are presented in subsequent subdivisions of this section. This entire treatment is intended to demystify the process of obtaining approximate values of critical exponents and other thermodynamic properties, without attempting to provide a working level of expertise in the methods. (For serious work with these approximations, a number of subtleties and advanced techniques recommend reference to specialized treatises, some of which are listed at appropriate points in this section.)

7.10.1 Series Expansions of System Properties. In the absence of an external B field, the first few terms of the partition function of Eq. (7.6.2) can be obtained by primitive counting of the closed graphs of r bonds on the lattice. Thus for the closed linear chain, $n(0) = 1$ and $n(N) = 1$ are the only nonzero terms, and $Z_N = (2 \cosh K)^N (1 + y^N)$, in agreement with Eq. (7.5.28). For a 2-d square lattice, $n(0) = 1$, $n(4) = N$, $n(6) = 2N$, since the 2×3 rectangle can be oriented in 2 different ways, and $n(8) = 7N + (1/2!)N(N - 5)$, where the second term of this expression counts the number of ways of placing two unit squares on the lattice with no coincident sides. More terms can be found in this manner, but typically the difficulty of finding the next term exceeds that of finding all terms up to that point, and accuracy becomes a problem. A number of sophisticated theorems and computer programs for counting various kinds of diagrams on a lattice have been developed, and their use is essential to efforts at extending the number of terms in these series representations. (Domb [1949a]; Domb [1949b]; Domb [1960]; Domb and Sykes [1961]; Guttmann [1989]; Oguchi [1950]; Oguchi [1951a]; Oguchi [1951b]; Sykes [1961])

Since terms of the sort already exhibited depend on N, and since N is required to become infinite in the thermodynamic limit, the single-site partition function is the appropriate choice, defined as

$$Z_1 = \lim_{N \to \infty} Z_N^{1/N}. \tag{7.10.1}$$

For the 2d square Ising lattice, the expression obtained thus far is

$$
\begin{aligned}
Z_1 &= \left[\left(2 \cosh^2 K \right)^N \left(1 + N x^4 + 2N x^6 + [7N + (1/2)N(N-5)]x^8 + \cdots \right) \right]^{1/N} \\
&= 2 \cosh^2 K \left[1 + x^4 + 2x^6 + 5x^8 + \cdots \right].
\end{aligned} \tag{7.10.2}
$$

For a simple cubic lattice, the corresponding expression is

$$Z_1 = 2 \cosh^3 K [1 + 3x^4 + 22x^6 + 192x^8 + 2046x^{10} + 24853x^{12} + \cdots, \tag{7.10.3}$$

and similar expressions are available for a variety of lattices, up to somewhat higher order than that of Eq. (7.10.3).

Since all thermodynamic information about a system is implicit in the partition function, it should be possible to extract this information, at least in series form, from the partition function of Eq. (7.10.1). For example, the single-site Helmholtz potential is $f_1 = -kT \ln Z_1$, and if this corresponds to F_B of Chapter 5, then $df_1 = -s_1 dT - \langle m \rangle \, dB_o$, where $\langle m \rangle = m \langle \sigma \rangle$ is the mean magnetic moment per site. The heat capacity at constant B_o, written as C_H, is given by

$$C_H = N_0 T \left(\frac{\partial s_1}{\partial T} \right)_{B_o} = N_0 \beta^2 \left(\frac{\partial^2 \ln Z_1}{\partial \beta^2} \right)_{B_o}. \tag{7.10.4}$$

The critical exponent α can be approximated from the series for C_H.

The isothermal susceptibility can also be found from df_1. First, the mean magnetic moment per spin is

$$\langle m \rangle = m \langle \sigma \rangle = - \left(\frac{\partial f_1}{\partial B} \right)_T = \frac{kT}{N} \left(\frac{\partial \ln Z_N}{\partial B} \right)_T, \tag{7.10.5}$$

where it is assumed that the external magnetic field terms in the partition function are included. The magnetization is $M = \mathcal{N} \langle m \rangle$, where $\mathcal{N}$ is the number of spins per unit volume. The isothermal susceptibility is

$$\chi_T = \mu_0 \left(\frac{\partial M}{\partial B} \right)_T = \mu_0 \mathcal{N} m \left(\frac{\partial \langle \sigma \rangle}{\partial B} \right)_T. \tag{7.10.6}$$

Usually in this context the susceptibility is normalized to the Curie value for a system of noninteracting spins, and examined in the limit $B = 0$. It is written here as

$$\bar{\chi}_T(T, B = 0) = \chi_T(T, B = 0)/\mu_0 \mathcal{N} m^2 \beta, \tag{7.10.7}$$

where the denominator provides the normalization. After the B-derivatives are taken, the normalization carried out, and B set to zero, the susceptibility is found to be

$$\bar{\chi}_T = Z_N^{-1} \sum_{\{\sigma_i = \pm 1\}} \sum_{n=1}^{N} \sigma_1 \sigma_n \prod_{i,j} e^{K \sigma_i \sigma_j} - \left(N Z_N^2 \right)^{-1} \left[\sum_{\{\sigma_i\}} \sum_{n=1}^{N} \sigma_n \prod_{i,j} e^{K \sigma_i \sigma_j} \right]^2$$

$$= \sum_{n=1}^{N} \left[\langle \sigma_1 \sigma_n \rangle - \langle \sigma_n \rangle^2 \right]. \tag{7.10.8}$$

This is the relationship used in the discussion of the Ginzburg criterion in Chapter 6, except that here the spin variable is the order parameter, whereas there in the Landau theory, the order parameter is represented in the continuum limit as a continuous variable. When the squared term is dropped because $[\langle \sigma \rangle]^2 = 0$ when $B = 0$, the remaining term can be expanded as a series.

In a one-dimensional closed chain, from Eq. (7.9.4),

$$\langle \sigma_1 \sigma_{n+1} \rangle = \frac{x^n + x^{N-n}}{1 + x^N},$$

and the final line of Eq. (7.10.8) then gives

$$\bar{\chi}_T = 1 + \sum_1^{N-1} \frac{x^n}{1+x^N} + \sum_1^{N-1} \frac{x^{N-n}}{1+x^N} = \frac{1+x}{1-x}, \tag{7.10.9}$$

in the limit of $N \to \infty$, where the initial 1 is the $n = 0$ term.

The graphs that contribute to $\bar{\chi}_T$ in one dimension are identifiable from the first line of Eq. (7.10.8), with the surviving term when $\langle \sigma_n \rangle$ is set equal to zero rewritten here as

$$\bar{\chi}_T = Z_N^{-1} (\cosh K)^N \sum_{\{\sigma_i\}} \sum_n \sigma_1 \sigma_{1+n} (1 + x\sigma_1\sigma_2)(1 + x\sigma_2\sigma_3)\cdots(1 + x\sigma_N\sigma_1), \tag{7.10.10}$$

which generates graphs of length 0 and length N for $n = 0$, giving the terms $1 + x^N$. For $0 < n \le N - 1$, the graphs that survive summation over both signs of all spins contribute the terms $x^n + x^{N-n}$. The sum of all the 1-d graphs represented by Eq. (7.10.8), omitting the factor 2^N, is

$$\sum = 1 + x^N + \sum_{n=1}^{N-1} \left(x^n + x^{N-n}\right)$$

$$= -\left(1 + x^N\right) + 2\sum_0^N x^n \to \frac{1+x}{1-x}, \tag{7.10.11}$$

where the final expression results when $N \to \infty$. The graphs that contribute to the sum of Eq. (7.10.11) are most easily described as lines on the chain, linking σ_1 with σ_{n+1}, going in either direction from site 1. All intermediate spins occur exactly twice in Eq. (7.10.10) and the open ends are then effectively closed by the factor $\sigma_1\sigma_{n+1}$. When the other factors required by Eq. (7.10.10) are included, the result of summing the graphs for which every spin appears an even number of times agrees with that of Eq. (7.10.9).

In two or more dimensions, the procedure is similar, and the graphs that contribute to the susceptibility arise from terms like those in Eq. (7.6.2) except for the factor $\sigma_1\sigma_n$ that determines which graphs survive the summation over spin signs. These turn out to be diagrams for which every spin is a termination point for an even number of x-bonds except spins σ_1 and σ_n. Such graphs may be broken lines that form a continuous path from σ_1 to σ_n, or they may include crossings and disjoint closed diagrams in other parts of the lattice, provided only that there exists a connection from site 1 to site n. For example, the first few terms of $\bar{\chi}$ for a simple square lattice are

$$\bar{\chi}_T = 1 + 4x + 12x^2 + 36x^3 + 100x^4 + 276x^5 + 740x^6 + 1942x^7 + \cdots. \tag{7.10.12}$$

Such series are known, typically up to powers of x around 30, for most common two- and three-dimensional lattices. (Domb [1974]) The longest one I have seen, with 54 terms, is attributed to B. G. Nickel for the square Ising lattice. (Guttmann [1989]) The susceptibility series are particularly nice to work with, because they contain all powers of x. This circumstance makes it possible to generate a significant number of terms in the series with much less effort than is required to produce the same number of nonzero terms in the partition function, and the approximation techniques are easier to use on such series.

The graphs generated by these high-temperature expansions illustrate the manner by which remote spins become correlated, even though the interactions are restricted to nearest neighbors. The rapid increase in the coefficients, for increasing powers of x, shows the multiplicity of connecting paths by which distant spins can interact indirectly.

7.10.2 The Asymptotic-Ratio Method. The thermodynamic properties of a system near the critical point are expected to be expressible in the form

$$A(x) = B(x)(1 - x/x_c)^{-q} + C(x), \qquad (7.10.13)$$

where $B(x)$ and $C(x)$ are regular, x is typically temperature dependent, and q is related to a critical exponent. Near x_c, the term $C(x)$ becomes unimportant, and $A(x)$ can be approximated as

$$A(x) \approx a_0 + a_1 x + a_2 x^2 + \cdots + a_n x^n + \cdots$$

$$\approx B(x)\left[1 + q(x/x_c) + \frac{q(q+1)x}{2!x_c} + \cdots + \frac{q(q+1)\cdots(q+n-1)x^n}{n!x_c^n} + \cdots\right]. (7.10.14)$$

The ratio of successive coefficients of x^n is

$$r_n = \frac{a_n}{a_{n-1}} = \frac{(q+n-1)}{nx_c}. \qquad (7.10.15)$$

A plot of r_n against $1/n$ is a straight line that intercepts the ordinate at $1/x_c$, with slope $(q-1)/x_c$. In practice, a plot of r_n against $1/n$ is expected to produce a curve that becomes an increasingly better approximation to a straight line as n increases, so that a reliable extrapolation to $1/n = 0$ can be drawn, to yield a value for the critical parameter x_c. The slope of this asymptotic straight line, $(q-1)/x_c$, together with the extrapolated estimate of x_c, yields an estimate of the exponent q, which is presumably simply related to a critical exponent.

As an example, the series for $\bar{\chi}_T$, obtained by counting graphs on a simple cubic lattice, is found to be

$$\bar{\chi}_T = 1 + 6x + 30x^2 + 150x^3 + 726x^4 + 3510x^5 + 16710x^6 + 79494x^7 + 375174x^8 +$$
$$+ 1769686x^9 + 8306862x^{10} + 38975286x^{11} + 182265822x^{12} + 852063588x^{13} +$$
$$+ 3973784886x^{14} + 18527532310x^{15} + 86228667894x^{16} + 401225391222x^{17} + \cdots \ (7.10.16)$$

From these terms, the ratios r_n of Eq. (7.10.15) can be calculated to yield approximate values for x_c and for $q = \gamma$. A plot of the numerical values of r_n against $1/n$ is nearly a straight line, with a least-squares fit to the data, omitting points for $n < 3$, of the form

$$r_n = 1.2060662/n + 4.5786079. \qquad (7.10.17)$$

The numerical results are thus $x_c = 1/4.5786079 = 0.218407$, and $\gamma = 1 + 1.2060662x_c = 1.263413$. The values obtained by more advanced methods give $x_c = 0.218140$, and $\gamma = 1.250$. The agreement is good enough to support the expectation that $\gamma = 5/4$, which is the accepted value, although there is apparently no requirement that critical exponents be rational numbers. Moreover, there are studies that indicate a slightly different value (Nickel [1982]).

Similar analyses for face-centered and body-centered cubic lattices lead to the same value of γ, but different values for x_c. Once the value of x_c has been determined, it can be used in the analysis of other quantities by supplying the constant term in the version of Eq. (7.10.15) that results from the appropriate r_n series. Graph-counting techniques permit series such as that of Eq. (7.10.16) to be written for systems with the dimensionality $d > 3$. A motivation for doing so is the observation that γ seems to approach the Landau value, $\gamma_L = 1 = 4/4$, as the dimensionality increases. For 2-d, $\gamma = 7/4$, for 3-d, $\gamma \approx 5/4$. An unexpected result is that for $d \geq 4$, even for nonintegral dimensionality, the critical exponents are found to be precisely the Landau values.

7.10.3 The Padé Approximant Method. When a series is not regular, in that some of the coefficients are either zero or negative, the asymptotic-ratio method of analysis fails to generate a result that can be easily interpreted, either because the ratios cannot be found or because the radius of convergence of the series is determined by a complex singularity closer to the origin than x_c. In such cases in particular, but as a useful working tool in general, the method of Padé approximants provides a powerful alternative. A given series,

$$A(x) = a_0 + a_1 x + a_2 x^2 + \cdots + a_L x^L, \tag{7.10.18}$$

is to be approximated by a ratio, called the M, N^{th} Padé approximant, written as

$$[M, N](x) = \frac{P_M(x)}{Q_N(x)} = \frac{p_0 + p_1 x + \cdots p_M x^M}{1 + q_1 x + \cdots q_N x^N}. \tag{7.10.19}$$

In the following development, M is always take equal to N. In order to determine the $2N+1$ coefficients in $[N, N](x)$, it is necessary to use $2N + 1$ coefficients in the known series for $A(x)$, which requires that $L \geq 2N$. The coefficients are determined by equating $[N, N](x)$ to the first $2N + 1$ terms of A, multiplying both sides of the equation by Q_N, and equating coefficients of like powers of x. Thus $[1, 1](x)$ is found from the equation

$$p_0 + p_1 x = (1 + q_1 x)(a_0 + a_1 x + a_2 x^2),$$

which gives

$$p_0 = a_0, \quad p_1 = \left(a_1^2 - a_0 a_2\right)/a_1, \quad q_1 = -a_2/a_1,$$

and allows $[1, 1](x)$ to be written

$$[1, 1](x) = \frac{\begin{vmatrix} a_1 & a_2 \\ a_0 x & a_0 + a_1 x \end{vmatrix}}{\begin{vmatrix} a_1 & a_2 \\ x & 1 \end{vmatrix}}$$

$$= a_0 + a_1 x + a_2 x^2 + \left(a_2^2/a_1\right) x^3. \tag{7.10.20}$$

This approximation differs from Eq. (7.10.18) only in third order. An obvious extension of this procedure leads to the expression

$$[n,n](x) = \cfrac{\begin{vmatrix} a_1 & a_2 & \cdots & a_{n+1} \\ a_2 & a_3 & \cdots & a_{n+2} \\ \vdots & \vdots & \cdots & \vdots \\ a_n & a_{n+1} & \cdots & a_{2n} \\ a_0 x^n & a_0 x^{n-1} + a_1 x^n & \cdots & a_0 + a_1 x + \cdots + a_n x^n \end{vmatrix}}{\begin{vmatrix} a_1 & a_2 & \cdots & a_{n+1} \\ a_2 & a_3 & \cdots & a_{n+2} \\ \vdots & \vdots & \cdots & \vdots \\ a_n & a_{n+1} & \cdots & a_{2n} \\ x^n & x^{n-1} & \cdots & 1 \end{vmatrix}}, \tag{7.10.21}$$

which differs from $A(x)$ only in terms of order x^{n+2}. The denominator of the approximant can be factored as

$$Q_n(x) = \prod_{s=1}^{n} (1 - x/x_s), \tag{7.10.22}$$

with the factors ordered such that

$$|x_1| \le |x_2| \le \cdots \le |x_n|. \tag{7.10.23}$$

The radius of convergence of the approximant is determined by the root $|x_1|$, the one closest to the origin in the complex plane. If this is a positive real root, it is expected to be an approximate value for x_c, the critical value.

When it is expected that a function represented by a series such as $A(x)$ in Eq. (7.10.18) behaves near the critical point like $A(x)$ in Eq. (7.10.13), it is useful to approximate A near the critical point as

$$A(x) \approx B(x_c)(1 - x/x_c)^{-q}[1 + \mathcal{O}(1 - x/x_c)]; \quad (x \to x_c^+). \tag{7.10.24}$$

The logarithmic derivative of $A(x)$ is

$$D(x) = \frac{d \ln A(x)}{dx} = \left(\frac{q}{x_c - x}\right)[1 + \mathcal{O}(x - x_c)], \tag{7.10.25}$$

which has a simple pole at x_c. The Padé approximant is known to represent simple poles exactly; therefore the successive approximants to the series for $D(x)$ should converge to give an accurate value of x_c and a good estimate of q. The approximant of Eq. (7.10.25) can be obtained either by rewriting the logarithmic derivative of Eq. (7.10.18) as a power series and using Eq. (7.10.21) with redefined coefficients, or by writing $[N,N](x)$ as in Eq. (7.10.20), crossmultiplying with the logarithmic derivative in the form A'/A, and identifying the coefficients in $[N, N]$ by equating coefficients of like powers of x. The results are identical.

As an example, the $[1,1]$ approximant to the logarithmic derivative of the series for $A(x)$ is

$$[1,1]_D(x) = \frac{1}{a_0} \frac{a_1 \left(2a_2 - a_1^2/a_0\right) + \left[\left(2a_2 - a_1^2/a_0\right)^2 - 3a_1 a_3 + 3a_1^2 a_2/a_0 - a_1^4/a_0^2\right] x}{2a_2 - a_1^2/a_0 - (3a_3 - 3a_1 a_2/a_0 + a_1^3/a_0^2) x}.$$

The use of the terms in Eq. (7.10.16) for the simple cubic lattice gives

$$[1,1]_D(x) = \frac{144 - 180x}{24 - 126x} \approx \frac{q}{x_c - x},$$

giving $x_c^{(1)} = 24/126 = 0.19047619$ and, with this value used for x in the numerator to give the residue, $q^{(1)} = \gamma^{(1)} = 0.870728299$. These values are not in good agreement with the better ones of Eq. (7.10.17), but successive orders converge rapidly. The approximant [4,4] gives $x_c^{(4)} = 0.218151$ and $\gamma^{(4)} = 1.25053$; similarly, [7,7] gives $x_c^{(7)} = 0.218138$ and $\gamma^{(7)} = 1.24984$. These results are to be compared to the usually accepted values of $x_c = 0.218140$ and $\gamma = 1.250 \pm 0.001$.

A number of techniques are available for finding one parameter, based on prior knowledge of another. For example, if q is already known with good accuracy, then the approximants for

$$[A(x)]^{1/q} \approx [B(x_c)]^{1/q} (1 - x/x_c)^{-1}$$

give estimates of x_c and $B(x_c)$ that are biased in that they depend upon the value of q. Similarly, if x_c is already known, the approximants can be written to give

$$(x - x_c)\frac{d}{dx} \ln A(x) = (x - x_c)D(x) \approx q.$$

This procedure is particularly useful in obtaining values for additional critical exponents after, say, both x_c and γ are known from the series expansion for $\bar{\chi}_T$. The use of several different approximants also provides a consistency check that is a measure of the reliability of the estimates. Several refinements of this sort are given in more extensive treatments of the subject (Baker, Jr. [1961]; Baker, Jr. [1965]; Baker, Jr. [1970]; Baker, Jr. [1975]; Baker, Jr. [1977]; Essam and Fisher [1963]; Gaunt and Guttmann [1974]; Graves-Morris [1973]; Guttmann [1989]). Computer programs and their results are also available(Guttmann [1989]; Mouritsen [1984]).

7.11 OTHER MODELS

Systems of interacting spins provide an easily visualized basis for a wide range of studies of phase transitions and critical phenomena. A one-to-one correspondence has already been indicated between the Ising model of magnetic interactions and the order-disorder system studied by Bragg and Williams. Another such correspondence exists between the Ising model and the so-called lattice gas, which is a fluid confined to a volume that is regarded as partitioned into molecule-size cells that can be either empty or occupied by no more than one molecule. (The cell is occupied if the center of a molecule is anywhere within it.) (Lee and Yang [1952]) The interaction energy of adjacent cells is $-J$ if both cells are occupied, and zero otherwise, with the added condition that the energy of a cell becomes infinite if it is occupied by more than one molecule. Correspondence can be established between Z_N of an Ising system and Z_G of a lattice gas. The gas exhibits local clustering that becomes systemwide at the critical point. It provides a good way to model the similarities of fluid and magnetic systems (Stanley [1971]). In this section, however, I want to consider more general systems.

Spin systems can be characterized by d, the number of spatial dimensions of the lattice, and by D, the dimensionality of the spin variable. The Ising system and its analogs can have any value for d, but the spins are one-dimensional, so $D = 1$. The Heisenberg model, proposed after Ising's original paper seemed to rule out phase transitions in such a simple system, invokes interactions of the form

$$\mathcal{H} = -J \sum_{nn} \mathbf{S}_i \cdot \mathbf{S}_j = -J \sum_{nn} (S_{ix}S_{jx} + S_{iy}S_{jy} + S_{iz}S_{jz}), \qquad (7.11.1)$$

with $D = 3$. Because the Heisenberg model is so much more difficult to solve than the Ising model, and because the critical exponents were found to differ from those of the Ising model, the X-Y model was proposed, for which the interacting spins are restricted to a plane and the interaction energy is

$$\mathcal{H} = -J \sum_{nn} \mathbf{S}_i \cdot \mathbf{S}_j = -J \sum_{nn} (S_{ix}S_{jx} + S_{iy}S_{jy}). \qquad (7.11.2)$$

Both of these models can have any spatial dimension d, but the X-Y model always has $D = 2$, and the Heisenberg model has $D = 3$. (An ironic outcome, in view of Heisenberg's proposal of the model to overcome the perceived deficiency of the Ising model, is that even in two dimensions, it does not lead to long-range order at any finite temperature.) One other characteristic of the Heisenberg and X-Y models is that the spins in the Hamiltonian are quantum operators, for which commutation rules must be obeyed. Thus there are at least four possible parameters that could determine the value of the critical exponents: spatial dimensionality d, spin dimensionality D, coordination number ν, and the choice of quantum or classical spin variables. (Brown and Luttinger [1955]; Mermin [1968]; Betts [1974]).

A linear open generalized Heisenberg chain has the Hamiltonian

$$\mathcal{H} = -\sum_{i=1}^{N}\sum_{j=1}^{N} \left[J_{ij}^{\parallel} S_{zi}S_{zj} + J_{ij}^{\perp}(S_{xi}S_{xj} + S_{yi}S_{yj}) \right] - m_s B \sum_{i=1}^{N} S_{zi}. \qquad (7.11.3)$$

The Ising system corresponds to $J_{ij}^{\perp} = 0$; the spatially isotropic Heisenberg system to $J_{ij}^{\perp} = J_{ij}^{\parallel}$. For the n-n Heisenberg model with $N + 1$ spins, the Hamiltonian becomes

$$\mathcal{H} = -2J_S \sum_{i=0}^{N} \mathbf{S}_i \cdot \mathbf{S}_j - m_s B \sum S_{zi}, \qquad (7.11.4)$$

which can be transformed by writing $2J_S S^2 \equiv J$; $m_s S \equiv m$; and $\mathbf{s}_i \equiv \mathbf{S}_i/S$. The resulting commutation relations are of the form

$$s_{xj}s_{yj} - s_{yj}s_{xj} = is_{zj}/S. \qquad (7.11.5)$$

In the limit of $S \to \infty$, the new spins commute, and a classical version of the Heisenberg model is obtained. (It is important to study both the classical and quantum versions of the same model in order to examine the differences, if any, in the critical exponents. It turns out that almost always there is no difference, for reasons that become more evident in the next chapter.) With $K = \beta J$ and $B = 0$, the partition function becomes

$$Z_N = \int \frac{d\Omega_0}{4\pi} \int \frac{d\Omega_1}{4\pi} \cdots \int \frac{d\Omega_N}{4\pi} \prod_{i=1}^{N} e^{K\mathbf{s}_i \cdot \mathbf{s}_{i-1}}$$

$$= \int \frac{d\Omega_0}{4\pi} \prod_{i=1}^{N} \int e^{K\cos\theta_i} \frac{\sin\theta_i \, d\theta_i}{2} = \left[\frac{\sinh K}{K} \right]^N. \qquad (7.11.6)$$

The energy per spin is

$$u_1 = \frac{-1}{N}\frac{\partial \ln Z_N}{\partial \beta} = -Ju(K), \qquad (7.11.7)$$

where

$$u(K) = \coth K - (1/K), \qquad (7.11.8)$$

the Langevin function. As in Eq. (7.10.8), the normalized susceptibility is given by

$$\begin{aligned}
\bar{\chi}_T &= \sum_{n=0}^{N}\langle s_{z0}s_{zn}\rangle \\
&= \frac{1}{3}\sum_{n=0}^{N}\langle \mathbf{s}_0 \cdot \mathbf{s}_n\rangle = \frac{1}{3}\sum_{n=0}^{N}u^n(K) = \frac{1}{3}\frac{1-u^{N+1}}{1-u}.
\end{aligned} \qquad (7.11.9)$$

Two tricks have been used: (1) in the absence of a B field, $\langle s_{z0}s_{zn}\rangle = \langle \mathbf{s}_0 \cdot \mathbf{s}_n\rangle/3$; and (2), $\langle \mathbf{s}_0 \cdot \mathbf{s}_n\rangle = u(K)\langle \mathbf{s}_0 \cdot \mathbf{s}_{n-1}\rangle$, a result that can be iterated. (Fisher [1964]; Joyce [1967])

The Hamiltonian of Eq. (7.11.4) can be generalized to an arbitrary spin dimensionality D, in the same classical limit. It is then called the D-vector model. (Stanley [1969]) In the absence of a B field, the Hamiltonian is

$$\mathcal{H} = -J\sum_{nn}\mathbf{s}_i \cdot \mathbf{s}_j, \qquad (7.11.10)$$

where each spin vector has D orthogonal components, with $\mathbf{s}_i \cdot \mathbf{s}_i = s_i^2 = \lambda$, where λ is usually taken to be either 1 or D. The classical spins can have any orientation; the usual partition-function sums are then replaced by integrals, and the partition function is written

$$Z_N(D, K) = \int d\mathbf{s}_1 \int d\mathbf{s}_2 \cdots \int d\mathbf{s}_N \prod_{i=1}^{N}[\delta(\lambda - s_i^2)]\exp\left[K\sum_{i=1}^{N-1}\mathbf{s}_i \cdot \mathbf{s}_{i+1}\right], \qquad (7.11.11)$$

where $K = \beta J$, and $\int d\mathbf{s}$ is an integral over a D-dimensional spin space of cartesian coordinates σ_j. If this is taken as an open chain, the integrals can be carried out sequentially, beginning with the one for s_N, here denoted as $\mathcal{Z}(D, K)$, and written as

$$\begin{aligned}
\mathcal{Z}(D, K) &= \int d\mathbf{s}_N \delta\left[\lambda - s_N^2\right]e^{K\mathbf{s}_N \cdot \mathbf{s}_{N-1}} \\
&= \frac{K}{2\pi i}\int_{-\infty}^{\infty} d\sigma_1 \cdots \int_{-\infty}^{\infty} d\sigma_D \int_{-i\infty}^{i\infty} dy \left[e^{yK(\lambda-\sum \sigma_j^2)}\right]e^{K\sum_n \sigma_n \tau_n}, \quad (7.11.12)
\end{aligned}$$

where the delta-function is written as a standard contour integral, and the τ's are the components of $\mathbf{s}_{N-1}$. The integrations can be completed by interchanging the order of integration, after insertion of the factor $\exp\left[Ka(\lambda - \sum \sigma_n^2)\right] = 1$, where a is a large positive real number, such that $(a+y)\sigma_n^2 - \tau_n\sigma_n > 0$. The result of these operations is, with $v = a+y$,

$$\mathcal{Z}(D, K) = \frac{K}{2\pi i}\int_{a-i\infty}^{a+i\infty} dv\, e^{Kv\lambda}\prod_{n=1}^{D}\int_{-\infty}^{\infty} d\sigma_n e^{-K(v\sigma_n^2 - \sigma_n\tau_n)}. \qquad (7.11.13)$$

The final integrals, given on p. 23, are

$$\int_{-\infty}^{\infty} d\sigma_n e^{-K(v\sigma_n^2 - \sigma_n \tau_n)} = \left(\frac{\pi}{2K}\right)^{1/2} e^{K\tau_n^2/4v}. \tag{7.11.14}$$

Observe that $\sum \tau_n^2 = \lambda$, which eliminates the τ's, and the expression simplifies to

$$\mathcal{Z}(D, K) = \frac{K}{2\pi i} \int_{a-\infty}^{a+\infty} dv\, e^{K\lambda(v+1/4v)} \left(\frac{\pi}{DK}\right)^{D/2}. \tag{7.11.15}$$

With the substitution $w = 2v$, the final expression becomes

$$\mathcal{Z}(D, K) = \frac{K}{2} \left(\frac{2\pi}{K}\right)^{D/2} \frac{1}{2\pi} \int_{2a-\infty}^{2a+\infty} dw\, e^{(K\lambda/2)(w+1/w)} w^{-D/2} = \frac{K}{2} \left(\frac{2\pi}{K}\right)^{D/2} I_{\frac{D}{2}-1}(\lambda K), \tag{7.11.16}$$

where I_r is the modified Bessel function of the first kind of order r.

After $N - 1$ successive integrations of this form, the remaining term, for s_1, is simply the volume of a D-dimensional sphere of radius $\lambda^{1/2}$; therefore the exact partition function for a $d = 1$ dimensional linear chain of D-dimensional classical spins is

$$Z_N(D, K) = \mathcal{Z}(D, 0)\,[\mathcal{Z}(D, K)]^{N-1} = (1 + \delta_{1,D})^N \frac{(\lambda\pi)^{D/2}}{\lambda\Gamma(D/2)} \left[\frac{K}{2}\left(\frac{2\pi}{K}\right)^{D/2} I_{\frac{D}{2}-1}(\lambda K)\right]^{N-1}. \tag{7.11.17}$$

Since $I_{1/2}(x) = (2/\pi x)^{1/2} \sinh x$ and $I_{-1/2}(x) = (2/\pi x)^{1/2} \cosh x$, the result reduces to that for a linear Ising chain when $D = 1$. The partition function, Eq. (7.11.17), contains all thermodynamic information, including the fact that there is no phase transition in one dimension for any choice of the spin dimension D.

The treatment of the D-vector model in higher spatial dimensions requires the use of series expansions and approximation techniques, the results of which are available. (Stanley [1974]) The critical exponents for this model are used in Chapter 8.

A number of other models are to be found in the literature. They are often variations, generalizations, and complications of the more primitive ones presented here. It is sometimes possible to transform a particular model into another that is its mathematical equivalent, although they may show little kinship at first glance. (Ashkin and Lamb, Jr. [1943]; Potts [1952]; Stephen and Mittag [1972]) A particularly rich and fruitful model is the one named after Renfrey Potts. For a well-written presentation of the various aspects of this model see the article by Wu. (Wu [1982])

7.12 COMMENTARY

I have always marveled at the ability of political cartoonists to create caricatures that are immediately recognizable by all. I am not so much impressed by the comic or grotesque aspects of the caricature as by the fact that a recognizable essence, captured with a few skillful strokes of the pen, can emphasize the unmistakable features of a subject. Yet consider the systems studied in this chapter. All of the models presented are caricatures of some underlying physical reality. They are too free of detail to be regarded as faithful representations of physical systems, and indeed if much more microscopic detail were included, the already complicated many-interacting-body systems would be completely intractable. Therefore our only hope of understanding these systems is that our caricatures are valid representations of the features we seek to comprehend. And, as with any such endeavors, they are not all equally successful. The good ones must emphasize the essential features, suppress nonessential detail, and provide us with a mathematicized model that is within our capability to solve. (This same analogy has been presented by Frenkel and reiterated by others.) (Fisher [1983]; Tamm [1962])

The early caricatures, the molecular field model and its analogs, are still useful, but they are so stark that they fail to include features that we now know are essential, and they typically do not provide reliable results. The failure of these models is attributable to the mathematical approximation methods used to solve them, and not to the underlying Hamiltonians, which are also caricatures. Thus the failure of these models could be regarded as a problem of second-order caricatures. When exact partition functions for the model Hamiltonians were achieved, their essential physics could be extracted, and the validity of the characterization assessed. (Bethe [1935]; Bragg and Williams [1934]) *The Ising model is the irreducible caricature of ferromagnetism.*

The exact solution of the one-dimensional Ising problem, with the result that there is no phase transition, pointed out the inadequacy of the molecular field theory, at least in one dimension, and prompted Heisenberg to propose his more-complicated $D = 3$ quantum-mechanical model. There are several good histories of the Ising model, from the original suggestion by Wilhelm Lenz, based on a paper he published in 1920, to his doctoral student Ernst Ising, who published his eponymic paper in 1925, and on to the exact solution in two dimensions, first by Onsager in 1942, and later by others with easier methods. (Brush [1967]; Brush [1983]; Domb [1960]; Hurst and Green [1960]; McCoy and Wu [1973]) Some of the more notable achievements should be mentioned. The mathematician Balthus van der Waerden showed in 1941 that the binary alloy problem of the Bragg-Williams type could be solved exactly, in three dimensions, by counting closed diagrams on a lattice. He obtained a few terms in the series that is equivalent to Eq. (7.6.2) by direct counting, and he set upper bounds for several more, but he did not have the sophisticated counting theorems of graph theory that enable this approach to count higher-order graphs with precision. (van der Waerden [1941]) At about the same time, H. A. Kramers and Gregory Wannier located the exact transition temperature for the two-dimensional Ising lattice, and showed that the partition function corresponded to the largest eigenvalue of a characteristic matrix. (Kramers and Wannier [1941]) (Their method was extended and generalized by others.) (Ashkin and Lamb, Jr. [1943]; Potts [1952]) Also at this time, Elliott Montroll showed that the phase transition occurred when the largest eigenvalue of the transfer matrix becomes degenerate, and his analysis effectively anticipated the idea of block spins that is central to Chapter 8. (Montroll [1941]) Lars Onsager's 1942 exact solution of the two-dimensional Ising problem, first published in 1944, was so complicated that few people at the time understood it. His student, Bruria Kaufman, improved, extended, and clarified the calculations by use of spinor notation, but they still remained abstruse. A calculation by Chen Ning Yang on spontaneous magnetization obliterated any illusions of clarity that Kaufman may have achieved (Kaufman [1949]; Kaufman and Onsager [1949]; Onsager [1944]; Yang [1952]). (After reading the Onsager papers and seeing Onsager write his formula, giving the critical exponent for spontaneous magnetization, as a conference remark, without proof, Elliott Montroll commented that the practice of announcing one's results in the form of a cryptogram—popular in the days of Galileo and Newton—had Onsager as the only modern who operates in this tradition.) (Montroll, Potts and Ward [1963])

The combinatorial breakthrough by Mark Kac and John C. Ward led the way to several order-of-magnitude improvements in the ease of solving the Ising problem in two dimensions, and from that point, the number of papers on the subject becomes large. (Fisher [1961]; Glasser [1970]; Green and Hurst [1964]; Hurst and Green [1960]; Kac and Ward [1952]; Kasteleyn [1963]; Kasteleyn [1967]; McCoy and Wu [1973]; Montroll [1966]; Montroll, Potts and Ward [1963]; Potts and Ward [1955]; Sherman [1960]; Vdovichenko [1965a]; Vdovichenko [1965b]) An interesting and fundamental paper by Michael Fisher not only presented the Fisher cities of Fig. 7.18, but also solved problems of dimer coverages of finite lattices without periodic boundaries, such as chess boards.

Even the imposing algebraic demands of the Onsager-Kaufman calculations were softened by more recent developments. (Schultz, Mattis and Lieb [1964]; Thompson [1965]) Several other problems of significance have been solved by these modern versions of the transfer-matrix method. They include an exact calculation of the residual entropy of square ice, (Lieb [1967a]), the exact solution of the Slater KDP model of a ferroelectric in two dimensions, with an electric field (!), (Lieb [1967b]), and an exact solution of the so-called Rhys F model of an antiferroelectric, also in two dimensions and with an applied field. This model exhibits a strange phase transition such that the Helmholtz potential and all of its derivatives are continuous at T_c (Lieb [1967c]). These problems have also been studied by graphical methods and series expansions. (Nagle [1974])

Analytic solutions for the Ising model in three dimensions, and for the model in two dimensions in an external B field, have remained beyond the abilities of everyone who has tried. The Heisenberg, X-Y, and D-vector models are not analytically solved even in two dimensions with $B = 0$. Excellent progress has been made in extracting the properties of these systems by the use of series-generating graphical methods and sophisticated handling of the resulting series, and, as shown in Chapter 8, by the concepts and methods of universality. Nevertheless, exact analytic solutions would represent not only an advance in our physical understanding of these systems and their phase transitions, but also significant advance in mathematical understanding.

One further model deserves mention: the spherical model. (Berlin and Kac [1952]) It removes the restriction on individual Ising spins that $\sigma_i^2 = 1$, and replaces it for a system of N spins by the condition $\sum \sigma_i^2 = N$, which is also satisfied by spins in the Ising model. The spherical model can be solved exactly in three dimensions, and its critical exponents are mentioned in Chapter 8. It is equivalent to the D-vector model of the previous section, with the spin dimension $D \to \infty$. (Stanley [1968])

Problems, puzzles, and surprises still remain as intrinsic features of this rich and incompletely explored subject. The insights gained from the realization that a variety of phenomena could be modeled in terms of systems of interacting spins have led to the exciting concept of universality, explored in Chapter 8. But an underlying feature of the usual statistical treatment of systems with an inherent dynamic behavior is that the sense of turmoil at the atomic level is missing. As in the dog-flea problem of Chapter 1, the equilibrium state is presented in terms of a set of time-independent probabilities, from which variances can be calculated, but the mere knowledge of this equilibrium probability set is insufficient to convey even the simple dynamics of fleas independently hopping from one dog to the other. The dynamic processes that take place in a system of interacting particles near the critical point are far more complex and far more interesting than those of flea hops. A necessarily abbreviated examination of some of these processes is to be found in Chapter 10. I mention this aspect of the problem in order to extol the amazing ability of an equilibrium theory to describe phase transitions, exhibit divergencies in response functions, and provide numerical values for critical exponents, all without any overt recognition of the active competitive processes taking place at the atomic level.

7.13 REFERENCES

The principal references for Chapter 7 are here. The complete reference list is to be found in the Bibliography, at the end of the book.

BAKER, JR., G. A. [1975], *Essentials of Padé Approximants*. Academic Press, New York, NY, 1975.

DOMB, C. [1960], On the Theory of Cooperative Phenomena in Crystals. *Advances in Physics* **9** (1960), 245–361.

FISHER, M. E. [1961], Statistical Mechanics of Dimers on a Plane Lattice. *Physical Review* **124** (1961), 1664–1672.

GREEN, H. S. AND HURST, C. A. [1964], *Order-Disorder Phenomena*. Interscience Publishers, A division of John Wiley & Sons, London, 1964.

KASTELEYN, P. W. [1963], Dimer Statistics and Phase Transitions. *Journal of Mathematical Physics* **4** (1963), 287–293.

MONTROLL, E. W. [1964], *Lattice Statistics*. In *Applied Combinatorial Mathematics*, E. F. Beckenbach, Ed. John Wiley & Sons, New York, NY, 1964, 96–143.

MONTROLL, E. W., POTTS, R. B. AND WARD, J. C. [1963], Correlations and Spontaneous Magnetization of the Two-Dimensional Ising Model. *Journal of Mathematical Physics* **4** (1963), 308–322.

NAGLE, J. F. [1968], The One-Dimensional KDP Model in Statistical Mechanics. *American Journal of Physics* **36** (1968), 1114–1117.

ONSAGER, L. [1944], Crystal Statistics I. A Two-dimensional Model of an Order-Disorder Transition. *Physical Review* **65** (1944), 117–149.

SCHULTZ, T. D., MATTIS, D. C. AND LIEB, E. H. [1964], Two-Dimensional Ising Model as a Soluble Problem of Many Fermions. *Reviews of Modern Physics* **36** (1964), 856–871.

7.14 PROBLEMS

1. Consider a lattice made up of alternating A and B sites, such that each spin on an A site interacts only with spins on neighboring B sites, and in such a way that the interaction energy is least when the spins are antiparallel. This is a model of antiferromagnetism. The Hamiltonian of the system can be written

$$H = \frac{1}{2} \sum_{(\sigma_a, \sigma_b)} J_{ij}\sigma_{ai}\sigma_{bj} - mB \sum_i (\sigma_{ai} + \sigma_{bi})$$

where $J_{ij} = 0$ unless σ_{ai} and σ_{bj} are nearest neighbors (each site has ν nearest neighbors on the other type of site), and $J_{ij} = J$ for the nearest neighbor interaction. Calculate the average value of each spin by the molecular field approximation. With $B = 0$, is there a phase transition? Calculate the isothermal susceptibility in the limit of small molecular field.

2. Try to modify the Bragg-Williams model of ferromagnetism to include a short-range order parameter s, such that

$$N_{++} = \gamma N_+^2 (1 + s), \quad N_{--} = \gamma N_-^2 (1 + s)$$

and

$$N_{+-} = N_{-+} = \gamma N_+ N_- (1 - s).$$

Evaluate γ from the condition that the total number of n.n. pairs is $N\nu/2$. Write an expression for the total energy, the usual expression for S (or a better one, if you can devise it), and calculate $\langle \sigma \rangle$ when $B = 0$. What does s do to T_c? Is this behavior of T_c vs. s physically reasonable?

3. Calculate $\langle \sigma_i \sigma_j \rangle_{nn}$ for the Bragg-Williams model of ferromagnetism.

4. An attempt to improve the molecular-field calculation of a phase transition is as follows: A representative spin, σ_1, interacts with ν nearest neighbors, $\sigma_{21}, \sigma_{22}, \cdots, \sigma_{2\nu}$, which in turn each interact with $\nu - 1$ other nearest neighbors of spin $\langle \sigma \rangle$. The energy of the system is given by

$$\mathcal{E} = -J[\sigma_1 + (\nu - 1)\langle \sigma \rangle](\sigma_{21} + \sigma_{22} + \cdots + \sigma_{2\nu}).$$

a. Find a general expression for the partition function of this subsystem, expressed in terms of ν and σ_1, since the sum over the values of σ_1 cannot be carried out unless ν is known.
b. Write the partition function explicitly for $\nu = 2, 3,$ and 4.
c. Use the terms in the unsummed version of the partition function to obtain an expression for $\langle \sigma_1 \rangle$. Carry out the required sums to obtain a general expression for $\langle \sigma_1 \rangle$, which will still contain unsummed σ_1's.
d. Find $\langle \sigma_1 \rangle$ for $\nu = 2, 3,$ and 4.
e. Set $\langle \sigma_1 \rangle = \langle \sigma \rangle$ to obtain an equation that can be solved for $\langle \sigma \rangle$. Do this explicitly for $\nu = 2, 3,$ and 4.
f. Find kT_c/J for $\nu = 2, 3,$ and 4. Compare the results with those of the simple Weiss theory, and, for $\nu = 4$, with the exact Ising-model result.
g. Nearest-neighbor correlations can be found by calculating $\langle \sigma_1 \sigma_{21} \rangle$. Do this for $\nu = 2, 3,$ and 4. In particular, show that this model, unlike that of Weiss, gives non-zero correlations for $T > T_c$.

5. Solve the Nagle model of KDP, when ϵ_2 and ϵ_3 are infinite, by the transfer matrix method.

6. Solve exactly the open linear Ising chain with nn and the next nn interactions; *i.e.*

$$\mathcal{H} = -J_1 \sum \sigma_i \sigma_{i+1} - J_2 {\sum}' \sigma_i \sigma_{i+2}.$$

7. Calculate the number of dimer coverages of a 4×4 square lattice.

8. Calculate the exact partition function for a linear double Ising chain, with $B = 0$, with $J_1 =$ n.n. coupling constant along the chain, $J_2 =$ coupling constant linking adjacent elements in the two parallel chains:
a. by the method of Eqs. (7.5.8)-(7.5.24), for the open chain.
b. by the transfer matrix method for a closed-loop chain.
c. compare $\lim_{N \to \infty} (\ln Z_N / N)$ given by parts 8.a and 8.b.
d. Is there a phase transition?

9. Consider a linear chain of molecules of the form H_3R, where R is an unspecified trivalent ion. The proton bonds in this chain are assumed to satisfy an ice condition, in that any R^{3-} ion is assigned an infinite energy unless it has exactly three n.n. protons near it in the chain. In analogy to the Slater model of KDP, if the three near protons are on the same side of R, the energy of that H_3R group is set equal to ϵ_1; if two protons are on one side and one on the other the energy of the group has the value ϵ_2.

a. Obtain the partition function for an N-molecule ring.

b. Calculate $U(T)$.

c. Show that as $N \to \infty$, there is a phase transition, find the transition temperature, and evaluate the average energy per molecule below and above the transition temperature.

d. Calculate the latent heat per particle.

10. Derive the critical exponents for the molecular-field model of Section 7.3.

7.15 SOLUTIONS

1. In the molecular-field model, each σ_a is regarded as surrounded by n.n.'s with spins $\langle \sigma_b \rangle$, and similarly for the σ_b. The single-spin partition function of a σ_a is

$$Z_{1a} = \sum_{\sigma_a = \pm 1} \exp\left[-\beta(J\nu \langle \sigma_b \rangle \sigma_a - mB\sigma_a)\right] = 2\cosh\left[-\beta(J\nu \langle \sigma_b \rangle - mB)\right],$$

and for σ_b,

$$Z_{1b} = \sum_{\sigma_b = \pm 1} \exp\left[-\beta(J\nu \langle \sigma_a \rangle \sigma_b - mB\sigma_b)\right] = 2\cosh\left[-\beta(J\nu \langle \sigma_a \rangle - mB)\right].$$

The average energy of each is

$$\langle \sigma_a \rangle = -mB_{ea} \langle \sigma_a \rangle = -\frac{\partial \ln Z_{1a}}{\partial \beta} = (J\nu \langle \sigma_a \rangle - mB)\tanh\left[-\beta(J\nu \langle \sigma_a \rangle - mB)\right];$$

$$\langle \sigma_b \rangle = -mB_{eb} \langle \sigma_b \rangle = -\frac{\partial \ln Z_{1b}}{\partial \beta} = (J\nu \langle \sigma_b \rangle - mB)\tanh\left[-\beta(J\nu \langle \sigma_b \rangle - mB)\right];$$

where $B_{ea} = B - J\nu \langle \sigma_b \rangle / m$, and $B_{eb} = B - J\nu \langle \sigma_a \rangle / m$. The average spins are seen to be

$$\langle \sigma_a \rangle = -\tanh\left[\beta(J\nu \langle \sigma_b \rangle - mB)\right];$$

$$\langle \sigma_b \rangle = -\tanh\left[\beta(J\nu \langle \sigma_a \rangle - mB)\right].$$

For $B = 0$, with $J\nu = kT_c$, these become

$$\langle \sigma_a \rangle = -\tanh\left(\frac{T_c}{T} \langle \sigma_b \rangle\right), \text{ and } \langle \sigma_b \rangle = -\tanh\left(\frac{T_c}{T} \langle \sigma_a \rangle\right).$$

The two equations may be plotted in $\langle \sigma_a \rangle$-$\langle \sigma_b \rangle$ space for $T > T_c$, $T = T_c$, and $T < T_c$ to display the solutions. For $T > T_c$, the only intersection is at the origin, so that $\langle \sigma_a \rangle = \langle \sigma_b \rangle = 0$ prevails; for $T = T_c$, the curves are tangent at the origin, signaling an incipient phase transition, and for $T < T_c$, the curves intersect in the second and fourth quadrants, with $\langle \sigma_a \rangle = -\langle \sigma_b \rangle$, so that $\langle \sigma_a \rangle + \langle \sigma_b \rangle = 0$, and there is no net magnetic moment.

In general, with $B \neq 0$, the net molar magnetic moment is $M = (N_o/2)m(\langle \sigma_a \rangle + \langle \sigma_b \rangle)$. The isothermal molar susceptibility is

$$\chi_T = \left(\frac{\partial M}{\partial B} \right)_T = \frac{(N_o/2)m^2\beta \left[2 - \langle \sigma_a \rangle^2 - \langle \sigma_b \rangle^2 - 2\beta J\nu \left(1 - (1 - \langle \sigma_a \rangle^2)(1 - \langle \sigma_b \rangle^2) \right) \right]}{1 - (\beta J\nu)^2 (1 - \langle \sigma_a \rangle^2)(1 - \langle \sigma_b \rangle^2)}.$$

For $J\nu = kT_c$, and $\langle \sigma_a \rangle$, $\langle \sigma_b \rangle \ll 1$, this becomes

$$\chi_T = N_o m^2 / k(T + T_c),$$

corresponding to Néel's law for antiferromagnetic materials.

2. With the given expressions for N_{++}, etc., and with $N_+ = (N/2)(1+x)$, $N_- = (N/2)(1-x)$, the condition on n.n. pairs is given by $\gamma = \nu/2N(1 + sx^2)$. The total energy then becomes

$$E = -\frac{J\nu N(s + x^2)}{2(1 + sx^2)} - mNBx.$$

The expression for S is as given in Eq. (7.4.5), and the path from there to Eq. (7.4.19) gives, for this problem with $B = 0$,

$$x = \langle \sigma \rangle = \tanh \beta \left(\frac{J\nu x(1 - s^2)}{(1 + sx^2)^2} \right).$$

Here the critical temperature becomes $T_c = J\nu(1 - s^2)/k$, which is seen to be lowered by the effect of n.n. interactions, as in the exact solution of the 2-d Ising model. This behavior is physically reasonable, since local clumps of like n.n. pairs are not as easily ordered long-range as are spins that can be individually aligned by the average molecular field. The heat capacity follows from the procedure leading to Eq. (7.3.10), here given by

$$C = \frac{RT_c^2}{T^2} \frac{x^2(1 - x^2)(1 - s^2)}{(1 + sx^2)[(1 + sx^2)^3 + (T_c/T)(1 - x^2)(3sx^2 - 1)]}.$$

When $s = 0$, this reduces to the result of Eq. (7.3.10); for $T > T_c$, $x = 0$, and the denominator of C becomes $T^2 - TT_c$, which behaves, so $C = 0$ for $T > T_c$, just as in the B-W model.

3. For the B-W model, each spin has all of its nearest neighbors with spins of value $\langle \sigma \rangle$. Therefore $\langle \sigma_i \sigma_i \rangle_{nn} = \langle \sigma \langle \sigma \rangle \rangle = \langle \sigma \rangle^2$. The correlation coefficient is

$$r_c = \frac{\langle \sigma_i \sigma_j \rangle_{nn} - \langle \sigma \rangle^2}{\langle \sigma^2 \rangle - \langle \sigma \rangle^2} = 0,$$

indicating the complete absence of correlated n.n. behavior, a well-known weakness of the B-W model.

4.

$$\mathcal{E} = -J(\sigma_1 + (\nu - 1)\langle \sigma \rangle)(\sigma_{21} + \sigma_{22} + \cdots + \sigma_{2\nu}).$$

$$Z = \sum_{\{\sigma_i\}} e^{\beta J(\sigma_1 + (\nu-1)\langle \sigma \rangle)(\sigma_{21} + \cdots + \sigma_{2\nu})}$$

$$= \sum_{\{\sigma_i\}} e^{K\sigma_1\sigma_{21}} \cdots e^{K\sigma_1\sigma_{2\nu}} e^{K(\nu-1)\langle \sigma \rangle \sigma_{21}} \cdots e^{K(\nu-1)\langle \sigma \rangle \sigma_{2\nu}},$$

where $K = \beta J$, $x = \tanh \beta J = \tanh K$, and $y = \tanh(\nu - 1) \langle \sigma \rangle K$.

a.

$$Z_\nu = (\cosh K)^\nu (\cosh(\nu - 1) \langle \sigma \rangle K)^\nu \sum_{\{\sigma_i\}} (1 + \sigma_1 \sigma_{21} x) \cdots (1 + \sigma_1 \sigma_{2\nu} x)(1 + \sigma_{21} y) \cdots (1 + \sigma_{2\nu} y).$$

Sum over σ_{2i} in $(1 + \sigma_1 \sigma_{2i})(1 + \sigma_{2i} y)$, to get

$$Z_\nu = (2 \cosh K)^\nu (\cosh(\nu - 1) \langle \sigma \rangle K)^\nu \sum_{\sigma_1 = \pm 1} (1 + \sigma_1 xy)^\nu = 2^{\nu+1} (\cosh K)^\nu (\cosh(\nu - 1) \langle \sigma \rangle K)^\nu f_\nu(xy),$$

with 4.b,

$$f_2(xy) = 1 + (xy)^2; \quad f_3(xy) = 1 + 3(xy)2; \quad f_4(xy) = 1 + 6(xy)^2 + (xy)^4;$$

$$f_5(xy) = 1 + 10(xy)^2 + 5(xy)^4; \quad f_6(xy) = 1 + 15(xy)^2 + 15(xy)^4 + (xy)^6;$$

etc., with $f_\nu(1) = 2^{\nu-1}$.

c. and d.

$$\langle \sigma_1 \rangle = Z_{-1} \sum_{\{\sigma_i\}} \sigma_1 e^{K(\sigma_1 + (\nu-1)\langle \sigma \rangle)(\sigma_{21} + \cdots + \sigma_{2\nu})} = \frac{1}{2 f_\nu(xy)} \sum_{\sigma_1 = \pm 1} \sigma_1 (1 + \sigma_1 xy)^\nu = \frac{g_\nu(xy)}{f_\nu(xy)},$$

with the g-functions given by

$$g_2(xy) = 2xy; \quad g_3(xy) = 3xy + (xy)^2; \quad g_4(xy) = 4xy[1 + (xy)^2];$$

$$g_5(xy) = 5xy + 10(xy)^3 + (xy)^5; \quad g_6(xy) = 6xy + 20(xy)^3 + 6(xy)^5;$$

etc., with $g_\nu(1) = 2^{\nu-1}$.

e. Set $\langle \sigma_1 \rangle = \langle \sigma \rangle$. At c, $\langle \sigma \rangle \approx 0$, so that no term higher than first power in y need be considered in g, and every $f \approx 1$. Thus, $\langle \sigma_1 \rangle = \langle \sigma \rangle \approx (\nu x_c)(\nu - 1) K_c \langle \sigma \rangle$, or $1 = \nu(\nu - 1) K_c \tanh K_c$.

f. Numerical solution for $\nu = 2$, 3, and 4 gives

$$\nu = 2 : \quad K_c = 0.77170; \quad kT_c = 1.2958J$$
$$\nu = 3 : \quad K_c = 0.41934; \quad kT_c = 2.3813J$$
$$\nu = 4 : \quad K_c = 0.29275; \quad kT_c = 3.4159J.$$

For the Weiss model, $kT_c = \nu J$, and for the square Ising lattice with $\nu = 4$, $kT_c = 2.263J$.

g. By the same procedure as was used to find $\langle \sigma_1 \rangle$, the correlation function is calculated to be

$$\langle \sigma_1 \sigma_{12} \rangle = \frac{1}{2 f_\nu(xy)} \sum_{\sigma_1 = \pm 1} (x + \sigma y)(1 + \sigma_1 xy)^{\nu-1} = \frac{h_\nu(x, y)}{f_\nu(xy)},$$

where

$$h_2(x, y) = x(1 + y^2); \quad h_3(x, y) = x(1 + 2y^2 + xy^2); \quad h_4(x, y) = x(1 + 3y^2 + 2xy^2 + x^2 y^3 + x^2 y^4),$$

and $h_\nu(1, 1) = 2^{\nu-1}$. At T_c, $y = 0$, so $f_\nu(xy) = 1$ for all ν, and $h_\nu(x, y) = x$. Therefore, for all ν at T_c,

$$\langle \sigma_1 \sigma_{21} \rangle_c = x_c = \tanh K_c,$$

which varies with ν. For $T > T_c$, $y = 0$, and

$$\langle \sigma_1 \sigma_{12} \rangle = \tanh K,$$

which varies with ν and T, but is nonvanishing for all finite T. Thus short-range correlations exist for $T > T_c$, as in the exact calculation. (This is the Bethe-Peierls-Weiss model.) (Weiss [1950])

5. For the Nagle KDP model, with ϵ_2 and ϵ_3 infinite, the transfer matrix, Eq. (7.5.43), becomes

$$P = \begin{bmatrix} 1 & 0 & 0 & 0 \\ 0 & x & x & 0 \\ 0 & x & x & 0 \\ 0 & 0 & 0 & 1 \end{bmatrix}.$$

The eigenvalues are 0, 1, 1, and $2x$, with the largest either 1 or $2x$. The phase transition occurs at $2x = 1$, or $e^{-\beta\epsilon_1} = 0.5$, as given in Eq. (7.5.40). This result illustrates a characteristic behavior of the eigenvalues of the transfer matrix at a phase transition: the largest becomes degenerate at T_c, and the competing pair exchange order at the phase transition. (Lassettre and Howe [1941]; Montroll [1941]; Stanley et al. [1970])

6. For the open linear Ising chain with nn and next nn interactions, let $\tau_i = \sigma_i \sigma_{i+1}$ for $1 \le i \le N-1$, $\tau_N = \sigma_N$. Then $\sigma_i \sigma_{i+2} = \sigma_i \sigma_{i+1} \sigma_{i+1} \sigma_{i+2} = \tau_i \tau_{i+1}$, and

$$Z_N = \sum_{\{\tau_i = \pm 1\}} \exp \left[K_1 \sum_{i=1}^{N-1} \tau_i + K_2 \sum_{i=1}^{N-2} \tau_i \tau_{i+1} \right]$$

$$= (\cosh K_1)^{N-1} (\cosh K_2)^{N-2} \sum_{\{\tau_i\}} \prod_i (1 + \tau_i z)(1 + \tau_i \tau_{i+1} y),$$

where $z = \tanh K_1$, and $y = \tanh K_2$. This is equivalent to Eq. (7.5.8) and the final solution, equivalent to Eq. (7.5.24), is

$$\lim_{N \to \infty} (\ln Z_N / N) = \ln (2 \cosh K_1 \cosh K_2) + \ln \lambda_1,$$

where the larger eigenvalue is

$$\lambda_1 = (1/2) \left[1 + y + \left((1-y)^2 + 4yz^2 \right)^{1/2} \right].$$

7. Number the points 1 to 4, 5 to 8, 9 to 12, and 13 to 16. Make the x-arrows all point from left to right; the y-arrows point up for odd-numbered columns and down for even-numbered ones, as in Fig. 7.9a. The Pfaffian is given by $|D| = Pf^2$, where the matrix $D = I_4 \otimes X + J_4 \otimes Y$, with

$$X = \begin{bmatrix} 0 & x & 0 & 0 \\ -x & 0 & x & 0 \\ 0 & -x & 0 & x \\ 0 & 0 & -x & 0 \end{bmatrix}, \quad Y = \begin{bmatrix} y & 0 & 0 & 0 \\ 0 & -y & 0 & 0 \\ 0 & 0 & y & 0 \\ 0 & 0 & 0 & -y \end{bmatrix},$$

and the special matrix J_4 is given by $X = xJ_4$. By means of the transformation $T_4 J_4 T_4^{-1} = \Lambda_4$, the matrix D is transformed to the matrix D', with the same determinant, given by

$$D' = I_4 \otimes X + \Lambda_4 \otimes Y,$$

and the determinant D' is found to be

$$|D'| = \prod_{n=1}^{4} \begin{vmatrix} y\lambda_n & x & 0 & 0 \\ -x & -y\lambda_n & x & 0 \\ 0 & -x & y\lambda_n & x \\ 0 & 0 & -x & -y\lambda_n \end{vmatrix}$$

$$= \prod_{n=1}^{4} \left[y^4 \lambda_n^4 - 3x^2 y^2 \lambda_n^2 + x^4 \right].$$

The eigenvalues λ_n of the matrix J_4 are found to be the roots of the equation $\lambda^4 + 3\lambda^2 + 1 = 0$, or

$$\lambda^2 = -\frac{3}{2} \pm \frac{1}{2}\sqrt{5},$$

so that each root occurs twice. With these results, the Pfaffian is seen to be

$$Pf = \sqrt{|D'|} = y^8 + 9y^6 x^2 + 16 y^4 x^4 + 9y^2 x^6 + x^8,$$

and the 36 dimer coverages can be put in correspondence with the terms of the Pfaffian, as was done in Fig. 7.9.

8. Number the spins in the chain σ_{1n} for the first linear chain and σ_{2n} for its companion. The partition function is

$$Z_N = \sum_{\{\sigma_i = \pm 1\}} \prod_i \exp\left[K_1 \left(\sigma_{1i}\sigma_{1,i+1} + \sigma_{2i}\sigma_{2,i+1} \right) + K_2 \sigma_{1i}\sigma_{2i} \right].$$

a. By direct expansion, this partition function may be written

$$Z_N = \left(\cosh^2 K_1 \right)^{N-1} \left(\cosh K_2 \right)^N \sum_{\{\sigma_i = \pm 1\}} \left(1 + x\sigma_{11}\sigma_{12} \right) \left(1 + x\sigma_{21}\sigma_{22} \right) \left(1 + y\sigma_{11}\sigma_{21} \right) [\cdots]$$

$$= \left(\cosh^2 K_1 \right)^{N-1} \left(\cosh K_2 \right)^N \sum_{\{\sigma_i\}} 4 \left(1 + x^2 y\sigma_{12}\sigma_{22} \right) [\cdots]$$

$$= \left(\cosh^2 K_1 \right)^{N-1} \left(\cosh K_2 \right)^N (4)^{n-1} \sum_{\{\sigma_i\}} \left(a_n + b_n \sigma_{1n}\sigma_{2n} \right) \left(1 + x\sigma_{1n}\sigma_{1,n+1} \right)$$

$$\times \left(1 + x\sigma_{2n}\sigma_{2,n+1} \right) \left(1 + y\sigma_{1n}\sigma_{2n} \right) [\cdots]$$

$$= \left(\cosh^2 K_1 \right)^{N-1} \left(\cosh K_2 \right)^N (4)^n \sum_{\{\sigma_i\}} \left(a_{n+1} + b_{n+1}\sigma_{1,n+1}\sigma_{2,n+1} \right) [\cdots]$$

where inactive factors have been indicated by $[\cdots]$. A matrix equation for the coefficients a_n and b_n follows from the final two steps After summation over σ_{1n} and σ_{2n}, it is found to be

$$\begin{bmatrix} a_{n+1} \\ b_{n+1} \end{bmatrix} = \begin{bmatrix} 1 & y \\ x^2 y & x^2 \end{bmatrix} \begin{bmatrix} a_n \\ b_n \end{bmatrix},$$

or, since $a_0 = 1$ and $b_0 = 0$, this equation becomes

$$\begin{bmatrix} a_N \\ b_N \end{bmatrix} = \begin{bmatrix} 1 & y \\ x^2 y & x^2 \end{bmatrix}^N \begin{bmatrix} 1 \\ 0 \end{bmatrix} = A^N \begin{bmatrix} 1 \\ 0 \end{bmatrix}.$$

The matrix A may be written

$$A = \begin{bmatrix} 1 & y \\ x^2 y & x^2 \end{bmatrix} = \begin{bmatrix} 1 & 0 \\ 0 & x^2 \end{bmatrix} \begin{bmatrix} 1 & y \\ y & 1 \end{bmatrix} = BC.$$

As in Eq. (7.5.15), find the square root of B and use it as shown in Eq. (7.5.18) to rewrite A^N. Then follow the procedure leading to Eq. (7.5.24) to get

$$\lambda_1 = \frac{1}{2} \left[1 + x^2 + \sqrt{(1 - x^2)^2 + 4x^2 y^2} \right]$$

and

$$\lim_{N \to \infty} \left(\frac{\ln Z_N}{N} \right) = \ln \left[4 \cosh^2 K_1 \cosh K_2 \; \lambda_1 \right]$$

$$= \ln \left[2 \cosh 2K_1 \cosh K_2 \left(1 + \sqrt{1 + \sinh^2 2K_1 \tanh^2 K_2} \right) \right].$$

b. The partition function can be symmetrized for the transfer-matrix method, to read

$$Z_N = \sum_{\{\sigma_i\}} \prod_i \exp \left[K_1 \left(\sigma_{1i} \sigma_{1,i+1} + \sigma_{2i} \sigma_{2,i+1} \right) + (K_2/2) \left(\sigma_{1i} \sigma_{2i} + \sigma_{1,i+1} \sigma_{2,i+1} \right) \right]$$

$$= \mathrm{Tr} P^N,$$

where the transfer matrix is

$$P = \begin{bmatrix} A & 1 & 1 & C \\ 1 & B & D & 1 \\ 1 & D & B & 1 \\ C & 1 & 1 & A \end{bmatrix},$$

with $A = \exp(2K_1 + K_2)$, $B = \exp(2K_1 - K_2)$, $C = \exp(-2K_1 + K_2)$, $D = \exp(-2K_1 - K_2)$. Since $\mathrm{Tr} P^N = \sum \lambda_i^N$, where the λ_i are the eigenvalues of P, then in the limit of large N,

$$\lim_{N \to \infty} \left(\frac{\ln Z_N}{N} \right) = \ln \lambda_1,$$

where λ_1 is the largest eigenvalue of P, and not the same as in part a. The eigenvalue equation is

$$(\lambda - A + C)(\lambda - B + D)[\lambda^2 - (A + B + C + D)\lambda + AD + BC + AB + CD - 4] = 0,$$

giving the largest eigenvalue as

$$\lambda_1 = \frac{A + B + C + D}{2} + \frac{1}{2} \sqrt{(A - B + C - D)^2 + 16}$$

$$= 2 \cosh K_2 \left[\cosh 2K_1 + \sqrt{1 + \sinh^2 2K_1 \tanh^2 K_2} \right].$$

c. Evidently 8.a and 8.b agree exactly in the limit of large N.

d. By inspection, there is no phase transition.

9.a. Let $x = e^{-\beta \epsilon_1}$. The transfer matrix is

	lll	llr	lrl	rll	rlr	rrl	lrr	rrr
lll	1	0	0	0	0	0	0	0
llr	0	x	x	x	0	0	0	0
lrl	0	x	x	x	0	0	0	0
rll	0	x	x	x	0	0	0	0
rlr	0	0	0	0	x	x	x	0
rrl	0	0	0	0	x	x	x	0
lrr	0	0	0	0	x	x	x	0
rrr	0	0	0	0	0	0	0	1

The eigenvalues of the transfer matrix satisfy the equation

$$(1 - \lambda)^2 \left[(x - \lambda)^3 + 2x^3 - 3x^2(x - \lambda)\right]^2 = 0,$$

with roots

$$\lambda = 0, \ 0, \ 0, \ 0, \ 1, \ 1, \ 3x, \ 3x.$$

The partition function is given by

$$\ln Z_1 = \lim_{N \to \infty} \left[\frac{\ln Z_N}{N}\right] = \ln \lambda_1,$$

where λ_1 is the largest eigenvalue.

b. The energy associated with each ion is

$$u_1 = -\frac{\partial \ln Z_1}{\partial \beta}.$$

c. There is a phase transition at $3x_c = 1$, or $x_c = 1/3$. For $T < \epsilon_1/k \ln 3$, the energy is zero. For $T > \epsilon_1/k \ln 3$, the energy per ion is ϵ_1. Therefore the latent heat per ion is also ϵ_1. This result illustrates that degeneracy of the largest eigenvalue is not sufficient for a phase transition, because degenerate sets of identical eigenvalues can arise as a consequence of symmetry. The requirement is that eigenvalues that are not identical (and therefore equal at all temperatures) attain equal values at a particular x, and the role of *largest eigenvalue* transfers from one of them to a distinctly different one at T_c.

10. The results of Section 3 show, with little effort, that the critical exponents are the classical Landau ones.

8. *RENORMALIZATION*

8.1 INTRODUCTION

Critical exponents were examined in Chapter 6 for systems that were considered to be continuous. A formal procedure called rescaling was introduced in Section 6 of that chapter, in terms of generalized homogeneous functions. By this physically mysterious rescaling process, the critical exponents of thermodynamics could be expressed in terms of just two independent parameters, with no more than a hint of the underlying physics. In Chapter 7, phase transitions were examined for systems that were made up of spins on a lattice, coupled to nearest neighbors. The order parameter $\langle \eta \rangle$ in Chapter 6 is described as a continuum field, since it is defined at every point of the continuous space occupied by the system, whereas the order parameter of Chapter 7 is called a lattice field, since it is defined only on the lattice points of the system. It was found that critical exponents for the Ising model, calculated exactly in two dimensions and well approximated by Padé or other methods in three dimensions, do not agree with the Landau values, but they do obey the equalities derived in Chapter 6 from the scaling hypothesis. The molecular field approximation, which exhibits a phase transition in one dimension where an exact calculation shows none, agrees with the Landau exponents. In this chapter, an analysis of the physics based on the rescaling of a characteristic length offers a key to the understanding of these behaviors.

Two characteristic lengths, the lattice spacing a, and the correlation length ξ, are the natural choices for expressing the spatial dependence of the lattice field $\langle \eta \rangle = \langle \sigma \rangle$ of Chapter 7, since the system size is usually chosen to be infinite. The physical argument is made that if a and ξ are quite different, a phenomenon that is characterized by one of them may be entirely independent of the other. Well-known examples of this kind of scaling fill the literature of hydrodynamics. (Birkhoff [1960]) As an introductory nonthermal example, consider a cylinder of radius R and height $3R$, to be filled to the top with uniform steel balls of radius $a \ll R$, and an identical cylinder to be filled with steel balls of radius $b = \sigma a$, where σ is a constant, say of order 10, but with $b \ll R$. How do the masses of the two filled cylinders compare? (Nearly every student I have asked has replied that the cylinder filled with the smaller balls has the larger mass, because they pack better.) Evidently the total mass of the balls in a cylinder may be written as $M(r, R, \rho)$, where r is the radius of the balls, either a or b, their density is ρ, and R is the length scale of the cylinder. Denote the volume of a ball of radius r by V_r, and the number of balls of radius r in a cylinder of radius R and height $3R$ by $N(r, R)$. The effective density of the cylinder full of balls of radius r is $\rho(r)$, assuming a constant density of the ball material. Then the mass of the balls of radius a may be written

$$M(a, R, \rho) = N(a, R)V_a \rho = \rho(a)V(R),$$

where $V(R)$ is the volume of the cylinder. A similar equation applies to the balls of radius b, and the problem is reduced to a comparison of the effective densities. Here a scaling operation is employed, in this case called a similarity transformation. The dimension R is rescaled to $R' = (b/a)R$ for the cylinder containing the balls of radius b. This new cylinder can be seen as a one-to-one mapping of the cylinder containing the smaller balls, such that $N(b, R') = N(a, R)$. But note that $V_b = (b/a)^3 V_a$, and $V(R') = (b/a)^3 V(R)$. Therefore the mass of the contents of the rescaled cylinder may be written

$$M(b, R', \rho) = N(b, R')V_b\rho = N(a, R)(b/a)^3 V_a\rho = \rho(b)(b/a)^3 V(R).$$

From this and the previous equation, it follows that $\rho(b) = \rho(a)$. To the extent that edge effects can be neglected, then, the mass of the balls held in a container of characteristic dimension much greater than the ball radius is completely independent of that radius. This example is a crude analog of rescaling of critical phenomena, with the ball radius corresponding to lattice spacing, and the cylinder dimension to correlation length.

8.2 KADANOFF BLOCK SPINS

One of the earliest examples of rescaling as a technique in the study of critical phenomena was explained by Kadanoff et al. [1967]. In a lattice of spacing a, near the critical point, blocks of spins, L lattice points on a side, are regarded as the interacting entities, instead of the bare spins *per se*. Here the number L is chosen such that $a < La \ll \xi$. The number of spins in a lattice of d dimensions is L^d; they are assumed to be highly correlated, since the block size is much less than the correlation length. The key assumption is that each of these block spins can be treated as a single rescaled spin of magnitude $L^d \langle \sigma \rangle$ in a new Ising model with lattice spacing La, where $\langle \sigma \rangle$ is here taken to be the local average spin of the block. Since the correlation length ξ has not changed, its measure in terms of the new lattice spacing is reduced to $\xi' = \xi/L$, and the new lattice seems to have a higher value for $|\tau|$, the reduced temperature, and the new ordering field h' is presumably increased to induce the same system-wide average of the order parameter at a higher reduced temperature. Assume that

$$\tau' = L^x\tau, \text{ and } h' = L^y h, \tag{8.2.1}$$

with x and y positive. The singular part of the Gibbs potential per spin is taken to be the same function in the rescaled system as in the original, or

$$L^d g_s(\tau, h) = g_s(L^x\tau, L^y h). \tag{8.2.2}$$

The definition $\lambda = L^d$ permits Eq. (8.2.2) to be written

$$\lambda g_s(\tau, h) = g_s(\lambda^{x/d}\tau, \lambda^{y/d}h) = g_s(\lambda^p\tau, \lambda^q h), \tag{8.2.3}$$

in agreement with Eq. (6.6.5), with $p = x/d$ and $q = y/d$. All of Sections 6.2 and 6.3 follow from this rescaling.

The correlation length behaves as $\xi'(\tau', h') = L^{-1}\xi(\tau, h)$. With $h = 0$, this becomes $\xi'(L^x\tau, 0) = L^{-1}\xi(\tau, 0)$. But since $\xi(\tau, 0) \sim \tau^{-\nu}$, it follows that

$$(L^x\tau)^{-\nu} \sim L^{-1}\tau^{-\nu}, \tag{8.2.4}$$

or $x = 1/\nu$, and $p = 1/\nu d$. Then, from Eq. (6.6.10),

$$\alpha = 2 - 1/p = 2 - \nu d, \text{ or } \nu d = 2 - \alpha, \tag{8.2.5}$$

an independent confirmation of the hyperscaling relation, Eq. (6.7.29).

The underlying assumption of this physically plausible analysis, Eq. (8.2.2), is unproved at this point, but it offers an intelligible basis for the phenomenally successful derivations of the critical-exponent equalities of Chapter 6, and even for the hyperscaling relation that introduces the dimensionality dependence of these exponents. A revised version of the block-spin picture is developed by means of the procedures identified with the *renormalization group*, which are examined throughout the remainder of this chapter.

8.3 PRELIMINARIES IN ONE DIMENSION

Just as much of the methodology for exact Ising-model calculations was introduced by means of one-dimensional systems, so it is for renormalization. The concepts, simple implementations, and even most of the complications that arise in higher dimensions, can be exposed in terms of 1-d systems. Several techniques of calculation are used, all of which illustrate the concept of renormalization. In the picture of the previous section, blocks of spins were assumed to interact as single rescaled spins, with the same form of the Hamiltonian, partition function, and thermodynamic functions as in the original system. In this section, effective spin blocks (one dimensional, at first) are generated by the mathematical procedure of summing over selected spins, so that they are eliminated from the partition function, and rescaling the spacing of the reduced lattice so that it is recomposed to look as much as possible like the original lattice. The coupling constants of the recomposed lattice are thereby calculated directly, in terms of those of the original lattice. Kadanoff called this process *decimation*, but I dislike both the retributive connotation of his term and its implied removal of every tenth spin. The term I use, *recomposition*, means putting together in a new form. It consists of the two steps: *elimination* and *rescaling*, after which the lattice is said to be recomposed or renormalized.

8.3.1 Kadanoff Recomposition of 1-d Ising Chain.
The partition function of a single Ising chain with nearest-neighbor interactions and no external ordering field is given by Eq. (7.5.27), here rewritten with a slightly modified numbering system as

$$
\begin{aligned}
Z_N &= \left(A_0(x)\cosh K\right)^N \sum_{\{\sigma_i = \pm 1\}} \prod_{i=1}^{N} \left(1 + \sigma_{i-1}\sigma_i x\right) \\
&= \left(\frac{A_0(x)}{\sqrt{1-x^2}}\right)^N \sum_{\{\sigma_i\}} (1 + \sigma_0\sigma_1 x)(1 + \sigma_1\sigma_2 x)(1 + \sigma_2\sigma_3 x)\cdots \\
&= \left(\frac{2A_0^2(x)}{1-x^2}\right)^{N/2} \sum_{\{\sigma_i\}} (1 + \sigma_0\sigma_2 x^2)(1 + \sigma_2\sigma_4 x^2)(1 + \sigma_4\sigma_6 x^2)\cdots \\
&= \left(\frac{A_1(x_1)}{\sqrt{1-x_1^2}}\right)^{N/2} \sum_{\{\sigma_i\}} (1 + \sigma_0\sigma_1 x_1)(1 + \sigma_1\sigma_2 x_1)(1 + \sigma_2\sigma_3 x_1)\cdots, \qquad (8.3.1)
\end{aligned}
$$

where $A_0(x) = 1$ in the original equation, the spin numbers now run from 0 to $N-1$, with $\sigma_N = \sigma_0$, and $\cosh K$ has been expressed in terms of $x = \tanh K$. The third line was obtained by summation over the spins with odd indices, so that these spins are eliminated, and the resultant partition function, which retains the same value, has the appearance of one that corresponds to a new lattice with half as many spins, spaced twice as far apart. The new value of the bond strength x is $x_1 = x^2$, as written in the final formulation. The new lattice spacing is implicitly $a_1 = 2a$. In the recomposed system, the spins should be renumbered as consecutive integers, so $2i \rightarrow i$, as in the final equation. The transformation of the parameters in the partition function is thus

$$A_1(x_1) = 2A_0^2(x)\sqrt{\frac{1+x_1}{1-x_1}}, \quad \text{with } x_1 = x^2. \tag{8.3.2}$$

This process can be iterated, giving after the nth recomposition,

$$A_n(x_n) = 2A_{n-1}^2(x_{n-1})\sqrt{\frac{1+x_n}{1-x_n}}, \quad \text{with } x_n = x_{n-1}^2. \tag{8.3.3}$$

Now define a function $f_1(x)$ as

$$f_1(x) = \lim_{N \to \infty} \left[\frac{1}{N} \ln \left(\frac{1}{\sqrt{1-x^2}} \right)^N \sum_{\{\sigma_i\}} \prod_{i=1}^{N}(1 + \sigma_{i-1}\sigma_i x) \right]. \tag{8.3.4}$$

With this notation, the partition function per original spin can be written

$$\ln Z_1 = \lim_{N \to \infty} \left[\frac{1}{N} \ln Z_N \right] = \ln A_0(x) + f_1(x)$$
$$= (1/2)[\ln A_1(x_1) + f_1(x_1)] = (1/2^n)[\ln A_n(x_n) + f_1(x_n)]. \tag{8.3.5}$$

The recursion relation for the function $f_1(x_n)$ follows immediately as

$$f_1(x_n) = 2f_1(x_{n-1}) - \ln \left(2\sqrt{\frac{1+x_{n-1}^2}{1-x_{n-1}^2}} \right). \tag{8.3.6}$$

Note that the bond strength iterates as $x_n = x_{n-1}^2$, and $x_n = \tanh K_n \leq 1$, so that successive recompositions lead to ever weakening of the bond strength and continued reduction of the correlation length. The calculation in 1-d is reversible; it is interesting therefore to follow the iterative process backwards, from a lattice of arbitrarily weak coupling, where the approximate value of $f_1(x_n)$ is

$$f_i(x_n) = \ln 2 + \mathcal{O}(x_n) \approx \ln 2, \tag{8.3.7}$$

since the only significant contribution from the sum over spin signs is 1 for each sign. The backwards iteration is

$$f_1(x_{n-1}) = (1/2)[f_1(x_n) + \ln \left(\sqrt{\frac{1+x_n}{1-x_n}} \right). \tag{8.3.8}$$

x'	$x = \sqrt{x'}$	$f_1(x')$	$f_1(x)$	exact $f_1(x)$
0.01	0.1	$\ln 2 = 0.6931$	0.698147	0.698172
0.1	0.316227766	0.698147	0.745815	0.745827
0.316227766	0.562341325	0.745815	0.883206	0.883212
0.562341325	0.749894209	0.883206	1.106302	1.106305
0.749894209	0.865964323	1.106302	1.386081	1.386083
0.865964323	0.930572040	1.386081	1.6979710	1.6979718
0.930572040	0.964661619	1.6979710	2.02687960	2.0268700
0.964661619	0.982171889	2.02687960	2.36453981	1.36454001
0.982171889	0.991045856	2.36453981	2.706636486	2.706636584
0.991045856	0.995512860	2.706636486	3.050966563	3.050966612

Table 8.1 Renormalization-group sequence for one dimension.

Table 8.1 shows a sequence of these numerical iterations, starting at the value $x' = 0.01$, where for this table, x' is used as the value of x_n on the right-hand side of Eq. (8.3.8), and x is used as the value of x_{n-1} on the left. Then the new value of x is used as x' in the next row of the table, and continued for several iterations. The exact value, in the last column, follows directly from the exact solution of Eq. (8.3.5), which gives $f_1(x) = \ln Z_1(x) - \ln A_0(x) = \ln Z_1(x)$, since $A_0 = 1$, and from the known exact solution of Chapter 7, $\ln Z_1 = \ln(2\cosh K) = \ln 2/\sqrt{1 - x^2}$. Two characteristics of this iteration are: (1) the value of $f_1(x)$ based on the initial approximation agrees better with the exact value as the iteration proceeds, and (2) the end point is at $x = 1$. If the exact initial value of $f_1(x') = \ln[2/\sqrt{1 - (.01)^2}] = 0.69319783$, had been used instead of the approximate one, then at every stage, the last two columns of the table would agree. If the iteration were taken in the forward direction, instead of the reverse, x' would have gone to 0 as the end point, with f_1 converging exactly to $f_1 = \ln 2$. (Chandler [1987])

The relationship of correlation lengths in successive approximations is $\xi_n = \xi_{n-1}/2$. At the critical point, where the correlation length becomes infinite, recomposing should not change this infinite value, and at infinite temperature, where the correlation length vanishes, there is also no change in apparent correlation length. Thus, if successive values of the correlation length are set equal, such that $\xi_{n-1} = \xi_n$, the only two possible solutions of these two equations are $\xi_n = 0$ and $\xi_n = \infty$. The same procedure of equating successive values of x in Eq. (8.3.6) gives $x_n = x_{n-1}^2 = x_n^2$, with the solutions $x_n = 0$ or $x_n = 1$. Thus the iteration process exhibits two fixed points: at $x = 0$, the bond energy is infinitely weaker than kT, so that the spins are not coupled, and at $x = 1$, for which $T = 0$, there is no thermal stirring to destroy correlations. Neither of these conditions is particularly interesting, but the 1-d Ising chain is already known to have no critical temperature above $T = 0$. Renormalization is not expected to give any result in disagreement with exactly known results calculated by other means.

8.3.2 Transfer-Matrix Renormalization. The same problem can be treated by a method based on the transfer-matrix calculations of Section 7.5.2. This time an external ordering field will be included. The Hamiltonian may be written as

$$\beta\mathcal{H} = -\sum_{i=1}^{N}\left[K\sigma_{i-1}\sigma_i + (h/2)(\sigma_{i-1} + \sigma_i) + C\right], \tag{8.3.9}$$

where C is an added constant that shifts the energy level, if convenient, but has no effect on thermodynamic behavior. It is used in this calculation as a means of preserving the form of the partition function after renormalization. In accordance with Eq. (7.5.30) and (7.5.31), the partition function may be written

$$Z_N = \sum_{\{\sigma_i=\pm1\}} \langle\sigma_0|P|\sigma_1\rangle\langle\sigma_1|P|\sigma_2\rangle\langle\sigma_2|P|\sigma_3\rangle\cdots|P|\sigma_0\rangle, \qquad (8.3.10)$$

where the transfer matrix P is

$$P = e^C \begin{pmatrix} e^{K+h} & e^{-K} \\ e^{-K} & e^{K-h} \end{pmatrix}. \qquad (8.3.11)$$

Instead of carrying out all the indicated matrix products to find the trace of P^N, only the products obtained by summation over the odd spins are produced, leading to a reduced expression for the partition function in which all odd spins have been eliminated. One of these partial-trace eliminations is seen to be

$$\langle\sigma_0|P|\sigma_1\rangle\langle\sigma_1|P|\sigma_2\rangle = \langle\sigma_0|P^2|\sigma_2\rangle = \langle\sigma_0|P_1|\sigma_2\rangle, \qquad (8.3.12)$$

where $P_1 = P^2$, which is then required to be of the same form as P, or

$$P_1 = e^{C_1} \begin{pmatrix} e^{K_1+h_1} & e^{-K_1} \\ e^{-K_1} & e^{K_1-h_1} \end{pmatrix}. \qquad (8.3.13)$$

Identification of terms in the defined relation $P^2 = P_1$ gives the equations

$$e^{2h_1} = e^{2h}\frac{\cosh(2K+h)}{\cosh(2K-h)} \text{ or } h_1 = h + \ln\sqrt{\frac{\cosh(2K+h)}{\cosh(2K-h)}},$$
$$e^{4K_1} = \frac{\cosh(2K+h)\cosh(2K-h)}{\cosh^2 h} \text{ or } 4K_1 = \ln\frac{\cosh(2K+h)\cosh(2K-h)}{\cosh^2 h},$$
$$e^{4C_1} = 16e^{8C}\cosh^2 h\cosh(2K+h)\cosh(2K-h) \text{ or }$$
$$4C_1 = 8C + \ln\left[\cosh^2 h\cosh(2K+h)\cosh(2K-h)\right]. \qquad (8.3.14)$$

Note that $|h_1| > |h|$, in agreement with Eq. (8.2.1).

This procedure can be iterated, as before. With C, K, and h renamed C_n, K_n, and h_n, and a similar transformation of the index 1 to $n+1$, Eq. (8.3.14) provides a set of mappings of points in the three-dimensional C-K-h space, starting from any arbitrary point but proceeding in a predetermined manner through a sequence of points to a fixed point or cluster point of the sequence. At each step, Eq. (8.3.10) could have its remaining spins renumbered as part of the recomposition, although this exercise is not necessary, once it is understood to be part of the formalism. Several further developments are needed before the Kadanoff scaling of Eq. (8.2.1) can be claimed as derived.

Agreement with the procedure of the previous section is obtained when C and h are set equal to zero, and appropriate identities are used, based on $x = \tanh K$. In general, $C_n = \ln A_n$, as given by Eq. (8.3.3).

8.3.3 The Nagle Linear Model of KDP. Although phase transitions are not usually possible in 1-d systems, the Nagle model of KDP, discussed in Section 7.5.3 exhibits such a transition because some of the possible configurational energies are infinite. The system is discussed there in some detail, and solved with all energies finite by the transfer matrix method, with the transfer matrix given in Eq. (7.5.43). Problem 7.5 asks for the solution of this model with the appropriate energies infinite by the transfer-matrix method, and the solution gives the transfer matrix as

$$P = \begin{pmatrix} 1 & 0 & 0 & 0 \\ 0 & x & x & 0 \\ 0 & x & x & 0 \\ 0 & 0 & 0 & 1 \end{pmatrix}. \tag{8.3.15}$$

The partition function can be written like that of Eq. (8.3.10), except that each spin index takes on four values, corresponding to those in Eq. (7.5.43). Recomposition by elimination of the odd-numbered indices, as in the previous subsection, again leads to the relation $P_1 = P^2$, with

$$P_1 = \begin{pmatrix} 1 & 0 & 0 & 0 \\ 0 & x_1 & x_1 & 0 \\ 0 & x_1 & x_1 & 0 \\ 0 & 0 & 0 & 1 \end{pmatrix}, \tag{8.3.16}$$

where the renormalized value of x is $x_1 = 2x^2$, as may be seen by multiplication of the matrix P by itself. In general, $x_n = 2x_{n-1}^2$. At a stationary point, then, when $x_n = x_{n-1}$, these two equations combine to give

$$x_n = 2x_n^2. \tag{8.3.17}$$

The solutions are seen to be $x_n = 0$ and $x_n = 1/2$, a nontrivial value, which was found as the critical value in Eq. (7.5.43). The transformation $x_{n+1} = 2x_n^2$ maps the point $x_n \neq 1/2$ onto a point that is farther from the critical value than was x_n, no matter which side of the critical value is chosen as the starting point. But the inverse transformation, here written $x' = \sqrt{x/2}$ converges to the critical value from any starting point on the line $0 < x < 1$. This is the paradigm of the behavior sought in renormalization-group theory. In reality, such clean, direct, unambiguous, and exactly solvable systems are virtually unknown in two or more dimensions.

8.3.4 Other Examples in One Dimension. Some of the complication that arises in two or more dimensions can be illustrated by models in one dimension with several different pathways by which spins may interact. As a first of these examples, consider the double chain of problem 7.8, where the partition function is

$$Z_N = \left(\cosh^2 K_1\right)^{N-1} \left(\cosh K_2\right)^N \sum_{\sigma_{ij}=\pm 1} \prod_{i=0}^{N-1} (1 + \sigma_{i0}\sigma_{i1}y)(1 + \sigma_{i0}\sigma_{i+1,0}x)(\sigma_{i1}\sigma_{i+1,1}x),$$

$$= C_N \sum_{\sigma_{ij}=\pm 1} \prod_{i=0}^{N-1} (1 + \sigma_{i0}\sigma_{i1}y)(1 + \sigma_{i0}\sigma_{i+1,0}x)(\sigma_{i1}\sigma_{i+1,1}x), \tag{8.3.18}$$

where $x = \tanh K_1$, $y = \tanh K_2$, and $K_i = \beta J_i$. Consider the group of spins labeled 00, 01, 10, 11, 20, 21; these are the ones in Eq. (8.3.18) for $i = 0$, 1, or 2, and no term with $i > 2$

need be considered in this initial reduction procedure. The product of the first three terms in Eq. (8.3.18), written out with the recognition that $\sigma_{ij}^2 = 1$, is

$$(1 + y\sigma_{00}\sigma_{01})(1 + x\sigma_{00}\sigma_{10})(1 + x\sigma_{01}\sigma_{11}) = 1 + y\sigma_{00}\sigma_{01} + x(\sigma_{00}\sigma_{10} + \sigma_{01}\sigma_{11}) +$$
$$+ xy(\sigma_{01}\sigma_{10} + \sigma_{00}\sigma_{11}) + x^2 y\sigma_{10}\sigma_{11} + x^2\sigma_{00}\sigma_{01}\sigma_{10}\sigma_{11}. \tag{8.3.19}$$

Each of these terms can be represented by a diagram on the lattice. The term $y\sigma_{00}\sigma_{01}$ is the y bond connecting the two spins, and no other bonds present. The term $x^2 y\sigma_{10}\sigma_{11}$ is the connection from σ_{10} to σ_{11} by way of σ_{00} and σ_{01}. (Remember that the direct connection is not among the three factors used so far.) The four-spin term is merely the two x bonds, each connecting a pair of spins. Next, the factor $(1 + y\sigma_{10}\sigma_{11})(1 + x\sigma_{10}\sigma_{20})$ is multiplied by the eight terms already found, and the result summed over $\sigma_{10} = \pm 1$. Aside from the factor 2 that always multiplies the surviving terms, the result is a collection of terms representing connections involving the spins 00, 01, 11, and 02, since 10 has been summed out. A similar procedure with the remaining factor that contains 11, which is $1 + \sigma_{11}\sigma_{21}$, and a final multiplication by the only other factor in Eq. (8.3.19) gives the terms

$$(1 + 2x^2 y^2 + x^4 y^2) + (\sigma_{00}\sigma_{01} + \sigma_{20}\sigma_{21})(y + x^2 y + x^4 y + x^2 y^3) + (\sigma_{00}\sigma_{20} + \sigma_{01}\sigma_{21})(x^2 + 3x^2 y^2)$$
$$+ (\sigma_{00}\sigma_{21} + \sigma_{01}\sigma_{20})(3x^2 y + x^2 y^3) + \sigma_{00}\sigma_{01}\sigma_{20}\sigma_{21}(x^4 + y^2 + 2x^2 y^2).$$

The spinless term represents the possible closed diagrams of 0, 4, and 6 bonds that can be drawn on the first six points of the double chain. The other terms represent open diagrams connecting the specified spins. Note that each x, y factor is the sum of four diagrams, and all x^n (or y^n) factors in a single term have n of the same parity (all even or all odd). Every such group of six spins that begins with $\sigma_{2i,j}$ can be summed in this way. The next step in the renormalization operation is to note that the i index in σ_{ij} is now always even, and since all odd-i spins have been summed out, the i index can be relabeled $i/2$. Each summation over an odd-i spin produced a factor 2, which appears outside the summation symbol. Also, each spinless term in the reduced expression for the partition function can be factored out, giving the renormalized partition function as

$$Z_N = 2^N C_N (1 + 2x^2 y^2 + x^4 y^2)^{N/2} \sum_{\sigma_{ij} = \pm 1} \prod_{i=0}^{(N/2)-1} \left[1 + (\sigma_{i0}\sigma_{i1} + \sigma_{i+1,0}\sigma_{i+1,1})y' + \right.$$
$$+ (\sigma_{i0}\sigma_{i+1,0} + \sigma_{i1}\sigma_{i+1,1})x' + (\sigma_{i0}\sigma_{i+1,1} + \sigma_{i1}\sigma_{i+1,0})z') + (\sigma_{i0}\sigma_{i1}\sigma_{i+1,0}\sigma_{i+1,1})w' \right]$$
$$= C_{N'} \sum_{\sigma_{ij} = \pm 1} \prod_{i=0}^{N'-1} \left[1 + (\sigma_{i0}\sigma_{i1} + \sigma_{i+1,0}\sigma_{i+1,1})y' + (\sigma_{i0}\sigma_{i+1,0} + \sigma_{i1}\sigma_{i+1,1})x' + \right.$$
$$+ (\sigma_{i0}\sigma_{i+1,1} + \sigma_{i1}\sigma_{i+1,0})z') + (\sigma_{i0}\sigma_{i1}\sigma_{i+1,0}\sigma_{i+1,1})w' \right], \tag{8.3.20}$$

where $N' = N/2$, and, for example,

$$x' = \frac{x^2 + 3x^2 y^2}{1 + 2x^2 y^2 + x^4 y^2}.$$

In this case, the renormalized partition function contains more coupling constants than did the original. In addition to nearest-neighbor bonds along and across the chain, there is now an effective bond along each diagonal, written as z', and a four-spin bond, w'. Along this double chain, further reductions and renormalizations do not continue to generate new terms; each renormalized partition function can be stated in the form of Eq. (8.3.20), with five independent equations relating the new coupling constants and C' to the old. Problems 3 and 4, at the end of this chapter, provide two additional examples of 1-d chains that can be reduced to the form of Eq. (8.3.20), but with much more complicated definitions of the renormalized coupling parameters in terms of the original ones.

8.4 GENERALIZATION

For Ising models in one dimension, each recomposition reduced the number of spins from N to $N' = N/2$, and the correlation length from ξ to $\xi' = \xi/2$. More generally, elimination could reduce the number of spins to $N' = N/b$ and the correlation length to $\xi' = \xi/b$, if some other grouping is used to eliminate spins. Extension to a lattice in d dimensions reduces the number of spins to $N' = N/b^d$. The parameters C, K, and h were transformed as given by Eq. (8.3.14). These transformations may be generalized as

$$C' = bC + \mathcal{R}_c(K, h); \quad K' = \mathcal{R}_k(K, h); \quad \text{and} \quad h' = \mathcal{R}_h(K, h). \qquad (8.4.1)$$

The Hamiltonian is renormalized as

$$\mathcal{H}'(N', K', h') = \mathcal{R}_b(N, K, h), \qquad (8.4.2)$$

and the partition function, which retains its value, obeys

$$Z_{N'}(\mathcal{H}') = Z_N(\mathcal{H}). \qquad (8.4.3)$$

The Helmholtz potential per spin, in units of kT, may be written

$$f(\mathcal{H}) = -\frac{1}{N} \ln Z_N(\mathcal{H})$$

$$= -\left(\frac{N'}{N}\right) \frac{1}{N'} \ln Z_{N'}(\mathcal{H}') = \frac{1}{b^d} f(\mathcal{H}'), \qquad (8.4.4)$$

which is equivalent to the renormalization

$$f(K', h', C') = b^d f(K, h, C). \qquad (8.4.5)$$

The temperature is contained in the parameter K. Before any renormalization, the temperature is $T = (1/K)(k/J)$, so $1/K$ measures the temperature in units of k/J. Therefore $1/K'$ may be taken as a measure of temperature after renormalization, and the temperature transformation may be written

$$T' = \mathcal{R}_T(T, h). \qquad (8.4.6)$$

In these calculations, the temperature transformation is typically used for two purposes: (a) to find the critical temperature, and (b) to discover the critical behavior of the system. For the former, the ideal procedure is that displayed in the discussion above of linear KDP. The critical point is a fixed point of the transformation, so Eq. (8.4.6) becomes

$$T'_c = T_c = \mathcal{R}_T(T_c, h_c). \qquad (8.4.7)$$

This equation may be used to relate the original temperature to that of the first renormalization, as in Eq. (8.3.17), which gave $x_c = 2x_c^2$, with the nontrivial solution $x_c = 1/2 = \exp(-\beta_c \epsilon_1)$. Thus the critical temperature can be found without ambiguity. For the second purpose, the temperature appears in critical-exponent relations, such as $\chi_T \sim \tau^{-\gamma}$, where numerical details are not made explicit, and any measure of T that exhibits the correct qualitative behavior is satisfactory.

At the critical point, for which $h_c = 0$, as shown below, the correlation length becomes infinite. Its transformation equation,

$$\xi(T') = \xi(T)/b, \tag{8.4.8}$$

shows that at a fixed point given by $T'^* = T^*$, it obeys $b\xi(T^*) = \xi(T^*)$, with the only possible solutions being $\xi(T^*) = 0$, or $\xi(T^*) = \infty$. The former corresponds to zero correlation length at $T^* = \infty$, where neighboring spins are effectively uncoupled by thermal agitation, or at $T^* = 0$ where there is perfect alignment, but the correlation coefficient $\langle \eta(\mathbf{r})\eta(0) \rangle - \langle \eta(\mathbf{r}) \rangle \langle \eta(0) \rangle$ vanishes. (This statement provides the answer to the question proposed in Problem 6.6.) The infinite correlation length confirms that the nontrivial fixed point is the critical point.

In the transformations of Eq. (8.4.1), the constant C has no effect on critical behavior, and it does not enter the renormalizations of K or h. Near the critical point the transformations may be written

$$\tau' = \mathcal{R}_\tau(\tau, h), \quad \text{and} \quad h' = \mathcal{R}_h(\tau, h). \tag{8.4.9}$$

In general, these are nonlinear transformations, as may be seen by the examples already worked out. They are assumed to be continuous and differentiable, so that for small departures from the critical point, linear approximations of these transformations can be used. In this region, the linear approximation may be written

$$\begin{pmatrix} \tau' \\ h' \end{pmatrix} = \begin{pmatrix} \left(\frac{\partial \tau'}{\partial \tau}\right)_c & \left(\frac{\partial \tau'}{\partial h}\right)_c \\ \left(\frac{\partial h'}{\partial \tau}\right)_c & \left(\frac{\partial h'}{\partial h}\right)_c \end{pmatrix} \begin{pmatrix} \tau \\ h \end{pmatrix}, \tag{8.4.10}$$

where the derivatives are evaluated at the critical point. Because the behavior of the system is symmetric with respect to the sign of h, it follows that the transformation matrix in this case must be diagonal. Each transformation can be written

$$\tau' = \Lambda_\tau(b)\tau, \quad \text{and} \quad h' = \Lambda_h(b)h, \tag{8.4.11}$$

where $\Lambda_x = (\partial x'/\partial x)_c$, written to emphasize the scaling dependence of the transformation. An iteration of the transformation gives, say for τ,

$$\tau'' = \Lambda_\tau(b)\tau' = \Lambda_\tau^2(b)\tau. \tag{8.4.12}$$

But since two successive b-scalings are equivalent to one b^2-scaling, a functional relationship becomes evident:

$$\Lambda^2(b) = \Lambda(b^2). \tag{8.4.13}$$

This is a particular case of Eq. (6.6.2), with the only possible solution for differentiable functions given by

$$\Lambda_\tau(b) = b^{\lambda_1}, \quad \text{and} \quad \Lambda_h(b) = b^{\lambda_2}. \tag{8.4.14}$$

Thus the scaling of τ and h are found to be $\tau' = b^{\lambda_1}\tau$ and $h' = b^{\lambda_2}h$. The singular part of Eq. (8.4.5) may now be written

$$f_s(\tau, h) = b^{-d} f_s(b^{\lambda_1}\tau, b^{\lambda_2}h), \tag{8.4.15}$$

or, to return to the notation of Chapter 6, Section 6.6.2, define $b^d = \lambda$, $b^{\lambda_1} = \lambda^{\lambda_1/d}$, $b^{\lambda_2} = \lambda^{\lambda_2/d}$, $\lambda_1/d = p$, and $\lambda_2/d = q$. The Helmholtz potential can be expressed in terms of the Gibbs potential through a Legendre transformation. This treatment recovers the scaling equations of Chapter 6, as well as the Kadanoff block-spin consequences of this chapter.

More generally, the renormalization transformation is much more complicated than that of Eq. (8.4.5), because numerous coupling constants appear. As shown in the one-dimensional models of the previous section and of Problems 8.4 and 8.5, four K's are found, leading to a 5×5 square matrix and five-element column matrices to replace those of Eq. (8.4.10). The scaling behavior of each such constant should show up in a generalization of Eq. (8.4.15), written as

$$f_s(\tau, h, K_3, K_4, \ldots) = b^{-d} f_s(b^{\lambda_1}\tau, b^{\lambda_2}h, b^{\lambda_3}K_3, \ldots), \tag{8.4.16}$$

where the additional λ's are the analogs of those in Eq. (8.4.14). As in the scaling analysis of Chapter 6, the first independent variable on the right-hand side can be equated to unity, to set $b = \tau^{-\lambda_1}$. The scaling equation then becomes

$$f_s(\tau, h, K_3, \ldots) = \tau^{d/\lambda_1} f_s(1, h\tau^{\lambda_2/\lambda_1}, K_3\tau^{\lambda_3/\lambda_1}, \ldots). \tag{8.4.17}$$

Since $\lambda_1/d = p$, as used in Chapter 6, it follows that $d/\lambda_1 = 2 - \alpha$. Also, the gap exponent, Δ is found from these results to be $\Delta = q/p = \lambda_2/\lambda_1$, so that the h term becomes h/τ^Δ. The remaining parameters are of the form $K_j/\tau^{\lambda_j/\lambda_1}$. Their importance near the critical point depends upon the sign of λ_j/λ_1; if this sign is positive as $\tau \to 0$, then the rescaled K_j parameter becomes large, and it is then said to be a *relevant* parameter. Similarly, when λ_j/λ_1 is negative, the rescaled K_j approaches zero with τ, and K_j is described as *irrelevant*. In examining critical phenomena, it is usually possible to ignore the irrelevant parameters. (Fisher [1974]; Fisher [1983]) Successive renormalizations result in the flow of points such as (τ, h) toward the so-called *fixed points* of the transformation. But as already seen in the discussion of Eq. (8.4.8) and of the Nagle model of KDP, the critical point is a fixed point of the transformation such that successive renormalizations move neighboring points *away* from it. They move toward other fixed points that are usually uninteresting, and often called *trivial*, such as those at $T = 0$ or $T = \infty$. A relevant parameter is carried rapidly away from criticality by successive renormalizations; irrelevant parameters ultimately vanish, and their fields can be ignored. When more than two parameters are relevant, the fixed point represents a multicritical point.

For qualitative understanding of the critical behavior of Ising models, it is often sufficient to carry along no more than an easily manageable number of couplings, as evinced by the success of Kadanoff's block-spin model, but for accurate determination of the nontrivial fixed point(s), many more couplings are necessary. In an analysis of a 2-d Ising model, Wilson carried 217 coupling constants, most of which arise from the Kadanoff reductions, and found it necessary to monitor successive generations of couplings to ascertain the importance of still others beyond the 217 selected ones. (Wilson [1975]) His calculation produced quite accurate numbers for the critical values of the original coupling constants. Such calculations are complicated and subject to potentially serious numerical errors. There is no guarantee that fixed points can be found, especially since correlation lengths near the critical point are thousands of lattice spacings, and a Kadanoff reduction to a block of such dimensions must then involve some 10^{3d} spins, with all possible couplings. (Some of these couplings may be small enough to ignore, unless the number of them is so large that they become important. The inclusion or omission of any one coupling constant can seriously alter the results of the computation.)

Even though the scaling laws and critical-exponent relations are given a good physical basis in this version of renormalization, actual calculations are quite difficult because of the many possible paths that contribute to a distant coupling. Symmetries must be preserved, while attempts are made to discard some of the less-significant terms, in order to attain expressions of reduced awkwardness. The process never, of itself, becomes a representation of a system in the continuum limit.

8.5 WILSON RENORMALIZATION

Of the many contributions Kenneth Wilson has made to the understanding of critical phenomena, a procedure I regard as one of the most useful is that for transforming the lattice fields of Chapter 7, and of the previous section, into continuum fields that exhibit the long-wave-length fluctuations characteristic of critical behavior. The system of interacting spins, each at its lattice position r_j, can be fourier transformed to the continuous order parameter η used in Chapter 6, but with essential refinements. (Wilson [1973]; Wilson [1975]; Wilson [1983]; Wilson and Kogut [1974])

In accordance with the usual practice in this book, the smoothing operation is presented first in one dimension. The techniques so developed are then extended to d dimensions, when necessary. On a 1-d lattice of spacing a, choose a system of N spins and length $L = Na$, where $L \gg a$, and for some mathematical operations, the number of lattice points, $N = L/a$, is allowed to become infinite. For convenience, model the system as a closed chain, as in Chapter 7, or regard it as having periodic boundary conditions. Designate the position of each spin by its dimensionless spin coordinate n, so that its distance from spin 0 is na. Fourier transform the spins, to get

$$S(k_m) = \frac{1}{\sqrt{N}} \sum_{n=0}^{N-1} \sigma_n e^{ink_m}, \tag{8.5.1}$$

where the dimensionless variable k_m is defined as $k_m = 2\pi ma/L = 2\pi m/N$, and $m = 0, 1, 2, \ldots N-1$, with $k_N \equiv k_0$. The exact fourier inverse of Eq. (8.5.1) is

$$s_n = \frac{1}{\sqrt{N}} \sum_{m=0}^{N-1} S(k_m) e^{-ink_m} = \frac{1}{N} \sum_{m=0}^{N-1} e^{-ink_m} \sum_{n'=0}^{N-1} \sigma_{n'} e^{in'k_m}$$

$$= \frac{1}{N} \sum_{n'=0}^{N-1} \sigma_{n'} \sum_{m=0}^{N-1} e^{ik_m(n'-n)} = \frac{1}{N} \sum_{n'=0}^{N-1} \sigma_{n'} N\delta_{n'n} = \sigma_n, \tag{8.5.2}$$

as required. For smoothing, n is replaced by a continuous variable, x, equivalent to the ordinary coordinate, but measured in units of the lattice spacing a, and s_n becomes $s(x)$, a continuous function of x, as shown.

$$s_n = \frac{1}{\sqrt{N}} \sum_{m=0}^{N-1} S(k_m) e^{-ink_m} \rightarrow \left(\frac{1}{\sqrt{N}}\right) \sum_{m=0}^{N-1} S(k_m) e^{-ixk_m}$$

$$= \frac{1}{N} \sum_{m=0}^{N-1} \sum_{n=0}^{N-1} \sigma_n e^{ik_m(n-x)} = \frac{1}{N} \sum_{n=0}^{N-1} \sigma_n \frac{1 - e^{2\pi i(n-x)}}{1 - e^{2\pi i(n-x)/N}}. \tag{8.5.3}$$

Thus each spin is replaced by a spin density, $s_n(x)$, given by

$$s_n(x) = \frac{\sigma_n}{N} \sum_{m=0}^{N-1} e^{ik_m(n-x)} = \frac{\sigma_n}{N} \frac{1 - e^{2\pi i(n-x)}}{1 - e^{2\pi i(n-x)/N}}. \tag{8.5.4}$$

This expression peaks to $s_n(n) = \sigma_n$ at $x = n$, drops to zero at $x = n \pm 1$, and oscillates with decreasing amplitude as x increases its distance from n. (It is called a wavelet; there is considerable recent literature on wavelets. (Daubechies [1990])) This is the sharpest resolution possible when the shortest allowed wavelength equals the lattice spacing. The total spin is conserved in this kind of smoothing, since

$$\int_0^N dx \sum_{n=0}^{N-1} s_n(x) = \int_0^N dx \, s(x) = \frac{1}{N} \sum_{n=0}^{N-1} \sigma_n \sum_{m=0}^{N-1} e^{ink_m} \frac{e^{-iNk_m} - 1}{-ik_m}$$

$$= \frac{1}{N} \sum_{n=0}^{N-1} \sum_{m=0}^{N-1} e^{ink_m} N \delta_{0m} = \sum_{n=0}^{N-1} \sigma_n, \tag{8.5.5}$$

as required. The total spin function $s(x)$, defined in this equation, is the order parameter of Landau theory, now written as

$$\eta(x) \equiv s(x) = \frac{1}{N} \sum_{n=0}^{N-1} \sigma_n \sum_{m=0}^{N-1} e^{ik_m(n-x)}. \tag{8.5.6}$$

The order parameter and its fourier transform are now written as the pair

$$\eta(x) = \frac{1}{\sqrt{N}} \sum_{m=0}^{N-1} \tilde{\eta}(k_m) e^{-ixk_m}, \quad \text{and} \quad \tilde{\eta}(k_m) = \frac{1}{\sqrt{N}} \sum_{n=0}^{N-1} \sigma_n e^{ink_m} = \frac{1}{\sqrt{N}} \int_0^N dx \, \eta(x) e^{ixk_m}. \tag{8.5.7}$$

The Landau-Ginzburg expression for the Gibbs potential that was used in Chapter 6, Section 7.1, is written here for one dimension as

$$G = \int_0^N dx \left[g_0 - h(x)\eta(x) + \tfrac{1}{2} g_2 \eta^2 + \tfrac{1}{4} g_4 \eta^4 + \frac{c}{2} (\nabla \eta)^2 \right]. \tag{8.5.8}$$

Their theory permits short-range fluctuations in η, but it does not recognize the long-range fluctuations that are important in critical phenomena. As a means of examining the effects of range, the order parameter is written as $\eta(x) = \eta_0(x) + \eta_1(x)$, where η_0 contains only the long-range fourier components, and η_1 the short-range ones. An arbitrary wavelength boundary is chosen as l, where $1 < l < N$. Recall that $N = L/a$, and define $N_1 = L/la = N/l$. Define the two parts of the order parameter as

$$\eta_0(x) = \frac{1}{\sqrt{N}} \sum_{m=0}^{N_1-1} \tilde{\eta}(k_m) e^{-ixk_m} \quad \text{and}$$

$$\eta_1(x) = \frac{1}{\sqrt{N}} \sum_{m=N_1}^{N-1} \tilde{\eta}(k_m) e^{-ixk_m} \tag{8.5.9}$$

Independent of the choice of l, a basic difference in the two parts is that

$$\int_0^N dx \, \eta_0(x) = \frac{1}{N} \sum_{m=0}^{N_1-1} \sum_{n=0}^{N-1} \sigma_n \int_0^N dx \, e^{ik_m(n-x)} = \frac{1}{N} \sum_{n=0}^{N-1} \sigma_n \sum_{m=0}^{N_1-1} e^{ink_m} \frac{e^{-iNk_m} - 1}{-ik_m}$$

$$= \sum_{n=0}^{N-1} \sigma_n = N \langle \sigma \rangle = \sqrt{N} \tilde{\eta}(0), \quad \text{and}$$

$$\int_0^N dx \, \eta_1(x) = \frac{1}{N} \sum_{m=N_1}^{N-1} \sum_{n=0}^{N-1} \sigma_n \int_0^N dx \, e^{ik_m(n-x)}$$

$$= \frac{1}{N} \sum_{n=0}^{N-1} \sigma_n \sum_{m=N_1}^{N-1} e^{ink_m} \frac{e^{-iNk_m} - 1}{-ik_m} = 0. \tag{8.5.10}$$

(Only the $m = 0$ component has a nonzero average value.) One further useful feature of the separation, independent of l, is

$$\int_0^N dx \, \eta_0(x)\eta_1(x) = \frac{1}{N} \sum_{m=0}^{N_1-1} \tilde{\eta}(k_m) \sum_{m'=N_1}^{N-1} \tilde{\eta}(k_{m'}) \int_0^N dx \, e^{-ix(k_m+k_{m'})} = 0, \tag{8.5.11}$$

since $k_m \pm k_{m'} \neq 0$.

The short-range modes, which are not regarded as important in critical phenomena, may be averaged out statistically, while the long-range modes are retained in the Landau-Ginzburg format. It is customary in this analysis to set the constant c in Eq. (8.5.8) as $c = 1$, and to absorb the factor $\beta_b = 1/k_B T_b$ into the other constants, where T_b is the heat-bath temperature. Restated expressions for the various terms in Eq. (8.5.8) are as follows.

$$\int_0^N dx \, \eta(x) = N \langle \eta \rangle = \sqrt{N} \tilde{\eta}(0); \tag{8.5.12}$$

$$\int_0^N dx \, \eta^2(x) = N \langle \eta_0^2 \rangle + \int_0^N dx \, \eta_1^2(x)$$

$$= N\eta_0^2 + \int_0^N dx \, \eta_1^2(x), \tag{8.5.13}$$

where η_0 is written henceforth as an imprecisely defined average of the long-range modes, and

$$\int_0^N dx\, \eta_1^2(x) = \frac{1}{N} \int_0^N dx \sum_{m=N_1}^{N-1} \tilde{\eta}(k_m) \sum_{m'=N_1}^{N-1} \tilde{\eta}(km')e^{-ix(k_m+k_{m'})}$$

$$= \frac{1}{N} \sum_{m=N_1}^{N-1} \tilde{\eta}(k_m) \sum_{m'=N_1}^{N-1} \tilde{\eta}(k_{m'}) N \delta_{-m}^{m'}$$

$$= \sum_{m=N_1}^{N-1} \tilde{\eta}(k_m)\tilde{\eta}^*(k_m) = \sum_{m=N_1}^{N-1} |\tilde{\eta}_1(k)|^2, \qquad (8.5.14)$$

Similarly, when it is noted that $\nabla\eta \to d\eta/dx$, the short-range contribution from $(\nabla\eta)^2$ becomes

$$\int_0^N dx\, (\nabla\eta_1)^2 = \frac{-1}{N} \int_0^N dx \sum_{m=N_1}^{N-1} k_m \tilde{\eta}(k_m) \sum_{m'=N_1}^{N-1} k_{m'} \tilde{\eta}(k_{m'})e^{-ix(k_m+k'_m)}$$

$$= \sum_{m=N_1}^{N-1} k_m^2 |\tilde{\eta}(k_m)|^2. \qquad (8.5.15)$$

The fourth-order term is carried only to second order in η_1, on the basis that fourth-order averages of the short-range part of the order parameter are small. The approximations are that any $\int_0^N dx$ of a term containing an odd power of $\eta_1(x)$ is zero, and that there is negligible mixing of the even-powered terms such as $\eta_0^2(x)\eta_1^2(x)$, so that

$$\int_0^N dx\, \eta_0^2(x)\eta_1^2(x) = \frac{1}{N} \int_0^N dx\, \eta_0^2(x) \sum_{m=N_1}^{N-1} \tilde{\eta}(k_m) \sum_{m'=N_1}^{N-1} \tilde{\eta}(k'_m)e^{-ix(K_m+k'_m)}$$

$$\approx \eta_0^2 \sum_{m=N_1}^{N-1} |\tilde{\eta}(k_m)|^2, \qquad (8.5.16)$$

where again the definition of η_0 is imprecise. (If $l = 1$, so that $N_1 = N$, then $\eta_0 = \sqrt{N}\tilde{\eta}(0)$, no matter how it arises, but for $l > 1$, other long-range terms enter the average, and the precise definition varies with the averaging process. For the purposes of this analysis, there seems little to be gained by the proliferation of these long-range averages, although a more careful treatment should recognize their differences.) The fourth-order term is approximated as

$$\int_0^N dx\, \eta^4(x) \approx \int_0^N dx\, \left(\eta_0^4 + 4\eta_0^3\eta_1 + 6\eta_0^2\eta_1^2\right) = N\eta_0^4 + 6N\eta_0^2 \langle\eta_1^2\rangle$$

$$= N\eta_0^4 + 6N\eta_0^2 \sum_{m=N_1}^{N-1} |\tilde{\eta}_1(k_m)|^2. \qquad (8.5.17)$$

The entire equation for the Gibbs potential can be restated as

$$\beta_b G(\eta, \tau, h) = \beta_b G_0 - N h \eta_0 + \int_0^N dx \left[(1/2)(\nabla \eta(x))^2 + (1/2)g_2 \eta^2(x) + (1/4)g_4 \eta^4(x) \right]$$

$$= \beta_b G_0 - N h \eta_0 + \frac{N}{2} \left[(\nabla \eta_0)^2 + g_2 \eta_0^2 + (1/2)g_4 \eta_0^4 \right]$$

$$+ \sum_{m=N_1}^{N-1} \left[(1/2)k_m^2 + (1/2)g_2 + (3/2)g_4 \eta_0^2 \right] |\bar{\eta}_1(k_m)|^2, \tag{8.5.18}$$

where G_0 contains the nonfluctuating terms that are otherwise not displayed, and the $\nabla \eta_0$ term is retained, at present for cosmetic purposes, even though η_0 appears to be defined as a constant. The various fourier components are regarded as capable of independent fluctuation; they increase the number of possible states of the system, and therefore the entropy, as well as providing modes for the storage of energy. For each value of k_m, the fourier component $\bar{\eta}(k_m)$ could, in principle, take any value in the complex $\bar{\eta}$ plane. Therefore, because of the inclusion of the fluctuating modes, the G of Eq. (8.5.18) is not the equilibrium value as specified in Chapter 2. Equilibrium must be sought in terms of minimizing the Gibbs potential, subject to constant $T = T_b$ and B_0 (or h), as in Problem 13 of Chapter 3. Any particular Gibbs potential may be written symbolically as $G = U - TS - \mathcal{M}B_o = G_0 - N h \eta_0 + \sum_i [p_i E_i + k T_b p_i \ln p_i]$, where p_i represents the probability of the energy term E_i, and the entropy is written in terms of the probabilities. The equilibrium expression for G is found by the usual variational technique, enhanced in detail but not in principle.

The varying quantity represented by E_i is the Landau-Ginzburg-Wilson energy written as the coefficient of $\bar{\eta}(k_m)$ in Eq. (8.5.18). For ease of notation, write $A(k_m) \equiv (1/2)[k_m^2 + g_2 + 3g_4 \eta_0^2]$. Write $\bar{\eta}(k_m) = y_m + i z_m$, so that $|\bar{\eta}(k_m)|^2 = y_m^2 + z_m^2$, with y_m and z_m regarded as independent. A probability density $\rho(y_m)$ or $\rho(z_m)$ (they turn out to be the same function) must be assigned to each variable y_m or z_m (*i.e.*, for every possible value of m), such that

$$I_y \equiv \int_{-\infty}^{\infty} dy_m \, \rho(y_m) = 1, \tag{8.5.19}$$

and equivalently for z_m. The Gibbs potential to be varied is now written, where use is made of the now-imprecise definition of η_0,

$$\beta_b G = \beta_b G_0 - N h \eta_0 + \frac{N}{2} \left[(\nabla \eta_0)^2 + g_2 \eta_0^2 + 1/2 g_4 \eta_0^4 \right]$$

$$+ \sum_{m=N_1}^{N-1} \left[\int_{-\infty}^{\infty} dy_m \left[A(k_m) y_m^2 \rho(y_m) + \beta_b k_B T_b \rho(y_m) \ln a_1 \rho(y_m) \right] + \right.$$

$$\left. + \int_{-\infty}^{\infty} dz_m \left[A(k_m) z_m^2 \rho(z_m) + \beta_b k_B T_b \rho(z_m) \ln a_1 \rho(z_m) \right] \right], \tag{8.5.20}$$

where a_1 is a dimensional constant required because ρ is a probability density, and G is to be minimized, subject to Eq. (8.5.19), by variation of the probability densities. This variational procedure is written as

$$\delta(\beta_b G + \alpha_y I_y + \alpha_z I_z) = \sum_{m=N_1}^{N-1} \left[\int_{-\infty}^{\infty} dy_m \, \delta\rho(y_m) \left[A(k_m) y_m^2 + (\ln a_1 \rho(y_m) + 1) + \alpha_y \right] + \right.$$
$$\left. + \int_{-\infty}^{\infty} dz_m \, \delta\rho(z_m) \left[A(k_m) z_m^2 + (\ln a_1 \rho(z_m) + 1) + \alpha_z \right] \right] = 0, \qquad (8.5.21)$$

where the α's are undetermined multipliers. For any fixed value of m, the coefficient of the variation $\delta\rho(y_m)$ must vanish, giving the required probability density as

$$a_1 \rho(y_m) = e^{(-1-\alpha_y)} e^{-A(k_m) y_m^2},$$

or

$$\rho(y_m) = Z_m^{-1} e^{-A(k_m) y_m^2}, \qquad (8.5.22)$$

where the partition function Z_m normalizes ρ, as is usual, and is given by

$$Z_m = \int_{-\infty}^{\infty} dy_m \, e^{-A(k_m) y_m^2} = \sqrt{\frac{\pi}{A(k_m)}}, \qquad (8.5.23)$$

and similarly for the $\rho(z_m)$ terms. The now-minimized Gibbs potential is restated as

$$\beta_b G(l) = \beta_b G_0 - Nh\eta_0 + (N/2) \left[(\nabla\eta_0)^2 + g_2\eta_0^2 + (1/2)g_4\eta_0^4 \right]$$
$$+ \sum_{m=N_1}^{N-1} \left[\int_{-\infty}^{\infty} dy_m \, \rho(y_m) \left[A(k_m) y_m^2 + (\ln a_1 - \ln Z_m - A(k_m) y_m^2) \right] + \right.$$
$$\left. + \int_{-\infty}^{\infty} dz_m \, \rho(z_m) \left[A(k_m) z_m^2 + (\ln a_1 - \ln Z_m - A(k_m) z_m^2) \right] \right]$$
$$= \beta_b G_0 - Nh\eta_0 + \frac{N}{2} \left[(\nabla\eta_0)^2 + g_2\eta_0^2 + (1/2)g_4\eta_0^4 \right] + \sum_{m=N_1}^{N-1} \left[\ln A(k_m) + \ln(a_1^2/\pi) \right]$$
$$\to \beta_b G_0 - Nh\eta_0 + (N/2) \left[(\nabla\eta_0)^2 + g_2\eta_0^2 + (1/2)g_4\eta_0^4 \right]$$
$$+ \frac{N}{2\pi} \int_{\Lambda}^{2\pi} dk \, \left[\ln A(k) + \ln(a_1^2/\pi) \right], \qquad (8.5.24)$$

where the sum has been replaced in the final version by an integral, for ease of further manipulations, and $\Lambda = 2\pi/l$. Recall from Chapter 6 that the correlation length is given by $\xi^2 = (g_2 + 3g_4\eta_0^2)^{-1}$, in present notation, and in Landau theory, $g_2 = g_{21}\tau$, where $\tau = (T - T_c)/T_c$.

Of interest at this point is a separate look at the averaged Landau-Ginzburg-Wilson energy, $\langle G_f(k_m) \rangle$, as determined by the probability densities that minimize the Gibbs potential. This energy, written in dimensionless form, is

$$\beta_b \langle G_f(k_m) \rangle = A(k_m) \left[\int_{-\infty}^{\infty} dy_m \, \rho(y_m) y_m^2 + \int_{-\infty}^{\infty} dz_m \, \rho(z_m) z_m^2 \right]$$
$$= A(k_m) \left[\frac{1}{2A(k_m)} + \frac{1}{2A(k_m)} \right] = 1, \qquad (8.5.25)$$

in units of $k_B T_b$. This is a classical calculation, and for approximations to second order in η_1, equipartition is expected, and every Landau-Ginzburg-Wilson mode has kT as its average energy. The mean-squared fluctuation amplitude of the mth spin wave is

$$\langle |\tilde{\eta}(k_m)|^2 \rangle = \langle y_m^2 \rangle + \langle z_m^2 \rangle = \frac{1}{k_m^2 + \xi^{-2}} \qquad (8.5.26)$$

This result shows that for any value of k_m, the fluctuation amplitude increases with the correlation length, as claimed, and that especially when ξ is large, the long-wave-length (small-k_m) amplitudes can become quite large.

As a step toward completion of the introduction to Wilson renormalization, it is necessary to generalize the final form of Eq. (8.5.24) to a space of d dimensions. Keep N as the total number of spins, and approximate the d-dimensional volume element in k space as if it were spherical, or $d^d k = B_d k^{d-1} dk$, where $B_d = 2\pi^{d/2}\Gamma(d/2)$ is the numerical coefficient in the d-dimensional volume element, as derived in Appendix F. The generalized expression to replace Eq. (8.5.24) is

$$\beta_b G(l) = \beta_b G_0 - N h \eta_0 + (N/2) \left[(\nabla \eta_0)^2 + g_2 \eta_0^2 + (1/2) g_4 \eta_0^4 \right]$$
$$+ \frac{N B_d}{2\pi} \int_\Lambda^{2\pi} k^{d-1} dk \left[\ln A(k) + \ln(a_1^2/\pi) \right], \qquad (8.5.27)$$

One of Wilson's methods of interpretation is to determine the behavior of the coefficients g_2 and g_4 as a consequence of a small change in l. The first step is to calculate $\partial \beta_b G / \partial l$ from Eq. (8.5.27) as

$$\frac{\partial \beta_b G}{\partial l} = \frac{-2\pi}{l^2} \frac{\partial \beta_b G}{\partial \Lambda} = \frac{N B_d}{2\pi} \left(\frac{-2\pi}{l^2} \right) (-\Lambda^{d-1}) \left[\ln A(\Lambda) + \ln(a_1^2/\pi) \right]. \qquad (8.5.28)$$

The first-order expansion of the Gibbs potential gives

$$\beta_b G(l + dl) = \beta_b G(l) + \frac{N B_d (2\pi)^{d-1} dl}{l^{d+1}} \left[\ln A(\Lambda) + \ln(a_1^2/\pi) \right]. \qquad (8.5.29)$$

The next step is to expand the expression for $\ln A(\Lambda)$, in the approximation that the maximum wave length, $l = 2\pi/\Lambda$, is appreciably less than the correlation length, ξ, or $\Lambda \xi > 1$, as follows:

$$\ln A(\Lambda) = \ln \left(\frac{\Lambda^2}{2} \right) + \ln \left(1 + \frac{g_2 + 3 g_4 \eta_0^2}{\Lambda^2} \right)$$
$$\approx \ln \left(\frac{\Lambda^2}{2} \right) + \frac{g_2 + 3 g_4 \eta_0^2}{\Lambda^2} - \frac{(g_2 + 3 g_4 \eta_0^2)^2}{2\Lambda^4}. \qquad (8.5.30)$$

The l-dependencies of the coefficients can be determined by comparison of the terms in η_0^2 and in η_0^4. These yield

$$g_2(l + dl) = g_2(l) + \frac{2 B_d (2\pi)^{d-1}}{l^{d-1}} \left[\left(\frac{3 g_4}{4\pi^2} \right) - \left(\frac{3 g_2 g_4 l^2}{8\pi^4} \right) \right] dl, \quad \text{and}$$

$$g_4(l + dl) = g_4(l) - \frac{2 B_d (2\pi)^{d-1}}{l^{d-3}} \left(\frac{9 g_4^2}{16\pi^4} \right) dl. \qquad (8.5.31)$$

From these expansions, the first-order differential equations for the l-dependent coefficients are seen to be

$$\frac{dg_2}{dl} = \frac{2B_d(2\pi)^{d-1}}{l^{d-1}} \left(\frac{3g_4}{4\pi^2} - \frac{3g_2 g_4 l^2}{8\pi^4} \right), \quad \text{and}$$

$$\frac{dg_4}{dl} = -\frac{2B_d(2\pi)^{d-1}}{l^{d-3}} \left(\frac{9g_4^2}{16\pi^4} \right). \tag{8.5.32}$$

Write $\epsilon = 4 - d$, and define the numerical constant in the g_4 equation as $A_4(d)$, so that it can be written

$$\frac{dg_4}{dl} = -A_4(d)l^{\epsilon-1}g_4^2. \tag{8.5.33}$$

It may be integrated directly, to give

$$g_4(l) = \frac{\epsilon g_4(1)}{A_4(d)g_4(1)(l^\epsilon - 1) + \epsilon}. \tag{8.5.34}$$

For $\epsilon > 0$ and $l \gg 1$, the solution becomes

$$g_4(l) \approx \frac{\epsilon l^{-\epsilon}}{A_4(d)}, \tag{8.5.35}$$

which is also a solution to Eq. (8.5.33). Note that the Landau coefficient g_4 is a function of the cutoff length l, and it decreases to a small value as l becomes large. For $\epsilon \to 0$,

$$g_4(l) \to \frac{g_4(1)}{A_4(d)g_4(1)\ln l + 1}, \tag{8.5.36}$$

which depends only slightly on l, in keeping with expectations from the Landau theory. The use of the large-l solution for g_4 in the differential equation for g_2 gives

$$\frac{dg_2}{dl} = \frac{3B_d(2\pi)^{d-1}g_4(l)}{4\pi^2 l^{d-1}} \left(1 - \frac{g_2(l)l^2}{2\pi^2} \right) \approx \frac{3B_d(2\pi)^{d-3}\epsilon l^{-3}}{A_4(d)} \left(1 - \frac{g_2 l^2}{2\pi^2} \right). \tag{8.5.37}$$

Temporarily change variable to $y_2 = g_2/2\pi^2$, and use the definition of the constant $A_4(d)$ to rewrite the equation as

$$\frac{dy_2}{dl} = \frac{\epsilon}{3}l^{-3}(1 - y_2 l^2). \tag{8.5.38}$$

Its solution is

$$y_2(l) = \frac{c_2}{2\pi^2}l^{-\epsilon/3} - \frac{\epsilon}{6}\left(\frac{1}{1 - \epsilon/6} \right)l^{-2}, \tag{8.5.39}$$

where c_2 is an arbitrary constant that includes the factor τ, for agreement with Landau theory. For large l, the second term is assumed to be negligible, so that g_2 becomes

$$g_2(l) \approx c_2 l^{-\epsilon/3} \sim l^{-\epsilon/3}\tau. \tag{8.5.40}$$

To examine the consequences of the wave-length cutoff on critical exponents, choose l to be some fixed fraction of ξ, the correlation length, so that $l \sim \xi$, but small enough that the expansion of the logarithm in Eq. (8.5.30) remains valid. Recall that $\xi = (g_2 + 3g_4\eta_0^2)^{-1/2}$, and that for large l, g_4 is assumed to be negligible, as indicated by Eq. (8.5.35). Therefore, near the critical point, $\xi \approx g_2^{-1/2} \sim \tau^{-1/2} l^{\epsilon/6}$. But if $l \sim \xi$, then

$$\xi \sim \tau^{-1/2} \xi^{\epsilon/6}, \quad \text{or} \quad \xi \sim \tau^{3/(\epsilon-6)}. \tag{8.5.41}$$

The critical exponent for ξ is ν, and the relation $\xi \sim \tau^{-\nu}$ leads to the ϵ-dependent expression for ν:

$$\nu = \frac{3}{6-\epsilon} = \frac{1}{2}\left(\frac{1}{1-\epsilon/6}\right) \approx \frac{1}{2} + \frac{\epsilon}{12}, \tag{8.5.42}$$

where the final approximation, to first order in ϵ, is often listed as the only one consistent with the other approximations used in its derivation. Note that for $d = 4$, or $\epsilon = 0$, the value $\nu = 1/2$ of Landau-Ginzburg theory is obtained, again confirming that $d = 4$ is the marginal dimension for the validity of Landau theory. The experimental values for ν in 3-d are in the range $0.6 \leq \nu \leq 0.7$, and the numerical estimates for the 3-d Ising model are $\nu = 0.64 \pm 0.005$. The first expression in Eq. (8.5.42) gives $\nu = 0.6$, and the first-order approximation gives $\nu = 0.583$. These values are not in perfect agreement with the comparison data, but they are an improvement over the theory of Chapter 6. The exact calculation for the 2-d Ising model gives $\nu = 1$, whereas Eq. (8.5.42) yields $\nu = 0.75$ and $\nu = 0.67$.

The Landau expression for the equilibrium value of the order parameter with $h = 0$ is $\eta_0^2 \sim g_2/g_4$, or, since $\eta_0 \sim \tau^\beta$,

$$\eta_0^2 \sim \frac{g_2}{g_4} \sim \frac{l^{-\epsilon/3}\tau}{l^{-\epsilon}}$$

$$\sim \xi^{2\epsilon/3}\tau \sim \tau^{1-2\nu\epsilon/3} \sim \tau^{2\beta}, \tag{8.5.43}$$

from which it follows that

$$\beta = \frac{1}{2} - \frac{\epsilon\nu}{3} = \frac{1}{2} - \frac{\epsilon}{6(1-\epsilon/6)} \approx \frac{1}{2} - \frac{\epsilon}{6}. \tag{8.5.44}$$

This result is again in agreement with Landau theory for $d = 4$, and it offers corrections for $d < 4$. The experimental values are $0.31 \leq \beta \leq 0.36$, and the numerical estimate for the 3-d Ising system is $\beta = 0.31 \pm 0.005$. The two values from Eq. (8.5.44) are $\beta = 0.3$ and $\beta = 0.33$, in quite good agreement for such a simplified version of the theory. For the 2-d Ising model, the two calculated values are $\beta = 0$ and $\beta = 0.167$, thus spanning the exact value $\beta = 0.125$.

Calculation of the other Landau-Ginzburg-Wilson critical exponents is left as an exercise. (See Problem 5.)

In order to complete the process of renormalization, the Gibbs potential must be rescaled to the cutoff length l. In the rescaled system, lengths are measured in units of la, rather than of a. The function to be rescaled is

$$\beta_B G = \beta_B G_0 - Nh\eta_0 + \frac{N}{2}((\nabla\eta_0)^2 + g_2\eta_0^2 + (1/2g_4\eta_0^4) +$$

$$+ \frac{NB_d}{2\pi}\int_\Lambda^{2\pi} dk\, k^{d-1}\left[\ln A(k) + \ln(a_1^2/\pi)\right], \tag{8.5.45}$$

where $\Lambda = 2\pi/l$. Neither the temperature nor the value of G changes under this rescaling transformation. The new spatial coordinate is $r_l = r/l$, with its consequence that $N_l = N/l^d$, the new number of units in the lattice. The next term to examine contains the gradient.

$$N\left(\frac{d\eta}{dx}\right)^2 = N_l\left(\frac{d\eta_l}{dx_l}\right)^2 = Nl^{-d}\left(l\frac{d\eta_l}{dx}\right)^2 = Nl^{2-d}\left(\frac{d\eta_l}{dx}\right)^2, \qquad (8.5.46)$$

which requires that $\eta_l = l^{-1+d/2}\eta$. This kind of analysis can be carried out term by term in Eq. (8.5.46) to obtain the complete recipe for rescaling:

$$r_l = r/l; \quad N_l = N/l^d; \quad \eta_l = \eta l^{-1+d/2};$$
$$h_l = hl^{1+d/2}; \quad g_{2l} = l^2 g_2; \quad g_{4l} = l^{4-d}g_4. \qquad (8.5.47)$$

This transformation empties the final integral of Eq. (8.5.45) and restores the Gibbs potential to its Landau-Ginzburg form. Continued iteration of this transformation is similar in effect to continued iteration of the Kadanoff reductions. The behavior of the asymptotic approximations of the constants g_2 and g_4, as given in Eq. (8.5.39) and Eq. (8.5.35), under the rescaling transformation is

$$g_{2l} = l^2 g_2 = cl^{2-\epsilon/3} - \frac{2\pi^2\epsilon}{3}\frac{1}{2-\epsilon/3}, \quad \text{and}$$

$$g_{4l} = l^\epsilon g_4 = \frac{g_4(1)}{a_4(d)}. \qquad (8.5.48)$$

At the critical point, the behavior of the Gibbs potential is expected to be independent of scale, and Eq. (8.5.48) unequivocally agrees with this expectation. The only l-dependent term, after scaling, contains the factor c, which is known to be proportional to τ, so that it becomes zero at $T = T_c$. Thus the critical point is a fixed point of the transformation, and for temperatures other than T_c, g_{2l} moves rapidly away from the fixed point. The resulting larger value of g_{2l} implies a reduction in correlation length, and thus a recession from criticality.

The set of operations consisting of the transformation from $\beta_B G(l)$ to $\beta_B G(l + \delta l)$, followed by rescaling, is called *the renormalization group*. Technically, it is usually a semigroup, since it lacks an inverse.

8.6 COMMENTARY

Since the time of Boltzmann, physicists have tried to provide a microscopic basis for macroscopic sciences such as thermodynamics and hydrodynamics. Such is the aim of statistical thermophysics. When the ever-present microscopic fluctuations can be averaged out, correctly formulated microscopic descriptions of physical systems lead to remarkably accurate and useful macroscopic theory. Successful averaging by the standard techniques requires that the fluctuations are localized, with correlation lengths of the order of the atomic or molecular spacing. To my knowledge, every system that can be so described has been treated in the literature of statistical mechanics. The really difficult phenomena to understand quantitatively are those that exhibit fluctuations at all wave lengths, from macroscopic to microscopic. These include turbulence and critical phase transitions. The importance of renormalization is that it offers a means of treating systems that exhibit these fluctuations on many length scales. Even more significantly, it offers a conceptual basis for thinking about such systems.

It is apparent that microscopic details cannot be of overriding importance in the interactions of macroscopic structures within the system. Kadanoff's spin blocks, and Niemeijer and van Leeuwen's majority-rule recomposition provided bold and imaginative instruction for investigating the effects of change of scale. (Niemeijer and Van Leeuwen [1976]) Wilson's renormalization, adapted from quantum field theory, but with considerable modification and enhancement, offers another, formally quite different, means to a similar end. Conceptually, the idea is to formulate the problem in terms of a Hamiltonian (or a partition function, or a Gibbs potential), average out, or otherwise consolidate, the components of the order parameter that are smaller than some arbitrary cutoff length l, and restate the problem in terms of a Hamiltonian that is rescaled so that its length unit is the cutoff length l. (It is expected that l is greater than the separation of the original interacting components.) The new Hamiltonian, sometimes after trivial generalizations, takes on the same form as the original, with the parameters redefined. The entire operation is called *renormalization*, and the set of renormalization transformations is called the renormalization group. The cutoff l can be considered continuous, and a plot of the renormalized constants, such as g_{4l} against g_{2l} in Eq. (8.5.48), parametric in l, produces a trajectory in the parameter space (in this case, g_{2l}-g_{4l} space) that should exhibit a fixed point. As already noted, this fixed point of the transformation is the critical point. (A complete plot of the behavior of the renormalized parameters requires the general solutions of the differential equations for g_2 and g_4; the one for g_2 was not given, since Eq. (8.5.37) used the near-critical approximation for g_4 instead of the more-complicated one for all values of l.)

The literature on the renormalization group is enormous. I have tried to present the conceptual basis of the approach and to provide a small sampling of the kinds of calculations that can be done, together with some of the results that are obtained. There are excellent treatments that offer much more detail. (Fisher [1983]; Ma [1976]; Patashinskii and Pokrovskii [1979]; Pfeuty and Toulouse [1977]; Wegner [1976a]; Wegner [1976b]; Wilson [1975]; Wilson and Kogut [1974])

Two important extensions of the theory have been omitted. The first is the inclusion of higher-dimensional spins, as mentioned in Section 7.11. The inclusion of the general spin n, (written as D in Chapter 7) leads to the following first-order expressions for the critical exponents:

$$\alpha = \frac{4 - n}{2(n + 8)}\epsilon,$$

$$\beta = \frac{1}{2} - \frac{3}{2(n + 8)}\epsilon,$$

$$\gamma = 1 + \frac{n + 2}{2(n + 8)}\epsilon,$$

$$\delta = 3 + \epsilon \qquad (8.6.1)$$

These agree with the first-order expressions already found for these exponents when $n = 1$. Expansions exist to higher order in both ϵ and n. It is usually true that expansion to order

ϵ^2 improves the results, but the ϵ^3 and higher exponents are not any better, and may be worse. (Fisher [1974])

The second important extension is to dynamic critical phenomena. Transport coefficients such as thermal and electrical conductivity, mobility, and viscosity, which are considered in Chapter 9, can be profoundly altered near the critical point. Treatment of transport phenomena of systems in the critical region by renormalization methods leads to a steady improvement in our understanding of the way these systems behave, and even in the way they approach equilibrium. The conceptual basis of the treatment is completely analogous to static renormalization, and calculations can be carried out at a wide range of levels of difficulty and exactness. (Hohenberg and Halperin [1977]; Ma [1976]; Patashinskii and Pokrovskii [1979])

The concept of universality, as developed in Chapter 6, is extended and clarified by renormalization theory. Universality classes are identified as dependent upon the dimensionality of the space, d, and that of the order parameter, n. Another determinant of the universality class, not mentioned here in any detail, relates to the symmetry of the system. Systems belonging to the same universality class have the same critical exponents, despite the possible lack of any obvious similarities in the underlying microscopic structures. In the neighborhood of a multicritical point, the system changes from one universality class to another, as a function of a parameter written as τ_2 in Chapter 6. This change of universality class is called *crossover*. Such a situation is identified in renormalization theory by the existence of more than two relevant variables. For states sufficiently close to a multicritical point, the critical exponents are determined by the universality class of the multicritical point. (Barber and Fisher [1973]; Bruce [1980]; Bruce and Cowley [1980])

The Wilson treatment of Landau-Ginzburg theory provides a relatively uncluttered caricature of the actions of the renormalization group in the theory of critical phenomena. A correct treatment is much more complicated. The long-range component of the order parameter, written as η_0 in my exposition, and described after Eq. (8.5.13) as an imprecisely defined average, must be treated with more care. The higher-order gradient terms must be considered. Either higher order terms in the Landau-Ginzburg expansion must be included, or many interaction parameters that arise in Kadanoff reductions must be used instead. It is too much to expect that phenomena as complicated as critical phase transitions in d dimensions could be characterized accurately by the behavior of two parameters, g_2 and g_4. It is remarkable that the crude treatment I have presented is so successful.

An excellent historical and technical summary of the subject is to be found in Wilson's Nobel Prize address. (Wilson [1983])

The elementary wave packets that I have called *wavelets*, used for spreading discrete spins over a continuum, are somewhat simpler in structure than those that have been polished to perfection. Nevertheless, they do provide a square-integrable orthonormal basis for the space, complete in the sense that any function that can be expressed as a fourier transform, with components of no shorter wavelength than the lattice spacing, can also be expressed as a sum of these wavelets. An easily pictured wavelet is of the form $\sin u/u$, where u is a suitable function of $\mathbf{r}$ (and perhaps t). Two of the problems illustrate their properties.

The renormalization group originated in quantum field theory, and the transfer of concepts to thermophysics has induced significant strides in the understanding of critical phenomena. A surprising reciprocity has developed, in that cosmological string theories lead to phase transitions with critical exponents. Some of the predictions of these theories have analogs in the theory of polymer chains, which are subject to experimental verification. Systems of the same universality class exhibit identifiably similar behavior, even when they seem to have little in common. (Duplantier and Saleur [1989b]; Flory [1969]; Freed [1987]; de Gennes [1979]; Raposo et al. [1991])

Near the critical point, the order parameter (say the spin orientation) shows long-range order. What are the shapes of the ordered domains? In a sea of positive spins, a bubble of negative spins may form, with characteristic dimension of the order of the correlation length. Within this bubble, there may form smaller bubbles of positive spins, and within them even smaller bubbles of negative spins, to all orders. This kind of fractal structure, for which there seems to be some evidence, is not at all like the almost-solid spin blocks that the Kadanoff picture describes, and the boundary energy seems too high to correspond to a Gibbs-potential minimum, but the entropy inherent in the fractal structure compensates. In a two-dimensional Ising lattice, the fractal dimension of the critical spin domain has been calculated to be $D_f = 187/96 \approx 1.95$. (Duplantier and Saleur [1989a]) Computer simulations of such lattices seem to be too coarse-grained to display this fractal structure over a sufficient range of scale lengths to make it evident, although they do show that the spin blocks are far from homogeneous. (Wilson [1979])

8.7 REFERENCES

A few of the important references are listed here; a complete list is to be found in the Bibliography at the end of this book.

BRUCE, A. D. [1980], Structural Phase Transitions. II Static Critical Behavior. *Advances in Physics* **29** (1980), 111–218.

FISHER, M. E. [1983], *Scaling, Universality, and Renormalization Group Theory.* In *Critical Phenomena*, F. J. W. Hahne, Ed. Springer-Verlag, New York–Heidelberg–Berlin, 1983, 1–139.

FISHER, M. E. [1974], The Renormalization Group in the Theory of Critical Behavior. *Reviews of Modern Physics* **46** (1974), 597–616.

KADANOFF, L. P., GÖTZE, W., HAMBLEN, D., HECHT, R., LEWIS, E. A. S., PALCIAUSKAS, V. V., RAYL, M., SWIFT, J., ASPNES, D. AND KANE, J. [1967], Static Phenomena Near Critical Points: Theory and Experiment. *Reviews of Modern Physics* **39** (1967), 395–431.

MA, S-K. [1976a], *Modern Theory of Critical Phenomena.* The Benjamin/Cummings Publishing Company, Inc., Reading, Massachusetts, 1976.

PATASHINSKII, A. Z. AND POKROVSKII, V. I. [1979], Pergamon Press, Oxford, England, 1979. *Fluctuation Theory of Phase Transitions.*

PFEUTY, P. AND TOULOUSE, G. [1977], *The Renormalization Group and Critical Phenomena.* John Wiley & Sons, New York, NY, 1977.

WEGNER, F. G. [1976a], *Critical Phenomena and Scale Invariance.* In *Critical Phenomena*, J. Brey and R. B. Jones, Eds. Springer-Verlag, New York–Heidelberg–Berlin, 1976, 1–29.

WEGNER, F. J. [1976b], *The Critical State, General Aspects.* In *Phase Transitions and Critical Phenomena, V. 6,* C. Domb and M. S. Green, Eds. Academic Press, New York, NY, 1976.

WILSON, K. G. [1975], The Renormalization Group: Critical Phenomena and the Kondo Problem. *Reviews of Modern Physics* **47** (1975), 773–840.

WILSON, K. G. [1983], The renormalization group and critical phenomena. *Reviews of Modern Physics* **55** (1983), 583–600.

WILSON, K. G. AND KOGUT, J. [1974], The Renormalization Group and the ε Expansion. *Physics Reports C* **12** (1974), 75–199.

8.8 PROBLEMS

1. For the linear H_3R chain of problem 7.9, determine the critical temperature by finding a nontrivial stationary point from renormalization of the transfer matrix, as was done for linear KDP in Section 8.3.3.

2. For the Nagle model of KDP, as in Section 8.3.3, renormalize the chain into blocks of four units, instead of the binary renormalization already exhibited. Show that the critical temperature found from the fixed point agrees with the result of Eq. (8.3.17). Examine numerically the convergence of the inverse transformation, starting at $x = 1$ and at $x = 0.01$, to the critical value.

3. Consider a double Ising chain, such as that of problem 7.8, with $h = 0$. Number the spins σ_{ij}, where $0 \leq i \leq N \to \infty$, and $j = 0$, 1. By listing the diagrams, write out a sample term in a fourfold Kadanoff reduction. (In the segment σ_{0j} to σ_{4j}, sum over all $i \neq 4n$. Then renumber by changing $4i$ to i.) Show that the resulting terms couple the same spin groups as those given in Eq. (8.3.20), and note that the effective coupling constants describe the multiplicity of paths by which coupling can be accomplished.

4. Use the diagram method to obtain the coupling constants for a triple Ising chain having only nearest-neighbor bonds, in its initial form. There are two cross bonds, labeled $y = \tanh K_2$, and three bonds along the chain, labeled $x = \tanh K_1$, for each link. Number the spins σ_{i0}, σ_{i1}, and σ_{i2}, where $0 \leq i \leq N - 1$. Renormalize the chain by summing out every spin σ_{ij} with i and/or j odd. This procedure will recompose the triple chain into a double chain, that remains as a double chain under continued renormalizations.

5. Use the Landau-theory result that $\gamma = 2\nu$, in addition to the expressions for ν and β found in Eq. (8.5.42) and (8.5.44) to find numerical values for the critical exponents α, δ, and γ in two and three dimensions.

6. Show that the wavelets defined in Eq. (8.5.6) constitute an orthonormal set.

7. The wavelet of Eq. (8.5.4) that replaces the spin σ_n was shown to spread the spin from $n - 1$ to $n + 1$. When the order parameter is split into two parts, in Eq. (8.5.9), the part of σ_n that is assigned to η_0 looks like the wavelet of Eq. (8.5.4), except that N becomes N_1. Show that the σ_n wavelet that is a part of η_0 now spreads from $n - l$ to $n + l$.

8.9 SOLUTIONS

1. The transfer matrix, from problem 7.9, is

$$
P = \begin{array}{c c} & \begin{array}{c c c c c c c c} lll & llr & lrl & rll & rlr & rrl & lrr & rrr \end{array} \\ \begin{array}{c} lll \\ llr \\ lrl \\ rll \\ rlr \\ rrl \\ lrr \\ rrr \end{array} & \begin{pmatrix} 1 & 0 & 0 & 0 & 0 & 0 & 0 & 0 \\ 0 & x & x & x & 0 & 0 & 0 & 0 \\ 0 & x & x & x & 0 & 0 & 0 & 0 \\ 0 & x & x & x & 0 & 0 & 0 & 0 \\ 0 & 0 & 0 & 0 & x & x & x & 0 \\ 0 & 0 & 0 & 0 & x & x & x & 0 \\ 0 & 0 & 0 & 0 & x & x & x & 0 \\ 0 & 0 & 0 & 0 & 0 & 0 & 0 & 1 \end{pmatrix} \end{array}
$$

The partition function is given by

$$
\begin{aligned}
Z_N &= \sum_{\{\sigma_i\}} \langle \sigma_0 | P | \sigma_1 \rangle \langle \sigma_1 | P | \sigma_2 \rangle \cdots \langle \sigma_{N-1} | P | \sigma_0 \rangle \\
&= \sum_{\{\sigma_i\}} \langle \sigma_0 | P^2 | \sigma_2 \rangle \cdots \rightarrow \sum_{\{\sigma_i\}} \langle \sigma_0 | P_1 | \sigma_1 \rangle \cdots ,
\end{aligned}
$$

where the final form is obtained by renumbering $\sigma_{2i} \rightarrow \sigma_i$, and $P^2 = P_1$ is just like P except that x is replaced by $3x^2$. (Note that the index σ_i takes on eight values, rather than the usual two.) The renormalization gives $x' = 3x^2$, and the stationary point is that for which $x' = x$, or $x = 3x^2$. The nontrivial stationary point occurs at $x_c = 1/3$, in agreement with the solution to problem 7.9.

2. As in Section 8.3.3, the partition function is

$$
Z_N = \sum_{\{\sigma_i\}} \langle \sigma_0 | P | \sigma_1 \rangle \langle \sigma_1 | P | \sigma_2 \rangle \cdots \langle \sigma_{N-1} | P | \sigma_0 \rangle ,
$$

with P given by

$$
P = \begin{pmatrix} 1 & 0 & 0 & 0 \\ 0 & x & x & 0 \\ 0 & x & x & 0 \\ 0 & 0 & 0 & 1 \end{pmatrix}.
$$

The partition function can be renormalized by summing over all σ_i's such that $i \neq 4n$, to get

$$
Z_N = \sum_{\{\sigma_i\}} \langle \sigma_0 | P^4 | \sigma_4 \rangle \langle \sigma_1 | P^4 | \sigma_8 \rangle \cdots \rightarrow \sum_{\{\sigma_i\}} \langle \sigma_0 | P' | \sigma_1 \rangle \langle \sigma_1 | P' | \sigma_2 \rangle \cdots ,
$$

where the indices have been renumbered $\sigma_{4i} \rightarrow \sigma_i$, and $P' = P^4$. The renormalization gives $x' = 8x^4$; stationary points occur at $x = 8x^4$, and the only real, nontrivial such point is at $x_c = 1/2$, as in Section 8.3.3. The inverse transformation is $x' = (x/8)^{1/4}$. The sequence starting with $x = 1$ is 1, 0.5946, 0.52213, 0.50544, 0.501356, 0.5003386, 0.5000846, 0.5000212, 0.5000053, 0.5000013, 0.5000003, and on to higher accuracy. The sequence starting with $x = 0.01$ is 0.01, 0.1880, 0.3915, 0.47035, 0.4924174, 0.4980935, 0.4995227, 0.4998806, 0.4999702, 0.4999925, 0.4999981, 0.4999995, again showing satisfactory convergence to $x_c = 1/2$.

3. In the style of Eq. (7.5.8), the partition function for the double Ising chain can be written

$$Z_N = (\cosh^2 K_1)^{N-1} (\cosh K_2)^N \sum_{\sigma_{ij}=\pm 1} \prod_{i=0}^{N-1} (1 + y\sigma_{i0}\sigma_{i1})(1 + x\sigma_{i0}\sigma_{i+1,0})(1 + x\sigma_{i1}\sigma_{i+1,1})$$

$$= (2^4 \cosh^2 K_1)^{N-1} (2^2 \cosh K_2)^N \sum_{\{\sigma_{ij}\}} \prod_{i=0}^{(N/4)-1} [(1 + 4x^2y^2 + 3x^4y^2 + 2x^6y^2 + 3x^4y^4 +$$

$$+ 2x^6y^4 + x^8y^2) +$$

$$+ (\sigma_{i0}\sigma_{i1} + \sigma_{i+1,0}\sigma_{i+1,1})(y + x^2y + x^4y + x^6y + x^8y + 3x^2y^3 + 4x^4y^3 + 3x^6y^3 + x^4y^5) +$$

$$+ (\sigma_{i0}\sigma_{i+1,0} + \sigma_{i1}\sigma_{i+1,1})(x^4 + 10x^4y^2 + 5x^4y^4) +$$

$$+ (\sigma_{i0}\sigma_{i+1,1} + \sigma_{i,1}\sigma_{i+1,0})(5x^4y + 10x^4y^3 + x^4y^5) +$$

$$+ (\sigma_{i0}\sigma_{i1}\sigma_{i+1,0}\sigma_{i+1,1})(y^2 + 2x^2y^2 + 6x^4y^2 + 2x^2y^4 + 2x^6y^2 + 2x^4y^4 + x^8)],$$

where the final expression results from the procedure explained below the figure.

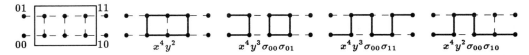

The left-hand diagram is the basic block being renormalized. All original spins within the solid rectangle are to be summed over, and the end spins, originally designated σ_{40} and σ_{41} have been renumbered. The other four diagrams represent the terms indicated, as examples of the counting procedure.

Note that every possible path connecting the spin pairs or clusters in each term is represented in the reduced expression for Z_N, which can be written in the form

$$Z_N = A(K_1, K_2) \sum_{\sigma_{ij}=\pm 1} \prod_{i=0}^{N'} [1 + (\sigma_{i0}\sigma_{i1} + \sigma_{i+1,0}\sigma_{i+1,1})y' + (\sigma_{i0}\sigma_{i+1,0} +$$

$$+ \sigma_{i1}\sigma_{i+1,1})x' + (\sigma_{i0}\sigma_{i+1,1} + \sigma_{i,1}\sigma_{i+1,0})z' + \sigma_{i0}\sigma_{i1}\sigma_{i+1,0}\sigma_{i+1,1}w'],$$

where the new coupling constants can be identified from the previous equation. Unlike reductions in two or more dimensions, this double-chain example does not become increasingly more complicated upon further reduction. All possible couplings are represented by the four parameters x', y', z', and w'. Further renormalizations redefine these constants in terms of those of the previous generation, but no more are required. Observe that there are sixteen paths necessary to define each of these coupling constants, not counting the additional sixteen paths that contribute to the spin-free term that is renormalized as 1.

4. The planar triple chain, when reduced by summation over every spin σ_{ij} such that i or j (or both) is odd, becomes equivalent to the double chain of the previous problem, with different values for the coupling constants x', y', z', and w'. The partition function is

$$Z_N = \left[(\cosh^3 K_1)^{N-1} (\cosh^2 K_2)^N \right] \sum_{\sigma_{ij}=\pm 1} \prod_{i=0}^{N-1} [(1 + \sigma_{i0}\sigma_{i1}y) \times$$

$$\times (1 + \sigma_{i1}\sigma_{i2}y)(1 + \sigma_{i0}\sigma_{i+1,0}x)(1 + \sigma_{i1}\sigma_{i+1,1}x)(1 + \sigma_{i2}\sigma_{i+1,2}x)].$$

After summation over all spins with i and/or j odd the renumbering of all surviving spins such that $i \to i/2$ and $j \to j/2$, the partition function becomes

$$Z_N = \left[\left(8 \cosh^3 K_1 \right)^{N-1} \left(4 \cosh^2 K_2 \right)^N \right] \sum_{\sigma_{ij} = \pm 1} \prod_{i=0}^{(N/2)-1} [(1 + 4x^2 y^2 + 2x^4 y^2 + 2x^2 y^4 + 7x^4 y^4) +$$

$$+ (\sigma_{i0}\sigma_{i+1,0} + \sigma_{i1}\sigma_{i+1,1})(x^2 + 3x^2 y^2 + 2x^4 y^2 + 5x^2 y^4 + 4x^4 y^4 + x^6 y^2) +$$

$$+ (\sigma_{i0}\sigma_{i1} + \sigma_{i+1,0}\sigma_{i+1,1})(y^2 + 3x^2 y^2 + 5x^4 y^2 + 2x^2 y^4 + 4x^4 y^4 + x^2 y^6) +$$

$$+ (\sigma_{i0}\sigma_{i+1,1} + \sigma_{i1}\sigma_{i+1,0})(6x^2 y^2 + 2x^4 y^2 + 2x^2 y^4 + 4x^4 y^4 + x^6 y^2 + x^2 y^6) +$$

$$+ \sigma_{i0}\sigma_{i1}\sigma_{i+1,0}\sigma_{i+1,1}(x^4 + y^4 + 6x^2 y^4 + 6x^4 y^2 + 2x^4 y^4)].$$

Renormalization is completed by factoring out the spinless term and defining new coupling constants. The renormalized partition function is

$$Z_N = A(K_1, K_2) \sum_{\sigma_{ij} = \pm 1} \prod_{i=0}^{N'} [1 + (\sigma_{i0}\sigma_{i1} + \sigma_{i+1,0}\sigma_{i+1,1})y' + (\sigma_{i0}\sigma_{i+1,0} +$$

$$+ \sigma_{i1}\sigma_{i+1,1})x' + (\sigma_{i0}\sigma_{i+1,1} + \sigma_{i,1}\sigma_{i+1,0})z' + \sigma_{i0}\sigma_{i1}\sigma_{i+1,0}\sigma_{i+1,1}w'],$$

which has the same form as that of problem 3. The new coupling constants can be read off, as required, from the previous equation. There is no further change in form of the partition function upon successive renormalizations.

5. With $\gamma = 2\nu$ and the known expression for β, the critical-exponent equalities of Chapter 6 can be used to obtain the others. It is easily seen that the parameters p and q in Section 6.6.3 are $p = (6 - \epsilon)/(12 - 3\epsilon)$, and $q = (6 - \epsilon)/(8 - 2\epsilon)$. From these, the remaining critical exponents are found to be

$$\alpha = 2 - (1/p) = \frac{\epsilon}{6(1 - \epsilon/6)} \approx \frac{\epsilon}{6}$$

$$\delta = \frac{q}{1 - q} = \frac{6 - \epsilon}{2 - \epsilon} = \frac{d + 2}{d - 2} \approx 3 + \epsilon \quad \text{and}$$

$$\gamma = (1 - 2q)/p = (1 - \epsilon/6)^{-1} \approx 1 + \epsilon/6.$$

Numerical values for $d = 3$, or $\epsilon = 1$ are (where $\approx$ denotes the first-order approximation): $\alpha = 0.2 \approx 0.167$, $\delta = 5 \approx 4$, and $\gamma = 1.2 \approx 1.167$. Numerical estimates for the 3-d Ising model are: $\alpha \approx 0.125$, $\delta \approx 5$, and $\gamma \approx 1.125$. For $d = 2$, or $\epsilon = 2$, the calculated values are $\alpha = 0.5 \approx 0.33$, $\delta = \infty \approx 5$, and $\gamma = 1.5 \approx 1.33$. Values for the 2-d Ising model are: $\alpha = 0$ (logarithmic singularity), $\delta = 15$, and $\gamma = 1.75$.

6. Write the wavelet in terms of its fourier components as

$$s_n(x) = \frac{1}{N} \sigma_n \sum_{m=0}^{N-1} e^{ik_m(n-x)},$$

and evaluate the integral

$$\int_0^N dx\, s_n(x)s_{n'}(x) = \frac{\sigma_n \sigma_n'}{N^2} \int_0^N dx \sum_{m=0}^{N-1}\sum_{m'=0}^{N-1} e^{ik_m n} e^{ik_{m'} n'} e^{-ix(k_m + k_{m'})}$$

$$= \frac{\sigma_n \sigma_{n'}}{N^2} \sum_{m=0}^{N-1}\sum_{m'=0}^{N-1} e^{ik_m n} e^{ik_{m'} n'} N\delta_{-m}^{m'}$$

$$= \frac{\sigma_n \sigma_{n'}}{N} \sum_{m=0}^{N-1} e^{ik_m(n-n')} = \sigma_n^2 \delta_n^{n'},$$

which establishes that the wavelets are orthonormal, since $\sigma_n^2 = 1$.

7. The wavelet $s_{n0}(x)$ that is part of η_0 is

$$s_{n0}(x) = \frac{\sigma_n}{N} \sum_{m=0}^{N_1} e^{ik_m(n-x)} = \frac{\sigma_N}{N}\frac{1 - e^{2\pi i(n-x)/l}}{1 - e^{2\pi i(n-x)/N}} \to \lim_{x \to n} = \frac{\sigma_n}{l}.$$

The first zero on either side of the central maximum is at $x = n \pm l$, defining the spread of the wavelet, and

$$\int_0^N dx\, s_{n0}(x) = \sigma_n.$$

When the system is rescaled, so that l becomes 1, the wavelet returns to its original height and spread.

9. *IRREVERSIBLE PROCESSES*

9.1 INTRODUCTION

Throughout much of its history, *thermodynamics* has been a misnomer. Practitioners in the field, from the time of Clausius, have regarded the phrase *nonequilibrium thermodynamics* as an oxymoron, in view of the prevailing attitude that there is no satisfactory way to define entropy or any of the intensive parameters such as temperature or pressure unless the system is in equilibrium. Boltzmann attempted to study the approach to equilibrium in terms of the evolution of a function $f(\mathbf{r}, \mathbf{p}, t)$ that represents the number of particles in a region $d\mathbf{r}d\mathbf{p} = d\mu$ of the six-dimensional phase space of the positions and momenta of the particles. His approach, while neither rigorous nor of general validity, was at least a good beginning that led to great strides in the development of the kinetic theory of gases. Boltzmann's interpretation of his $fd\mu$ as the actual number of particles in a volume element $d\mu$, and not as the probable number, based on the data at hand, is discussed in the Commentary section of Chapter 1. The dog-flea and wind-tree models treated in that chapter illustrate the difference in the points of view and in the basic interpretation of the equilibrium concept.

The range of problems requiring use of the methodology of thermodynamics and statistical thermophysics for systems not in equilibrium is enormous. Much, if not most, of the matter in the universe is not in a state of equilibrium with its surroundings. Some stars, during their entire histories, have never even come close to anything resembling equilibrium. More mundanely, meteorology studies a nonequilibrium system (yet the local temperatures and pressures are regularly measured and reported, without any quibble about their definitions); biology, the science of living systems, is often cited as a study of intrinsically nonequilibrium entities; steam power plants could not function without irreversibility; much of plasma physics, including massive efforts to solve the formidable problems of exploiting thermonuclear reactors as environmentally acceptable, nonpolluting energy sources, attempts to understand fiendishly uncooperative systems that must be held far from equilibrium with their surroundings for long enough periods of time to permit the release of useful amounts of energy. These and other evident examples show that many, perhaps most, of the current problems within the purview of statistical physics are those involving nonequilibrium systems.

I expect treatments of irreversible processes by methods based on information theory to supersede most of the methods that have been used to attack such problems. I could present the material in this chapter as if these new methods were already in place and widely accepted, ignoring much of the older literature. But since the aims of a textbook must include both guidance to the pertinent literature and introduction to expected future developments, insofar as possible, I have included discussions (admittedly sometimes sketchy) of the major methods that have been used to study nonequilibrium processes. Fluctuations and related phenomena are examined in Chapter 10.

9.2 IRREVERSIBLE THERMODYNAMICS

Three kinds of analysis of irreversible thermodynamic systems are given in this section. The first, based on the Bernoulli equation of elementary physics, considers a system in steady flow, with time-independent irreversible processes taking place in a central chamber, a black box with no diagram of its internal structure. The entering fluid is in equilibrium, with well-defined and directly measurable properties, as is the effluent. Thus the irreversible processes are treated by examining the changes in equilibrium states. This procedure is manifestly correct for all applicable systems, but it sometimes fails to provide details of interest about happenings inside the black box (Tribus [1961]).

The second, usually attributed to Onsager, attempts to extend the equations of thermostatics to nonequilibrium systems. Several assumptions are made, not all of which are universally valid or free from objection. Because of the apparent success of the theory and its widespread use, it is examined in some detail.

The third analysis, that of continuum mechanics, treats properly defined *thermodynamic* quantities in time-dependent equations, considering simple homogeneous systems of the kind examined in Chapter 2, and extending the analysis to include inhomogeneous systems. It treats the intensive parameters as locally defined fields, and the extensive parameters become densities or specific (per unit mass) quantities, which then become continuous fields subject to conservation laws. Local differentiability is assumed, so that local values of the intensive parameters can be defined in terms of derivatives.

9.2.1 The Generalized Bernoulli Equation.
Consider a fluid in a state of thermodynamic equilibrium, flowing at a uniform speed through an insulated tube of uniform size, past a point 1, as shown in Fig. 9.1. It enters a chamber R, which might be supplied with thermal energy at the uniform rate $\dot{Q}$, and it might have work done on it at the rate $\dot{W}$. No fluid is stored or created within the chamber. When it reaches the point 2 it is again in a state of equilibrium. (Transverse velocity gradients, as in viscous laminar flow, are disregarded in this sketch, but they are easily included when necessary.) Steady flow requires that

$$\Gamma = \rho_1 v_1 A_1 = \rho_2 v_2 A_2, \qquad (9.2.1)$$

where ρ is the density, v is the flow speed, and A is the area of the cross-section. The $\rho v A$

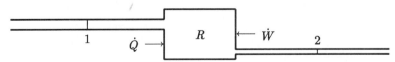

Figure 9.1 Generalized flow system. The fluid enters at point 1 and leaves at point 2, both in states of thermodynamic equilibrium, and in steady flow. Work and heat processes may take place in the chamber R, at rates $\dot{W}$ and $\dot{Q}$, as well as other irreversible processes.

product is the mass-flow rate, Γ. At points 1 and 2, the fluid has internal energy per unit mass (or per mole) u_i, kinetic flow-energy per unit mass $v_i^2/2$, potential energy per unit mass γ_i, and other specific properties, such as the entropy per unit mass, s_i. Therefore the energy per second carried into R by the fluid passing point 1 is $(u_1 + v_1^2/2 + \gamma_1)\Gamma$. Additional energy is supplied to the entering fluid at the rate $P_1 A_1 v_1$, the work per second done on the fluid by the pressure P_1 to move it past point 1. The energy balance equation for the flow system in steady state is thus seen to be

$$(u_1 + v_1^2/2 + \gamma_1)\Gamma + P_1 A_1 v_1 + \dot{Q} + \dot{W} = (u_2 + v_2^2/2 + \gamma_2)\Gamma + P_2 A_2 v_2. \tag{9.2.2}$$

Division by Γ gives

$$u_1 + v_1^2/2 + \gamma_1 + P_1/\rho_1 + \Gamma^{-1}(\dot{Q} + \dot{W}) = u_2 + v_2^2/2 + \gamma_2 + P_2/\rho_2, \tag{9.2.3}$$

or, since the specific enthalpy is $h = u + Pv = u + P/\rho$,

$$h_1 + v_1^2/2 + \gamma_1 + \Gamma^{-1}(\dot{Q} + \dot{W}) = h_2 + v_2^2/2 + \gamma_2. \tag{9.2.4}$$

When (1) the heat and work terms are zero, (2) $\gamma = g$, the acceleration due to gravity, and (3) $u_1 = u_2$, Eq. (9.2.4) becomes the usual Bernoulli equation of first courses in physics. When (1) the kinetic energy terms are negligible, (2) $\gamma_1 = \gamma_2$, and (3) the heat and work terms are zero, Eq. (9.2.4) becomes that of an ordinary throttling process, as represented by the Joule-Thomson porous-plug experiment. The black box could be a steam boiler or turbine, a heat exchanger in an air conditioner or steam power plant, a solar collector, or a variety of other obvious devices. Even more generally, it could be an ohmic resistor, with the fluid being an electron current, or one of several thermoelectric devices. There can be many channels into and out of the chamber, with conservation of mass and energy being the basic requirements. Chemical processes can then be included, with individual accountability of each atomic species. Since the equilibrium properties of the system are measurable at points along the input and output channels, the specific entropies are known, and the net rate of entropy production can be determined unambiguously. This kind of analysis is thus seen to be of general validity, and it has provided sound guidance for generations of engineers. Attempts to replace it by continuum theories, in which differences are replaced by gradients or divergencies, as appropriate, require details of what happens in microscopic black boxes, and these details often have not been carefully handled.

9.2.2 Onsager's Theory. In much the same way that thermostatic stability was studied in Chapter 2, Onsager set out to analyze the time dependence of thermodynamic systems in general. (Casimir [1945]; de Groot and Mazur [1962]; Kreuzer [1981]; Meixner [1973]; Onsager [1931a]; Onsager [1931b]) Consider a small subsystem of a large microcanonical system. It is prepared in a macroscopic state that is not in equilibrium with its surroundings, and constraints are released to allow it to approach equilibrium. Its entropy per mole is presumed to be expandable around the equilibrium value as

$$s = s_0(\{x_{0i}\}) + \sum_i \left(\frac{\partial s}{\partial x_i}\right)_0 (x_i - x_{0i}) + \frac{1}{2}\sum_{ij} \left(\frac{\partial^2 s}{\partial x_i \partial x_j}\right)_0 (x_i - x_{0i})(x_j - x_{0j}) + \cdots, \tag{9.2.5}$$

where $\{x_{0i}\}$ is the set of equilibrium extensive parameters, expressed in units per mole (internal energy per mole, for example), the x_i are the instantaneous values of these parameters for states that may be other than equilibrium. Since s is expected to be a maximum when the x_i have their equilibrium values, the first derivatives of the entropy, evaluated at equilibrium, must all be zero. For sufficiently small departures from equilibrium, Eq. (9.2.5) may be rewritten

$$\xi_s = -(1/2)\sum_{ij} G_{ij}\xi_i\xi_j, \tag{9.2.6}$$

where $\xi_s = s - s_0$, $\xi_i = x_i - x_{0i}$, and

$$G_{ij} = -\left(\frac{\partial^2 s}{\partial x_i \partial x_j}\right)_0 = G_{ji}. \tag{9.2.7}$$

Onsager defined quantities that he called *affinities* and I call *forcings*, as the generalized (and gerundized) forces that drive the system toward equilibrium. They may be written as

$$f_i \equiv \frac{\partial \xi_s}{\partial \xi_i} = -\sum_j G_{ij}\xi_j. \tag{9.2.8}$$

The assumption seems physically plausible in that the forcings all vanish at equilibrium, and the direction of a forcing changes sign with the displacements, the ξ_i. The time rate of change of the molar entropy displacement may now be written

$$\frac{d\xi_s}{dt} = \sum_i f_i\dot{\xi}_i. \tag{9.2.9}$$

Onsager's first major assumption was that the rate of change of a displacement is given in terms of the forcings as

$$\dot{\xi}_i = \sum_j L_{ij}f_j, \tag{9.2.10}$$

where the L_{ij} are called *L-coefficients* or *Onsager's phenomenological coefficients* or *kinetic coefficients*. This assumption is also plausible; the coefficients are taken to be unspecified functions of the intensive parameters of the surrounding equilibrium system. (Note that the per-mole parameters of the surrounding system are negligibly affected by changes that may occur in the subsystem, and the intensive parameters are similarly unaffected, so that there is no ambiguity in their definitions.) Note that Eq. (9.2.9) can be rewritten as

$$\dot{\xi}_s = \sum_{ij} L_{ij}f_if_j. \tag{9.2.11}$$

In matrix notation, the set of ξ_i's is written as a column matrix ξ, with similar treatment for the other parameters. Then $f = -G\xi$, and

$$\dot{\xi} = -LG\xi. \tag{9.2.12}$$

This matrix equation has the solution

$$\xi(t+\tau) = e^{-LG\tau}\xi(t), \quad \tau > 0. \tag{9.2.13}$$

The displacements are thus expected to decay exponentially to zero, but not necessarily at the same rate.

Onsager next assumed that the parameters of the subsystem fluctuate naturally, as a consequence of interactions with its surroundings, and that the probabilities associated with these fluctuations are those derivable for equilibrium systems. Thus for the entire microcanonical system, its equilibrium entropy is given by

$$S_0 = k \ln \Omega_0, \tag{9.2.14}$$

as in Chapter 3, where Ω_0 is the number of distinguishable microscopic configurations of the system that are possible, subject to the existing constraints. The probability of any macroscopic microcanonical configuration is proportional to the number Ω of distinguishable microscopic configurations, and the entropy is taken to be $S = k \ln \Omega$. The entropy of the total system, with the subsystem having molar entropy s_1, is

$$S = k \ln \Omega = (N - N_1)s_0 + N_1 s_1 = k \ln \Omega_0 + N_1 \xi_s. \tag{9.2.15}$$

From this, the probability for a particular nonequilibrium configuration is

$$p \sim \Omega = \Omega_0 \exp[(N_1/k)\xi_s] = \Omega_0 \exp[-(N_1/2k)\xi^T G\xi], \tag{9.2.16}$$

where ξ^T is the row-matrix transpose of ξ. In order to obtain a true probability set, it is necessary to divide the p of Eq. (9.2.16) by its integral over all possible values of the ξ_i. Since there is negligible contribution from points that are far from equilibrium, it is permissible to extend the limits of integration to $\pm\infty$. The integral to be evaluated, denoted by Y, is

$$Y = \int_{-\infty}^{\infty} \cdots \int_{-\infty}^{\infty} d\xi_1 \cdots d\xi_n \exp[-(N_1/2k)\xi^T G\xi]. \tag{9.2.17}$$

Since G is a real, symmetric matrix, it can be diagonalized by an orthogonal transformation, say T, such that $T^T = T^{-1}$, giving

$$TGT^{-1} = \Lambda. \tag{9.2.18}$$

Define a new matrix η, given by

$$\eta = T\xi\sqrt{N_1/k}, \quad \eta^T = \sqrt{N_1/k}\,\xi^T T^{-1}. \tag{9.2.19}$$

Since the Jacobian of the transformation is simply $(k/N_1)^{n/2}|\det T|$, the required integral becomes

$$Y = \left(\frac{k}{N_1}\right)^{n/2} \int_{-\infty}^{\infty} \cdots \int_{-\infty}^{\infty} d\eta_1 \cdots d\eta_n \exp[-(1/2)\eta^T \Lambda \eta] = \left(\frac{k}{N_1}\right)^{n/2} \prod_{i=1}^{n} \int_{-\infty}^{\infty} d\eta_i \, e^{-\lambda_i \eta_i^2/2}$$

$$= \left(\frac{k}{N_1}\right)^{n/2} \prod_{i=1}^{n} \sqrt{\frac{2\pi}{\lambda_i}} = \frac{(2\pi k/N_1)^{n/2}}{\sqrt{\det \Lambda}}. \tag{9.2.20}$$

Note that $\prod \lambda_i = \det \Lambda$, that $\det \Lambda = \det G$, and that this procedure was used in Section 1.9. Y can now be written

$$Y = (2\pi k/N_1)^{n/2}/\sqrt{\det G}. \tag{9.2.21}$$

The covariances $\langle \xi_q \xi_r \rangle$ can be calculated directly from the normalized probability:

$$\langle \xi_q \xi_r \rangle = Y^{-1} \int_{-\infty}^{\infty} \cdots \int_{-\infty}^{\infty} d\xi_1 \cdots d\xi_n (\xi_q \xi_r) \exp[-(1/2) \sum G_{ij} \xi_i \xi_j]$$

$$= \frac{-2k}{N_1} \frac{\partial \ln Y}{\partial G_{qr}} = \frac{k}{N_1} \frac{\partial \det \ln G}{\partial G_{qr}} = \frac{k}{N_1} \frac{g_{rq}}{\det G}, \tag{9.2.22}$$

where g_{rq} is the cofactor of G_{qr} in the expansion of $\det G$. The covariance matrix Ξ, with elements $\langle \xi_i \xi_j \rangle$, is evidently a multiple of the inverse of G, since

$$G\Xi = \Xi G = (k/N_1)I, \tag{9.2.23}$$

where I is the $n \times n$ unit matrix.

The correlations of displacements ξ_i with forcings f_j may be found from the results of Eq. (9.2.22). By use of Eq. (9.2.8), the correlation $\langle \xi_i f_j \rangle$ may be written

$$\langle \xi_i f_j \rangle = - \left\langle \xi_i \sum_k G_{jk} \xi_k \right\rangle = - \sum_k G_{jk} \Xi_{ki} = -(k/N_1)\delta_{ij}, \tag{9.2.24}$$

since the sum of products of one column with cofactors of another gives zero. Thus a displacement is found to be correlated only with its own forcing. The correlation of the forcings is found similarly:

$$\langle f_i f_j \rangle = \sum_{kl} G_{ik} G_{jl} \langle \xi_k \xi_l \rangle = \frac{k}{N_1} \sum_k G_{ik} \delta_{jk} = \frac{k}{N_1} G_{ij}, \tag{9.2.25}$$

and in matrix terms, $F = (k/N_1)G$, where F is the matrix with elements $\langle f_i f_j \rangle$.

The N_1 factor that appears in these results is a consequence of my choosing molar displacements. If densities had been used, N_1 would be replaced by V_1, and if the total entropy of the subsystem had been chosen, expressed in terms of total subsystem extensive variables, the factor would have been unity.

The next assumption about the behavior of the subsystem, called Onsager's *regression hypothesis*, is that every spontaneous fluctuation that appears macroscopically at time t as $\xi(t)$, despite the underlying diversity of its possible microstates, decays according to the deterministic expression of Eq. (9.2.13). A time-dependent covariance matrix can now be defined by use of Eq. (9.2.13), with elements

$$\langle \xi_i(t+\tau) \xi_j(t) \rangle = e^{-LG\tau} \langle \xi_i(t) \xi_j(t) \rangle, \tag{9.2.26}$$

or, in terms of the matrix Ξ,

$$\Xi(\tau) = e^{-LG\tau} \Xi(0) = \Xi(0) e^{-GL\tau}, \quad \tau > 0, \tag{9.2.27}$$

so that Ξ obeys the same differential equation as ξ. Two algebraic steps have been taken to obtain the last expression. First, recall that $\Xi(0) = (k/N_1)G^{-1}$, with both matrices symmetric. Then use the general matrix identity $f(LG)G^{-1} = G^{-1}f(GL)$. Onsager's hypothesis on the decay of spontaneous fluctuations is that they decay in exactly the same way as the regression of the autocorrelation functions, given in Eq. (9.2.26) when $i = j$. The assumption that $\Xi(\tau)$, as defined by its elements in Eq. (9.2.26), is independent of t implies that if t is replaced by $t - \tau$, the definition of $\Xi(\tau)$ should be unchanged. But the left-hand side of Eq. (9.2.26) becomes

$$\Xi_{ij}(\tau) = \langle \xi_i(t)\xi_j(t - \tau) \rangle = \langle \xi_j(t - \tau)\xi_i(t) \rangle = \Xi_{ji}(-\tau), \tag{9.2.28}$$

or $\Xi(-\tau) = \Xi^T(\tau)$, for $\tau > 0$. At this point, Onsager invoked his principle of *microscopic reversibility*, discussed subsequently, which says, in effect, that the equations governing the motions of the particles of the system are invariant under time reversal. Therefore the covariance matrix $\Xi(\tau)$ should equal its time-reversed counterpart, $\Xi(-\tau)$, which then requires, from the transpose of Eq. (9.2.27),

$$\Xi(\tau) = \Xi(-\tau) = \Xi(0)e^{-GL^T\tau} = e^{-L^T G\tau}\Xi(0). \tag{9.2.29}$$

Comparison with Eq. (9.2.27) shows that it is necessary that

$$L = L^T, \tag{9.2.30}$$

or $L_{ij} = L_{ji}$, the new and generally unexpected consequence of the Onsager theory upon which its usefulness rests.

The time-reversal invariance of the equations of motion is a familiar consequence of accelerations being second time derivatives. If there are force terms with first time derivatives, such as those with $\mathbf{v} \times \mathbf{B}$, or Coriolis terms, since time reversal changes the sign of $\mathbf{v}$, it is necessary to reverse $\mathbf{B}$ and the angular velocities responsible for the Coriolis forces. Any other quantities that must be reversed for time-reversal invariance are identifiable in the equations of motion. A custom to indicate this necessity is to write Eq. (9.2.30) as $L(\mathbf{B}) = L^T(-\mathbf{B})$, suppressing the other dependencies in L. As indicated in the commentary of Chapter 1, and in the discussions of the dog-flea and wind-tree models, the seeming incompatibility of microscopic reversibility and equilibration has caused much puzzlement and sometimes bitter controversy. Those of us who regard equilibrium as a state of information, rather than of dynamics, find these problems easier to resolve.

For most applications, a somewhat different formulation of the Onsager theory is useful. In this case, the subsystem is a fixed region of space, subject to flows of energy, electrical charges, particles of various kinds, *etc.*, and either allowed to relax to equilibrium or held in a steady state that requires continuous forcing. The development begins with an equation similar to Eq. (9.2.5), except this time the x_i's represent densities (per unit volume), rather than their molar equivalents. In this case, volume is not a parameter. But since a typical steady-state process may have temperature, pressure, and concentration gradients throughout the system, the local densities themselves, rather than their departures from some remote equilibrium value, are the parameters of the calculation. The entropy density is taken as the same function of the local densities as in equilibrium, which eliminates ambiguity in defining the entropy without an assumption of local thermodynamic equilibrium. The starting point is the equation

$$ds = \sum_i F_i dx_i, \tag{9.2.31}$$

where, as in Eq. (9.2.8), the F_i are given by

$$F_i = \frac{\partial s}{\partial x_i}. \tag{9.2.32}$$

Each of the densities is a conserved quantity that satisfies a continuity equation,

$$\frac{dx_i}{dt} = \frac{\partial x_i}{\partial t} + \nabla \cdot \mathbf{J}_i = 0, \tag{9.2.33}$$

where $\mathbf{J}_i$ is the flux of x_i. The entropy density is not, in general, conserved, but it should satisfy an equation that accounts for its production:

$$\frac{ds}{dt} = \frac{\partial s}{\partial t} + \nabla \cdot \mathbf{J}_s, \tag{9.2.34}$$

where the entropy flux $\mathbf{J}_s$ follows from Eq. (9.2.31); or

$$\mathbf{J}_s = \sum_i F_i \mathbf{J}_i. \tag{9.2.35}$$

The two terms on the right-hand side of Eq. (9.2.34) can be evaluated with the help of Eq. (9.2.31) and Eq. (9.2.35). The first gives

$$\frac{\partial s}{\partial t} = \sum_i F_i \frac{\partial x_i}{\partial t} = -\sum_i F_i \nabla \cdot \mathbf{J}_i, \tag{9.2.36}$$

by use of Eq. (9.2.33), and the second provides

$$\nabla \cdot \mathbf{J}_s = \sum_i \left(\nabla F_i \cdot \mathbf{J}_s + F_i \nabla \cdot \mathbf{J}_i \right). \tag{9.2.37}$$

The rate of production of entropy per unit volume is thus found to be, upon cancellation of the $\nabla \cdot \mathbf{J}_i$ terms,

$$\frac{ds}{dt} = \sum_i \nabla F_i \cdot \mathbf{J}_i \tag{9.2.38}$$

in analogy with Eq. (9.2.9). This time the forcings are seen to be the gradients of the quantities F_i, that act in place of the equilibrium intensive parameters associated with the entropy, and the local rates of change of the displacements are replaced by the corresponding currents. The Onsager assumption equivalent to Eq. (9.2.10) is

$$\mathbf{J}_i = \sum_j L_{ij} \nabla F_j, \tag{9.2.39}$$

so that Eq. (9.2.11) becomes, in this formulation,

$$\dot{s} = \sum_{ij} L_{ij} \nabla F_i \nabla F_j. \tag{9.2.40}$$

When $\dot{s}$ can be written in this form, the current densities of Eq. (9.2.39) are expected to be related through the reciprocal relations of Eq. (9.2.30), or $L_{ij}(\mathbf{B}) = L_{ji}(-\mathbf{B})$. When the system is sufficiently close to local thermodynamic equilibrium, the L-coefficients can be regarded as functions of the local intensive parameters. It is possible, and sometimes useful, to develop the Onsager theory to higher order, so that systems further from equilibrium can be followed. A good choice of starting point is Eq. (9.2.5), with higher-order terms retained. Another such choice is at Eq. (9.2.39), including higher terms such as $L_{ijk}\nabla F_j\nabla F_k$. Since the Onsager theory does not provide theoretical expressions for the L-coefficients, but merely reduces the number of independent ones by symmetry considerations based on microscopic reversibility, there is considerable freedom for exploration. For systems far from equilibrium, the L-coefficients are expected to depend upon quantities that are not well defined, and therefore their meaning becomes unclear. Many systems exist with memory, so that their evolutions are determined by other than present parameters, as was required in Eq. (9.2.12); in these cases, the Onsager development cannot be expected to be valid, and indeed it is not. Such systems are discussed in a subsequent section of this chapter. (For a well-documented criticism of Onsager theory, see Truesdell, 1984, but also see Meixner, 1973.)

9.2.3 Application of the Onsager Theory. As an example, the transport of heat and electrical charge in a wire subject to gradients in temperature and electric potential is treated both here and in Section 9.3, on the Boltzmann equation. The wire is considered to be uniform and aligned parallel to the y axis. In this approximation, both heat and electric current are regarded as carried by the electrons. The number density of the electrons, rather than their molar density, is used in the calculation, so the chemical potential is that per electron. In this case, Eq. (9.2.31) becomes

$$ds = \left(\frac{1}{T}\right) du' - \left(\frac{\mu}{T}\right) dn, \tag{9.2.41}$$

where μ is now called the *electrochemical potential*, and $u' = u + \gamma$ is necessary, where γ is the potential energy density of Eq. (9.2.2). (This energy term must be included in energy-flux calculations, as should be clear from Eq. (9.2.2).) In the absence of an electric potential difference, μ is the ordinary chemical potential, here written as μ_c, and in a metallic conductor taken to be the Fermi energy at the local temperature and electron density. The local electrical potential ϕ effectively adds the potential energy $-e\phi$ to μ_c, giving $\mu = \mu_c - e\phi$. Equation (9.2.35) relating the currents becomes

$$\mathbf{J}_s = \left(\frac{1}{T}\right) \mathbf{J}_u - \left(\frac{\mu}{T}\right) \mathbf{J}_n, \tag{9.2.42}$$

where $\mathbf{J}_u$ is now the flux of the enhanced internal energy density u'. The Onsager expressions for the conserved currents are

$$\mathbf{J}_u = L_{uu}\nabla\left(\frac{1}{T}\right) - L_{un}\nabla\left(\frac{\mu}{T}\right) \quad \text{and}$$

$$\mathbf{J}_n = L_{nu}\nabla\left(\frac{1}{T}\right) - L_{nn}\nabla\left(\frac{\mu}{T}\right). \tag{9.2.43}$$

For this calculation, it is easier to relate the currents to known properties if the internal-energy current is replaced by the thermal-energy flux, or *heat-flow vector*, $\mathbf{J}_q$, given by

$$\mathbf{J}_q = T\mathbf{J}_s = \mathbf{J}_u - \mu\mathbf{J}_n. \tag{9.2.44}$$

Equation (9.2.44) for $\mathbf{J}_q$ restricts usage of the term to equivalent expressions, even though other definitions are often encountered. With this replacement, Eq. (9.2.38) for the rate of entropy generation becomes

$$\dot{s} = \nabla\left(\frac{1}{T}\right)\cdot(\mathbf{J}_q + \mu\mathbf{J}_n) - \nabla\left(\frac{\mu}{T}\right)\cdot\mathbf{J}_n$$

$$= \nabla\left(\frac{1}{T}\right)\cdot\mathbf{J}_q - \frac{1}{T}\nabla\mu\cdot\mathbf{J}_n, \tag{9.2.45}$$

and the one-dimensional Onsager current densities are

$$J_q = -\left(\frac{L_{qq}}{T^2}\right)\frac{dT}{dy} - \left(\frac{L_{qn}}{T}\right)\frac{d\mu}{dy} \quad \text{and}$$

$$J_n = -\left(\frac{L_{nq}}{T^2}\right)\frac{dT}{dy} - \left(\frac{L_{nn}}{T}\right)\frac{d\mu}{dy}. \tag{9.2.46}$$

It may be seen by direct substitution that if $L_{un} = L_{nu}$, it then follows that $L_{qn} = L_{nq}$. When the temperature is uniform, and $\mu = \epsilon_f - e\phi$, the Fermi energy is independent of position, and $d\mu/dy = -ed\phi/dy = eE$. The particle current becomes

$$J_n = -j/e = -\sigma E/e = -L_{nn}eE/T, \tag{9.2.47}$$

where j is the electric current density, and σ is the conductivity. Therefore the Onsager coefficient L_{nn} is found to be $L_{nn} = \sigma T/e^2$. There is a heat current accompanying the electric current, given by $J_q = -L_{qn}eE/T$, with L_{qn} as yet undetermined.

The thermal conductivity, by contrast, is usually measured at zero electric current, not at zero electric field. When Eqs. (9.2.46) are solved with $J_n = 0$, the heat flux is found to be

$$J_q = -\left(\frac{L_{qq}L_{nn} - L_{qn}L_{nq}}{L_{nn}T^2}\right)\frac{dT}{dy} = -\kappa\frac{dT}{dy}, \tag{9.2.48}$$

where κ is the thermal conductivity. Thus there are two transport coefficients, σ and κ, that have been related to three independent L-coefficients. At least one other independent transport coefficient is needed, in order to determine the L's uniquely.

Note that even though the heat and particle currents are interrelated by equations such as (9.2.46), it is not possible in general to suppress the heat flux by application of an electric field in the absence of an electric current, or to suppress the electric current by a temperature gradient with zero heat flux. This statement is expressed in another way in Eq. (9.2.24).

Since $\mu = \epsilon_f - e\phi$, it is possible in principle to apply an electric field of such magnitude and direction as to make $d\mu/dy = 0$. In this case, the currents of Eq. (9.2.46) become $J_q = -(L_{qq}/T^2)dT/dy$, and $J_n = -(L_{nq}/T^2)dT/dy$. A measurement of, say, the electric current under these conditions provides a value for L_{nq}, from which it becomes possible to determine L_{qq}, and a measurement of J_q offers the opportunity for independent verification of the result.

The rate of entropy production by the currents of Eq. (9.2.46) is found from Eq. (9.2.38) to be

$$\dot{s} = \frac{L_{qq}}{T^4}\left(\frac{dT}{dy}\right)^2 + \frac{2L_{qn}}{T^3}\left(\frac{dT}{dy}\right)\left(\frac{d\mu}{dy}\right) + \frac{L_{nn}}{T^2}\left(\frac{d\mu}{dy}\right)^2. \tag{9.2.49}$$

In the absence of a temperature gradient, only the final term contributes, giving

$$\dot{s} = \frac{L_{nn}}{T^2}(eE)^2 = \frac{\sigma E^2}{T} = \frac{jE}{T}, \tag{9.2.50}$$

which is seen to be the Joule heating per unit volume divided by the temperature, in agreement with expectation. In order to maintain a time-independent temperature of the system, it is necessary to remember that there must be a heat flux out of the system that is not carried by the current J_q. It corresponds to the $\dot{Q}$ of Fig. 9.1. Note also that the dissipative, irreversible processes in this example must necessarily lead to the production of entropy. Therefore, it follows from Eq. (9.2.49) that $L_{qq} > 0$ and $L_{nn} > 0$. Also, in order for the quadratic form to be positive definite, it is necessary that $L_{qq}L_{nn} - L_{qn}L_{nq} \geq 0$, in agreement with the result of Eq. (9.2.48), since the heat conductivity is non-negative. The negative signs in Eq. (9.2.46), at least for the diagonal terms, are then a consequence of the necessity that a natural flow is downhill, for the appropriate gradient.

9.2.4 Continuum Thermodynamics. Nowhere did Gibbs imply that the concepts of entropy and temperature should be restricted to states of equilibrium, or to homogeneous systems. In his scientific papers (Gibbs [1961]), p. 39, he asserts that when a body is not in a state of equilibrium, it nevertheless has a volume, energy, and entropy that are the sums of the volumes, energies, and entropies of its parts. A number of books and articles, often concerned primarily with chemical reactions, treat time-dependent phenomena in terms of thermodynamics, with emphasis on the Onsager theory of the previous subsection. (Fitts [1962]; de Groot and Mazur [1962]; Kreuzer [1981]) In this section, I call attention to the approach called *Rational Thermodynamics*, the methodology of which was developed by practitioners of a discipline called *Rational Mechanics*. The workers in rational thermodynamics since about 1960 have advanced their subject impressively by emphasis on careful definitions and sound mathematical procedures. (Coleman and Mizel [1963]; Coleman and Noll [1963]; Coleman and Noll [1964]; Coleman and Owen [1974]; Noll [1973]; Serrin [1979]; Truesdell [1984]; Truesdell [1986]; Truesdell and Toupin [1960]) Their work is extensive and deserving of careful study. But because the technical language of their writings differs from that learned by most physicists, as modern Italian differs from classical Latin, some effort is required to gain a true translation. I offer only a small sample here, since the space required to present the subject adequately would fill another book the size of this one. My demonstration that translation is possible should encourage others to attempt it. (This segment is based on Truesdell [1984].)

First, for a simple system, one can express the Clausius-Planck inequality as

$$T\frac{dS}{dt} \geq Q, \tag{9.2.51}$$

where Q is called the *heating*; it is evidently the rate of adding thermal energy to the system. (I have omitted some careful definitions, and I have changed the notation to resemble that used in this book, in an effort to circumvent barriers that could preclude further study; the original source is recommended.) The rate of change of the internal energy is given by

$$\dot{U} = Q + W, \tag{9.2.52}$$

where W, called the *working*, is the rate at which work is done on the system. The Helmholtz potential is $F = U - TS$, and its time derivative is

$$\dot{F} = \dot{U} - S\dot{T} - T\dot{S} = W + Q - S\dot{T} - T\dot{S},$$

or

$$\dot{F} - W + S\dot{T} = Q - T\dot{S} \leq 0, \tag{9.2.53}$$

where Eq. (9.2.51) has been used. This result is called *the reduced dissipation inequality*.

As an illustration of the use of this result, consider a simple system composed of a material without memory. The governing equations for such a system (called *equations of state*) are

$$
\begin{aligned}
W &= -\sum P_i \dot{X}_i, \\
F &= F(T, \{X_i\}), \quad \text{and} \\
S &= S(T, \{X_i\}),
\end{aligned}
\tag{9.2.54}
$$

where the P_i are generalized pressures, and the X_i are extensive parameters. When these expressions are used in the reduced dissipation inequality, the result is

$$\left(\frac{\partial F}{\partial T} + S \right) \dot{T} + \sum \left(\frac{\partial F}{\partial X_i} + P_i \right) \dot{X}_i \leq 0. \tag{9.2.55}$$

For any state of the system, as determined by T and the set $\{X_i\}$, the coefficients of $\dot{T}$ and each $\dot{X}_i$ are fixed, but the temperature and extensive parameters can have positive or negative rates of change, independently. Thus it is possible to choose the signs of the rates so as to violate the inequality, unless the coefficients in parentheses are independently zero. It follows therefore that

$$S = -\frac{\partial F}{\partial T}, \quad \text{and} \quad P_i = -\frac{\partial F}{\partial X_i}. \tag{9.2.56}$$

Thus the equality in Eq. (9.2.53) always holds for systems described by Eq. (9.2.54), and the Helmholtz potential is a thermodynamic potential, from which the entropy and the generalized pressures can be determined. In such a system, all processes are reversible, and a process is adiabatic if and only if it is isentropic. Evidently, even in this introductory example, the reduction of classical thermodynamics to statements about equilibrium is neither essential nor advantageous.

From this beginning, the extensive parameters S and $\{X_i\}$ are redefined in terms of the local specific entropy and other specific quantities, such that, for example,

$$S = \int \eta \, dM, \quad \text{and} \quad U = \int \epsilon \, dM, \tag{9.2.57}$$

where η is the specific entropy, ϵ the specific internal energy, and the mass M is a non-negative measure, defined over the measurable subsets of the body, and assumed to be a continuous function of volume. (The entropy density thus becomes a field, and the practitioners of continuum thermodynamics in general must be careful to avoid reducing the entropy to merely one of many fields, thereby bedimming its fundamental position.)

The time derivatives of the system properties are written without the usual reluctance exhibited by main-line thermodynamicists to consider temperature or specific entropy as dynamic variables. Rational thermodynamicists treat wave propagation in dissipative media, diffusion, transport phenomena, and chemical reactions, all without recourse to Onsager theory. They develop constitutive relations that imply the symmetries, when valid, that the Onsager theory obtains by methods they regard as specious. (Truesdell [1984]) They study nonlinear phenomena, multiphase transport of fluids, and the thermodynamics of materials with memory. They work in the continuum limit, as mentioned in Chapter 2, but they often are able to provide mathematical descriptions of the observed phenomenology. They make no appeal to the underlying microscopic behavior of the system, and therefore a further presentation of their methods is not justified within the context of this book. Nevertheless, thermophysics is the poorer for its failure to synthesize a discipline that includes the methods and results of rational thermodynamics.

9.3 THE BOLTZMANN EQUATION

For his pioneering study of the approach to equilibrium of a system of particles, *e.g.* a gas, Boltzmann chose to describe his system in terms of a function denoted by $f(\mathbf{r}, \mathbf{v}, t)$, such that the number of particles in the region of his 6-dimensional μ-space in the infinitesimal region with coordinates between $\mathbf{r}$ and $\mathbf{r} + d\mathbf{r}$ and velocities between $\mathbf{v}$ and $\mathbf{v} + d\mathbf{v}$ at time t is given by

$$dn(t) = f(\mathbf{r}, \mathbf{v}, t)d\mu, \qquad (9.3.1)$$

where $d\mu = d^3 r d^3 v$. He could have normalized his function to be a probability density by dividing by the total number of particles in the system, but he chose not to. Most work with the Boltzmann equation follows his choice, and I shall accede to standard practice.

The number of particles in the system is a constant, N, given by

$$N = \int f d\mu, \qquad (9.3.2)$$

where the limits of integration throughout this section, unless otherwise noted, are the boundaries of the volume for $\mathbf{r}$ and $\pm\infty$ for each cartesian velocity component. The number density f satisfies a continuity equation, with the understanding that collision processes can cause the points representing individual particles to disappear from one particular volume element $d\mu$ and reappear elsewhere. (This is not quite an accurate statement, because if the particles are followed over short enough time intervals to allow the motions to be seen as continuous, they then move smoothly through μ-space, but typical collision times for air molecules under standard conditions are of the order of 10^{-10} s., and for time intervals longer by an order of magnitude, the motion seems discontinuous. Boltzmann thought of collisions as if atoms were like billiard balls, for which collisions result in instantaneous velocity changes.) The continuity equation for f is therefore written

$$\frac{\partial f}{\partial t} + \mathbf{v} \cdot \nabla_r f + \mathbf{a} \cdot \nabla_v f = \left(\frac{\partial f}{\partial t}\right)_c, \qquad (9.3.3)$$

where the right-hand side is the collision term, and the acceleration $\mathbf{a}$ is the analog in velocity space of the velocity in coordinate space. Accelerations are a consequence of fields in coordinate space, and not those caused by collisions. Boltzmann wrote the collision term as an integral over velocity space, as

$$\left(\frac{\partial f}{\partial t}\right)_c = \int \int \int \int [f(\mathbf{v}_1')f(\mathbf{v}') - f(\mathbf{v}_1)f(\mathbf{v})]\, \sigma' v_r d^3 v_1 d\Omega_1, \qquad (9.3.4)$$

where the first term sums the collisions of particles with velocities $\mathbf{v}_1'$ and $\mathbf{v}'$ that result in a particle with velocity $\mathbf{v}$; the second term counts the collisions of particles with velocities $\mathbf{v}_1$ and $\mathbf{v}$ that shift a particle's velocity from $\mathbf{v}$ to some other value; σ' is the differential cross-section; v_r is the relative speed of the particles before or after collision; and Ω_1 is the solid angle of the relative velocities before collision, $\mathbf{v}_1' - \mathbf{v}'$, with respect to those after collision, $\mathbf{v} - \mathbf{v}_1$. Thus the full Boltzmann equation is an integro-differential equation that can become even more complicated if higher-order collision terms are considered. I leave these complexities, insofar as possible, to discussions of kinetic theory, since it is not possible to treat them adequately in the space available, and since their details are not central to the conceptual understanding of irreversible processes.

The Boltzmann equation has led to much of our present insight and understanding of how microscopic processes in gases lead to macroscopic consequences. Some of its uses are presented, or at least caricatured, in subsequent subsections. It works well in dilute gases with short-range, two-body interactions, but it is intrinsically a single-particle equation. Systems with long-range forces and many-body interactions have been treated with much less success. Modifications of the Boltzmann equation that provide better results in such systems are discussed briefly, with the understanding that they are the subject matter of kinetic theory, and to treat the subject properly would require an entire book. Appropriate references are provided. (Brush [1972]; Liboff [1990b])

9.3.1 Collisional Invariants and Hydrodynamic Equations. The local particle density of a Boltzmann gas of identical particles is found by integration of f over velocity space, and a similar integration gives the momentum density $\mathbf{G}$ and translational kinetic energy density u_t, i.e.

$$n(\mathbf{r}, t) = \int d^3 v f(\mathbf{r}, \mathbf{v}, t),$$

$$\mathbf{G}(\mathbf{r}, t) = \int d^3 v f(\mathbf{r}, \mathbf{v}, t) m\mathbf{v}, \quad \text{and}$$

$$E(\mathbf{r}, t) = \int d^3 v f(\mathbf{r}, \mathbf{v}, t) u_t = \int d^3 v f(\mathbf{r}, \mathbf{v}, t) mv^2/2. \qquad (9.3.5)$$

The average momentum and translational kinetic energy per particle are evidently

$$\langle \mathbf{p}(\mathbf{r}, t) \rangle = \mathbf{G}/n, \quad \text{and}$$

$$\langle \epsilon_{tot}(\mathbf{r}, t) \rangle = E/n. \qquad (9.3.6)$$

The local average velocity is

$$\mathbf{w}(\mathbf{r}, t) \equiv \langle \mathbf{v}(\mathbf{r}, t) \rangle = \langle \mathbf{p}(\mathbf{r}, t) \rangle /m. \qquad (9.3.7)$$

The velocity $\mathbf{w}$ represents the local flow velocity of the fluid. Individual molecules have a distribution of velocities around the average, and their translational kinetic energies in a coordinate system that moves at the average velocity are regarded as thermal. The total velocity of a particle is now written

$$\mathbf{v} = \mathbf{w} + \mathbf{c}, \tag{9.3.8}$$

where $\mathbf{c}$ is often called the *peculiar velocity*, and is associated with the thermal motion. The average thermal translational kinetic energy per particle is

$$\epsilon_{th} = \frac{m}{2} \langle \mathbf{c}^2 \rangle = \frac{m}{2} (\langle v^2 \rangle - w^2); \tag{9.3.9}$$

therefore the total translational kinetic energy per particle, as seen in the laboratory, is

$$\epsilon_{tot} = \epsilon_{fl} + \epsilon_{th}, \tag{9.3.10}$$

where the flow kinetic energy is $\epsilon_{fl} = mw^2/2$.

Because of the conservation laws of mechanics, the average energy, momentum, and total number of particles in a Boltzmann gas are unchanged by collisions, provided certain processes are disallowed, such as ionizations, dissociations, radiative collisions, and chemical reactions. For a monatomic gas, the entire energy is translational kinetic, and energy conservation is easy to formulate. The Boltzmann equation can be used as a starting point for the derivation of time- and space-dependent densities that are carried by the fluid. The terms in the Boltzmann equation are multiplied by the functions already used in Eq. (9.3.5) and then integrated over velocity space. To do these integrations in general, write $\phi(\mathbf{v})$ as an unspecified collision-invariant function of $\mathbf{v}$ alone. The first of them is

$$\int d^3 v \phi(\mathbf{v}) \frac{\partial f}{\partial t} = \frac{\partial}{\partial t} \langle n\phi \rangle. \tag{9.3.11}$$

The second is

$$\int d^3 v \phi(\mathbf{v}) \mathbf{v} \cdot \nabla_r f = \nabla_r \cdot \langle n\mathbf{v}\phi \rangle, \tag{9.3.12}$$

and the third is

$$\int d^3 v \phi(\mathbf{v}) \mathbf{a} \cdot \nabla_v f = \int d^3 v \nabla_v \cdot (\mathbf{a}\phi f) - \int d^3 v (\mathbf{a} \cdot \nabla_v \phi) f - \int d^3 v \phi(\mathbf{v})(\nabla_v \cdot \mathbf{a}) f. \tag{9.3.13}$$

The final term in Eq. (9.3.13) vanishes for the usual case that there is no velocity-dependent component of the force parallel to $\mathbf{v}$, so that all terms like $\partial a_z / \partial v_z$ are zero. The first term on the right-hand side of Eq. (9.3.13) vanishes by the divergence theorem, since f is assumed to drop off with increasing v fast enough to reduce any fluxes at infinity to zero. Only the middle term remains. The resulting equation is seen to be

$$\frac{\partial}{\partial t} \langle n\phi \rangle + \nabla_r \cdot \langle n\mathbf{v}\phi \rangle - \langle n\mathbf{a} \cdot \nabla_v \phi \rangle = 0. \tag{9.3.14}$$

To find the fluid equation for the particle density, the function used is simply $\phi = 1$, giving

$$\frac{\partial n}{\partial t} + \nabla_r \cdot (n\mathbf{w}) = 0, \tag{9.3.15}$$

which is the equation of continuity for particles. If $\phi = m$, the continuity equation for mass results; it is written

$$\frac{\partial \rho}{\partial t} + \nabla_r \cdot \mathbf{J}_m = 0, \tag{9.3.16}$$

where ρ is the density, and $\mathbf{J}_m = \rho \mathbf{w}$, the mass flux vector. Similarly, the equation for momentum density is found to be

$$\frac{\partial \mathbf{G}}{\partial t} + \nabla_r \cdot (\mathbf{G}\mathbf{w}) + \nabla_r \cdot \widehat{\mathbf{P}} = \langle nm\mathbf{a} \rangle, \tag{9.3.17}$$

where the second and third terms come from the ∇_r term of Eq. (9.3.14). The dyad, or second-order tensor, written as $\mathbf{G}\mathbf{w}$ could also be written as a 3×3 matrix with elements $G_i w_j$, and the pressure tensor $\widehat{\mathbf{P}}$ has components $P_{ij} = \langle mnc_ic_j \rangle$. Its terms represent, in the reference frame of the fluid, the averaged rate of transport per unit area of momentum component mc_i as a consequence of the velocity c_j. If orthogonal components of the peculiar velocity are uncorrelated, as in a typical gas with no turbulence or vorticity, the pressure tensor is diagonal. If the fluid is isotropic, the diagonal elements are equal, and the pressure tensor reduces to a scalar, with value $\langle mnc^2 \rangle /3$, since $\langle c^2 \rangle = 3 \langle c_i^2 \rangle$.

Equation (9.3.17) is often rewritten by expanding the terms and eliminating two of them by means of Eq. (9.3.15). The expansion is

$$m\mathbf{w}\frac{\partial n}{\partial t} + mn\frac{\partial \mathbf{w}}{\partial t} + m\mathbf{w}\nabla \cdot (n\mathbf{w}) + mn\mathbf{w} \cdot \nabla\mathbf{w} + \nabla \cdot \widehat{\mathbf{P}} = \langle nm\mathbf{a} \rangle. \tag{9.3.18}$$

The first and third terms add to zero because of Eq. (9.3.15). The total time derivative, resulting from local changes and from those occurring as a consequence of fluid transport, often called the *convective derivative*, is written

$$\frac{D}{Dt} \equiv \frac{\partial}{\partial t} + \mathbf{w} \cdot \nabla. \tag{9.3.19}$$

With this notation, Eq. (9.3.18) becomes

$$nm\frac{D\mathbf{w}}{Dt} = -\nabla \cdot \widehat{\mathbf{P}} + \langle nm\mathbf{a} \rangle, \tag{9.3.20}$$

which is $\mathbf{F} = m\mathbf{a}$ for a volume element of the Boltzmann fluid. The divergence of the pressure tensor becomes the ordinary pressure gradient for scalar pressure, and the other term is the body force density attributable to local fields, but not collisions. A similar calculation for which $\phi = mv^2/2$ in Eq. (9.3.14) reduces, after the elimination of terms that sum to zero by use of Eq. (9.3.15) and Eq. (9.3.20), to

$$\frac{D}{Dt}\left(\frac{1}{2}nm\langle c^2 \rangle\right) = -\nabla_r \cdot \mathbf{J}_k - \widehat{\mathbf{P}} : \widehat{\mathbf{\Lambda}}, \tag{9.3.21}$$

where the vector $\mathbf{J}_k = \langle (m/2)nc^2\mathbf{c} \rangle$ is usually written in kinetic theory as $\mathbf{J}_q$, and called the *heat-flux vector* since it represents the flow of energy stored in the microscopic motions of the system. However, since $\mathbf{J}_q$ has already been preempted by Eq. (9.2.44), another notation must be used if the Boltzmann-equation results are to be compared with those of Onsager theory. The symmetric rate-of-strain tensor Λ has components

$$\Lambda_{ij} = \frac{1}{2}\left\langle \frac{\partial w_i}{\partial r_j} + \frac{\partial w_j}{\partial r_i} \right\rangle. \tag{9.3.22}$$

The double dot symbolizes the scalar product of the two tensors, and Eq. (9.3.21) attributes the rate of change of the (translational) thermal energy density to two sources: (1) the divergence of the heat flux, and (2) the rate of deformation (compression, for example) of the volume in response to ambient stress. Note that the work done by external fields does not contribute to the local heating. (See Section 9.10.2 for a different aspect of this point.)

Although these derivations are based on the Boltzmann equation, with all its limitations to low densities and short-range interactions, the resulting hydrodynamic equations are quite generally valid, since they represent basic conservation laws of physics. Calculation of the transport coefficients, which relate the fluxes to the forcings (thermal and electrical conductivity, diffusivity, *etc.*), requires that an appropriate nonequilibrium Boltzmann function be used, and the expressions for these coefficients are dependent upon the actual Boltzmann function used in the calculations. It is in the derivation of correct phenomenological and constitutive relations, and not in the derivation of the conservation laws, that the limitations of the Boltzmann equation become apparent.

9.3.2 The Relaxation-Time Approximation. Boltzmann developed methods for treating the collision term by following in detail the conservation of momentum and energy, but I regard this procedure as so much a part of kinetic theory, and so well developed there that I shall be content to supply references (Cohen [1962]; Liboff [1990b]; McLennan [1989]). The treatment used in this section is based on the result of Boltzmann's calculations, that from any initial distribution of points in μ-space, the function f evolves to a time-independent equilibrium state. Since this is the basic concept required for extracting useful physics from the Boltzmann equation, the simplest way to represent this concept is to assume that in the absence of external fields or spatial inhomogeneities, the Boltzmann equation can be written

$$\frac{\partial f}{\partial t} = -\frac{f(t) - f_0}{\tau}, \tag{9.3.23}$$

where τ is called the *relaxation time*, f_0 is the equilibrium Boltzmann function, and the collision term is caricatured by the right-hand side. Since f_0 is time-independent, the f on the left-hand side may be replaced by $f - f_0$, to give a solution

$$f(t) - f_0 = (f(0) - f_0)\, e^{-t/\tau}, \tag{9.3.24}$$

which yields as $f(t)$ a function that decays monotonously to the equilibrium function f_0. This form of the collision term can now be used in the calculation of more-interesting phenomena.

The relaxation-time version of the Boltzmann equation to be used here is

$$\left(\frac{\partial}{\partial t} + \mathbf{v}\cdot\nabla_r + \mathbf{a}\cdot\nabla_v\right)f = -\left(\frac{f - f_0}{\tau}\right). \tag{9.3.25}$$

For steady-state transport problems, f is not a function of time. The required Boltzmann function is approximated by solving Eq. (9.3.25) for the f on the right-hand side, giving

$$f(\mathbf{r}, \mathbf{v}) = f_0(\mathbf{r}, \mathbf{v}) - \tau \left(\mathbf{v} \cdot \nabla_r + \mathbf{a} \cdot \nabla_v \right) f(\mathbf{r}, \mathbf{v}), \tag{9.3.26}$$

which is seen to be a solution for f in terms of itself. The usual procedure is to consider near-equilibrium steady states, for which, on the right-hand side, $f = f_0$ is a suitable first approximation. The process can be iterated, if deemed useful, to obtain higher approximations. As an illustration, consider a steady-state electron gas subject to a temperature gradient and to an electric field, both in the y direction, in accord with that already examined in the previous section in terms of Onsager theory. In this case, Eq. (9.3.26) becomes

$$f = f_0 - \tau \left(v_y \frac{\partial f_0}{\partial y} - \frac{eE_y}{m} \frac{\partial f_0}{\partial v_y} \right). \tag{9.3.27}$$

The initial Boltzmann function can be taken as locally Maxwellian, for a dilute electron gas (or for a dilute gas of monatomic ions, with $-e$ replaced by the appropriate charge), or as an ideal fermion gas for a metallic conductor. These are the usual choices for a gas of noninteracting particles. For the Maxwell gas, f_0 is found from Eq. (3.3.39) to be

$$f_0 = n(y) \left(\frac{m}{2\pi kT(y)} \right)^{3/2} e^{-mv^2/2kT(y)}, \tag{9.3.28}$$

where $n(y)$ is the local number density. Unless otherwise stated, the number density is taken to be independent of position, and written simply n.

The integrals needed for calculation of the fluxes for a Maxwellian gas can be found from two basic integrals:

$$F_1(\alpha) = \int_{-\infty}^{\infty} dx \, e^{-\alpha x^2} = \sqrt{\pi/\alpha}, \tag{9.3.29}$$

and

$$F_2(\alpha) = \int_{-\infty}^{\infty} dx \, x e^{-\alpha x^2} = 0. \tag{9.3.30}$$

Differentiation r times of both expressions for $F_1(\alpha)$ with respect to α leads to

$$(-)^r \int_{-\infty}^{\infty} dx \, x^{2r} e^{-\alpha x^2} = (-)^r \pi^{1/2} \frac{d^r}{d\alpha^r} \alpha^{-1/2}, \tag{9.3.31}$$

and the same procedure applied to $F_2(\alpha)$, or the mere observation that x to an odd power is an odd function, leads to zero for all odd powers of x in the integrand. With these tools in hand, and the assumption that the relaxation time τ is independent of $\mathbf{v}$, the particle flux and kinetic-energy flux are found to be

$$J_n = \int_{-\infty}^{\infty} d^3v \, v_y f = -\frac{n\tau}{m} \left[k\frac{dT}{dy} + eE_y \right], \tag{9.3.32}$$

and

$$J_k = \int_{-\infty}^{\infty} d^3v \, (\tfrac{1}{2}mv^2)v_y f = -\frac{nkT\tau}{m} \left[5k\frac{dT}{dy} + \frac{5}{2}eE_y \right]. \tag{9.3.33}$$

From these results, it is usual in kinetic theory to consider J_k as the heat flux, deliberately disregarding any contribution from a background lattice or gas, and to proceed directly with the calculation of the electrical and thermal conductivities. The electrical conductivity is normally defined for systems with zero temperature gradient. The electric current density is found from the particle current density to be

$$J_e = -eJ_n = \frac{ne^2\tau}{m} E_y = \sigma E_y, \tag{9.3.34}$$

which gives for the electrical conductivity σ,

$$\sigma = ne^2\tau/m. \tag{9.3.35}$$

The thermal conductivity is defined for the condition that $J_n = 0$, which establishes that $eE_y = -kdT/dy$. The assumption that $J_k = J_q$ then gives

$$J_k = -\frac{5}{2}\frac{nk^2T\tau}{m}\frac{dT}{dy} = -\kappa\frac{dT}{dy}, \tag{9.3.36}$$

so that the thermal conductivity κ is given as

$$\kappa = 5nk^2T\tau/2m. \tag{9.3.37}$$

These results are correct, despite the fact that J_k is not the proper expression for the heat flux, as defined by Eq. (9.2.44), but the reason will become evident. The motivation for using the proper J_q is that the expected Onsager symmetry $L_{qn} = L_{nq}$ is not found if J_k is taken to be the expression for J_q.

According to the discussion following Eq. (9.2.41), the internal energy per particle is given by $u' = u - e\phi$, where u is the normal internal energy of the ideal gas, and its flux is J_k. The electric potential energy, which is usually not considered since it remains constant, must be accounted for in flow processes. Therefore the internal energy flux must be written

$$J_u = J_k - e\phi J_n, \tag{9.3.38}$$

since every particle carries the potential energy $-e\phi$. The heat flux, defined in Eq. (9.2.44), is

$$J_q = J_u - \mu J_n = J_k - \mu_c J_n, \tag{9.3.39}$$

since $\mu = \mu_c - e\phi$, where μ_c is the ordinary chemical potential. In the absence of particle current, as is usual in the calculation of the thermal conductivity, only J_k contributes to the thermal current, and the result of Eq. (9.3.37) is seen to be correct.

The Onsager coefficients may be obtained from the current densities J_n of Eq. (9.3.32) and the proper J_q, which has become

$$J_q = -\frac{n\tau}{m}\left[\left(5k^2T - k\mu_c\right)\frac{dT}{dy} + \left(\frac{5}{2}kT - \mu_c\right)eE_y\right]. \tag{9.3.40}$$

(It is an easy exercise to show formally that this expression for the heat flux leads to the result of Eq. (9.3.37) for the thermal conductivity.)

The Onsager expressions for these currents, Eq. (9.2.46), must be rewritten in terms of the temperature gradient and the electric field for comparison. The chemical potential for a monatomic ideal gas is found from the partition function to be

$$\mu_c = kT \ln n\lambda^3, \tag{9.3.41}$$

where λ is the thermal wave length. Note that

$$\frac{d\mu_c}{dy} = \left(\frac{\partial \mu_c}{\partial T}\right)_n \frac{dT}{dy} = \left(\frac{\mu_c}{T} - \frac{3k}{2}\right)\frac{dT}{dy}. \tag{9.3.42}$$

With this help, the fluxes can be written in two ways:

$$\begin{aligned}
J_q &= -\frac{n\tau}{m}\left[\left(5k^2T - k\mu_c\right)\frac{dT}{dy} + \left(\tfrac{5}{2}\,kT - \mu_c\right)eE_y\right] \\
&= -\left[\frac{L_{qq}}{T^2} + \frac{L_{qn}}{T}\left(\frac{\mu_c}{T} - \frac{3k}{2}\right)\right]\frac{dT}{dy} - \frac{L_{qn}}{T}eE_y,
\end{aligned} \tag{9.3.43}$$

and

$$\begin{aligned}
J_n &= -\frac{n\tau}{m}\left[k\frac{dT}{dy} + eE_y\right] \\
&= -\left[\frac{L_{nq}}{T^2} + \frac{L_{nn}}{T^2}\left(\mu_c - \tfrac{3}{2}\,kT\right)\right]\frac{dT}{dy} - \frac{L_{nn}}{T}eE_y.
\end{aligned} \tag{9.3.44}$$

The identification of coefficients in the two pairs of equations gives as the Onsager coefficients

$$\begin{aligned}
L_{nn} &= nT\tau/m, \\
L_{qn} = L_{nq} &= \frac{nT\tau}{m}\left(\tfrac{5}{2}\,kT - \mu_c\right), \\
L_{qq} &= \frac{nT\tau}{m}\left[\tfrac{35}{4}\,(kT)^2 - 5kT\mu_c + \mu_c^2\right].
\end{aligned} \tag{9.3.45}$$

The elementary calculation based on the simplest approximation of a steady-state nonequilibrium solution to the Boltzmann equation leads to physically acceptable expressions for the transport coefficients σ and κ; it gives direct solutions for the values of all four Onsager coefficients; and the required symmetry condition is manifestly satisfied. The one adjustable parameter, τ, and the argument that the mass m could be an effective mass m^*, allow for accurate fitting of the theoretical expressions for σ and κ to measured values, but the ratio $\kappa/\sigma = 5k^2T/2e^2$ offers no opportunity for manipulation. It says that universally the ratio of thermal to electrical conductivity is independent of any property of the material, depending only on T and the physical constants e and k. This is certainly in the spirit of the Wiedemann-Franz law, proposed on the basis of observation that the best electrical conductors are also the best thermal conductors. But since this result would seem to depend upon the assumption that the conducting electron gas obeys Maxwell-Boltzmann statistics, a recalculation is in order based on the statistics for noninteracting fermions.

The equilibrium Boltzmann function for noninteracting electrons that obey fermion statistics is found from Eq. (4.5.1) to be

$$f_0(\mathbf{v}) = 2\left(\frac{m}{h}\right)^3 \frac{1}{e^{\beta(\epsilon - \mu_c)} + 1}, \tag{9.3.46}$$

where $\epsilon = mv^2/2$, and the appropriate conversions have been made to velocity space. The Fermi energy has been written as μ_c, in agreement with the notation used in Eq. (9.3.39). The perturbed Boltzmann function is written as in Eq. (9.3.27), with the required derivatives of f_0 obtained from Eq. (9.3.46). Note that

$$\frac{\partial f_0}{\partial y} = \frac{\partial f_0}{\partial T}\frac{dT}{dy} + \frac{\partial f_0}{\partial \mu_c}\frac{d\mu_c}{dy}, \tag{9.3.47}$$

which can be rewritten by use of the identities:

$$\frac{\partial f_0}{\partial \mu_c} = -\frac{\partial f_0}{\partial \epsilon}, \quad \text{and} \quad \frac{\partial f_0}{\partial T} = -\frac{\epsilon - \mu_c}{T}\frac{\partial f_0}{\partial \epsilon}. \tag{9.3.48}$$

With these substitutions, Eq. (9.3.27) becomes

$$\begin{aligned}
f &= f_0 + \tau v_y \left[\frac{\epsilon - \mu_c}{T}\frac{dT}{dy} + eE_y + \frac{d\mu_c}{dy}\right]\frac{\partial f_0}{\partial \epsilon} \\
&= f_0 + \tau v_y \left[\frac{\epsilon - \mu_c}{T}\frac{dT}{dy} + \frac{d\mu}{dy}\right]\frac{\partial f_0}{\partial \epsilon},
\end{aligned} \tag{9.3.49}$$

where the relation $\mu = \mu_c - e\phi$ has been used. The particle current density can now be found, as

$$\begin{aligned}
J_n &= \int d^3v\, v_y f = \tau \int d^3v\, v_y^2 \left[\frac{\epsilon - \mu_c}{T}\frac{dT}{dy} + \frac{d\mu}{dy}\right]\frac{\partial f_0}{\partial \epsilon} \\
&= \frac{4\pi}{3}\int_0^\infty dv\, v^4 \left[\frac{\epsilon - \mu_c}{T}\frac{dT}{dy} + \frac{d\mu}{dy}\right]\frac{\partial f_0}{\partial \epsilon},
\end{aligned} \tag{9.3.50}$$

where the the v_y axis has been chosen as the polar axis, and the angle variables have been integrated out. The remaining term in v is replaced by means of $\epsilon = mv^2/2$ to give $v^4 dv = (2\sqrt{2}\epsilon^{3/2}/m^{5/2})d\epsilon$, and Eq. (9.3.50) is rewritten as

$$\begin{aligned}
J_n &= \frac{8\sqrt{2}\pi\tau}{3m^{5/2}}\int_0^\infty d\epsilon\, \epsilon^{3/2}\left[\frac{\epsilon - \mu_c}{T}\frac{dT}{dy} + \frac{d\mu}{dy}\right]\frac{\partial f_0}{\partial \epsilon} \\
&= -\frac{8\sqrt{2}\pi\tau}{3m^{5/2}}\int_0^\infty d\epsilon\left[\left(\frac{5}{2}\epsilon^{3/2} - \frac{3}{2}\mu_c\epsilon^{1/2}\right)\frac{1}{T}\frac{dT}{dy} + \frac{3}{2}\epsilon^{1/2}\frac{d\mu}{dy}\right]f_0 \\
&= -\frac{16\sqrt{2}\pi\tau m^{3/2}}{3mh^3}\left\{\mu_0^{5/2}\left(1 + \frac{5}{8}\left(\frac{\pi kT}{\mu_0}\right)^2\right) - \mu_0\left(1 - \frac{1}{12}\left(\frac{\pi kT}{\mu_0}\right)^2\right) \times \right. \\
&\quad \left. \mu_0^{3/2}\left(1 + \frac{1}{8}\left(\frac{\pi kT}{\mu_0}\right)^2\right)\frac{1}{T}\frac{dT}{dy} + \mu_0^{3/2}\left(1 + \frac{1}{8}\left(\frac{\pi kT}{\mu_0}\right)^2\right)\frac{d\mu}{dy}\right\} \\
&= -\frac{n\tau}{m}\left[\frac{7}{12}\frac{\pi^2 k^2 T}{\mu_0}\frac{dT}{dy} + \left(1 + \frac{1}{8}\left(\frac{\pi kT}{\mu_0}\right)^2\right)\frac{d\mu}{dy}\right],
\end{aligned} \tag{9.3.51}$$

where the first step is an integration by parts, in the second it is necessary to use second-order approximations of Fermi integrals from Appendix B and the expansion to second order of μ_c in terms of μ_0, Eq. (4.5.5), and finally the factor $\mu_0^{3/2}$ is replaced by its equivalent from Eq. (4.5.4). In order to obtain J_n in the same form as Eq. (9.3.32), replace μ by $\mu_c - e\phi$, and find $d\mu_c/dT$ from Eq. (4.5.5), to obtain

$$J_n = -\frac{n\tau}{m}\left[\frac{5\pi^2 k^2 T}{12\mu_0}\frac{dT}{dy} + \left(1 + \frac{1}{8}\left(\frac{\pi kT}{\mu_0}\right)^2\right)eE_y\right]. \tag{9.3.52}$$

(The seemingly negligible second-order term is retained in the coefficient of E_0 because it is needed, in a sense, for J_q.) An equivalent calculation for J_k gives

$$\begin{aligned}
J_k &= -\frac{n\tau\mu_0}{m}\left[\frac{11\pi^2 k^2 T}{12\mu_0}\frac{dT}{dy} + \left(1 + \frac{5}{8}\left(\frac{\pi kT}{\mu_0}\right)^2\right)\frac{d\mu}{dy}\right] \\
&= -\frac{n\tau\mu_0}{m}\left[\frac{3\pi^2 k^2 T}{4\mu_0}\frac{dT}{dy} + \left(1 + \frac{5}{8}\left(\frac{\pi kT}{\mu_0}\right)^2\right)eE_y\right].
\end{aligned} \tag{9.3.53}$$

The heat flux $J_q = J_k - \mu_c J_n$ is now seen to be

$$J_q = -\frac{n\tau}{m}\left[\frac{\pi^2 k^2 T}{3}\frac{dT}{dy} + \frac{7\pi^2 k^2 T^2}{12\mu_0}eE_y\right], \tag{9.3.54}$$

and the contribution of the second term, which came from the difference in the second-order corrections to J_k and $\mu_c J_n$, is nearly always negligible. The electric conductivity is

$$\sigma = ne^2\tau/m, \tag{9.3.55}$$

as in Eq. (9.3.35), but with the difference that the velocity dependence of τ is not in question, since the Fermi integrands contribute significantly only quite near the Fermi energy. For this reason, τ was not included as a factor of the integrands.

The thermal conductivity κ is seen to be

$$\kappa = n\tau\pi^2 k^2 T/3m, \tag{9.3.56}$$

and the Wiedemann-Franz ratio is

$$\frac{\kappa}{\sigma} = \frac{\pi^2 k^2 T}{3e^2}, \tag{9.3.57}$$

independent of the material. The constant numerical value of $\kappa/\sigma T$, and its reasonably good numerical agreement with experiment, for many metals, is a remarkable result. It is sometimes argued that such a result is inevitable, by dimensional analysis, since the only possible combination of constants that survives dimensionally is $(k/e)^2$. But such an argument is invalid, because the quantity $(k/e)^2(kT/\mu_0)^n$ has the same dimensions for any n. The absence of the Fermi energy in the ratio is a consequence of the coincidence that although the heat capacity per electron is much smaller than that of a Boltzmann particle, the electrons at the Fermi level move so much faster than Boltzmann particles at the same temperature that they more than compensate for carrying such small amounts of thermal energy.

9.3.3 More Transport Coefficients. Expressions for three additional transport coefficients are derived in this subsection, and others are mentioned. The first, called *mobility*, is in general the ratio of the mean particle velocity to the applied forcing. In the present context, it is the ratio of the charged-particle velocity to the applied electric field. It is usually denoted by μ, but since that letter is already overworked, I use b, from its German name, *Beweglichkeit*. The defining equation is

$$w = \langle v \rangle = bE, \tag{9.3.58}$$

where more generally w and E can be vectors, b becomes a tensor, and bE becomes a dot product. The existence of such an equation implies that w is a limiting speed, dependent for its limitation on collisions of the particle with the medium through which it moves.

The electric conductivity relates the current density to the field by

$$J_e = \sigma E, \tag{9.3.59}$$

and the current density may be written

$$J_e = n q_e w, \tag{9.3.60}$$

where q_e is the particle's electric charge, and n is the number of such particles per unit volume. Use of Eq. (9.3.58) and Eq. (9.3.59) leads to

$$n q_e b E = \sigma E, \tag{9.3.61}$$

so that the mobility is found to be

$$b = \sigma / n q_e. \tag{9.3.62}$$

The conductivity, rewritten here for a general q_e, is given by Eq. (9.3.35) as $\sigma = n q_e^2 \tau / m$, so that the mobility becomes

$$b = q_e \tau / m. \tag{9.3.63}$$

It takes the sign of the charge, thereby giving the direction of the so-called *drift velocity* with respect to the electric field.

The second additional transport coefficient is the *diffusivity*, D, that relates a particle current density to a concentration gradient, or

$$\mathbf{J}_n = -D \nabla n. \tag{9.3.64}$$

This expression, known as Fick's Law, is an analog of Ohm's law, rewritten from Eq. (9.3.59) as $\mathbf{J}_e = -\sigma \nabla \phi$. For a Maxwell-Boltzmann ideal gas with a concentration gradient, the initial Boltzmann function is

$$f_0(\mathbf{r}, \mathbf{v}) = n(\mathbf{r}) \left(\frac{m}{2\pi kT} \right)^{3/2} e^{mv^2/2kT}, \tag{9.3.65}$$

and the perturbed function f, from Eq. (9.3.26), is

$$f(\mathbf{r}, \mathbf{v}) = f_0 \left[1 - \tau \mathbf{v} \cdot \nabla_r \ln n(\mathbf{r}) \right], \tag{9.3.66}$$

in the absence of applied forces or temperature gradients. The particle current density is thus

$$\mathbf{J}_n = \int d^3v\, \mathbf{v}f = -\tau \int d^3v\, f_0 \mathbf{v}\mathbf{v} \cdot \nabla_r \ln n = -\frac{kT\tau}{m}\nabla_r n. \qquad (9.3.67)$$

Comparison with Eq. (9.3.64) provides an expression for the diffusivity:

$$D = kT\tau/m. \qquad (9.3.68)$$

The ratio of diffusivity to mobility is seen to be

$$D/b = kT/e, \qquad (9.3.69)$$

a result usually known as the Einstein relation, because he rediscovered it in the course of his work on Brownian motion, discussed in Chapter 10. It was first presented by Walther Nernst during studies of charge transport by ions in solution. (Einstein [1905]; Fürth [1956]; Nernst [1884]) It is an example of the fluctuation-dissipation theorem, discussed in Chapter 10.

For a system of tagged particles diffusing through a medium of equivalent but nontagged particles, the continuity equation for the tagged particles is

$$\frac{\partial n}{\partial t} = -\nabla \cdot \mathbf{J}_n = D\nabla^2 n. \qquad (9.3.70)$$

If all N_0 of the tagged particles are considered to start at the origin at time $t = 0$, so that $n(0,t) = N_0\delta(\mathbf{r})$, then the solution to Eq. (9.3.70) is well known to be

$$n(r,t) = \frac{N_0}{(4\pi Dt)^{3/2}}e^{-r^2/4Dt}. \qquad (9.3.71)$$

This solution represents an irreversible spreading of the initial concentration of particles to an ever-more-nearly uniform distribution. The probability per unit volume that one of the particles is at a distance r from the origin at time t is just $p(r,t) = n(r,t)/N_0$, and the mean displacement from the origin of a particle at time t is

$$\langle r(t)\rangle = \int_0^\infty d^3r\, r p(r,t) = 4\sqrt{Dt/\pi}. \qquad (9.3.72)$$

The more-commonly-given result is the mean squared displacement,

$$\langle r^2(t)\rangle = \int_0^\infty d^3r\, r^2 p(r,t) = 6Dt. \qquad (9.3.73)$$

A cartesian component of the mean displacement is, for example, $\langle y(t)\rangle = 0$, but the corresponding mean squared displacement is $\langle y^2(t)\rangle = 1/3\,\langle r^2(t)\rangle$. There is thus no satisfactory way to associate a velocity with the average motion of a diffusing particle.

The viscosity, ζ, is defined from the stress tensor of Eq. (9.3.17) by the relation

$$P_{zy} = -\zeta\frac{\partial w_y}{\partial z}, \qquad (9.3.74)$$

where the stress P_{zy} is the net rate of transport per unit area of y-momentum in the $+z$ direction. As a consequence of a z dependence of the fluid velocity component w_y, particles that move from a region just below a particular plane perpendicular to the z axis to a region just above that plane because of their peculiar velocities do not transport y-momentum upward at the same rate as do the particles that cross the plane in the opposite direction. Suppose, for example, that the only nonzero component of $\mathbf{w}$ is w_y, given by

$$w_y = Az. \tag{9.3.75}$$

At a particular z level, w_y is fixed, but the peculiar velocity c_y of any particle crossing the plane is correlated with the component c_z of that same particle, because particles with $c_z > 0$ come from below the plane, where $w_y = A(z - \delta z)$, and those with $c_z < 0$ come from above the plane, where $w_y = A(z + \delta z)$. Thus P_{yz} represents a pressure-like force per unit area on the plane at z, but unlike the scalar pressure, the force is tangent to the plane.

The relaxation-time approximation for the Boltzmann function, in the absence of gradients in temperature and particle density, and with $\mathbf{a} = 0$, is

$$
\begin{aligned}
f &= f_0 - \tau v_z \frac{\partial f_0}{\partial z} = f_0 - \tau v_z \frac{\partial f_0}{\partial w_y} \frac{\partial w_y}{\partial z} \\
&= f_0 \left(1 - \tau A v_z m c_y / kT \right),
\end{aligned}
\tag{9.3.76}
$$

where Eq. (9.3.75) has been used. The tensor component P_{yz} can be evaluated as

$$
\begin{aligned}
P_{yz} &= \int d^3v\, m c_y c_z f = P_{zy} = -\frac{\tau A m^2}{kT} \int d^3v\, c_y^2 c_z v_z f_0 \\
&= -\frac{\tau A m^2 n}{kT} \left(\frac{m}{2\pi kT} \right)^{3/2} \sqrt{\frac{2\pi kT}{m}} \left(\frac{\pi}{4} \right) \left[\left(\frac{2\pi kT}{m} \right)^{3/2} \right]^2 \\
&= -\tau A n kT = -\tau n kT \frac{\partial w_y}{\partial z}.
\end{aligned}
\tag{9.3.77}
$$

The coefficient of viscosity is thus found to be

$$\zeta = nkT\tau, \tag{9.3.78}$$

in the relaxation-time approximation with constant τ. The thermal conductivity from Eq. (9.3.37) can be written

$$\kappa = \frac{5}{2} \frac{nk^2 T\tau}{m} = nc_p kT\tau, \tag{9.3.79}$$

where the specific heat is $c_p = C_p/N_0 m = 5k/2m$. The ratio of thermal conductivity to viscosity is thus given by

$$\kappa/\zeta = c_p, \tag{9.3.80}$$

a result that at first seems counterintuitive.

Other transport coefficients relate thermal, electric, and magnetic effects, such as those named for Peltier, Seebeck, Thompson, Nernst, Hall, Righi-Leduc, and Ettinghausen. Some of these, especially the ones that relate magnetic fields to other transport phenomena, are often found to be of practical importance in laboratory procedures. They can be understood conceptually by means of the relaxation-time approximation to the Boltzmann equation, and sometimes by more-elementary means. (Bridgman [1961]; Kramers and Wannier [1941]) They can also be examined by means of the Onsager relations. (Desloge [1966]) Detailed, quantitative treatments of these effects are to be found in books on condensed-matter physics. (Blakemore [1988]; Callaway [1974]; Kittel [1986])

9.4 THE HYDRODYNAMIC EQUATIONS

Although the general derivation of the hydrodynamic equations from the Boltzmann equation has led to Eqs. (9.3.15), (9.3.20) and (9.3.21), it is useful to examine these results less generally, first with $f = f_0$, the zeroth-order approximation, and then with the first-order relaxation-time approximation given by

$$f_1 = f_0 - \tau \left(\frac{\partial}{\partial t} + \mathbf{v} \cdot \nabla_r + \mathbf{a} \cdot \nabla_v \right) f_0, \tag{9.4.1}$$

where

$$f_0(\mathbf{r}, \mathbf{v}, t) = n(\mathbf{r}, t) \left(\frac{m}{2\pi k T(\mathbf{r}, t)} \right)^{3/2} \exp[-mc^2/2kT(\mathbf{r}, t)], \tag{9.4.2}$$

with $\mathbf{c} = \mathbf{v} - \mathbf{w}(\mathbf{r}, t)$, and functional dependencies as indicated. Thus the spatial and temporal dependencies of f are determined implicitly by the presumably observable velocity, temperature, and particle-density fields. The zero-order Boltzmann function is thus the analog of $n(\mathbf{r}, t)$ times a single-particle probability density determined by information-theoretic methods that maximize the entropy subject to existing constraints. This use of observable macroscopic quantities to parameterize the statistical function is fundamental to the development of equilibrium statistical mechanics as presented, and it provides a foundation for entropy-maximum-based nonequilibrium statistical mechanics as developed later in this chapter. The technique of expressing the Boltzmann function in terms of its dependencies on n, T, and $\mathbf{w}$ is known as the Hilbert-Enskog method. (Chapman and Cowling [1970]; Enskog [1917]; Hilbert [1912]; Hirschfelder, Curtis and Bird [1966])

The population density, momentum density, and kinetic-energy density are collisional invariants, as already noted. With the Boltzmann operator denoted by $\mathcal{B}$, where

$$\mathcal{B} = \frac{\partial}{\partial t} + \mathbf{v} \cdot \nabla_r + \mathbf{a} \cdot \nabla_v, \tag{9.4.3}$$

the quantities that lead to the hydrodynamic equations are obtained from

$$\int d^3 v \phi(\mathbf{v}) \mathcal{B} f = \frac{\partial \langle n\phi \rangle}{\partial t} + \nabla_r \cdot \langle n\phi\mathbf{v} \rangle - \langle n\mathbf{a} \cdot \nabla_v \phi \rangle = 0, \tag{9.4.4}$$

where, as before, ϕ is chosen to yield an equation for the required density, and $f = f_0$ or $f = f_1$. The derivatives needed in the zeroth-order calculation are:

$$\frac{\partial f_0}{\partial t} = \frac{\partial f_0}{\partial n}\frac{\partial n}{\partial t} + \frac{\partial f_0}{\partial T}\frac{\partial T}{\partial t} + \sum_i \frac{\partial f_0}{\partial w_i}\frac{\partial w_i}{\partial t},$$

$$\mathbf{v} \cdot \nabla_r f_0 = \mathbf{v} \cdot \left(\nabla_r n \frac{\partial f_0}{\partial n} + \nabla_r T \frac{\partial f_0}{\partial T} + \sum_i \nabla_r w_i \frac{\partial f_0}{\partial w_i} \right),$$

$$\mathbf{a} \cdot \nabla_v f_0 = -\frac{m}{kT}\mathbf{c} \cdot \mathbf{a} f_0. \tag{9.4.5}$$

The derivatives of f_0 appearing in these equations are:

$$\frac{\partial f_0}{\partial n} = \frac{f_0}{n}; \quad \frac{\partial f_0}{\partial w_i} = \frac{mc_i}{kT} f_0;$$

$$\frac{\partial f_0}{\partial T} = \frac{1}{T}\left(\frac{mc^2}{2kT} - \frac{3}{2}\right); \quad \text{and} \quad \frac{\partial f_0}{\partial v_i} = -\frac{mc_i}{kT} f_0. \tag{9.4.6}$$

With the help of these derivatives, the averages indicated in Eq. (9.4.4) can be evaluated directly. When $\phi = 1$, use of Eq. (9.3.12) in Eq. (9.4.4) leads to the continuity equation,

$$\frac{\partial n}{\partial t} + \nabla_r (n\mathbf{w}) = 0, \tag{9.4.7}$$

as before. When $\phi = m\mathbf{v}$, the momentum conservation equation is obtained. Elimination of redundant terms by means of Eq. (9.4.7), and use of the vector identity $\mathbf{w} \cdot \nabla_r w^2 = 2\mathbf{w} \cdot (\mathbf{w} \cdot \nabla_r)\mathbf{w}$, reduces the momentum equation, originally given by Eq. (9.3.20), to

$$mn\frac{D\mathbf{w}}{Dt} = -\nabla P + nm\langle \mathbf{a}\rangle, \tag{9.4.8}$$

where $P = nkT$, the scalar pressure. (Only the diagonal terms of the pressure tensor differ from zero when $f = f_0$.) Similarly, the energy equation that was written more generally as Eq. (9.3.21) can be reduced to

$$\frac{DT}{Dt} + \frac{2T}{3}\nabla_r \cdot \mathbf{w} = 0. \tag{9.4.9}$$

The set Eq. (9.4.7), Eq. (9.4.8), and Eq. (9.4.9) are usually called the Euler equations in hydrodynamics.

When rewritten after multiplication by $3nk/2$, the Eq. (9.4.9) becomes easier to interpret:

$$n\frac{D(3kT/2)}{Dt} + P\nabla_r \cdot \mathbf{w} = 0. \tag{9.4.10}$$

The first term accounts for the time rate of change of the translational kinetic energy per unit volume as the fluid moves along, and the second term is the negative of the rate of work done by pressure as a consequence of change in volume during flow. Thus there is no energy flow across the boundaries of a moving fluid element as a consequence of a temperature gradient, even though one may exist, and the flow must be adiabatic. This conclusion, for all its evident lack of generality, can be derived formally from Eq. (9.4.10) and Eq. (9.4.7). From the former,

$$\frac{3n}{2kT}\frac{DkT}{Dt} = -n\nabla_r \cdot \mathbf{w},$$

and from the latter,

$$\frac{Dn}{Dt} = -n\nabla_r \cdot \mathbf{w}.$$

Since the right-hand sides are identical, the left-hand sides must be equal, or

$$\frac{D}{Dt}\left(\frac{n}{(kT)^{3/2}}\right) = 0. \tag{9.4.11}$$

The constancy of $n/(kT)^{3/2}$ during the flow leads directly to the equation for an adiabat in P-T space for a monatomic gas:

$$PT^{-5/2} = \text{constant},\tag{9.4.12}$$

from which other equations for adiabats may be derived with the help of the ideal-gas law.

One further interesting consequence of the Euler equations is found from Eq. (9.4.8), rewritten for steady, irrotational flow. A different version of the vector identity used in its derivation is $2(\mathbf{w} \cdot \nabla_r)\mathbf{w} = \nabla_r w^2 - 2\mathbf{w} \times (\nabla_r \times \mathbf{w})$, where only the first term survives for irrotational flow. With the partial time derivative also zero, Eq. (9.4.8) becomes

$$\nabla_r \left(\tfrac{1}{2}w^2 + \phi^* \right) + \frac{1}{\rho}\nabla_r P = 0,\tag{9.4.13}$$

where ϕ^* is the potential energy per unit mass, and the density is $\rho = mn$. This is a version of the Bernoulli equation. For an isothermal fluid, Eq. (9.4.13) becomes

$$\nabla_r \left(\tfrac{1}{2}w^2 + \phi^* + \frac{kT}{m}\ln\rho \right) = 0,$$

or, upon integration,

$$\rho = \rho_0 \exp\left[-\beta \left(\tfrac{1}{2}mw^2 + m\phi^* \right) \right],\tag{9.4.14}$$

a slightly generalized expression of the barometric equation, where $\beta = 1/kT$.

The next approximation in the relaxation-time hierarchy repeats the calculations of the conservation laws, this time with f_1 of Eq. (9.3.58) as the Boltzmann function. Use of the derivatives already given in Eq. (9.4.5) and Eq. (9.4.6) leads directly to a working expression for f_1:

$$f_1 = f_0 - \tau f_0 \left[\frac{1}{n}Dn + \frac{1}{T}\left(\frac{mc^2}{2kT} - \frac{3}{2} \right) DT - \frac{m}{kT}\mathbf{c}\cdot(D\mathbf{w} + \mathbf{a}) \right],\tag{9.4.15}$$

where

$$D \equiv \frac{\partial}{\partial t} + \mathbf{v}\cdot\nabla_r.$$

Note that f_1 has terms that are proportional to the spatial variations of n, T, and $\mathbf{w}$. These terms lead to corrections of two of the conservation laws; the continuity equation remains as it was. The calculations are much like those for zeroth order, or the more-general ones that produced Eq. (9.3.20) and Eq. (9.3.21). For a true first-order approximation to the conservation laws, it is customary to regard the macroscopic fields as varying slowly in space and time, so derivatives of n, T, and $\mathbf{w}$ are considered to be of first order. The transport coefficients κ, ζ, and when applicable, any mobilities b, are also regarded as first order, and their derivatives can be dropped. The diagonal elements of the pressure tensor continue to be denoted by P, and the terms of the momentum equation that contain off-diagonal elements are written as viscous forces, involving spatial derivatives of the flow velocity. With these simplifications, the first-order conservation laws, called the *Navier-Stokes* equations, become:

$$\frac{\partial n}{\partial t} + \nabla_r \cdot (n\mathbf{w}) = 0.\tag{9.4.16}$$

$$mn\frac{D\mathbf{w}}{Dt} = -\nabla_r P - \zeta\left(\nabla_r^2\mathbf{w} + \tfrac{1}{3}\nabla_r(\nabla_r\cdot\mathbf{w})\right) + mn\mathbf{a}. \qquad (9.4.17)$$

$$n\frac{D(\tfrac{3}{2}kT)}{Dt} + P\nabla_r\mathbf{w} = \kappa\nabla_r^2 T. \qquad (9.4.18)$$

The $\mathbf{F} = m\mathbf{a}$ equation now contains viscosity, and the energy conservation equation includes heat transfer. These Navier-Stokes equations are evidently in better agreement with the behavior of real fluids than were the Euler equations, and the next order of approximation, the Burnett equations, which contain higher derivatives of the observable fields, are expected to offer further improvement (an expectation that does not always turn out to be verified). Treatments of the higher approximations, as well as calculations using the full integro-differential formulation of the Boltzmann equation, are to be found in more-specialized writings. (Burnett [1935a]; Burnett [1935b]; Chapman and Cowling [1970]; Cohen [1962]; Cohen and Thiring [1973]; Grad [1958]; McLennan [1989])

9.5 THE VLASOV EQUATIONS

In the study of fully ionized plasmas, the ions and electrons rarely interact in the billiard-ball-like collisions envisioned by Boltzmann. Instead, the long-range interactions may be attributed to the electric and magnetic fields that arise from the distribution of charge and current within the plasma. Each species of particle is described by a so-called collisionless Boltzmann equation, or Vlasov equation,

$$\left(\frac{\partial}{\partial t} + \mathbf{v}\cdot\nabla_r + \mathbf{a}_i\cdot\nabla_v\right)f_i(\mathbf{r},\mathbf{v},t) = 0, \qquad (9.5.1)$$

where the acceleration is

$$\mathbf{a}_i = (\mathbf{E} + \mathbf{v}\times\mathbf{B})\,q_i/m_i. \qquad (9.5.2)$$

The local number density of each species is

$$n_i(\mathbf{r},t) = \int d^3v\, f_i(\mathbf{r},\mathbf{v},t), \qquad (9.5.3)$$

and the electric charge density is

$$\rho_e = \sum_i q_i n_i(\mathbf{r},t), \qquad (9.5.4)$$

where q_i is the charge of the ith species. Similarly, the local electric current density of each species is

$$\mathbf{J}_i = q_i\int d^3v\,\mathbf{v} f_i(\mathbf{r},\mathbf{v},t) = q_i n_i\mathbf{w}_i, \qquad (9.5.5)$$

and the net local electric current density is

$$\mathbf{J} = \sum_i \mathbf{J}_i. \qquad (9.5.6)$$

The electric field in Eq. (9.5.2) is the superposition of any E arising from an external source and the Coulomb and induced fields attributable to the charges and current variations within the plasma. The magnetic field is the superposition of any **B** arising from an external source and the B-fields that originate within the plasma as a consequence of conduction and displacement currents.

When Boltzmann collisions can be neglected with sufficient accuracy, the Vlasov description of the plasma can be an excellent approximation. On first inspection, it seems to offer a coupled set of differential equations, from which a closed solution should be forthcoming for any specified initial state and boundary conditions. There is no *ad hoc* assumption of a stochastic nature, such as the collision-time approximation or the original Boltzmann probalistic treatment of collisions. Although the function f is itself closely related to a probability density, its evolution as expressed by the Vlasov equation seems deterministic.

A useful starting point for many plasma problems is that of charge neutrality everywhere, so that there is no internal Coulomb field. But even in this case, local fluctuations that cause departures from neutrality are sources of electric fields that affect the motion of the particles. The plasma medium turns out to be extremely responsive. Waves propagate from the site of a fluctuation, and energy can be fed into such waves as they travel, causing a variety of instabilities. The Vlasov plasma is nonlinear, and the equations for a nonlinear system can be quite intractable. Linearized approximations to these equations are often used in elementary treatments. They begin with Maxwellian f-functions, not always at the same ion and electron temperatures, and they lead to an impressive spectrum of well-behaved wave modes. But the full consequences of the nonlinear equations are far from being fully explored. I provide the interested reader with a number of references to this extensive and complicated field. (Balescu [1967]; Davidson [1972]; Montgomery and Tidman [1964]; Spitzer [1956]; Stix [1962])

9.6 THE MASTER EQUATION

The Boltzmann equation and its relatives treat nonequilibrium processes in terms of the statistical behavior of molecules. The statistical unit is the molecule, and the average behavior is that of a suitably large number of individual molecules. The Gibbs view, in contrast, is that an entire thermodynamic system is the statistical unit, and the average behavior is that of a suitably large number of thermodynamic systems, regarded by him as an ensemble. Information theory can accommodate either point of view, with the probability sets that yield the entropy and the macroscopic observables of the system chosen to pertain to either statistical entity. A different approach to the calculation of probability sets of either type is to be found in the *master equation*, proposed by Wolfgang Pauli in 1928. (Pauli [1928]) It was intended to provide a set of equations, soundly based in quantum mechanics, leading to a rigorous derivation of the macroscopic equations of thermodynamics from the microscopic equations of a system of particles. He offered a simple derivation for such systems, but it was based on an assumption of random phases that is not always justifiable. The Pauli master equation is nothing more than the equation for the time evolution of a probability set as a consequence of Markov processes. It may be written

$$\dot{p}(n,t) = -\sum_m \left[W_{nm} p(n,t) - W_{mn} p(m,t) \right], \tag{9.6.1}$$

where $p(n,t)$ is the probability that the system is in state n at time t, and $W_{mn}\delta t$ is the probability that a system in state m will make a transition to state n in the time interval δt.

The processes are Markovian because they depend only on the state at time t, and not on that at other times, past or future. The transition probabilities can be written as a matrix,

$$W = (W_{mn}). \tag{9.6.2}$$

The states ϕ_n of the system are assumed to be a complete set of eigenstates of a Hamiltonian $\mathcal{H}_0$, whereas the Hamiltonian for the system is

$$\mathcal{H} = \mathcal{H}_0 + \mathcal{H}_1, \tag{9.6.3}$$

and $\mathcal{H}_1$ is a small, perhaps incompletely known, term that may represent internal interactions not included in $\mathcal{H}_0$ or the inevitable influence of the outside world, even on what is supposed to be a closed system. The significance of $\mathcal{H}_1$ is that it is responsible for inducing transitions between the eigenstates of $\mathcal{H}_0$.

The basic problem that thwarts all attempts to derive the macroscopic, irreversible equations of thermodynamics solely from the microscopic ones of classical or quantum mechanics is that the equations of motion in Hamiltonian mechanics are time-reversible. This problem is examined in Chapter 1, where the dog-flea and wind-tree models, proposed by the Ehrenfests, are treated by equations that are equivalent to Eq. (9.6.1). The transition probabilities, always proportional in those examples to the number of entities capable of making the transition, are given explicitly, and the states are labelled by their occupation numbers. As noted there, the Ehrenfests were concerned about the irreversible exponential decay of the system to equilibrium, according to their analysis, despite the evident time-reversibility of the underlying microscopic processes. My treatment of these problems by means of the master equation leads to a set of time-dependent probabilities that represents our knowledge of the system as it evolves in time away from the state of last observation. It is clear and intuitively acceptable that as our knowledge of the system becomes less and less current, our description of it becomes, if anything, less sharp. For the two Ehrenfest models, the master-equation descriptions of the systems evolve to time-independent sets of probabilities, each of which provides a perfect description of the state of equilibrium in terms of our knowledge and expectations. The general solution to the master equation is also clearly not time reversible, since the equation is first order in time. (van Hove [1962]).

For the wind-tree model, the Ehrenfests calculated transition probabilities in terms of the geometry of the tree molecules and the speed of the wind particles. The result was that the probability of a wind molecule in one of its four possible single-particle states making a transition to either one of the two accessible states in time δt is $\alpha \delta t$, where the α was related by them to the physics of the model. Similarly, the Pauli equation uses transition probabilities derived from the physics of the problem under consideration. In both cases, the stochastic assertion that leads to irreversible behavior is that the transitions cannot be described deterministically. This is not to say that there is anything fuzzy or indefinite about any particular transition. Presumably each transition occurs during a suitably brief time interval. The indeterminism arises, not because the process itself is inherently blurry, but because we are unable to predict its time of occurrence. (As in radioactive decay, we do not have the ability to predict the precise time of a particular nuclear process, and it has always seemed odd to me that similar inabilities, for whatever reason, are often excluded from the theoretical formalism.) Thus it is the evolution of our state of knowledge, as contained in the time-dependent probability set, and not in the microscopic motions of the particles themselves, that leads to equilibrium. In a gas system, for example, we usually describe equilibrium in terms of a very few macroscopic parameters, such as T, P, and N. The expected values of these few parameters, as calculated from the time-evolved probability set described by the Pauli equation, are usually in excellent agreement with the observed ones, in cases where the Pauli equation can be solved. (The amusing experiments of Kohlrausch and Schrödinger, in an attempt to confirm the Ehrenfests' analysis of the dog-flea model, provide data in agreement with the master-equation treatment offered in Chapter 1.)

In a quantum system, the indices m and n are usually taken to designate particular eigenstates, of which there is assumed to be a complete orthonormal set. The eigenstates may be those of an unperturbed Hamiltonian, and the transition probabilities are calculated by standard methods in terms of the matrix elements of a perturbation. They are regarded as known. It is usual, although not always necessary, to describe the states as interconnected, which means that it is possible to go from any state to any other, although not necessarily in one step. This implies that the system does not comprise two or more noninteracting subsystems. The states are then said to form an irreducible Markov chain.

The expectation value of an observable that corresponds to an operator $\mathcal{A}$, when the system is in state n, is designated a_n, and also regarded as known. The statistical expectation value of $\mathcal{A}$ is $\langle a(t) \rangle$, and that of any function of the operator, $f(\mathcal{A})$ is

$$\langle f[a(t)] \rangle = \sum_n f[a_n]p(n,t). \tag{9.6.4}$$

Even when the master equation cannot be solved in closed form, it is possible to extract useful expressions for the expected time variation of the observables. In order to do this for the operator $\mathcal{A}$, first define the quantities

$$J_r(a_n) = \sum_m (a_m - a_n)^r W_{nm}, \tag{9.6.5}$$

known as transition moments, or jump moments, and presumed calculable. The time-derivative of $\langle a \rangle$ is then seen to be

$$\langle \dot{a} \rangle = \sum_n a_n \dot{p}(n,t) = \sum_n \sum_m [a_n W_{mn} p(m,t) - a_n W_{nm} p(n,t)]$$

$$= \sum_n J_1(a_n)p(n,t) = \langle J_1[a(t)] \rangle, \tag{9.6.6}$$

where the indices were interchanged in one summation. A similar calculation gives $\langle \dot{a}^2 \rangle$ as

$$\langle \dot{a}^2 \rangle = \sum_n [J_2(a_n) + 2a_n J_1(a_n)]p(n,t) = \langle J_2[a(t)] \rangle + 2\langle a(t)J_1[a(t)] \rangle. \tag{9.6.7}$$

Since the $p(n,t)$ are not known unless the master equation has been solved, these expressions and others of a similar nature seem useless. But note that the temporal behavior of one or more observables may be known well enough to provide input data for the determination of the probability set by means of information theory. Equations of the form

$$\langle X_r(t) \rangle = \sum_n X_{rn} p(n,t) \quad \text{for } t_1 \le t \le t_2, \tag{9.6.8}$$

that can be used as input allow the use of an information-theoretic procedure of entropy maximization, subject to the known constraints, to obtain the least-biased estimate of a set of time-dependent probabilities that are in agreement with macroscopic observations. These probabilities can then be used to calculate the expected time dependencies of other observables. The rationale is essentially the same as that used in thermostatics. We maximize the information entropy subject to what we know, even though the input information may be time dependent, because by so doing, we obtain the best possible set of probabilities without implying extraneous assumptions. It is not clear at this point that the information entropy, here denoted by S_I, is identifiable as the thermodynamic entropy, S_t, but the probabilities that are determined by S_I are nevertheless the best approximation to the true set, in the absence of a closed solution of the master equation, or of an approximate one that not only reproduces the macroscopic observations, but also uses at least some of the physics inherent in the master equation.

The entropy-maximization calculation proceeds as in Chapter 1, now with Lagrange multipliers that depend on time, $\lambda_r(t)$. The variation that yields the probability set for the specified time interval is

$$\delta \sum_n \left[p(n,t) \ln p(n,t) + \left(\lambda_0'(t) + \sum_r \lambda_r(t) X_{rn} p(n,t) \right) \right] = 0, \tag{9.6.9}$$

which leads to

$$p(n,t) = \exp \left[-\lambda_0(t) - \sum_r \lambda_r(t) X_{rn} \right], \quad \text{for } t_1 \le t \le t_2. \tag{9.6.10}$$

Since the probabilities sum to unity, the time-dependent partition function is seen to be

$$Z(\{\lambda_r(t)\}) = e^{\lambda_0(t)} = \sum_n \exp \left[-\sum_r \lambda_r(t) X_{rn} \right], \tag{9.6.11}$$

and the λ's are evaluated by means of the set of equations that relate the input data, here denoted by $\bar{X}_r$, and the expectation values $\langle X_r \rangle$, or

$$\bar{X}_q = \langle X_q \rangle = - \left(\frac{\partial \ln Z}{\partial \lambda_q} \right)_{\lambda_{r \ne q}}. \tag{9.6.12}$$

There are, of course, exactly as many equations as λ_r's.

Other procedures exist for finding approximate solutions to the master equation. It should be noted that the one given here maximizes $S_I(t)$ for every t in the interval of input data, and any other approximate solution based only on the same input data must either agree with this one or exhibit a smaller information entropy. In the latter case, the approximant implies more knowledge of the system than appears to be justified by the available data, and it is therefore unacceptable. If, however, the approximant makes better use of the relationships among the probabilities than does the entropy-maximum procedure, it could give an approximate entropy that is correctly diminished. The information entropy S_I^*, found from the time-dependent probability set obtained by exact solution of the master equation, is the best definition of thermodynamic entropy available for the system so described. It gives the minimum information entropy that may be found validly, at every time within the interval covered by the solution to the master equation, and it is compatible with the expectation value of every macroscopic thermodynamic variable that can be computed from the probability set. These observations may be expressed by the the relation

$$S_I(t) \geq S_I^*(t) = S(t),\tag{9.6.13}$$

where $S(t)$ is the correct expression for the time-dependent thermodynamic entropy, for thermal systems. (The Ehrenfest models of Chapter 1, for example, lead to time-dependent entropies that satisfy the inequality part of (9.6.13), but the set of fleas on two dogs and the system of specularly reflected, single-speed wind molecules are not usually regarded as thermal systems.) An information entropy approximation to the solution of the radioactive-decay problem of Chapter 1, given the expectation that $\langle n(t) \rangle = Ne^{-\alpha t}$, is found to be of the form

$$p(n,t) = \exp[-\lambda_0(t) - n\lambda_1(t)],\tag{9.6.14}$$

whereas the exact solution to Eq. (1.5.4) was found in Eq. (1.5.5) to be

$$p(n,t) = \binom{N}{n}\left(e^{-\alpha t}\right)^n \left(1 - e^{-\alpha t}\right)^{N-n}.\tag{9.6.15}$$

In Eq. (9.6.14), the multiplier $\lambda_0(t)$ is determined in terms of $\lambda_1(t)$ because $\sum p(n,t) = 1$ for all t. There is evidently no simple algebraic solution for $\lambda_1(t)$ that will give results that agree with the exact expression for $p(n,t)$, and we are confronted with the reality that it is possible to derive an information-theoretic maximum-entropy estimate of a probability set that is in serious disagreement with the correct set. (See Problem (12.), Chapter 1, for a similar kind of problem but with the equilibrium probability set from information theory in complete agreement with the exact solution of the master equation.) The expectation value of $n(t)$, obtained from Eq. (9.6.14), can be equated to the input value to give

$$\frac{\langle n(t) \rangle}{N} = e^{-\alpha t} = x\left[\frac{x^N(1-x) - (1-x^N)/N}{(1-x)(1-x^{N+1})}\right],\tag{9.6.16}$$

where $x = \exp[-\lambda_1(t)]$. In principle, Eq. (9.6.16) can be solved for $\lambda_1(t)$ in terms of αt, and the result used in Eq. (9.6.14) to give the required probabilities. But this procedure seems much too difficult to be of any practical value.

9.7 ENTROPY MAXIMIZATION, INFORMATION ENTROPY, AND EQUILIBRIUM

There is little disagreement with the long-standing result, independent of information theory, that a proper expression for the equilibrium entropy of a thermal system is given by

$$S = -k\sum_i p_i \ln p_i,\tag{9.7.1}$$

where p_i is the probability of the ith state of the system. (It is assumed that the system has been described as having a denumerable set of possible states, to each of which there can be assigned a probability p_i.) I think the principal controversy between those who support and those who reject the approach of information theory is in the method of determining the set of probabilities. At no point is the dispute more clearly expressed than in the context of the master equation. At equilibrium, Eq. (9.6.1) says

$$\dot{p}_n = 0 = -\sum_m \left[W_{nm}p_n - W_{mn}p_m \right], \tag{9.7.2}$$

where the equilibrium probabilities are assumed to be independent of time, and the transition probabilities are calculable from quantum mechanics. It follows that the probability set can be determined exactly, in principle, from the solution to Eq. (9.7.2), and that therefore the entropy must exist as an absolute quantity, dependent only upon our quantum-mechanical determination of the set of coefficients of the p's, the transition probabilities. It can have nothing to do with our knowledge of the state of the system, or indeed on any input from the laboratory. The transition probabilities depend upon matrix elements of interaction terms in the Hamiltonian. So how does a temperature arise?

9.7.1 Statistical Descriptions of Equilibrium.

Although Eq. (9.7.2) is a necessary and sufficient condition for equilibrium, several other relationships among the probabilities and transition probabilities are encountered in the literature, at least as sufficient conditions. These include the principles of microscopic reversibility and detailed balance, the ergodic hypothesis, and a less-familiar hypothesis relating the transition probabilities, here called the W-hypothesis. Most of these conditions are invoked for isolated systems, of the type called *microcanonical* in Chapter 3. They are described as follows, in accordance with a lucid paper by Thomsen [1953]. (1) *Detailed Balance* —(**D**) is the condition that transitions between any two states m and n take place at equilibrium at the same rate in both directions, or

$$(\textbf{D}) \quad W_{mn}p_m = W_{nm}p_n. \tag{9.7.3}$$

(2) *Microscopic Reversibility*—(**M**) is the condition that the transition probabilities between any two states are the same in both directions, or

$$(\textbf{M}) \quad W_{mn} = W_{nm}. \tag{9.7.4}$$

(3) *The Ergodic Hypothesis*—(**E**) is the assumption, rather than the derivation from information theory, that for a microcanonical system in equilibrium, all states that correspond to the energy of the system are equally probable, or

$$(\textbf{E}) \quad p_n = 1/\Omega, \tag{9.7.5}$$

where Ω is the number of microstates with the same energy as the macrostate. (4) *The W-Hypothesis*—(**W**) is the condition that the sum of all elements in any row of the transition matrix W of Eq. (9.6.2) is equal to the sum of all the elements in the corresponding column, or

$$(\textbf{W}) \quad \sum_n W_{mn} = \sum_n W_{nm}. \tag{9.7.6}$$

Two lemmas follow immediately from $(\mathbf{W})$, denoted by $(\mathbf{W}_1)$ and $(\mathbf{W}_2)$. For any factor that depends only on the index m, it is evident that if $(\mathbf{W})$ holds, then

$$(\mathbf{W}_1) \quad \sum_n f_m W_{mn} = \sum_n f_m W_{nm},$$

since the factor f_m could just as well be outside the summation. If $(\mathbf{W}_1)$ is then summed with respect to m, and the indices interchanged on the right-hand side, one obtains

$$(\mathbf{W}_2) \quad \sum_{mn} f_m W_{mn} = \sum_{mn} f_n W_{mn}.$$

Thomsen presents theorems showing the logical relationships among these four principles, and also with the so-called *Second Law of Thermodynamics*, $\dot{S} \geq 0$, which he denotes as $(\mathbf{S})$. Most of them are derived by inspection; others are left as exercises. He concludes that the principles $(\mathbf{W})$, $(\mathbf{E})$, and $(\mathbf{S})$ are equivalent, or

$$(\mathbf{W}) \rightleftharpoons (\mathbf{E}) \rightleftharpoons (\mathbf{S}), \tag{9.7.7}$$

and that microscopic reversibility is equivalent to the joint invocation of the ergodic hypothesis and detailed balance, or

$$(\mathbf{M}) \rightleftharpoons [(\mathbf{E}) + (\mathbf{D})]. \tag{9.7.8}$$

He then shows that neither detailed balance $(\mathbf{D})$ nor the W-hypothesis $(\mathbf{W})$ implies microscopic reversibility.

A standard result of quantum mechanics is that the transition probabilities are proportional to the squared absolute values of the matrix elements of the perturbation causing the transition, or, since the Hamiltonian is hermitian,

$$\begin{aligned} W_{mn} &\sim |\langle m|\mathcal{H}_1|n\rangle|^2 = \langle m|\mathcal{H}_1|n\rangle^* \langle m|\mathcal{H}_1|n\rangle \\ &= \langle n|\mathcal{H}_1|m\rangle \langle m|\mathcal{H}_1|n\rangle = \langle m|\mathcal{H}_1|n\rangle \langle n|\mathcal{H}_1|m\rangle, \end{aligned} \tag{9.7.9}$$

which leads to the conclusion that $W_{mn} = W_{nm}$, since the coefficients of the squared matrix elements are the same in both directions. Thus it follows that microscopic reversibility is always to be expected for systems that can be described by the master equation. (This statement ignores the presence of magnetic fields, Coriolis forces, and the other possible situations that restrict microscopic reversibility. It can be made to include them when necessary, but for simplicity the extra words are to be regarded as implicit. It should be noted that for terms in the Hamiltonian such as $\mathbf{B} \cdot \mathbf{S}$, where $\mathbf{S}$ is a spin vector, both factors change sign under time reversal, resulting in no net change.) If, as seems to be true for microcanonical systems that can be described by the master equation, microscopic reversibility restricts the possible transition matrix W to being symmetric, implying the ergodic hypothesis $\mathbf{E}$, then the equilibrium probability set is given by Eq. (9.7.5), in agreement with the microcanonical set derived by means of information theory in Chapter 3. Therefore, for these equilibrium systems, there cannot be a probability set obtained by direct solution of the master equation that differs from that derived by maximizing the information entropy, subject to the constraint that all allowed states of the system have the same energy.

This kind of analysis can be extended to cover any kind of system; it should be sufficient to outline the method as applied to a canonical system, based on the careful work of Richard T. Cox. (Cox [1950]; Cox [1955]) Regard the canonical system as one of two loosely coupled parts of a microcanonical system, the other of which is very much larger, and is considered to be a heat bath at temperature T. The two systems together are assumed to be describable by a master equation, and therefore satisfy **M**, with all consequent implications. The Hamiltonian for the combined system is

$$\mathcal{H} = \mathcal{H}_a + \mathcal{H}_b + \mathcal{H}_1, \tag{9.7.10}$$

where the index a refers to the canonical system, b to the heat bath, and $\mathcal{H}_1$ is the weak interaction term that induces transitions between the states of each system. The energies of system a are denoted by ϵ_i, and those of system b, the heat bath, by E_n. Since the complete system is microcanonical, every allowed transition must conserve energy, or

$$\epsilon_i + E_m = \epsilon_j + E_n, \tag{9.7.11}$$

for any allowed transition $(i \to j) + (m \to n)$. The transition probability for this combined process, written as $W_{im,jn}$, must satisfy

$$W_{im,jn} = W_{jn,im}, \tag{9.7.12}$$

because of microscopic reversibility. It is assumed that the temperature of the heat bath is not significantly altered by interaction with the canonical system. For the heat bath at equilibrium, write $U_b = \sum p'_n E_n$, $S_b = -k \sum p'_n \ln p'_n$, and $F_b = U_b - TS_b$, where the p'_n are the probabilities at equilibrium associated with the heat bath. It then follows that

$$\ln p'_n = (F_b - E_n)/kT. \tag{9.7.13}$$

To verify this, write

$$S_b = -k \sum_n p'_n \ln p'_n = (U_b - F_b)/T, \tag{9.7.14}$$

as a direct consequence of Eq. (9.7.13).

The canonical system is taken to be initially in some state that is not in equilibrium with the heat bath. Its probability set is assumed to evolve in accordance with a master equation, with the possibility that transitions are not required to conserve energy, because there is interaction with the heat bath. Thus it is no longer necessary either that $W_{ij} = 0$ if $\epsilon_i \neq \epsilon_j$, or that $W_{ij} = W_{ji}$. In order to obtain expressions for these transition probabilities, it is necessary to sum over the heat-bath states. A transition from state i to state j in the canonical system can be any one of the transitions satisfying Eq. (9.7.11). The transition probability $W_{ij}(mn)$ for that process is given by

$$W_{ij}(mn) = p'_m W_{im,jn}, \tag{9.7.15}$$

where p'_m is the probability that the heat bath is in its state m. The total transition probability W_{ij} is the sum over m and n of all possible transitions like that of Eq. (9.7.15), or

$$W_{ij} = \sum_m \sum_n p'_m W_{im,jn}. \tag{9.7.16}$$

For the inverse transition, a similar expression is

$$W_{ji} = \sum_m \sum_n p'_n W_{jn,im} = \sum_m \sum_n p'_n W_{im,jn}, \tag{9.7.17}$$

where Eq. (9.7.12) has been used. The probabilities for heat-bath states are already known from Eq. (9.7.13), allowing the transition probabilities to be written

$$W_{ij} = e^{\beta F_b} \sum_m \sum_n e^{-\beta E_m} W_{im,jn}, \quad \text{and}$$

$$W_{ji} = e^{\beta F_b} \sum_m \sum_n e^{-\beta E_n} W_{im,jn}. \tag{9.7.18}$$

As suggested by Eq. (9.7.11), multiplication of the first of these equations by $\exp(-\beta\epsilon_i)$ and the second by $\exp(-\beta\epsilon_j)$ gives

$$w_{ij} \equiv W_{ij} e^{-\beta\epsilon_i} = e^{\beta F_b} \sum_m \sum_n W_{im,jn} e^{-\beta(\epsilon_i + E_m)}, \quad \text{and}$$

$$w_{ji} \equiv W_{ji} e^{-\beta\epsilon_j} = e^{\beta F_b} \sum_m \sum_n W_{im,jn} e^{-\beta(\epsilon_j + E_n)}. \tag{9.7.19}$$

The right-hand sides of these two equations are equal, so that $w_{ij} = w_{ji}$. The transition probabilities for the canonical system thus become

$$W_{ij} = e^{\beta\epsilon_i} w_{ij}, \quad \text{and}$$

$$W_{ji} = e^{\beta\epsilon_j} w_{ji} = e^{\beta\epsilon_j} w_{ij}. \tag{9.7.20}$$

The master equation for the canonical system therefore becomes

$$\dot{p}_i(t) = -\sum_j \left[e^{\beta\epsilon_i} p_i(t) - e^{\beta\epsilon_j} p_j(t) \right] w_{ij}, \tag{9.7.21}$$

and when the system is in equilibrium with the heat bath, it is necessary that

$$\frac{p_i}{p_j} = \frac{e^{-\beta\epsilon_i}}{e^{-\beta\epsilon_j}}. \tag{9.7.22}$$

This result is equivalent to the statement that

$$p_i = e^{-\beta\epsilon_i}/Z, \tag{9.7.23}$$

where

$$Z = \sum_i e^{-\beta\epsilon_i}, \tag{9.7.24}$$

in complete agreement with the information-theoretic results of Chapter 3. By this and similar arguments, it follows that the equilibrium probability set derived by the maximum-information-entropy formalism is identical to the one that is obtained from the master equation, at least for the standard microcanonical, canonical, and grand canonical systems. Therefore any subjectivity in the entropy, the basis for most of the rejection of the information-theoretic formulation of equilibrium statistical thermophysics, must correspond exactly to the subjectivity in specifying the states of the system to which the probabilities of the set $\{p_n\}$ apply.

9.7.2 Master Equation Treatment of Magnetic Resonance. A two-level spin system, such as that of Section 3.5.2, provides a good example of the time-evolution of a probability set as described by the master equation. The spins are regarded as so isolated that each spin can be treated as a slightly perturbed canonical system. The Hamiltonian is taken to be

$$\mathcal{H} = -\mu B \sigma + \mathcal{H}_i + \mathcal{H}_\omega, \tag{9.7.25}$$

where the spin variable is $\sigma = \pm 1$, $\mathcal{H}_i$ is the weak interaction of the spin with other spins and with the components of the lattice, and $\mathcal{H}_\omega = f(\mu B_\omega \sigma)$ is the contribution of an applied microwave magnetic field of angular frequency ω. The two possible energies of the unperturbed states are

$$\epsilon_\sigma = -\sigma \mu B, \tag{9.7.26}$$

with corresponding probabilities $p_\sigma(t)$. The master equation for this system is

$$\dot{p}_+ = -p_+ \left(W_{+-} + w \right) + p_- \left(W_{-+} + w \right)$$
$$\dot{p}_- = -p_- \left(W_{-+} + w \right) + p_+ \left(W_{+-} + w \right), \tag{9.7.27}$$

where the W's are the transition probabilities that derive from $\mathcal{H}_i$, and w is the transition probability induced by $\mathcal{H}_\omega$. It is expected that w will be negligible unless the applied frequency is very near the resonance value, for which $\hbar \omega = |\epsilon_- - \epsilon_+| = 2\mu B$. The W's satisfy the relation of Eq. (9.7.20), or

$$\frac{W_{+-}}{W_{-+}} = \frac{\exp\left[-\beta\mu B\right]}{\exp\left[\beta\mu B\right]} = \exp\left[-2\beta\mu B\right]. \tag{9.7.28}$$

For notational convenience, write $p_\sigma = (1/2)(1 + \sigma\alpha)$, use $x = \beta\mu B$, and define $W = e^x W_{+-} = e^{-x} W_{-+}$. Then the difference of the two rates in Eq. (9.7.27) becomes

$$\dot{p}_+ - \dot{p}_- = \dot{\alpha} = 2W \sinh x - 2\alpha \left(w + W \cosh x \right). \tag{9.7.29}$$

The solution is

$$\alpha(t) = \alpha(0)e^{-2yt} + \frac{W}{y} \sinh x \left(1 - e^{-2yt} \right), \tag{9.7.30}$$

where $y = w + W \cosh x$. In general, the value of α decays in time to either an equilibrium or a steady-state value given by

$$\alpha \to \frac{W \sinh x}{w + W \cosh x}. \tag{9.7.31}$$

When $w = 0$, in the absence of microwave stirring, this reduces to $\alpha = \tanh x$, and the equilibrium probabilities are in agreement with Eq. (3.5.20). The decay of α to its equilibrium value shows how the small interaction part of the Hamiltonian contributes to the process of equilibration. The possibility of negative temperatures, as discussed in Chapter 3, requires that the interaction term be weak enough to preclude the rapid decay of α to its normal equilibrium value. The presence of the microwave field reduces the absolute value of α, thereby simulating an increased temperature. For the kind of system for which negative temperatures are observed, W is so weak that the microwave field effectively obliterates any alignment of the spins by the applied steady magnetic field; *i.e.* for $w \gg W \sinh x$, the spins behave as if the temperature were infinite.

To treat the same problem by means of the information-entropy formalism, suppose that $\langle \sigma(t) \rangle$ is known from experiment or some other theory, such as the one based on the master equation. Note that

$$\langle \sigma(t) \rangle = \sum_\sigma \sigma p_\sigma(t) = \alpha(t), \tag{9.7.32}$$

which is known from Eq. (9.7.30). Maximization of the information entropy subject to the constraint of Eq. (9.7.32) leads to the probability set

$$p_\sigma(t) = e^{\sigma \lambda_1(t)}/Z(t), \tag{9.7.33}$$

where $Z(t) = 2 \cosh \lambda_1(t)$, arrived at by the requirement that $\sum p_\sigma = 1$ at all times. From these probabilities, another expression is found for the expectation value of the spin, and it is set equal to that of Eq. (9.7.32) in order to evaluate λ_1:

$$\langle \sigma(t) \rangle = \alpha(t) = \tanh \lambda_1(t). \tag{9.7.34}$$

It is therefore possible to choose $\lambda_1(t)$ so that the information entropy produces a probability set that corresponds to any known function that is used as input, including that from the solution of the master equation. The only disagreement that could arise between the presumably correct solution of the master equation and that derived by means of information theory must have as its source an incorrect expression for $\alpha(t)$ in Eq. (9.7.34). In this case, the discrepancy could motivate further examination both of the transition probabilities in the master equation and of the input data for $\langle \sigma(t) \rangle$, if there is a difference.

Other examples of the use of master equations to study nonequilibrium properties of systems can be found in studies of shock waves and related phenomena. (Clarke and McChesney [1976]; Zel'dovich and Raizer [1966])

9.8 THE LIOUVILLE EQUATION

In Chapter 1, Section 6, the classical Liouville function ρ_N was described as the probability density in the $6N$-dimensional Γ-space introduced by Willard Gibbs. It was shown there that as a consequence of the conservation of ρ_N, its total time derivative is always zero, and ρ_N satisfies the Liouville theorem

$$\frac{d\rho_N}{dt} = \frac{\partial \rho_N}{\partial t} + \{\rho_N, \mathcal{H}\} = 0, \tag{9.8.1}$$

where the classical Poisson bracket is, because of the equations of motion,

$$\{\rho_N, \mathcal{H}\} = \sum_i \left(\dot{\mathbf{r}}_i \cdot \nabla_{ri} + \dot{\mathbf{p}}_i \cdot \nabla_{pi} \right) \rho_N. \tag{9.8.2}$$

The Gibbs entropy is then written

$$S = -k \int \rho_N \ln \left(h^{3N} \rho_N \right) d\Gamma_N. \tag{9.8.3}$$

The conventional treatment of the Gibbs entropy concludes that it is not equivalent to the thermodynamic entropy, at least for systems that are away from equilibrium, because of Eq. (9.8.1). If the Gibbs entropy does not evolve, it cannot represent the thermodynamic entropy, which is expected to increase with time to a maximum that is taken to represent equilibrium, subject to the existing constraints.

A similar kind of derivation is sometimes given for the von Neumann statistical operator of quantum systems, introduced in Section 8 of Chapter 1. The statistical operator ρ is often said to satisfy

$$\frac{d\rho}{dt} = \frac{\partial \rho}{\partial t} + \frac{1}{i\hbar} \left(\rho \mathcal{H} - \mathcal{H} \rho \right) = 0, \tag{9.8.4}$$

which is then written

$$\frac{\partial \rho}{\partial t} = \frac{1}{i\hbar} \left[\mathcal{H}, \rho \right], \tag{9.8.5}$$

since the quantum Poisson bracket is $1/i\hbar$ times the commutator. The entropy, given as

$$S = \mathrm{Tr} \left(\rho \ln \rho \right), \tag{9.8.6}$$

is thus claimed to be independent of time, and thus cannot represent the thermodynamic entropy away from equilibrium. (I make no such claim; see Section 9 for a further treatment.)

Despite the evident flaws found in this kind of treatment of them, the Liouville function and its relatives have been found useful in several contexts. (Liboff [1990b]) Reduced density functions, derived by integration over most of the coordinates and momenta, can be persuaded to evolve reasonably. The treatment of harmonic-oscillator chains in Chapter 1 shows how a reduced density function can develop in time to the classical equilibrium density, with the correct entropy. Further discussion of this point appears in a later subsection.

9.8.1 Classical Systems. The Liouville function for a closed classical system is so fettered by the Liouville theorem, Eq. (9.8.1), that it is of little use *per se* in the study of nonequilibrium systems. It is often stated, in the manner of Æsop's grapeless fox, that the Liouville function contains more information than is of interest, and it surely implies more than is usually known. Nevertheless, its properties can be examined in preparation for the next subsection.

The μ-space of the Boltzmann equation is six-dimensional, with a point specified by three position coordinates and three velocity components. The volume element is $d\mu = d^3r d^3v$. The Γ-space volume element is customarily expressed in terms of $3N$ coordinate components and a like number of momentum components, as $d\Gamma = \prod_i (m^3 d\mu_i)$. The normalization of ρ_N is given by

$$\int \rho_N d\Gamma = 1, \tag{9.8.7}$$

where the spatial integrations are over the volume of the container or the limits allowed by confining potential energies, and the cartesian momentum integrations are from $-\infty$ to $+\infty$. It is convenient to write $m^3 d\mu_i = dX_i$, where the subscript refers to the particle. The statistical average of any function $A(\{X_i\})$ is seen to be

$$\langle A \rangle = \int A(\{X_i\})\rho_N d\Gamma. \tag{9.8.8}$$

The time derivative of any function A is already known to be

$$\frac{dA}{dt} = \frac{\partial A}{\partial t} + \{A, \mathcal{H}\}. \tag{9.8.9}$$

Therefore, even if A and ρ_N are not explicitly time dependent, the time derivative $\langle \dot{A} \rangle$ is not necessarily zero. It is sufficient, when there is no explicit time dependence, for the Poisson bracket $\{A, \mathcal{H}\}$ to vanish, in which case A is said to be a constant of the motion.

9.8.2 The BBGKY Hierarchy. For many systems, the particles interact with short-range forces, such that most interactions involve only two molecules at a time. For such systems, it is useful to derive reduced probability densities from the full Liouville function, and to derive the so-called equations of motion for these reduced densities directly from the Liouville equation. It will turn out that the equation of motion for a single particle (corresponding to the Boltzmann equation) involves the two-particle probability density; the equation of motion for the two-particle density involves the three-particle density; and on to higher densities in the same way. This system of equations, with slight modification, is known as the BBGKY hierarchy (for Bogoliubov, Born, Green, Kirkwood, and Yvon, who proposed it). (Bogoliubov [1962]; Born and Green [1949]; Cohen [1962]; Kirkwood [1946]; Liboff [1990b]; McLennan [1989]; Yvon [1935])

A reduced probability density can be defined from the Liouville function as

$$\rho_s(X_1 \ldots X_s; t) = \int \rho_N(X_1 \ldots X_N; t) dX_{s+1} \ldots dX_N, \quad 0 < s \leq N. \tag{9.8.10}$$

The meaning of a reduced density is that $\rho(X_1 \ldots X_s; t) dX_1 \ldots dX_s$ is the probability that at time t, particle 1 is in the interval dX_1 around the point designated as X_1, particle 2 is similarly placed with respect to X_2, *etc.* It follows from Eq. (9.8.7) that every ρ_s is normalized, or

$$\int \rho_s(X_1 \ldots X_s; t) dX_1 \ldots dX_s = 1. \tag{9.8.11}$$

The slight modification that is usually found in BBGKY theory is that the reduced functions are normalized differently; they are defined as

$$F_s(X_1 \ldots X_s; t) = V^s \rho_s(X_1 \ldots X_s; t), \quad 0 < s < N, \tag{9.8.12}$$

where V is the volume of the container. The functions F_s are symmetric in the X_i, and the s particles can be any s of the original N. The average of any single-particle function, such as the momentum $\mathbf{p}_1$, can be found from F_1, since

$$\langle \mathbf{p}_1(t) \rangle = \int \mathbf{p}_1 \rho_N d\Gamma = \frac{1}{V} \int \mathbf{p}_1 F_1(X_1; t) dX_1. \tag{9.8.13}$$

The average over the system for a function that is the sum of single-particle functions is

$$\langle A_1(t) \rangle = \frac{N}{V} \int a_1 F_1 \, dX_1, \tag{9.8.14}$$

where

$$A_1 = \sum_{i=1}^{N} a_i, \tag{9.8.15}$$

and each a_i is dependent only on the coordinates and momenta of the ith particle. Similarly, a two-particle function such as the potential energy of interaction $\phi(|\mathbf{r}_i - \mathbf{r}_j|)$ can be averaged by means of F_2. Suppose such a symmetric two-body function is

$$A_2 = \frac{1}{2} \sum_{i \neq j} a_{ij}. \tag{9.8.16}$$

Then the time-dependent expectation value of A_2 is

$$\langle A_2(t) \rangle = \frac{N(N-1)}{2V^2} \int a_{12} F_2 \, dX_1 \, dX_2, \tag{9.8.17}$$

where the combinatorial factor counts the number of ways of finding one particle at position 1 and the other at 2. This procedure can be continued as needed.

The Liouville equation, Eq. (9.8.1), can be written

$$\frac{\partial \rho_N}{\partial t} = i\mathcal{L}_N \rho_N, \tag{9.8.18}$$

where the Liouville operator for classical systems is defined as

$$i\mathcal{L}_N = \sum_j \left(\frac{\partial \mathcal{H}_N}{\partial \mathbf{r}_j} \frac{\partial}{\partial \mathbf{p}_j} - \frac{\partial \mathcal{H}_N}{\partial \mathbf{p}_j} \frac{\partial}{\partial \mathbf{r}_j} \right) = \{ \mathcal{H}_N, \ \}. \tag{9.8.19}$$

(A notation for the Poisson bracket that differs from that of Eq. (9.8.2) has been used in this expression. The use of i in the definition of $\mathcal{L}$ is standard practice, intended to emphasize the similarities of the Liouville and Schrödinger equations. It is often useful in formal manipulations of the Liouville equation, such as writing the solution symbolically as

$$\rho(\{X_i\}; t) = e^{i\mathcal{L}_N t} \rho(\{X_i\}; 0). \tag{9.8.20}$$

In quantum mechanics, $\mathcal{L}$ becomes a *superoperator* (an operator that operates on operators to produce other operators) defined by the equation

$$\mathcal{L}\mathcal{A} = -\hbar^{-1}[\mathcal{H}, \mathcal{A}], \tag{9.8.21}$$

where $\mathcal{A}$ is any operator. (Some authors define $\mathcal{L}$ as the negative of the one used here.) This kind of result is examined again subsequently.)

The Hamiltonian for the system can be written

$$\mathcal{H}_N = \sum_{i=1}^{N} \frac{p_i^2}{2m} + \sum_i \phi_i + \frac{1}{2}\sum_{i \neq j} \phi_{ij}, \tag{9.8.22}$$

where $\phi_i = \phi_1(\mathbf{r}_i)$, the single-particle potential energy of the ith particle, and $\phi_{ij} = \phi_2(|\mathbf{r}_i - \mathbf{r}_j|)$, the two-particle interaction potential energy, assumed to be repulsive in this treatment. The Liouville operator thus becomes

$$\mathcal{L}_N = -i\sum_{i=1}^{N}\left[\left(\frac{\partial \phi_i}{\partial \mathbf{r}_i}\cdot\frac{\partial}{\partial \mathbf{p}_i} + \sum_{i<j}^{s}\Theta_{ij}\right) - \frac{\mathbf{p}_i}{m}\cdot\frac{\partial}{\partial \mathbf{r}_i}\right],$$
$$= \mathcal{L}_s + \mathcal{L}_{N-s} + \mathcal{L}'_s, \tag{9.8.23}$$

where Θ_{ij} is defined to be

$$\Theta_{ij} = \frac{\partial \phi_{ij}}{\partial \mathbf{r}_i}\cdot\frac{\partial}{\partial \mathbf{p}_i} + \frac{\partial \phi_{ij}}{\partial \mathbf{r}_j}\cdot\frac{\partial}{\partial \mathbf{p}_j}, \tag{9.8.24}$$

and in anticipation of the subsequent calculations, the Liouville operator has been partitioned. For the terms $\mathcal{L}_s$ and $\mathcal{L}_{N-s}$, the indices i and j are limited to the set $1 \leq \{i,j\} \leq s$ and $s+1 \leq \{i,j\} \leq N$, respectively. The only kind of term in the interaction Liouville operator, $\mathcal{L}'_s$ has $1 \leq i \leq s$ <u>and</u> $s+1 \leq j \leq N$, so that

$$\mathcal{L}'_s = -i\sum_{i=1}^{s}\sum_{j=s+1}^{N}\Theta_{ij}. \tag{9.8.25}$$

Because of Eq. (9.8.18), the equations of motion for the BBGKY functions can be derived without extensive manipulation. Note that

$$0 = V^s\int dX_{s+1}\dots dX_N\left(\frac{\partial}{\partial t} - i\mathcal{L}_N\right)\rho_N = V^s\int dX_{s+1}\dots dX_N\left(\frac{\partial}{\partial t} - i\mathcal{L}_s - i\mathcal{L}_{N-s} - i\mathcal{L}'_s\right)\rho_N. \tag{9.8.26}$$

The $N-s$ terms integrate to zero, because they are all gradient terms that are presumed to vanish when integrated and evaluated either at coordinate or momentum boundaries. At the momentum boundaries, the density function goes to zero; at the coordinate boundaries, either the density function goes to zero at the extremes of the coordinates because of the potential energy term ϕ_i, or the particles are confined in a box that may be taken as a specularly reflecting cube, such that ρ_N is unchanged by reversal of any normal component of velocity. The integration contains such terms as

$$\int dx_i \int dp_i \; p_i\frac{\partial \rho_N}{\partial x_i} = \int dp_i \; p_i\rho_N|_x,$$

where the integrated term is to be evaluated at the boundary. Either $\rho_N = 0$ at the limit of the x_i excursion because of the potential-energy term, or the probability density at the wall is an even function of p_i, and the subsequent p_i integral yields zero because the integrand is odd. Thus all integrals of the $N-s$ terms vanish.

The remaining parts of Eq. (9.8.26) can be rewritten to give

$$\left(\frac{\partial}{\partial t} - i\mathcal{L}_s\right) F_s(X_1 \ldots x_s; t) = iV^s \int dX_{s+1} \ldots dX_N \mathcal{L}'_s \rho_N$$

$$= V^s \int dX_{s+1} \ldots dX_N \sum_{i=1}^{s} \sum_{j=s+1}^{N} \Theta_{ij} \rho_N$$

$$= V^s(N-s) \int dX_{s+1} \sum_{i=1}^{s} \Theta_{i,s+1} \int dX_{s+2} \ldots dX_N \rho_N$$

$$= \frac{N-s}{V} \int dX_{s+1} \sum_{i=1}^{s} \Theta_{i,s+1} F_{s+1}. \tag{9.8.27}$$

Note that the only surviving terms on the right-hand side of the equation for F_s involve the operators $\Theta_{i,s+1}$, defined in Eq. (9.8.24), acting on the next higher function, F_{s+1}. These terms symbolize the interactions of the particles $1 \le i \le s$ in the reduced set with the representative outside particle, numbered $s+1$, and the result multiplied by $N-s$, the number of such particles. The source term in what would otherwise be a continuity equation for any reduced BBGKY function comes from the next-higher cluster of two-body interactions. Thus the source term for $s = 1$, which is effectively the Boltzmann equation, is constructed of two-body interactions; the $s = 2$ source term comes from coupled pairs of simultaneous two-body interactions; *etc.* If the particles do not interact, the right-hand sides all vanish, and $F_s(X_1 \ldots X_s; t) = F_1(X_1; t) \times \cdots \times F_1(X_s; t)$.

In the thermodynamic limit, where $V \to \infty$ with $N/V = 1/v$, the constant particle number density, the BBGKY hierarchy becomes

$$\left(\frac{\partial}{\partial t} - i\mathcal{L}_s\right) F_s(X_1 \ldots x_s; t) = \frac{1}{v} \int dX_{s+1} \sum_{i=1}^{s} \Theta_{i,s+1} F_{s+1}. \tag{9.8.28}$$

The most-used BBGKY functions are F_1 and F_2; the equations for them are:

$$\left(\frac{\partial}{\partial t} - i\mathcal{L}_1\right) F_1(X_1; t) = \frac{1}{v} \int dX_2 \Theta_{12} F_2(X_1, X_2; t), \text{ and}$$

$$\left(\frac{\partial}{\partial t} - i\mathcal{L}_2\right) F_2(X_1, X_2; t) = \frac{1}{v} \int dX_3 (\Theta_{13} + \Theta_{23}) F_3(X_1, X_2, X_3; t). \tag{9.8.29}$$

The left-hand sides of these equations are the total time derivatives of the functions as the probability-like fluid streams through its $6s$-dimensional space. These total time derivatives would vanish, as in the original Liouville equation, were it not for the right-hand sides. Entropies defined in terms of F_1 or F_2 evolve in time as entropies are expected to do, and for sufficiently dilute systems of gases, the ideal-gas equilibrium entropy is approached. But the problem remains that these are coupled equations, and the success of any calculation that uses them depends upon the approximations that are invoked to truncate the hierarchy.

The various treatments of the BBGKY hierarchy form a substantial part of any modern course in kinetic theory, and I cannot begin to cover them in this book. My purpose in the above discussion is to introduce the reader to an important area of physics that every physicist should at least know about. Constructive understanding requires further study, with the given references providing a good start. (Cohen [1962]; Liboff [1990b]; McLennan [1989])

9.8.3 Evolution in Time of the Classical Gibbs Entropy. It is a fiction to claim that because we know the classical Hamiltonian, the entropy defined by it does not evolve in time. First, even if we can write the complete Hamiltonian, we do not know the values of the pertinent physical constants to infinite accuracy. Therefore our knowledge is fuzzy, not only because of the uncertainty of the initial conditions (which the Liouville theorem preserves), but also because of our imprecise knowledge of the physical constants that appear in the Hamiltonian. As shown in Problem 11, Chapter 1, an isolated single harmonic oscillator ultimately must be described as thermalized, because all of the phase information that goes into a kinetic description becomes blurred by the imprecision of our knowledge of the frequency. This is not pedantic nit-picking! We cannot know the frequency with infinite precision, even classically. Therefore we cannot, without repeated observation, keep track of the phase of the oscillator. Without the phase, the dynamic description is lost, and the only description of the system that remains for our use is a thermal one. The details are given in the solution. The entropy of this system is known at all times in terms of the covariance matrix, W, as shown in Section 1.9. In this case, the matrix is 2×2, with the elements derived in Problem 11, Chapter 1, section (f) . The entropy is, in the notation of that problem,

$$S = k \left(1 + \ln(\sqrt{\det W}/\hbar)\right), \tag{9.8.30}$$

where W is calculated to be

$$W = \begin{pmatrix} \langle y_1^2 \rangle - \langle y_1 \rangle^2 & \langle y_1 y_2 \rangle - \langle y_1 \rangle \langle y_2 \rangle \\ \langle y_1 y_2 \rangle - \langle y_1 \rangle \langle y_2 \rangle & \langle y_2^2 \rangle - \langle y_2 \rangle^2 \end{pmatrix}$$
$$= \begin{pmatrix} \alpha^2 \cos^2 \omega t + \delta^2 \sin^2 \omega t & (\delta^2 - \alpha^2) \sin \omega t \cos \omega t \\ (\delta^2 - \alpha^2) \sin \omega t \cos \omega t & \alpha^2 \sin^2 \omega t + \delta^2 \cos^2 \omega t \end{pmatrix}, \tag{9.8.31}$$

At $t = 0$, and for all later times, this unblurred Gibbs entropy is seen to be

$$S(0) = S(t) = k[1 + \ln(\alpha\delta/\hbar)], \tag{9.8.32}$$

which is the entropy attributable to the initial uncertainty of our knowledge of position and momentum, carried forever by our assumed exact knowledge of the Hamiltonian. It corresponds precisely to uncertainties attributable to thermal motion, as discussed in Chapter 1. The initially known dynamic kinetic and potential energies of the oscillator do not appear in the entropy, as should be expected. But in a description of the system that better approximates physical reality, as t increases, imprecision of phase information acts as a mechanism for enlarging the error bars around the position and momentum coordinates in the phase space, with the end result that all phase information is lost. The ω-averaged Liouville function evolves to a time-independent two-dimensional Gaussian probability density, with the entire initial kinetic, potential, and thermal energies lumped into a single entity that appears as its variance. This variance may be related to the temperature by

$$(\alpha^2 + \delta^2 + u_1^2 + u_2^2) = kT. \tag{9.8.33}$$

The final entropy may then be written

$$S(\infty) = k[1 - \ln \beta \hbar \omega_0],$$

(9.8.34)

where $\beta = 1/kT$, with kT given by Eq. (9.8.33). The imprecision of our knowledge of the oscillator frequency was represented by the Gaussian expression

$$\rho_\omega = \frac{\exp\left[-(\omega - \omega_0)^2/2\omega_1^2\right]}{\omega_1\sqrt{2\pi}},$$

(9.8.35)

and the average of any function $f(\omega)$ is given by

$$\bar{f} = \int_{-\infty}^{\infty} f(\omega)\rho_\omega d\omega.$$

(9.8.36)

As an illustration, the behavior of one of the matrix elements in Eq. (9.8.31) was found to be

$$\langle \bar{y}_1^2 \rangle - \langle \bar{y}_1 \rangle^2 = \frac{\alpha^2 + \delta^2}{2} + \left[\frac{\alpha^2 - \delta^2}{2}\right] e^{-2\omega_1^2 t^2} \cos 2\omega_0 t +$$

$$+ \frac{(1 - e^{-\omega_1^2 t^2})}{2}\left[(u_1^2 + u_2^2) - (u_1^2 - u_2^2)e^{-\omega_1^2 t^2}\cos 2\omega_0 t - 2u_1 u_2 e^{-\omega_1^2 t^2}\sin 2\omega_0 t\right], (9.8.37)$$

and the others are to be found in Problem 11, Chapter 1. It may be seen that the rate of phase-information loss is proportional to $\exp[-(\omega_1 t)^2]$, where ω_1^2 is the variance of ω. Clearly, the smaller the variance, the more slowly the dynamic information is lost, as expected, but for any non-zero variance, the t^2 time dependence inexorably erases the phase information that permits us to describe the behavior of the oscillator by means of mechanical variables. Furthermore this loss of ability to track phase extends from $t = 0$ symmetrically into the past or into the future. (Another look at this example is provided as a problem at the end of this chapter.)

It should be noted that the simple harmonic oscillator presents the least-complicated nontrivial isolated system possible, one for which there are no uncertainties caused by collisions or attributable to outside influences from a heat bath or elsewhere. Yet even this barest caricature of a microcanonical system, when faced with the fact that our knowledge cannot be infinitely precise, evolves symmetrically in either direction of time, away from the time of observation, to a state that must be described as thermalized. A classical physical system is subject to at least four sources of uncertainty, in addition to imprecision of the known Hamiltonian, that hasten the blurring of any calculated dynamic description based on initial observation: (1) there are doubtless interactions within the system among its constituents that are either unknown or regarded as negligible (such as induced polarization effects and gravitation) that cause a steady, progressive erosion in the quality of phase information; (2) there are outside influences that cannot be screened out, especially gravitational forces and cosmic rays (recall the calculation by Borel, mentioned in Section 2.3.1, that the motion of one gram of matter through a distance of one centimeter at a point 4 light years away would make it impossible to compute the trajectories of molecules in a macroscopic container of gas at standard conditions for more than one microsecond.); (3) interactions with a heat bath (this is the thermalizing agent invoked in Section 1.7 to bring about equilibration of a system of coupled harmonic oscillators; and (4) chaos (recall the quotations from Joseph Ford in Section 1.9.7). Only (3) is expected to produce any significant change in the energy of the system; the others always lead to thermalization of the system's energy through loss of phase information, usually in short enough times to render untenable any claim that the proper Liouville function does not evolve in time. By *proper Liouville function* I mean the Γ-space density that represents our knowledge of the system, based on the equations of motion, the initial observations, and the inevitable broadening of our uncertainties because of our lack of omniscience and omnipotence. In the section on Predictive Statistical Mechanics, a time-dependent density function is offered that is expected to be an acceptable replacement for the conceptually incomplete Liouville function that cannot evolve.

The formal solution to the Liouville equation for N particles, Eq. (9.8.20), exhibits the properties needed to extend the single-oscillator example of this section. The Liouville operator, even in its classical version, is hermitian, and therefore has real eigenvalues. The eigenvalue expansion of Eq. (9.8.20) is a series of oscillatory terms, each with a frequency corresponding to one of the eigenvalues of the Liouville operator. These frequencies are not known to infinite accuracy; they are uncertain in the same way as that of the single-oscillator example. They are then subject to frequency averaging, with the result that all phase information that the Liouville function pretends to preserve is ultimately so blurred that the system becomes describable only in thermal terms. There is no need to plod through the formalism of this result for present purposes. It is enough to recognize that we cannot know the parameters of an N-body problem well enough to describe it in dynamic terms forever, based on an observation only of the initial conditions.

9.9 THE STATISTICAL OPERATOR

The von Neumann statistical operator for quantum systems is introduced in Chapter 1, and used for equilibrium calculations in Chapter 3. It will be examined in more detail in this section. A many-body system is described in quantum mechanics by a state vector in a Hilbert space, satisfying a Schrödinger equation

$$i\hbar \frac{d\,|n_0\rangle}{dt} = \mathcal{H}_0\,|n_0\rangle\,, \tag{9.9.1}$$

where $\mathcal{H}_0$ is the Hamiltonian, taken here to have no explicit time dependence, and $|n_0\rangle$ is one of a complete, orthonormal set of state vectors, all of which satisfy Eq. (9.9.1). The formal time dependence of any one of these state vectors can be obtained directly from Eq. (9.9.1) as

$$|n_0(t)\rangle = e^{-i\mathcal{H}_0 t/\hbar}\,|n_0(0)\rangle \equiv U_0(t)\,|n_0(0)\rangle\,, \tag{9.9.2}$$

where the unitary operator $U_0(t)$ can be regarded as transforming the time-independent kets $|n_0(0)\rangle$ to the usual time-dependent ones. It is sometimes convenient to describe the time-evolution of a system in terms of that of its state vector as referred to a fixed basis in the Hilbert space. For this purpose, the complete set $\{|n_0(0)\rangle\}$ is used, and the resulting description of the system is called the Heisenberg picture, whereas that in terms of the set $\{|n_0(t)\rangle\}$ is called the Schrödinger picture. The inverse transformation is $|n_0(0)\rangle = U_0^\dagger\,|n_0(t)\rangle$, and the bra vectors are $\langle n_0(0)| = \langle n_0(t)|\,U_0(t)$, with $U_0^\dagger(t) = U_0^{-1}(t)$.

The statistical operator has been defined as

$$\rho_0(t) = \sum_n |n_0(t)\rangle \, p_n \, \langle n_0(t)|, \tag{9.9.3}$$

where the summation index is properly n_0, and the p_n are the probabilities assigned to the particular states, as explained in Chapters 1 and 3. For a system in equilibrium, the probabilities are taken to be real, non-negative constants. In terms of the Heisenberg basis, Eq. (9.9.3) can be rewritten as

$$\rho_0(t) = e^{-i\mathcal{H}_0 t/\hbar} \left(\sum_n |n_0(0)\rangle \, p_n \, \langle n_0(0)| \right) e^{\mathcal{H}_0 t/\hbar} = U_0(t)\rho_0(0)U_0^\dagger(t), \tag{9.9.4}$$

where $\rho_0(0)$ is the time-independent initial statistical operator, and the unitary matrices govern the evolution of ρ. The time derivative of the statistical operator, the von Neumann equation, follows directly as

$$i\hbar \frac{d\rho_0}{dt} = [\mathcal{H}_0, \rho_0] = -\hbar \mathcal{L}_0 \rho_0, \tag{9.9.5}$$

where the total time derivative should be noted, as well as the contrast with the Liouville equation, $[\ ,\ ]$ is the commutator, and $\mathcal{L}_0$ is the Liouville superoperator of Eq. (9.8.21). The p_n are regarded for equilibrium as being not explicitly time dependent. The solution of Eq. (9.9.5) is often written

$$\rho_0(t) = e^{i\mathcal{L}_0 t} \rho_0(0). \tag{9.9.6}$$

Any operator $\mathcal{F}(t)$ may be specified by its matrix elements as

$$\mathcal{F}_{mn} = \langle m_0(t)|\mathcal{F}|n_0(t)\rangle = \langle m_0(0)|U_0^\dagger \mathcal{F} U_0|n_0(0)\rangle = \langle m_0(0)|\mathcal{F}_H|n_0(0)\rangle, \tag{9.9.7}$$

where $\mathcal{F}_H = U_0^\dagger \mathcal{F} U_0$ is the operator in the Heisenberg picture. The time derivative of the matrix element is then given by

$$i\hbar \frac{d\mathcal{F}_{mn}}{dt} = \langle m_0(0)| - \mathcal{H}_0 \mathcal{F}_H + \mathcal{F}_H \mathcal{H}_0 + i\hbar U_0^\dagger \frac{\partial \mathcal{F}}{\partial t} U_0 |n_0(0)\rangle. \tag{9.9.8}$$

This result permits the equivalent of Eq. (9.9.5) for operators to be written

$$i\hbar \frac{d\mathcal{F}_H}{dt} = [\mathcal{F}_H, \mathcal{H}_0] + i\hbar \frac{d\mathcal{F}}{dt}, \tag{9.9.9}$$

where the final term accounts for explicit time dependence of the Schrödinger operator. The general solution of this equation is

$$\mathcal{F}_H(t) = U_0^\dagger(t)\mathcal{F}(t)U_0(t), \tag{9.9.10}$$

where $\mathcal{F}(t)$ is the original Schrödinger operator. Note that the Heisenberg operator would evolve in time even if there were no explicit time dependence in $\mathcal{F}$.

For systems that are subject to a time-dependent perturbation, the evolution of the statistical operator leads to theories of irreversible phenomena. Suppose the Hamiltonian has the structure

$$\mathcal{H} = \mathcal{H}_0 + \lambda\mathcal{H}_1(t), \tag{9.9.11}$$

where λ is a constant for keeping track of orders of magnitude, and $\mathcal{H}_1(t) = 0$ for $t < 0$. The system is assumed to be in equilibrium at $t = 0$, at which time the perturbation is turned on. At $t = 0$, the density operator ρ_0 satisfies Eq. (9.9.4), with $\mathcal{H} = \mathcal{H}_0$. The Volterra integral equation of the second kind for $\rho(t)$, which also must satisfy a similar equation, is found to be

$$\rho(t) = \rho_0(t) - \frac{i\lambda}{\hbar}\int_0^t e^{-i(t-t')\mathcal{H}_0/\hbar}[\mathcal{H}_1(t'), \rho(t')]e^{i(t-t')\mathcal{H}_0/\hbar}\,dt'. \tag{9.9.12}$$

To verify this solution, note that from Eq. (9.9.12)

$$\begin{aligned}
\frac{d\rho(t)}{dt} &= \frac{d\rho_0}{dt} - \frac{i\lambda}{\hbar}[\mathcal{H}_1(t), \rho(t)] + \frac{1}{i\hbar}\left(\mathcal{H}_0\frac{i\lambda}{\hbar}\int_0^t \cdots - \frac{i\lambda}{\hbar}\int_0^t \cdots \mathcal{H}_0\right) \\
&= \frac{1}{i\hbar}[\mathcal{H}_0, \rho_0] + \frac{1}{i\hbar}[\lambda\mathcal{H}_1, \rho(t)] + \frac{1}{i\hbar}[\mathcal{H}_0, \rho(t) - \rho_0] \\
&= \frac{1}{i\hbar}[\mathcal{H}, \rho(t)] = i\mathcal{L}\rho. \tag{9.9.13}
\end{aligned}$$

As shown in Eq. (1.8.5), the expectation value of an operator $\mathcal{F}$ is given by

$$\langle\mathcal{F}\rangle = \text{Tr}(\rho\mathcal{F}). \tag{9.9.14}$$

For the Hamiltonian of Eq. (9.9.11), a Volterra integral-equation solution of Eq. (9.9.9) is similar to that of Eq. (9.9.12) for ρ:

$$\mathcal{F}_H(t) = \mathcal{F}_{0H} + \frac{i\lambda}{\hbar}\int_0^t dt'\, e^{-i(t-t')\mathcal{H}_0/\hbar}[\mathcal{H}_1(t'), \mathcal{F}(t')]e^{i(t-t')\mathcal{H}_0/\hbar}, \tag{9.9.15}$$

where $\mathcal{F}_{0H}$ satisfies Eq. (9.9.9). The verification follows the details given for ρ in Eq. (9.9.13), but it must be observed that the explicit time dependence of $\mathcal{F}$ does not depend upon the Hamiltonian. These formal solutions are exact for the specified kind of perturbation that is turned on at $t = 0$, but in each case, the required operator is given in terms of itself. It is usually not possible to extract closed-form expressions for the operators from the integral equations, but the format of these equations is ideal for the iterative approximations of perturbation theory. These approximations are often presented through diagram techniques similar to those invented by Feynman for quantum electrodynamics. (Feynman [1949]; Fujita [1966]; Montroll [1960]; Montroll [1962]; Montroll and Ward [1959]) A direct application of the basic integral equations is offered in a subsequent subsection.

9.9.1 Time Evolution of the Statistical Operator. The statistical operator found in Eq. (9.9.12) can be represented as a generalization of ρ_0 as given in Eq. (9.9.3), but with time-dependent probabilities and off-diagonal terms, since it is permissible to describe the state of the system in terms of any set of basis vectors. Therefore write the basis vectors as $|n_0(0)\rangle \equiv |n\rangle$, and represent $\rho(t)$ as

$$\rho(t) = \sum_{m,n} |m_0(t)\rangle\, P_{mn}(t)\, \langle n_0(t)| = U_0(t) \left(\sum_{m,n} |m\rangle\, P_{mn}(t)\, \langle n| \right) U_0^\dagger(t). \qquad (9.9.16)$$

It is convenient to explore the behavior of the core operator of this expression, here denoted by $\rho_I(t)$ because of its resemblance to operators in the interaction picture, and defined to be

$$\rho_I(t) = U_0^\dagger(t)\rho(t)U_0(t) = \sum_{m,n} |m\rangle\, P_{mn}(t)\, \langle n| . \qquad (9.9.17)$$

With $\rho_0 = Z_0^{-1} \exp(-\beta \mathcal{H}_0)$, where $Z_0 = \mathrm{Tr}\exp(-\beta\mathcal{H}_0)$, which commutes with U_0, an expression for ρ_I is found from Eq. (9.9.12) as

$$\begin{aligned}
\rho_I(t) &= \rho_0 - \frac{i\lambda}{\hbar} \int_0^t dt'\, U_0^\dagger(t')\, [\mathcal{H}_1(t'), \rho(t')]\, U_0(t') \\
&= \rho_0 - \frac{i\lambda}{\hbar} \int_0^t dt'\, U_0^\dagger(t')\, \Big[\mathcal{H}_1, U_0\rho_I U_0^\dagger\Big]_{t'}\, U_0(t') \\
&= \rho_0 - \frac{i\lambda}{\hbar} \int_0^t dt'\, [\mathcal{H}_I(t'), \rho_I(t')] ,
\end{aligned} \qquad (9.9.18)$$

which is equivalent to

$$i\hbar\dot\rho_I = [\mathcal{H}_I, \rho_I], \qquad (9.9.19)$$

where $\mathcal{H}_I = U_0^\dagger \mathcal{H}_1 U_0$. The generalized probabilities that express the time development of ρ_I are the matrix elements $\langle m|\rho_I|n\rangle$, given by

$$P_{mn}(t) = \langle m|\rho_I(t)|n\rangle = p_n\delta_{mn} - \frac{i\lambda}{\hbar} \int_0^t dt'\, \langle m|[\mathcal{H}_I(t'), \rho_I(t')]|n\rangle , \qquad (9.9.20)$$

and their time derivatives are

$$\begin{aligned}
\dot P_{mn} &= -\frac{i\lambda}{\hbar} \sum_s \Big(\langle m|\mathcal{H}_I(t)|s\rangle\, \langle s|\rho_I(t)|n\rangle - \langle m|\rho_I(t)|s\rangle\, \langle s|\mathcal{H}_I|n\rangle \Big) \\
&= -\frac{i\lambda}{\hbar} \sum_s \Big(\langle m|\mathcal{H}_I(t)|s\rangle\, P_{sn}(t) - P_{ms}(t)\, \langle s|\mathcal{H}_I(t)|n\rangle \Big) .
\end{aligned} \qquad (9.9.21)$$

This is an exact generalized master equation, in which the coefficients of the P_{mn} are presumably calculable without approximation as the matrix elements of the perturbing Hamiltonian, with no requirement that these terms be small. The usual terms in the master equation are the diagonal ones in the transition matrix. They are found to be

$$\dot P_n \equiv \dot P_{nn} = -\frac{i\lambda}{\hbar} \sum_s \Big(\langle n|\mathcal{H}_I(t)|s\rangle\, \langle s|\rho_I(t)|n\rangle - \langle n|\rho_I(t)|s\rangle\, \langle s|\mathcal{H}_I|n\rangle \Big) . \qquad (9.9.22)$$

But the $P_{mn}(t)$ in general are dependent upon history, as shown explicitly in Eq. (9.9.20), and it is evidently not usually possible to reduce Eq. (9.9.22) to a Pauli master equation of the Markov type.

The time evolution of the statistical operator can be written quite generally, from Eq. (9.9.13), as $\rho(t) = U(t, t_0)\rho(t_0)U^\dagger(t, t_0)$, where

$$i\hbar \frac{dU(t, t_0)}{dt} = \mathcal{H}U(t, t_0). \qquad (9.9.23)$$

When $\mathcal{H} = \mathcal{H}_0$, the solution is

$$U(t, t_0) = U_0(t, t_0) = e^{-i\mathcal{H}_0(t - t_0)/\hbar}.$$

The operator $U(t, t_0)$ is unitary, and the propagation of $\rho(t)$ forward in time from a presumably known initial value $\rho(t_0)$ is as deterministic as that of the classical probability density ρ_N of Eq. (9.8.1), which is governed by the Liouville theorem. The unitary transformation cannot lead to equilibration; it represents the propagation of the states at t_0 to dynamically allowed successor states at the times t. The initial information entropy, $S_I(t_0) = -k\mathrm{Tr}\rho(t_0)\ln\rho(t_0)$, which equals the thermodynamic entropy if the initial state is one of equilibrium, evolves unchanged under the unitary transformation, since every eigenvalue of $\rho(t)$ is a constant of the motion. This statement is true for every system that obeys the Schrödinger equation. Yet if the system actually does evolve, after some unspecified time, to a new state that is regarded as one of equilibrium, the new statistical operator ρ' should be obtained by the maximization of the entropy, $S' = \mathrm{Tr}\rho'\ln\rho'$, subject to the new constraints. A fundamental tenet of thermodynamics, and one that is easily verified by example at the level of a first course in physics, is that the natural evolution of the system is in the direction of increasing entropy, subject to the existing constraints. It is therefore expected that $S' > S_I(t_0)$. What, then, is wrong?

I have offered several arguments, with examples, to the effect that the Hamiltonian of Eq. (9.9.23) is never known well enough to permit the predicted long-time evolution of ρ to be taken seriously, and I have shown how the uncertainties allow ρ to evolve from one equilibrium value to another. These uncertainties, at least in a closed system, introduce no new energy; they simply provide mechanisms for the loss of phase information, thereby vastly enlarging the possible set of microstates that can evolve from the initial set. If the original entropy can be written $S(t_0) = k\ln\Omega_0$, where Ω_0 is the number of microstates compatible with the initial macrostate, the unitary transformation of Eq. (9.9.23) simply propagates this set of states forward in time, without allowing their number to increase; hence the entropy cannot increase. But the action of the Hamiltonian on these original states allows their energy (and other macroscopic parameters) to change, so that, for example, $\langle X(t)\rangle = \mathrm{Tr}(\rho(t)X)$ presumably evolves correctly in time, where X is the operator corresponding to some macroscopic observable. This is equivalent to the use of Eq. (9.8.20) in Eq. (9.8.8) to obtain classical time-dependent expectation values. The evolved states under the unitary transformation are representative of the action of the Hamiltonian. If the macroscopic energy of the system has increased, these evolved microstates have the same increases of energy. However, the *total* number of equally probable final states Ω_f having this increased energy is expected to exceed Ω_0, typically by a large number, since the additional energy offers more ways for the macroscopically indistinguishable microstates to exist. Their ratio is $\Omega_f/\Omega_0 = \exp[(S' - S(t_0))/k]$. (See problem 11.) Therefore, even though the statistical operator that evolves by way of Eq. (9.9.23) does not become the correct one for equilibrium, it propagates forward a presumably unbiased sampling of states representative of the totality of equilibrium final states. Averages and predictions of most macroscopic observables, based on this sampling, are assumed to be indistinguishable from those of equilibrium theory (much more so, for example, than the samplings by pollsters that forecast the actual outcome of an election, although the sampling ratio may be many orders of magnitude smaller). For this reason, transport properties calculated by means of $\rho(t)$, the unitary transformation of $\rho(t_0)$, are regarded as valid, and the techniques for deriving these properties are worth study.

A general expression for $U(t, t_0)$ in closed form is usually difficult or impossible to find, although it may be written as

$$U(t, t_0) = \mathcal{T} \exp\left[-(i/\hbar) \int_{t_0}^{t} \mathcal{H}(t')dt'\right],$$

where $\mathcal{T}$ is the time-ordering operator of quantum field theory. An integral-equation expression for $U(t, t_0)$ is

$$U(t, t_0) = U_0(t, t_0)\left[1 - \frac{i\lambda}{\hbar} \int_{t_0}^{t} dt'\, U_0^\dagger(t', t_0)\mathcal{H}_1(t')U(t', t_0)\right], \qquad (9.9.24)$$

following Eq. (9.9.12), with $\mathcal{H}$ given in Eq. (9.9.11). This expression can be iterated to produce an expansion for $U(t, t_0)$.

Application of the time-evolved statistical operator in the calculation of physical properties almost always involves some method of approximation, using a master equation, quantum analogs of Boltzmann and Vlasov equations, or linear response theory and its iterations, as examined next.

9.9.2 The Green-Kubo Formalism. The derivation of transport coefficients by means of linear response theory was published by M. S. Green in 1951 and developed systematically by Ryogo Kubo, beginning in 1957. (Green [1951]; Kubo [1957]; Kubo, Yokota and Nakajima [1957]) The abbreviated derivation presented here does not seem to depend very much on the underlying fluctuations that are responsible for the behavior of the system. Another, more detailed, treatment of the subject is offered in Chapter 10.

In linear response theory, which is a first-order perturbation theory, $\rho(t')$ in Eq. (9.9.12) is replaced by $\rho_0(t')$ when necessary. In this derivation, the system is regarded as being in equilibrium for times $t < 0$, with ρ_0 independent of time. The calculation given here is for the electrical conductivity, representative of transport processes that are driven by externally applied forcings, rather than of those such as diffusion or thermal conduction that are driven by internal inhomogeneities. The perturbation, as yet not specifically electrical, is taken to be

$$\lambda\mathcal{H}_1(t) = -E_0 e^{i\omega t}\mathcal{R}, \qquad (9.9.25)$$

where E_0 is the c-number amplitude of the time dependence of the perturbation, which is switched on at $t = 0$, and $\mathcal{R}$ is an operator that has no explicit time dependence. The expression for ρ, from Eq. (9.9.12), is

$$\rho(t) = \rho_0 + \frac{i}{\hbar} \int_0^t dt' e^{-i(t-t')\mathcal{H}_0/\hbar} E_0 e^{i\omega t'} [\mathcal{R}, \rho(t')] e^{i(t-t')\mathcal{H}_0/\hbar}$$

$$= \rho_0(t) + \frac{i}{\hbar} \int_0^t d\tau e^{-i\tau \mathcal{H}_0/\hbar} E_0 e^{i\omega(t-\tau)} [\mathcal{R}, \rho(t-\tau)] e^{i\tau \mathcal{H}_0/\hbar}, \qquad (9.9.26)$$

where the substitution $t - t' = \tau$ was used. By means of Eq. (9.9.4), and the zeroth-order approximation of a factor in what is already a first-order term, $\rho(t - \tau)$ may be written

$$\rho(t - \tau) \approx \rho_0(t - \tau) = U_0^\dagger(\tau)\rho_0(t)U_0(\tau) \approx U_0^\dagger(\tau)\rho(t)U_0(\tau). \qquad (9.9.27)$$

Take the operator $\mathcal{F}$ to represent a generalized current, a transport of some property that flows in response to the perturbation. The expectation value of this current is given by Eq. (9.9.14); it is further expected that the response is nil in the absence of the perturbation, or $\mathrm{Tr}(\rho_0 \mathcal{F}) = 0$. After the perturbation is switched on, the response is not expected to attain its steady-state value instantaneously, and there may be transient responses to the perturbation other than at turn-on time. Calculations of the transport coefficients are usually for steady-state behavior, for which the time dependence of the perturbation may be a sum of sinusoidal oscillations, each at constant amplitude. One of these was already chosen in Eq. (9.9.25), and the response up to time t is

$$\langle \mathcal{F}(t) \rangle_\omega = \frac{iE_0}{\hbar} e^{i\omega t} \int_0^t d\tau e^{-i\omega\tau} \mathrm{Tr}\left(\mathcal{F} e^{-i\tau \mathcal{H}_0/\hbar} [\mathcal{R}, e^{i\tau \mathcal{H}_0/\hbar} \rho(t) e^{-i\tau \mathcal{H}_0/\hbar}] e^{i\tau \mathcal{H}_0/\hbar} \right).$$

If, for this linear-response-theory approximation, the statistical operator is taken to be $\rho(t) = \rho_0 = Z_0^{-1} \exp(-\beta\mathcal{H}_0)$, its accompanying factors cancel, since they commute with $\mathcal{H}_0$. The trace operation is invariant under cyclic permutation of the terms subject to the operation; in this case, then, the expectation value can be rewritten as

$$\langle \mathcal{F}(t) \rangle_\omega = \frac{i}{\hbar} E_0 e^{i\omega t} \int_0^t d\tau e^{-i\omega\tau} \mathrm{Tr}\left(e^{i\tau \mathcal{H}_0/\hbar} \mathcal{F} e^{-i\tau \mathcal{H}_0/\hbar} [\mathcal{R}, \rho_0] \right)$$

$$= \frac{i}{\hbar} E_0 e^{i\omega t} \int_0^t d\tau e^{-i\omega\tau} \mathrm{Tr}\left(\mathcal{F}_H(\tau) [\mathcal{R}, \rho_0] \right), \qquad (9.9.28)$$

where Eq. (9.9.10) has been used.

The commutator can be evaluated by means of an identity derived by Kubo. Note that

$$[\mathcal{R}, \rho_0] = Z_0^{-1} [\mathcal{R}, e^{-\beta\mathcal{H}_0}], \qquad (9.9.29)$$

where Z_0 is the equilibrium partition function of the unperturbed system. By formal application of Eq. (9.9.10), it is seen that

$$e^{\lambda\mathcal{H}_0} \mathcal{R}(t_1) e^{-\lambda\mathcal{H}_0} = \mathcal{R}^{(0)}(t_1 - i\lambda\hbar), \qquad (9.9.30)$$

where t_1 is an arbitrary and irrelevant time, since $\mathcal{R}$ is not explicitly time dependent, and $\mathcal{R}^{(0)}$ is the Heisenberg operator when $\mathcal{H}_1 = 0$. The derivative with respect to λ of Eq. (9.9.30) gives

$$\frac{d}{d\lambda}\left(e^{\lambda\mathcal{H}_0}\mathcal{R}(t_1)e^{-\lambda\mathcal{H}_0}\right) = -i\hbar\dot{\mathcal{R}}^{(0)}(t_1 - i\lambda\hbar) = \left[\mathcal{H}_0, \mathcal{R}^{(0)}(t_1 - i\lambda\hbar)\right], \qquad (9.9.31)$$

where $\dot{\mathcal{R}}$ is the derivative of $\mathcal{R}$ with respect to its argument. Integration of Eq. (9.9.31) gives

$$\int_0^\beta d\left(e^{\lambda\mathcal{H}_0}\mathcal{R}(t_1)e^{-\lambda\mathcal{H}_0}\right) = e^{\beta\mathcal{H}_0}\mathcal{R}(t_1)e^{-\beta\mathcal{H}_0} - \mathcal{R}(t_1) = -i\hbar\int_0^\beta d\lambda\dot{\mathcal{R}}^{(0)}(t_1 - i\lambda\hbar)$$

$$= \int_0^\beta d\lambda\left[\mathcal{H}_0, \mathcal{R}(t_1 - i\lambda\hbar)\right]. \qquad (9.9.32)$$

Multiplication from the left by $\exp(-\beta\mathcal{H}_0)$ gives, with the use of Eq. (9.9.29),

$$\left[\mathcal{R}(t_1), e^{-\beta\mathcal{H}_0}\right] = -i\hbar e^{-\beta\mathcal{H}_0}\int_0^\beta d\lambda\dot{\mathcal{R}}^{(0)}(t_1 - i\hbar\lambda) = Z_0\left[\mathcal{R}(t_1), \rho_0\right]. \qquad (9.9.33)$$

With these results, Eq. (9.9.28) can be expressed as

$$\langle\mathcal{F}(t)\rangle = E_0 e^{i\omega t}\int_0^t d\tau e^{-i\omega\tau}\mathrm{Tr}\left(Z_0^{-1}\mathcal{F}_H(\tau)e^{-\beta\mathcal{H}_0}\int_0^\beta d\lambda\dot{\mathcal{R}}^{(0)}(t_1 - i\hbar\lambda)\right). \qquad (9.9.34)$$

When E_0 represents the amplitude of an externally applied electric field, then $\mathcal{F} = \mathcal{J}$, the electric current density, given by

$$\mathcal{J} = V^{-1}\sum_V\sum_i q_i v_i = V^{-1}\sum_V\sum_i q_i p_i/m_i, \qquad (9.9.35)$$

where the ith species of charge carrier has charge q_i, mass m_i, and momentum p_i, and the sum is taken over the entire volume of the system. If the system is a homogeneous, isotropic rectangular parallelopiped with the face at $x = 0$ at potential $\phi(0) = 0$ and the face at $x = L$ at potential $\phi(L) = \phi_L(t)$, then the potential at any x is $\phi(x) = (x/L)\phi_L$. The electric field is $E_x(t) = -\phi_L/L$, and the potential can be written $\phi(x, t) = -xE_x(t)$. The perturbation term of Eq. (9.9.25) thus becomes the added electrostatic energy:

$$-E_0 e^{i\omega t}\mathcal{R} = \sum q_i\phi_i = \sum q_i x_i\phi_L(t)/L = -\sum q_i x_i E_0 e^{i\omega t}, \qquad (9.9.36)$$

thus identifying the operator $\mathcal{R}$ as

$$\mathcal{R} = \sum q_i x_i. \qquad (9.9.37)$$

The time derivative of $\mathcal{R}$, needed in Eq. (9.9.34), is seen to be

$$\dot{\mathcal{R}} = \sum q_i\dot{x}_i = V\mathcal{J}, \qquad (9.9.38)$$

where $\mathcal{J}$ is that of Eq. (9.9.35). With these identifications, Eq. (9.9.34) becomes

$$\left\langle \mathcal{J}(t) \right\rangle = E_0 e^{i\omega t} V \int_0^t d\tau e^{-i\omega\tau} \text{Tr} \left(\mathcal{J}(\tau) Z_0^{-1} e^{-\beta\mathcal{H}_0} \int_0^\beta d\lambda \mathcal{J}^{(0)}(t_1 - i\beta\lambda) \right)$$

$$= E_0 e^{i\omega t} V \int_0^t d\tau e^{-i\omega\tau} \int_0^\beta d\lambda \left\langle \mathcal{J}(\tau)\mathcal{J}^{(0)}(t_1 - i\hbar\lambda) \right\rangle_0, \qquad (9.9.39)$$

where the final subscript indicates that the average is the trace with respect to the initial statistical operator. The Kubo expression for the electrical conductivity is found from the steady-state version of Eq. (9.9.39), taken with $t \to \infty$ and $t_1 = 0$, the time at which the electric field is switched on. For an infinite time limit, it is necessary to include a factor $\exp(\alpha t)$ in the perturbation, and then take the limit as $\alpha \to 0$. The result is

$$\sigma = \lim_{\alpha \to 0} \frac{\left\langle \mathcal{J}(t) \right\rangle_{av.}}{E_0 e^{(\alpha+i\omega)t}} = V \lim_{\alpha \to 0} \int_0^\infty d\tau e^{-(\alpha+i\omega)\tau} \int_0^\beta d\lambda \left\langle \mathcal{J}(\tau)\mathcal{J}^{(0)}(-i\hbar\lambda) \right\rangle_0. \qquad (9.9.40)$$

More discussion of the relationship of transport coefficients to fluctuations is found in Chapter 10. In particular, the physical significance of correlation functions that invoke an imaginary time parameter needs to be examined. The commentary and other parts of Chapter 10 contain references and criticisms pertinent to the Kubo relations.

9.10 PREDICTIVE STATISTICAL MECHANICS

A conceptual barrier to the comfortable development of nonequilibrium statistical mechanics is that the Liouville and (perhaps) the von Neumann equations claim that the density operator ρ is constant in time. Therefore the entropy, which is defined as a function only of ρ, cannot evolve to an equilibrium value. We know that something is wrong with this state of affairs, and a justification for the standard calculations is given in Section 9.9.1, based on the idea that a representative sample set of microstates is propagated forward by the unitary transformation of an initial statistical operator, and this sample provides a good-enough, although perhaps minuscule, representation of the time-evolved system to generate accurate values for its macroscopic properties. This idea was stated without proof, since there is none, as what seems to be the only plausible way to reconcile the Liouville theorem and its quantum counterpart with the fact of macroscopic equilibration. (Jaynes [1985]) Nevertheless, a formalism has not yet emerged that (1) allows ρ to evolve to equilibrium, (2) is generally accepted, and (3) is clearly correct. In Chapter 1, several systems were shown to evolve in time, as described by the probability sets representing our knowledge of their states. Also, in the treatment there of a chain of coupled harmonic oscillators, the N-oscillator system was shown to evolve in time to equilibrium with the rest of the infinite chain, regarded as a heat bath. The formalism there used a reduced Liouville function, as does the BBGKY formalism of this chapter. I have also discussed in this chapter the single, isolated, microcanonical oscillator of Problem 11, Chapter 1, for which imprecision in the value of the frequency leads to loss of phase information and thus to the necessity for a thermal description of its behavior. This kind of imprecision, which is always with us, provides part of the conceptual basis for discarding as a gross oversimplification the temporal invariance of the Liouville function and its quantum counterpart. (Recall that the Liouville function and its analogs carry forward all details of the initial conditions, and these usually are not relevant in the description of evolved equilibrium states.) But these mechanisms, while of some fundamental interest, are at best too awkward to be useful as means toward the calculation of transport properties or as procedures for predicting the behavior of macroscopic systems from the suitably averaged properties of the possible microstates. A concise and elegant general formalism for carrying out the evolutionary blurring of the statistical operator is not to be found among the *ad hoc* procedures exhibited so far. For this purpose, I return to the methods based on information theory, as introduced into thermodynamics by Jaynes. In this section, I introduce a formalism that seems to be the best candidate for the needed objective, although it has probably not reached its final form. (Grandy, Jr. [1980]; Grandy, Jr. [1987]; Hobson [1971]; Jaynes [1979]; Jaynes [1980]; Robertson [1966]; Robertson [1967]; Robertson [1970]; Robertson [1979]; Robertson and Mitchell [1971]; Zubarev [1974]; Zubarev and Kalashnikov [1970])

9.10.1 The Time-Dependent Maximum-Entropy Formalism. Generalization of the formalism, based on information theory, to systems that can be time dependent and spatially nonuniform does not differ, in principle, from the conceptual structure of equilibrium thermodynamics as presented in the first three chapters. The information entropy of the system is given by

$$S_I = -k \sum_n p_n \ln p_n,$$

$$S_I = -k \int d\Gamma \, \rho(\Gamma) \ln h^{3N} \rho(\Gamma), \quad \text{or}$$

$$S_I = -k \text{Tr} \rho \ln \rho, \tag{9.10.1}$$

depending upon the context. The information entropy is maximized, subject to constraints of the form $\bar{X}_r = \langle X_r \rangle = \sum p_n X_{rn}$, or its equivalent, where $\bar{X}_r$ is the measured (or otherwise known) value of the rth observable, $\langle X_r \rangle$ is the expected value of the observable as determined by the probability set $\{p_n\}$, and X_{rn} is the value this observable would have if the system were in its nth state. The one additional constraint is that $\sum p_n = 1$, or its equivalent. For equilibrium systems, the information entropy was shown by plausibility arguments, and then postulated, to be equal to the thermodynamic entropy. The conceptual basis for this formulation was provided by Gibbs, as discussed in the Commentaries of the first three chapters. It was exploited in the context of thermodynamics by Callen. The explicit formulation of the theory in terms of the Shannon measure of information entropy was, and continues to be, predominantly the work of Jaynes and his colleagues. The underlying concept is that maximization of the information entropy, subject to the known constraints, is tantamount to finding the least-biased, and therefore the most general, probabilistic description of the system. To do otherwise is to claim more knowledge than is warranted by the data at hand. (Jaynes [1980]; Jaynes [1982b])

The same concept provides a sound basis for the study of nonequilibrium phenomena. There is no claim that the information entropy represents the correct expression for the thermodynamic entropy of a system that is not in equilibrium (or that it does not). The units are chosen to be those of thermodynamic entropy, since the two are equal at equilibrium. The information entropy is regarded as a measure of our ignorance of the system, which may vary from place to place and with time. To maximize our ignorance, subject to what is known, means that we select the most broadly based probability set possible, and thereby predict the behavior that can happen in the greatest number of ways, consistent with the observations. Predictions made with this probability set are expected either to confirm further measurements, say of other observables, or to disagree. In the latter case, the new data provide additional constraints, by which the probability set can be refined.

The maximum-information-entropy formalism for time-dependent systems has already been introduced, beginning with Eq. (9.6.8). The constraints consist of known or measured values, $\bar{X}_r(t)$, of the observables X_r in the time interval $t_1 \leq t \leq t_2$. The probability set that maximizes S_I, subject to the constraints, is

$$p(n,t) = Z^{-1} \exp\left[-\sum_r \lambda_r(t) X_{rn}\right], \tag{9.10.2}$$

where X_{rn} is, as before, the value of X_r when the system is in state n, and $Z(\{\lambda_r\}) = \sum_n \exp[-\sum_r \lambda_r(t) X_{rn}]$. The multipliers $\lambda_r(t)$ were determined by setting

$$\bar{X}_j(t) = \langle X_j(t)\rangle = \sum_n p(n,t) X_{jn} = -\frac{\partial \ln Z}{\partial \lambda_j(t)}. \tag{9.10.3}$$

The information entropy is

$$S_I = -k \sum_n p(n,t) \ln p(n,t) = k \sum_r \lambda_r(t) \langle X_r(t)\rangle + k \ln Z, \tag{9.10.4}$$

and its differential is

$$dS_I = k \sum_r \langle X_r\rangle \, d\lambda_r + k \sum_r \lambda_r d\langle X_r\rangle + k \sum_r \frac{\partial \ln Z}{\partial \lambda_r} d\lambda_r = k \sum_r \lambda_r d\langle X_r\rangle, \tag{9.10.5}$$

showing that $S_I = S_I(\{\langle X_r(t)\rangle\})$.

These results have been compared with, and found to be equivalent to, those derived directly from the master equation, and they suffer from the same lack of generality. They are memoryless. The master equation describes a Markov process, and this solution for the probability set produces an entropy that is a function of t, but not of prior history. Thus it fails to utilize all of the available information, and this limits its validity to systems without the fading-memory effects that typify the normal behavior of even the simplest systems. (Recall the gradual fading of the initial data in the harmonic-oscillator chain of Chapter 1.)

For ease of notation, suppose there is a single observable, X, that can be represented by the operator $\mathcal{X}$. In the Heisenberg picture, it can be written as

$$\mathcal{X}(t) = U^\dagger(t,t_0)\mathcal{X}(t_0)U(t,t_0), \tag{9.10.6}$$

where $\mathcal{X}(t_0)$ is identical to the Schrödinger operator at $t = t_0$. The expectation value of the observable, at $t = t_0$, is given by

$$\langle X(t_0)\rangle = \mathrm{Tr}\left(\rho_0 \mathcal{X}(t_0)\right), \tag{9.10.7}$$

where the statistical operator at $t = t_0$ is $\rho_0 = \rho(t_0)$, given by

$$\rho_0 = Z_0^{-1} e^{-\lambda_0 \mathcal{X}(t_0)}; \quad \text{and} \quad Z_0 = \text{Tr} e^{-\lambda_0 \mathcal{X}(t_0)}. \tag{9.10.8}$$

The expectation value of some unconstrained operator $\mathcal{A}$, at $t = t_0$, is $\langle A_0 \rangle = \text{Tr}(\rho_0 \mathcal{A}(t_0))$, just as it would be in equilibrium theory.

Suppose that at $t = t_1 > t_0$, another expectation value is found. It is expected to be obtained in the usual way:

$$\langle X(t_1) \rangle = \text{Tr}(\rho_1 \mathcal{X}(t_1)), \tag{9.10.9}$$

but now $\rho_1 = \rho(t_1)$ retains the information in ρ_0 and incorporates that of Eq. (9.10.9) to define a new Lagrange multiplier, so that the statistical operator becomes

$$\rho_1 = Z_1^{-1} e^{-[\lambda_0 \mathcal{X}(t_0) + \lambda_1 \mathcal{X}(t_1)]}, \quad \text{and} \quad Z_1 = \text{Tr} e^{-[\lambda_0 \mathcal{X}(t_0) + \lambda_1 \mathcal{X}(t_1)]}. \tag{9.10.10}$$

Expectation values of unconstrained operators can be found at two times, t_0 and t_1. Evidently this procedure can be continued, for a succession of times for which $t_{i+1} > t_i$, such that at time t_n, the statistical operator is

$$\rho_n = Z_n^{-1} \exp\left[-\sum_{i=0}^{n} \lambda_i \mathcal{X}(t_i) \right]; \quad \text{and} \quad Z_n = \text{Tr} \exp\left[-\sum_{i=0}^{n} \lambda_i \mathcal{X}(t_i) \right]. \tag{9.10.11}$$

In the limit of very large n over a finite total time interval, with equal divisions of that interval, the sum can be replaced by an integral, and the information is then regarded as continuous. The statistical operator becomes

$$\rho = Z^{-1} \exp[-\int_{t_0}^{t_f} dt \lambda(t) \mathcal{X}(t)]; \quad t_0 \leq t \leq t_f. \tag{9.10.12}$$

The partition functional $Z[\lambda(t)]$ is the trace of the exponential expression in Eq. (9.10.12), and the function $\lambda(t)$ is found by identifying the expectation values with the input data, or

$$\langle X(t) \rangle = \text{Tr}[\rho \mathcal{X}(t)] = -\frac{\delta \ln Z}{\delta \lambda(t)}, \tag{9.10.13}$$

where the δ's indicate functional differentiation. For times later than t_f, the statistical operator can be propagated forward by means of the unitary transformation of Eq. (9.9.23), so that all of the information in the data-gathering interval continues to influence predictions of expectation values. Because of the invariance of the trace under cyclic permutation, Eq. (9.10.13) can be written for any time t as

$$\langle X(t) \rangle = \text{Tr}[\rho \mathcal{X}(t)] = \text{Tr}[\rho(t) \mathcal{X}], \tag{9.10.14}$$

with either the ρ of Eq. (9.10.12) and the operator $\mathcal{X}(t)$ given by (9.10.6), or $\rho(t)$ so transformed, with $\mathcal{X}$ simply the Schrödinger operator.

The quantity X might be measurable as a function of position throughout the system, as well as one of time. For each position $\mathbf{r}_j$, there is a Lagrange multiplier $\lambda_j(t)$, so that the statistical operator becomes

$$\rho = Z^{-1}\exp[-\int_{t_0}^{t_f} dt \sum_j \lambda_j(t)\mathcal{X}(\mathbf{r}_j,t)], \qquad (9.10.15)$$

where $\sum_j$ covers the set of positions for which X is measured. For a dense such set, the sum is replaced by an integral over the volume of the system.

A number of observables may provide input data for the determination of ρ, leading to a generalization of Eq. (9.10.15):

$$\rho = Z^{-1}\exp\left[-\sum_i \int_{R_i} dt\, d^3r\, \lambda_i(\mathbf{r},t)\mathcal{X}_i(\mathbf{r},t)\right], \qquad (9.10.16)$$

where the space-time region of observation of the ith observable, R_i, can be different for every i. The entropy is maximized subject to the existing constraints provided by the input data, and expectation values are predicted by means of the usual recipe given in Eq. (9.10.14).

In principle, the theory now utilizes all information that can be incorporated in the input data, including memory effects and spatial dependencies of a defining set of observables. The entropy has been maximized, subject to the given information. There is no need to require that the observables be independent, since the Lagrange multipliers for any redundant data disappear from the equations, and the redundant data therefore do not disproportionally weight the statistical operator. Any contradictory information causes the procedure for identification of the Lagrange multipliers to fail, so that the statistical operator cannot be found. The resulting predictions may not be correct, but they are the best possible, subject to what is known. The partition functional is

$$Z(\{\lambda_i(\mathbf{r},t)\}) = \mathrm{Tr}\exp\left[-\sum_i \int_{R_i} dt\, d^3r\, \lambda_i(\mathbf{r},t)\mathcal{X}_i(\mathbf{r},t)\right]. \qquad (9.10.17)$$

The Lagrange multipliers are determined, as always, by the requirement that the expectation values of the observables, as calculated by use of the statistical operator, agree with the input data over the range of observation, or

$$\langle\mathcal{X}_i(\mathbf{r},t)\rangle = -\frac{\delta\ln Z}{\delta\lambda_i(\mathbf{r},t)} = \bar{X}_i(\mathbf{r},t)$$
$$= \mathrm{Tr}[\rho\mathcal{X}_i(\mathbf{r},t)]; \quad \mathbf{r},t \text{ in } R_i. \qquad (9.10.18)$$

The set $\{\bar{X}_i\}$ is the input data, and the δ's denote functional differentiation. Prediction of any other quantity, represented by the operator $\mathcal{Y}$, is effected by evaluating

$$\langle\mathcal{Y}(\mathbf{r},t)\rangle = \mathrm{Tr}[\rho\mathcal{Y}(\mathbf{r},t)] = \mathrm{Tr}[\rho(t)\mathcal{Y}], \qquad (9.10.19)$$

where, in the final form, $\rho(t)$ is the time-evolved statistical operator, and $\mathcal{Y}$ is the time-independent Schrödinger operator. Then, since $U(t,t_0)$ of Eq. (9.9.23) is usually approximated by iteration of Eq. (9.9.24), all required expectation values can be found with the same approximate form of the time-evolved statistical operator. (The warning of the previous section, that the time-evolved statistical operator propagates forward in time the deterministic histories of the states at $t = t_0$, which may be a ridiculously small sampling of the states that become accessible at some later time, should be kept in mind, but I am not aware of an example that predictions based on this sampling suffer inaccuracies attributable to sample bias. One reason for this is that linear response theory typically treats only small departures from equilibrium, for which the sampling should almost always be adequate, except, perhaps, in the neighborhood of a critical point.)

9.10.2 Generalized Heating and Mitchell's Thermal Sources.

In most treatments of irreversible phenomena, especially that of response theory, the system is regarded as driven by forcings that can be represented as terms in a Hamiltonian. The coupled-oscillator system of Chapter 1 described the heat bath in mechanical terms. It is evident that such a specific mechanistic description of the heating process is inappropriate, and it should not be necessary. The consequences to the system of adding thermal energy should be independent of the nature of the heat source, subject of course to obvious constraints on energy density and heating rate. Furthermore, for thermal sources, detailed knowledge of the Hamiltonian of the source is usually missing. The prevailing attitude, seldom formally expressed, is that since the details of the source are not important, then we should be free to invoke any energy-transfer mechanism that may be convenient, provided the source is regarded as thermalized. This was the procedure used in Chapter 1, and it provided a good model of the evolution of a system to equilibrium. The heating mechanism used there is the exchange by mechanical coupling of the kinetic and potential energies of the oscillators in the chain, but thermalized by averaging over the probability density of the heat-bath variables. Mitchell has devised an elegant generalization of this kind of procedure, leading to deepened understanding of thermal sources. (Grandy, Jr. [1987]; Jaynes [1985]; Mitchell [1967])

Suppose some thermodynamic property such as the energy of the system, $\langle E \rangle = U$, but not limited to that, is denoted by A. In its ith state, this property has the value A_i. The probability set that maximizes the entropy, subject to the known constraints, is $\{p_i\}$, and the expectation value of A is $\langle A \rangle = \sum_i p_i A_i$. If an infinitesimal process takes place that results in a change in A, it may be written

$$\delta \langle A \rangle = \sum_i p_i \delta A_i + \sum_i A_i \delta p_i. \qquad (9.10.20)$$

The term $\sum p_i \delta A_i = \langle \delta A \rangle$ is the expectation value of the change in $\langle A \rangle$. For example, if $A_i = E_i$, then δE_i is the chance in the energy of the ith energy state, as discussed in Chapter 2, as a consequence of work processes such as a change in volume or magnetization, and δA_i represents a generalization of this work, so that $\langle \delta A \rangle = \delta W_A$ is the expected value of the generalized work done on the system. Then, in agreement with the meaning of thermal energy transfer, as introduced in Chapter 2 as a mode of energy transfer not identifiable as macroscopic work, the remaining term is $\sum A_i \delta p_i = \delta Q_A$, the generalized heating, corresponding to a change in the probabilities, but not in the values A_i of the property in the possible states of the system. Thus

$$\delta \langle A \rangle = \delta W_A + \delta Q_A, \qquad (9.10.21)$$

is a generalization of usual first law of thermodynamics, which this becomes when $A_i = E_i$. The entropy, given by $S = -k \sum p_i \ln p_i$, is changed only by changes in the $\{p_i\}$, which are associated with the heating term, δQ_A. This is the kind of result over which one may exclaim, "Aha!"

The equilibrium probabilities of a system may be determined by the maximization of the information entropy subject to known expectation values of several properties, A, B, C, with the probability of the ith state given by

$$p_i = Z^{-1} \exp\left(-\sum_j \lambda_j A_{ji}\right), \tag{9.10.22}$$

where the $\{A_j\}$ are the physical properties A, B, etc., that correspond to the input data set $\{A_j'\}$, A_{ji} is the value of A_j when the system is in its ith state, and $Z = \sum_i \exp\left(-\sum_j \lambda_j A_{ji}\right)$. The Lagrange multipliers, or potentials, $\{\lambda_j\}$, are evaluated by the requirement that $\langle A_n \rangle = A_n'$, where

$$\langle A_n \rangle = -\frac{\partial \ln Z}{\partial \lambda_n}. \tag{9.10.23}$$

There are exactly enough of these equations to evaluate all of the potentials. Note also that the covariances of the macroscopic properties are found directly from Z:

$$\begin{aligned} K_{mn} &= -\frac{\partial \langle A_m \rangle}{\partial \lambda_n} = -\frac{\partial \langle A_n \rangle}{\partial \lambda_m} \\ &= K_{nm} = \langle A_m A_n \rangle - \langle A_m \rangle \langle A_n \rangle. \end{aligned} \tag{9.10.24}$$

For a system initially at equilibrium, the covariances are defined with respect to the equilibrium probability set. The covariance of two macroscopic properties is given by

$$K_{AB} = \langle AB \rangle_0 - \langle A \rangle_0 \langle B \rangle_0, \tag{9.10.25}$$

where $\langle \cdots \rangle_0$ denotes the average with respect to the equilibrium probabilities. If a heating δQ_A results in a positive $\delta \langle A \rangle$, and if $K_{AB} > 0$, then $\delta \langle B \rangle > 0$ is also expected, but there is as yet no quantitative relationship. If, for some property C, such that A and C are not coupled and therefore $K_{AC} = 0$, the heating δQ_A then results in no change in the expectation value of C, or $\delta \langle C \rangle = 0$. Since this should be true for any such C, construct a property $C = B - xA$, where x is to be determined, such that $\delta \langle C \rangle = \delta \langle B \rangle - x\delta \langle A \rangle = 0$, or $x = \delta \langle B \rangle / \delta \langle A \rangle$. But $K_{AC} = K_{AB} - xK_{AA} = 0$, or $x = K_{AB}/K_{AA}$, and

$$\delta \langle B \rangle = x\delta \langle A \rangle = (K_{AB}/K_{AA})\delta \langle A \rangle = (K_{AB}/K_{AA})\delta Q_A, \tag{9.10.26}$$

since $\delta \langle A \rangle = \delta Q_A$ in the absence of a work term. This is a general result for the effect of a small generalized heating source δQ_A on some other property B.

Suppose the only two properties that determine the equilibrium state are A and B, and a small heating, δQ_A, operates to change the system to a new equilibrium state. The changes are given by

$$\delta \langle A \rangle = \frac{\partial \langle A \rangle}{\partial \lambda_A}\delta\lambda_A + \frac{\partial \langle A \rangle}{\partial \lambda_B}\delta\lambda_B = -(K_{AA}\delta\lambda_A + K_{AB}\delta\lambda_B)$$

$$\delta \langle B \rangle = \frac{\partial \langle B \rangle}{\partial \lambda_A}\delta\lambda_A + \frac{\partial \langle B \rangle}{\partial \lambda_B}\delta\lambda_B = -(K_{BA}\delta\lambda_A + K_{BB}\delta\lambda_B), \quad \text{or} \tag{9.10.27}$$

$$-\begin{pmatrix} \delta\langle A\rangle \\ \delta\langle B\rangle \end{pmatrix} = \begin{pmatrix} K_{AA} & K_{AB} \\ K_{BA} & K_{BB} \end{pmatrix} \begin{pmatrix} \delta\lambda_A \\ \delta\lambda_B \end{pmatrix},$$ (9.10.28)

where $\delta\langle A\rangle = \delta Q_A$. The Gibbs stability criterion, given as Eq. (2.11.1), is written here in general terms as

$$\sum_j \delta\lambda_j \delta Q_j = \sum_j \delta\lambda_j(\delta\langle A_j\rangle - \langle\delta A_j\rangle) \le 0.$$ (9.10.29)

For the two-property system under consideration, this criterion becomes

$$\delta\lambda_A \delta Q_A + \delta\lambda_B \delta\langle B\rangle \le 0,$$ (9.10.30)

since $\delta W_B = 0 = \langle\delta B\rangle$. When Eq. (9.10.28) is used to expand the heating and response terms, the result is a quadratic form that can be written as

$$-\begin{pmatrix} \delta\lambda_A & \delta\lambda_B \end{pmatrix} \begin{pmatrix} K_{AA} & K_{AB} \\ K_{BA} & K_{BB} \end{pmatrix} \begin{pmatrix} \delta\lambda_A \\ \delta\lambda_B \end{pmatrix} \le 0,$$ (9.10.31)

which says that the covariance matrix is positive definite. Therefore Eq. (9.10.28) can be inverted, giving $\delta\lambda_A$ and $\delta\lambda_B$ uniquely in terms of $\delta\langle A\rangle$ and $\delta\langle B\rangle$. More generally, for any number of constraining parameters A_i', the Gibbs convexity relation is equivalent to the assertion that the covariance matrix is positive definite.

If $\delta\lambda_B = 0$, Eq. (9.10.28) says that $\delta\langle A\rangle = -K_{AA}\delta\lambda_A = \delta Q_A$, or $\delta\lambda_A = -\delta Q_A/K_{AA}$. It then says that $\delta\langle B\rangle = -K_{BA}\delta\lambda_A = (K_{BA}/K_{AA})\delta Q_A$, the same as that found in Eq. (9.10.26) as caused by δQ_A, with no constraint on B. Because the inversion of Eq. (9.10.28) is unique, it is established that, in general, a heating δQ_A does not change the potentials conjugate to the other system properties. This is one of Mitchell's results. He showed that as a consequence of a heating δQ_A, with the property B unconstrained, the following statements are equivalent:

1. Properties C, such that $K_{CA} = 0$, are unchanged.

2. Potentials λ_B of unconstrained parameters are unchanged.

3. The changes in the $\{p_i\}$ resulting from δQ_A automatically remaximize the entropy, giving the new entropy the value it would have with $A' + \delta Q_A = \langle A\rangle$, and no constraint from B.

The first statement is evident and intuitively obvious; the second is established in the previous paragraph; and the third follows by direct calculation. Nevertheless, despite the demonstrated equivalence of these three statements, each seems to be saying something totally different and unrelated to the others.

Another of Mitchell's results follows when a third property, C, is introduced explicitly, so that Eq. (9.10.28) becomes

$$-\begin{pmatrix} \delta\langle A\rangle \\ \delta\langle B\rangle \\ \delta\langle C\rangle \end{pmatrix} = \begin{pmatrix} K_{AA} & K_{AB} & K_{AC} \\ K_{BA} & K_{BB} & K_{BC} \\ K_{CA} & K_{CB} & K_{CC} \end{pmatrix} \begin{pmatrix} \delta\lambda_A \\ \delta\lambda_B \\ \delta\lambda_C \end{pmatrix}$$ (9.10.32)

When the only cause of change is the heating δQ_A, with B and C unconstrained, the changes in the expectation values are $\delta\langle A\rangle = \delta Q_A$, $\delta\langle B\rangle = (K_{BA}/K_{AA})\delta Q_A$, and $\delta\langle C\rangle = $

$(K_{CA}/K_{AA})\delta Q_A$, with $\delta\lambda_B = 0 = \delta\lambda_C$. But if C is kept at its original value, so that $\delta\langle C\rangle = 0$, there results the constraint equation $\delta\langle C\rangle = 0 = K_{CA}\delta\lambda_A + K_{CB}\delta\lambda_B + K_{CC}\delta\lambda_C$, or

$$\delta\lambda_C = -(K_{CA}/K_{CC})\delta\lambda_A - (K_{CB}/K_{CC})\delta\lambda_B, \tag{9.10.33}$$

Since B is unconstrained, it has been established that $\delta\lambda_B = 0$, and, from Eq. (9.10.32), $\delta\lambda_C = -(K_{CA}\ K_{CC})\delta\lambda_A$. Direct substitution of this result in the other two equations of Eq. (9.10.32) gives

$$-\delta\langle A\rangle = \left(\frac{K_{AA}K_{CC} - K_{AC}K_{CA}}{K_{CC}}\right)\delta\lambda_A = K'_{AA}\delta\lambda_A = -\delta Q_A, \quad \text{and}$$

$$-\delta\langle B\rangle = \left(\frac{K_{BA}K_{CC} - K_{BC}K_{CA}}{K_{CC}}\right)\delta\lambda_A = -\left(\frac{K_{BA}K_{CC} - K_{BC}K_{CA}}{K_{AA}K_{CC} - K_{AC}K_{CA}}\right)\delta Q_A. \tag{9.10.34}$$

The A-C correlation coefficient is

$$R_{AC} = K_{AC}/\sqrt{K_{AA}K_{CC}}. \tag{9.10.35}$$

Its use permits $\delta\langle B\rangle$ of Eq. (9.10.34) to be written

$$\delta\langle B\rangle = \left(\frac{K_{BA}}{K_{AA}} - \frac{K_{BC}K_{CA}}{K_{AA}K_{CC}}\right)\delta Q'_A, \tag{9.10.36}$$

where $\delta Q'_A$, called the *renormalized source strength*, is given by

$$\delta Q'_A = \frac{\delta Q_A}{1 - R_{AC}^2}. \tag{9.10.37}$$

The first term in $\delta\langle B\rangle$ is the usual response of Eq. (9.10.26), but this time to the renormalized source strength, $\delta Q'_A$. The second term can be written $(K_{BC}/K_{CC})\delta Q'_C$, where $\delta Q'_C = -(K_{CA}/K_{AA})\delta Q'_A$, and $\delta Q'_C$ is the effective heating attributable to the constraint that C is constant. It is the negative of the response δQ_C if C were unconstrained.

Thus the constraint on C has modified the consequences of the heating source δQ_A by changing its apparent strength as it affects B, and it has introduced a virtual source $\delta Q'_C$ that acts to nullify any change in C that would otherwise have been produced by $\delta Q'_A$. These results, and the linear equations leading to them, are the exact analogs of well-known theorems in electrical network theory, in and their extension to time- and space-dependent systems, they become the analogs of reciprocity theorems in field theories. These extensions are documented by the authors referred to in the opening paragraph of this section.

9.11 COMMENTARY

In this chapter, already much too long, I have introduced a number of standpoints from which nonequilibrium phenomena can be viewed, with abbreviated glimpses of each landscape. The subject is vast and active. I am not aware of a book that covers the entire subject well enough to please even a majority of those who have contributed to it, although there are a number of good samplings of the broad field and specialized treatments of particular approaches. (Balescu [1975]; Cox [1955]; Grandy, Jr. [1987]; Kreuzer [1981]; Liboff [1990b]; Lindblad

[1983]; McLennan [1989]; Prigogine [1962]) I have provided a reference list at the end of the book that contains titles and authors of books and other references that I have found useful as background for this book, even though I may not have referred to some of them in the text. These offer a convenient starting point for further study.

A good introduction to the important work of Truesdell and his school of rational thermodynamics is Truesdell [1984]. The reformulation of thermodynamics to include arbitrary time variations and fading-memory effects is impressive. I look forward to the evolution of a common language that will encourage the interaction of the rest of us with the practitioners of this linguistically obscure but unmistakably important phenomenological theory that, among other things, exposed the limitations of the Onsager relations.

Linear response theory was introduced, but since fluctuations are an important feature in the development of the subject, it is continued in Chapter 10.

My coverage of kinetic theory is severely limited by the lack of space, and the competition of other topics. I have caricatured the subject by means of the relaxation-time approximation, thereby omitting any serious treatment of collisions. As in most of statistical thermophysics, the macroscopic phenomenology is often insensitive to microscopic details, especially when the underlying microscopic conservation laws leak through the statistical treatment to yield the macroscopic laws of motion required as output. Kinetic theory is often treated entirely as a classical discipline, in the tradition of Boltzmann. In its quantum formulation, transport phenomena, especially in the presence of a magnetic field, are more-easily treated by means of the quantum Boltzmann equation than by linear response theory. (van Vliet and Vasilopoulos [1988])

The proof offered in Section 9.7 that the Gibbs-Jaynes equilibrium entropy, obtained by maximizing the information entropy subject to known constraints, agrees exactly with the equilibrium entropy calculated from the master-equation equilibrium-probability set provides a different view of the issue of subjectivity. The objection is that the information entropy is subjective, depending on an individual's knowledge of the system; therefore it cannot represent an objective, measurable property such as thermodynamic equilibrium entropy. What seems to have emerged is that the subjective element of the calculation lies in the assertion that one knows the microstates of the system, and not in the method of assigning probabilities to those states. Once the states are agreed upon, the master-equation and information-theoretic prescriptions for calculating their equilibrium probabilities lead to identical results. I have never seen or heard an objection to the master equation based on the fact that the states to which it attributes probabilities are subjectively determined.

The failure of the Gibbs entropy to evolve to a new equilibrium value is even now sometimes presented as another argument against the Gibbs-Jaynes formulation of thermodynamics. I have discussed and resolved this problem in Section 9.9.1, where it was shown that the quantum time-evolution operator carries forward a fixed set of Ω initial states. These states, and their classical analogs that are propagated by the Liouville operator, maintain their fixed entropy, $S = k \ln \Omega$, while allowing other macroscopic properties to evolve under the influence of the time-dependent Hamiltonian. The classical problem is discussed in Section 9.8.3, where it is shown that inherent uncertainties in our knowledge of any system inevitably lead to a blurring of the phase information that the dynamics professes to preserve, thereby leading to thermalization. As an example, a single, isolated classical harmonic oscillator was shown to lose all phase information, at constant energy, and thus become describable only in thermal terms. Under the action of the Liouville operator, the orbit is restricted to a small volume of the accessible phase space, determined by the initial uncertainties in position and momentum. But the undeniable uncertainty in the frequency, since it cannot be known with infinite accuracy, gradually spreads these uncertainties in such a way that the entire constant-energy volume of the oscillator's phase space is uniformly represented in the evolved probability density. The entropy has evolved to a new maximum value, representative of thermal equilibrium, with the oscillator's entire energy having become thermal. (Note that no coarse-graining of the phase space is required.) Once this kind of blurring is known to exist, presumably for all systems, both classical and quantum, it is no longer necessary, for many purposes, to invoke a blurring operator and complicate the analysis with its effects. When a new state of equilibrium is reached, the information entropy can be remaximized to obtain the correct thermodynamic entropy and all other equilibrium expectation values. For transport processes and other phenomena involving persistent nonequilibrium states, expectation values can be obtained as averages over the evolved initial states. The statistical operator used for this average is the deterministic unitary transformation of the known initial statistical operator. If memory effects need be considered, the use of the statistical operator given in Eq. (9.10.16) is essential.

To display an operator that induces thermalization (or, at least, growth of the information entropy commensurate with the actual rate of blurring) requires more knowledge than is usually available about such things as the rate of loss of phase information attributable to external magnetic, acoustic, and gravitational fields, ignored isotopic mass differences, the presence of impurities, and the limited accuracy of data on atomic masses, electric and magnetic moments, and transition times between states. Nevertheless, in the absence of such knowledge, it is evident that the constancy of the Gibbs entropy (which implies that the evolutionary history of the initial states is completely determined by the known Hamiltonian) is an artifact of the formalism. (Jaynes [1982a]; Liboff [1974]; Liboff [1990a]) The artifact is harmless in many calculations, but the underlying physics should be understood. In particular, for systems that are chaotic, turbulent, or near a critical point, the deterministic evolution of the initial states by a known unitary transformation seems to me to be not only dubious, but also not likely to provide states that continue to be representative of the entire evolved system.

The treatment of irreversible processes driven by thermal sources, as introduced by Mitchell and outlined in Section 9.10.2, has been overlooked by most authors who write about nonequilibrium statistical thermophysics, except for Grandy. I am not aware of many solved problems based on Mitchell's work, although a few examples are to be found. (Grandy, Jr. [1987]; Jaynes [1985]; Mitchell [1967]) Nevertheless, I am convinced that the theory of thermally driven systems is destined to become a major part of the methodology of nonequilibrium statistical mechanics.

I have not included a discussion of reduced statistical operators, or projection operators. An elementary example of these is shown in Chap.1, for the system of coupled harmonic oscillators. There a reduced probability density is obtained for the system by a procedure equivalent to averaging over all other oscillators in the chain. The BBGKY hierarchy of this chapter is similar in method. The quantum procedure is somewhat more complicated (Agarwal [1971]; Zwanzig [1961]), and it may yield non-positive-definite reduced density operators for systems that exhibit non-markovian behavior. (Seke [1990])

I do not have the space to review the extensive work on information entropy and statistical mechanics being produced in Argentina. (Duering et al. [1985]) I am certain that there are other such efforts that I have neglected, and I apologize.

Further aspects of time-dependent and irreversible processes, including the influence of fluctuations, are found in Chapter 10.

9.12 REFERENCES

The principal references are listed here. The complete reference list is to be found in the bibliography at the end of the book.

BALESCU, R. [1975], *Equilibrium and Nonequilibrium Statistical Mechanics*. John Wiley & Sons, New York, NY, 1975.

COHEN, E. G. D. [1962], *The Boltzmann Equation and its Generalization to Higher Densities*. In *Fundamental Problems in Statistical Mechanics*, E. G. D. Cohen, Ed. North-Holland, Amsterdam, 1962, 110–156.

COX, R. T. [1955], *Statistical Mechanics of Irreversible Change*. Johns Hopkins Univ. Press, Baltimore, MD, 1955.

GIBBS, J. W. [1961], *The Scientific Papers of J. Willard Gibbs, V1*. Dover, Mineola, NY, 1961.

GRAD, H. [1958], *Principles of the Kinetic Theory of Gases*. In *Handbuch der Physik*, S. Flugge, Ed. vol. XII, Springer-Verlag, New York–Heidelberg–Berlin, 1958.

GRANDY, JR., W. T. [1980], Principle of Maximum Entropy and Irreversible Processes. *Physics Reports* **62** (1980), 175–266.

GRANDY, JR., W. T. [1987], *Foundations of Statistical Mechanics V.II:Nonequilibrium Phenomena*. D. Reidel, Dordrecht–Boston–London, 1987.

JAYNES, E. T. [1979], *Where Do We Stand On Maximum Entropy?* In *The Maximum Entropy Formalism*, R. D. Levine and M. Tribus, Eds. MIT Press, Cambridge, MA, 1979, A Conference held at MIT, 2-4 May 1978.

JAYNES, E. T. [1980] *The Minimum Entropy Production Principle*. In *Annual Reviews of Physical Chemistry*. vol. 31, 1980, 579–601.

JAYNES, E. T. [1982a], *Gibbs vs. Boltzmann Entropies*. In *Papers on Probability, Statistics, and Statistical Physics*, R. D. Rosenkrantz, Ed. D. Reidel, Dordrecht–Boston–London, 1982, Reprinted from Am. J. Phys. **33**,391-398 (1965).

KREUZER, H. J. [1981], *Nonequilibrium Thermodynamics and its Statistical Foundations*. Oxford University Press, London, 1981.

KUBO, R. [1957], Statistical Mechanical Theory of Irreversible Processes I. General Theory and Simple Applications to Magnetic and Conduction Problems. *Journal of the Physical Society of Japan* **12** (1957), 570–586.

KUBO, R., YOKOTA, M. AND NAKAJIMA, S. [1957], Statistical Mechanical Theory of Irreversible Processes II. Response to Thermal Disturbance. *Journal of the Physical Society of Japan* **12** (1957), 1203–1211.

LIBOFF, R. L. [1990b], *Kinetic Theory: Classical, Quantum, and Relativistic Descriptions.* Prentice Hall, Englewood Cliffs, NJ, 1990.

MONTROLL, E. W. [1962], *Some Remarks on the Integral Equations of Statistical Mechanics.* In *Fundamental Problems in Statistical Mechanics*, E. G. D. Cohen, Ed. North-Holland, Amsterdam, 1962, 223–249.

THOMSEN, J. S. [1953], Logical Relations among the Principles of Statistical Mechanics and Thermodynamics. *Physical Review* **91** (1953), 1263–1266.

TRIBUS, M. [1961], *Thermostatics and Thermodynamics.* D. van Nostrand Co., Inc., New York, 1961.

TRUESDELL, C. A. [1984], *Rational Thermodynamics, 2nd Ed..* Springer-Verlag, New York–Heidelberg–Berlin, 1984.

ZWANZIG, R. W. [1961], *Statistical Mechanics of Irreversibility.* In *Lectures in Theoretical Physics*, V. III, W. E. Brittin, B. W. Downs and J. Downs, Eds. Interscience Publishers, A division of John Wiley & Sons, London, 1961, 106–141.

9.13 PROBLEMS

1. Find the Onsager coefficients for combined particle and thermal transport in one dimension, in the presence of a temperature gradient and a parallel external electric field, for a fermion gas of noninteracting electrons. Show that the expected symmetry follows from the direct calculation, to the lowest nontrivial order.

2. Show that an apparent generalization of the governing equations of rational thermodynamics, Eq. (9.2.54), to

$$W = W_0(T, \{X_i\}) - \sum P_i \dot{X}_i,$$
$$F = F_0(T, \{X_i\}) - \sum C_i \dot{X}_i,$$
$$S = S_0(T, \{X_i\}) - \sum D_i \dot{X}_i,$$

after application of the reduced dissipation inequality and recognition that $W_0 = 0$ at equilibrium, reduces to the original set of governing equations.

3. Show that the equations of motion of a classical mechanical system can be written as $\dot{\mathbf{f}} = -i\mathcal{L}_n \mathbf{f}$, where $\mathcal{L}_n$ is the classical Liouville operator of Eq. (9.8.19) and $\mathbf{f} = \{\mathbf{p}_1, \mathbf{r}_1, \ldots, \mathbf{p}_n, \mathbf{r}_n\}$. Find the formal solution of the equation for $\mathbf{f}$, when $\mathcal{L}_n$ is not explicitly dependent on t.

4. Suppose for a classical system the Hamiltonian is $\mathcal{H} = \mathcal{H}_0 + \lambda \mathcal{H}_1(t)$, as in Eq. (9.9.5), with $\mathcal{H}_0$ independent of t, and the small perturbation $\lambda \mathcal{H}_1(t)$ turned on at $t = 0$. Suppose further that $\mathcal{L} = \mathcal{L}_0 + \lambda \mathcal{L}_1(t)$, where the Liouville operator is derived in the usual way from the Hamiltonian. Show that the integral equations satisfied by the Liouville function and any dynamic variable $A(t)$, including $\mathbf{f}(t)$ from Problem 3, are given by

$$\rho(t) = \rho_0(t) + i\lambda \int_0^t dt' e^{i(t-t')\mathcal{L}_0} \mathcal{L}_1(t')\rho(t'), \quad \text{and}$$
$$A(t) = A_0(t) - i\lambda \int_0^t dt' e^{i(t-t')\mathcal{L}_0} \mathcal{L}_1(t')A(t'),$$

where $\dot{\rho}_0 = i\mathcal{L}_0\rho_0$ and $\dot{A}_0 = -i\mathcal{L}_0 A_0$. (These are the classical analogs of Eq. (9.9.12) and Eq. (9.9.15).)

5. Use the definitions given in Eq. (9.7.3) to Eq. (9.7.6) for the principles of microscopic reversibility (**M**), detailed balance (**D**), the ergodic hypothesis (**E**), the (**W**) hypothesis, and the so-called *Second Law of Thermodynamics*, $\dot{S} \geq 0$, (denoted by (**S**)) to show that (**E**) $\rightleftharpoons$ (**W**) $\rightleftharpoons$ (**S**) by showing that (a) (**E**) $\rightarrow$ (**W**), (b) (**W**) $\rightarrow$ (**S**), and (c) (**S**) $\rightarrow$ (**E**).

6. With the definitions and proofs of Problem 5, show that (**M**) $\rightleftharpoons$ ((**E**) + (**D**)), as claimed in Eq. (9.7.8).

7. Show by counterexample that (a) (**D**) $\xrightarrow{\text{not}}$ (**M**), and (b) (**W**) $\xrightarrow{\text{not}}$ (**M**). Choose a three-level system for simplicity.

8. A 2-d gaussian probability density is written as

$$\rho(x_1, x_2) = \left(\frac{1}{2\pi\sigma\tau(1-\gamma^2)^{1/2}}\right) \exp\left[\left(\frac{-1}{2(1-\gamma^2)}\right)\left(\frac{x_1^2}{\sigma^2} + \frac{x_2^2}{\tau^2} - \frac{2\gamma x_1 x_2}{\sigma\tau}\right)\right],$$

where it is directly verifiable that $\sigma^2 = \langle x_1^2 \rangle$, $\tau^2 = \langle x_2^2 \rangle$, and $\langle x_1 x_2 \rangle = \sigma\tau\gamma$, where γ is the correlation function. a. Find the entropy $S/k = -\int d^2x\, \rho \ln h'\rho$ that corresponds to the given ρ, where h' is an appropriate constant. b. Show that this entropy agrees with that calculated by another method for the microcanonical simple harmonic oscillator of problem 11, Chapter 1, and choose a dimensionally correct constant h'.

9. In a one-dimensional repeatable experiment to measure properties of uniform microscopic spheres suspended in a liquid, it is observed that the average displacement of a sphere from its starting point is summarized by $\langle x(t) \rangle = 0$, and the mean squared displacement by $\langle x^2(t) \rangle = at = 2Dt$, where the constant a is expressed in terms of the diffusivity D, in accord with expectations from Eq. (9.3.73). Use the experimental results in the formalism of information theory to derive an expression for the probability density $\rho(x, t)$, and relate the result to Eq. (9.3.71).

10. Solve the relaxation-time approximation to the Boltzmann equation for the Boltzmann function $f(\mathbf{r}, \mathbf{v}, t) = f_0(\mathbf{v}) + f_1(\mathbf{v}, t)$, for a uniform system of electrons, subject to a spatially uniform, time-dependent electric field given by $\mathbf{E} = i E_0 \exp(i\omega t)$, switched on at $t = 0$. The system is taken to be at equilibrium before the field is turned on, with $f = f_0$, and the relaxation time is assumed to be independent of speed. Calculate the frequency-dependent current density (a) when f_0 is maxwellian, and (b) when f_0 is for noninteracting fermions. Is the reactive part of the steady-state current inductive or capacitive?

11. One kilogram of water, initially at equilibrium at $20°$C. is heated to $100°$C. and left in an insulated container, again at equilibrium. Calculate the ratio Ω_f/Ω_0 of the number of final microstates to initial microstates of the system. (This problem is to illustrate the enormous increase in the number of microstates of a macroscopic thermodynamic system as a consequence of a finite entropy increase. Most of these final states are excluded from the subspace of initial states that are propagated by Eq. (9.9.23) from the initial state.)

9.14 SOLUTIONS

1. The flux densities J_n and J_q are given by Eqs. (9.3.52) and (9.3.54). The Onsager equations must be rewritten from Eq. (9.2.46), as was done in Eq. (9.3.43) and (9.3.44), except for fermions.

The proper Onsager equations then are found to be

$$J_n = -\left[\frac{L_{nq}}{T^2} - \frac{L_{nn}\mu_0}{6T^2}\left(\frac{\pi kT}{\mu_0}\right)^2\right]\frac{dT}{dy} - \frac{L_{nn}}{T}eE_y, \quad \text{and}$$

$$J_q = -\left[\frac{L_{qq}}{T^2} - \frac{L_{qn}\mu_0}{6T^2}\left(\frac{\pi kT}{\mu_0}\right)^2\right]\frac{dT}{dy} - \frac{L_{qn}}{T}eE_y.$$

From these and the expressions derived from the Boltzmann equation, the Onsager coefficients may be read off as

$$L_{nn} = \frac{nT\tau}{m}; \quad L_{qn} = \frac{7}{12}\frac{nT\tau\mu_0}{m}\left(\frac{\pi kT}{\mu_0}\right)^2;$$

$$L_{nq} = \frac{L_{nn}\mu_0}{6}\left(\frac{\pi kT}{\mu_0}\right)^2 + \frac{5}{12}\frac{nT\tau\mu_0}{m}\left(\frac{\pi kT}{\mu_0}\right)^2 = L_{qn}; \quad \text{and}$$

$$L_{qq} = \frac{L_{qn}\mu_0}{6}\left(\frac{\pi kT}{\mu_0}\right)^2 + \frac{1}{3}\frac{nT\mu_0^2\tau}{m}\left(\frac{\pi kT}{\mu_0}\right)^2 \approx \frac{n\pi^2 k^2 T^3\tau}{3m}.$$

2. From the reduced dissipation inequality, Eq. (9.2.53), use of the supposed generalizations gives

$$\left[\frac{\partial F_0}{\partial T} + S_0 - \left(\sum \dot{C}_i \dot{X}_i - \sum D_i \dot{X}_i\right)\right]\dot{T} - W_0 + \sum\left(\frac{\partial F_0}{\partial X_i} + P_i\right)$$

$$- \sum_{i,j}\frac{\partial C_i}{\partial X_i}\dot{X}_i\dot{X}_j - \sum C_i\ddot{X}_i \leq 0.$$

Since W_0 is a function of system properties that are defined without reference to the state of equilibrium, it must be identically zero. Since any $\ddot{X}_i$ can have either sign, while all other terms in the inequality are zero, it is necessary that $C_i = 0$ for all i. Similar arguments require that $D_i = 0$, reducing the supposed generalization to the original equations.

3. By direct operation with $\mathcal{L}_n$, it may be seen that

$$-i\mathcal{L}_n \mathbf{f} = \left\{-\frac{\partial\mathcal{H}}{\partial \mathbf{r}_1}, \frac{\partial\mathcal{H}}{\partial \mathbf{p}_1}, \ldots, -\frac{\partial\mathcal{H}}{\partial \mathbf{r}_n}, \frac{\partial\mathcal{H}}{\partial \mathbf{p}_n}\right\} = \dot{\mathbf{f}}.$$

By identifying components on both sides of the equation, one obtains the classical equations of motion as derived from the Hamiltonian. The formal solution for $\mathbf{f}$ is

$$\mathbf{f}(t) = e^{-i\mathcal{L}_n t}\mathbf{f}(0).$$

4. The equation to be solved is

$$\dot{\rho} = i\mathcal{L}\rho = i(\mathcal{L}_0 + \lambda\mathcal{L}_1)\rho.$$

Direct differentiation of the proposed expression for ρ gives

$$\dot{\rho} = \dot{\rho}_0 + i\lambda\mathcal{L}_1(t)\rho(t) + i\mathcal{L}_0\left(i\lambda\int_0^t dt' e^{i(t-t')\mathcal{L}_0}\mathcal{L}_1(t')\rho(t')\right)$$

$$= i\mathcal{L}_0\rho_0 + i\lambda\mathcal{L}_1(t)\rho(t) + i\mathcal{L}_0[\rho(t) - \rho_0(t)] = i\mathcal{L}\rho(t),$$

as required. A similar procedure shows that $\dot{A} = -i\mathcal{L}A(t)$.

5. **a.** The master equation says $\dot{p}_i = \sum_j (p_j w_{ji} - p_i w_{ij})$. At equilibrium, $\dot{p}_i = 0$ for all i, giving $0 = \sum_j (p_j w_{ji} - p_i w_{ij})$. Since $(\mathbf{E}) \rightarrow p_i = 1/\Omega$ for all i, the p_i's factor out, leaving $\sum_j w_{ij} = \sum_j w_{ji}$.

b. Since $\dot{S} = -k \sum_i \dot{p}_i \ln p_i$, it follows that

$$-\frac{\dot{S}}{k} = \sum_{i,j} (p_j w_{ji} - p_i w_{ij}) \ln p_i$$

$$= \sum_{ij} (p_j w_{ji} \ln p_i - p_j w_{ji} \ln p_j + p_j w_{ji} \ln p_j - p_i w_{ij} \ln p_i),$$

where the extra terms are equal, and could cancel. Note that the final sum is equivalent to the third sum with the indices interchanged, so they cancel, leaving

$$-\frac{\dot{S}}{k} = \sum_{i,j} (p_j w_{ji} \ln p_i - p_j w_{ji} \ln p_j) = \sum_{i,j} (p_j w_{ji} \ln(p_i/p_j))$$

$$\leq \sum_{i,j} p_j w_{ji} \left(\frac{p_i}{p_j} - 1\right) = \sum_{i,j} w_{ji}(p_i - p_j) = \sum_i p_i \sum_j (w_{ji} - w_{ij}) = 0,$$

by $(\mathbf{W})$, where the inequality $\ln x \leq x - 1$ has been used, and the last step came from an interchange of indices in the second summation. The result is that $\dot{S} \geq 0$, which shows that $(\mathbf{W}) \rightarrow (\mathbf{S})$, as required.

c. Given $\dot{S} \geq 0$, it follows that at equilibrium the entropy must be a maximum. For a microcanonical system, it was shown in Chapter 3 that maximization of the entropy requires that $p_i = 1/\Omega$ for all i. Therefore $(\mathbf{S}) \rightarrow (\mathbf{E})$, and the circuit is complete.

6. Note that $(\mathbf{M})$ is a special case of $(\mathbf{W})$, so $(\mathbf{M}) \rightarrow (\mathbf{W}) \rightarrow (\mathbf{E})$, from Problem (5.). Also, it is evident that $(\mathbf{M}) + (\mathbf{E}) \rightarrow (\mathbf{D})$, so $(\mathbf{M}) \rightarrow (\mathbf{D})$. Another inspection shows that $(\mathbf{E}) + (\mathbf{D}) \rightarrow (\mathbf{M})$, verifying Eq. (9.7.8).

7.**a.** For a 3-level system, pick three arbitrary, unequal probabilities, such as $p_1 = 1/2$, $p_2 = 1/3$, and $p_3 = 1/6$. Then choose the transition probabilities w_{ij} to satisfy $(\mathbf{D})$, or $p_i w_{ij} = p_j w_{ji}$. These conditions lead to a transition matrix proportional to

$$W = \begin{pmatrix} 0 & 2 & 1 \\ 3 & 0 & 1 \\ 3 & 2 & 0 \end{pmatrix}$$

which is not symmetric. Thus $(\mathbf{D})$ does not imply $(\mathbf{M})$.

b. An evident transition matrix that satisfies $(\mathbf{W})$ but not $(\mathbf{M})$ is

$$W = \begin{pmatrix} 0 & 1 & 0 \\ 0 & 0 & 1 \\ 1 & 0 & 0 \end{pmatrix}.$$

Here the transitions can take place from 1 to 2, from 2 to 3, and from 3 to 1, all with equal transition probabilities, but without detailed balance, even at equilibrium. A similar situation occurs with the wind-tree model of Chapter 1.

8.a. With $\theta \equiv (1 - \gamma^2)$, the entropy is given by

$$S/k = - \int_{-\infty}^{\infty} dx_1 \int_{-\infty}^{\infty} dx_2 \rho \ln h' \rho$$

$$= \left(\frac{1}{2\pi\sigma\tau\theta^{1/2}} \right) \left(\frac{1}{2\theta} \right) \int_{-\infty}^{\infty} dx_1 \int_{-\infty}^{\infty} dx_2 \exp\left[-\frac{1}{2\theta} \left(\frac{x_1^2}{\sigma^2} + \frac{x_2^2}{\tau^2} - \frac{2\gamma x_1 x_2}{\sigma\tau} \right) \right] \times$$

$$\times \left[\ln\left(\frac{2\pi\sigma\tau\sqrt{1-\gamma^2}}{h'} \right) + \left(\frac{x_1^2}{\sigma^2} + \frac{x_2^2}{\tau^2} - \frac{2\gamma x_1 x_2}{\sigma\tau} \right) \right].$$

By use of

$$\int_{-\infty}^{\infty} dx e^{-ax^2+bx} = \sqrt{\frac{\pi}{a}} e^{-b^2/4a}$$

and its derivatives with respect to a and b, the required integrations are direct. They yield

$$S/k = 1 + \ln\left[\left(\frac{2\pi}{h'} \right) \sigma\tau\sqrt{1-\gamma^2} \right].$$

b. For the oscillator of Problem 11, Chapter 1, note that h' has already been identified as $h' = \hbar\omega_0$. The matrix W of that problem can be rewritten in terms of the identifications given for σ, τ, and γ as

$$W = \begin{pmatrix} \langle (y_1')^2 \rangle & \langle y_1' y_2' \rangle \\ \langle y_1' y_2' \rangle & \langle (y_2')^2 \rangle \end{pmatrix} = \begin{pmatrix} \sigma^2 & \sigma\tau\gamma \\ \sigma\tau\gamma & \tau^2 \end{pmatrix}$$

and $\det W = \sigma^2\tau^2(1 - \gamma^2)$. The entropy from Eq. (1.7.53) is $S/k = 1 + \ln[(\det W)^{1/2}/\hbar] = 1 + \ln[\sigma\tau\sqrt{1-\gamma^2}/\hbar\omega_0]$. When σ^2, τ^2, and $\sigma\tau\gamma$ are identified with the ω-averaged quantities found in part (g) of problem 11, Chapter 1, the information entropy of the isolated harmonic oscillator is seen to evolve to the time-independent equilibrium value given in Eq. (9.8.34).

9. The information entropy is to be maximized, subject to constraints; the conditions to be satisfied are:

$$S/k = - \int_{-\infty}^{\infty} dx \, \rho \ln h' \rho, \quad \text{maximized subject to}$$

$$1 = \int_{-\infty}^{\infty} dx \, \rho,$$

$$0 = \int_{-\infty}^{\infty} dx \, \rho \, x \ (= \langle x(t) \rangle),$$

$$2Dt = \int_{-\infty}^{\infty} dx \, \rho \, x^2 \ (= \langle x^2(t) \rangle).$$

The well-known variational procedure explained in Chapter 1 yields

$$\rho = Z^{-1} e^{-\lambda_1 x - \lambda_2 x^2},$$

where the λ's are undetermined (time-dependent) multipliers, and

$$Z = \int_{-\infty}^{\infty} dx \, e^{-\lambda_1 x - \lambda_2 x^2} = \sqrt{\frac{\pi}{\lambda_2}} e^{\lambda_1^2/4\lambda_2}.$$

The use of this expression for ρ in the constraint equations gives $\lambda_1 = 0$ and $\lambda_2 = 1/4Dt$. The final expression for ρ is thus seen to be

$$\rho(x,t) = \frac{1}{\sqrt{4\pi Dt}} e^{-x^2/4Dt},$$

a one-dimensional version of Eq. (9.3.71).

10. The Boltzmann equation may be written

$$\frac{\partial f}{\partial t} - \frac{eE(t)}{m}\frac{\partial f}{\partial v_x} = -\frac{f - f_0}{\tau_r},$$

where τ_r is the relaxation time. With $f = f_0 + f_1$, the equation for the first-order term is

$$\frac{\partial f_1}{\partial t} + \frac{f_1}{\tau_r} = \frac{eE(t)}{m}\frac{\partial f_0}{\partial v_x}.$$

The general solution is

$$f_1(\mathbf{v},t) = \frac{e}{m}\frac{\partial f_0}{\partial v_x}\int_{-\infty}^{t} dt'\, e^{-(t-t')/\tau_r} E(t') = \frac{eE_0 e^{i\omega t}}{m}\frac{\partial f_0}{\partial v_x}\int_0^t d\tau\, e^{-(i\omega+1/\tau_r)\tau}$$

$$= \frac{eE_0}{m}\frac{\partial f_0}{\partial v_x}\left(\frac{e^{i\omega t} - e^{-t/\tau_r}}{i\omega + 1/\tau_r}\right),$$

where the substitution $\tau = t - t'$ was used in the integration. The current density is

$$j_e(t) = -e\int d^3v\, v_x f(\mathbf{v},t) = -\frac{e^2\tau_r}{m}\left(\frac{E_0\left(e^{i\omega t} - e^{t/\tau_r}\right)}{1 + i\omega\tau_r}\right)\int d^3v\, v_x \frac{\partial f_0}{\partial v_x}$$

$$= \frac{ne^2\tau_r}{m}\left(\frac{E_0\left(e^{i\omega t} - e^{-t/\tau_r}\right)}{1 + i\omega\tau_r}\right),$$

after integration by parts, independent of the specific form of f_0. For steady-state behavior, the factor $1 + i\omega\tau_r$ in the denominator shows that the reactance is inductive, both because of the positive sign and because of the direct proportionality to ω.

11. For a microcanonical system, the entropy is given by $S = k\ln\Omega$, where Ω is the number of equally probable microstates that could reproduce the given macrostate. Therefore, $\Omega_f/\Omega_0 = \exp[(S_f - S_0)/k]$. The entropy change of the water is

$$\frac{S_f - S_0}{k} = \frac{1}{k}\int_{T_0}^{T_f}\frac{MC}{T}dT,$$

where $M = 1$kg is the mass and $C = 4180$J/kgK is the approximate specific heat capacity of water. With $k = 1.38 \times 10^{-23}$ J/K, the result is $(S_f - S_0)/k = 7.31 \times 10^{25}$, and $\Omega_f/\Omega_0 = \exp[7.31 \times 10^{25}]$, a respectably large number!

10. FLUCTUATIONS

10.1 INTRODUCTION

Fluctuations are the underlying phenomena of thermodynamics. In disciplines such as classical mechanics and optics, variational principles are often mysteriously invoked to assert that a set of material particles or photons "choose" a space-time evolution that minimizes some property such as the action or the travel time, implying that in some arcane manner such a system has the opportunity to explore these alternatives before (or in the course of) its manifest behavior. The Feynman path-integral method, presented in Chapter 3, exploits and partially demystifies the processes involved. For thermodynamic systems, however, the microscopic activities by which equilibrium is maintained or equilibration attained are the physical fluctuations attributable to the motions of the particles comprising them.

Thermodynamic response functions and stability conditions are shown in Chapters 2, 4, and 6 to be related to certain variances that are often called fluctuations, and even in the harmonic-oscillator system of Chapter 1, the temperature emerged from a covariance matrix. These examples have hinted at the fundamental importance of fluctuations, without offering any study of the dynamics of the fluctuations *per se*, except for the simple model systems of the Ehrenfests, treated in Chapter 1 by means of master equations. The Onsager relations and the Kubo formula of Chapter 9 depend for their microscopic roots on the kinds of fluctuation phenomena examined in this chapter.

10.2 BROWNIAN MOTION

The phenomenon of Brownian motion, first described by the botanist Robert Brown in 1828, during his studies that began with the erratic migration of pollen grains suspended in water. It was studied systematically by Einstein and, later, by Smoluchowski. (Einstein [1905]; Smoluchowski [1916]) Brown described the motion of a small object such as a pollen grain, a soot particle of the sort found in the suspension called India ink, or even a fragment of the Sphinx, by observing and recording its position at regular time intervals. He then drew straight lines connecting successive positions, finding what appeared to be a jagged and aimless path through the accessible space. A continuous observation would have shown an almost imperceptible jiggling of the particle, as a consequence of its many collisions with the molecules of the fluid surrounding it, but an almost smooth meandering from one observation point to the next. Nevertheless, Brownian motion is sometimes described as if the particle actually followed the straight lines between observation points. From this description, Brownian motion is then modelled as a Markovian random walk. In its full generality, the random walker takes successive steps of arbitrary length in randomly chosen directions; a common restriction limits the steps to fixed length; and an interesting and useful further restriction confines the walker to points on a periodic lattice. Most of the conclusions about the consequences of Brownian motion are not particularly sensitive to the details of the model, but there are unexpected dependencies on the dimensionality of the space available for the random walk. The literature on random walks, with a wide variety of restrictions, is extensive, and only a brief introduction can be given here. (Barber and Ninham [1970]; Montroll [1964]; Montroll [1967]; Spitzer [1956])

10.2.1 Markovian Random Walks on Lattices. A Markovian random walk is one for which the walker's next step is not influenced by his previous steps. The first example is that of a one-dimensional random walk on the integers. The walker starts at the origin and steps in the $+x$ direction with probability p, or in the $-x$ direction with probability $q = 1 - p$. The probability that the walker will be at position s at step t, $P_t(s)$, is given in terms of his probable positions at step $t - 1$ as

$$P_t(s) = qP_{t-1}(s+1) + pP_{t-1}(s-1). \tag{10.2.1}$$

Define the generating function

$$G(\phi,t) = \sum_{s=-\infty}^{\infty} P_t(s)e^{is\phi}, \tag{10.2.2}$$

and note that this is the characteristic function for the discrete probability set. A difference equation for G may be found from Eq. (10.2.1):

$$G(\phi,t) = \sum_{s=-\infty}^{\infty} [qP_{t-1}(s+1) + pP_{t-1}(s-1)] = (qe^{-i\phi} + pe^{i\phi})G(\phi,t-1)$$
$$= (qe^{-i\phi} + pe^{i\phi})^t, \tag{10.2.3}$$

where the final step is an obvious solution of the difference equation. The coefficient of $\exp(is\phi)$ in the generating function is the probability $P_t(s)$ that the walker will be at s after t steps. To extract $P_t(s)$ from the generating function, recall that

$$\frac{1}{2\pi}\int_{-\pi}^{\pi} e^{-ir\phi}d\phi = \delta_0^r,$$

so that

$$P_t(s) = \frac{1}{2\pi}\int_{-\pi}^{\pi} d\phi e^{-is\phi}G(\phi,t). \tag{10.2.4}$$

When $p = q = 1/2$,

$$G(\phi,t) = (\cos\phi)^t, \tag{10.2.5}$$

and

$$P_t(s) = \frac{1}{2\pi}\int_{-\pi}^{\pi} d\phi \cos^t\phi\, e^{-is\phi} = \frac{1}{2^{t+1}}\binom{t}{(t-s)/2}\left[1 + (-1)^{t+s}\right], \tag{10.2.6}$$

where the final factor assures that the walker is at an even/odd integer after an even/odd number of steps. Note that $P_t(-s) = P_t(s)$ for $p = q = 1/2$, as expected. The generating function for the probability that the walker is at s after any number of steps is

$$U_1(z,s) = \sum_{n=0}^{\infty} z^n P_n(s) = \frac{1}{2\pi}\sum_{n=0}^{\infty}\int_{-\pi}^{\pi} d\phi e^{-is\phi}z^n\cos^n\phi; \quad |z| < 1$$
$$= \frac{1}{2\pi}\int_{-\pi}^{\pi}\frac{d\phi e^{-is\phi}}{1 - x\cos\phi} = \frac{(1 - \sqrt{1-z^2})^{|s|}}{z^{|s|}\sqrt{1-z^2}}, \tag{10.2.7}$$

a result that will be used in subsequent derivations.

For a random walk on an unbounded 2-*d* lattice of integers, starting at $(0,0)$, choose p_1, q_1 as the probabilities for a step in the $\pm x_1$ direction, and p_2, q_2 as those for the $\pm x_2$ direction, such that $p_1 + q_1 + p_2 + q_2 = 1$. The coefficient of $\exp[i(s_1\phi_1 + s_2\phi_2)]$ in the generating function

$$G(\phi_1, \phi_2, t) = \left(p_1 e^{i\phi_1} + q_1 e^{-i\phi_1} + p_2 e^{-i\phi_2} + q_2 e^{-i\phi_2}\right)^t \tag{10.2.8}$$

is the probability $P_t(s)$ that the walker will be at $s = is_1 + js_2$, after having taken t steps. For $p_i = q_i = 1/4$, the generating function becomes

$$G(\phi_1, \phi_2, t) = [(1/2)(\cos\phi_1 + \cos\phi_2)]^t. \tag{10.2.9}$$

The probability in either case is given by

$$P_t(s) = \frac{1}{(2\pi)^2} \int_{-\pi}^{\pi} d\phi_1 \int_{-\pi}^{\pi} d\phi_2 G(\phi_1, \phi_2, t) e^{-i\phi \cdot s}. \tag{10.2.10}$$

For equal probabilities, the generating function corresponding to Eq. (10.2.7) is

$$U_2(z, s) = \sum_{n=0}^{\infty} z^n P_n(s) = \frac{1}{(2\pi)^2} \int_{-\pi}^{\pi} \int_{-\pi}^{\pi} \frac{d^2\phi e^{-is\cdot\phi}}{1 - (z/2)(\cos\phi_1 + \cos\phi_2)}. \tag{10.2.11}$$

The generalization to d dimensions, with all probabilities equal to $1/2d$, and the lattices not bounded, is straightforward. The results are

$$P_t(s) = \frac{1}{(2\pi)^d} \int_{-\pi}^{\pi} \cdots d \cdots \int_{-\pi}^{\pi} d^d\phi e^{-is\cdot\phi} [(1/d)(\cos\phi_1 + \cdots + \cos\phi_d)]^t, \tag{10.2.12}$$

and

$$U_d(z, s) = \frac{1}{(s\pi)^d} \int_{-\pi}^{\pi} \cdots \int_{-\pi}^{\pi} d^d\phi \frac{\exp(-is\cdot\phi)}{1 - (z/d)(\cos\phi_1 + \cdots + \cos\phi_d)}. \tag{10.2.13}$$

For more exotic walks, a general, systematic procedure for obtaining $U_d(z, s)$ is required. For this purpose, define f_j as the probability that any event ϵ may occur for the first time on the jth trial, and u_j as the probability that ϵ occurs on the jth trial, whether or not it has previously occurred. For formalistic convenience, set $u_0 \equiv 1$ and $f_0 \equiv 0$. Construct the generating functions

$$(a) \quad U(z) = u_0 z^0 + u_1 z^1 + \cdots u_n z^n + \cdots \quad \text{and}$$
$$(b) \quad F(z) = f_1 z^1 + f_2 z^2 + \cdots, \tag{10.2.14}$$

and note that $u_n = u_0 f_n + u_1 f_{n-1} + \cdots + u_{n-1} f_1$, since, if the event occurs after step n, it must have occurred first at some step k and then recurred in $n - k$ steps. When this equation is multiplied by z^n and summed over n, the result is

$$\sum_{n=0}^{\infty} z^n u_n = U(z) = \sum_{n=0}^{\infty} z^n \sum_{k=0}^{n-1} u_k f_{n-k}$$
$$= 1 + u_o f_1 z + (u_0 f_2 + u_1 f_1)z^2 + (u_0 f_3 + u_1 f_2 + u_2 f_1)z^3 + \cdots$$
$$= 1 + u_0(f_1 z + f_2 z^2 + f_3 z^3 + \cdots) + u_1 z(f_1 z + f_2 z^2 + \cdots) + \cdots$$
$$= 1 + F(z)U(z), \tag{10.2.15}$$

which leads to

$$U(z) = \frac{1}{1 - F(z)}, \quad \text{and} \quad F(z) = 1 - \frac{1}{U(z)}. \tag{10.2.16}$$

The probability that ϵ ever occurs is $F(1) = 1 - 1/U(1)$. Thus, if $U(1)$ is infinite, the event ϵ is a certainty, and if $U(1)$ is finite, there is then a finite probability that ϵ will never occur. Note that $U_d(z, \mathbf{s})$ generates the probabilities that a walker will be at $\mathbf{s}$ at any step n, so that $U_d(1, \mathbf{s})$ is exactly the function $U(1)$ for the event that the walker arrives at $\mathbf{s}$ at some time during the walk. As an example, it is easily seen that a random walker on the integers in one dimension is certain to return to the origin, since $U_1(z, 0) = 1/\sqrt{1 - z^2}$ from Eq. (10.2.7), and $U_1(1, 0) = \infty$. This same result can be found more laboriously by adding the probabilities, *e.g.* $U_1(1, 0) = 1 + 1/2 + 3/8 + 5/16 + \cdots$, which diverges.

The probability that a random walker, starting at the origin, escapes is seen to be $1 - F(1)$. Thus the probability of escape in one dimension with equally probable steps in either direction is zero. More generally, when $p > q$, it is left as an exercise to show that $F_1(1, 0) = 2q < 1$, and the probability of escape is $1 - 2q > 0$. In two dimensions, the probability of escape for symmetric walks is again zero, but in three and more dimensions, escape is possible.

These calculations may be generalized to include steps of arbitrary length and direction. Let $p(\mathbf{s})$ represent the probability of a step from $\mathbf{s}'$ to $\mathbf{s}' + \mathbf{s} = \mathbf{s}''$, regarded as independent of $\mathbf{s}'$. Then, for a denumerable set of possible finite steps, the lattice version of a Green function equation for P_t is

$$P_{t+1}(\mathbf{s}) = \sum_{\mathbf{s}'} P_t(\mathbf{s}') p(\mathbf{s} - \mathbf{s}'). \tag{10.2.17}$$

Define the generating functions

$$(a) \quad \Pi_t(\phi) = \sum_{\mathbf{s}} P_t(\mathbf{s}) e^{i\phi \cdot \mathbf{s}}, \quad \text{and}$$

$$(b) \quad \lambda(\phi) = \sum_{\mathbf{s}} p(\mathbf{s}) e^{i\phi \cdot \mathbf{s}}. \tag{10.2.18}$$

Multiply Eq. (10.2.17) by $\exp(i\phi \cdot \mathbf{s})$ and sum on $\mathbf{s}$ to get

$$\sum_{\mathbf{s}} P_{t+1}(\mathbf{s}) e^{i\phi \cdot \mathbf{s}} = \Pi_{t+1}(\phi) = \sum_{\mathbf{s}} \sum_{\mathbf{s}'} p(\mathbf{s} - \mathbf{s}') P_t(\mathbf{s}') e^{i\phi \cdot (\mathbf{s} - \mathbf{s}' + \mathbf{s}')}, \quad \text{or}$$

$$\Pi_{t+1}(\phi) = \lambda(\phi) \Pi_t(\phi). \tag{10.2.19}$$

Note that if the walker is known to have started at the origin, then $\Pi_0(\phi) = 1$, from Eq. (10.2.18), so that Eq. (10.2.19) becomes

$$\Pi_t(\phi) = [\lambda(\phi)]^t. \tag{10.2.20}$$

The d-dimensional Fourier inversion of Eq. (10.2.18)(a) gives

$$P_t(\mathbf{s}) = \frac{1}{(2\pi)^d} \int_{-\pi}^{\pi} \cdots \int_{-\pi}^{\pi} d^d\phi \, [\lambda(\phi)]^t \, e^{-i\mathbf{s} \cdot \phi}. \tag{10.2.21}$$

From this result, the generating function $U_d(z, \mathbf{s})$ is found to be

$$U_d(z, \mathbf{s}) = \sum_{t=0}^{\infty} z^t P_t(\mathbf{s}) = \frac{1}{(2\pi)^d} \int_{-\pi}^{\pi} \cdots \int_{-\pi}^{\pi} d^d\phi \frac{\exp(-i\mathbf{s} \cdot \phi)}{1 - z\lambda(\phi)}, \qquad (10.2.22)$$

which is an evident generalization of Eq. (10.2.13).

For a simple cubic lattice with all probabilities equal, $\lambda_s(\phi)$ has already been given in Eq. (10.2.13) as $\lambda_s(\phi) = (1/3)(\cos\phi_1 + \cos\phi_2 + \cos\phi_3)$. For a face-centered cubic lattice, Eq. (10.2.18)(b) gives

$$\lambda_f(\phi) = \frac{1}{12} \left[e^{i(\phi_1+\phi_2)} + e^{i(\phi_1-\phi_2)} + e^{i(-\phi_1+\phi_2)} + e^{i(-\phi_1-\phi_2)} + \cdots \right],$$

where the other terms are similar pairwise combinations of ϕ_3 with either ϕ_1 or ϕ_2. The combined terms can be reduced to

$$\lambda_f(\phi) = (1/3)(\cos\phi_1 \cos\phi_2 + \cos\phi_2 \cos\phi_3 + \cos\phi_3 \cos\phi_1). \quad \text{(fcc)}$$

For a body-centered cubic lattice, with equally probable steps from a lattice point to any one of the eight adjacent cube centers, use of Eq. (10.2.18)(b) leads to

$$\lambda_b(\phi) = \cos\phi_1 \cos\phi_2 \cos\phi_3.$$

From these results, and the choice $\mathbf{s} = 0$, the U_3 functions are

$$U_3^{(s)} = \frac{1}{(2\pi)^3} \int_{-\pi}^{\pi}\int_{-\pi}^{\pi}\int_{-\pi}^{\pi} \frac{d\phi_1 d\phi_2 d\phi_3}{1 - (z/3)(\cos\phi_1 + \cos\phi_2 + \cos\phi_3)}, \quad \text{(sc)}$$

$$U_3^{(f)} = \frac{1}{(2\pi)^3} \int_{-\pi}^{\pi}\int_{-\pi}^{\pi}\int_{-\pi}^{\pi} \frac{d\phi_1 d\phi_2 d\phi_3}{1 - (z/3)(\cos\phi_1 \cos\phi_2 + \cos\phi_2 \cos\phi_3 + \cos\phi_3 \cos\phi_1)}, \quad \text{(fcc)}$$

$$U_3^{(b)} = \frac{1}{(2\pi)^3} \int_{-\pi}^{\pi}\int_{-\pi}^{\pi}\int_{-\pi}^{\pi} \frac{d\phi_1 d\phi_2 d\phi_3}{1 - z \cos\phi_1 \cos\phi_2 \cos\phi_3}, \quad \text{(bcc)} \qquad (10.2.23)$$

These famous and imposing integrals were first obtained by W. F. van Peype when he was a doctoral student of H. A. Kramers, in 1938. He evaluated them numerically (van Peype [1938]), but both he and Kramers were unable to integrate them in closed form. Kramers showed them to R. H. Fowler, who passed them to G. H. Hardy, all to no avail. (Hardy could not offer them to Ramanujan, who had died in 1920.) They soon became widely known, and they were first integrated in closed form by G. N. Watson (1939). (Barber and Ninham [1970]) For $z = 1$, the values are found to be

$$U_3^{(s)}(1,0) = \frac{4}{3\pi^2} \left[18 + 12\sqrt{2} - 10\sqrt{3} - 7\sqrt{6} \right] K^2[(2 - \sqrt{3})(\sqrt{3} - \sqrt{2})]$$

$$= 4\left(\frac{\sqrt{6}}{\pi^2}\right) \Gamma(1/24)\Gamma(5/24)\Gamma(7/24)\Gamma(11/24) = 1.5163860591$$

$$U_3^{(f)}(1,0) = \frac{9[\Gamma(1/3)]^6}{2^{14/3}\pi^4} = 1.3446610732$$

$$U_3^{(b)}(1,0) = \frac{[\Gamma(1/4)]^4}{4\pi^3} = 1.3932039297. \qquad (10.2.24)$$

(The sc expression in terms of gamma functions was derived by Glasser and Zucker (1977).) The probabilities of return to the origin, calculated from these results and Eq. (10.2.16), $F(1) = 1 - 1/U(1)$, are

$$F_3^{(s)}(1,0) = 0.340537330, \quad \text{for the simple cubic;}$$
$$F_3^{(f)}(1,0) = 0.256318237, \quad \text{for fcc;}$$
$$F_3^{(b)}(1,0) = 0.282229985, \quad \text{for bcc.} \tag{10.2.25}$$

The number of nearest neighbors for the sc is 6; for the bcc is 8; and for the fcc is 12. The probability of return, as expected, diminishes with the number of choices, but in all cases in three dimensions, return to the origin is by no means a certainty, as it is in one or two dimensions.

The mean value of a component s_i of the random-walker's position $\mathbf{s}$, after t steps, is

$$\langle s_i \rangle_t = \sum_{\mathbf{s}} s_i P_t(\mathbf{s}) = \lim_{\phi \to 0} \left(-i \frac{\partial \Pi_t(\phi)}{\partial \phi_i} \right) = -it \left[\lambda(0) \right]^{t-1} \left(\frac{\partial \lambda}{\partial \phi_i} \right)_{\phi=0}$$
$$= -it \sum_{\mathbf{s}} is_i p(\mathbf{s}) = t \sum_{\mathbf{s}} s_i p(\mathbf{s}) = t\mu_i, \tag{10.2.26}$$

since $\lambda(0) = 1$, where μ_i is the mean of the ith component of a step, or one of the first moments. In one dimension, with $p(1) = p$; $p(-1) = q = 1 - p$, it follows that $p(s) = 0$ for $s \neq \pm 1$, and $\langle s \rangle_t = t(p - q)$.

The second moments are found similarly, as shown here in general:

$$\langle s_j s_k \rangle_t = \sum_{\mathbf{s}} s_j s_k P_t(\mathbf{s}) = -\lim_{\phi \to 0} \frac{\partial^2 \Pi_t(\phi)}{\partial \phi_j \partial \phi_k}$$
$$= -t \lim_{\phi \to 0} \frac{\partial}{\partial \phi_j} \left[\lambda^{t-1}(\phi) \frac{\partial \lambda(\phi)}{\partial \phi_k} \right]$$
$$= -t(t-1)\lambda^{t-2}(0) \left(\frac{\partial \lambda(\phi)}{\partial \phi_j} \frac{\partial \lambda(\phi)}{\partial \phi_k} \right)_{\phi=0} - t\lambda^{t-1}(0) \left(\frac{\partial^2 \lambda(\phi)}{\partial \phi_j \partial \phi_k} \right)_{\phi=0}$$
$$= t(t-1) \left[\left(\sum_{\mathbf{s}} s_j p(\mathbf{s}) \right) \left(\sum_{\mathbf{s}} s_k p(\mathbf{s}) \right) \right] + t \sum_{\mathbf{s}} s_j s_k p(\mathbf{s})$$
$$= t\mu_{jk} + t(t-1)\mu_j \mu_k. \tag{10.2.27}$$

It follows immediately from Eq. (10.2.26) and Eq. (10.2.27) that the covariance is

$$\langle s_j s_k \rangle_t - \langle s_j \rangle_t \langle s_k \rangle_t = t \left(\mu_{jk} - \mu_j \mu_k \right). \tag{10.2.28}$$

For symmetric random walks, with $\mu_j = 0$ for all j, (and therefore $\langle s_j \rangle = 0$ for all j), the only nonzero second moments are the squared distances, or

$$\langle s_i s_j \rangle_t = t\mu_{jk}\delta_{jk}, \tag{10.2.29}$$

showing that the squared displacement is proportional to t, quite generally. For a simple cubic lattice, $\mu_{11} = \mu_{22} = \mu_{33} = 1/3$, giving $\langle s_i^2 \rangle_t = t/3$, and

$$\langle s^2 \rangle_t = \langle s_1^2 \rangle_t + \langle s_2^2 \rangle_t + \langle s_3^2 \rangle_t = t. \qquad (10.2.30)$$

Here t is the number of steps, which can be proportional to the time for a regular step rate. Thus the mean squared displacement of a random walker is proportional to the number of steps, in qualitative agreement with the displacement of a diffusing particle, as given in Eq. (9.3.72). Brown's original diagrams showing apparently random displacements of microscopic particles during successive time intervals suggested the random-walk model of diffusion, and the agreement so far indicates that the model is useful.

The probability of the random walker's return to the origin was found to depend upon the dimensionality of the lattice. In particular, in one dimension the return to the origin was shown to be a certainty. But even in this case, the full story has not yet been told. A further pertinent question asks for the average time between successive returns. The function that generates first returns to the origin in one dimension is, from Eq. (10.2.14)(b), Eq. (10.2.16), and the paragraph that follows it,

$$F_1(z,0) = 1 - 1/U_1(z,0) = \sum_{t=1}^{\infty} z^t f_t(0). \qquad (10.2.31)$$

The mean return time is

$$\begin{aligned} \langle t \rangle &= \sum_{t=1}^{\infty} t f_t(0) = \lim_{z \to 1} \frac{\partial F_1(z,0)}{\partial z} \\ &= \lim_{z \to 1} \frac{\partial}{\partial z} \left[1 - \sqrt{1 - z^2} \right] = \lim_{z \to 1} \frac{z}{\sqrt{1 - z^2}} = \infty. \end{aligned} \qquad (10.2.32)$$

Thus, even though return to the origin is a certainty in one dimension, the time interval, on average, between successive returns is infinite. In two or more dimensions, the result applies even more strongly.

For a 1-d walk of a large number of steps, with $p = q = 1/2$, the distribution of probable positions becomes approximately Gaussian. To show this, note that the characteristic function of Eq. (10.2.5) is sharply peaked around $\phi = 0$. With the number of steps now written as N in place of t, the logarithm of the characteristic function may be expanded around the origin as

$$\begin{aligned} \ln G_N(\phi) = N \ln \cos \phi &= N \left[\ln 1 + \phi \left(\frac{-\sin \phi}{\cos \phi} \right)_{\phi=0} + \frac{\phi^2}{2} \left(-1 - \frac{\sin^2 \phi}{\cos^2 \phi} \right)_{\phi=0} + \mathcal{O}(\phi^3) \right] \\ &= N \left[-\frac{\phi^2}{2} + \mathcal{O}(\phi^3) \right] \approx -\phi^2 N/2. \end{aligned} \qquad (10.2.33)$$

The expansion near the origin ignores the peaks of $(\cos \phi)^N$ near $\phi = \pm\pi$. These can be accounted for by restricting the expansion of Eq. (10.2.33) to $-\pi/2 \le \phi \le \pi/2$ and incorporating the final factor of Eq. (10.2.6), which requires s and N to have the same parity. The characteristic function is then $G_N(\phi) \approx \exp(-\phi^2 N/2)$, for ϕ near the origin, and the probability that the walker is at s at step N is approximated by

$$P_N(s) \approx \left(\frac{1+(-1)^{N+s}}{2}\right) \frac{2}{2\pi} \int_{-\pi/2}^{\pi/2} d\phi\, e^{-\phi^2 N/2} e^{-is\phi} \approx \left(\frac{1+(-1)^{N+s}}{2\pi}\right) \int_{-\infty}^{\infty} d\phi\, e^{-\phi^2 N/2} e^{-is\phi}$$

$$= \left(\frac{1+(-1)^{N+s}}{2\pi}\right) \left(\frac{2\pi}{N}\right)^{1/2} e^{-s^2/2N} = \frac{2}{\sqrt{2\pi N}} e^{-s^2/2N}, \qquad (10.2.34)$$

where the change in limits of integration has negligible effect for large N, and the final expression is valid for the nonzero probabilities for which the parities of N and s are equal. If the lattice spacing is a, the points of nonzero probability for a fixed N are separated by a distance $\delta x = 2a$. The occupied lattice points, say for N even, are at $x = 2as$ for all integers s. In the continuum limit of very small a, a displacement of s steps is equivalent to a distance $x = 2sa$, with x now regarded as a continuous variable. If the steps are taken at a regular frequency ν, then the number of steps in time t is $N = \nu t$. With these substitutions, and the recognition that $P_N(s)\delta s \to P_t(x)(dx/2a)$, with $\delta s = 2$, the probability that the walker's displacement lies in the interval dx around x at time t becomes

$$P_t(x)dx = \frac{1}{\sqrt{8\pi a^2 \nu t}} e^{-x^2/8\nu a^2 t} dx = \frac{1}{\sqrt{4\pi Dt}} e^{-x^2/4Dt}, \qquad (10.2.35)$$

where the diffusivity is written as $D = 2\nu a^2$. The probability density $P_t(x)$ satisfies the one-dimensional diffusion equation, similar to Eq. (9.3.70). The mean squared displacement is found to be

$$\langle x^2(t) \rangle = 4a^2 \nu t = 2Dt, \qquad (10.2.36)$$

as expected for diffusion in one dimension. It is thus seen that random walk provides an interesting caricature of diffusion, and of Brownian motion.

Suppose that, in addition to diffusion, there is a steady particle drift. Then x can be replaced by $x - vt$, and the probability density of Eq. (10.2.35) becomes

$$P_t(x) = \frac{1}{\sqrt{4\pi Dt}} e^{-(x-vt)^2/4Dt}. \qquad (10.2.37)$$

The differential equation satisfied by $P = P_t(x)$ is easily found to be

$$\frac{\partial P}{\partial t} = -v\frac{\partial P}{\partial x} + D\frac{\partial^2 P}{\partial x^2}. \qquad (10.2.38)$$

This and the ordinary diffusion equation belong to a family of equations, known as *Fokker-Planck Equations*, that frequently arise in the contexts of diffusive transport, Brownian motion, and fluctuations. A more general form for these equations is

$$\frac{\partial P}{\partial t} = -\frac{\partial}{\partial x}\left[F(x)P\right] + \frac{1}{2}\frac{\partial^2}{\partial x^2}\left[G(x)P\right], \qquad (10.2.39)$$

where $F(x)$ and $G(x)$ are functions of position that emerge in the context of the particular problem. Fokker-Planck equations, and the Smoluchowski equation, which is closely related, are often solved with the help of Laplace- or Fourier-transform methods. (Chandrasekhar [1943]; Kac [1947]; Uhlenbeck and Ornstein [1930]; Wang and Uhlenbeck [1945])

The Gaussian probability density of Eq. (10.2.35) is an example of the central limit theorem, which proclaims the result that in the continuum limit, the joint probability density of a set of n similarly distributed random variables becomes Gaussian in the limit $n \to \infty$. (Cramér [1946])

10.3 CORRELATIONS

For a different view of Brownian motion, consider the Langevin picture of a classical microscopic Brownian particle of mass M suspended in a medium of thermalized molecules (of masses much less than M), subject to their never-ending bombardment. The net force exerted on the Brownian particle by the molecules of the medium is regarded primarily as a random variable that consists of two parts: the first is an average force that opposes the motion, in recognition of the highway-traffic effect that the particle experiences more oncoming than overtaking encounters (or collisions), and the second is the remaining purely random force, $\mathbf{f}(t)$, such that its ensemble average, or expectation value, is zero. (Doob [1942]) The equation of motion for the Brownian particle is

$$\dot{\mathbf{p}} = -\zeta \mathbf{p} + \mathbf{f}(t), \tag{10.3.1}$$

where $-\zeta \mathbf{p}$ is the instantaneous average force, with ζ itself some measure of the collision frequency, and $\langle \mathbf{f}(t) \rangle = 0$. By inspection, the solution to Eq. (10.3.1) is

$$\mathbf{p}(t) = \mathbf{p}(0)e^{-\zeta t} + e^{-\zeta t} \int_0^t dt' e^{\zeta t'} \mathbf{f}(t'). \tag{10.3.2}$$

The expectation value of the particle's momentum is

$$\langle \mathbf{p}(t) \rangle = \langle \mathbf{p}(0) \rangle e^{-\zeta t}, \tag{10.3.3}$$

since the expectation value of the final term is zero. (Here it is assumed that the order of the operations, differentiation or integration and averaging, is interchangeable.) Thus the particle's expected momentum decays to zero from any initial value.

Much more information can be coaxed from Eq. (10.3.2) by straightforward manipulations. Note that $\mathbf{p} = M\mathbf{v} = m d\mathbf{r}/dt$. Then use the identities

$$\mathbf{p} \cdot \mathbf{r} = \frac{M}{2} \frac{dr^2}{dt}, \quad \text{and} \quad \dot{\mathbf{p}} \cdot \mathbf{r} = \frac{M}{2} \left[\frac{d^2 r^2}{dt^2} - 2v^2 \right] \tag{10.3.4}$$

to write

$$M \frac{d^2 \langle r^2 \rangle}{dt^2} - 2M \langle v^2 \rangle = -\zeta M \frac{d \langle r^2 \rangle}{dt}, \tag{10.3.5}$$

since $\langle \mathbf{r} \cdot \mathbf{f} \rangle = 0$. If the particle's mean squared speed is related by its classical energy expectation value to the temperature, or $\langle v^2 \rangle = 3kT/M$, Eq. (10.3.5) is readily solved to give

$$\langle r^2(t) \rangle = \frac{6kT}{M\zeta} \left[t - \frac{1}{\zeta} \left(1 - e^{-\zeta t} \right) \right]. \tag{10.3.6}$$

For $\zeta t \ll 1$, the mean squared displacement is $\langle r^2(t) \rangle = (3kT/M)t^2 = \langle v^2 \rangle t^2$, or $r_{rms}(t) = v_{rms}t$, showing that a characteristic instantaneous speed of the particle is just that expected from its equilibrium kinetic energy. But the direction of motion changes randomly, and for $\zeta t \gg 1$, it follows from Eq. (10.3.6) that

$$\langle r^2(t) \rangle = 6kTt/M\zeta = 6Dt, \tag{10.3.7}$$

where D is the diffusivity, given here by $D = kT/M\zeta$, and the final equality is imported from Eq. (9.3.72), the standard result from Fick's Law for diffusion in three dimensions.

From the kinematic equation of motion, if $\mathbf{r}(0) = 0$,

$$\mathbf{r}(t) = \frac{1}{m} \int_0^t dt' \mathbf{p}(t').$$

Another way to write the mean-square displacement is thus

$$\langle r^2(t) \rangle = \frac{1}{m^2} \left\langle \int_0^t d\mathbf{p}(t') \cdot \int_0^t dt'' \mathbf{p}(t'') \right\rangle = \frac{1}{m^2} \int_0^t dt' \int_0^t dt'' \langle \mathbf{p}(t') \cdot \mathbf{p}(t'') \rangle, \tag{10.3.8}$$

from which it follows that

$$\begin{aligned}
\frac{d\langle r^2 \rangle}{dt} &= \frac{2}{m^2} \int_0^t dt' \langle \mathbf{p}(t) \cdot \mathbf{p}(t') \rangle = \frac{2}{m^2} \int_0^t dt' \langle \mathbf{p}(t-t') \cdot \mathbf{p}(0) \rangle \\
&= \frac{-2}{m^2} \int_t^0 d\tau \langle \mathbf{p}(\tau) \cdot \mathbf{p}(0) \rangle = \frac{2}{m^2} \int_0^t d\tau \langle \mathbf{p}(\tau) \cdot \mathbf{p}(0) \rangle \\
&= 2 \int_0^t d\tau \langle \mathbf{v}(0) \cdot \mathbf{v}(\tau) \rangle = \int_{-\infty}^{\infty} d\tau \phi_v(\tau) \to 6D, \tag{10.3.9}
\end{aligned}$$

where $\phi_v(\tau) = \langle \mathbf{v}(0) \cdot \mathbf{v}(\tau) \rangle$, the velocity autocorrelation function, and the final expression comes from the time derivative of Eq. (10.3.7), in the limit as $t \to \infty$. The result, given by

$$D = \frac{1}{3} \int_0^{\infty} dt \langle \mathbf{v}(0) \cdot \mathbf{v}(t) \rangle = \frac{1}{6} \int_{-\infty}^{\infty} dt\, \phi_v(t), \tag{10.3.10}$$

which relates a transport coefficient, the diffusivity, to the time-integral of an autocorrelation function, is an example of a *Green-Kubo* formula. Other such formulas are to be found in Sec. 10.5. Note that $\phi(t) = \phi(-t)$.

If an additional constant force $\mathbf{F}$ is imposed on the particle, then for $\zeta t \gg 1$, the expectation value of the momentum found from Eq. (10.3.2) with $\mathbf{f} \to \mathbf{f} + \mathbf{F}$ is seen to be

$$\langle \mathbf{p}(t) \rangle = \langle \mathbf{p}(0) \rangle e^{-\zeta t} + e^{-\zeta t}\mathbf{F} \left(e^{\zeta t} - 1 \right)/\zeta \to \mathbf{F}/\zeta. \tag{10.3.11}$$

The mobility b is given by $\mathbf{v} = b\mathbf{F}$, from which $b = 1/M\zeta$, or the frequency of Eq. (10.3.1) is identified as

$$\zeta = 1/Mb. \tag{10.3.12}$$

The diffusivity becomes $D = kTb$, as known from the Nernst-Einstein relation of Chap. 9.

The mean squared momentum is obtained from Eq. (10.3.2) as

$$\langle p^2(t) \rangle = \langle p^2(0) \rangle e^{-2\zeta t} + 2e^{-2\zeta t} \int_0^t dt' e^{\zeta t} \langle \mathbf{p}(0) \cdot \mathbf{f}(t') \rangle + e^{-2\zeta t} \int_0^t dt' \int_0^t dt'' e^{\zeta(t'+t'')} \langle \mathbf{f}(t') \cdot \mathbf{f}(t'') \rangle$$

$$= \langle p^2(0) \rangle e^{-2\zeta t} + 3e^{-2\zeta t} \int_0^t dt' \int_0^t dt'' e^{\zeta(t'+t'')} \phi(t'' - t'), \qquad (10.3.13)$$

where the second term in the first line vanishes because $\langle \mathbf{p}(0) \cdot \mathbf{f}(t') \rangle = \langle \mathbf{p}(0) \rangle \cdot \langle \mathbf{f}(t') \rangle = 0$. The autocorrelation function of the random force is defined as $\phi_f(t'' - t') = \langle f_i(t') f_i(t'') \rangle$ for $i = x$, y, or z, and the factor 3 comes from the sum of the three terms in the dot product. The random force that leads to Brownian motion is regarded as stationary, in that the autocorrelation function depends only on the time difference, and not on either t' or t'' alone. For this reason, the autocorrelation function is an even function of its argument, since $\phi_f(\tau) = \langle f_i(t) f_i(t+\tau) \rangle = \langle f_i(t-\tau) f_i(t) \rangle$, and the order of terms in the final product can be interchanged to give $\phi_f(\tau) = \phi_f(-\tau)$. Other properties of the autocorrelation function follow from

$$0 \le \left\langle \left[f(t) - f(t+\tau) \right]^2 \right\rangle = \langle f^2(t) \rangle + \langle f^2(t+\tau) \rangle - 2\langle f(t)f(t+\tau) \rangle = 2\left[\phi_f(0) - \phi_f(\tau) \right],$$

or $0 \le \phi_f(0) \ge \phi_f(\tau)$. For $\tau \gg 1/\zeta$, the forces are not correlated, and

$$\lim_{\tau \gg 1/\zeta} \phi_f(\tau) = \langle f_i(t) \rangle \langle f_i(t+\tau) \rangle = 0. \qquad (10.3.14)$$

A change of variables in Eq. (10.3.13) to $t' + t'' = 2u$ and $t'' - t' = v$ permits the integral to be rewritten, since the Jacobian of the transformation is $\partial(t', t'')/\partial(u, v) = 1$,

$$\int_0^t dt' \int_0^t dt'' e^{\zeta(t'+t'')} \phi_f(t'' - t') = \int_{-t}^t dv \int_{|v|}^{t-|v|} du \, e^{2\zeta u} \phi_f(v)$$

$$= \frac{e^{2\zeta t}}{\zeta} \int_0^t dv \, \phi_f(v) \left(e^{-2\zeta v} - e^{2\zeta(v-t)} \right), \qquad (10.3.15)$$

where the symmetry of $\phi(v)$ has been used to change the lower limit of the v integral. The use of this result in Eq. (10.3.13) gives

$$\langle p^2(t) \rangle = \langle p^2(0) \rangle e^{-2\zeta t} + \frac{3}{\zeta} \int_0^t dv \, \phi_f(v) \left(e^{-2\zeta v} - e^{2\zeta(v-t)} \right). \qquad (10.3.16)$$

In the limit of infinite t, the left-hand side is expected to equal its thermal equilibrium value, and the second term in the integral to vanish. The result in the limit is thus

$$\langle p^2(t \to \infty) \rangle = 3MkT = \frac{3}{\zeta} \int_0^\infty dv \, \phi_f(v) e^{-2\zeta v},$$

or

$$MkT\zeta = \int_0^\infty dv \, \phi_f(v) e^{-2\zeta v}. \qquad (10.3.17)$$

This equation relates ζ to the autocorrelation function of the random force. Note that this is not a Laplace transform that permits $\phi_f(v)$ to be determined by inversion, since ζ is a fixed parameter given by Eq. (10.3.12), not a variable. As an example, if $\phi_f(v) = \phi_f(0)\exp(-\alpha\zeta|v|)$, where $\phi_f(0) = \langle f^2(0)\rangle$ and α is an undetermined numerical factor, Eq. (10.3.17) gives $MkT\zeta = \phi_f(0)/(2+\alpha)\zeta$, or $\phi_f(0) = (2+\alpha)kT/Mb^2$, where Eq. (10.3.12) has been used.

The autocorrelation function for the random forces acting on a Brownian particle is expected to build up and decay in a much shorter time interval than the decay time of the particle's initial conditions, because these forces are attributed to short-range-force collisions of molecules with the relatively ponderous Brownian particle. Therefore the numerical factor α in the assumed form of ϕ_f is expected to satisfy $\alpha \gg 1$. With this assumption, Eq. (10.3.16) becomes

$$\langle p^2(t)\rangle = \langle p^2(0)\rangle e^{-2\zeta t} + 3MkT\zeta(2+\alpha)\int_0^t dv\left[e^{-\zeta(2+\alpha)v} - e^{-2\zeta t}e^{\zeta(2-\alpha)v}\right]$$

$$= \langle p^2(0)\rangle e^{-2\zeta t} + 3MkT\left(1 - e^{-\zeta(2+\alpha)t}\right) + 3MkT\left(\frac{\alpha+2}{\alpha-2}\right)\left(e^{-\zeta\alpha t} - e^{-2\zeta t}\right)$$

$$\approx \langle p^2(0)\rangle e^{-2\zeta t} + 3MkT\left(1 - e^{-\zeta(2+\alpha)t}\right) + 3MkT\left(e^{-\zeta\alpha t} - e^{-2\zeta t}\right)$$

$$= \langle p^2(0)\rangle e^{-2\zeta t} + 3MkT\left(1 - e^{-2\zeta t}\right)\left(1 + e^{-\alpha\zeta t}\right)$$

$$\approx \langle p^2(0)\rangle e^{-2\zeta t} + 3MkT\left(1 - e^{-2\zeta t}\right), \tag{10.3.18}$$

where the approximations are $(\alpha+1)/(\alpha-1) \approx 1$, and $\alpha\zeta t \gg 1$, both of which are valid for the expected large values of α over most time intervals of interest. Two features of this result are noteworthy: (1) if $\langle p^2(0)\rangle = 3MkT$, the equilibrium value, then for all time, $\langle p^2(t)\rangle = 3MkT$, and (2) all trace of the assumed autocorrelation function has disappeared from the final result, implying that if $\phi_f(\tau)$ differs from zero only over a very short time, such that $|\tau| \ll 1/\zeta$, then the form of $\phi_f(\tau)$ is not of great importance.

The mean squared velocity can be found from Eq. (10.3.18) as

$$\langle v^2(t)\rangle = \langle v^2(0)\rangle e^{-2\zeta t} + \frac{3kT}{M}\left(1 - e^{-2\zeta t}\right). \tag{10.3.19}$$

This result can be used in Eq. (10.3.5) to obtain a more detailed expression for $\langle r^2(t)\rangle$ than that given by Eq. (10.3.6). This calculation is left as a exercise. (See Problem 5.)

10.4 SPECTRAL DENSITY AND AUTOCORRELATION FUNCTIONS

In agreement with the definition introduced in the previous section, the autocorrelation function associated with a stationary random function of time, $f(t)$, is

$$\phi(\tau) = \langle f(t)f(t+\tau)\rangle. \tag{10.4.1}$$

It is sometimes necessary, in order to avoid convergence problems, to approach the definition of Eq. (10.4.1) as a limiting process. For this purpose, define a segment of $f(t)$ as

$$f_{Ta}(t) = f(t) \quad \text{for} \quad a \le t \le a + T,$$
$$f_{Ta}(t) = 0, \quad \text{otherwise.} \tag{10.4.2}$$

Define the Fourier transform of $f_{Ta}(t)$ as

$$F_{Ta}(\omega) = \frac{1}{\sqrt{2\pi}} \int_{-\infty}^{\infty} dt\, e^{-i\omega t} f_{Ta}(t). \qquad (10.4.3)$$

Its Fourier inverse is

$$f_{Ta}(t) = \frac{1}{\sqrt{2\pi}} \int_{-\infty}^{\infty} d\omega\, F_{Ta}(\omega) e^{i\omega t}. \qquad (10.4.4)$$

The Fourier transform of $f_{Ta}(t+\tau)$ is seen to be

$$\frac{1}{\sqrt{2\pi}} \int_{-\infty}^{\infty} dt\, e^{-i\omega t} f_{Ta}(t+\tau) = \frac{1}{\sqrt{2\pi}} \int_{-\infty}^{\infty} dt\, e^{-i\omega(t-\tau)} f_{Ta}(t) = e^{i\omega\tau} F_{Ta}(\omega), \qquad (10.4.5)$$

and its inverse follows directly from Eq. (10.4.4) as

$$f_{Ta}(t+\tau) = \frac{1}{\sqrt{2\pi}} \int_{-\infty}^{\infty} d\omega\, e^{i\omega(t+\tau)} F_{Ta}(\omega). \qquad (10.4.6)$$

The kind of product needed in the autocorrelation function is

$$f_{Ta}(t) f_{Ta}(t+\tau) = \frac{1}{2\pi} \int_{-\infty}^{\infty} d\omega \int_{-\infty}^{\infty} d\omega'\, e^{i\omega(t+\tau)} e^{i\omega' t} F_{Ta}(\omega) F_{Ta}(\omega'). \qquad (10.4.7)$$

The infinite time integral of this product is

$$\begin{aligned}
\int_{-\infty}^{\infty} dt\, f_{Ta}(t) f_{Ta}(t+\tau) &= \frac{1}{2\pi} \int_{-\infty}^{\infty} dt \int_{-\infty}^{\infty} d\omega \int_{-\infty}^{\infty} d\omega'\, e^{i\omega(t+\tau)} e^{i\omega' t} F_{Ta}(\omega) F_{Ta}(\omega') \int_{-\infty}^{\infty} dt\, e^{i(\omega+\omega')t} \\
&= \frac{1}{2\pi} \int_{-\infty}^{\infty} d\omega \int_{-\infty}^{\infty} d\omega'\, F_{Ta}(\omega) F_{Ta}(\omega') e^{i\omega\tau} \\
&= \int_{-\infty}^{\infty} d\omega \int_{-\infty}^{\infty} d\omega'\, \delta(\omega+\omega') e^{i\omega\tau} F_{Ta}(\omega) F_{Ta}(\omega') \\
&= \int_{-\infty}^{\infty} d\omega /, F_{Ta}(\omega) F_{Ta}(-\omega) e^{i\omega\tau} = \int_{-\infty}^{\infty} d\omega\, |F_{Ta}(\omega)|^2 e^{i\omega\tau}, \qquad (10.4.8)
\end{aligned}$$

since

$$\delta(\omega+\omega') = \frac{1}{2\pi} \int_{-\infty}^{\infty} dt\, e^{i(\omega+\omega')t}.$$

Define $f(t)$ as a stationary random function if $\langle f_{Ta}(t) f_{Ta}(t+\tau) \rangle$ depends only on τ, and not on a. For such a function, define an interval autocorrelation function, $\phi_T(\tau)$, as

$$\phi_T(\tau) = \frac{1}{T} \int_{-\infty}^{\infty} dt\, f_T(t) f_T(t+\tau), \qquad (10.4.9)$$

and observe that the autocorrelation function is given by

$$\begin{aligned}
\phi(\tau) = \lim_{T\to\infty} \phi_T(\tau) &= \lim_{T\to\infty} \frac{1}{T} \int_{-\infty}^{\infty} dt\, f_T(t) f_T(t+\tau) \\
&= \lim_{T\to\infty} \frac{1}{T} \int_{-\infty}^{\infty} d\omega\, |F_T(\omega)|^2 e^{i\omega t}. \qquad (10.4.10)
\end{aligned}$$

Define a function called the *spectral density*, $G(\omega)$ (sometimes called the *power spectrum*), as

$$G(\omega) \equiv \lim_{T \to \infty} \frac{\sqrt{2\pi}}{T} |F_T(\omega)|^2. \tag{10.4.11}$$

This definition permits the autocorrelation function to be written

$$\phi(\tau) = \frac{1}{\sqrt{2\pi}} \int_{-\infty}^{\infty} d\omega \, G(\omega) e^{i\omega t} = \sqrt{\frac{2}{\pi}} \int_0^\infty d\omega \, G(\omega) \cos \omega\tau, \tag{10.4.12}$$

where the second form is valid because $G(\omega)$ is an even function of its argument. The Fourier inverse of Eq. (10.4.12) is

$$G(\omega) = \frac{1}{\sqrt{2\pi}} \int_{-\infty}^{\infty} d\tau \, \phi(\tau) e^{-i\omega\tau} = \sqrt{\frac{2}{\pi}} \int_0^\infty d\tau \, \phi(\tau) \cos \omega\tau. \tag{10.4.13}$$

The result that the spectral density and the autocorrelation function are Fourier cosine transforms of each other is known as the *Wiener-Khinchine Theorem*.

As an example, suppose the autocorrelation function is $\phi(\tau) = A \exp(-\alpha\tau)$. Then the spectral density is

$$G(\omega) = \sqrt{\frac{2}{\pi}} \int_0^\infty d\tau \, A e^{-\alpha\tau} \cos \omega\tau = A \sqrt{\frac{2}{\pi}} \frac{\alpha}{\alpha^2 + \omega^2}, \tag{10.4.14}$$

which is called a Lorentzian spectrum. It is readily seen that $G(\omega) = G(0)/2$ for $\omega = \pm\alpha$, so a measure of the frequency range required to represent the autocorrelation function is $\Delta\omega = 2\alpha$. The larger the value of α, the shorter the correlation time, and the higher the frequency range required for representation of the random variable.

10.4.1 The Nyquist Theorem. It follows directly from Eq. (10.3.2) that the velocity autocorrelation function for particles that obey the Langevin equation is

$$\phi_v(t) = \langle v_x(0) v_x(t) \rangle = \langle v_x^2(0) \rangle e^{-\zeta t} = \frac{kT}{m} e^{-\zeta t}, \tag{10.4.15}$$

where the final expression is valid for Boltzmann particles. The corresponding spectral density is found from Eq. (10.4.14) to be

$$G_v(\omega) = \sqrt{\frac{2}{\pi}} \frac{kT}{m} \frac{\zeta}{\zeta^2 + \omega^2}. \tag{10.4.16}$$

The velocity fluctuations of electrons moving in a resistor give rise to fluctuations in the voltage across the resistor, known as *Johnson noise*, after the discoverer. (Johnson [1928]; Nyquist [1928]) The relationship of current to voltage for a resistor is given by Ohm's law, written as

$$V = IR = RAj = -RAne \langle v_x \rangle, \tag{10.4.17}$$

where j is the current density, A is the conductor cross-sectional area, and n is the electron density. Write $n \langle v_x \rangle = (Al)^{-1} \sum v_{xi}$, where the sum is over all electrons in the resistor. Use this expression to rewrite Eq. (10.4.17) as

$$V = -\frac{Re}{l} \sum_i v_{xi} = \sum_i V_i, \qquad (10.4.18)$$

where the random voltage produced by the ith electron is $V_i = -Rev_{xi}/l$. (The resistor is here considered to be in equilibrium with a heat bath at temperature T, but not connected electrically in a closed circuit, so this calculation yields the open-circuit voltage produced in the resistor by the thermal fluctuations of its electrons.) The autocorrelation function of the noise voltage is

$$\langle V_i(0)V_i(t) \rangle = \left(\frac{Re}{l} \right)^2 \langle v_{ix}(0)v_{ix}(t) \rangle = \left(\frac{Re}{l} \right)^2 \left(\frac{kT}{m} \right) e^{-\zeta t}, \qquad (10.4.19)$$

where Eq. (10.4.15) has been used, and the electrons have been assumed to be Maxwell-Boltzmann particles. (It turns out that the result is independent of the velocity distribution of the electrons.) The spectral density, from Eq. (10.4.16) is

$$G_V(\omega) = \left(\frac{Re}{l} \right)^2 \left(\frac{kT}{m} \right) \sqrt{\frac{2}{\pi}} \frac{\zeta}{\zeta^2 + \omega^2} \rightarrow \left(\frac{Re}{l} \right)^2 \left(\frac{kT}{m} \right) \sqrt{\frac{2}{\pi}} \left(\frac{1}{\zeta} \right), \qquad (10.4.20)$$

where the final expression applies when $\omega \ll \zeta$. The expectation value of the squared fluctuation voltage produced by one electron is

$$\langle V_i^2 \rangle = \sqrt{\frac{2}{\pi}} \int_{\omega_1}^{\omega_2} d\omega \, G_V(\omega) = \frac{2}{\pi} \left(\frac{Re}{l} \right)^2 \left(\frac{kT}{m} \right) \left(\frac{\Delta \omega}{\zeta} \right), \qquad (10.4.21)$$

where $\Delta \omega = \omega_2 - \omega_1$, and $\omega_1 < \omega_2 \ll \zeta$. If the electron velocity fluctuations are independent (no collective oscillations, for example), then the mean squared voltage fluctuation is given by the Nyquist theorem as

$$\langle V^2 \rangle = nAl \langle V_i^2 \rangle = \frac{2kTR^2}{\pi} \left(\frac{ne^2 A}{lm\zeta} \right) \Delta \omega = 4kTR\Delta\nu, \qquad (10.4.22)$$

where $\nu = \omega/2\pi$, and Eq. (9.3.55) has been used to write the resistance in terms of the conductivity, as

$$R = \frac{l}{\sigma A} = \frac{lm\zeta}{ne^2 A}. \qquad (10.4.23)$$

This is an example of how a transport coefficient (R) is related in a fundamental way to the *equilibrium* thermal fluctuations within the system. A few more such examples are offered in the next subsection, and a more general derivation of the fluctuation-dissipation theorem is given in the section on response theory.

It should be noted that the open-circuit voltage across the resistor is not the fundamental fluctuating quantity, independent of the circuit; the current is. The mean-squared current fluctuations are written as

$$\langle I^2 \rangle = \frac{\langle V^2 \rangle}{R^2} = \left(\frac{4kT}{R} \right) \Delta\nu = 4kTG\Delta\nu, \qquad (10.4.24)$$

where G is the conductance. (The impedance is $Z = R + iX$, where X is the reactance; the admittance is $Y = 1/Z = G - iB$, where B is the elastance.) If the resistor R is in parallel with a general, unspecified impedance, $Z' = R' + iX'$, such that they are in thermal equilibrium with each other at constant temperature, then there can be no net transfer of energy from one to the other. The resistor R acts as a generator of Johnson noise, producing a current in the circuit given by $I_R = V_R/(R + Z')$, where V_R is the noise voltage of Eq. (10.4.22). A narrow-band passive filter in the circuit allows only a very limited frequency range for the exchange of this noise energy, in order that the reactance can be treated as constant. (This is a mathematical convenience that easily can be dispensed with, but at the cost of readability.) The resulting voltage across the impedance is $(V_{Z'})_R = Z'I_R$. The average power delivered to Z' is

$$P' = \langle |I_R|^2 \rangle R' = \langle V_R^2 \rangle R'/|R + Z'|^2. \tag{10.4.25}$$

Since the impedance receives power *from* R, it must supply an equivalent amount of power *to* R. Its circuit current can be written $I_{Z'} = V_{Z'}/(R + Z')$, and the power it delivers to R is

$$P = \langle |I_{Z'}|^2 \rangle R = \langle V_{Z'}^2 \rangle R/|R + Z'|^2. \tag{10.4.26}$$

When these are equal, the result is

$$\langle V_{Z'}^2 \rangle R = \langle V_R^2 \rangle R', \quad \text{or} \quad \langle V_{Z'}^2 \rangle = \langle V_R^2 \rangle (R'/R) = 4R'kT\Delta\nu, \tag{10.4.27}$$

for any narrow band $\Delta\nu$. It is therefore true over the entire spectrum, with the possibility that $R' = R'(\nu)$. (When the impedance is nonreactive, the restriction to a narrow pass band can be lifted.) For example, if Z' is a metallic conductor with fermion charge carriers, instead of the Maxwell-Boltzmann particles used in the original development, the result is still valid. This verifies the earlier claim that the Nyquist theorem is independent of the statistics of the charge carriers. The theorem has been verified experimentally for metallic conductors, p- and n-type semiconductors, and ionic conductors such as solutions of electrolytes.

Johnson noise is not the only noise source in some resistors. Various kinds of composition resistors, widely used in electronic circuitry, are often made of a ceramic mixture containing finely ground particles of a conductor such as carbon (in the form of lamp black). These resistors generate a current-dependent noise, orders of magnitude greater than Johnson noise, and with an entirely different spectral density. It is attributable to the nature of the percolation-like conduction paths, from grain to grain, and the electric-field dependence of the contacts within the resistive medium. It is not intrinsically a thermal phenomenon, but this type of resistor usually exhibits a negative temperature coefficient of resistance.

10.4.2 Other Fluctuation-Dissipation Relations. Examples that relate transport coefficients to the equilibrium fluctuations of a system are numerous and frequently encountered. Only two of these are identified here. From Eq. (10.3.12), the mobility of a Brownian particle is related to the reciprocal decay time of the fluctuating force attributable to the bombarding molecules of the medium by $1/b = M\zeta$. From Eq. (10.3.17), $M\zeta$ is given by

$$M\zeta = \frac{1}{kT} \int_0^\infty d\tau\, \phi_f(\tau)e^{-2\zeta\tau},$$

where $\phi_f(\tau) = \langle f_i(0)f_i(\tau)\rangle$. The resulting fluctuation-dissipation relation is

$$\frac{1}{b} = \frac{1}{kT}\int_0^\infty d\tau\,\phi_f(\tau)e^{-2\zeta\tau}. \qquad (10.4.28)$$

Since the diffusivity is derivable from the mobility by the Nernst-Einstein relation, there is automatically another such fluctuation-dissipation expression:

$$\frac{1}{D} = \frac{1}{(kT)^2}\int_0^\infty d\tau\,\phi_f(\tau)e^{-2\zeta\tau}, \qquad (10.4.29)$$

which is equivalent to Eq. (10.3.17). Typically, the autocorrelation function of the random forces, $\phi_f(t)$, decays in a much shorter time than $1/2\zeta$, so that $\phi_f(t)\exp(-2\zeta t) \approx \phi_f(t)$, and (10.4.29), for example, becomes

$$\frac{1}{D} = \frac{1}{2(kT)^2}\int_{-\infty}^\infty d\tau\,\phi_f(\tau), \qquad (10.4.30)$$

where the past-future symmetry of $\phi_f(t)$ has been invoked. Note that $\phi_f(t)$ is the autocorrelation function for any component of the random force, and $\langle \mathbf{f}(0)\cdot\mathbf{f}(t)\rangle = 3\phi_f(t)$.

10.5 MORE RESPONSE THEORY

Elementary derivations of response coefficients are commonplace in undergraduate physics courses. For example, Ohm's law is made plausible, and an expression for the conductivity of a metal, in terms of electron mass, charge, number density, and an ill-defined collision time, is derived early in the study of electricity. The conductivity σ is a response coefficient, and the current density is then written

$$J = \sigma E, \qquad (10.5.1)$$

where E is taken to be the applied electric field at zero frequency. This expression, as sophistication develops, is generalized to include nonlinear effects at zero frequency and inertial or other reactive effects at nonzero frequencies. The nonlinear current density at zero frequency is typically written in one of two ways:

$$\begin{aligned} J &= \sigma_1 E + \sigma_2 E^2 + \cdots \\ &= \sigma(E)E, \end{aligned} \qquad (10.5.2)$$

where the σ_n in the first expression are proportional to $(d^n J/dE^n)_{E=0}$, and the second expression is more ornamental than useful, since the parameter of significance is dJ/dE, rather than J/E. For a time-dependent generalization, the current density may be written

$$J(t) = \int_{t_0}^t dt_1\,E(t_1)\phi_1(t-t_1) + \int_{t_0}^t dt_1\int_{t_0}^{t_1} dt_2\,E(t_1)E(t_2)\phi_2(t_1-t_2,t-t_2) + \cdots, \qquad (10.5.3)$$

where it is assumed that the field is turned on at $t = t_0$, and the quantities ϕ_1 and ϕ_2 are the first-order and second-order response functions. (Note that the first term is the most general form for which $J(\lambda E, t) = \lambda J(E, t)$.) It is possible, in terms of the Green-Kubo formalism and its generalizations, to derive exact, quantum-mechanically correct formal expressions for response functions and response coefficients. As already noted in Sec. 9.9.1, in practice the formal expressions are of little practical value, since exact solutions for the unitary evolution operator $U(t, t_0)$ are not usually to be found. Nevertheless, the exact theory can be presented efficiently and offered both as a challenge and as a point of departure for the standard approximations of linear response theory and its relatives. (Bernard and Callen [1959]; Grandy, Jr. [1980]; Grandy, Jr. [1987]; Green [1961]; Kreuzer [1981]; Kubo [1957]; Kubo [1965]; Kubo [1966]; Kubo, Yokota and Nakajima [1957]; Montroll [1961]; Montroll [1962]; Peterson [1967]; van Vliet [1978]; Zwanzig [1960]; Zwanzig [1961]; Zwanzig [1965])

The Hamiltonian for the system is taken to be

$$\mathcal{H}_t = \mathcal{H}_0 - \sum_j \mathcal{A}_j F_j(t); \quad t \geq t_0, \tag{10.5.4}$$

where $\mathcal{H}_0$ is the Hamiltonian of the system before the external forcings are coupled to the system at $t = t_0$. The operators $\mathcal{A}_j$ are Schrödinger operators, not explicitly time dependent, that couple the forcings to the system, as does $\mathcal{R}$ in Eq. (9.9.25), and the $F_j(t)$ are, like the E_0 factor in that equation, dimensional c-numbers that are switched on at time $t = t_0$. A response to this forcing corresponding to an operator $\mathcal{B}_i$ is

$$\langle \mathcal{B}_i \rangle_t - \langle \mathcal{B}_i \rangle_0 = \mathrm{Tr}\big(\mathcal{B}_i \rho(t)\big), \tag{10.5.5}$$

where $\rho(t)$ is the time-dependent statistical operator that satisfies

$$i\hbar d\rho(t)/dt = [\mathcal{H}_t, \rho(t)]. \tag{10.5.6}$$

For efficient development of the theory, $\langle \mathcal{B}_i \rangle_0$ is ignored as unnecessary clutter, and $\rho(t) = U(t, t_0)\rho(t_0)U^\dagger(t, t_0)$ is regarded as being carried forward in time by the unitary transformation, where $\rho(t_0)$ is the statistical operator at the onset of the external forcings, determined from Eq. (10.5.6) with $\mathcal{H}_t \to \mathcal{H}_0$. For most applications, $\mathcal{H}_0$ is a time-independent Hamiltonian, and $\rho(t_0)$ is the equilibrium statistical operator, but such a restriction is not necessary in principle. The response of a nonequilibrium system can be written in this formalism in the same way, with $\rho(t_0)$ found by a procedure such as that outlined in Sec. 9.10.1 and given in Eq. (9.10.16). The unitary operator $U(t, t_0)$ satisfies Eq. (9.9.23), written here as

$$i\hbar dU(t, t_0)/dt = \mathcal{H}_t U(t, t_0). \tag{10.5.7}$$

Its solution, from Eq. (9.9.24), is

$$U(t, t_0) = U_0(t, t_0) \left[1 - \frac{i}{\hbar} \int_{t_0}^{t} dt' \, U_0^\dagger(t', t_0)\mathcal{H}_1(t')U(t', t_0) \right], \tag{10.5.8}$$

where $\mathcal{H}_1(t) = \sum \mathcal{A}_j F_j(t)$. The response functions (or functionals) and response coefficients, in both linear and nonlinear response theory, are obtained from the (functional) derivatives of the responses, $\langle \mathcal{B}_i \rangle_t$, with respect to the forcings, $F_j(t)$, for all orders. For example,

$$\sigma_{ij}(F_1, F_2, \cdots, F_n; t) \equiv \delta \langle \mathcal{B}_i \rangle_t / \delta F_j(t), \quad \text{and}$$
$$\sigma_{ijk}(F_1, F_2, \cdots, F_n; t) \equiv \delta^2 \langle \mathcal{B}_i \rangle_t / \delta F_j(t)\delta F_k(t). \tag{10.5.9}$$

The operator $\mathcal{B}_i$ is not intrinsically a function of the forcings; the derivatives of Eq. (10.5.9) find their F_j-dependencies implicit in the unitary operator $U(t,t_0)$. One of the required derivatives of the unitary operator is

$$\frac{\delta U(t,t_0)}{\delta F_m(t)} = \frac{-i}{\hbar}U(t,t_0)\int_{t_0}^t dt'\, U^\dagger(t',t_0)\frac{\delta\mathcal{H}_{t'}}{\delta F_m(t')}U(t',t_0). \qquad (10.5.10)$$

This assertion may be verified by noting that both sides vanish at $t=t_0$, and showing that both satisfy the same differential equation; the first is

$$i\hbar\frac{d}{dt}\left(\frac{\delta U(t,t_0)}{\delta F_m(t)}\right) = \frac{\delta}{\delta F_m(t)}\left(i\hbar\frac{dU(t,t_0)}{dt}\right) = \frac{\delta}{\delta F_m(t)}\left(\mathcal{H}_t U(t,t_0)\right), \qquad (10.5.11)$$

and the second is

$$i\hbar\frac{d}{dt}\left(\frac{-i}{\hbar}U(t,t_0)\int_{t_0}^t dt'\, U^\dagger(t',t_0)\frac{\delta\mathcal{H}_{t'}}{\delta F_m(t')}U(t',t_0)\right)$$

$$= \mathcal{H}_t\frac{\delta U(t,t_0)}{\delta F_m(t)} + U(t,t_0)U^\dagger(t,t_0)\frac{\delta\mathcal{H}_t}{\delta F_m(t)}U(t,t_0)$$

$$= \frac{\delta}{\delta F_m(t)}\left(\mathcal{H}_t U(t,t_0)\right). \qquad (10.5.12)$$

The Hermitian conjugate of Eq. (10.5.10) is

$$\frac{\delta U^\dagger(t,t_0)}{\delta F_m} = \frac{i}{\hbar}\left(\int_{t_0}^t dt'\, U^\dagger(t',t_0)\frac{\delta\mathcal{H}_{t'}}{\delta F_m(t')}U(t',t_0)\right)U(t,t_0). \qquad (10.5.13)$$

Since the trace operation of Eq. (10.5.5) is invariant under cyclic permutations of its operand, the time dependence can be transferred to $\mathcal{B}_i$, which becomes a Heisenberg operator:

$$\langle\mathcal{B}_i\rangle_t = \mathrm{Tr}\big(\mathcal{B}_i\rho(t)\big) = \mathrm{Tr}\big(\mathcal{B}_i(t,t_0)\rho(t_0)\big), \qquad (10.5.14)$$

where $\rho(t_0)$ has already been specified, and the Heisenberg form of the operator is

$$\mathcal{B}_i(t,t_0) = U^\dagger(t,t_0)\mathcal{B}_i U(t,t_0). \qquad (10.5.15)$$

In the middle expression of Eq. (10.5.14), the operator is averaged over a representative set of states that have been carried forward exactly by $U(t,t_0)$ from t_0; in the final expression, the effect of the operator, averaged over the initial states, is propagated forward by the exact unitary transformation. The results are equivalent. The functional derivative of this operator with respect to F_m is found by the use of Eqs. (10.5.10) and (10.5.13) to be

$$\frac{\delta\mathcal{B}_i(t,t_0)}{\delta F_m(t)} = \frac{\delta U^\dagger(t,t_0)}{\delta F_m(t)}\mathcal{B}_i U(t,t_0) + U^\dagger(t,t_0)\mathcal{B}_i\frac{\delta U(t,t_0)}{\delta F_m(t)}$$

$$= \frac{i}{\hbar}\left(\int_{t_0}^t dt'\, U^\dagger(t',t_0)\frac{\delta\mathcal{H}_{t'}}{\delta F_m(t')}U(t',t_0)\right)U^\dagger(t,t_0)\mathcal{B}_i U(t,t_0)+$$

$$-\frac{i}{\hbar}U^\dagger(t,t_0)\mathcal{B}_i U(t,t_0)\int_{t_0}^t dt'\, U^\dagger(t',t_0)\frac{\delta\mathcal{H}_{t'}}{\delta F_m(t')}U(t',t_0)$$

$$= \frac{i}{\hbar}\int_{t_0}^t dt'\,[-\mathcal{A}_m(t',t_0),\mathcal{B}_i(t,t_0)] \qquad (10.5.16)$$

since $\delta \mathcal{H}_t / \delta F_m(t) = -\mathcal{A}_m$, and $\mathcal{A}_m$ is transformed into a Heisenberg operator. The first-order generalized response coefficient is thus found to be

$$\sigma_{ij}(F_1, F_2, \cdots, F_n; t) = \text{Tr}\left(\rho(t_0)\frac{\delta\mathcal{B}_i(t, t_0)}{\delta F_j(t)}\right)$$

$$= \frac{-i}{\hbar}\int_{t_0}^t dt' \left\langle [\mathcal{A}_j(t', t_0), \mathcal{B}_i(t, t_0)]\right\rangle_0. \qquad (10.5.17)$$

The integrand may be written

$$\Phi_{ij}(t, t'; t_0) = \frac{1}{i\hbar}\left\langle [\mathcal{A}_j(t', t_0), \mathcal{B}_i(t, t_0)]\right\rangle_0, \qquad (10.5.18)$$

thus defining a set of *response functionals*. The total response of the observable represented by the operator $\mathcal{B}_i$ to the set of n forcings turned on at $t = t_0$ becomes

$$\left\langle \mathcal{B}_i \right\rangle_t = \sum_{j=1}^n \int_{t_0}^t dt' \, F_j(t')\Phi ij(t, t'; t_0). \qquad (10.5.19)$$

This expression is the complete nonlinear response, calculated without approximations, and valid for any magnitude of the external forcings. The response functionals of Eq. (10.5.18) are identical in form to those derived by means of the Kubo theory; they differ in that the Heisenberg operators in Eq. (10.5.18) are defined with respect to the complete Hamiltonian, $\mathcal{H}_t$, whereas Kubo's not only use $\mathcal{H}_0$, but also require it to represent an equilibrium system. Although the general result can be derived with no more effort than is demanded for the Kubo derivation, and its formal simplicity is displayed for all to behold, to my knowledge no constructive use has been made of it. Therein lies the challenge.

At some point, as the theory goes, approximations must be invoked. The basic approximation is equivalent to the replacement, in Eq. (10.5.8), of the final $U(t', t_0)$ by $U_0(t', t_0)$, since the major problem in the use of exact nonlinear response theory lies in the practical impossibility, at our present state of analytical competence, of finding a useful expression for $U(t, t_0)$.

The times that appear in these response functions and their relatives can be redistributed in various ways. Even though the exact solution for $U(t, t_0)$ is not usually known, some of its general properties are independent of the details of its structure. The two required basic properties are

$$U(t, t_0) = U(t, t')U(t', t_0); \quad t_0 \le t' \le t, \quad \text{and}$$
$$U^\dagger(t, t_0) = U^{-1}(t, t_0) = U(t_0, t). \qquad (10.5.20)$$

By use of these, and the invariance of the trace under cyclic permutation, the response functional of Eq. (10.5.18) can be written

$$\Phi_{ij}(t, t'; t_0) = \frac{1}{i\hbar}\left\langle [\mathcal{A}_j(t_0), \mathcal{B}_i(t, t')]\right\rangle_0. \qquad (10.5.21)$$

The physical meaning of these response functionals, which is evident from Eq. (10.5.19), is less intuitive than that of their analogs in linear response theory. There, the response functions give the response at time t to a δ-function forcing at time t'.

The only difference between Φ_{ij} of Eq. (10.5.21) and that of linear response theory is that in Eq. (10.5.21), $\mathcal{B}(t, t') = U^\dagger(t, t')\mathcal{B}U(t, t')$, whereas that of linear response theory is $\mathcal{B}_{lrt}(t, t') \equiv \mathcal{B}^{(0)} = U_0^\dagger(t, t')\mathcal{B}U_0(t, t')$. Thus $\mathcal{B}$ is propagated not only by $\mathcal{H}_0$, but also by the nonzero forcings that are turned on at $t = t_0$ in $\mathcal{H}$, whereas in linear response theory only the (equilibrium) Hamiltonian $\mathcal{H}_0$ propagates $\mathcal{B}^{(0)}$. Both forms have memory. In linear response theory, it is assumed that the forcings are so small that no serious error is made by the use of $\mathcal{H}_0$. Both forms have expectation values found by the use of $\text{Tr}(\rho_0 \ldots)$. This is the analog of the procedure used with the coupled-oscillator chain in Chap. 1. The evolution of each particle is carried forward by the dynamics, and given in terms of the initial positions and momenta of all of the particles in the chain. Statistical averaging can then be done over the original initial conditions, as they appear in the evolved expressions for positions and momenta, or by propagating forward the initial probability density for use in averaging over the time-dependent positions and momenta *per se*. These are equivalent (unless the probability density or statistical operator is reduced, as was done in Chap. 1), and in their equivalence lies the explanation of the Liouville theorem.

Even though it is usually necessary to revert to linear response theory or some other approximation, it is instructive to continue the development of the exact theory. At any point, the comparable result of linear response theory can be recovered by a process equivalent to the replacement of the exact Heisenberg operator $\mathcal{B}_i(t, t')$ in Eq. (10.5.21) by its zero-forcing counterpart $\mathcal{B}_i^{(0)}(t, t')$, given by

$$\mathcal{B}_0^{(0)}(t, t') = U_0^\dagger(t, t')\mathcal{B}_i U_0(t, t'). \tag{10.5.22}$$

When the forcings are constants, turned on abruptly at $t = t_0$, then the exact unitary transformation becomes

$$U(t, t_0) = e^{-i(t-t_0)\mathcal{H}/\hbar} = U(t - t_0), \tag{10.5.23}$$

with $\mathcal{H} = \mathcal{H}_0 - \sum F_j \mathcal{A}_j$, for $t \geq t_0$. As seen in general from problem 7, but derived easily for $U(t - t_0)$ as written in Eq. (10.5.23), the exact response coefficient of Eq. (10.5.17) becomes

$$\sigma_{ij}(F_1, F_2, \cdots, F_n; t) = \frac{1}{i\hbar} \int_{t_0}^{t} dt' \, \text{Tr}\left(\rho(t)[\mathcal{A}_j(t' - t), \mathcal{B}_i]\right), \tag{10.5.24}$$

where

$$\rho(t) = e^{-(t-t_0)\mathcal{H}/\hbar} \rho(t_0) e^{(t-t_0)\mathcal{H}/\hbar}. \tag{10.5.25}$$

Let $t - t' = \tau$, so that $dt' = -d\tau$, and rewrite this equation as

$$\sigma_{ij}(F_1, F_2, \cdots, F_n; t) = \frac{1}{i\hbar} \int_{0}^{t-t_0} d\tau \, \text{Tr}\left(\rho(t)[\mathcal{A}_j(-\tau), \mathcal{B}_i]\right). \tag{10.5.26}$$

This form is sometimes useful in the calculation of zero-frequency electric polarization, magnetization by a static magnetic field, and electric current in a conductor that is in good contact with a heat bath, so that heating by ohmic dissipation can be removed in order to maintain a steady state. As indicated, the response to the sudden turn-on of henceforth-static forcings is time-dependent, and Eq. (10.5.26) in principle exhibits this time dependence. The actual response to such forcings is

$$\langle \mathcal{B}_i \rangle_t = \sum_{j=1}^{n} \frac{F_j}{i\hbar} \int_0^{t-t_0} d\tau \, \mathrm{Tr} \left(\rho(t) [\mathcal{A}_j(-\tau), \mathcal{B}_i] \right), \qquad (10.5.27)$$

and the coefficients of the F_j are seen to be field-dependent quantities of the type given in the second form of Eq. (10.5.8), since the unitary transformations of the $\mathcal{A}_j$ are effected by the operator of Eq. (10.5.23), which is simple but exact.

A sometimes-useful trick, originated by Kubo and already used in the first look at linear response theory in Chap. 9, allows the transformation of an operator $\mathcal{A}$ to one involving $\dot{\mathcal{A}}$. This procedure is of advantage when the laboratory observable is not the operator found in the Hamiltonian, but its time derivative. As an example, in the problem of electric conductivity, the electrostatic energy term added to $\mathcal{H}_0$ is given in terms of the operator $\mathcal{R}$ of Eq. (9.9.37) as

$$\mathcal{H}_1 = -E_0 \mathcal{R} = \sum q_i \phi_i = \sum q_i x_i \phi_L / L, \qquad (10.5.28)$$

where ϕ_L is the potential difference across the sample, and L is its length. The operator $\mathcal{R} = \sum q_i x_i$ is not always a useful observable, but its time derivative, given by Eq. (9.9.35), is related to the current density operator by

$$\mathcal{J}_x = V^{-1} \sum_V \sum_i q_i v_i = V^{-1} \sum_V \dot{\mathcal{R}}, \qquad (10.5.29)$$

where V is the volume of the sample. It is thus useful to know about the Kubo transformation.

The expectation values written in Eq. (10.5.21) are understood as

$$\langle [\mathcal{A}, \mathcal{B}(t,t')] \rangle_0 = \mathrm{Tr} \left(\rho_0 [\mathcal{A}, \mathcal{B}(t,t')] \right). \qquad (10.5.30)$$

When ρ_0 is an equilibrium canonical statistical operator, (10.5.30) can be written

$$\langle [\mathcal{A}, \mathcal{B}(t,t')] \rangle_0 = Z_0^{-1} \mathrm{Tr} \left(e^{-\beta \mathcal{H}_0} [\mathcal{A}, \mathcal{B}(t,t')] \right). \qquad (10.5.31)$$

The equilibrium operator U_0 is already known to be $U_0(t,t_0) = \exp(-i(t-t_0)\mathcal{H}_0/\hbar)$, so that $U_0^\dagger(t,t_0)\mathcal{A}(t_0)U_0(t,t_0) = \mathcal{A}^{(0)}(t)$. In this formalism, one may then write

$$e^{\lambda \mathcal{H}_0} \mathcal{A}(t_1) e^{-\lambda \mathcal{H}_0} = \mathcal{A}^{(0)}(t_1 - i\hbar\lambda), \qquad (10.5.32)$$

where t_1 is a dummy argument that will soon vanish, and $t \to -i\lambda\hbar$. By use of Eq. (9.9.9), another formal manipulation gives

$$\frac{d}{d\lambda} \left(e^{\lambda \mathcal{H}_0} \mathcal{A}(t_1) e^{-\lambda \mathcal{H}_0} \right) = \left[\mathcal{H}_0, e^{\lambda \mathcal{H}_0} \mathcal{A}(t_1) e^{-\lambda \mathcal{H}_0} \right] = -i\hbar \dot{\mathcal{A}}^{(0)}(t_1 - i\lambda\hbar), \qquad (10.5.33)$$

where $\dot{\mathcal{A}}$ is the derivative of $\mathcal{A}$ with respect to its argument. Integration of this equation gives

$$\int_0^\beta d\left(e^{\lambda\mathcal{H}_0}\mathcal{A}(t_1)e^{-\lambda\mathcal{H}_0}\right) = e^{\beta\mathcal{H}_0}\mathcal{A}(t_1)e^{-\beta\mathcal{H}_0} - \mathcal{A}(t_1) = -i\hbar\int_0^\beta d\lambda\,\dot{\mathcal{A}}^{(0)}(t_1 - i\lambda\hbar), \quad (10.5.34)$$

which can be rewritten

$$\left[\mathcal{A}(t_1), e^{-\beta\mathcal{H}_0}\right] = -i\hbar e^{-\beta\mathcal{H}_0}\int_0^\beta d\lambda\,\dot{\mathcal{A}}^{(0)}(t_1 - i\lambda\hbar). \quad (10.5.35)$$

Because of its invariance under cyclic permutations of its terms, the trace of Eq. (10.5.31) can be written as

$$\mathrm{Tr}\left(e^{-\beta\mathcal{H}_0}(\mathcal{A}\mathcal{B} - \mathcal{B}\mathcal{A})\right) = -\mathrm{Tr}\left([\mathcal{A}, e^{-\beta\mathcal{H}_0}]\mathcal{B}\right),$$

and the original expectation value of Eq. (10.5.30) becomes

$$\begin{aligned}
\langle[\mathcal{A}, \mathcal{B}(t, t')]\rangle_0 &= -Z_0^{-1}\mathrm{Tr}\left([\mathcal{A}(t_1), e^{-\beta\mathcal{H}_0}]\mathcal{B}(t, t')\right) \\
&= i\hbar Z_0^{-1}\mathrm{Tr}\left(e^{-\beta\mathcal{H}_0}\int_0^\beta d\lambda\,\dot{\mathcal{A}}^{(0)}(t_1 - i\lambda\hbar)\,\mathcal{B}(t, t')\right) \\
&= i\hbar\int_0^\beta d\lambda\,\left\langle\dot{\mathcal{A}}^{(0)}(t_1 - i\lambda\hbar)\mathcal{B}(t, t')\right\rangle_0 \\
&= i\hbar\int_0^\beta d\lambda\,\left\langle e^{\lambda\mathcal{H}_0}\dot{\mathcal{A}}(t_1)e^{-\lambda\mathcal{H}_0}\mathcal{B}(t, t')\right\rangle_0.
\end{aligned} \quad (10.5.36)$$

The dummy argument t_1 can be dropped from the result at this point, and the remaining puzzle is the interpretation of the imaginary-time operator $\dot{\mathcal{A}}(-i\lambda\hbar)$. The same pseudo-time is seen in Sec. 3.5.4 on Feynman path integrals, where t becomes $-i\hbar\beta$. The expectation value $\left\langle\dot{\mathcal{A}}^{(0)}(-i\hbar\lambda)\mathcal{B}\right\rangle$ turns out to be related to a standard correlation function. Largely for this reason, the principal treatment of response functions appears in this chapter rather than the previous one.

As an example, the zero-frequency current density that arises from the sudden turn-on at $t = t_0$ of a potential difference across a homogeneous, isotropic sample, in good contact with a heat bath, is calculated exactly, for comparison with the linear-response-theory result of Chap. 9. Let $\mathcal{B}_i$ become $\mathcal{J}_x$, the current-density operator, and $F_j\mathcal{A}_j$ the potential-energy operator of the charges in the sample, or F_j becomes E_x and $\mathcal{A}_j$ becomes $\sum q_n x_n$ over the sample. Then $\dot{\mathcal{A}}_j = \sum q_n\dot{x}_n = V\mathcal{J}_x$, where V is the volume. From Eqs.(10.5.19) and (10.5.21), the expectation value of the current density is, with $t_0 = 0$,

$$\begin{aligned}
J_x = \langle\mathcal{J}_x\rangle_t &= \frac{E_x}{i\hbar}\int_0^t dt'\,\left\langle\left[\sum q_n x_n, \mathcal{J}_x(t')\right]\right\rangle_0 \\
&= VE_x\int_0^t dt'\int_0^\beta d\lambda\,\left\langle\mathcal{J}_x^{(0)}(-i\lambda\hbar), \mathcal{J}_x(t')\right\rangle_0,
\end{aligned} \quad (10.5.37)$$

where Eq. (10.5.36) has been used to rewrite the trace. This is an exact, nonlinear result, since $\mathcal{J}_x(t')$ is the exactly transformed current density that includes the effects of the internal forcing, E_x. The zero-frequency conductivity is given by

$$\sigma_{xx}(E_x, t) = \frac{J_x}{E_x} = V \int_0^t dt' \int_0^\beta d\lambda \left\langle \mathcal{J}_x^{(0)}(-i\lambda\hbar), \mathcal{J}_x(t') \right\rangle_0, \qquad (10.5.38)$$

which is dependent upon both the magnitude of the electric field and the time. As $t \to \infty$ it is expected that the steady-state conductivity will continue to be field dependent, but the time dependence should disappear. The result is just that of linear response theory if $\mathcal{J}_x(t')$ becomes $\mathcal{J}_x^{(0)}(t')$, and it is therefore in agreement with Eq. (9.9.40), with $\omega = 0$.

10.5.1 Linear Response Theory. For most purposes, response-theory calculations are based on an approximation to the solution for $U(t, t_0)$ of Eq. (10.5.8), given by

$$U(t, t_0) \approx U_0(t, t_0) \left[1 - \frac{i}{\hbar} \int_{t_0}^t dt' \, U_0^\dagger(t', t_0) \mathcal{H}_1(t') U_0(t', t_0) \right], \qquad (10.5.39)$$

where the final U of the exact theory has been changed to $U_0 = \exp[-i(t - t_0)\mathcal{H}_0/\hbar]$, and $\mathcal{H}_0$ is always required to be a time-independent Hamiltonian. The time $t = t_0$ is such that the system has attained equilibrium, and the statistical operator has the canonical value $\rho_0 = \exp(-\beta\mathcal{H}_0)/Z_0$. The response functionals of Eq. (10.5.21) become

$$\Phi_{ij}(t, t'; t_0) \approx \frac{1}{i\hbar} \left\langle [\mathcal{A}_j(t_0), \mathcal{B}_i^{(0)}(t, t')] \right\rangle_0$$

$$= \frac{1}{i\hbar} \left\langle \left[\mathcal{A}_j(t_0), \left(1 - \frac{i}{\hbar} \sum_j \int_{t'}^t dt_1 F_j(t_1) U_0^\dagger(t_1, t') \mathcal{A}_j U_0(t_1, t') \right) U_0^\dagger(t, t') \mathcal{B}_i \right. \right.$$

$$\left. \left. \times U_0(t, t') \left(1 + \frac{i}{\hbar} \sum_j \int_{t'}^t dt_1 F_j(t_1) U_0^\dagger(t_1, t') \mathcal{A}_j U_0(t_1, t') \right) \right] \right\rangle_0. \qquad (10.5.40)$$

After straightforward, but tedious, algebra, or by use of superoperator techniques (Peterson [1967]), the response functionals can be expanded as

$$\Phi_{ij}(t, t') = \frac{1}{i\hbar} \left\langle \left[\mathcal{A}_j, \mathcal{B}_i^{(0)}(t - t') \right] \right\rangle_0 +$$

$$+ \left(\frac{1}{i\hbar} \right)^2 \sum_k \int_{t'}^t dt_1 F_k(t_1) \left\langle \left[\mathcal{A}_j, \left[\mathcal{A}_k, \mathcal{B}_i^{(0)}(t - t') \right] \right] \right\rangle_0 + \cdots$$

$$= \phi_{ij}(t - t') + \sum_k \int_{t'}^t dt_1 \, F_k(t_1) \phi_{ijk}(t - t', t_1 - t') + \cdots, \qquad (10.5.41)$$

where the first- and second-order *linear response functions* are defined by the terms of this equation. The total expected response is now written

$$\langle \mathcal{B}_i(t) \rangle_0 \approx \sum_j \int_{t_0}^t dt' F_j(t') \phi_{ij}(t - t') + \sum_{j,k} \int_{t_0}^t dt_1 F_k(t_1) \int_{t_0}^{t_1} dt_2 F_j(t_2) \phi_{ijk}(t - t_2, t_1 - t_2) + \cdots.$$

$$(10.5.42)$$

The linear response functions are the zero-forcing coefficients that are common in such linear phenomenological equations as Ohm's law. There is much discussion in the literature about the conditions for applicability of linear response theory, or indeed of this form of nonlinear response theory. Some authors who prefer the kinetic-theory approach claim that there is too little physics in response theory. They regard the use of the Boltzmann equation and its relatives, including the quantum modifications, as the only physically meaningful way to account for details of collisions in the calculated response functions. Others object that even if linear response theory is valid, there are no means provided by the theory to delimit its range of validity. For how large a forcing can the response be calculated by linearized theory? A counterargument for the kinetic theorists is that the details of interaction of the microscopic constituents of a system often disappear from the calculated expressions for macroscopic observables. Furthermore, response theory is derived directly from rather basic quantum mechanics, and to deny its validity is to deny its basis. The objection that linear response theory may indeed be valid, but only for such small forcings that it is useless for most purposes, is one that must be considered seriously.

The first-order response of the operator $\mathcal{B}$ to a single forcing $\mathcal{A}F(t)$ is given in terms of the linear response function,

$$\phi_{BA}(t - t') = \frac{1}{i\hbar} \left\langle \left[\mathcal{A}, \mathcal{B}^{(0)}(t - t') \right] \right\rangle_0, \tag{10.5.43}$$

by the first term in Eq. (10.5.42):

$$\langle \mathcal{B}(t) \rangle_0 = \int_{t_0}^{t} dt' \phi_{BA}(t - t') F(t'). \tag{10.5.44}$$

Assume that the forcing is turned on at $t_0 = -\infty$; change variable to $\tau = t - t'$, so that $dt' = -d\tau$; and rewrite Eq. (10.5.44) as

$$\langle \mathcal{B}(t) \rangle_0 = -\int_{\infty}^{0} d\tau \, \phi_{BA}(\tau) F(t - \tau) = \int_{0}^{\infty} d\tau \, \phi_{BA}(\tau) F(t - \tau). \tag{10.5.45}$$

Let $F(t - \tau) = \delta(t - \tau)$. Then $\langle \mathcal{B}(t) \rangle_0 = \phi_{BA}(t)$, and the linear response function is seen to be the response of the system to unit-impulse forcing.

The symmetry of the first-order response function is easily found to be

$$\phi_{AB}(\tau) = -\phi_{BA}(-\tau). \tag{10.5.46}$$

The verification of this assertion is left as an exercise. (See problem 7.)

If the forcing is turned on at $t_0 = 0$ in Eq. (10.5.44), the Laplace- and Fourier-transform convolution theorems can be used. For a set of functions $\{f_i(t)\}$, their Laplace transforms are defined as

$$\bar{F}_i(s) = \int_{0}^{\infty} dt \, e^{-st} f_i(t). \tag{10.5.47}$$

The convolution theorem for Laplace transforms states that if

$$f_3(t) = \int_{0}^{t} dt' \, f_2(t - t') f_1(t'), \tag{10.5.48}$$

then the Laplace transform of f_3 is

$$\bar{F}_3(s) = \bar{F}_1(s)\bar{F}_2(s).$$ (10.5.49)

If an operator that does not depend explicitly on the time is inserted anywhere as part of the integrand of Eq. (10.5.48), its only effect is to appear in a corresponding position in Eq. (10.5.49). A similar theorem holds for Fourier transforms, and both are useful in linear response theory.

The Laplace transform of Eq. (10.5.44), with $t_0 = 0$, is

$$\int_0^\infty dt\, e^{-st} \left\langle \mathcal{B}(t) \right\rangle_0 \equiv \bar{B}(s) = \bar{\phi}_{BA}(s)\bar{F}(s),$$ (10.5.50)

where $\bar{\phi}_{BA}(s)$ is sometimes called the *generalized susceptibility*, and is defined as

$$\bar{\phi}_{BA}(s) \equiv \int_0^\infty dt\, e^{-st}\phi_{BA}(t) = \frac{1}{i\hbar}\int_0^\infty dt\, e^{-st} \left\langle \left[\mathcal{A}, \mathcal{B}^{(0)}(t) \right] \right\rangle_0.$$ (10.5.51)

For a known linear-response function $\phi_{BA}(t)$ and a specified time-dependence of the forcing, $F(t)$, the Laplace transform $\bar{B}(s)$, from Eq. (10.5.50), can be inverted to produce the required time-dependent response, $\left\langle \mathcal{B}(t) \right\rangle_0$.

An interesting and important result of linear response theory describes the relaxation of the system after a constant forcing of long duration is abruptly turned off. Suppose that in Eq. (10.5.44), $F(t) = F_0$, a constant for $t_0 \leq 0$, and $F(t) = 0$ for $t > 0$. Set $t_0 = -\infty$ and write Eq. (10.5.44) as

$$\left\langle \mathcal{B}(t) \right\rangle = \int_{-\infty}^t dt'\, \phi_{BA}(t - t')F(t') = F_0 \int_{-\infty}^0 dt'\, \phi_{BA}(t - t')$$
$$= -F_0 \int_\infty^t d\tau\, \phi_{BA}(\tau) = F_0 \int_t^\infty d\tau\, \phi_{BA}(\tau),$$ (10.5.52)

where $\tau = t - t'$. The relaxation function, $\Psi_{BA}(t)$ is defined as

$$\Psi_{BA}(t) = \int_t^\infty dt'\, \phi_{BA}(t') = \frac{1}{i\hbar}\int_t^\infty dt' \left\langle \left[\mathcal{A}, \mathcal{B}^{(0)}(t') \right] \right\rangle_0.$$ (10.5.53)

If $\phi_{BA}(t) \to 0$ as $t \to \infty$, then it follows that

$$\frac{d\Psi_{BA}(t)}{dt} = -\phi_{BA}(t) = -\frac{1}{i\hbar} \left\langle \left[\mathcal{A}, \mathcal{B}^{(0)}(t) \right] \right\rangle.$$ (10.5.54)

The relaxation function not only implies the response function; it contains additional information that is sometimes useful.

10.6 THE KRAMERS-KRONIG RELATIONS

The response of a system to a forcing, as calculated by linear response theory, is not intrinsically causal. The calculated response may begin before the forcing is turned on.

As a preliminary illustration of how this unphysical situation occurs, I invoke a well-known example. The high-frequency conductivity of electrons in a fully ionized plasma is sometimes derived on the assumption that the ions are too massive to be responsive, and only the electrons have low-enough inertia to contribute to the current. Regard the electrons as independent and noninteracting classical particles. The equation of motion of one of these is

$$m\ddot{x} = -eE_0 e^{-i\omega t},$$

(10.6.1)

with its first integral seen by inspection to be

$$\dot{x} = (-ieE_0/m\omega)e^{-i\omega t} + \dot{x}_0.$$

(10.6.2)

The current density is given by

$$j = -ne\dot{x} - ne\left\langle \dot{x}_0 \right\rangle = \frac{ine^2 E_0}{m\omega}e^{-i\omega t} = \sigma E_0 e^{-i\omega t},$$

(10.6.3)

where n in the electron number density, and $\left\langle \dot{x}_0 \right\rangle = 0$. The frequency-dependent conductivity is found to be

$$\sigma(\omega) = ine^2/m\omega.$$

(10.6.4)

The current density caused by an impulse field at $t = 0$, $E_0\delta(t)$, is given by

$$j(t) = \frac{1}{\sqrt{2\pi}} \int_{-\infty}^{\infty} d\omega \, \sigma(\omega)\tilde{E}(\omega)e^{-i\omega t},$$

(10.6.5)

where the Fourier transform of the impulse field is

$$\tilde{E}(\omega) = \frac{1}{\sqrt{2\pi}} \int_{-\infty}^{\infty} dt \, e^{i\omega t}E(t) = \frac{E_0}{\sqrt{2\pi}},$$

(10.6.6)

which is independent of ω. The current density thus becomes

$$\begin{aligned} j(t) &= \frac{E_0}{2\pi} \int_{-\infty}^{\infty} d\omega \left(\frac{ine^2}{m\omega}\right) e^{-i\omega t} = \frac{ine^2}{2\pi m\omega} \int_{-\infty}^{\infty} d\omega \, \frac{e^{-i\omega t}}{\omega} \\ &= -ne^2 E_0/2m, \quad t < 0 \\ &= +ne^2 E_0/2m, \quad t > 0. \end{aligned}$$

(10.6.7)

The improper integral may be evaluated as follows: Note that $e^{-i\omega t} = \cos\omega t - i\sin\omega t$, and that since the cosine integrand is odd it integrates to zero. The sine integrand is even, and can be written as twice the integral over only positive values of ω. Express it as a limit,

$$\int_{-\infty}^{\infty} d\omega \, \frac{e^{-i\omega t}}{\omega} = -2i \lim_{\alpha \to 0} \int_{0}^{\infty} d\omega \, e^{-\alpha\omega} \frac{\sin\omega t}{\omega} \equiv I(\alpha)$$

(10.6.8)

Find $dI/d\alpha$, which is easily integrated by elementary methods, to give $dI/d\alpha = 2it/(\alpha^2+t^2)$. This expression can be integrated directly, to give $I(\alpha) = 2i[\tan^{-1}(\alpha/t) - C]$. Since $I(\alpha = \infty) = 0$, the constant is found to be $C = (\pi/2)\mathrm{sgn}\,t$, and the required integral is $I(0) = -i\pi \text{ sgn } t$, leading to the results of Eq. (10.6.7). The appearance of current for times $t < 0$ violates causality, and corrective measures are necessary. The conductivity can be written

$$\sigma(\omega) = \frac{ne^2}{2m}\left[\pi\delta(\omega) + \frac{i}{\omega}\right],\tag{10.6.9}$$

so that the current density becomes

$$\begin{aligned}
j(t) &= \frac{ne^2 E_0}{2\pi m}\int_{-\infty}^{\infty} d\omega\, e^{-i\omega t}\left[\pi\delta(\omega) + \frac{i}{\omega}\right]\\
&= \frac{ne^2 E_0}{2\pi m}\left[\pi - \pi\right] = 0;\quad t < 0\\
&= \frac{ne^2 E_0}{2\pi m}\left[\pi + \pi\right] = \frac{ne^2 E_0}{m};\quad t > 0.
\end{aligned}\tag{10.6.10}$$

The chosen conductivity now satisfies causality, and it agrees with the calculated expression for $\omega > 0$. A general procedure is evidently necessary for assuring that the calculated transport coefficients and susceptibilities are causal functions.

The first-order response of the operator $\mathcal{B}$ to a single forcing, from Eq. (10.5.43), with $t_0 = 0$ is

$$\langle\mathcal{B}(t)\rangle = \int_0^t dt'\, \phi(t - t')F(t'),\tag{10.6.11}$$

where subscripts that are not needed for present purposes have been dropped. For the situation that $F(t) = 0$ for $t < 0$, and the response function similarly obeys $\phi(t) = 0$ for $t < 0$, Eq. (10.6.11) can be written

$$\langle\mathcal{B}(t)\rangle = \int_{-\infty}^{\infty} dt'\, \phi(t - t')F(t'),\tag{10.6.12}$$

because the integrand vanishes outside the limits specified originally. With these new limits, the Fourier convolution theorem gives

$$\tilde{B}(\omega) = \chi(\omega)\tilde{F}(\omega),\tag{10.6.13}$$

where

$$\begin{aligned}
\tilde{B}(\omega) &= \int_{-\infty}^{\infty} dt\, \langle\mathcal{B}(t)\rangle\, e^{i\omega t} = \int_0^{\infty} dt\, \langle\mathcal{B}(t)\rangle\, e^{i\omega t},\\
\chi(\omega) &= \tilde{\phi}(\omega) = \int_{-\infty}^{\infty} dt\, \phi(t)e^{i\omega t},\quad\text{and}\\
\tilde{F}(\omega) &= \int_{-\infty}^{\infty} dt\, F(t)e^{i\omega t}.
\end{aligned}\tag{10.6.14}$$

The response function $\phi(t)$ is real; therefore the susceptibility χ can be written

$$\chi(\omega) = \lim_{\alpha\to\infty}\int_0^{\infty} dt\, e^{(i\omega - \alpha)t}\phi(t) = \chi'(\omega) + i\chi''(\omega),\tag{10.6.15}$$

where χ' and χ'' are real, and the α-limiting process is implied henceforth, when needed. The real and imaginary parts of the susceptibility are given by

$$\chi'(\omega) = \int_0^\infty \cos \omega t \, \phi(t); \tag{10.6.16}$$

and

$$\chi''(\omega) = \int_0^\infty \sin \omega t \, \phi(t). \tag{10.6.17}$$

Each of these equations is a Fourier transform, and each can be inverted, to give a fruitful basis for further manipulation. If $\chi'(\omega)$ exists as a useful property, it must be analytic, regular, continuous, and bounded in the upper half-plane. With these limitations, the inverse Fourier transforms can be written

$$\phi(t) = \frac{2}{\pi} \int_0^\infty d\omega \, \chi'(\omega) \cos \omega t = \frac{2}{\pi} \int_0^\infty d\omega \, \chi''(\omega) \sin \omega t. \tag{10.6.18}$$

The use of the first of these in Eq. (10.6.17) yields

$$\begin{aligned}
\chi''(\omega) &= \frac{2}{\pi} \lim_{T \to \infty} \int_0^T dt \, \sin \omega t \int_0^\infty du \, \chi'(\omega) \cos ut \\
&= \frac{1}{\pi} \lim_{T \to \infty} \int_0^\infty du \, \chi'(u) \int_0^T dt \, [\sin(\omega + u)t + \sin(\omega - u)t] \\
&= \frac{-1}{\pi} \lim_{T \to \infty} \int_0^\infty du \, \chi'(u) \left[\frac{\cos(\omega + u)t}{\omega + u} + \frac{\cos(\omega - u)t}{\omega - u} \right]_{t=0}^T \\
&= \frac{1}{\pi} P \int_{-\infty}^\infty du \, \frac{\chi'(u)}{\omega - u}, \tag{10.6.19}
\end{aligned}$$

where, in the next to the last line, the Riemann-Lebesgue lemma has been used to eliminate the infinite-upper-limit terms in the third line by considering their integrals with respect to u, as $T \to \infty$. Since both χ' and χ'' are real, the principal value is required. (The principal value is defined by

$$P \int_{-\infty}^\infty \frac{dy \, f(y)}{x - y} \equiv \lim_{\epsilon \to 0} \left[\int_{-\infty}^{x-\epsilon} + \int_{x+\epsilon}^\infty \right] \frac{dy \, f(y)}{x - y}. \tag{10.6.20}$$

It is usually evaluated by closing the contour and replacing the path through the pole by half the sum of two paths, one of which detours around the pole in the upper half plane, and the other in the lower. Another technique uses the trick employed in Eq. (10.6.8), after a change of variable. See Problem 8.) A similar calculation gives the complementary relation; both are displayed here as the celebrated Kramers-Kronig relations.

$$\chi'(\omega) = \frac{-1}{\pi} P \int_{-\infty}^\infty du \, \frac{\chi''(u)}{\omega - u}, \quad \text{and}$$

$$\chi''(\omega) = \frac{1}{\pi} P \int_{-\infty}^\infty du \, \frac{\chi'(u)}{\omega - u}. \tag{10.6.21}$$

These constitute a Hilbert-transform pair, sometimes called *dispersion relations*. Their derivation was made possible because the causality condition, that $F(t) = 0$ for $t < 0$ implies that $\phi(t) = 0$ for $t < 0$, allows the integration limits in Eq. (10.6.12) to become infinite, so that the Fourier convolution theorem can be invoked. Therefore these Hilbert transforms are regarded as the sought-for conditions that the calculated susceptibilities be causal.

Similar conditions are available for the admittances. The first-order response of the operator $\mathcal{B}$, from Eq. (10.5.44), by way of Eq. (10.6.12), is written

$$\langle \mathcal{B}(t) \rangle = \int_{-\infty}^{\infty} dt' \, \phi_{BA}(t - t')F(t') = \frac{1}{2\pi} \int_{-\infty}^{\infty} d\omega \, e^{-i\omega t}\chi(\omega)\tilde{F}(\omega). \qquad (10.6.22)$$

Its time derivative is

$$\left\langle \dot{\mathcal{B}}(t) \right\rangle \equiv \int_{-\infty}^{\infty} dt' \, \phi_{\dot{B}A}(t - t')F(t') = \frac{1}{2\pi} \int_{-\infty}^{\infty} d\omega \, e^{-i\omega t}[-i\omega\chi(\omega)]\tilde{F}(\omega)$$

$$= \frac{1}{2\pi} \int_{-\infty}^{\infty} d\omega \, e^{-i\omega t}Y(\omega)\tilde{F}(\omega), \qquad (10.6.23)$$

where $Y(\omega)$ is usually called the *admittance* for the particular process.

As a specific example, suppose $\mathcal{B} = \sum q_i x_i$, where the q_i represent electric charges, and the x_i are the positions of these charges, measured, say, from their equilibrium positions, and $F(t)$ is the electric field, turned on at $t = 0$. The response $\mathcal{B}(t)$ is the total electric dipole moment, or the polarization, depending in how the system is defined. The time derivative $\dot{\mathcal{B}} = \sum q_i \dot{x}_i$, which is then a current or current density. The admittance, or one of its relatives, is the response function for the current, whereas the susceptance, or one of its relatives, is that for the polarization. (As to the proper names for these functions, the familiar *conductance* and *conductivity* pair, corresponding to *resistance* and *resistivity*, provide some guidance.) I do not distinguish the terms accurately in a general treatment, since the ultimate proper usage depends upon the definitions of the operators and the system.

As is evident from Eq. (10.6.23), the admittance and the susceptance are related:

$$Y(\omega) = -i\omega\chi(\omega) = Y'(\omega) + iY''(\omega) = \omega\chi''(\omega) - i\omega\chi'(\omega). \qquad (10.6.24)$$

The Kramers-Kronig relations can be rewritten in terms of the Y's as:

$$\chi'(\omega) = -\frac{Y''(\omega)}{\omega} = \frac{1}{\pi}P\int_{-\infty}^{\infty} \frac{du\, Y'(u)}{u(\omega - u)} \quad \text{and}$$

$$\chi''(\omega) = \frac{Y'(\omega)}{\omega} = -\frac{1}{\pi}P\int_{-\infty}^{\infty} \frac{du\, Y''(u)}{u(\omega - u)}. \qquad (10.6.25)$$

(There are frequent variations in the signs of these relations; these arise from the choice of sign in the Fourier-transform pair, and sometimes in the choice of sign for the imaginary component of the admittance.)

10.7 FLUCTUATION-DISSIPATION THEOREMS

Two puzzling aspects of formal linear response theory remain to be resolved. First, the response function, as given in Eq. (10.5.43),

$$\phi_{BA}(t - t') = (1/i\hbar)\left\langle [\mathcal{A}, \mathcal{B}^{(0)}(t - t')]\right\rangle_0,$$

is expressed as the trace of a commutator, which is not usually an observable quantity. Second, the Kubo transform that replaced the commutator by a correlation function, Eq. (10.5.36),

$$\left\langle [\mathcal{A}, \mathcal{B}^{(0)}(t - t')]\right\rangle_0 = i\hbar \int_0^\beta d\lambda \left\langle \dot{\mathcal{A}}^{(0)}(-i\lambda\hbar)\mathcal{B}^{(0)}(t - t')\right\rangle_0,$$

involves imaginary time. A more physical expression for the response function, such as those of Eq. (10.4.19) for the resistance in terms of a voltage autocorrelation function or of Eq. (10.4.30) for the diffusivity, is needed. Specifically, the objective of this section is to use an ordinary correlation function, defined as

$$\psi_{BA}(t) = \left\langle \{\dot{\mathcal{A}}(0)\mathcal{B}(t)\}\right\rangle_0 \equiv \frac{1}{2}\left\langle \dot{\mathcal{A}}(0)\mathcal{B}(t) + \mathcal{B}(t)\dot{\mathcal{A}}(0)\right\rangle_0, \tag{10.7.1}$$

and to find a real decaying kernel $\Gamma(t)$, such that the response function can be written

$$\phi_{BA}(t) = \int_{-\infty}^\infty dt'\Gamma(t - t')\psi_{BA}(t'). \tag{10.7.2}$$

If such a function exists, the Fourier convolution theorem says

$$\tilde{\phi}_{BA}(\omega) = \chi_{BA}(\omega) = \tilde{\Gamma}(\omega)\tilde{\psi}_{BA}(\omega). \tag{10.7.3}$$

It is customary to write

$$\tilde{\Gamma}(\omega) = 1/E_\beta(\omega) = \int_{-\infty}^\infty dt\, e^{i\omega t}\Gamma(t). \tag{10.7.4}$$

For the continuation of this analysis, it is assumed that $\left\langle \dot{\mathcal{A}}^{(0)}(-i\lambda\hbar)\mathcal{B}^{(0)}(t)\right\rangle_0$ is analytic in the strip $-\beta\hbar \le \Im t = y \le \beta\hbar$, where, for integration in the complex plane, $t = x + iy$. It is further assumed that $\lim_{x \to \pm\infty}\left\langle \dot{\mathcal{A}}\mathcal{B}(t)\right\rangle = 0$. From Eq. (10.6.14), write

$$\begin{aligned}
\chi_{AB}(\omega) &= \int_{-\infty}^\infty dt\, e^{i\omega t}\int_0^\beta d\lambda \left\langle \dot{\mathcal{A}}^{(0)}(-i\lambda\hbar)\mathcal{B}^{(0)}(t)\right\rangle_0 \\
&= \int_0^\beta d\lambda\, e^{\lambda\hbar\omega}\int_{-\infty}^\infty dt \left\langle e^{i\omega(t+i\lambda\hbar)}\dot{\mathcal{A}}(0)\mathcal{B}^{(0)}(t + i\lambda\hbar)\right\rangle_0 \\
&= \int_0^\beta d\lambda\, e^{\lambda\hbar\omega}\int_{-\infty+i\lambda\hbar}^{\infty+i\lambda\hbar} dt \left\langle e^{i\omega t}\dot{\mathcal{A}}(0)\mathcal{B}^{(0)}(t)\right\rangle_0 \\
&= \frac{e^{\beta\hbar\omega} - 1}{\hbar\omega}\int_{-\infty}^\infty dt\, e^{i\omega t}\left\langle \dot{\mathcal{A}}(0)\mathcal{B}^{(0)}(t)\right\rangle_0, \tag{10.7.5}
\end{aligned}$$

where in the second step, the correlation is regarded as stationary in time, so that it depends only on the difference of the two arguments. In the third step, an obvious change of variable allows the time integral to be rewritten, and the analyticity and limit assumptions about the integrand permit the path of integration to be returned to the real axis. Thus the Fourier transform of the first half of the symmetrized correlation function $\psi_{BA}(t)$ has been related to $\chi_{BA}(\omega)$. (This result, as it stands, is sometimes presented as a fluctuation-dissipation theorem.) The next set of manipulations, starting with the final expression in Eq. (10.7.5), produces the other half of $\psi_{BA}(t)$. The various steps, such as the use of the cyclic invariance of the trace, are either obvious or sufficiently similar to those of Eq. (10.7.5) to require no further comment.

$$
\begin{aligned}
\chi_{BA}(\omega) &= \frac{e^{\beta\hbar\omega} - 1}{\hbar\omega Z_0} \int_{-\infty}^{\infty} dt\, e^{i\omega t} \mathrm{Tr}\left(\left(e^{-\beta\mathcal{H}_0} \dot{A}(0) e^{\beta\mathcal{H}_0} \right) \left(e^{-\beta\mathcal{H}_0} \mathcal{B}^{(0)}(t) \right) \right) \\
&= \frac{e^{\beta\hbar\omega} - 1}{\hbar\omega Z_0} \int_{-\infty}^{\infty} dt\, e^{i\omega t} \mathrm{Tr}\left(e^{-\beta\mathcal{H}_0} \mathcal{B}^{(0)}(t) \dot{A}^{(0)}(i\beta\hbar) \right) \\
&= \frac{e^{\beta\hbar\omega} - 1}{\hbar\omega} \int_{-\infty}^{\infty} dt\, e^{i\omega(t-i\beta\hbar)} \left\langle \mathcal{B}^{(0)}(t - i\beta\hbar) \dot{A}(0) \right\rangle_0 e^{-\beta\hbar\omega} \\
&= \frac{1 - e^{-\beta\hbar\omega}}{\hbar\omega} \int_{-\infty-i\beta\hbar}^{\infty-i\beta\hbar} dt\, e^{i\omega t} \left\langle \mathcal{B}^{(0)}(t) \dot{A}(0) \right\rangle_0 \\
&= \frac{1 - e^{-\beta\hbar\omega}}{\hbar\omega} \int_{-\infty}^{\infty} dt\, e^{i\omega t} \left\langle \mathcal{B}^{(0)}(t) \dot{A}(0) \right\rangle_0 .
\end{aligned}
\tag{10.7.6}
$$

The required Fourier transform is extracted from the final expressions in these two equations as

$$
\begin{aligned}
\tilde{\psi}_{BA}(\omega) &= \frac{1}{2} \int_{-\infty}^{\infty} dt\, e^{i\omega t} \left\langle \dot{A}(0)\mathcal{B}^{(0)}(t) + \mathcal{B}^{(0)}(t)\dot{A}(0) \right\rangle_0 \\
&= \frac{\hbar\omega}{2} \left[\frac{1}{e^{\beta\hbar\omega} - 1} + \frac{1}{1 - e^{-\beta\hbar\omega}} \right] \chi_{BA}(\omega).
\end{aligned}
\tag{10.7.7}
$$

This result permits Eq. (10.7.3) to become

$$
\begin{aligned}
\tilde{\psi}_{BA}(\omega) &= \left(\frac{\hbar\omega}{2} \right) \coth\left(\frac{\beta\hbar\omega}{2} \right) \chi_{BA}(\omega) \\
&= \chi_{BA}(\omega)/\bar{\Gamma}(\omega) = \chi_{BA}(\omega) E_\beta(\omega),
\end{aligned}
\tag{10.7.8}
$$

where the universal, system-independent function $E_\beta(\omega)$ has been identified as

$$
E_\beta(\omega) = \left(\frac{\hbar\omega}{2} \right) \coth\left(\frac{\beta\hbar\omega}{2} \right).
\tag{10.7.9}
$$

This is the expression found in Chapter 3 for the expected energy of a single harmonic oscillator, in terms of its frequency and $T = 1/k\beta$. It represents the energy associated with spontaneous equilibrium fluctuations. From (10.7.9) and Eq. (10.7.4), the function $\Gamma(t)$ may be written: (Kubo [1957])

$$
\Gamma(t) = \frac{1}{2\pi} \int_{-\infty}^{\infty} d\omega\, e^{-i\omega t} \left(\frac{\tanh(\beta\hbar\omega/2)}{(\hbar\omega/2)} \right) = \frac{2}{\pi\hbar} \ln \coth\left(\frac{\pi|t|}{2\beta\hbar} \right).
\tag{10.7.10}
$$

In the classical system, as $\hbar \to 0$, the kernel function becomes $\Gamma(t) = \beta\delta(t)$, and the susceptance is given by

$$\chi_{BA}(\omega) = \beta \int_{-\infty}^{\infty} e^{i\omega t}\psi_{BA}(t) = \beta\tilde{\psi}_{BA}(\omega). \qquad (10.7.11)$$

Both here and in the general result of Eq. (10.7.8), susceptance is determined by the Fourier transform of an ordinary (even if symmetrized) correlation function of equilibrium fluctuations. This correlation function is a measure of the ever-present spontaneous fluctuations that are excited thermally within the system. The response function $\phi_{BA}(t)$, which measures the behavior of the system when perturbed from its equilibrium state, is thus fundamentally related to the behavior of the system in response to its own internally produced fluctuations, as Onsager said in the context of thermodynamics. This is the general content of fluctuation-dissipation theorems. The response functions, which may lead to dissipation of energy in response to a forcing, are completely expressed by the equilibrium correlation functions of internal thermal fluctuations, which are not dissipative.

The admittance is given in terms of the susceptance by Eq. (10.6.24), so that

$$\tilde{\psi}_{BA}(\omega) = \frac{i}{\omega}E_\beta(\omega)Y_{BA}(\omega) = \frac{i}{\omega}E_\beta(\omega)\left[Y'_{BA}(\omega) + iY''(\omega)\right]. \qquad (10.7.12)$$

The real and imaginary parts of $\tilde{\psi}_{BA}(\omega)$ are related to those of the admittance as

$$\tilde{\psi}'_{BA}(\omega) = -\left(\frac{\hbar}{2}\coth\frac{\beta\hbar\omega}{2}\right)Y''_{BA}(\omega) \quad \text{and}$$

$$\tilde{\psi}''_{BA}(\omega) = \left(\frac{\hbar}{2}\coth\frac{\beta\hbar\omega}{2}\right)Y'_{BA}(\omega). \qquad (10.7.13)$$

10.8 COMMENTARY

The yin and yang of statistical physics are the suppression by averaging of microscopic phenomena to extract macroscopic parameters and the re-emergence of these microscopic phenomena as fluctuations, retaining their underlying reality and their fundamental importance. As an early example, recall the identification of temperature that was found in Chapter 4 for noninteracting fermions, bosons, and Boltzmann particles:

$$kT = (2/3)\langle\Delta(NE)\rangle/\langle N\rangle, \qquad (10.8.1)$$

where $\langle\Delta(NE)\rangle$ is the E-N covariance in a grand-canonical system. (One small modification, found necessary for bosons at $T < T_c$, is that $\langle N\rangle$ excludes the ground-state population.) The covariance measures the fluctuation in a macroscopic variable, which was calculated from the (three quite different) partition functions that assign probabilities to the microscopic states. In general, the macroscopic properties of the three ideal gases are so dissimilar that it is not valid to employ the elementary description of temperature as a measure of the mean kinetic energy of a particle. But to describe the temperature as a measure of a covariance, a property of the equilibrium fluctuations within the system, has general validity.

10.8.1 Random Walks. Just as lattices are useful for the study of interacting particles, even though details of the lattice structure disappear from the calculation of critical exponents, so are they useful in the modeling of physical processes in terms of random walks. Brownian motion and elementary aspects of diffusion are introduced in this chapter, but the many applications to such diverse areas as molecular biology, economics, chemical kinetics, dynamic critical phenomena, epidemiology, population dynamics, cosmology, numerical analysis, and polymer science have necessarily been neglected. I have presented a few techniques that use generating functions and Fourier transforms as an introduction to the theory of random walks, and I have displayed Watson's three integrals, not only as pure art, both beautiful and awe-inspiring, but also as an indication of the difficulty to be expected in the analytic solution of such problems as the three-dimensional Ising system. (Fortunately, a few people are still to be found, such as M. L. Glasser, who are skillful in the evaluation of the kinds of integrals that arise in lattice walks and other problems that can be modeled in this context.) A more recent and quite readable survey of several applications of the theory of random walks, by Raposo *et al.* is recommended for further reading (Barber and Ninham [1970]; Glasser and Zucker [1977]; Montroll [1967]; Raposo et al. [1991]).

10.8.2 Fluctuation-Dissipation Theorems. Johnson noise in resistors and the Nyquist explanation of it were reported in adjoining articles in the Physical Review (Johnson [1928]; Nyquist [1928]). Like the much earlier Nernst-Einstein relation, in its derivation, a transport coefficient is expressed in terms of the equilibrium fluctuations of some system parameter. The first general theory of this kind, to the best of my knowledge, is that of Callen and Welton, which is still worth reading because of its directness (Callen and Welton [1951]). These fluctuation-dissipation theorems are now derived much more efficiently and generally, but much less transparently, by way of linear transport theory. It is at first surprising that the transport coefficients, which usually imply dissipation and entropy production, are related to dissipationless equilibrium fluctuations by a theory that allows no production of entropy, and no dissipation.

The equilibrium correlation functions measure the relaxation of these fluctuations, implying the existence of a relaxation time such as that introduced in the Langevin theory of diffusion. But no relaxation occurs in a finite system, because the Heisenberg operators are harmonic in time, and the response function $\phi_{BA}(t)$ is oscillatory. This statement is true, despite the formal existence of the relaxation function of Eq. (10.5.15). Localized disturbances propagate through the medium in the manner of acoustic waves, which are examples of an adiabatic linear response to the kind of external forcing that can be represented as a term in the Hamiltonian. For this reason, even though transport coefficients are given formally in linear response theory in terms of correlation functions, these functions are obtained from linear response theory only after the introduction of a means to allow dissipation.

A number of techniques exist for introducing entropy growth and dissipation into linear response theory, some of which have been mentioned in Chapter 9. The use of a reduced statistical operator, which is equivalent to the treatment of the coupled harmonic oscillator system of Chapter 1, selects part of the overall (often infinite) system, as the canonical system to be studied, and reduces the statistical operator by integration over the outside (heat bath) variables in the classical case, or by taking an equivalent partial trace in a quantum system. By this procedure, the system can evolve to a state of equilibrium with the heat bath by a Hamiltonian process. The final equilibrium behavior of the same system, but now in equilibrium with a heat bath that is not described by a Hamiltonian, can be obtained by remaximization of the entropy, subject to the constraint provided by the heat bath. The Mitchell theory of thermal sources, described in Chapter 9, offers another route, with great conceptual advantage is some cases. Other means of allowing entropy growth involve coarse-graining, either in classical phase space, or in the quantum energy levels. My discussion in Chapter 9 of the evolution of the classical Gibbs entropy as a consequence of imprecise knowledge of the Hamiltonian, for a variety of reasons given there, can be carried over to quantum systems. The harmonic behavior of the Heisenberg operators, with frequencies given by $\omega_{rs} = |E_r - E_s|/\hbar$, becomes blurred in time as a consequence of the same kinds of imprecision that I invoked to allow the classical Gibbs entropy to evolve, with similar consequences. The off-diagonal elements of the statistical operator decay in time with exponents proportional to $-t^2$, resulting in a diagonal statistical operator that can be converted to a transition matrix for a master equation. (It is sometimes necessary to group an energy continuum into a countable set of discrete but approximate energy levels, in order to write the master equation for a corresponding set of probabilities. This is a kind of coarse graining.)

In the van Hove derivation of the master equation, the perturbation of the equilibrium Hamiltonian is kept small, so that in Eq. (9.9.11), $\lambda \to 0$. The transition probabilities, given by Wigner's golden rule, allow the master equation of Eq. (9.9.22) to be written

$$\dot{P}_n = \frac{2\pi\lambda^2}{\hbar} \sum_s{}' \delta(E_n - E_s) \left|\langle n|\mathcal{H}_I(t)|s\rangle\right|^2 (P_s - P_n).$$ (10.8.2)

In the van Hove limit, where $\lambda \to 0$ and $t/\tau_r \to \infty$, such that $\lambda^2 t = \mathcal{O}1$, with τ_r as the longest relaxation time that is found from the weak-interaction limit of the Liouville operator, the density operator becomes diagonal, and the evolution of the system can be described by Eq. (10.8.2) (van Hove [1960]; van Hove [1962]; van Vliet [1978]; van Vliet [1979]).

The Pauli master equation seems to offer a plausible means of following the temporal evolution of a nonequilibrium system, except that it represents a Markov process, with no memory. For systems that are effectively memoryless, it usually provides an easier way to extract information about a system than does linear response theory *per se*. Van Vliet has written a excellent survey of linear response theory, including a generalized inhomogeneous master equation that incorporates streaming terms.

10.8.3 Criticism of Linear Response Theory. The linearization of response theory means that the Heisenberg operators are carried forward in time by the unperturbed Hamiltonian, and the statistical average is over the initial unperturbed states. This procedure must have the effect of restricting the applicability of the theory to perturbations with some unspecified limiting magnitude, or for some limited time after the onset of the perturbation. The van Hove limit states such a restriction, but it arises in the context of a master-equation formulation of response theory, rather than as a general criterion.

Van Kampen objected to linear response theory on the grounds that it is a mere exercise in mathematical manipulation, rather than a theory with physical content, describing the macroscopic responses to microscopic interactions. This objection has already been answered by the statement that it is firmly based in quantum mechanics and uses valid perturbation methods. The only open question, as I see it, is the time-amplitude limit of applicability of the derived results. Van Kampen estimates this as follows, for the calculation of electric conductivity: An electron acted upon by a field moves a distance $x = (eE/2m*)t^2$, which he claims should be small compared to the diameter of the scatterers in order for the response to be linear. (Here, $m*$ is the effective mass, and I would rephrase his criterion to require x to be of the order of a mean free path, but this is not my objection.) He estimates the limiting value for x to be 100Å, he supposes that $m* = 0.01m$, where m is the electron mass, and he takes an observation time, for which he expects linear response theory to be valid, of 1 s. His criterion in general is $t^2 E \leq 2xm*/e$, and his choices of values give $t^2 E = 10^{-18}$ Vs^2/cm, or $E \leq 10^{-18}$ V/cm. Van Vliet and others have argued that the electron-electron and electron-phonon collision times in a metal at room temperature are in the range 10^{-13}-10^{-12} s, and that these scatterings determine the time for which linear response theory must work, since each electron effectively starts anew after a few scatterings. Then if the appropriate value for the time of validity is taken as ten times the electron-electron or electron-phonon collision time, the limiting electric field becomes $E_{max} = 100$-1000 V/cm, a much more reasonable value. It should be noted, however, that these collisions are not normally included explicitly in the Hamiltonian that carries the Heisenberg operators (although they could be regarded as implicit in $\mathcal{H}_0$), and the inclusion of their effects based on physical arguments is supplementary to the theory (van Kampen [1971]; van Vliet [1988]; van Vliet [1978]).

A major conceptual problem with response theory is the absence of dissipation. The unitary transformation of the Heisenberg operators, including the statistical operator, produces no entropy, and no dissipation. The fluctuation-dissipation theorems relate response functions to equilibrium correlations, but ohmic heating is not forthcoming. The linear-response-theory description of an electric heater allows the energy of the heater to increase as a consequence of work done on the system by forcing a current through it, but the work is regarded as causing energy to be stored in the vibrational modes of the heater molecules, without any production of entropy. This is merely a restatement of the fact that the transfer of electrical energy to a system, even in equilibrium thermodynamics, is regarded as work. The *de facto* loss of information that allows the system to be regarded as thermalized is accomplished by a new maximization of the entropy, subject to the known constraints, as discussed in the Commentary section of Chapter 9.

The use of models has long been a means by which understanding is achieved. The Boltzmann dog-flea and wind-tree models that I treated in Chapter 1, together with the harmonic oscillator models that also appear there, are systems that can be analyzed in a variety of formalisms, including those resembling the Boltzmann equation, the master equation, the Liouville equation, and some of the techniques of information theory. The comparison of the results of different approaches provides insight that can be transferred to systems of more interest in physics. In addition to the papers mentioned in the Commentary section of Chapter 1, I recommend two papers by Max Dresden that offer explicit comparisons of the results obtained by different approaches, for several models. He finds behavior in some models that is not at all in agreement with the usual intuitive picture of how systems evolve to equilibrium. There continues to be a need for models having completely specified Hamiltonians that can be solved exactly, with special emphasis on the possibility that some such models should be designed to provide limits on the behavior of real physical systems (Dresden [1962]; Dresden and Feiock [1972]).

I regret that space does not allow treatments of time-dependent critical fluctuations and the kinds of statistical fluctuations that appear in chaotic systems. Both topics are expected to grow in importance as they become better understood.

10.9 REFERENCES

The principal references are listed here. A complete listing is to be found in the Bibliography at the end of the book.

BARBER, M. N. AND NINHAM, B. W. [1970], *Random and Restricted Walks*. Gordon and Breach, New York, NY, 1970.

BERNARD, W. AND CALLEN, H. B. [1959], Irreversible Thermodynamics of Nonlinear Processes and Noise in Driven Systems. *Reviews of Modern Physics* **31** (1959), 1017–1044.

CHANDRASEKHAR, S. [1943], Stochastic Problems in Physics and Astronomy. *Reviews of Modern Physics* **15** (1943), 1–89.

GRANDY, JR., W. T. [1987], *Foundations of Statistical Mechanics V.II: Nonequilibrium Phenomena*. D. Reidel, Dordrecht–Boston–London, 1987.

KREUZER, H. J. [1981], *Nonequilibrium Thermodynamics and its Statistical Foundations*. Oxford University Press, London, 1981.

KUBO, R. [1957], Statistical Mechanical Theory of Irreversible Processes I. General Theory and Simple Applications to Magnetic and Conduction Problems. *Journal of the Physical Society of Japan* **12** (1957), 570–586.

KUBO, R. [1965], *Linear Response Theory of Irreversible Processes*. In *Statistical Mechanics of Equilibrium and Non-Equilibrium*, J. Meixner, Ed. North-Holland, Amsterdam, 1965, 81–99.

MONTROLL, E. W. [1962], *Some Remarks on the Integral Equations of Statistical Mechanics*. In *Fundamental Problems in Statistical Mechanics*, E. G. D. Cohen, Ed. North-Holland, Amsterdam, 1962, 223–249.

MONTROLL, E. W. [1967], *Stochastic Processes and Chemical Kinetics*. In *Energetics in Metallurgical Phenomena, V. III*, W. M. Mueller, Ed. Gordon and Breach, New York, NY, 1967, 123–187.

PETERSON, R. L. [1967], Formal Theory of Nonlinear Response. *Reviews of Modern Physics* **39** (1967), 69–77.

UHLENBECK, G. E. AND ORNSTEIN, L. S. [1930], On the Theory of Brownian Motion. *Physical Review* **36** (1930), 823–841.

VAN VLIET, K. M. [1978], Linear Response Theory Revisited. I. The Many-Body van Hove Limit. *Journal of Mathematical Physics* **19** (1978), 1345–1370.

VAN VLIET, K. M. [1979], Linear Response Theory Revisited. II. The Master Equation Approach. *Journal of Mathematical Physics* **20** (1979), 2573–2595.

ZWANZIG, R. W. [1961], *Statistical Mechanics of Irreversibility*. In *Lectures in Theoretical Physics, V. III*, W. E. Brittin, B. W. Downs and J. Downs, Eds. Interscience Publishers, A division of John Wiley & Sons, London, 1961, 106–141.

10.10 PROBLEMS

1. For a 1-d random walker on the integers, starting at the origin, calculate the probability that the walker returns to the origin for the first time after $2n$ steps, when the one-step probabilities are equal.

2. For a 1-d random walk on the integers, starting at the origin, calculate the probability that the walker will never return to the origin when the one-step probabilities to right and left, p and q, are unequal.

3. On the plane $x = 0$, N_0 tagged molecules are released to diffuse in the surrounding medium. A parallel plane at $x = x_0$ absorbs every one of these molecules that reaches it. Use the Fokker-Planck equation with $v = 0$ to find $P_t(x)$. Assume that the behavior of the molecules can be described as a one-dimensional diffusion, and that $x_0 = -a$. Find the number of molecules absorbed as a function of time. How many of these molecules are never absorbed? How long does it take for half of the molecules to be absorbed?

4.a. Suppose the correlation time is so short that the autocorrelation function in Eq. (10.4.26) can be represented as a delta function, $\phi_f(v) = \phi_0 \delta(v)$. Show that the approximate result for $\langle p^2(t) \rangle$ obtained in Eq. (10.4.26) becomes exact for this choice of ϕ_f.
b. Now suppose that $\phi_f(v) = \phi_0$ for $-a \leq v \leq a$ and zero otherwise. Show that for $t \gg a$, the result of Eq. (10.4.26) is valid.

5. Use the almost exact result for $\langle v^2(t) \rangle$ given in Eq. (10.4.29) in the exact differential equation for $\langle r^2(t) \rangle$, Eq. (10.3.5), to find the almost exact solution, to supersede the cruder approximation of Eq. (10.3.6).

6. From Eq. (10.3.2), find an expression for the velocity autocorrelation function required in Eq. (10.3.9), and use it to calculate an expression for the diffusivity D. The result should agree with that following Eq. (10.3.7).

7. Verify that $\phi_{AB}(\tau) = -\phi_{BA}(-\tau)$, as asserted in Eq. (10.5.46).

8. Verify the internal consistency of the Hilbert-transform pair that appears as the Kramers-Kronig relations of Eq. (10.6.21) by taking the Fourier transform of the second of these, as given in Eq. (10.6.18), to obtain a causally correct expression for $\phi(t)$.

10.11 SOLUTIONS

1. From Eq. (10.2.6), the probability that the walker will be at the origin after $2n$ steps is

$$P_{2n}(0) = \frac{1}{4^n} \binom{2n}{n}.$$

From Eq. (10.2.7), the generating function that the walker is at the origin after any number of steps is

$$U_1(z, 0) = 1/\sqrt{1 - z^2}.$$

From Eq. (10.2.16), the generating function for the probabilities of first-time occurrences is

$$F(z) = 1 - 1/U(z) = 1 - \sqrt{1 - z^2},$$

where $U(z) = U_1(z)$ from the previous equation. Direct binomial expansion of this expression gives

$$F(z) = \sum_{n=1}^{\infty} f_n z^n = \sum_{n=1}^{\infty} \frac{z^{2n}}{4^n(2n-1)} \binom{2n}{n} = \sum_{n=1}^{\infty} \frac{P_{2n}}{2n-1} z^{2n}.$$

Therefore the probability that the random walker returns to the origin for the first time after $2n$ steps is $P_{2n}/(2n-1)$, or one in $2n-1$ returns to the origin after $2n$ steps will be for the first time.

2. From Eq. (10.2.31), the probability that the walker will never return to the origin is $1/U_1(1,0)$; Eq. (10.2.7) gives $U_1(z,s)$ as

$$U_1(z,s) = \sum_{n=0}^{\infty} z^n P_n(s) = \sum_{n=0}^{\infty} \frac{z^n}{2\pi} \int_{-\pi}^{\pi} d\phi\, e^{-is\phi} G(\phi,n)$$

$$= \frac{1}{2\pi} \int_{-\pi}^{\pi} d\phi\, e^{-is\phi} \sum_{n=0}^{\infty} z^n \left(qe^{-i\phi} + pe^{i\phi}\right)^n = \frac{1}{2\pi} \int_{-\pi}^{\pi} \frac{d\phi\, e^{-is\phi}}{1 - z(pe^{i\phi} + qe^{-i\phi})},$$

where $G(\phi,n)$ was obtained from Eq. (10.2.3). The substitution $\zeta = \exp(i\phi)$ and the choice $s = 0$ give the required generating function as

$$U_1(z,0) = \left(\frac{-1}{pz}\right) \frac{1}{2\pi i} \oint \frac{d\zeta}{\zeta^2 - (\zeta/pz) + (q/p)}.$$

The contour integral is most easily evaluated by choosing $z < 1$, carrying out the integration, and then allowing z to approach 1. For the choice $q \neq p$, the unit circle encloses only one pole of the denominator, since the other pole is at $\zeta = q/p > 1$, for $q > p$, as $z \to 1$. By this approach, the value of the integral is found to be

$$U_1(1,0) = 1/|(q-p)|,$$

and the probability that the walker never returns to the origin is $|q - p|$.

3. The solution to the Fokker-Planck equation, Eq. (10.2.37), with $P_t(x_0) = 0$ for all t is easily seen to be

$$P_t(x) = \frac{1}{\sqrt{4\pi Dt}} \left[e^{-x^2/4Dt} - e^{-(2x_0-x)^2/4Dt} \right].$$

With the coordinate system chosen so that the origin is between $x = x_0$ and $x = \infty$, the fraction of unabsorbed molecules, $\theta(t)$, is

$$\theta(t) = \int_{x_0}^{\infty} dx\, P_t(x) = \frac{1}{\sqrt{4\pi Dt}} \left[\int_{x_0}^{\infty} dx\, e^{-x^2/4Dt} - \int_{x_0}^{\infty} dx\, e^{-(2x_0-x)^2/4Dt} \right]$$

$$= \frac{1}{\sqrt{4\pi Dt}} \left[\int_{x_0}^{\infty} dx\, e^{-x^2/4Dt} - \int_{-x_0}^{\infty} dy\, e^{y^2/4Dt} \right]$$

$$= \frac{1}{\sqrt{4\pi Dt}} \left[\int_{-a}^{\infty} - \int_{a}^{\infty} \right] dx\, e^{-x^2/4Dt} = \frac{1}{\sqrt{4\pi Dt}} \int_{-a}^{a} dx\, e^{-x^2/4Dt},$$

where the substitutions $x_0 = -a$ and $y = x - 2x_0$ have been used. The final integral can be written as twice the integral over half of the interval. Now write $z^2 = x^2/4Dt$, and get $N(t) = N_0\theta(t)$ as

$$N(t) = \frac{2N_0}{\sqrt{\pi}} \int_0^{a/\sqrt{4Dt}} dz\, e^{-z^2} = N_0 \mathrm{erf}(a/\sqrt{4Dt}).$$

As a check, note that $N(0) = N_0$, $N(\infty) = 0$. Therefore, eventually all of the molecules are absorbed. The number absorbed by time t is $N_a(t) = N_0 - N(t)$, or

$$N_a(t) = N_0 \left[1 - \mathrm{erf}(a/\sqrt{4Dt}) \right].$$

When $N_a(t) = N_0/2$, the argument of the error function is ≈ 0.497, and $t_{1/2} \approx a^2/0.998D \approx a^2/D$.

4.a. With $\phi_f(v) = \phi_0 \delta(v)$, integration of Eq. (10.3.16) gives

$$\left\langle p^2(t) \right\rangle = \left\langle p^2(0) \right\rangle e^{-2\zeta t} + \frac{3\phi_0}{\zeta} \left[1 - e^{-2\zeta t} \right].$$

But $\left\langle p^2(\infty) \right\rangle = 3MkT = 3\phi_0/\zeta$, from which Eq. (10.4.28) follows directly.

b. For the given ϕ_f, integration of Eq. (10.3.16) gives

$$\left\langle p^2(t) \right\rangle = \left\langle p^2(0) \right\rangle e^{-2\zeta t} + \frac{3\phi_0}{\zeta^2} \left[1 - e^{-2a\zeta} - e^{-2\zeta t} \left(e^{2a\zeta} - 1 \right) \right]$$

$$= \left\langle p^2(0) \right\rangle e^{-2\zeta t} + \frac{3\phi_0}{\zeta^2} \left(1 - e^{-2a\zeta} \right) \left(1 - e^{-2\zeta(t-a)} \right).$$

In this case, $\left\langle p^2(\infty) \right\rangle = 3MkT = (3\phi_0/\zeta^2)[1 - \exp(-2\zeta a)]$, from which

$$\left\langle p^2(t) \right\rangle = \left\langle p^2(0) \right\rangle e^{-2\zeta t} + 3MkT \left(1 - e^{-2\zeta(t-a)} \right),$$

which agrees with Eq. (10.4.26) for $t \gg a$.

5. Let $\left\langle r^2(t) \right\rangle = y$ in Eq. (10.3.5), which then becomes $d^2y/dt^2 + \zeta dy/dt = 2 \left\langle v^2(t) \right\rangle$. The initial conditions are $y(0) = 0$ and $\dot{y}(0) = 0$. The first time integral is elementary, leaving a first-order equation for y of the form $\dot{y} = -\zeta y + B(t) - B(0)$, where $B(t)$ is obtained by integration of $2 \left\langle v^2(t) \right\rangle$. The solution for y is seen to be

$$y(t) = e^{-\zeta t} \int_0^t dt' e^{\zeta t'} \left[\frac{\left\langle v^2(0) \right\rangle}{\zeta} \left(1 - e^{-\zeta t} \right) + \frac{6kTt'}{M} - \frac{3kT}{M\zeta} \left(1 - e^{-2\zeta t'} \right) \right]$$

$$= \frac{\left\langle v^2(0) \right\rangle}{\zeta^2} \left(1 - e^{-\zeta t} \right)^2 + \frac{6kTt}{M\zeta} - 3kTM\zeta^2 \left(3 - e^{-\zeta t} \right) \left(1 - e^{-\zeta t} \right) = \left\langle r^2(t) \right\rangle.$$

6. Since $\left\langle \mathbf{p}(0) \cdot \mathbf{f}(t) \right\rangle = 0$, the momentum autocorrelation function becomes

$$\phi_p(t) = \left\langle \mathbf{p}(0) \cdot \mathbf{p}(t) \right\rangle = \left\langle p^2(0) \right\rangle e^{-\zeta t} = \left\langle p^2 \right\rangle e^{-\zeta t},$$

thus resembling Eq. (10.3.3), since the system average squared momentum is regarded as independent of time. From this expression, the diffusivity is given by Eq. (10.3.10) as

$$D = \frac{\left\langle v^2 \right\rangle}{3} \int_0^\infty d\tau \, e^{-\zeta \tau} = \frac{\left\langle v^2 \right\rangle}{3\zeta}.$$

If $\left\langle v^2 \right\rangle = 3kT/M$, then $D = kT/M\zeta$, as before.

7. The proof is as follows:

$$\phi_{AB}(\tau) = \frac{1}{i\hbar} \left\langle \left[\mathcal{B}, \mathcal{A}^{(0)}(\tau) \right] \right\rangle_0 = \frac{1}{i\hbar} \left\langle \left[\mathcal{B}, e^{i\tau\mathcal{H}_0/\hbar} \mathcal{A} e^{-i\tau\mathcal{H}_0/\hbar} \right] \right\rangle_0$$

$$= \frac{1}{i\hbar} \left\langle \mathcal{B} e^{i\tau\mathcal{H}_0/\hbar} \mathcal{A} e^{-i\tau\mathcal{H}_0/\hbar} - e^{i\tau\mathcal{H}_0/\hbar} \mathcal{A} e^{-i\tau\mathcal{H}_0/\hbar} \mathcal{B} \right\rangle_0$$

$$= \frac{1}{i\hbar} \left\langle e^{-i\tau\mathcal{H}_0/\hbar} \mathcal{B} e^{i\tau\mathcal{H}_0/\hbar} \mathcal{A} - \mathcal{A} e^{-i\tau\mathcal{H}_0/\hbar} \mathcal{B} e^{i\tau\mathcal{H}_0/\hbar} \right\rangle_0$$

$$= -\frac{1}{i\hbar} \left\langle \left[\mathcal{A}, \mathcal{B}^{(0)}(-\tau) \right] \right\rangle_0 = -\phi_{BA}(-\tau). \quad \text{Q.E.D.}$$

8. From Eq. (10.6.18) and the recognition that the integrands are even functions of ω, the second one may be rewritten

$$\phi(t) = \frac{i}{\pi} \int_{-\infty}^{\infty} d\omega\, \chi''(\omega) e^{-i\omega t}.$$

Use of the second of Eq. (10.6.21) for χ'' gives

$$\phi(t) = \frac{i}{\pi} \int_{-\infty}^{\infty} d\omega\, e^{-i\omega t} \frac{1}{\pi} P \int_{-\infty}^{\infty} \frac{du\,\chi'(u)}{\omega - u} = \frac{i}{\pi^2} P \int_{-\infty}^{\infty} \frac{dy}{y} e^{-ity} \int_{-\infty}^{\infty} du\,\chi'(u) e^{-itu}$$

$$= \frac{1}{\pi}\phi(t) P \int_{-\infty}^{\infty} \frac{dy}{y} \sin ty = \phi(t)\operatorname{sgn} t.$$

The trick shown in Eq. (10.6.8) has been used, after the change of variable to $y = \omega - u$. When $t > 0$, this result says $\phi(t) = \phi(t)$, as it should. For $t < 0$, it says $\phi(t) = -\phi(t)$, which is satisfied iff $\phi(t) = 0$, the causality condition. Thus the Kramers-Kronig relation gives self-consistent results that impose causality on the response function.

A.   GENERATING FUNCTIONS

A.1 INTRODUCTION

Generating functions are useful mathematical entities that contain or imply, usually in a compact and economical form, a wealth of mathematical information. The partition function is an example of a generating function; in addition to being a compact repository of thermodynamic information, the canonical partition function can generate the microcanonical partition functions for all energies, and the grand partition function can generate the canonical partition function for any population N of the system. Some authors use the term *generating function* instead of *partition function* (Heer [1972]).

In this appendix, generating functions are introduced by a well-known example, which serves as a vehicle for demonstrating some of their power and convenience, as well as typical operations involving them.

The generating functions used in this treatment are of the form

$$G(x,t) = \sum_n f_n(x)t^n, \tag{A.1.1}$$

where the $f_n(x)$ are typically a set of mathematical functions of a variable x, with the function index n covering a range of values. In the examples of this appendix, the index is an integer, but it need not be so limited. The range of n is typically from 0-∞ or from $-\infty$-∞. The generating function presumably can be obtained in a compact form that has the series expansion given in Eq. (A.1.1). The functions $f_n(x)$ are then obtained by finding the coefficients of the t^n in the series expansion, but usually *not* by expanding the generating function in a series.

A.2 AN EXAMPLE

The functions $f_n(x)$ are sometimes known to be related, for example by a recursion relation. In this example, since the recursion relation is that for Bessel and related functions, $J_n(x)$ will be used instead of $f_n(x)$. I start with one recursion relation and one boundary condition, to obtain a handbookfull of Bessel-function relationships and formulas. The recursion relation is

$$J_n'(x) = (1/2)[J_{n-1}(x) - J_{n+1}(x)], \tag{A.2.1}$$

and the generating function is

$$G(x,t) = \sum_{n=-\infty}^{\infty} J_n(x)t^n. \tag{A.2.2}$$

The x-derivative of G gives

$$\frac{\partial G}{\partial x} = \sum_{-\infty}^{\infty} J_n'(x)t^n = \frac{1}{2}\sum_{-\infty}^{\infty} t^n[J_{n-1}(x) - J_{n+1}(x)]$$
$$= (1/2)(t - 1/t)G(x,t). \tag{A.2.3}$$

525

This first-order differential equation is easily solved, with the boundary condition $G(0,t) = 1$, to give

$$G(x,t) = e^{(x/2)(t-1/t)}. \tag{A.2.4}$$

Thus the function $J_n(x)$, which happens to be the Bessel function that is usually so denoted, is the coefficient of t^n in the series expansion of a known, simple function. If the right-hand side of Eq. (A.2.4) is written as the product of the series expansions for $e^{xt/2}$ and $e^{-x/2t}$, and the terms collected for each t^n, the coefficients are the power-series expressions for $J_n(x)$.

Note that various summation formulas follow directly: as examples,

$$G(x,1) = \sum_{-\infty}^{\infty} J_n(x) = 1,$$

$$G(x+y,t) = G(x,t)G(y,t), \tag{A.2.5}$$

from which $J_n(x+y)$ is found to be

$$J_n(x+y) = \sum_{r=-\infty}^{\infty} J_r(x)J_{n-r}(y). \tag{A.2.6}$$

From the t derivative of G, it follows that

$$t\frac{\partial G}{\partial t} = \sum_n nt^n J_n(x) = \frac{x}{2}\left(t + \frac{1}{t}\right)\sum_n t^n J_n(x),$$

which, by equating like powers of t, gives the other usual recursion relation for Bessel functions:

$$J_n(x) = (x/2n)[J_{n-1}(x) + J_{n+1}(x)]. \tag{A.2.7}$$

From the two recursion relations, the Bessel differential equation follows directly:

$$\frac{1}{x}\frac{d}{dx}(xJ_n'(x)) + \left(1 - \frac{n^2}{x^2}\right)J_n(x) = 0. \tag{A.2.8}$$

The substitution $t = e^{i\theta}$ gives

$$\begin{aligned}G(z,t) = e^{iz\sin\theta} &= \cos(z\sin\theta) + i\sin(z\sin\theta)\\ &= \sum_n J_n(z)e^{in\theta} = \sum_n J_n(z)[\cos n\theta + i\sin n\theta],\end{aligned} \tag{A.2.9}$$

leading to the useful formulas

$$\cos(z\sin\theta) = \sum_{n=-\infty}^{\infty} J_n(z)\cos n\theta$$

$$\sin(z\sin\theta) = \sum_{n=-\infty}^{\infty} J_n(z)\sin n\theta, \tag{A.2.10}$$

from which a collection of sums can be obtained by selecting particular values for θ.

A frequently used procedure for extracting a particular $f_n(x)$ from a known $G(x,t)$ is by the use of contour integration. The Cauchy theorem applied to Eq. (A.1.1) gives

$$f_r(x) = \frac{1}{2\pi i} \oint \frac{G(x,z)}{z^{r+1}}.$$

It is often convenient to carry out the integration by means of the substitution $z = e^{i\theta}$. Applied to Eq. (A.2.9), this technique gives

$$J_n(x) = \frac{1}{2\pi} \int_0^{2\pi} d\theta \, e^{i(x\sin\theta - n\theta)}$$
$$= \frac{1}{\pi} \int_0^{\pi} d\theta \, \cos(x\sin\theta - n\theta). \qquad (A.2.11)$$

A number of other such relationships can be developed from this start, but it is not my purpose to exhibit an exhaustive catalog of Bessel-function properties. Those already presented, with the evident ease of derivation, serve to illustrate the power and usefulness of the generating function as a mathematical tool.

A.3 APPLICATION TO PROBABILITIES

The dog-flea problem in Chapter 1 presents a set of coupled time-dependent probabilities, related by Eq. (1.5.4). The solution to this equation can be obtained in a variety of ways, particularly since the fleas act independently, but it provides a good example of the use of generating functions in this kind of calculation, and not every such equation is so trivially solved by other techniques. The basic differential equation for the probabilities, rewritten in time units for which $\alpha = 1$, is

$$\dot{p}_n(t) = -np_n(t) + (n+1)p_{n+1}(t), \qquad (A.3.1)$$

with the assumed initial condition that all N fleas are on dog 1 at $t = 0$, so $p_N(0) = 1$ and $p_r(0) = 0$ for $r \neq N$. A suitable generating function is

$$G(x,t) = \sum_{n=0}^{\infty} p_n(t)x^n. \qquad (A.3.2)$$

The partial derivative of G with respect to t gives

$$\frac{\partial G(x,t)}{\partial t} = -\sum_n np_n(t)x^n + \sum_n (n+1)p_{n+1}(t)x^n$$
$$= -x\frac{\partial G}{\partial x} + \frac{\partial G}{\partial x} = (1-x)\frac{\partial G}{\partial x}. \qquad (A.3.3)$$

where Eq. (A.3.1) has been used, and the sums are then rewritten as evident partial derivatives with respect to x. Since the sum of the probabilities is always 1, it follows that $G(1,t) = 1$; the initial conditions for the probabilities give the other necessary boundary condition, $G(x,0) = x^N$. With these conditions, the first-order partial differential equation for G can be easily solved to give

$$G(x,t) = \left[1 - (1-x)e^{-t}\right]^N,\tag{A.3.4}$$

which can be expanded as a power series in x as

$$G(x,t) = \sum_{n=0}^{\infty} \binom{N}{n} \left(1 - e^{-t}\right)^{N-n} \left(e^{-t}\right)^n x^n.\tag{A.3.5}$$

From this result, the probabilities can be read off as

$$p_n(t) = \binom{N}{n} \left(1 - e^{-t}\right)^{N-n} e^{-nt},\tag{A.3.6}$$

in agreement with Eq. (1.5.5).

This kind of calculation is often useful in the study of equilibration as a consequence of transitions of a system from one state to another when the transition probabilities are known.

A.4 OTHER USEFUL GENERATING FUNCTIONS

A number of reference books on the functions that appear in mathematical physics provide lists of generating functions (Abramowitz and Stegun [1964]; Gradshteyn and Ryzhik [1980]). A few of the ones used in this book are listed here.

1. The number of ways of choosing m objects from a collection of n distinguishable objects, independent of order, is $\binom{n}{m}$. The generating function is the binomial expansion

$$(1 + x)^n = \sum_{m=0}^{n} \binom{n}{m} x^m.\tag{A.4.1}$$

2. The Legendre polynomials are generated by

$$\frac{1}{\sqrt{1 - 2xt + t^2}} = \sum_{n=0}^{\infty} P_n(x)t^n.\tag{A.4.2}$$

Recursion relations and the differential equation for Legendre functions are readily obtained from the generating function.

3. The Bernoulli polynomials, or particular values of them, are encountered in several calculations in this book. They are generated by the function

$$\frac{te^{xt}}{e^t - 1} = \sum_{n=0}^{\infty} \frac{B_n(x)}{n!} t^n, \quad |t| < 2\pi,\tag{A.4.3}$$

The Bernoulli polynomials are also generated by the statement that $B_n(x)$ is a polynomial of degree n in x, and its moving average is x^n, or

$$x^n = \int_x^{x+1} dt\, B_n(t). \qquad (A.4.4)$$

This statement allows all of the coefficients in $B_n(x)$ to be determined uniquely, for $n \geq 0$. The Bernoulli numbers are $B_n = B_n(0)$. An even more efficient way to obtain the Bernoulli numbers is to recognize that

$$B_n(x) = \sum \binom{n}{k} B_{n-k} x^l, \qquad (A.4.5)$$

so that a determination of $B_n(x)$ by means of Eq. (A.4.4) yields values of all the B_k in the range $0 \leq k \leq n$.

The first few Bernoulli numbers are: $B_0 = 1$, $B_1 = -1/2$, $B_2 = 1/6$, $B_4 = -1/30$, $B_6 = 1/42$, $B_8 = -1/30$, $B_{10} = 5/66$, $B_{12} = -671/2730$. ($B_{2n+1} = 0$ for $n > 0$.) Others are to be found in standard mathematical tables.

The first five Bernoulli polynomials are:

$$
\begin{aligned}
B_0(x) &= 1 \\
B_1(x) &= -(1/2) + x \\
B_2(x) &= (1/6) - x + x^2 \\
B_3(x) &= (1/2)x - (3/2)x^2 \\
B_4(x) &= -(1/30) + x^2 - 2x^3 + x^4.
\end{aligned} \qquad (A.4.6)
$$

Chapter 10 includes a number of calculations, using generating functions, in the theory of random walks.

A.5 REFERENCES

ABRAMOWITZ, M. AND STEGUN, I. A., Eds. [1964], *Handbook of Mathematical Functions.* U. S. National Bureau of Standards, Washington, D. C., 1964.

GRADSHTEYN, I. S. AND RYZHIK, I. M. [1980], *Table of Integrals, Series, and Products.* Academic Press, New York, NY, 1980.

HEER, C. V. [1972], *Statistical Mechanics, Kinetic Theory, and Stochastic Processes.* Academic Press, New York, NY, 1972.

B. FERMI AND BOSE INTEGRALS

B.1 FERMI INTEGRALS

In the study of noninteracting fermions, integrals commonly encountered are of the form

$$\mathcal{F}_\phi(\beta, \mu) = \int_0^\infty \frac{d\epsilon \, \phi(\epsilon)}{e^{\beta(\epsilon - \mu)} + 1},$$

$$= \mu \int_0^\infty \frac{dx \, \phi(\mu x)}{1 + e^{s(x-1)}}, \qquad (B.1.1)$$

where $s = \beta\mu$, $x = \epsilon/\mu$, and $\phi(\epsilon)$ is a function of the energy, ϵ, of a single fermion. Since these integrals usually do not permit evaluation in closed form, it is the purpose of this section to develop suitable approximations to them for systems that are highly degenerate. For these systems, the characteristic thermal energy per particle, kT, is much less than the chemical potential, μ, which is also referred to as the Fermi energy, ϵ_F. Therefore the approximation requires the validity of the inequality $\beta\mu \gg 1$.

It is convenient to define the Fermi function, $F(x)$, as

$$F(x) = \left(1 + e^{s(x-1)}\right)^{-1}, \qquad (B.1.2)$$

and to write its first derivative as

$$F'(x) = -s e^{s(x-1)} \left(1 + e^{s(x-1)}\right)^{-2},$$

$$= \frac{-s}{(1 + e^{s(x-1)})(1 + e^{s(1-x)})}. \qquad (B.1.3)$$

The general Fermi integral of Eq. (B.1.1) can now be written

$$\mathcal{F}_\phi(\beta, \mu) = \mu \int_0^\infty dx \, \phi(\mu x) F(x). \qquad (B.1.4)$$

Integration by parts yields

$$\mathcal{F}_\phi(\beta, \mu) = -\int_0^\infty dx \, \Phi(\mu x) F'(x), \qquad (B.1.5)$$

where

$$\Phi(\epsilon) = \int_0^\epsilon d\epsilon' \phi(\epsilon'),$$

which is equivalent to

$$\Phi(\mu x) = \mu \int_0^x dx' \phi(\mu x'),$$

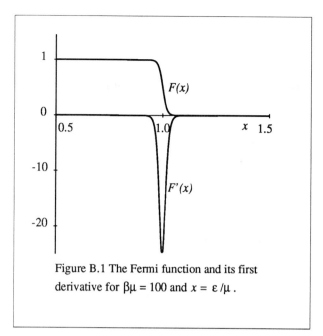

Figure B.1 The Fermi function and its first derivative for $\beta\mu = 100$ and $x = \epsilon/\mu$.

where the integrated term in Eq. (B.1.5) has vanished at the upper limit because of $F(x)$ and at the lower limit because of $\Phi(\epsilon)$. The basis of the integration procedure is the behavior of the Fermi function and its derivative, as shown in Fig. B.1.

Note that $F'(x)$ differs significantly from zero only near $x = 1$. It is therefore useful to expand the function $\Phi(\mu x)$ about this point and integrate the resulting power series term by term. A change of variables will permit this procedure to be carried out directly. Let $y = s(x - 1)$, so that Eq. (B.1.5) may be written

$$\mathcal{F}_\phi(\beta, \mu) = \int_{-s}^{\infty} \frac{dy\, \Phi(\mu + ykT)}{(1 + e^y)(1 + e^{-y})},$$
$$\approx \int_{-\infty}^{\infty} \frac{dy\, \Phi(\mu + ykT)}{(1 + e^y)(1 + e^{-y})}, \qquad (B.1.6)$$

where the extension of the lower limit of integration is in accord with the approximation procedure, as indicated by the behavior of $F'(x)$. The function $\Phi(\mu + ykT)$ is expanded in powers of y to give

$$\Phi(\mu + ykT) = \sum_{n=0}^{\infty} \frac{(ykT)^n}{n!} \Phi^{(n)}(\mu). \qquad (B.1.7)$$

Insertion of this expression in Eq. (B.1.6) gives

$$\mathcal{F}_\phi(\beta, \mu) \approx \sum_{n=0}^{\infty} \frac{(kT)^n}{n!} \Phi^{(n)}(\mu) \int_{-\infty}^{\infty} \frac{dy\, y^n}{(1 + e^y)(1 + e^{-y})},$$

$$= 2 \sum_{n=0}^{\infty} \frac{(kT)^{2n}}{(2n)!} \Phi^{(2n)}(\mu) \int_0^{\infty} \frac{dy \, y^{2n}}{(1+e^y)(1+e^{-y})},$$

where only the even values of n need be considered, since the denominator of the integrand is an even function of y. The integral may be rewritten as

$$\mathcal{F}_{\phi}(\beta, \mu) = \sum_{n=0}^{\infty} \frac{(kT)^{2n}}{(2n)!} \Phi^{(2n)}(\mu) I_{2n}. \qquad (B.1.8)$$

The integrals I_{2n} are convertible into numbers, as follows:

$$
\begin{aligned}
I_{2n} &\equiv 2 \int_0^{\infty} \frac{dy \, y^{2n}}{(1+e^y)(1+e^{-y})}, \\
&= 2 \int_0^{\infty} \frac{dy \, y^{2n} e^y}{(1+e^y)^2}, \\
&= \frac{-2y^{2n}}{1+e^y} \Big|_0^{\infty} + 4n \int_0^{\infty} \frac{dy \, y^{2n-1}}{1+e^y}, \\
&= \delta_{0n} + 4n \int_0^{\infty} \frac{dy \, y^{2n-1} e^{-y}}{1+e^{-y}}, \\
&= \delta_{0n} - 4n \int_0^{\infty} dy \, y^{2n-1} \sum_{k=1}^{\infty} (-)^k e^{-ky}, \\
&= \delta_{0n} - 4n\Gamma(2n) \sum_{k=1}^{\infty} \frac{(-)^k}{k^{2n}}, \\
&= 2(2n)! \left[1 - \frac{1}{2^{2n}} + \frac{1}{3^{2n}} - \frac{1}{4^{2n}} + \cdots \right], \quad (n \neq 0) \\
&= 2(2n)! \zeta(2n)[1 - 2^{1-2n}], \quad (n \neq 0). \qquad (B.1.9)
\end{aligned}
$$

The first three of these are:

$$I_0 = 1, \quad I_2 = \pi^2/3, \quad \text{and} \quad I_4 = 7\pi^4/15.$$

It is now possible to write the approximate expression for $\mathcal{F}_{\phi}$ as

$$\mathcal{F}_{\phi}(\beta, \mu) = \Phi(\mu) + \frac{(\pi kT)^2}{6} \Phi''(\mu) + \frac{7(\pi kT)^4}{360} \Phi^{iv}(\mu) + \cdots, \qquad (B.1.10)$$

where it is not usual to require even this many terms, since high-order calculations for idealized systems are seldom of any experimental interest.

Another technique for evaluating the integrals I_n makes use of a generating-function technique. Note that

$$I_n \equiv \int_{-\infty}^{\infty} \frac{x^n \, dx}{(e^x + 1)(e^{-x} + 1)}.$$

Write the generating function $J(k)$ as

$$J(k) = \sum_0^\infty (ik)^m I_m/m! = \int_{-\infty}^\infty \frac{e^{ikx}\,dx}{(e^x+1)(e^{-x}+1)}.\qquad (B.1.11)$$

Use the Cauchy integral formula to integrate around a contour in the complex plane, first along the real axis from $-R$ to R, next from R to $R+2i\pi$, then along the line $y=2i\pi$ to $-R+2i\pi$, and finally close to $-R$. In the limit as $R\to\infty$, the integral along the real axis becomes $J(k)$, the two end integrals in the y-direction vanish, and the integral along the line $y=2i\pi$ becomes $-e^{-2k\pi}J(k)$. The only pole within the contour is on the imaginary axis at $z=i\pi$. Shift the variable x in Eq. (B.1.11) to $z+i\pi$, to get, for small z,

$$\oint \frac{e^{ik(z+i\pi)}\,dz}{(e^{z+i\pi}+1)(e^{-z-i\pi}+1)} = e^{-k\pi}\oint \frac{e^{ikz}\,dz}{(-z)(z)} = -e^{-k\pi}\oint \frac{e^{ikz}\,dz}{z^2}.\qquad (B.1.12)$$

Recall that a consequence of Cauchy's formula is

$$f^{(n)}(a) = \frac{n!}{2\pi i}\oint \frac{f(z)dz}{(z-a)^{n+1}},$$

provided $f(z)$ is properly analytic and well-behaved within and on the contour. It therefore follows that the expression in Eq. (B.1.12) has the value $2\pi k e^{-\pi k}$, and the contour integration yields the result

$$J(k)\big[1 - e^{-2k\pi}\big] = 2k\pi e^{-k\pi},$$

or

$$J(k) = \frac{k\pi}{\sinh k\pi}.\qquad (B.1.13)$$

Now all that is necessary to obtain any integral I_n is to refer to Eq. (B.1.11), and obtain the coefficient of k^n in the generating function. The easiest way to do this is to recognize that the generating function for the Bernoulli polynomials, from Appendix A, is

$$\frac{te^{zt}}{e^t-1} = \sum_{n=0}^\infty \frac{B_n(z)t^n}{n!},$$

and the expression $x/\sinh x$ can be written

$$\frac{x}{\sinh x} = \frac{2xe^x}{e^{2x}-1} = \frac{te^{t/2}}{e^t-1},$$

where $t=2x$. This result permits $J(k)$ to be written as

$$J(k) = \sum_{n=0}^\infty \frac{B_n(1/2)}{n!}(2\pi k)^n.$$

From these results, it follows that for all odd values of n, $I_n=0$, since $B_n(1/2)=0$ for all odd n. Also, since $B_0(1/2)=1$, $B_2(1/2)=-1/12$, and $B_4(1/2)=7/240$, the values of the first three nonzero integrals I_n are seen to be those given above. A general expression for $B_n(1/2)$ is

$$B_n(1/2) = -(1 - 2^{1-n})B_n,$$

where B_n are the Bernoulli numbers, given in Appendix A.

The most frequent functions integrated by this procedure are of the form

$$\phi(\epsilon) = \epsilon^{\nu-1},$$

where ν is a real, positive number. In this case, the Fermi integrals may be written as

$$\begin{aligned}
\mathcal{F}_\nu(\beta, \mu) &\equiv \int_0^\infty \frac{d\epsilon\, \epsilon^{\nu-1}}{e^{\beta(\epsilon-\mu)} + 1}, \\
&= \frac{1}{\nu} \sum_{n=0}^\infty \frac{(kT)^{2n}}{(2n)!} I_{2n} \Phi^{(2n)}(\mu), \\
&= \frac{\mu^\nu}{\nu} \left[1 + \left(\frac{\nu(\nu-1)}{3 \times 2!} \right) \left(\frac{\pi}{\beta\mu} \right)^2 + \right. \\
&\qquad \left. + \left(\frac{7\nu(\nu-1)(\nu-2)(\nu-3)}{15 \times 4!} \right) \left(\frac{\pi}{\beta\mu} \right)^4 + \cdots \right],
\end{aligned} \qquad (B.1.14)$$

where it is clear that the approximation is a power-series expansion in the small parameter $\pi/\beta\mu$. A useful extension of Eq. (B.1.14) is an expression for the μ-derivative of $\mathcal{F}$,

$$\left(\frac{\partial \mathcal{F}_\nu(\beta, \mu)}{\partial \mu} \right)_T = (\nu - 1) \mathcal{F}_{\nu-1}(\beta, \mu). \qquad (B.1.15)$$

This result is used in several derivations in Chapter 4. Generalizations of Eq. (B.1.14) for parafermi integrals are given in the final section of this appendix.

B.2 BOSE INTEGRALS

The corresponding integrations for noninteracting bosons are much less troublesome, because in these integrals, the chemical potential, μ, is negative, making $z < 1$. The integrals to be evaluated are written as

$$\begin{aligned}
\mathcal{G}_\phi(\beta, z) &= \int_0^\infty \frac{d\epsilon\, \phi(\epsilon)}{e^{\beta(\epsilon-\mu)} - 1} = \int_0^\infty \frac{d\epsilon\, z e^{-\beta\epsilon} \phi(\epsilon)}{1 - z e^{-\beta\epsilon}}, \\
&= \int_0^\infty d\epsilon\, \phi(\epsilon) \sum_{n=1}^\infty (z e^{-\beta\epsilon})^n, \\
&= \sum_{n=1}^\infty z^n \int_0^\infty d\epsilon\, \phi(\epsilon) e^{-n\beta\epsilon} = \sum_{n=1}^\infty z^n \overline{\phi}(n\beta), \qquad (B.2.1)
\end{aligned}$$

where $\overline{\phi}(s)$ is the Laplace transform of $\phi(\epsilon)$; i.e.,

$$\overline{\phi}(s) = \int_0^\infty dx\, \phi(x) e^{-sx}.$$

This expansion is valid for all functions $\phi(\epsilon)$ that behave well enough to allow the Laplace transform to exist, and tables of these transforms are readily available.

For $\phi(\epsilon) = \epsilon^{\nu-1}$, the Laplace transform is $\overline{\phi}(n\beta) = \Gamma(\nu)/(n\beta)^\nu$, and Eq. (B.2.1) becomes

$$
\begin{aligned}
\mathcal{G}_\nu(\beta, z) &\equiv \int_0^\infty \frac{d\epsilon\, \epsilon^{\nu-1}}{e^{\beta(\epsilon-\mu)} - 1}, \\
&= \frac{\Gamma(\nu)}{\beta^\nu} \sum_{n=1}^\infty \frac{z^n}{n^\nu}, \\
&= \frac{\Gamma(\nu)}{\beta^\nu} g_\nu(z),
\end{aligned}
\tag{B.2.2}
$$

where the g-functions are defined as

$$
g_\nu(z) = \sum_{n=1}^\infty \frac{z^n}{n^\nu}.
\tag{B.2.3}
$$

The two principal properties of the g-functions are:
1. $g_\nu(1) = \zeta(\nu)$, the Riemann zeta function, and
2. $g_\nu'(z) = g_{\nu-1}(z)/z$.

It is sometimes convenient to write these functions in terms of μ, rather than z. In this case the functions are renamed as $G_\nu(\beta\mu) = g_\nu(z(\beta\mu))$.

B.3 PARAFERMION INTEGRALS

When the number of occupants of any single-particle state cannot exceed the integer p, the parafermi function corresponding to $F(x)$ in Eq. (B.1.2) is found to be

$$
F_p(x) = \frac{1}{e^{s(x-1)} - 1} - \frac{p+1}{e^{s(p+1)(x-1)} - 1},
\tag{B.3.1}
$$

where the function $F_1(x)$ is identically the $F(x)$ of Eq. (B.1.2). The function behaves qualitatively in the same way as $F(x)$ for any finite p, and the method of approximating the integrals is essentially the same. If Fig. B.1 were a plot of the function $F_p(x)$, the ordinate value 1 should be replaced by p; the derivative exhibits its characteristic localized departure from near zero at the same point; and the representative integral to be evaluated is, corresponding to Eq. (B.1.4),

$$
\begin{aligned}
\mathcal{F}_\phi(\beta, \mu; p) &= \mu \int_0^\infty dx\, \Phi(\mu x) F_p'(x) \\
&= \sum_{n=0}^\infty \frac{(kT)^n}{(2n)!} \Phi^{(2n)}(\mu) I_{2n}(p),
\end{aligned}
\tag{B.3.2}
$$

where $I_{2n}(p)$, generalizing the result of Eq. (B.1.9), is given by

$$
I_{2n}(p) = p\delta_{0n} + 2(2n)!\zeta(2n)\left(1 - (p+1)^{1-2n}\right).
\tag{B.3.3}
$$

When $\phi(\epsilon) = \epsilon^{\nu-1}$, the parafermi integrals corresponding to those of Eq. (B.1.14) become

$$\mathcal{F}_\nu(\beta,\mu;p) = \frac{\mu^\nu}{\nu}\left[p + \frac{\Gamma(\nu+1)I_2(p)}{2!\Gamma(\nu-1)}\left(\frac{kT}{\mu}\right)^2 + \frac{\Gamma(\nu+1)I_4(p)}{4!\Gamma(\nu-3)}\left(\frac{kT}{\mu}\right)^4 + \cdots\right]$$

$$= \frac{p\mu^\nu}{\nu}\left[1 + \frac{\Gamma(\nu+1)}{3(p+1)\Gamma(\nu-1)}\left(\frac{\pi}{\beta\mu}\right)^2 + \right.$$

$$\left. + \frac{\Gamma(\nu+1)(p^2+3p+3)}{45(p+1)^3\Gamma(\nu-3)}\left(\frac{\pi}{\beta\mu}\right)^4 + \cdots\right]. \tag{B.3.4}$$

The relationship for the μ-derivative, given in Eq. (B.1.15), is unchanged. With these extensions of the approximation, calculations for parafermion properties become no more difficult than those for fermions. (Vieira and Tsallis [1987])

B.4 REFERENCES

VIEIRA, M. C. de S. AND TSALLIS, C. [1987], D-dimensional Ideal Gas in Parastatistics: Thermodynamic Properties. *J. Stat. Phys.* **48** (1987), 97–120.

C. *COMBINATORIAL MISCELLANEA*

C.1 INTRODUCTION

In statistical procedures, it is often necessary to count the number of ways some collective event can occur. Some of these ways are regarded as combinations, some as permutations, and some as neither. If there are three objects, a, b, and c, the *permutations* of these objects are: abc, acb, bac, bca, cab, and cba. The one *combination* of these objects is the collection of all three, in any order. The permutations of them taken two at a time are: ab, ac, ba, ca, bc, and cb. The combinations of them taken two at a time are: ab, ac, and bc. There may be a pile of each kind of object, so that there can be more than one of a kind in a collection. The permutations of three objects taken from these piles, with repeats allowed, but without treating the interchange of like objects as a different permutation, are: aaa, aab, aac, aba, aca, baa, caa, similar sets containing multiple occurrences of b and of c, and the original set of six permutations with no repeats, totalling 27 possible permutations. Of these 27, there are 10 different combinations, of which three have all objects alike, six have two alike, and one has all three different. This appendix provides some of the necessary combinatorial formulas, and some procedures for generating others when needed.

C.2 BASIC COMBINATORIAL FORMULAS

(1) The number of permutations of N identifiable individuals, also called the number of permutations of N such individuals taken N at a time, is

$$_N P_N = N!, \qquad (C.2.1)$$

since the first one can be chosen in N ways, the second in $N-1$ ways, and on down to the the last one, which can be chosen in only one way. I use the term *identifiable individuals* to distinguish this category from two other such categories that must be considered: *nonidentifiable individuals*, such as the objects in pile a of the previous section or the fleas on one of the dogs in Chapter 1, and *nonidentifiable nonindividuals*, or simply *nonindividuals* for collective entities that are discussed in the next section. (2) The number of permutations of N identifiable individuals taken M at a time is

$$_N P_M = N \times (N-1) \times (N-2) \times \cdots \times (N-M+1) = N!/(N-M)!, \qquad (C.2.2)$$

since the argument given in (1) now terminates after M selections. (3) The number of combinations of these same N individuals, taken M at a time simply says that the $M!$ permutations within each group of individuals are not regarded as different, so that the required number of combinations is

$$_N C_M = \frac{N!}{M!(N-M)!} = \binom{N}{M}. \qquad (C.2.3)$$

(4) The number of ways that N individuals, identifiable or not, can be separated into two groups, with M of them in one group and $N-M$ in the other, is exactly $_N C_M$, since taking

M of them, in any order, for one group leaves $N - M$ of them as a remaining group. (5) The number of permutations of M kinds of things (nonidentifiable individuals) taken N at a time, with repeats allowed, is M^N, since the first may be selected in M ways, the second may also be selected in M ways, and so for all N choices. This is the same as the number of ways of placing N identifiable individuals in M numbered boxes, with no restriction on the number per box. Note that in both cases, there is no requirement that N be greater or smaller than M. (6) The number of ways that N identifiable individuals can be placed in M numbered boxes, with no more than one individual per box, and $N \le M$, is $_M P_N$, since the first individual has M choices, the second has $M - 1$, and on down to the last having $M - N + 1$. (7) The number of ways that N nonidentifiable individuals can be placed in M numbered boxes, with no limitation on the number per box, which is the same as finding the number of combinations of M kinds of nonidentifiable individuals taken N at a time, is not simply related to the permutations found in (5). To analyze this problem, consider the array

$$A_1 x x x A_2 A_3 x A_4 x x x x A_5 x x \cdots A_M x x x,$$

where the $\{A_i\}$ represent the numbered boxes, and each nonidentifiable individual in such a box is designated by an x in the space immediately following it. There are $M + N$ objects in the array. The first must always be a box (an A), and there are M ways to choose it. The remaining $M + N - 1$ items may be chosen in any order. There are $(M + N - 1)!$ permutations of these symbols, making a total of $M(M + N - 1)!$ permissible permutations of the elements of the array. To count combinations, however, note that the $N!$ permutations of the N nonidentifiable individuals in a particular array all represent the same combination, so that the number of permutations must be divided by $N!$. Furthermore, if the boxes, each with its contents intact, are shuffled in position, the reordered array is regarded as representing the same combination of m_1 objects in box A_1, m_2 objects in A_2, *etc.* There are $M!$ such permutations. Therefore the number of combinations of M kinds of nonidentifiable individuals taken N at a time is

$$\frac{M(M + N - 1)!}{M! \, N!} = \frac{(M + N - 1)!}{(M - 1)! \, N!}. \tag{C.2.4}$$

This result is used in the microcanonical treatment of quantum oscillators in Chapter 3.

Although similar procedures could be used to obtain other combinatorial formulas, a different approach is introduced in the next section.

C.3 PROBABILITIES AND COMBINATORIAL FACTORS

Since combinatorial formulas are used in this book mostly for the purpose of obtaining probabilities, it is of some interest to show how the two are related. As a vehicle for this section, I refer to the Ehrenfests' dog-flea model, discussed in Chapter 1, and here generalized.

If two events are not correlated, the probability of both of them occurring is given by $p_1 p_2$, where p_1 is the probability that the first event occurs, and p_2 is that of the second event. Suppose two flealess dogs enter a room infested with fleas, which are regarded as individuals that may be identifiable. Flea 1 will select dog A as its host with probability p_1, and dog B with probability q_1, with similar probabilities for all N fleas. If there are only two fleas, the joint probability-generating function is

$$(p_1 + q_1)(p_2 + q_2) = p_1 p_2 + p_1 q_2 + p_2 q_1 + q_1 q_2. \tag{C.3.1}$$

The four terms represent, respectively, the probability that both fleas select dog A, that flea 1 selects dog A and flea 2 selects dog B, *etc.* The coefficients of these terms (which are all 1) are the numbers of ways that the individual, identifiable fleas could be distributed on the two dogs. If the fleas are indistinguishable, and therefore have identical probability sets, then Eq. (C.3.1) can be written

$$(p+q)^2 = p^2 + 2pq + q^2. \tag{C.3.2}$$

Again the coefficients represent the numbers of ways that the two individual, nonidentifiable fleas could be distributed on the two dogs. There is one way that both fleas could be on dog A, similarly for B, and 2 ways that one flea could be on each dog. If there are N such individual, nonidentifiable fleas, Eq. (C.3.2) generalizes to become

$$(p+q)^N = \sum_{k=0}^{N} \binom{N}{k} p^k q^{N-k}, \tag{C.3.3}$$

the binomial distribution. This is the generating function both for the number of ways that k of N fleas can be on dog A and for the actual probability $P_N(k)$ that such an event will occur. If the dogs are equally attractive, so that $p = q = 1/2$, Eq. (C.3.3) becomes

$$1 = \frac{1}{2^N} \sum_{k=0}^{N} \binom{N}{k}, \tag{C.3.4}$$

and the probabilities are given by

$$P_N(k) = 2^{-N} \binom{N}{k}. \tag{C.3.5}$$

If there are M dogs, and the probability that any one of the N identical individual fleas chooses dog n is p_n, where $1 \leq n \leq M$, and $\sum p_n = 1$, the collective probability-generating function is

$$\left(\sum_{n=1}^{M} p_n \right)^N = \sum_{\{m_n\}} (N; m_1, m_2, \cdots, m_N) p_1^{m_1} p_2^{m_2} \cdots p_M^{m_M}, \tag{C.3.6}$$

where the multinomial coefficient is

$$(N; m_1, m_2, \cdots, m_N) = \frac{N!}{m_1! m_2! \cdots m_M!}, \tag{C.3.7}$$

and $\sum m_n = N$. The multinomial coefficient counts the number of ways that N identical individuals (fleas) can be assigned to M containers (dogs), with specifically m_1 fleas on dog 1, m_2 fleas on dog 2, *etc.* If the dogs are equally attractive, so that $p_n = 1/M$ for all n, the probability of finding the fleas distributed according to the set $\{m_n\}$ is

$$P_N(\{m_n\}) = M^{-N}(N; \{m_n\}). \tag{C.3.8}$$

These multinomial distributions represent equilibrium states of the generalized dog-flea problem. It is always possible, in principle, to establish any arbitrary initial distribution of fleas, stand clear, and let them evolve by Markovian random hops.

C.4 COMBINING NONINDIVIDUALS

The problem here can be explained in terms that are remote from physics. Suppose the manager of a motel separately registers three guests, Alpher, Bethe, and Gamow, and assigns them, in arbitrary order, to rooms 1, 2, and 3. Even though the manager regards the guests, for most purposes, as nonidentifiable, so that a room is either occupied by a generalized customer or else it is empty, there remains the vague realization that there are six ways to place the three guests in the three rooms, because each occupant is an individual, even if not regarded as identifiable. Sometime later, the three guests meet and recognize each other as old friends. They then ask that the internal doors connecting their rooms be unlocked, so that they can reorganize their quarters into a suite, to be occupied by all three. When this is done, the innkeeper realizes that each occupant is now distributed over three rooms, and that now there is only one possible arrangement of the three guests, such that all of them occupy all of the rooms. No longer is Alpher in room 1, Bethe in room 2, and Gamow in room 3, or some other permutation. All three occupy one suite, and any one could leave the suite by any one of the three outside doors.

This is the situation in the quantum theory of noninteracting particles. A single such ideal gas particle in a closed volume can be described by a Schrödinger equation, the solution of which is a set of eigenfunctions, each representing an eigenstate of the single-particle system. When two identical particles of this kind are in the closed volume, each could be in its own single-particle eigenstate, but quantum symmetry conditions, discussed in Chapter 4, require the two-particle state to be described as a composite state with both particles sharing both eigenstates. The same symmetry requirements exist for N such particles. They exist in one way, in the analog of a suite of states, and not each in its individual eigenstate. When the number of particles is of the order of the number of available states, the possibility exists that more than one particle could occupy the same single-particle state, and in the case of fermions, this multiple occupancy is forbidden. The formalism for handling such cases is well known, and the statistical consequences of the quantum symmetry requirements are covered in Chapter 4. The point of this section is that in the other extreme, when there are orders of magnitude more available single-particle states than there are particles, so that the probability is negligible that any two particles are in the same single-particle state, the right words need to be said to produce an accurate counting of many-body states. It is usually stated, in agreement with the innkeeper before the separate rooms were interconnected, that three rooms were occupied by three undifferentiated individuals, and since the 3! permutations of the occupants all result in the same state of occupancy, they may as well be regarded as one state. But this argument is wrong statistically, because there are indeed six ways to permute the occupants of the three rooms. The correct statement is that a single N-body state is a suite of N one-body states, with every particle occupying the entire suite.

As in the previous section, the number of ways for N particles to occupy M one-body states, with m_n particles in each state, is given by Eq. (C.3.7). When $M \gg N$, the probability of multiple occupancy is negligible, so that for almost all m_n, either $m_n = 0$ or $m_n = 1$, and the value of the multinomial coefficient is $N!$. But this is the number of ways that N individuals can be distributed, and when the individuals form a collective entity, or the states form a suite, and there is no longer individual occupancy of the single-particle states, the multinomial coefficient overstates the statistical weight of a configuration by a factor of $N!$. The remedy is to divide by the offending $N!$, giving the properly weighted probability, from Eq. (C.3.6) as

$$P_N(\{m_n\}) = (N!)^{-1}(N; m_1, m_2, \cdots, m_N) p_1^{m_1} p_2^{m_2} \cdots p_M^{m_M} = \prod_{n=1}^{M} (p_n)^{m_n}, \qquad (C.4.1)$$

where $\sum m_n = N$, and every m_n is expected to be either 1 or 0. The single-particle probabilities are found in Chapter 3 to be

$$p_n = Z^{-1} e^{-\beta \epsilon_n}. \qquad (C.4.2)$$

The single-particle partition function, as written in Chapter 3, is

$$Z_1 = \sum_{n=1}^{M} e^{-\beta \epsilon_n} = \sum_{n=1}^{M} a_n, \qquad (C.4.3)$$

where M is the total number of single-particle states. For unlike, noninteracting systems, the joint partition function is shown there to be the product of the partition functions of the systems before they are combined. For like systems, this same rule would lead to

$$Z_N = Z_1^N = \sum_{\{m_n\}} (N; m_1, m_2, \cdots, m_N) a_1^{m_1} a_2^{m_2} \cdots a_M^{m_M}, \qquad (C.4.4)$$

where the multinomial coefficient counts the number of ways the N-particle system can have m_1 particles in state 1, m_2 in state 2, *etc.* For quantum N-particle states where the particles meld into a suite of single-particle states, as just explained, and $M \gg n$, so that the probability of multiple occupancy is negligible, the multinomial coefficient overweights almost all of the states by a factor of $N!$. Therefore, the correct N-particle partition function for such systems is

$$Z_N = Z_1^N / N!, \qquad (C.4.5)$$

where the famous $N!$ divisor, originally inserted by Gibbs because it was needed for physical reasons, and then justified by specious argument, appears as a consequence of the nature of quantum N-particle states.

Further discussion of this treatment appears in Chapter 3.

D.   JACOBIANS

A frequent situation in which Jacobians are encountered is the following: Suppose $x = \phi(u,v)$ and $y = \psi(u,v)$. The necessary and sufficient condition that these equations can be solved for u and v is that

$$J\left(\frac{x,y}{u,v}\right) \equiv \frac{\partial(x,y)}{\partial(u,v)} = \begin{vmatrix} \frac{\partial x}{\partial u} & \frac{\partial x}{\partial v} \\ \frac{\partial y}{\partial u} & \frac{\partial y}{\partial v} \end{vmatrix} \neq 0, \qquad (D.0.1)$$

where the Jacobian is defined as the given determinant. The necessary and sufficient condition that a functional relationship exist between x and y, so that they are not independent, is that the Jacobian vanish. The same formalism can be extended to any number of variables.

In another context, as a result of a coordinate transformation, the new volume element is given in terms of the old as

$$dx\,dy\,\cdots dz = \frac{\partial(x,y,\cdots z)}{\partial(u,v,\cdots w)}\,du\,dv\,\cdots dw. \qquad (D.0.2)$$

Jacobians have a number of useful properties, all of which follow directly from the definition. These are listed below for convenient reference.

$$\frac{\partial(u,v,\cdots,w)}{\partial(x,y,\cdots,z)} = \frac{\partial(u,v,\cdots,w)}{\partial(r,s,\cdots,t)}\frac{\partial(r,s,\cdots,t)}{\partial(x,y,\cdots,z)}. \qquad (D.0.3)$$

$$\frac{\partial(u,v,\cdots,w)}{\partial(x,y,\cdots,z)} = \left(\frac{\partial(x,y,\cdots,z)}{\partial(u,v,\cdots,w)}\right)^{-1}. \qquad (D.0.4)$$

$$\frac{\partial(u,y,\cdots,z)}{\partial(x,y,\cdots,z)} = \frac{\partial u}{\partial x}. \qquad (D.0.5)$$

These relationships are particularly useful in the expression of the derivatives that occur in thermodynamics in terms of known or measurable parameters and response functions.

E. SUMMATION TECHNIQUES

E.1 THE EULER-MACLAURIN SUMMATION FORMULA

The summation formula treated in this section was submitted for publication by Euler in 1732, but Stirling told him that Maclaurin had already obtained it. Euler's article finally came out in 1738, and I am not aware that Maclaurin ever published his version of the formula.

The problem addressed by them is that of finding a general and efficient procedure for the summation of such series as

$$F(n) = \sum_{k=0}^{n} f(k). \tag{E.1.1}$$

I show here a symbolic derivation of the formula, with necessary and sufficient conditions for its applicability omitted, but usually evident. Note that

$$F(n+1) - F(n) = f(n+1). \tag{E.1.2}$$

Expand $F(x+1)$ in a Taylor series, as

$$F(x+1) = F(x) + F'(x) + \frac{1}{2}F''(x) + \frac{1}{3!}F'''(x) + \cdots$$

$$= e^D F(x); \quad \text{where} \quad D = \frac{d}{dx}. \tag{E.1.3}$$

Similarly, $f(x+1) = e^D f(x)$, giving $(e^D - 1)F(x) = e^D f(x)$. This final expression can be rewritten to become

$$F(x) = \frac{e^D}{e^D - 1} f(x) = \left(1 + \frac{1}{e^D - 1}\right) f(x)$$

$$= f(x) + \frac{1}{D} \sum_{k=0}^{\infty} \frac{B_k}{k!} D^k f(x)$$

$$= f(x) + \int f(x)\, dx - \frac{1}{2}f(x) + \sum_{k=1}^{\infty} \frac{B_{2k}}{(2k)!} f^{(2k-1)}(x), \tag{E.1.4}$$

where D^{-1} symbolizes integration, the B_k are the Bernoulli numbers, given in Appendix A, and the series containing them comes from the generating function for Bernoulli polynomials, also given in Appendix A. When this result is expressed again in terms of the original series, it may be written in either of two forms, depending on what is done with the first $f(x)$ in the final expression. These forms are

$$F(n) = \int_0^n dx\, f(x) + \frac{1}{2}\left[f(n) + f(0)\right] + \sum_{k=1}^{\infty} \frac{B_{2k}}{(2k)!} \left[f^{(2k-1)}(n) - f^{(2k-1)}(0)\right], \tag{E.1.5}$$

and

$$F(n) - f(n) = F(n-1) = \int_0^n dx\, f(x) - \frac{1}{2}\left[f(n) - f(0)\right] +$$

$$+ \sum_{k=1}^{\infty} \frac{B_{2k}}{(2k)!}\left[f^{(2k-1)}(n) - f^{(2k-1)}(0)\right], \qquad (E.1.6)$$

At first, it may appear that there is no particular advantage in replacing one series to be summed by another, especially if the replacement is an infinite series. The advantage, if any, can only lie in the rate of convergence, and the expectation that the integral alone is a good approximation is the basis for the assumption, not always valid, that the remaining terms are small, and the series of corrections will converge to the required accuracy after only a few terms. Several examples are provided, to illustrate Euler-Maclaurin summation techniques.

As a first example, consider the well known zeta function defined by a series:

$$\zeta(2) = \sum_{k=1}^{\infty} k^{-2} = \frac{\pi^2}{6} = 1.644934067. \qquad (E.1.7)$$

An efficient way to use Eq. (E.1.5) to evaluate the series is to write

$$F_\zeta = \sum_{k=1}^{9} k^{-2} + \sum_{k=0}^{\infty}(k+10)^{-2}$$

$$= 1.539767731 + \int_0^{\infty} \frac{dx}{(x+10)^2} + \frac{1}{2}f(0) - \sum_{k=0}^{\infty}\frac{B_{2k}}{(2k)!}f^{(2k-1)}(0)$$

$$= 1.539767731 + 0.1 + 0.005 + 1.663356817 \times 10^{-4} = 1.6449340667, \qquad (E.1.8)$$

where $f(x) = (x+10)^{-2}$, so that $f^{(n)}(x) = (-)^n(n+1)!(x+10)^{-n-2}$, and

$$-\sum_{k=1}^{\infty}\frac{B_{2k}}{(2k)!}f^{(2k-1)}(0) = \sum_{k=1}^{\infty}\frac{B_{2k}}{10^{2k+1}}$$

$$= \frac{10^{-3}}{6} - \frac{10^{-5}}{30} + \frac{10^{-7}}{42} - \frac{10^{-9}}{30} + \frac{5\times 10^{-11}}{66} + \cdots \qquad (E.1.9)$$

As can be seen in Eq. (E.1.9), successive terms in the Euler-Maclaurin series are smaller by two orders of magnitude, and the series converges much more rapidly than does the original series definition of the zeta function. The higher-order zeta functions converge even more rapidly, and for those of odd order, for which simple expressions in terms of π do not exist, this technique provides excellent numerical values.

As a second example, I demonstrate the use of Eq. (E.1.6) to find an expression equivalent to the Stirling formula for $\ln n!$.

$$\ln n! = \sum_{k=1}^{n}\ln n = \sum_{k=0}^{n-1}\ln(k+1). \qquad (E.1.10)$$

Write $f(x) = \ln(x + 1)$, and note that $f^{(k)}(x) = (-)^{k+1}(k-1)!(x+1)^{-k}$. The Euler-Maclaurin summation gives

$$\ln n! = (n+1)\ln(n+1) - n - \frac{1}{2}\ln(n+1) + \sum_{k=1}^{\infty} \frac{B_{2k}}{(2k)(2k-1)} \left[\frac{1}{(n+1)^{2k-1}} - 1 \right]$$

$$= (n+1/2)\ln(n+1) - n + \frac{1}{12}\left[\frac{1}{n+1} - 1 \right] - \frac{1}{360}\left[\frac{1}{(n+1)^2} - 1 \right] + \cdots (E.1.11)$$

This result is equivalent to that of the Stirling formula, and the approximation used for most purposes in statistical physics for large n,

$$\ln n! \approx n\ln n - n, \qquad (E.1.12)$$

is clearly insensitive to the fine structure of Eq. (E.1.11).

As a final example, consider the function

$$A(\alpha) = \sum_{0}^{\infty} e^{-\alpha n^2}. \qquad (E.1.13)$$

The E-M summation of this series contains only two terms! Here, $f(x) = e^{-\alpha x^2}$, an even function of x such that it and its derivatives to all orders vanish at infinity. Therefore only the derivatives at $x = 0$ need be considered in the Bernoulli-number sum of Eq. (E.1.5). These are all of odd order, and all such derivatives vanish at $x = 0$. Thus the E-M formula for $A(\alpha)$ is

$$A(\alpha) = \int_0^{\infty} dx\, e^{-\alpha x^2} + \frac{1}{2}f(0) = \frac{1}{2}\left[\sqrt{\frac{\pi}{\alpha}} + 1 \right]. \qquad (E.1.14)$$

For all $\alpha \geq 0$, it is evident that $A(\alpha) \geq 1$, when the original expression of Eq. (E.1.13) is examined. Without any numerical detail, the E-M result of Eq. (E.1.14) is clearly not valid for $\alpha \geq \pi$, since it then gives $A(\alpha) \leq 1$. For such values of α, direct summation of the original series converges rapidly enough for many purposes. For small α, where many terms in the series are needed to yield good numerical values, the E-M formula is useful.

E.2 THE POISSON SUMMATION FORMULA

It is sometimes possible to use a different summation formula, originated by Poisson, to supplement or improve that of Euler and Maclaurin. Suppose $f(x)$ is a continuous function for all real x, defined for $-\infty \leq x \leq \infty$. Construct the periodic function

$$g(x) = \sum_{k=-\infty}^{\infty} f(x+k), \qquad (E.2.1)$$

and assume that the series converges in every finite x interval. Note that $g(x) = g(x+1)$, and do what every physicist does with a periodic function: Expand it in a Fourier series.

$$g(x) = \sum_{n=-\infty}^{\infty} a_n e^{2\pi i n x}; \quad a_m = \int_0^1 dx\, g(x) e^{-2\pi i m x}, \qquad (E.2.2)$$

which may be rewritten as

$$a_m = \sum_{k=-\infty}^{\infty} \int_0^1 dx\, e^{-2\pi imx} f(x+k) = \sum_{k=-\infty}^{\infty} \int_n^{n+1} dx\, e^{-2\pi imx} f(x) = \int_{-\infty}^{\infty} dx\, e^{-2\pi imx} f(x).$$
$$(E.2.3)$$

On the assumption that these manipulations are all valid, it follows that

$$\sum_{k=-\infty}^{\infty} f(x+k) = \sum_{k=-\infty}^{\infty} e^{2\pi ikx} \int_{-\infty}^{\infty} dx'\, e^{-2\pi ikx'} f(x'). \qquad (E.2.4)$$

To apply this formula to the summation of a series, set $x=0$, to get

$$\sum_{n=-\infty}^{\infty} f(n) = \sum_{k=-\infty}^{\infty} \int_{-\infty}^{\infty} dy\, e^{-2\pi iky} f(y). \qquad (E.2.5)$$

This is the Poisson summation formula. In addition to the requirements that $f(x)$ be continuous for all real x, and that the series of Eq. (E.2.1) converge in every finite x interval, the two other sufficient conditions for the validity of the Poisson derivation are that the integral of $f(x)$ be absolutely convergent over the entire line, and that the sum of the $|a_n|$ also converges, or

$$\int_{-\infty}^{\infty} dx\, |f(x)| < \infty; \quad \text{and} \quad \sum_{n=-\infty}^{\infty} |a_n| < \infty. \qquad (E.2.6)$$

Under these conditions, Eq. (E.2.4) is valid for all x.

As an example of the use of the Poisson summation formula, I return to the third example in the E-M section. Write $B(\alpha)$ as

$$B(\alpha) = \sum_{k=-\infty}^{\infty} e^{-\alpha k^2}, \qquad (E.2.7)$$

which differs from $A(\alpha)$ of Eq. (E.1.13) in the summation limits. The Poisson formula, Eq. (E.2.5), says

$$B(\alpha) = \sum_{k=-\infty}^{\infty} \int_{-\infty}^{\infty} dy\, e^{-2\pi iky} e^{-\alpha y^2} = \sqrt{\frac{\pi}{\alpha}} \sum_{n=-\infty}^{\infty} e^{-n^2\pi^2/\alpha}. \qquad (E.2.8)$$

From the definitions of A and B, it follows that

$$A(\alpha) = \frac{1}{2}\left(B(\alpha)+1\right)$$
$$= \frac{1}{2}\left(\sqrt{\frac{\pi}{\alpha}}+1\right) + \sqrt{\frac{\pi}{\alpha}}\left(e^{-\pi^2/\alpha}+e^{-4\pi^2/\alpha}+\cdots\right), \qquad (E.2.9)$$

where the correction to the E-M value of Eq. (E.1.14) is shown as the final major term. In this way, the accuracy of the E-M expression can be assessed. When $\alpha = 0.01$, the first Poisson correction is $10\sqrt{\pi}e^{-100\pi^2} \approx 10^{-426}$, and the subsequent corrections are even smaller! For $\alpha = 1/4$, the first Poisson correction is 2.5×10^{-17}, which is negligible for most purposes. But for $\alpha = \pi$, the first Poisson correction is 0.04321391827, and the first three add to give a total correction of 0.04321740561. For $\alpha = \pi$, the Poisson formula is an identity, $A(\alpha) = A(\alpha)$. The correction terms are exactly the terms of the original series. It becomes evident that for $\alpha > \pi$, the series is best summed directly, rather than by means of either of the summation formulas given in this appendix.

The two summation formulas can produce remarkably accurate results under the proper circumstances, and unacceptably poor results in others. Use them with caution.

F. N-SPACE

F.1 SPHERICAL SURFACES AND VOLUMES IN N-SPACE

The volume or surface area of a sphere in n dimensions is often required for the calculations given in this book. The volume can be written

$$V_n = B_n r^n = \frac{A_n}{n} r^n, \qquad (F.1.1)$$

where the coefficients A_n and B_n are to be determined. An easy way to do this is by means of the well known n-dimensional gaussian integral

$$
\begin{aligned}
I(n) &= \int_{-\infty}^{\infty} dx_1 \cdots \int_{-\infty}^{\infty} dx_n e^{-(x_1^2 + \cdots + x_n^2)} = \left(\sqrt{\pi}\right)^n \\
&= \int_0^{\infty} A_n r^{n-1} e^{-r^2} dr, \qquad (F.1.2)
\end{aligned}
$$

where the last step replaces the cartesian volume element $dx_1 \cdots dx_N$ by the spherical one, $dV_n = A_n r^{n-1} dr$. The final integral can be converted to a gamma function by the substitution $r^2 = y$; it thus becomes

$$I(n) = \frac{A_n}{2} \int_0^{\infty} dy\, y^{(n/2)-1} e^{-y^2} = \frac{A_n}{2} \Gamma(n/2). \qquad (F.1.3)$$

Thus A_n is seen to be

$$A_n = \frac{2\pi^{n/2}}{\Gamma(n/2)}, \qquad (F.1.4)$$

and

$$V_n = B_n r^n = \frac{\pi^{n/2}}{\Gamma(n/2 + 1)}. \qquad (F.1.5)$$

As a check, note that $\Gamma(1/2) = \sqrt{\pi}$; $\Gamma(n+1) = n\Gamma(n)$; and, for integers, $\Gamma(n+1) = n!$. Then $A_2 = 2\pi/\Gamma(1) = 2\pi$, and $A_3 = 2\pi^{3/2}/(1/2)\sqrt{\pi} = 4\pi$.

An important ratio is the fractional increase in volume per fractional increase in radius, given by

$$\frac{dV_n}{V_n} = \frac{A_n r^{n-1} dr}{A_n r^n / n} = \frac{n\, dr}{r}. \qquad (F.1.6)$$

For n of the order of the Avogadro number, the fractional volume change is a huge multiple of the fractional change in radius, and essentially the entire volume is in a thin surface layer.

G. NUMERICAL CONSTANTS

G.1 PHYSICAL CONSTANTS

Constant	Symbol	Value (as of 1986*)
Avogadro constant	N_0	$6.02212367 \times 10^{23}$ mol^{-1}
Boltzmann constant	k	1.380657×10^{-23} J/K
Planck constant	h	$6.6260754 \times 10^{-34}$ Js
Universal gas constant	R	8.314510 J/mol K
Gravitational constant	G	6.67260×10^{-11} m^3/s^{2}kg
Speed of light in vacuum	c	2.99792458×10^8 m/s
Elementary charge	e	$1.60217738 \times 10^{-19}$ C
Unified atomic mass unit	u	$1.6605402 \times 10^{-27}$ kg
Electron mass	m_e	$5.48579902 \times 10^{-4}$ u
Neutron mass	m_n	1.008664704 u
Hydrogen mass	m_p	1.007825035 u
Permittivity constant	ϵ_0	$8.85418781762 \times 10^{-12}$ F/m
Permeability constant	μ_0	$1.25663706143 \times 10^{-6}$ H/m
Electron magnetic moment	μ_e	$9.2847700 \times 10^{-24}$ J/T
Bohr magneton	μ_B	$9.2740154 \times 10^{-24}$ J/T
Mass of the sun	$M_\odot$	1.991×10^{30} kg
Radius of the sun	$R_\odot$	6.96×10^8 m
Mass of the earth	M_E	5.98×10^{24} kg
Radius of the earth	R_E	6.378×10^6 m

*Selected mostly from tabulated values in *Symbols, Units and Nomenclature (IUPAP)*, prepared by E. R. Cohen and P. Giacomo (1986)

G.2 MATHEMATICAL CONSTANTS

$\zeta(2)$	$\pi^2/6$	$\zeta(3)$	1.202
$\zeta(4)$	$\pi^4/90$	$\zeta(5)$	1.037
$\zeta(6)$	$\pi^6/945$	$\zeta(7)$	1.008
$\zeta(8)$	$\pi^8/9450$	$\zeta(3/2)$	2.612
$\zeta(5/2)$	1.037	$\zeta(7/2)$	1.127
$\Gamma(1/2)$	$\sqrt{\pi}$	$\Gamma(n+1)$	$n\Gamma(n)$

BIBLIOGRAPHY

ABRAMOWITZ, M. AND STEGUN, I. A., Eds. [1964], *Handbook of Mathematical Functions*. U. S. National Bureau of Standards, Washington, D. C., 1964.

ACETTA, F. S. AND GLEISER, M. [1987], Thermodynamics of Higher Dimensional Black Holes. *Annals of Physics* **176** (1987), 278–300.

ADKINS, C. J. [1983], *Equilibrium Thermodynamics, 3rd ed.*. Cambridge University Press, Cambridge, England, 1983.

AGARWAL, G. S. [1971], Entropy, the Wigner Distribution, and the Approach to Equilibrium of a System of Coupled Harmonic Oscillators. *Physical Review A* **3** (1971), 828–831.

AHARONY, A. [1983], *Multicritical Points*. In *Critical Phenomena*, F. J. W. Hahne, Ed. Springer-Verlag, New York–Heidelberg–Berlin, 1983, 209–258.

ALLIS, W. P. [1956], *Motion of Ions and Electrons*. In *Handbuch der Physik*. vol. XXI, Springer-Verlag, New York–Heidelberg–Berlin, 1956.

ALS-NIELSEN, J. AND BIRGENEAU, R. J. [1977], Mean Field Theory, the Ginzburg Criterion, and Marginal Dimensionality of Phase Transitions. *American Journal of Physics* **45** (1977), 554–560.

ANDERSON, P. W. [1950], Two Comments on the Validity of the P. R. Weiss Theory of Ferromagnetism. *Phys. Rev.* **80** (1950), 922–923.

ANDREWS, T. [1869], On the Continuity of the Gaseous and Liquid States of Matter [Bakerian Lecture]. *Philosophical Transactions of the Royal Society of London* **159** (1869), 575–590.

ARNOL'D, V. I. AND AVEZ, A. [1968], *Ergodic Problems of Classical Mechanics*. W.A. Benjamin, Inc., New York, Amsterdam, 1968.

ARNOL'D, V. I. [1986], *Catastrophy Theory, 2nd rev. ed.*. Springer-Verlag, New York–Heidelberg–Berlin, 1986.

ASHKIN, J. AND LAMB, JR., W. E. [1943], Statistics of Two-Dimensional Lattices with Four Components. *Physical Review* **64** (1943), 178–184.

BAIERLEIN, R. [1971], *Atoms and Information Theory*. W. H. Freeman and Company, San Francisco, CA, 1971.

BAKER, JR., G. A. [1961], Applications of the Padé Approximant Method to the Investigation of Some Magnetic Properties of the Ising Model. *Physical Review* **124** (1961), 768–774.

BAKER, JR., G. A. [1965], The Theory and Applications of the Padé Approximant Method. *Advances in Physics* **1** (1965), 1–58.

BAKER, JR., G. A. [1970], *The Padé Approximant Method and Some Related Generalizations*. In *The Padé Approximant in Theoretical Physics*, G. A. Baker, Jr. and J. L. Gammel, Eds. Academic Press, New York, NY, 1970, 1.

BAKER, JR., G. A. [1975], *Essentials of Padé Approximants*. Academic Press, New York, NY, 1975.

BAKER, JR., G. A. [1977], *The Application of Padé Approximants to Physical Phenomena*. In *Padé and Rational Approximations, Theory and Applications*, E. B. Saff and R. S. Varga, Eds. Academic Press, New York, NY, 1977, 323–338.

BAKER, JR., G. A. [1984], *Critical Point Statistical Mechanics and Quantum Field Theory*. In *Phase Transitions and Critical Phenomena, V. 9*, C. Domb and J. L. Lebowitz, Eds. Academic Press, New York, NY, 1984, 233–311.

BALESCU, R. [1961], *General Theory of Nonequilibrium Phenomena*. In *Lectures in Theoretical Physics, V. III*, W. E. Brittin, B. W. Downs and J. Downs, Eds. Interscience Publishers, A division of John Wiley & Sons, London, 1961, 382–444.

BALESCU, R. [1967], *Statistical Mechanics of Charged Particles*. John Wiley & Sons, New York, NY, 1967.

BALESCU, R. [1968], Some Comments on Irreversibility. *Physics Letters* **27A** (1968), 249–250.

BALESCU, R. [1975], *Equilibrium and Nonequilibrium Statistical Mechanics*. John Wiley & Sons, New York, NY, 1975.

BARBER, M. N. AND FISHER, M. E. [1973], Critical Phenomena in Systems of Finite Thickness 1. The Spherical Model. *Annals of Physics* **77** (1973), 1–78.

BARBER, M. N. AND NINHAM, B. W. [1970], *Random and Restricted Walks*. Gordon and Breach, New York, NY, 1970.

BEDFORD, T. AND SWIFT, J., Eds. [1988], *New Directions in Dynamical Systems*. Cambridge University Press, Cambridge, England, 1988.

BELINFANTE, F. J. [1939a], Undor Calculus and Charge Conjugation. *Physica* **6** (1939), 849–869.

BELINFANTE, F. J. [1939b], Undor Equation of the Meson Field. *Physica* **6** (1939), 870–886.

BELINFANTE, F. J. [1939c], Spin of Mesons. *Physica* **6** (1939), 887–898.

BERLIN, T. H. AND KAC, M. [1952], The Spherical Model of a Ferromagnet. *Physical Review* **86** (1952), 821–835.

BERNAL, J. D. AND FOWLER, JR., R. H. [1933], A Theory of Water and Ionic Solution, with Particular Reference to Hydrogen and Hydroxyl Ions. *Journal of Chemical Physics* **1** (1933), 515–548.

BERNARD, W. AND CALLEN, H. B. [1959], Irreversible Thermodynamics of Nonlinear Processes and Noise in Driven Systems. *Reviews of Modern Physics* **31** (1959), 1017–1044.

BETHE, H. A. [1935], Statistical Superlattices. *Proc. Royal Society (London) A* **150** (1935), 552–575.

BETTS, D. D. [1974], X-Y Model. In *Phase Transitions and Critical Phenomena, V. 3*, C. Domb and M. S. Green, Eds. Academic Press, New York, NY, 1974, 569–652.

BINDER, K. [1987], Theory of First-order Phase Transitions. *Reports on Progress in Physics* **50** (1987), 783–860.

BIRKHOFF, G. [1960], *Hydrodynamics; A Study in Logic, Fact, and Similitude*. Princeton Univ. Press, Princeton, NJ, 1960.

BJERRUM, N. [1951], Structure and Properties of Ice. *K. Danske Vidensk. Selsk., Mat.-fis. Medd.* **27** (1951), 1–56.

BLAKEMORE, J. S. [1988], *Solid State Physics, 2nd Rev. Ed.*. Cambridge University Press, Cambridge, England, 1988.

BLATT, J. M. [1959], An Alternative Approach to the Ergodic Problem. *Progress in Theoretical Physics* **22** (1959), 745–756.

BOGOLIUBOV, N. N. [1962], *Problems of a Dynamical Theory in Statistical Physics*. In *Studies in Statistical Mechanics V.1*, G. E. Uhlenbeck and J. de Boer, Eds. North-Holland, Amsterdam, 1962.

BOGOLUBOV, N. N. AND BOGOLUBOV, JR., N. N. [1982], *Introduction to Quantum Statistical Mechanics*. World Scientific Publishing Co., Singapore, 1982.

BOHR, N. [1909], Application of the Electron Theory to Explain the Physical Properties of Metals. M. Sc. Examination Paper, Copenhagen, 1909.

BOHR, N. [1911], Studies on the Electron Theory of Metals. Dissertation, Copenhagen, 1911.

BOLTZMANN, L. [1872], Weitere Studien über Wärmegleichgewicht unter Gasmolekülen (H-Theorem). *Wein Ber.* **66** (1872).

BOREL, É. [1913], La Mécanique statistique et l'irréversibilité. *Journal de Physique* **3, 5th series** (1913), 189–196.

BOREL, É. [1914], *Introduction Géometrique á Quelques Théories Physiques*. Gauthier-Villars, Paris, 1914, Reprint of [Borel 1913].

BORN, M. [1964], *The Natural Philosophy of Cause and Chance*. Dover, Mineola, NY, 1964.

BORN, M. AND GREEN, H. S. [1949], *A General Theory of Liquids*. Cambridge University Press, Cambridge, England, 1949.

BOWERS, R. G. [1975], Some Fundamental Problems in Thermodynamics. *Il Nuovo Cimento* **29B** (1975), 303–321.

BOYER, T. H. [1974a], Concerning the Zero-point Energy of a Continuum. *Am. J. Phys.* **42** (1974), 518–519.

BOYER, T. H. [1974b], Van der Waals Forces and Zero-point Energy for Dielectric and Permeable Materials. *Phys. Rev. A* **9** (1974), 2078–2084.

BRAGG, W. L. AND WILLIAMS, E. J. [1934], The Effect of Thermal Agitation on the Atomic Arrangement in Alloys. *Proceedings of the Royal Society (London)* **A145** (1934), 699–730.

BRIDGMAN, P. W. [1961], *The Thermodynamics of Electrical Phenomena in Metals*. Dover, Mineola, NY, 1961.

BRILLOUIN, L. [1956], *Science and Information Theory*. Academic Press, New York, NY, 1956.

BRILLOUIN, L. [1962], *Science and Information Theory, 2nd ed.*. Academic Press, New York, NY, 1962.

BROUT, R. [1965], *Phase Transitions*. W. A. Benjamin, Inc., New York, Amsterdam, 1965.

BROWN, H. A. AND LUTTINGER, J. M. [1955], Ferromagnetic and Antiferromagnetic Curie Temperatures. *Physical Review* **100** (1955), 685–692.

BROWN, W. F. [1953], Magnetic Energy Formulas and Their Relation to Magnetization Theory. *Reviews of Modern Physics* **25** (1953), 131–135.

BRUCE, A. D. [1980], Structural Phase Transitions. II Static Critical Behavior. *Advances in Physics* **29** (1980), 111–218.

BRUCE, A. D. AND COWLEY, R. B. [1980], Structural Phase Transitions. III Critical Dynamics. *Advances in Physics* **29** (1980), 219–321.

BRUSH, S. G. [1967], History of the Lenz-Ising Model. *Reviews of Modern Physics* **39** (1967), 883–893.

BRUSH, S. G. [1972], *Kinetic Theory*. Pergamon Press, Oxford, England, 1972.

BRUSH, S. G. [1976], *The Kind of Motion We Call Heat*. North-Holland, Amsterdam, 1976.

BRUSH, S. G. [1983], *Statistical Physics and the Atomic Theory of Matter*. Princeton Univ. Press, Princeton, NJ, 1983.

BUCHDAHL, H. A. [1966], *The Concepts of Classical Thermodynamics*. Cambridge University Press, Cambridge, England, 1966.

BURNETT, D. [1935a], The Distribution of Velocities in a Slightly Non-Uniform Gas. *Proceedings of the London Mathematical Society* **39**, **Series 2** (1935), 385–400.

BURNETT, D. [1935b], The Distribution of Molecular Velocities and the Mean Motion in a Non-Uniform Gas. *Proceedings of the London Mathematical Society* **40** (1935), 382–435.

CALDIROLA, P., Ed. [1961], *Ergodic Theories*. Academic Press, New York, NY, 1961.

CALLAWAY, J. [1974], *Quantum Theory of the Solid State, V1 and V2*. Academic Press, New York, NY, 1974.

CALLEN, H. B. [1985], *Thermodynamics and an Introduction to Thermostatistics*. John Wiley & Sons, New York, NY, 1985.

CALLEN, H. B. AND WELTON, T. A. [1951], Irreversibility and Generalized Noise. *Physical Review* **83** (1951), 34–40.

CARATHÉODORY, C. [1909], Untersuchungen über die Grundlagen der Thermodynamik. *Mathematische Annalen* **67** (1909), 355–386.

CARNAP, R. [1977], *Two Essays on Entropy*. University of California Press, Berkeley, Los Angeles, London, 1977.

CASATI, G., FORD, J., VIVALDI, F. AND VISSCHER, W. M. [1984], One-Dimensional Classical Many-Body System Having a Normal Thermal Conductivity. *Physical Review Letters* **52** (1984), 1861–1864.

CASIMIR, H. B. G. [1945], On Onsager's Principle of Microscopic Reversibility. *Reviews of Modern Physics* **17** (1945), 343–350.

CHANDLER, D. [1987], *Introduction to Modern Statistical Mechanics*. Oxford University Press, London, 1987.

CHANDRASEKHAR, S. [1939], *An Introduction to the Study of Stellar Structure*. Dover, Mineola, NY, 1939.

CHANDRASEKHAR, S. [1943], Stochastic Problems in Physics and Astronomy. *Reviews of Modern Physics* **15** (1943), 1–89.

CHANDRASEKHAR, S. [1978], *Why are the stars as they are?* In *Physics and Astrophysics of Neutron Stars and Black Holes*, R. Giaconni and R. Ruffini, Eds. International School of Physics "Enrico Fermi," North-Holland, Amsterdam, 1978.

CHANDRASEKHAR, S. [1984] On Stars, Their Evolution and Stability. *Reviews of Modern Physics* **56** (1984), 137–147.

CHAPMAN, S. AND COWLING, T. G. [1970], *The Mathematical Theory of Non-Uniform Gases, 3rd ed..* Cambridge University Press, Cambridge, England, 1970.

CHEN, Y.-H., WILCZEK, F., WITTEN, E. AND HALPERIN, B. I. [1989], On Anyon Superconductivity. *Int. J. Mod. Phys.* **B3** (1989), 1001–1067.

CHENG, H. AND WU, T. T. [1967], Theory of Toeplitz Determinants and Spin Correlations of the Two-Dimensional Ising Model. III. *Physical Review* **164** (1967), 719–735.

CLARKE, J. F. AND McCHESNEY, M. [1976], *Dynamics of Relaxing Gases*. Butterworths, London, 1976.

COHEN, E. G. D. [1962], *The Boltzmann Equation and its Generalization to Higher Densities*. In *Fundamental Problems in Statistical Mechanics*, E. G. D. Cohen, Ed. North-Holland, Amsterdam, 1962, 110–156.

COHEN, E. G. D. [1966], *On the Statistical Mechanics of Moderately Dense Gases Not in Equilibrium*. In *Lectures in Theoretical Physics, Vol. VIIIA*. University of Colorado Press, Boulder, 1966.

COHEN, E. G. D. AND THIRING, W., Eds. [1973], *The Boltzmann Equation: Theory and Application*. Springer-Verlag, New York–Heidelberg–Berlin, 1973.

COLEMAN, B. D. AND MIZEL, V. J. [1963], Thermodynamics and Departure from Fourier's Law of Heat Conduction. *Archive for Rational Mechanics and Analysis* **13** (1963), 245–261.

COLEMAN, B. D. AND NOLL, W. [1963], The Thermodynamics of Elastic Materials with Heat Conduction. *Archive for Rational Mechanics and Analysis* **13** (1963), 167–178.

COLEMAN, B. D. AND NOLL, W. [1964], Thermodynamics of Materials with Memory. *Archive for Rational Mechanics and Analysis* **17** (1964), 1–46.

COLEMAN, B. D. AND OWEN, D. R. [1974], A Mathematical Foundation for Thermodynamics. *Archive for Rational Mechanics and Analysis* **54** (1974), 1–104.

COMPAGNER, A. [1989], Thermodynamics as a Continuum Limit of Statistical Mechanics. *American Journal of Physics* **57** (1989), 106–117.

COULON, C., ROUX, D. AND BELLOCQ, A. M. [1991], Second-Order Symmetric-Antisymmetric Phase Transition of Randomly Connected Membranes. *Physical Review Letters* **66** (1991), 1709–1712.

COWLEY, R. A. [1980], Structural Phase Transitions I. Landau Theory. *Advances in Physics* **29** (1980), 1–110.

COX, R. T. [1946], Probability, Frequency, and Reasonable Expectation. *American Journal of Physics* **17** (1946), 1–13.

COX, R. T. [1950], The Statistical Method of Gibbs in Irreversible Change. *Reviews of Modern Physics* **22** (1950), 238–248.

Cox, R. T. [1955], *Statistical Mechanics of Irreversible Change*. Johns Hopkins Univ. Press, Baltimore, MD, 1955.

Cox, R. T. [1961], *The Algebra of Probable Inference*. Johns Hopkins Univ. Press, Baltimore, MD, 1961.

Cox, R. T. [1979], *Of Inference and Inquiry*. In *The Maximum Entropy Formalism*, R. D. Levine and M. Tribus, Eds. MIT Press, Cambridge, MA, 1979, A Conference held at MIT, 2-4 May 1978.

Cramér, H. [1946], *Mathematical Methods of Statistics*. Princeton Univ. Press, Princeton, NJ, 1946.

Crawford, J. D., Lott, J. and Rieke, H. [1982], Connection Between Nonlinear Response and Statistical Behavior. *American Journal of Physics* 54 (1982), 363–373.

Crutchfield, J. P., Farmer, J. D. and Packard, N. H. [December 1986], Chaos. *Scientific American* 255 (December 1986), 46–58.

Daubechies, I. [1990], The Wavelet Transform, Time-Frequency Localization and Signal Analysis. *IEEE Transactions on Information Theory* 36 (1990), 961–1005.

David, F. [1989], 0(n) Gauge Models and Self-Avoiding Random Surfaces in Three Dimensions. *Europhysics Letters* 9 (1989), 575–580.

Davidson, R. C. [1972], *Methods in Nonlinear Plasma Theory*. Academic Press, New York, NY, 1972.

Day, W. A. [1969], A Theory of Thermodynamic Materials with Memory. *Archive for Rational Mechanics and Analysis* 34 (1969), 85–96.

Day, W. A. [1987], A Comment on a Formulation of the Second Law of Thermodynamics. *Archive for Rational Mechanics and Analysis* 98 (1987), 211–227.

Day, W. A. [1988], *A Commentary on Thermodynamics*. Springer-Verlag, New York–Heidelberg–Berlin, 1988.

Debye, P. [1912], Theory of Specific Heats. *Annalen der Physik* 39 (1912), 789–839.

Denbigh, K. G. and Denbigh, J. S. [1985], *Entropy in Relation to Incomplete Knowledge*. Cambridge University Press, Cambridge, England, 1985.

Desloge, E. A. [1966], *Statistical Physics*. Holt, Rinehart and Winston, Inc., New York, 1966.

Domb, C. [1949a], Order-Disorder Statistics. II. *Proc. Royal Soc. (London)* A 199 (1949), 199–221.

Domb, C. [1949b], Order-Disorder Statistics. I. *Proc. Royal Soc. (London)* A 196 (1949), 36–50.

Domb, C. [1960], On the Theory of Cooperative Phenomena in Crystals. *Advances in Physics* 9 (1960), 245–361.

Domb, C. [1974], *Ising Model*. In *Phase Transitions and Critical Phenomena, V.3*, C. Domb and M. S. Green, Eds. Academic Press, New York, NY, 1974, 357–484.

Domb, C. and Sykes, M. F. [1957], On the Susceptibility of a Ferromagnetic Above the Curie Point. *Proceedings of the Royal Society (London)* A 240 (1957), 214–228.

Domb, C. and Sykes, M. F. [1961], Use of Series Expansions for the Ising Model Susceptibility and Excluded Volume Problem. *J. Math. Physics* 2 (1961), 63–67.

Doob, J. L. [1942], The Brownian Movement and Stochastic Equations. *Annals of Mathematics* 43 (1942), 351–369.

Dresden, M. [1961], Recent Developments in the Theory of Transport and Galvanometric Phenomena. *Reviews of Modern Physics* 33 (1961), 265–342.

Dresden, M. [1962], *A Study of Models in Non-Equilibrium Statistical Mechanics*. In *Studies in Statistical Mechanics, V. I*, J. De Boer and G. E. Uhlenbeck, Eds. North-Holland, Amsterdam, 1962, 299–343.

Dresden, M. and Feiock, F. [1972], Models in Nonequilibrium Statistical Mechanics. *Journal of Statistical Physics* 4 (1972), 111–173.

Duering, E., Otero, D., Plastino, A. and Proto, A. N. [1985], Information Theory and the Linear Response Approach. *Physical Review A* 32 (1985), 2455.

DUHEM, P. [1886], *Le Potentiel Thermodynamique et ses Applications.* A. Hermann, Paris, 1886.

DUPLANTIER, B. AND SALEUR, H. [1989a], Stability of the Polymer Θ Point in Two Dimensions. *Physical Review Letters* **62** (1989), 1368–1371, 2641.

DUPLANTIER, B. AND SALEUR, H. [1989b], Exact Fractal Dimension of 2D Ising Clusters. *Physical Review Letters* **63** (1989), 2536–2537.

ECKMAN, J. -P. AND RUELLE, D. [1985], Ergodic Theory of Chaos and Strange Attractors. *Reviews of Modern Physics* **57** (1985), 617–656, Addendum, p. 1115.

EHRENBERG, W. [November 1967], Maxwell's Demon. *Scientific American* **217** (November 1967), 103–110.

EHRENFEST, P. AND EHRENFEST, T. [1959], *The Conceptual Foundations of the Statistical Approach in Mechanics.* Cornell University Press, Ithaca, 1959, English translation of the article published by B.G. Teubner (Leipzig, 1912) as No. 6 of Vol. IV 2 II of the *Encyklopädie der mathematischen Wissenschaften.*.

EINSTEIN, A. [1905], Über die von molekular-kinetische Theorie der Wärme geforderte Bewegung von in ruhenden Flüßigkeiten suspendierten Teilchen. *Annalen der Physik* **17** (1905), 549–560.

EINSTEIN, A. [1906a], New Determination of Molecular Dimensions. *Annalen der Physik* **19** (1906), 289–306.

EINSTEIN, A. [1906b], Zur Theorie der Brownschen Bewegung. *Annalen der Physik* **19** (1906), 371–381.

ENSKOG, D. [1917], *Kinetische Theorie der Vorgänge in mässig verdünten Gasen.* In *Dissertation.* Uppsala, 1917.

ESSAM, J. W. AND FISHER, M. E. [1963], Padé Approximant Studies of the Lattice Gas and Ising Ferromagnet below the Critical Point. *Journal of Chemical Physics* **38** (1963), 802–812.

FALK, H. [1970], Inequalities of J. W. Gibbs. *American Journal of Physics* **38** (1970), 858–869.

FALKOFF, D. L. [1961], *Master Equations and H-Theorems.* In *Ergodic Theories,* P. Caldirola, Ed. Academic Press, New York, NY, 1961, 170–176.

FERMI, E. [1956], *Thermodynamics.* Dover, Mineola, NY, 1956.

FERMI, E., PASTA, J. AND ULAM, S. [1955], *Studies of Non Linear Problems.* In *The Collected Papers of Enrico Fermi.* vol. 2, 1955, 977–988, Ulam's commentary is interesting..

FEYNMAN, R. P. [1949], The Theory of Positrons. *Physical Review* **76** (1949), 749–759.

FEYNMAN, R. P. [1972], *Statistical Mechanics.* W. A. Benjamin, Inc., New York, Amsterdam, 1972.

FEYNMAN, R. P. AND HIBBS, A. R. [1965], *Quantum Mechanics and Path Integrals.* McGraw-Hill, New York, NY, 1965.

FISHER, M. E. [1961], Statistical Mechanics of Dimers on a Plane Lattice. *Physical Review* **124** (1961), 1664–1672.

FISHER, M. E. [1964a], Correlation Functions and the Critical Region of Simple Fluids. *Journal of Mathematical Physics* **5** (1964), 944–962.

FISHER, M. E. [1964b], Magnetism in Infinite Systems—The Heisenberg Model for Infinite Spin. *American Journal of Physics* **32** (1964), 343–346.

FISHER, M. E. [1974], The Renormalization Group in the Theory of Critical Behavior. *Reviews of Modern Physics* **46** (1974), 597–616.

FISHER, M. E. [1983], *Scaling, Universality, and Renormalization Group Theory.* In *Critical Phenomena,* F. J. W. Hahne, Ed. Springer-Verlag, New York–Heidelberg–Berlin, 1983, 1–139.

FITTS, D. D. [1962], *Nonequilibrium Thermodynamics.* McGraw-Hill, New York, NY, 1962.

FLORY, P. J. [1969], *Statistical Mechanics of Chain Molecules.* Interscience Publishers, A division of John Wiley & Sons, London, 1969.

FLORY, P. J. [1971], *Principles of Polymer Chemistry.* Cornell University Press, Ithaca, NY, 1971.

FOMIN, S. V., KORNFELD, I. P. AND SINAI, I. G. [1982], *Ergodic Theory*. Springer Verlag, Berlin New York, 1982.

FORD, G. W., KAC, M. AND MAZUR, P. [1965], Statistical Mechanics of Assemblies of Coupled Oscillators. *Journal of Mathematical Physics* **6** (1965), 504–515.

FORD, J. [1970], Irreversibility in Non-linear Oscillator Systems. *Pure and Applied Chemistry* **22** (1970), 401–408.

FORD, J. [1983], How Random is a Coin Toss? *Physics Today* **36, No. 4** (1983), 40–47.

FORD, J. [1986], *Chaos: Solving the Unsolvable, Predicting the Unpredictable!* In *Chaotic Dynamics and Fractals*, M. F. Barnsley and S. G. Demko, Eds. Academic Press, New York, NY, 1986.

FORD, J. AND LUNSFORD, G. H. [1970] Stochastic Behavior of Resonant Nearly Linear Oscillator Systems in the Limit of Zero Nonlinear Coupling. *Physical Review A* **1** (1970), 59–70.

FORTE, S. [1992], Quantum Mechanics and Field Theory with Fractional Spin and Statistics. *Reviews of Modern Physics* **64** (1992), 193–236.

FREED, K. F. [1987], *Renormalization Group Theory of Macromolecules*. John Wiley & Sons, New York, NY, 1987.

FUJITA, S. [1966], *Introduction to Non-Equilibrium Statistical Mechanics*. W. B. Saunders Co., Philadelphia, 1966.

FÜRTH, R., Ed. [1956], *Investigations on the Theory of Brownian Movement*. Dutton, New York, 1956.

GABOR, D. [1950], Communication Theory and Physics. *Philosophical Magazine* **41** (1950), 1161–1187.

GABOR, D. [1961], *Light and Information*. In *Progress in Optics*, E. Wolf, Ed. vol. 1, North-Holland, Amsterdam, 1961.

GAL-OR, B. [1974a], *Statistical Mechanics, Fallacy and a New School of Thermodynamics*. In *Modern Developments in Thermodynamics*, B. Gal-Or, Ed. John Wiley & Sons, New York, NY, 1974.

GAL-OR, B. [1974b], *On a Paradox-Free Definition of Time and Irreversibility*. In *Modern Developments in Thermodynamics*, B. Gal-Or, Ed. John Wiley & Sons, New York, NY, 1974.

GAUNT, D. S. AND GUTTMANN, A. J. [1974], *Asymptotic Analysis of Coefficients*. In *Phase Transitions and Critical Phenomena*, V. 3, C. Domb and M. S. Green, Eds. Academic Press, New York, NY, 1974, 181–243.

DE GENNES, P. G. [1979], *Scaling Concepts in Polymer Physics*. Cornell University Press, Ithaca, NY, 1979.

GIBBS, J. W. [1876, 1878], On the Equilibrium of Heterogeneous Substances. *Transactions of the Connecticut Academy* **3** (1876, 1878), 108–248, 343-524.

GIBBS, J. W. [1960], *Elementary Principles in Statistical Mechanics*. Dover, Mineola, NY, 1960, Reprint of the original 1902 edition.

GIBBS, J. W. [1961], *The Scientific Papers of J. Willard Gibbs, V1*. Dover, Mineola, NY, 1961.

GINZBURG, V. L. [1961], Some Remarks on Phase Transitions of the Second Kind and the Macroscopic Theory of Ferromagnetic Materials. *Soviet Physics-Solid State* **2** (1961), 1824–1834.

GLASSER, M. L. [1970], Exact Partition Function for the Two-Dimensional Ising Model. *Am. J. Phys.* **38** (1970), 1033–1036.

GLASSER, M. L. AND ZUCKER, I. J. [1977], Extended Watson Integrals for the Cubic Lattices. *Proceedings of the National Academy of Science, U. S. A.* **74** (1977), 1800–1801.

GRAD, H. [1958], *Principles of the Kinetic Theory of Gases*. In *Handbuch der Physik*, S. Flugge, Ed. vol. XII, Springer-Verlag, New York–Heidelberg–Berlin, 1958.

GRAD, H. [1961], The Many Faces of Entropy. *Communications of Pure and Applied Mathematics* **14** (1961), 323–354.

GRADSHTEYN, I. S. AND RYZHIK, I. M. [1980], *Table of Integrals, Series, and Products*. Academic Press, New York, NY, 1980.

GRANDY, JR., W. T. [1980], Principle of Maximum Entropy and Irreversible Processes. *Physics Reports* **62** (1980), 175–266.

GRANDY, JR., W. T. [1987a], *Foundations of Statistical Mechanics V.I: Equilibrium Theory*. D. Reidel, Dordrecht–Boston–London, 1987.

GRANDY, JR., W. T. [1987b], *Foundations of Statistical Mechanics V.II:Nonequilibrium Phenomena*. D. Reidel, Dordrecht–Boston–London, 1987.

GRAVES-MORRIS, P. R., Ed. [1973], *Padé Approximants and their Applications*. Academic Press, New York, NY, 1973.

GREEN, H. S. [1961], Theories of Transport in Fluids. *Journal of Mathematical Physics* **2** (1961), 344–348.

GREEN, H. S. [1972], Parastatistics, Leptons, and the Neutrino Theory of Light. *Progress of Theoretical Physics* **47** (1972), 1400–1409.

GREEN, H. S. AND HURST, C. A. [1964], *Order-Disorder Phenomena*. Interscience Publishers, A division of John Wiley & Sons, London, 1964.

GREEN, M. S. [1951], Brownian Motion in a Gas of Noninteracting Molecules. *Journal of Chemical Physics* **19** (1951), 1036–1046.

GREEN, M. S. [1976], *Invariance Properties of the Renormalization Group*. In *Critical Phenomena*, J. Brey and R. B. Jones, Eds. Springer-Verlag, New York–Heidelberg–Berlin, 1976, 30–43.

GRIFFITHS, D. J. [1987], *Introduction to Electrodynamics*. Prentice Hall, Englewood Cliffs, NJ, 1987.

GRIFFITHS, R. B. [1964], A Proof that the Free Energy of a Spin System is Extensive. *Journal of Mathematical Physics* **5** (1964), 1215–1222.

GRIFFITHS, R. B. [1965a], Thermodynamic Inequality near the Critical Point for Ferromagnets and Fluids. *Physical Review Letters* **14** (1965), 623–624.

GRIFFITHS, R. B. [1965b], Ferromagnets and Simple Fluids near the Critical Point: Some Thermodynamic Inequalities. *Journal of Chemical Physics* **43** (1965), 1958–1968.

GRIFFITHS, R. B. [1970], Thermodynamics Near the Two-Fluid Critical Mixing Point in He^3-He^4.. *Physical Review Letters* **24** (1970), 715–717.

GRIFFITHS, R. B. [1973], Proposal on Notation at Tricritical Points. *Physical Review* **B7** (1973), 545–551.

DE GROOT, S. R. AND MAZUR, P. [1962], *Non-Equilibrium Thermodynamics*. North-Holland, Amsterdam, 1962.

GRÜNBAUM, A. [1974], *Is the Coarse-Grained Entropy an Anthropomorphism?* In *Modern Developments in Thermodynamics*, B. Gal-Or, Ed. John Wiley & Sons, New York, NY, 1974.

GUGGENHEIM, E. A. [1936] The Thermodynamics of Magnetization. *Proceedings of the Royal Society* **A155** (1936), 70–101.

GUGGENHEIM, E. A. [1945a], Notes on Magnetic Energy. *Physical Review* **68** (1945), 273–276.

GUGGENHEIM, E. A. [1945b], The Principle of Corresponding States. *Journal of Chemical Physics* **13** (1945), 253–261.

GUGGENHEIM, E. A. [1967], *Thermodynamics, 5th ed.*. North-Holland, Amsterdam, 1967.

GURTIN, M. E. AND WILLIAMS, W. O. [1967], An Axiomatic Foundation for Continuum Thermodynamics. *Archive for Rational Mechanics and Analysis* **26** (1967), 83–117.

GURTIN, M. E. [1973], Thermodynamics and the Energy Criterion for Stability. *Archive for Rational Mechanics and Analysis* **52** (1973), 93–103.

GURTIN, M. E. [1975], Thermodynamics and Stability. *Archive for Rational Mechanics and Analysis* **59** (1975), 63–96.

GUTTMANN, A. J. [1989], *Asymptotic Analysis of Power-Series Expansions*. In *Phase Transitions and Critical Phenomena*, C. Domb and J. L. Lebowitz, Eds. vol. 13, Academic Press, New York, NY, 1989, 1–234.

TER HAAR, D. [1954], *Elements of Statistical Mechanics*. Rinehart and Company, Inc., New York, 1954.

TER HAAR, D. AND WERGELAND, H. [1966], *Elements of Thermodynamics*. Addison Wesley, Reading, MA, 1966.

DE HAAS, W. D. AND VAN ALPHEN, P. M. [1930a], Dependence of the Susceptibility of Diamagnetic Metals upon the Field. *Proc. K. Akad. Amsterdam* **33** (1930), 1106–1118.

DE HAAS, W. D. AND VAN ALPHEN, P. M. [1930b], Relation of the Susceptibility of Diamagnetic Metals to the Field. *Proc. K. Akad. Amsterdam* **33** (1930), 680–682.

HALMOS, P. R. [1956], *Lectures on Ergodic Theory*. Chelsea, New York, NY, 1956.

HECHT, C. E. [1990], *Statistical Thermodynamics and Kinetic Theory*. W. H. Freeman and Company, San Francisco, CA, 1990.

HEER, C. V. [1972], *Statistical Mechanics, Kinetic Theory, and Stochastic Processes*. Academic Press, New York, NY, 1972.

HEINE, V. [1956], The Thermodynamics of Bodies in Static Electromagnetic Fields. *Proceedings of the Cambridge Philosophical Society* **52** (1956), 546–552.

HELLEMAN, R. H. G. [1980], *Self-generated Chaotic Behavior in Nonlinear Mechanics*. In *Fundamental Problems in Statistical Mechanics V*, E. G. D. Cohen, Ed. North-Holland, Amsterdam, 1980.

HEMMER, P. C. [1959], Dynamic and Stochastic Types of Motion in the Linear Chain. Norges Tekniske Høgskole, Dissertation, Trondheim, Norway, 1959.

HERZBERG, G. [1939], *Molecular Spectra and Molecular Structure I: Diatomic Molecules*. Prentice Hall, Englewood Cliffs, NJ, 1939.

HERZBERG, G. [1945], *Infrared and Raman Spectra*. D. van Nostrand Co., Inc., New York, 1945.

HILBERT, D. [1912], Begründung der kinetischen Gastheorie. *Mathematische Annalen* **72** (1912), 562–577.

HIRSCHFELDER, J. O., CURTIS, C. F. AND BIRD, R. B. [1954, 1966], *The Molecular Theory of Gases and Liquids*. John Wiley & Sons, New York, NY, 1954, 1966.

HOBSON, A. [1966a], Irreversibility and Information in Mechanical Systems. *The Journal of Chemical Physics* **45** (1966), 1352–1357.

HOBSON, A. [1966b], Irreversibility in Simple Systems. *American Journal of Physics* **34** (1966), 411–416.

HOBSON, A. [1968a], The Status of Irreversibility in Statistical Mechanics. *Physics Letters* **26A** (1968), 649–650.

HOBSON, A. [1968b], Further Comments on Irreversibility. *Physics Letters* **28A** (1968), 183–184.

HOBSON, A. [1971], *Concepts in Statistical Mechanics*. Gordon and Breach, New York, NY, 1971.

HOBSON, A. [1972], The Interpretation of Inductive Probabilities. *Journal of Statistical Physics* **6** (1972), 189–193.

HOBSON, A. AND CHENG, B. [1973], A Comparison of Shannon and Kullback Information Measures. *Journal of Statistical Physics* **7** (1973), 301–310.

HOBSON, A. AND LOOMIS, D. N. [1968], Exact Classical Nonequilibrium Statistical-Mechanical Analysis of the Finite Ideal Gas. *Physical Review* **173** (1968), 285–295.

HOHENBERG, P. C. AND HALPERIN, B. I. [1977], Theory of Dynamic Critical Phenomena. *Reviews of Modern Physics* **49** (1977), 435–479.

HOLDEN, A. V., Ed. [1986], *Chaos*. Princeton Univ. Press, Princeton, NJ, 1986.

VAN HOVE, L. [1960], *Lectures on Statistical Mechanics of Nonequilibrium Phenomena*. In *The Theory of Neutral and Ionized Gases*, C. De Witt and J-F. Detœuf, Eds. Wiley, 1960, 149–183.

VAN HOVE, L. [1962], *Master Equation and Approach to Equilibrium for Quantum Systems*. In *Fundamental Problems in Statistical Mechanics*, E. G. D. Cohen, Ed. North-Holland, Amsterdam, 1962, 157–172.

HUERTA, M. A. AND ROBERTSON, H. S. [1969], Entropy, Information Theory, and the Approach to Equilibrium of Coupled Harmonic Oscillator Systems. *Journal of Statistical Physics* **1** (1969), 393–414.

HUERTA, M. A. AND ROBERTSON, H. S. [1971], Approach to Equilibrium of Coupled Harmonic Oscillator Systems. II. *Journal of Statistical Physics* **3** (1971), 171–189.

HUERTA, M. A., ROBERTSON, H. S. AND NEARING, J. C. [1971], Exact Equilibration of Harmonically Bound Oscillator Chains. *Journal of Mathematical Physics* **12** (1971), 2305–2311.

HURST, C. A. AND GREEN, H. S. [1960], New Solution of the Ising Problem for a Rectangular Lattice. *Journal of Chemical Physics* **33** (1960), 1059–1062.

ISING, E. [1925], Beitrag zur Theorie des Ferromagnetismus. *Zeitschrift für Physik* **31** (1925), 253–258.

JACKSON, J. D. [1976], *Classical Electrodynamics, 2nd ed.*. John Wiley & Sons, New York, NY, 1976.

JAYNES, E. T. [1957a], Information Theory and Statistical Mechanics. *Physical Review* **106** (1957), 620–630.

JAYNES, E. T. [1957b], Information Theory and Statistical Mechanics II. *Physical Review* **108** (1957), 171–190.

JAYNES, E. T. [1979], *Where Do We Stand On Maximum Entropy?* In *The Maximum Entropy Formalism*, R. D. Levine and M. Tribus, Eds. MIT Press, Cambridge, MA, 1979, A Conference held at MIT, 2-4 May 1978.

JAYNES, E. T. [1980] *The Minimum Entropy Production Principle*. In *Annual Reviews of Physical Chemistry*. vol. 31, 1980, 579–601.

JAYNES, E. T. [1982a], *Brandeis Lectures*. In *Papers on Probability, Statistics, and Statistical Physics*, R. D. Rosenkrantz, Ed. D. Reidel, Dordrecht–Boston–London, 1982.

JAYNES, E. T. [1982b], In *Papers on Probability, Statistics, and Statistical Physics*, R. D. Rosenkrantz, Ed. D. Reidel, Dordrecht–Boston–London, 1982.

JAYNES, E. T. [1982c], *Gibbs vs. Boltzmann Entropies*. In *Papers on Probability, Statistics, and Statistical Physics*, R. D. Rosenkrantz, Ed. D. Reidel, Dordrecht–Boston–London, 1982, Reprinted from Am. J. Phys. **33**, 391-398 (1965).

JAYNES, E. T. [1984], *Bayesian Methods: General Background*. In *Maximum Entropy and Bayesian Methods in Applied Statistics*, J. H. Justice, Ed. Cambridge University Press, Cambridge, England, 1984.

JAYNES, E. T. [1985], *Generalized Scattering*. In *Maximum-Entropy and Bayesian Methods in Inverse Problems*, C. R. Smith and W. T. Grandy, Eds. D. Reidel, Dordrecht–Boston–London, 1985, 377–398.

JOHNSON, J. B. [1928], Thermal Agitation of Electricity in Conductors. *Physical Review* **32** (1928), 97–109.

JUMARIE, G. M. [1986], *Subjectivity, Information, Systems*. Gordon and Breach, New York, NY, 1986.

KAC, M. [1947], Random Walks and the Theory of Brownian Motion. *American Mathematical Monthly* **54** (1947), 369–391.

KAC, M. [1959], *Probability and Related Topics in Physical Sciences*. Interscience Publishers, A division of John Wiley & Sons, London, 1959.

KAC, M. AND WARD, J. C. [1952], A Combinatorial Solution of the Two-Dimensional Ising Problem. *Physical Review* **88** (1952), 1332–1337.

KADANOFF, L. P. [1966], Scaling Laws for Ising Models near T_c. *Physics* **26** (1966), 263–272.

KADANOFF, L. P. [1969], Mode-Mode Coupling near the Critical Point. *Journal of the Physical Society of Japan (Supplement)* **26** (1969), 122–126.

KADANOFF, L. P. AND BAYM, G. [1971], *Quantum Statistical Mechanics: Green's Function Methods in Equilibrium and Nonequilibrium Problems*. W. A. Benjamin, Inc., New York, Amsterdam, 1971.

KADANOFF, L. P. [1976], *Scaling, Universality, and Operator Algebras*. In *Phase Transitions and Critical Phenomena*, V. 5a, C. Domb and M. S. Green, Eds. Academic Press, New York, NY, 1976, 1–34.

KADANOFF, L. P., GÖTZE, W., HAMBLEN, D., HECHT, R., LEWIS, E. A. S., PALCIAUSKAS, V. V., RAYL, M., SWIFT, J., ASPNES, D. AND KANE, J. [1967], Static Phenomena Near Critical Points: Theory and Experiment. *Reviews of Modern Physics* **39** (1967), 395–431.

VAN KAMPEN, N. G. [1962], *Fundamental Problems in Statistical Mechanics of Irreversible Processes.* In *Fundamental Problems in Statistical Mechanics,* E. G. D. Cohen, Ed. North-Holland, Amsterdam, 1962, 173–201.

VAN KAMPEN, N. G. [1971], The Case Against Linear Response Theory. *Physica Norvegica* **5** (1971), 279–284.

VAN KAMPEN, N. G. [1984], *The Gibbs Paradox.* In *Essays in Theoretical Physics In Honour of Dirk ter Haar,* W. E. Parry, Ed. Pergamon Press, Oxford, England, 1984.

KAPUR, J. N. [1989], *Maximum-Entropy Models in Science and Engineering.* John Wiley & Sons, New York, NY, 1989.

KAPUR, J. N. AND KESAVAN, H. K. [1987], *Generalized Maximum Entropy Principle (with Applications).* Sandford Education Press, Waterloo, Ontario, Canada, 1987.

KASTELEYN, P. W. [1962], The Statistics of Dimers on a Lattice. *Physica* **27** (1962), 1209–1225.

KASTELEYN, P. W. [1963], Dimer Statistics and Phase Transitions. *Journal of Mathematical Physics* **4** (1963), 287–293.

KASTELEYN, P. W. [1967], *Graph Theory and Crystal Physics.* In *Graph Theory and Theoretical Physics,* F. Harary, Ed. Academic Press, New York, NY, 1967, 43–110.

KATZ, A. [1967], *Principles of Statistical Mechanics: The Information Theory Approach.* W. H. Freeman and Company, San Francisco, CA, 1967.

KAUFMAN, B. [1949], Crystal Statistics II. Partition Function Evaluated by Spinor Analysis. *Physical Review* **76** (1949), 1232–1243.

KAUFMAN, B. AND ONSAGER, L. [1949], Crystal Statistics III. Short-range Order in a Binary Ising Lattice. *Physical Review* **76** (1949), 1244–1252.

KESAVAN, H. K. AND KAPUR, J. N. [1989], The Generalized Maximum Entropy Principle. *IEEE Transactions on Systems, Man, and Cybernetics* **19** (1989), 1042–1052.

KHINCHIN, A. I. [1957], Dover, Mineola, NY, 1957. *Mathematical Foundations of Information Theory.*

KIRKWOOD, J. G. [1946], The Statistical Mechanical Theory of Transport Processes. I General Theory. *Journal of Chemical Physics* **14** (1946), 180–201, errata 347.

KIRKWOOD, J. G. [1947], The Statistical Mechanical Theory of Transport Processes. II Transport in Gases. *Journal of Chemical Physics* **15** (1947), 72–76, errata 155.

KITCHENS, B. [1981], *Ergodic Theory and Dynamical Systems.* Cambridge University Press, Cambridge, England, 1981.

KITTEL, C. [1986], *Introduction to Solid State Physics, 6th Ed..* John Wiley & Sons, New York, NY, 1986.

KLEIN, M. J. AND TISZA, L. [1949], Theory of Critical Fluctuations. *Physical Review* **76** (1949), 1861–1868.

KOHLRAUSCH, K. W. F. AND SCHRÖDINGER, E. [1926], The Ehrenfest Model of the H-Curve. *Physikalische Zeitschrift* **27** (1926), 306–313.

KRAMERS, H. A. AND WANNIER, G. H. [1941], Statistics of the 2-dimensional Ferromagnet. I-II. *Physical Review* **60** (1941), 252–263-278.

KREUZER, H. J. [1981], *Nonequilibrium Thermodynamics and its Statistical Foundations.* Oxford University Press, London, 1981.

KUBO, R. [1957], Statistical Mechanical Theory of Irreversible Processes I. General Theory and Simple Applications to Magnetic and Conduction Problems. *Journal of the Physical Society of Japan* **12** (1957), 570–586.

KUBO, R. [1965], *Linear Response Theory of Irreversible Processes.* In *Statistical Mechanics of Equilibrium and Non-Equilibrium,* J. Meixner, Ed. North-Holland, Amsterdam, 1965, 81–99.

KUBO, R. [1966], The Fluctuation-Dissipation Theorem. *Reports on Progress in Physics* **29** (1966), 255–284.

KUBO, R. AND NAGAMIYA, T. [1969], *Solid State Physics*. McGraw-Hill, New York, NY, 1969.

KUBO, R., YOKOTA, M. AND NAKAJIMA, S. [1957], Statistical Mechanical Theory of Irreversible Processes II. Response to Thermal Disturbance. *Journal of the Physical Society of Japan* **12** (1957), 1203–1211.

KULLBACK, S. [1968], *Information Theory and Statistics*. Dover, Mineola, NY, 1968.

LANDAU, L. D. [1930], Paramagnetism of Metals. *Zeitschrift für Physik* **64** (1930), 629–637.

LANDAU, L. D. AND LIFSHITZ, E. M. [1960], *Electrodynamics of Continuous Media*. Addison Wesley, Reading, MA, 1960.

LANDAU, L. D. AND LIFSHITZ, E. M. [1969], *Statistical Physics, 2nd. ed.*. Addison Wesley, Reading, MA, 1969.

LANDSBERG, P. T. [1961], *Thermodynamics*. Interscience Publishers, A division of John Wiley & Sons, London, 1961.

LANDSBERG, P. T., Ed. [1970], *International Conference on Thermodynamics* Cardiff, U.K. 1-4 April, 1970. Butterworth's Scientific Publications, London, 1970.

LANDSBERG, P. T., Ed. [1971], *Problems in Thermodynamics*. Pion, London, 1971.

LANDSBERG, P. T. [1984], Is Equilibrium Always an Entropy Maximum? *Journal of Statistical Physics* **35** (1984), 159–169.

LANDSBERG, P. T. AND TRANAH, D. T. [1980a], Thermodynamics of Non-extensive Entropies I. *Collective Phenomena* **3** (1980), 73–80.

LANDSBERG, P. T. AND TRANAH, D. T. [1980b], Thermodynamics of Non-extensive Entropies II. *Collective Phenomena* **3** (1980), 81–88.

LAWRIE, I. AND SARBACH, S. [1984], *Theory of Tricritical Points*. In *Phase Transitions and Critical Phenomena, V9*, C. Domb and J. L. Lebowitz, Eds. Academic Press, New York, NY, 1984, 1–161.

LEE, T. D. AND YANG, C. N. [1952], Statistical Theory of Equations of State and Phase Transitions. II. Lattice Gas and Ising Model. *Physical Review* **87** (1952), 410–419.

VAN LEEUWEN, J. H. [1919], Magnetism. Dissertation, Leyden, 1919.

VAN LEEUWEN, J. H. [1921], Problems of the Electronic Theory of Magnetism. *J. de Physique (6)* **2** (1921), 361–367.

LEFF, H. S. AND REX, A. F., Eds. [1990], *Maxwell's Demon: Entropy, Information, Computing*. Princeton Univ. Press, Princeton, NJ, 1990.

LEINAAS, J. M. AND MYRHEIM, J. [1977], On the Theory of Identical Particles. *Il Nuovo Cimento* **37B** (1977), 1–23.

LEUPOLD, H. A. [1969], Notes on the Thermodynamics of Paramagnets. *American Journal of Physics* **37** (1969), 1047–1054.

LEVINE, R. D. [1980], An Information Theoretical Approach to Inversion Problems. *Journal of Physics A* **13** (1980), 91–108.

LEWIS, G. N. [1907], Outline of a New System of Thermodynamic Chemistry. *Proc. Am. Acad.* **43** (1907), 259–293.

LEWIS, G. N. AND RANDALL, M. [1923], *Thermodynamics*. McGraw-Hill, New York, NY, 1923.

LIBOFF, R. A. [1974], Gibbs vs. Shannon Entropies. *Journal of Statistical Physics* **11** (1974), 343–357.

LIBOFF, R. A. [1990a], Microscopic Reversibility and Gibbs Entropy. *Journal of Non-Equilibrium Thermodynamics* **15** (1990), 151–155.

LIBOFF, R. L. [1990b], *Kinetic Theory: Classical, Quantum, and Relativistic Descriptions*. Prentice Hall, Englewood Cliffs, NJ, 1990.

LIEB, E. H. [1967a], Exact Solution of the *F* Model of an Antiferroelectric. *Physical Review Letters* **18** (1967), 1046–1048.

LIEB, E. H. [1967b], Residual Entropy of Square Ice. *Physical Review* **162** (1967), 162–172.

LIEB, E. H. [1967c], Exact Solution of the Two-Dimensional Slater KDP Model of a Ferroelectric. *Physical Review Letters* **19** (1967), 108–110.

LIEB, E. H. [1967d], Solution of the Dimer Problem by the Transfer Matrix Method. *Journal of Mathematical Physics* **8** (1967), 2339–2341.

LINDBLAD, G. [1983], *Non-Equilibrium Entropy and Irreversibility*. D. Reidel, Dordrecht–Boston–London, 1983.

LINN, S. L. AND ROBERTSON, H. S. [1984], Thermal Energy Transport in Harmonic Systems. *Journal of the Physics and Chemistry of Solids* **45** (1984), 133–140.

LIVENS, G. H. [1945], On the Energy and Mechanical Relations of the Magnetic Field. *Philosophical Magazine* **36** (1945), 1–20.

LIVENS, G. H. [1947], Note on Magnetic Energy. *Physical Review* **71** (1947), 58–63.

LIVENS, G. H. [1948], The Thermodynamics of Magnetization. *Proceedings of the Cambridge Philosophical Society* **44** (1948), 534–545.

MA, S-K. [1976a], *Scale Transformations in Dynamic Models*. In *Critical Phenomena*, J. Brey and R. B. Jones, Eds. Springer-Verlag, New York–Heidelberg–Berlin, 1976, 43–78.

MA, S-K. [1976b], *Modern Theory of Critical Phenomena*. The Benjamin/Cummings Publishing Company, Inc., Reading, Massachusetts, 1976.

MA, S-K. [1981], Calculation of Entropy from Data of Motion. *Journal of Statistical Physics* **26** (1981), 221–240.

MA, S-K. [1985], *Statistical Mechanics*. World Scientific, Philadelphia, Singapore, 1985.

MACKEY, M. C. [1989], The Dynamic Origin of Increasing Entropy. *Reviews of Modern Physics* **61** (1989), 981–1015.

MAN, C-S. [1986], *New Perspectives in Thermodynamics*. Springer-Verlag, New York–Heidelberg–Berlin, 1986.

MARADUDIN, A. A., MONTROLL, E. W. AND WEISS, G. H. [1963], *Theory of Lattice Dynamics in the Harmonic Approximation*. Academic Press, New York, NY, 1963.

MATTIS, D. C. [1965], *The Theory of Magnetism*. Harper and Row, N.Y., 1965.

MAXWELL, J. C. [1871], *Theory of Heat*. Longmans, London, 1871.

MAYER-KRESS, G., Ed. [1986], *Dimensions and Entropies in Chaotic Systems: Quantification of Complex Behavior*. Synergetics, vol. 32, Springer-Verlag, New York–Heidelberg–Berlin, 1986.

MAZUR, P. [1962], *Statistical Considerations on the Basis of Non-Equilibrium Thermodynamics*. In *Fundamental Problems in Statistical Mechanics*, E. G. D. Cohen, Ed. North-Holland, Amsterdam, 1962, 203–229.

MAZUR, P. [1965], *On a Statistical Mechanical Theory of Brownian Motion*. In *Statistical Mechanics of Equilibrium and Non-Equilibrium*, J. Meixner, Ed. North-Holland, Amsterdam, 1965, 52–68.

McCoy, B. M. AND WU, T. T. [1967a], Theory of Toeplitz Determinants and the Spin Correlation of the Two-Dimensional Ising Model, II. *Physical Review* **155** (1967), 438–542.

McCoy, B. M. AND WU, T. T. [1967b], Theory of Toeplitz Determinants and the Spin Correlation of the Two-Dimensional Ising Model, IV. *Physical Review* **162** (1967), 436–475.

McCoy, B. M. AND WU, T. T. [1973], *The Two-Dimensional Ising Model*. Harvard University Press, Cambridge, 1973.

McLENNAN, J. A. [1989], *Introduction to Non-Equilibrium Statistical Mechanics*. Prentice Hall, Englewood Cliffs, NJ, 1989.

MEIXNER, J. [1973], Consistency of the Onsager-Casimir Reciprocal Relations. *Advances in Molecular Relaxation Phenomena* **5** (1973), 319–331.

MERMIN, N. D. [1968], Crystalline Order in Two Dimensions. *Physical Review* **176** (1968), 250–254.

MERMIN, N. D. AND WAGNER, H. [1966], Absence of Ferromagnetism or Antiferromagnetism in One- or Two-Dimensiomal Isotropic Heisenberg Models. *Physical Review Letters* **17** (1966), 1133–1136.

MILLER, K. S. [1964], *Multidimensional Gaussian Distributions*. John Wiley & Sons, New York, NY, 1964.

MINC, H. [1978], *Permanents*. Addison Wesley, Reading, MA, 1978.

MITCHELL, W. C. [1967], Statistical Mechanics of Thermally Driven Systems. Washington University, PhD Dissertation, St. Louis, MO, 1967.

MONTGOMERY, D. C. AND TIDMAN, D. A. [1964], *Plasma Kinetic Theory*. McGraw-Hill, New York, NY, 1964.

MONTROLL, E. W. [1941], Statistical Mechanics of Nearest-neighbor Systems. *Journal of Chemical Physics* **9** (1941), 706–721.

MONTROLL, E. W. [1960], *Topics on Statistical Mechanics of Interacting Particles*. In *The Theory of Neutral and Ionized Gases*, C. De Witt and J. F. Detœuf, Eds. John Wiley & Sons, New York, NY, 1960, 2–148.

MONTROLL, E. W. [1961], *Nonequilibrium Statistical Mechanics*. In *Lectures in Theoretical Physics, V. III*, W. E. Brittin, B. W. Downs and J. Downs, Eds. Interscience Publishers, A division of John Wiley & Sons, London, 1961, 221–325.

MONTROLL, E. W. [1962], *Some Remarks on the Integral Equations of Statistical Mechanics*. In *Fundamental Problems in Statistical Mechanics*, E. G. D. Cohen, Ed. North-Holland, Amsterdam, 1962, 223–249.

MONTROLL, E. W. [1964], *Lattice Statistics*. In *Applied Combinatorial Mathematics*, E. F. Beckenbach, Ed. John Wiley & Sons, New York, NY, 1964, 96–143.

MONTROLL, E. W. [1966], *Lectures on the Ising Model of Phase Transitions*. In *Statistical Physics, Phase Transitions, and Superfluidity*. vol. 11, Brandeis University Summer Institute in Theoretical Physics, Gordon and Breach, New York, NY, 1966.

MONTROLL, E. W. [1967], *Stochastic Processes and Chemical Kinetics*. In *Energetics in Metallurgical Phenomena, V. III*, W. M. Mueller, Ed. Gordon and Breach, New York, NY, 1967, 123–187.

MONTROLL, E. W., POTTS, R. B. AND WARD, J. C. [1963], Correlations and Spontaneous Magnetization of the Two-Dimensional Ising Model. *Journal of Mathematical Physics* **4** (1963), 308–322.

MONTROLL, E. W. AND SHLESINGER, M. F. [1983], Maximum Entropy Formalism, Fractals, Scaling Phenomena, and 1/f Noise: A Tale of Tails. *Journal of Statistical Physics* **32** (1983), 209–230.

MONTROLL, E. W. AND WARD, J. C. [1959], Quantum Statistics of Interacting Particles. II Cluster Integral Development of Transport Coefficients. Presented at *Physica* (1959).

MORSE, P. M. [1969], *Thermal Physics*. Benjamin/Cummings, Reading, Mass., 1969.

NAGLE, J. F. [1966], Lattice Statistics of Hydrogen-Bonded Crystals. II. *J. Math. Physics* **7** (1966), 1492–1496.

NAGLE, J. F. [1968], The One-Dimensional KDP Model in Statistical Mechanics. *American Journal of Physics* **36** (1968), 1114–1117.

NAGLE, J. F. [1974], *Ferroelectric Models*. In *Phase Transitions and Critical Phenomena*, C. Domb and M. S. Green, Eds. vol. 3, Academic Press, New York, NY, 1974, 653–666.

NE'EMAN, Y. [1974], *Time Reversal Asymmetry at the Fundamental Level—and its Reflection on the problem of the Arrow of Time*. In *Modern Developments in Thermodynamics*, B. Gal-Or, Ed. John Wiley & Sons, New York, NY, 1974.

NERNST, W. [1884], *Zeitschrift für Physicalische Chemie* **9** (1884), 613.

NICKEL, B. G. [1982], In *Phase Transitions, Cargese 1980*, M. Levy, J. -C. le Gouillou and J. Zinn-Justin, Eds. Plenum Press, New York, NY, 1982.

NIEMEIJER, T. AND VAN LEEUWEN, J. M. J. [1976], *Renormalization Theory for Ising-like Spin Systems.* In *Phase Transitions and Critical Phenomena, V.6*, C. Domb and M. S. Green, Eds. 1976, 425–505.

NOLL, W. [1973], Lectures on the Foundations of Continuum Mechanics and Thermodynamics. *Archive for Rational Mechanics and Analysis* **52** (1973), 62–92.

NORTHCOTE, R. S. AND POTTS, R. B. [1964], Energy Sharing and Equilibrium for Nonlinear Systems. *Journal of Mathematical Physics* **5** (1964), 383–398.

NYQUIST, H. [1928], Thermal Agitation of Electric Charge in Conductors. *Physical Review* **32** (1928), 110–114.

O'NEILL, E. L. [1963], *Introduction to Statistical Optics.* Addison Wesley, Reading, MA, 1963.

OGUCHI, T. [1950], Statistics of the Three-Dimensional Ferromagnet. I. *J. Phys. Soc. Japan* **5** (1950), 75–81.

OGUCHI, T. [1951a], Statistics of the Three-Dimensional Ferromagnet. II. *J. Phys. Soc. Japan* **6** (1951), 27–31.

OGUCHI, T. [1951b], Statistics of the Three-Dimensional Ferromagnet. III. *J. Phys. Soc. Japan* **6** (1951), 31–35.

OHNUKI, Y. AND KAMEFUCHI, S. [1982], *Quantum Field Theory and Parastatistics.* Springer-Verlag, New York–Heidelberg–Berlin, 1982.

ONSAGER, L. [1931b], Reciprocal Relations in Irreversible Processes. I. *Physical Review* **37** (1931), 405–426.

ONSAGER, L. [1931a], Reciprocal Relations in Irreversible Processes. II. *Physical Review* **38** (1931), 2265–2279.

ONSAGER, L. [1944], Crystal Statistics I. A Two-dimensional Model of an Order-Disorder Transition. *Physical Review* **65** (1944), 117–149.

ORNSTEIN, L. S. AND ZERNIKE, F. [1914], Accidental Deviations of Density and Opalescence at the Critical Point of a Single Substance. *Proc. Amst. Akad. Sci.* **17** (1914), 793–806.

ORNSTEIN, L. S. AND ZERNIKE, F. [1917], Influence of Accidental Deviations of Density on the Equation of State. *Proc. Amst. Akad. Sci* **19** (1917), 1312–1315.

ORNSTEIN, L. S. AND ZERNIKE, F. [1918], The Linear Dimensions of Density Variations. *Physik. Zeits.* **19** (1918), 134–137.

ORNSTEIN, L. S. AND ZERNIKE, F. [1926], Molecular Scattering of Light at the Critical State. *Physik. Zeits.* **27** (1926), 761–763.

PARSEGIAN, A. [1987], Book review of Denbigh and Denbigh [1985]. *Journal of Statistical Physics* **46** (1987), 431–432.

PATASHINSKII, A. Z. AND POKROVSKII, V. I. [1979], Pergamon Press, Oxford, England, 1979. *Fluctuation Theory of Phase Transitions.*

PATERA, R. P. AND ROBERTSON, H. S. [1980], Information Theory and Solar Energy Collection. *Applied Optics* **19** (1980), 2403–2407.

PATHRIA, R. K. [1972], *Statistical Mechanics.* Pergamon Press, Oxford, England, 1972.

PAULI, W. [1927], Gas Degeneration and Paramagnetism. *Zeitschrift für Physik* **41** (1927), 81–102.

PAULI, W. [1928], In *Probleme der Modernen Physik (Festschrift zum 60. Geburtstage A. Sommerfelds)*, P. Debye, Ed. Hirzel, Leipzig, 1928, 30.

PAULI, W. [1940], Connection Between Spin and Statistics. *Physical Review* **58** (1940), 716–722.

PAULI, W. [1950], On the Connection between Spin and Statistics. *Prog. Theor. Phys.* **5** (1950), 526–536.

PAULI, W. AND BELINFANTE, F. J. [1940], Statistics of Elementary Particles. *Physica* **7** (1940), 177–192.

PENROSE, O. [1970], *Foundations of Statistical Mechanics; a Deductive Treatment.* Pergamon Press, Oxford, England, 1970.

PENROSE, O. AND LAWRENCE, A. [1974], *The Direction of Time and the Microcanonical Ensemble.* In *Modern Developments in Thermodynamics*, B. Gal-Or, Ed. John Wiley & Sons, New York, NY, 1974.

PETERSON, R. L. [1967], Formal Theory of Nonlinear Response. *Reviews of Modern Physics* **39** (1967), 69–77.

VAN PEYPE, W. F. [1938], Zur Theorie der magnetischen Anisotropie Kubischen Kristalle beim absoluten Nullpunkt. *Physica* 5 (1938), 465–483.

PFEUTY, P. AND TOULOUSE, G. [1977], *The Renormalization Group and Critical Phenomena*. John Wiley & Sons, New York, NY, 1977.

PHILLIPSON, P. E. [1971], State Function Formulation of Elementary Relaxation Processes. *American Journal of Physics* 39 (1971), 373–387.

PHILLIPSON, P. E. [1974], Effects of Long Range Interactions in Harmonically Coupled Systems. I. Equilibrium Fluctuations and Diffusion. *Journal of Mathematical Physics* 15 (1974), 2127–2147.

PLANCK, M. [1923], *Theorie der Wärmestrahlung, 5th ed.*. Barth, Leipzig, 1923.

PLANCK, M. [1932], *Theorie der Wärme, 2nd ed.*. Hirzel, Leipzig, 1932.

PLANCK, M. [1945], *Treatise on Thermodynamics*. Dover, Mineola, NY, 1945.

PLISCHKE, M. AND BERGERSEN, B. [1989], *Equilibrium Statistical Mechanics*. Prentice Hall, Englewood Cliffs, NJ, 1989.

POTTS, R. B. [1952], Some Generalized Order-Disorder Transformations. *Proc. Camb. Philos. Soc.* 48 (1952), 106–109.

POTTS, R. B. [1955a], Combinatorial Studies of the Triangular Ising Lattice, pt. 3. *Proc. Phys. Soc. London* A68 (1955), 145–148.

POTTS, R. B. [1955b], Spontaneous Magnetization of a Triangular Lattice. *Physical Review* 88 (1955), 352.

POTTS, R. B. AND WARD, J. C. [1955], The Combinatorial Method and the Two-Dimensional Ising Model. *Progress in Theoretical Physics* 13 (1955), 38–46.

POUNDSTONE, W. [1985], *The Recursive Universe*. William Morrow and Co., New York, 1985.

PRICE, H. [1991], Review Article (Denbigh and Denbigh 1985). *British Journal of Philosophy of Science* 42 (1991), 111–144.

PRIGOGINE, I. [1962], *Non-Equilibrium Statistical Mechanics*. Interscience Publishers, A division of John Wiley & Sons, London, 1962.

PRIGOGINE, I. [1967], *Introduction to Thermodynamics of Irreversible Processes*. Interscience Publishers, A division of John Wiley & Sons, London, 1967.

PRIGOGINE, I. AND STENGERS, I. [1984], *Order Out of Chaos*. Bantam Books, Toronto, New York, London, Sydney, 1984.

PRIVOROTSKII, I. [1976], *Thermodynamic Theory of Domain Structures*. John Wiley & Sons, New York, NY, 1976.

PYNN, R. AND SKJELTORP, A., Eds. [1983], *Multicritical Phenomena*. Plenum Press, New York, 1983.

RAHAL, V. [1989], A Note on the Fluctuations of Noninteracting Particles. *American Journal of Physics* 57 (1989).

RAPOSO, E. P., DE OLIVEIRA, S. M., NEMIROVSKY, A. M. AND COUTINHO-FILHO, M. D. [1991], Random Walks: A Pedestrian Approach to Polymers, Critical Phenomena, and Field Theory. *American Journal of Physics* 59 (1991), 633–645.

RASETTI, M. [1986], *Modern Methods in Equilibrium Statistical Mechanics*. World Scientific Publishing Co. Pte Ltd., Singapore, Philadelphia, 1986.

REICHL, L. E. [1980], *A Modern Course in Statistical Physics*. University of Texas Press, Austin, 1980.

ROBERTSON, B. [1966], Equations of Motion in Nonequilibrium Statistical Mechanics. *Physical Review* 144 (1966), 151–161.

ROBERTSON, B. [1967], Equations of Motion in Nonequilibrium Statistical Mechanics. II. Energy Transport. *Physical Review* 160 (1967), 175–183.

ROBERTSON, B. [1970], Quantum Statistical Mechanical Derivation of Generalized Hydrodynamic Equations. *Journal of Mathematical Physics* **11** (1970), 2482–2488.

ROBERTSON, B. [1979], *Applications of Maximum Entropy to Nonequilibrium Statistical Mechanics*. In *The Maximum Entropy Formalism*, R. D. Levine and M. Tribus, Eds. MIT Press, Cambridge, MA, 1979, A Conference held at MIT, 2-4 May 1978.

ROBERTSON, B. AND MITCHELL, W. C. [1971], Equations of Motion in Nonequilibrium Statistical Mechanics. III. Open Systems. *Journal of Mathematical Physics* **12** (1971), 563–568.

ROBERTSON, H. S. AND HUERTA, M. A. [1970a], Entropy Oscillation and the *H* Theorem for Finite Segments of Infinite Coupled Oscillator Chains. *Pure and Applied Chemistry* **22** (1970).

ROBERTSON, H. S. AND HUERTA, M. A. [1970b], Information Theory and the Approach to Equilibrium. *American Journal of Physics* **38** (1970), 619–630.

ROTHSTEIN, J. [1951], Information, Measurement, and Quantum Mechanics. *Science* **114** (1951), 171–175.

ROTHSTEIN, J. [1952], Information and Thermodynamics. *The Physical Review* **85** (1952), 135.

ROTHSTEIN, J. [1957], Nuclear Spin Echo Experiments and the Foundations of Statistical Mechanics. *American Journal of Physics* **25** (1957), 510–518.

ROTHSTEIN, J. [1959], Physical Demonology. *Methodos* **XI** (1959), 94–117.

ROTHSTEIN, J. [1979], *Generalized Entropy, Boundary Conditions, and Biology*. In *The Maximum Entropy Formalism*, R. D. Levine and M. Tribus, Eds. MIT Press, Cambridge, MA, 1979, A Conference held at MIT, 2-4 May 1978.

ROWLINSON, J. S. [1970], Probability, Information, and Entropy. *Nature* **225** (1970), 1196–1198.

RUELLE, D. [1988], *Chaotic Evolution and Strange Attractors*. Cambridge University Press, Cambridge, England, 1988.

RUSHBROOKE, G. S. [1963], On the Thermodynamics of the Critical Region for the Ising Model. *Journal of Chemical Physics* **39** (1963), 842–843.

RUSHBROOKE, G. S., BAKER, JR., G. A. AND WOOD, P. J. [1974], *Heisenberg Model*. In *Phase Transitions and Critical Phenomena, V. 3*, C. Domb and M. S. Green, Eds. Academic Press, New York, NY, 1974, 246–356.

SCHRÖDINGER, E. [1945], *What is Life?*. Cambridge University Press, Cambridge, England 1945.

SCHULTZ, T. D., MATTIS, D. C. AND LIEB, E. H. [1964], Two-Dimensional Ising Model as a Soluble Problem of Many Fermions. *Reviews of Modern Physics* **36** (1964), 856–871.

SEKE, J. [1990], Self-Consistent Projection-Operator Method for Describing the non-Markovian Time-Evolution of Subsystems. *Journal of Physics A: Math. Gen* **23** (1990), L61–L66.

SERRIN, J. [1979], Conceptual Analysis of the Classical Second Law of Thermodynamics. *Archiv for Rational Mechanics and Analysis* **70** (1979), 355–371.

SERRIN, J., Ed. [1986], *New Perspectives in Thermodynamics*. Springer-Verlag, New York–Heidelberg–Berlin, 1986.

SHANNON, C. E. [1948], The Mathematical Theory of Communication. *Bell System Technical Journal* **27** (1948), 379–424, 623-657.

SHANNON, C. E. [1949], Communication Theory in Secrecy Systems. *Bell System Technical Journal* **28** (1949), 656–715.

SHANNON, C. E. [1951], Prediction and Entropy in Printed English. *Bell System Technical Journal* **30** (1951), 50–64.

SHANNON, C. E. AND WEAVER, W. [1949], *The Mathematical Theory of Communication*. Univ. of Illinois Press, Urbana, 1949.

SHAPIRO, S. L. AND TEUKOLSKY, S. A. [1983], *Black Holes, White Dwarfs, and Neutron Stars*. John Wiley & Sons, New York, NY, 1983.

SHENKER, S. H. [1984], *Field Theories and Phase Transitions*. In *Recent Advances in Field Theory and Statistical Mechanics*, J-B. Zuber and R. Stora, Eds. North-Holland, Amsterdam, 1984, 1–38.

SHERMAN, S. [1960], Combinatorial Aspects of the Ising Model of Ferromagnetism I. A Conjecture of Feynman on Paths and Graphs. *Journal of Mathematical Physics* 1 (1960), 202–217.

SHERMAN, S. [1962], Combinatorial Aspects of the Ising Model of Ferromagnetism II. An Analogue to the Witt Identity. *Bulletin of the American Mathematical Society* 68 (1962), 225–229.

SINAI, I. G. [1976], *Introduction to Ergodic Theory*. Princeton Univ. Press, Princeton, NJ, 1976.

SINGH, J. [1966], *Great Ideas in Information Theory, Language, and Cybernetics*. Dover, Mineola, NY, 1966.

SKAGERSTAM, B-S. [1975], On the Notions of Entropy and Information. *Journal of Statistical Physics* 12 (1975), 449–462.

SMITH-WHITE, W. B. [1949], On the Mechanical Forces in Dielectrics. *Philosophical Magazine* 40 (1949), 466–479.

SMITH-WHITE, W. B. [1950], Energy in Electrostatics. *Nature* 166 (1950), 689–690.

SMOLUCHOWSKI, M. v. [1916], Drei Vorträge über Diffusion, Brownsche Bewegung, und Koagulation von Kolloidteilchen. *Physikalische Zeitung* 17 (1916), 557, 585.

SMORODINSKY, M. [1971], *Ergodic Theory, Entropy*. Springer-Verlag, New York–Heidelberg–Berlin, 1971.

SPITZER, L. [1956], *The Physics of Fully Ionized Gases*. Interscience Publishers, A division of John Wiley & Sons, London, 1956.

STANLEY, H. E. [1968], Spherical Model as the Limit of Infinite Spin Dimensionality. *Physical Review* 176 (1968), 718–722.

STANLEY, H. E. [1969], Exact Solution for a Linear Chain of Isotropically Interacting Classical Spins of Arbitrary Dimensionality. *Physical Review* 179 (1969), 570–577.

STANLEY, H. E. [1971], *Introduction to Phase Transitions and Critical Phenomena*. Oxford University Press, London, 1971.

STANLEY, H. E. [1974], *D-Vector Model*. In *Phase Transitions and Critical Phenomena*, C. Domb and M. S. Green, Eds. vol. 3, Academic Press, New York, NY, 1974, 485–567.

STANLEY, H. E., BLUME, M., MATSUMO, K. AND MILOŠEVIĆ, S. [1970], Eigenvalue Degeneracy as a Possible "Mathematical Mechanism" for Phase Transitions. *Journal of Applied Physics* 41 (1970), 1278–1282.

STANLEY, H. E. AND KAPLAN, T. A. [1967], On the Possible Phase Transitions in Two-Dimensional Heisenberg Models. *Journal of Applied Physics* 38 (1967), 975–977.

STEPHEN, M. J. AND MITTAG, L. [1972], Pseudo-Hamiltonians for the Potts Model at the Critical Point. *Physics Letters* 41A (1972), 357–358.

STIX, T. H. [1962], *The Theory of Plasma Waves*. McGraw-Hill, New York, NY, 1962.

SYKES, M. F. [1961], Some Counting Theorems in the Theory of the Ising Model and the Excluded Volume Problem. *J. Math. Physics* 2 (1961), 52–62.

SZILARD, L. [1929], Über die Entropieverminderung in Einem Thermodynamischen System bei Eingriffen Intelligenter Wessen. *Zeitschrift für Physik* 53 (1929), 840–856.

SZILARD, L. [1964], On the Decrease of Entropy in a Thermodynamic System by the Intervention of Intelligent Beings. *Behavioral Science* 9 (1964), 301.

TAMM, I. E. [1962], Yakov Il'ich Frenkel. *Soviet Physics Uspekhi* 5 (1962), 173–194.

THOMPSON, C. J. [1965], Algebraic Derivation of the Partition Function of a Two-Dimensional Ising Model. *Journal of Mathematical Physics* 6 (1965), 1392–1395.

THOMSEN, J. S. [1953], Logical Relations among the Principles of Statistical Mechanics and Thermodynamics. *Physical Review* 91 (1953), 1263–1266.

TISZA, L. [1966], *Generalized Thermodynamics*. MIT Press, Cambridge, MA, 1966.

TOLÉDANO, J-C. AND TOLÉDANO, P. [1987], *The Landau Theory of Phase Transitions*. World Scientific, Singapore, 1987.

TOLMAN, R. C. [1938], *The Principles of Statistical Mechanics*. Oxford University Press, London, 1938, Reprinted by Dover, New York (1979).

TRIBUS, M. [1961a], Information Theory as the Basis for Thermostatics and Thermodynamics. *Journal of Applied Mechanics* **28** (1961), 1–8.

TRIBUS, M. [1961b], *Thermostatics and Thermodynamics*. D. van Nostrand Co., Inc., New York, 1961.

TRIBUS, M. [1979], *Thirty Years of Information Theory*. In *The Maximum Entropy Formalism*, R. D. Levine and M. Tribus, Eds. MIT Press, Cambridge, MA, 1979, A Conference held at MIT, 2-4 May 1978.

TRIBUS, M. AND MOTRONI, H. [1972], Comments on the paper "Jaynes' Maximum Entropy Prescription and Probability Theory." *Journal of Statistical Physics* **4** (1972), 227–228.

TRIBUS, M., SHANNON, P. AND EVANS, R. E. [1966], Why Thermodynamics is a Logical Consequence of Information Theory? *American Institute of Chemical Engineers Journal* **12** (1966), 244–248.

TRUESDELL, C. A. [1961], *Ergodic Theory in Classical Stat. Mechanics*. In *Ergodic Theories*, P. Caldirola, Ed. Academic Press, New York, NY, 1961, 21–56.

TRUESDELL, C. A. [1969], *Rational Thermodynamics*. McGraw-Hill, New York, NY, 1969.

TRUESDELL, C. A. [1984], *Rational Thermodynamics, 2nd Ed.*. Springer-Verlag, New York–Heidelberg–Berlin, 1984.

TRUESDELL, C. A. [1986], *New Perspectives in Thermodynamics*. Springer-Verlag, New York–Heidelberg–Berlin, 1986.

TRUESDELL, C. A. AND BHARATHA, S. [1977], *The Concepts and Logic of Classical Thermodynamics as a Theory of Heat Engines*. Springer-Verlag, New York–Heidelberg–Berlin, 1977.

TRUESDELL, C. A. AND TOUPIN, R. A. [1960], *The Classical Field Theories*. In *Handbuch der Physik*, S. Flügge, Ed. vol. III/1, Springer-Verlag, New York–Heidelberg–Berlin, 1960, 226–793.

UHLENBECK, G. E. AND ORNSTEIN, L. S. [1930], On the Theory of Brownian Motion. *Physical Review* **36** (1930), 823–841.

VDOVICHENKO, N. V. [1965a], A Calculation of the Partition Function of a Plane Dipole Lattice. *Soviet Physics JETP* **20** (1965), 477–479.

VDOVICHENKO, N. V. [1965b], Spontaneous Magnetization of a Plane Dipole Lattice. *Soviet Physics JETP* **21** (1965), 350–352.

VIEIRA, M. C. de S. AND TSALLIS, C. [1987], D-dimensional Ideal Gas in Parastatistics: Thermodynamic Properties. *J. Stat. Phys.* **48** (1987), 97–120.

VAN VLECK, J. H. [1932], *The Theory of Electric and Magnetic Susceptibilities*. Oxford University Press, London, 1932.

VAN VLECK, J. H. [1978], Quantum Mechanics: The Key to Understanding Magnetism. *Science* **201** (1978), 113–120.

VAN VLIET, C. M. [1988], On van Kampen's Objections Against Linear Response Theory. *Journal of Statistical Physics* **53** (1988), 49–60.

VAN VLIET, C. M. AND VASILOPOULOS, P. [1988], The Quantum Boltzmann Equation and Some Applications to Transport Phenomena. *Journal of the Physics and Chemistry of Solids* **49** (1988), 639–649.

VAN VLIET, K. M. [1978], Linear Response Theory Revisited. I. The Many-Body van Hove Limit. *Journal of Mathematical Physics* **19** (1978), 1345–1370.

VAN VLIET, K. M. [1979], Linear Response Theory Revisited. II. The Master Equation Approach. *Journal of Mathematical Physics* **20** (1979), 2573–2595.

VAN DER WAERDEN, B. L. [1941], Die lange Reichwerte der regelmäßigen Atomanordnung in Mischkristallen. *Zeitschrift der Physik* **118** (1941), 473–488.

WALDRAM, J. R. [1985], *The Theory of Thermodynamics*. Cambridge University Press, Cambridge, England, 1985.

WALTERS, P. [1982], *An Introduction to Ergodic Theory*. Springer-Verlag, New York–Heidelberg–Berlin, 1982.

WANG, M. C. AND UHLENBECK, G. E. [1945], On the Theory of Brownian Motion II. *Reviews of Modern Physics* **17** (1945), 323–342.

WATANABE, S. [1969], *Knowing and Guessing*. John Wiley & Sons, New York, NY, 1969.

WATANABE, S. [1985], *Pattern Recognition: Human and Mechanical*. John Wiley & Sons, New York, NY, 1985.

WATSON, G. N. [1939], Three Triple Integrals. *Quarterly Journal of Mathematics, Oxford Ser. (1)* **10** (1939), 266.

WAX, N., Ed. [1954], *Selected Papers on Noise and Stochastic Processes*. Dover, Mineola, NY, 1954.

WEGNER, F. G. [1976a], *Critical Phenomena and Scale Invariance*. In *Critical Phenomena*, J. Brey and R. B. Jones, Eds. Springer-Verlag, New York–Heidelberg–Berlin, 1976, 1–29.

WEGNER, F. J. [1976b], *The Critical State, General Aspects*. In *Phase Transitions and Critical Phenomena*, *V. 6*, C. Domb and M. S. Green, Eds. Academic Press, New York, NY, 1976.

WEINBERGER, W. A. AND SCHNEIDER, W. G. [1952], On the Liquid-Vapor Coexistence Curve of Xenon in the Region of the Critical Temperature. *Canadian Journal of Physics* **30** (1952), 422–437.

WEISS, E. C., Ed. [1977], *The Many Faces of Information Science*. Westview Press for the American Association for the Advancement of Science, 1977, AAAS selected symposium; 3.

WEISS, P. R. [1950], The Application of the Bethe-Peierls Method to Ferromagnetism. *Phys. Rev.* **74** (1950), 1493–1504.

WERGELAND, H. [1962], *Fluctuations, Stochastic Processes, Brownian Motion*. In *Fundamental Problems in Statistical Mechanics*, E. G. D. Cohen, Ed. North-Holland, Amsterdam, 1962, 33–58.

WETMORE, F. E. W. AND LeRoy, D. J. [1951], *Principles of Phase Equilibria*. Dover, Mineola, NY, 1951.

WIDOM, B. [1973], *The Critical Point and Scaling Theory*. In *Proceedings of the van der Waals Conference on Statistical Mechanics*, C. Prins, Ed. North-Holland, Amsterdam, 1973, 107–118.

WIENER, N. [1948], *Cybernetics*. John Wiley & Sons, New York, NY, 1948.

WIENER, N. [1956], What is Information Theory? *IRE Trans. on Information Theory* **IT-2** (1956), 48.

WILCZEK, F. [1982], Quantum Mechanics of Fractional-Spin Particles. *Phys. Rev. Letters* **49** (1982), 957–959.

WILCZEK, F. [1990], *Fractional Statistics and Anyon Superconductivity*. World Scientific, Teaneck, NJ, 1990.

WILSON, K. G. [1973], *Critical Phenomena in 3.99 Dimensions*. In *Proceedings of the van der Waals Conference on Statistical Mechanics*, C. Prins, Ed. North-Holland, Amsterdam, 1973, 119–128.

WILSON, K. G. [1975], The Renormalization Group: Critical Phenomena and the Kondo Problem. *Reviews of Modern Physics* **47** (1975), 773–840.

WILSON, K. G. [1983], The Renormalization Group and Critical Phenomena. *Reviews of Modern Physics* **55** (1983), 583–600.

WILSON, K. G. [Aug. 1979], Problems in Physics with Many Scales of Length. *Scientific American* **241, No. 2** (Aug. 1979), 158–179.

WILSON, K. G. AND KOGUT, J. [1974], The Renormalization Group and the ε Expansion. *Physics Reports C* **12** (1974), 75–199.

WOODS, L. C. [1975], *The Thermodynamics of Fluid Systems*. Oxford University Press, London, 1975.

WU, F. Y. [1982], The Potts Model. *Reviews of Modern Physics* **54** (1982), 235–268, Errata in V.55, p. 315.

WU, T. T. [1966], Theory of Toeplitz Determinants and the Spin Correlations of the Two-Dimensional Ising Model. I.. *Physical Review* **149** (1966), 380–401.

WU, Y.-S. [1984], General Theory of Quantum Statistics in Two Dimensions. *Phys. Rev. Letters* **52** (1984), 2103–2106.

YANG, C. N. [1952], The Spontaneous Magnetization of a Two-Dimensional Ising Model. *Physical Review* **85** (1952), 808–816.

YANG, C. N. AND LEE, T. D. [1952], Statistical Theory of Equations of State and Phase Transitions. I. Theory of Condensation. *Physical Review* **87** (1952), 404–409.

YU, F. T. S. [1976], *Optics and Information Theory.* John Wiley & Sons, New York, NY, 1976.

YVON, J. [1935], *La Theorie Statistique des Fluides et l'Equation d'Etat.* Hermann et Cie, Paris, 1935.

YVON, J. [1969], *Correlations and Entropy in Classical Statistical Mechanics.* Pergamon Press, Oxford, England, 1969.

ZEL'DOVICH, Y. B. AND RAIZER, Y. P. [1966], *Physics of Shock Waves and High-Temperature Hydrodynamic Phenomena.* Academic Press, New York, NY, 1966.

ZEMANSKY, M. W. AND DITTMAN, R. H. [1981], *Heat and Thermodynamics.* McGraw-Hill, New York, NY, 1981.

ZERNIKE, F. [1916], The Clustering Tendency of the Molecules in the Critical State and the Extinction of Light Caused Thereby. *Proc. Amst. Akad. Sci.* **18** (1916), 1520–1527.

ZUBAREV, D. N. [1974], *Nonequilibrium Statistical Thermodynamics.* Consultants Bureau, New York, 1974.

ZUBAREV, D. N. AND KALASHNIKOV, V. P. [1970], Derivation of the Nonequilibrium Statistical Operator from the Extremum of Information Theory. *Physica* **46** (1970), 550–554.

ZUREK, W. H., Ed. [1990], *Complexity, Entropy, and the Physics of Information.* Addison Wesley, Reading, MA, 1990.

ZWANZIG, R. W. [1960], Ensemble Methods in the Theory of Irreversibility. *Journal of Chemical Physics* **33** (1960), 1338–1341.

ZWANZIG, R. W. [1961], *Statistical Mechanics of Irreversibility.* In *Lectures in Theoretical Physics, V. III,* W. E. Brittin, B. W. Downs and J. Downs, Eds. Interscience Publishers, A division of John Wiley & Sons, London, 1961, 106–141.

ZWANZIG, R. W. [1965], Time Correlation Function and Transport Coefficients in Statistical Mechanics. *Annual Reviews of Physical Chemistry* **16** (1965), 67–102.

INDEX